CANCER

CANCER

Prevention, Early Detection, Treatment and Recovery

SECOND EDITION

Edited by

Gary S. Stein
Kimberly P. Luebbers
University of Vermont Larner College of Medicine
149 Beaumont Avenue, HSRF 326
Burlington, VT 05405, USA

WILEY Blackwell

Registered Office
John Wiley & Sons, Inc., 111 River Street, Hoboken, NJ 07030, USA

Editorial Office
Boschstr. 12, 69469 Weinheim, Germany

For details of our global editorial offices, customer services, and more information about Wiley products visit us at www.wiley.com.

Wiley also publishes its books in a variety of electronic formats and by print-on-demand. Some content that appears in standard print versions of this book may not be available in other formats.

Library of Congress Cataloging-in-Publication data has been applied for
9781118962886 (paperback)

Cover Design: Wiley
Cover Images: © Sebastian Kaulitzki/Shutterstock, © bogdanhoda/Shutterstock, © Rido/Shutterstock, © Giovanni Cancemi/Shutterstock

Set in 11/13pt Times by SPi Global, Pondicherry, India
Printed and bound in Singapore by Markono Print Media Pte Ltd

10 9 8 7 6 5 4 3 2 1

CONTENTS

LIST OF CONTRIBUTORS

Thomas P. Ahern Assistant Professor, Department of Biochemistry and Department of Surgery, Larner College of Medicine, University of Vermont, Burlington, VT, USA

Christopher B. Allard Massachusetts General Hospital, Boston, MA, USA

Sarah R. Amend The James Buchanan Brady Urological Institute, Johns Hopkins University School of Medicine, Baltimore, MD, USA

Ogheneruona Apoe University of Vermont Medical Center, Burlington, VT, USA

Jesse N. Aronowitz Department of Radiation Oncology, University of Massachusetts Medical School; UMass Memorial Healthcare Systems, Worcester, MA, USA

Elizabeth Bannon Department of Radiation Oncology, University of Massachusetts Medical School; UMass Memorial Healthcare Systems, Worcester, MA, USA

Anant Bhave Larner College of Medicine, University of Vermont, Burlington, VT, USA

Maryann Bishop-Jodoin Department of Radiation Oncology, University of Massachusetts Medical School, Worcester, MA, USA. IROC Rhode Island, Lincoln, RI, USA

Michael L. Blute Massachusetts General Hospital, Boston, MA, USA

Bruce A. Bornstein Department of Radiation Oncology, University of Massachusetts Medical School; UMass Memorial Healthcare Systems, Worcester, MA, USA

Carla D. Bradford Department of Radiation Oncology, University of Massachusetts Medical School; UMass Memorial Healthcare Systems, Worcester, MA, USA

Leslie Bradford Maine Medical Center, Scarborough, ME, USA

Harry Bushe Department of Radiation Oncology, University of Massachusetts Medical School; UMass Memorial Healthcare Systems, Worcester, MA, USA

Frances E. Carr Professor, University of Vermont Cancer Center, Department of Pharmacology, Cancer Center, University of Vermont College of Medicine, Burlington, VT, USA

Maria G. Cicchetti Department of Radiation Oncology, University of Massachusetts Medical School; UMass Memorial Healthcare Systems, Worcester, MA, USA. IROC Rhode Island, Lincoln, RI, USA

Eric M. Clark Department of Surgery, University of Vermont, Burlington, VT, USA

Bernard F. Cole Professor, Department of Mathematics and Statistics, College of Engineering and Mathematical Sciences, University of Vermont, Burlington, VT, USA

Marion Couch Richard T. Miyamoto Professor and Chair, Department of Otolaryngology–Head and Neck Surgery, Indiana University School of Medicine, Indianapolis, IN, USA

David Cranmer President, Vermont Cancer Survivor Network, Montpelier, VT, USA

W. Christian Crannell Department of Surgery, University of Vermont, Burlington, VT, USA

Kristin DeStigter Larner College of Medicine, University of Vermont, Burlington, VT, USA

Linda Ding Department of Radiation Oncology, University of Massachusetts Medical School; UMass Memorial Healthcare Systems, Worcester, MA, USA

Liang Dong The James Buchanan Brady Urological Institute, Johns Hopkins University School of Medicine, Baltimore, MD, USA. Department of Urology, Renji Hospital, Shanghai Jiao Tong University, School of Medicine, Shanghai, China

Mai K. ElMallah Gene Therapy Center, University of Massachusetts Medical School, Worcester, MA, USA. Department of Pediatrics, University of Massachusetts Medical School, Worcester, MA, USA

Mark F. Evans Assistant Professor, Department of Pathology & Laboratory Medicine, University of Vermont, Burlington, VT, USA

Thomas J. FitzGerald Department Chair & Professor, Department of Radiation Oncology, University of Massachusetts Medical School; UMass Memorial Healthcare, Worcester, MA, USA. Principal Investigator, IROC Rhode Island, Lincoln, RI, USA

Terence R. Flotte Gene Therapy Center, University of Massachusetts Medical School, Worcester, MA, USA. Department of Pediatrics, University of Massachusetts Medical School, Worcester, MA, USA. Microbiology & Physiologic Systems, University of Massachusetts Medical School, Worcester, MA, USA

Garth Garrison Division of Pulmonary and Critical Care Medicine, University of Vermont Medical Center, Burlington, VT, USA

Robert H. Getzenberg Dr. Kiran C. Patel College of Allopathic Medicine, Nova Southeastern University, Fort Lauderdale, FL, USA

Jonathan M. Glanzman Department of Radiation Oncology, University of Massachusetts Medical School; UMass Memorial Healthcare Systems, Worcester, MA, USA

David J. Goff Department of Radiation Oncology, University of Massachusetts Medical School; UMass Memorial Healthcare Systems, Worcester, MA, USA

Solomon A. Graf Clinical Research Division, Fred Hutchinson Cancer Research Center, Seattle, WA, USA. Department of Medicine, University of Washington, Seattle, WA, USA. Veterans Affairs Puget Sound Health Care System, Seattle, WA, USA

Kristina Guyton Department of Surgery, University of Chicago Medicine, Chicago, IL, USA

Jean R. Harvey Associate Dean for Research, College of Agriculture and Life Sciences, Robert L. Bickford, Jr. Endowed Professor, Chair, Department of Nutrition and Food Sciences, University of Vermont, Vermont, VT, USA

Jessica L. Heath Assistant Professor of Pediatrics and Biochemistry, Pediatric Hematology-Oncology, University of Vermont, Burlington, VT, USA

Beth B. Herrick Department of Radiation Oncology, University of Massachusetts Medical School; UMass Memorial Healthcare Systems, Worcester, MA, USA

Sally Herschorn Larner College of Medicine, University of Vermont, Burlington, VT, USA

Jessica R. Hiatt Department of Radiation Oncology, University of Massachusetts Medical School; UMass Memorial Healthcare Systems, Worcester, MA, USA

Neil Hyman Department of Surgery, University of Chicago Medicine, Chicago, IL, USA

Ted James Department of Surgery, University of Vermont Medical Center, Burlington, VT, USA

Chris A. Jones Department of Surgery, University of Vermont, Burlington, VT, USA

Michael Kalfopolous Gene Therapy Center, University of Massachusetts Medical School, Worcester, MA, USA

Farrah B. Khan Division of Hematology and Oncology, University of Vermont Medical Center, Burlington, VT, USA

Sahn-Ho Kim Vattikuti Urology Institute, Henry Ford Health System, Detroit, MI, USA

C. Matthew Kinsey Division of Pulmonary and Critical Care Medicine, University of Vermont Medical Center, Burlington, VT, USA

I-Lin Kuo Department of Radiation Oncology, University of Massachusetts Medical School; UMass Memorial Healthcare Systems, Worcester, MA, USA

Jennifer LaFemina Division of Surgical, UMass Memorial Health Center, University of Massachusetts Medical School, Worcester, MA, USA

Benjamin Lange Larner College of Medicine, University of Vermont, Burlington, VT, USA

Fran Laurie Department of Radiation Oncology, University of Massachusetts Medical School, Worcester, MA, USA. IROC Rhode Island, Lincoln, RI, USA

Yuan-Chyuan Lo Department of Radiation Oncology, University of Massachusetts Medical School; UMass Memorial Healthcare Systems, Worcester, MA, USA

Kimberly P. Luebbers Assistant Dean for Clinical Research, Direct for the Office of Clinical Trials Research, University of Vermont Larner College of Medicine Office of Clinical Trials, Burlington, VT, USA

Jennifer L. May Outpatient Oncology Dietitian, The University of Vermont Cancer Center, VT, USA

Kathleen McBeth University of Vermont Medical Center, Burlington, VT, USA

Wendy McKinnon Certified Genetic Counselor, University of Vermont Cancer Center, Burlington, VT, USA

Janaki Moni, MD Department of Radiation Oncology, University of Massachusetts Medical School; UMass Memorial Healthcare Systems, Worcester, MA, USA. IROC Rhode Island, Lincoln, RI, USA

Jesse Moore Department of Surgery, University of Vermont, Burlington, VT, USA

Michael Moore Arilla Spence DeVault Professor of Otolaryngology–Head and Neck Surgery, Chief, Division of Head and Neck Surgery, Indiana University School of Medicine, Indianapolis, IN, USA

David C. Nalepinski Assistant Vice President for Community Oncology, Duke Cancer Institute, Durham, NC, USA

Mitchell Norotsky Department of Surgery, University of Vermont, Burlington, VT, USA

Amrita Pandit University of Vermont Medical Center, Burlington, VT, USA

Scott Perrapato University of Vermont Larner College of Medicine, Burlington, VT, USA

Kenneth J. Pienta The James Buchanan Brady Urological Institute, Johns Hopkins University School of Medicine, Baltimore, MD, USA

Richard S. Pieters Department of Radiation Oncology, University of Massachusetts Medical School; UMass Memorial Healthcare Systems, Worcester, MA, USA

Carlos A. Pino Professor of Anesthesiology and Pain Medicine, Director, Center for Pain Medicine, The Robert Larner, MD College of Medicine, The University of Vermont, Burlington, VT, USA

Mark K. Plante Division of Urology, Department of Surgery, University of Vermont Medical Center, Burlington, VT, USA

Kumar G. Prasad Clinical Instructor, Department of Otolaryngology–Head and Neck Surgery, Indiana University School of Medicine, Indianapolis, IN, USA.

Department of Otolaryngology–Head and Neck Surgery, Meritas Health, North Kansas City, MO, USA

G. Premveer Reddy Vattikuti Urology Institute, Henry Ford Health System, Detroit, MI, USA

Paul S. Rava Department of Radiation Oncology, University of Massachusetts Medical School; UMass Memorial Healthcare Systems, Worcester, MA, USA

Michaela Reagan Assistant Professor, Tufts University School of Medicine, Boston, MA, USA. Faculty Scientist I, Maine Medical Center Research Institute, Scarborough, ME, USA

Lawrence Recht Department of Neurology & Clinical Neurosciences, Stanford University School of Medicine, Stanford, CA, USA

Lekkala V. Reddy Vattikuti Urology Institute, Henry Ford Health System, Detroit, MI, USA

Judy R. Rees Associate Professor, Department of Epidemiology, Geisel School of Medicine, Dartmouth College, Hanover, NH, USA. Norris Cotton Cancer Center, Lebanon, NH, USA

Sean Reynolds Larner College of Medicine, University of Vermont, Burlington, VT, USA

Christopher Riberdy Department of Radiation Oncology, University of Massachusetts Medical School; UMass Memorial Healthcare Systems, Worcester, MA, USA

Alan Rosmarin Wolters Kluwer Health, Waltham, MA, USA

Allison Sacher Department of Radiation Oncology, University of Massachusetts Medical School; UMass Memorial Healthcare Systems, Worcester, MA, USA

Jonathan Saleeby Department of Radiation Oncology, University of Massachusetts Medical School; UMass Memorial Healthcare Systems, Worcester, MA, USA

Baddr Shakhsheer Department of Surgery, University of Chicago Medicine, Chicago, IL, USA

Shirin Sioshansi Department of Radiation Oncology, University of Massachusetts Medical School; UMass Memorial Healthcare Systems, Worcester, MA, USA

David W. Sobel Division of Urology, Department of Surgery, University of Vermont Medical Center, Burlington, VT, USA

Brian L. Sprague Departments of Biochemistry and Department of Surgery, Larner College of Medicine, University of Vermont, Burlington, VT, USA

Gary S. Stein Director, University of Vermont Cancer Center, Perelman Professor and Chair, Department of Biochemistry, Professor, Department of Surgery, University of Vermont Larner College of Medicine, Burlington, VT, USA

F. Marc Stewart Clinical Research Division, Fred Hutchinson Cancer Research Center, Seattle, WA, USA. Department of Medicine, University of Washington, Seattle, WA, USA

Celeste Straight UMass Memorial Health Care, Central Mass Ob/Gyn, Shrewsbury, MA, USA

Jill Sudhoff-Guerin Health Policy and Advocacy Consultant, The Vermont Medical Society, Montpelier, VT, USA

Bradley Switzer Division of Medical Oncology, UMass Memorial Health Center, University of Massachusetts Medical School, Worcester, MA, USA

Gabriella Szalayova University of Vermont Medical Center, Burlington, VT, USA

Phil Trabulsy University of Vermont, Burlington, VT, USA

Kenneth Ulin Department of Radiation Oncology, University of Massachusetts Medical School; UMass Memorial Healthcare Systems, Worcester, MA, USA

John M. Varlotto Department of Radiation Oncology, University of Massachusetts Medical School; UMass Memorial Healthcare Systems, Worcester, MA, USA

Harold James Wallace, III Division of Radiation Oncology, Larner College of Medicine, University of Vermont, Burlington, VT, USA

Tao Wang Department of Radiation Oncology, University of Massachusetts Medical School, Worcester, MA, USA

George J. Weiner Director, C.E. Block Chair of Cancer Research, Holden Comprehensive Cancer Center, The University of Iowa, Iowa City, IA, USA. Professor, Department of Internal Medicine, Carver College of Medicine, Iowa City, IA, USA. Professor, Department of Pharmaceutical Sciences and Experimental Therapeutics, College of Pharmacy, Iowa City, IA, USA

Giles F. Whalen Division of Surgical, UMass Memorial Health Center, University of Massachusetts Medical School, Worcester, MA, USA

Marie E. Wood University of Vermont Medical Center, Burlington, VT, USA

Richard C. Zieren The James Buchanan Brady Urological Institute, Johns Hopkins University School of Medicine, Baltimore, MD, USA. Department of Urology, Amsterdam UMC, University of Amsterdam, Amsterdam, The Netherlands

Susan Zweizig University of Massachusetts, UMass Memorial Health Care, Worcester, MA, USA

PREFACE

Cancer is a leading cause of death globally. The social and economic burdens are profound. Making a difference in cancer prevention, early detection, treatment, and survivorship is not an option but a requirement and obligation. This responsibility is shared by cancer scientists, physicians, and a broad spectrum of healthcare professionals as well as by cancer advocates, educators, thought leaders, and lawmakers who are committed to making cancer preventable and treatable. Every life lost to cancer is a powerful motivation to accelerate progress toward bringing cancer under control. We are moving forward with confidence that more than half of the people who encounter cancer are cured. Reinforcing progress, an increasing number of people are living with cancer, undergoing long-term treatments, and experiencing quality lives.

Collaborative, multidisciplinary partnerships of dedicated scientists and physician-investigators are providing unique opportunities to reduce the burden of cancer. There is increasing capacity and capability to abbreviate the translation of discoveries in laboratories to advances in cancer prevention, early detection, treatment with clinical trials, and survivorship. There is recognition for the necessity to rectify disparities that result from restrictive healthcare access in many rural and urban communities. Initiatives to expand treatment options and opportunities for cancer patients irrespective of social or economic limitations must be a priority. An investment that enhances capabilities in cancer prevention, early detection, targeted treatment, and survivorship should be obligatory. The return on the investment will be immediate, progressively increase, and make a sustainable difference in quality of life.

"Cancer: Prevention, Early Detection, Treatment and Recover" was developed to provide a patient-centric book that synthesizes current understanding of the biology and treatment of cancer. We have responded to the importance of readily accessible explanations that accurately, directly, and comprehensively explain the biomedical parameters of cancer. We were guided by the necessity for facts, perspectives, and resources to enable cancer patients and caregivers to responsibly access options for "informed" and "comfortable" decisions.

This book provides a unique and valued resource, presenting an overview of cancer with emphasis on navigating the emerging "roadmap" from cancer prevention to early detection, treatment, and recovery. There are comprehensive books focused on the mechanistic and clinical facets of cancer. "Cancer: Prevention, Early Detection, Treatment and Recovery" authoritatively outlines the challenges and opportunities associated with cancer biology, cancer research, and the spectrum of clinical considerations that are central to "experiencing cancer" as well as to studying, investigating, and treating cancer.

"Cancer: Prevention, Early Detection, Treatment and Recovery" is dedicated to Dr. Arthur B. Pardee, who has been instrumental in establishing the foundation for progress in the prevention and treatment of cancer. Arthur Pardee has, for many years, been making visionary and far reaching discoveries in cancer-compromised biological control that translate molecular and cellular changes during the onset and progression of cancer to novel strategies for cancer diagnosis and therapy. His contributions include the mechanistic understanding for control of gene expression, regulation of the cell cycle, and the coordinate control of genes that are responsible for physiological responsiveness and compromises associated with cancer onset and progression. Until his death on February 24, 2019, Arthur Pardee's discoveries consistently accelerated progress toward making cancer preventable and treatable. He will continue to be an inspirational role model for scientists and physician investigators throughout the international cancer research community.

Gary S. Stein

Kimberly P. Luebbers

I

CANCER AND THE CANCER EXPERIENCE

1

CANCER AND THE CANCER EXPERIENCE

Gary S. Stein[1] and Kimberly P. Luebbers[2]

[1] *University of Vermont Cancer Center, Department of Biochemistry and Surgery, University of Vermont Larner College of Medicine, Burlington, VT, USA*
[2] *Clinical Research, Clinical Trials Research, University of Vermont Larner College of Medicine Office of Clinical Trials, Burlington, VT, USA*

ENCOUNTERING CANCER

The impact and consequences of cancer are immense and far-reaching. Everyone is vulnerable. Many of us have experienced cancer and most of us have been close to someone encountering this challenging disease. The cumulative toll is extensive. The data is compelling. Cancer is the leading cause of death in many states as well as globally. The social and economic (financial) burdens are profound.

A cancer diagnosis is frightening, confusing, and generally unexpected, evoking thoughts and feelings that are unfamiliar and unsettling. The patient has concerns that come without warning or preparation. The patient may experience a roller coaster of signs, symptoms, and emotions that are difficult to understand and tame. Cancer often changes the lives of patients, families, and friends, derailing goals, wishes, hopes, and dreams. Plans must be put on hold and decisions are required without experience to draw on. Priorities instantly change. Survival becomes a reality with many perplexing unanswered/unanswerable questions.

Cancer: Prevention, Early Detection, Treatment and Recovery, Second Edition. Edited by Gary S. Stein and Kimberly P. Luebbers.
© 2019 John Wiley & Sons, Inc. Published 2019 by John Wiley & Sons, Inc.

CHALLENGES, OPPORTUNITIES AND COMMITMENTS

Despite the complex causes for cancer, the uncertainties associated with cancer risk and the requirements for unique strategies to advance cancer diagnosis and treatment, we can and should be optimistic that significant advances are being made toward reducing the burden of cancer. Making a difference in cancer prevention, early detection, treatment, and survivorship is not an option but a requirement and obligation. This responsibility is shared by dedicated scientists, physicians, a broad spectrum of healthcare professionals, cancer advocates, educators, thought leaders, and law makers who are committed to making cancer a preventable and treatable disease. And, while there is so much to be learned and so much to be accomplished we are making tremendous advances in understanding the biology and treatment of cancer – knowledge that can be captured to move toward putting cancer behind us as a life threatening disease. Numbers speak. While every life lost to cancer is powerful motivation to accelerate progress toward bringing cancer under control we are moving forward with confidence that more than half of the people who encounter cancer are cured. Reinforcing progress, an increasing number of people are living with cancer undergoing long-term treatments and experiencing (compatible with) quality lives.

WHAT IS CANCER?

Pursuit of the causes for cancer and discovery of strategies for treatment have been elusive. Cancer defies conventional definitions for a disease with unique, sometimes ambiguous and frequently challenging biology and pathology. The complexities of changes in cells, tissues, and organs that do not respond to traditional therapies are often unfamiliar and unpredictable. The elusiveness of cancer is an enigma to dedicated scientists at the forefront of studying cancer in the laboratory and to physicians who are committed to maximizing effectiveness of treatment for cancer patients. Sustained effectiveness for the treatment of cancer and the recurrence of disease have been and continue to be difficult problems. Cancer can be refractory to drugs, develop resistance, and acquire capabilities to circumvent therapeutic targets.

While there are many gaps in our knowledge about cancer (biology and pathology) there is emerging appreciation for the absence of a single cause. This is in striking contrast to polio, which is caused by an identifiable infectious agent and prevented by a specific vaccine. The lack of a shared treatment responsiveness for all tumors can be frustrating. However, it is widely recognized that the dominant unifying features of cancer are compromised control of cell proliferation with decline in specialized cell functions and abrogated (relinquished) competency for the responsiveness to environmental regulatory signals. Cancer cells generally exhibit unrestricted proliferation that may not be compensated by cell loss.

The onset as well as progression of cancer is associated with changes that include but are not restricted to: enhanced cell mobility and motility providing capability for metastasis, the development of secondary malignant growth at a distance from the primary tumor; decline in the patient's capability to enlist the immune system to recognize and reject tumor cells; a shift from aerobic (oxidative phosphorylation) to anaerobic (glycolysis) metabolism; recapitulated expression of fetal proteins in tumor cells; and, competency for tumor cells to populate and multiply in vital organs interfering with essential functions. Many of the modifications exhibited by tumor cells provide biomarkers, biochemical indicators for the presence of cancer and full effectiveness of cancer therapy. Others are *targets* for the identification and design of pharmaceuticals to treat cancer.

Acceptance of the absence for a single cause of cancer has been long coming. During the past century there have been an evolving series of infectious agents and environmental factors that have been held accountable for cancer. Genetic predisposition and recently epigenetic control, non-DNA-encoded cellular regulatory information that is retained and distributed to progeny cells during cell division, have been implicated as accountable for cancer. Viruses that include HPV, EBV, and HIV have been directly linked to cancer. Inflammation is associated with the onset and progression of tumors. And, a long list of cancer causing chemicals designated *carcinogens* has been identified. Compelling evidence has linked smoking, alcohol consumption, ultraviolet exposure, and obesity with cancer risk. Rather than assigning blame to a single cause for cancer, prevailing evidence that is widely understood and accepted supports a multifactorial basis for cancer with the realization that synergy between risk factors can exacerbate and accelerate cancer onset, progression, and recurrence of disease. The incidence of cancer increases with age, attributable to accumulation of acquired genome mutations, compromised capability to repair damaged DNA/chromosomes, genes, and compromised metabolic control.

Objective, scientifically based documentation for cancer risk has been leveraged through public awareness, effective advocacy and support from law makers into policies, recommendations and legislative mandates to reduce exposure to factors and practices that contribute to cancer. Regulations, guidelines, and persuasive public opinion is implementing cancer prevention, the benefits for quality of life are increasingly understood and accepted. The economic impact (consequences) is recognized and appreciated. Cancer prevention is becoming understood and accepted as a shared responsibility. With enhanced capabilities for treating cancer, emphasis on survivorship is becoming increasingly important.

CANCER AND THE PATIENT

The emotional impact of cancer cannot be captured in words. With so many unfamiliar considerations and perplexing decisions, it is difficult to navigate the cancer experience with confidence. What should I do? Who should I turn to? How can I have assurance my decisions are appropriate? There is so much information, most authoritative and some questionable. There are so many options. There are numerous opinions, most highly responsible but not all consistent. And it can be difficult to distinguish between authoritative and anecdotal reports and recommendations.

Confirmation of a Cancer Diagnosis

A cancer diagnosis may result from cancer screening; a mammogram, a colonoscopy, or a prostate cancer test. A routine physical examination or evaluation of a suspicious symptom can be the basis for diagnosing cancer. Regardless of the competency and experience of the diagnosing physician and the capabilities of the medical practice or health center where the diagnosis is made, the patient should not hesitate to seek a *second opinion* for multiple reasons. Obtaining a *second* or *third opinion* does not preclude treatment by the physician who provides the initial diagnosis. Confirmation of a diagnosis provides assurance that the full spectrum of diagnostic options are utilized or considered and that a comprehensive evaluation of treatment plans is performed. A second opinion should not be a challenge to confidence in the treating physician. Rather, a contributory diagnosis provides a valuable perspective and reassurances for decision-making by the patient and the doctor. Criteria for tumor diagnosis and staging are evolving and options for treatment are rapidly emerging. Seeking recommendations from experts is vitally important.

Specialized diagnostic procedures that can be informative for decision-making may be available at centers that specialize in treating certain tumors. With rapidly accruing developments in treatment options, not only rare tumors but also the prevalent types of cancer, it is advantageous to secure a comprehensive assessment of the treatment possibilities as well as an understanding of the strengths and limitations of the options. Decisions may be between medical, surgical, or radiation therapy or a combination of these treatments. Consultation to obtain multiple perspectives is becoming increasingly important for informed decision-making with an understanding of the strengths and limitations of options that are available.

Treatment Options

Medical treatments include conventional *cytotoxic chemotherapy* that is directed to selectively kill rapidly dividing cancer cells with doses, treatment regimens, and combinations of drugs that are designed to maximize specificity and minimize non-specific, *off-target* affects. Precision medicine is emerging with the objective to target the unique defect of a specific tumor and requires molecular diagnostic evaluation involving analysis of the tumor genome. Another emerging dimension to medical oncology is immunotherapy where the tumor is specifically targeted or the patient's immune system is rendered competent to reject the tumor by providing cues that the tumor is *non-self.*

Surgical options are expanding. Minimal invasive surgical procedures are being developed that are highly effective and reduce recovery time for the patient. With complex surgical cases a consideration is the experience of the surgeon with the specific procedure.

Radiation treatments are advancing rapidly with more focus on *treatment fields* to prevent/reduce tissue and organ complications. Specific modalities of radiation therapy can be beneficial for certain tumors and be advantageous for contained or disseminated (metastasized) cancers. There are advances in treatment plans that combine medical, surgical, and radiation therapy, both sequential and combined.

To treat or not to treat is an important decision for DCIS and prostate cancer. There is growing concern that these tumors may be excessively treated and guidelines are being developed for *active surveillance*, where the lesions are frequently reevaluated. The objective is to avoid unnecessary medical, surgical, or radiation intervention where consequences can include treatment/procedure-related complications including the risk for secondary cancers.

Where to Obtain Treatment?

It is important to engage the most knowledgeable and experienced physicians to establish and confirm a cancer diagnosis and contribute to development of a treatment plan. Academic cancer centers provide the most comprehensive opportunities for state-of-the-art cancer diagnosis and treatment. Multidisciplinary team approaches are utilized for diagnosis and decision-making. The most advanced capabilities, instrumentation and expertise are available for characterizing tumors and evaluating patients. Clinical trials are available to provide the most up-to-date options for tumor detection and therapy. Research at cancer centers translates laboratory discovery to novel options for treating cancer.

Community-based cancer centers offer cancer patients quality care and are affiliated with academic cancer centers, working collaboratively to provide patients with the advantage of treatment near home with access to academic cancer centers for specialized diagnoses or therapy. Community-based physicians who are affiliated with academic cancer centers have the ability to draw on the expertise and experience of the centers as well as participate in clinical trials. For highly specialized medical, surgical, or radiation therapy it may be advantageous to obtain treatment where unique capabilities are available.

Patient and Family-Centered Care

There is growing appreciation for the importance of family and friends for the well-being of the patient. It is recognized that involvement of family and friends should be a priority from the initial visit. Engagement of people close to the patient can expand availability of care givers. Unquestionably, involvement of family and friends makes a decisive difference in the quality of life for a cancer patient.

Assistance with care and emotional reinforcement are decisive factors. Understanding the cancer patient, the disease, and treatments enables families and friends to be maximally supportive with an accurate understanding of the challenges confronting the patients. Beyond the treatment, involvement of family and friends continues to be important for survivorship and should not be dismissed.

Shared Decision-Making

The options for treatment decisions and strategies for survivorship are extensive and complex. Turning the clock back, physicians provided treatment plans to patients and generally there was minimal dialog beyond clarification of expectations of patient compliance and the anticipated outcomes. Now there is emphasis on shared decision-making with discussions between the patient and physician on cancer detection, treatment options, recovery, and surviving cancer as a treatable disease. Cancer prevention is another area where there is a spectrum of choices and these can best be evaluated by considering consequences of options by physician–patient discussion. Engaging the patient and physician in shared decision-making provides the patient with in-depth understanding that will impact on patient compliance. The dialog will allow the physician to understand the thoughts and feelings of the patient.

PATHWAYS TO DISCOVERY

Building on Discoveries of the Aberrant Biology of Tumor Cells

Broad-based engagement of scientists and physician/investigators has provided an understanding of cancer biology that can translate to clinical applications. Initially, the focus was on a cause and emphasis was on a cellular change. There is now recognition for genetic, epigenetic, and environmental contributions to cancer risk, initiation, and progression. There is compelling documentation for multiple cancer-related alterations in control of cell proliferation and chromosome structure and function that are associated with altered metabolic control and compromised capabilities for responsiveness to regulatory signals and execution of activities required by specialized cells. Particularly significant is acceptance that cancer derails essential cellular processes. Cross-talk between parameters of control is altered and compensatory mechanisms are invoked during the onset and progression of cancer that are mechanistically informative and clinically relevant. There is increasing emphasis on multiple dimensions of regulation and integration of activities. There are emerging *success stories* where genetic, epigenetic, metabolic, cellular, and biochemical properties of cancer cells are informative for cancer diagnosis, decisions and options for treatment, monitoring therapeutic responsiveness, and monitoring relapse. Monitoring the structural and functional properties of cancer cells enhances the effectiveness of conventional chemotherapy, radiation therapy, immunotherapy, and cancer stem-cell-based interventions. With the development of genome-based medicine where aberrant genes and altered gene expression support refined diagnosis and targeted therapy it is increasingly important to consider cellular changes that occur with cancer and are responsive to treatment.

Embracing the *Value Added* from Collaboration

The complexity of cancer research, diagnosis, and treatment requires collaborative team approaches. Effectiveness in cancer investigation and patient care necessitates partnerships where complementary perspectives, expertise, and experience optimize capabilities.

Collaboration accelerates mechanistic, behavioral, clinical, and translational cancer research. Multidisciplinary approaches reinforce creativity and innovation as well as enhance the transition from discovery to clinical applications. Collaboration supports development of novel laboratory-based and behavioral strategies, data acquisition and analysis as well as integration of complex data sets including sophisticated informatics. Research initiatives that incorporate scientists and physician/investigators support pursuit of fundamental mechanisms within the context of clinical applications. Collaborative research is the longstanding approach in the pharmaceutical and biotechnology industries. Cancer centers, to accommodate expectations from the National Cancer Institute emphasize and incentivize multidisciplinary, inter-programmatic and inter-institutional collaboration. And educational institutions are supporting collaborative contributions for academic advancement.

Collaboration increases the effectiveness of clinical care with increasing development of disease-site-specific multidisciplinary clinics. To expedite diagnosis and development of a treatment plan with comprehensive clinical input, during the initial visit the patient is evaluated by a medical oncologist, surgical oncologist, and radiation oncologist and there is participation of a radiologist and pathologist. In addition to follow-up evaluations for immediate initiation of treatment, the patient is generally spared the anxiety associated with a series of sequential visits.

Developmental and Age-Dependent Considerations

There are striking age-dependent cancer treatment considerations that are becoming increasingly evident. Pediatric oncology is an expanding field and there is accruing understanding for drug selection, dose, and treatment regimens that are effective and compatible with patient tolerance and responses that change with advancing age. With pediatric patients there is growing evaluation of minimal chemotherapy doses that achieve maximal responses to protect young children from lifelong vulnerability to treatment-related complications. There are similar treatment and dose considerations for radiation treatment of pediatric and geriatric cancer patients. The changes in treatment responsiveness and dose tolerances that occur with the transition from pediatric to adolescence are important clinical considerations. Further understanding the optimal approaches for treating adolescent cancer patients will be important.

EMPHASIS ON CANCER PREVENTION AND SURVIVORSHIP

The advances in cancer diagnosis and therapy have been significant. Success stories include testicular cancer, breast cancer, skin cancer, GI (gastrointestinal) tumors, and leukemias. Equally important, there is an increasing emphasis on cancer prevention and survivorship that can advance progress toward making cancer preventable and treatable. Progress in cancer prevention reflects the combined effectiveness of education on the importance of healthy living, emphasizing diet and exercise and tobacco control. Policies and legislation that mandate warnings on tobacco products and prohibit the sale of tobacco and alcohol to minors and prohibiting smoking in many public areas has been influential. Identification of carcinogens that impact on cancer risk has had a positive effect, including scrutiny of inclusion in food products.

Early detection has been a driving force in alleviating the cancer burden. There has been development and refinement of technologies and assays for cancer screening with strong support of physicians, healthcare insurers, educators, law makers, and healthcare advocacy organizations. Breast cancer, prostate cancer, skin cancer, and recently lung cancer screening programs have made a difference in reducing the incidence of advanced stage cancer diagnoses. Acceptance of the importance for cancer screening is gaining momentum and will be an increasingly positive force for detecting cancer at a stage where response to treatment is effective. With an emerging pipeline of targeted drugs and immunotherapy, early detection is paving the way for cancer to be treated as a chronic and controllable illness. Living a high-quality life with cancer is an increasingly prevalent reality.

As a consequence of advances in cancer detection and treatment, survivorship is becoming increasingly important. Survivorship begins with a cancer diagnosis, continues through active surveillance or treatment, and continues thereafter to limit the risk of recurrence (relapse) and to support a productive and meaningful life.

Survivorship is not always intuitive. Guidance for strategies to live with and/or go beyond a cancer experience is available from multiple sources. Treating physicians, social workers, cancer support groups, and cancer advocacy organizations have programs that *reach out* to cancer patients and families to assist with adjustments in perspective and activities that are valuable for survivorship with maximal capabilities and minimal limitations. There is increasing emphasis on capturing opportunities for growing from a cancer experience – acquiring appreciation and understanding for priorities that should not (can) be taken for granted.

LESSONS LEARNED AND ENCOURAGING PROSPECTS

A century of dedicated commitment to exploring the deviant behavior of cancer cells, the pursuit of the elusive cause of cancer, and the development of innovative strategies to eliminate or neutralize unrestricted proliferation of tumors has yielded valuable understanding for the biology and treatment of cancer. The recent advocacy by Vice President Joe Biden for the Cancer Moonshot will support initiatives that translate discovery to transformative clinical applications with immediate, far-reaching, and long-lasting impact. Drawing on breakthroughs in understanding cancer and advances in cancer therapy and learning from approaches that encountered obstacles, we are poised to accelerate progress toward making cancer a preventable and treatable disease.

II

CURRENT UNDERSTANDING OF CANCER BIOLOGY

2

CANCER BIOLOGY AND PATHOLOGY

G. Premveer Reddy, Lekkala V. Reddy, and Sahn-Ho Kim

Vattikuti Urology Institute, Henry Ford Health System, Detroit, MI, USA

INTRODUCTION

Cancer has plagued mankind for many thousands of years. It is prevalent in all parts of the world and in all human races, as well as in all vertebrates. Worldwide there were 14.1 million new cases and 8.2 million cancer-related deaths in 2012 (http://globocan.iarc.fr/Pages/fact_sheets_cancer.aspx), and in the United States there were approximately 1.7 million new cases and 0.6 million cancer-related deaths in 2016 [1]. Despite recent advances in therapy, cancer remains the second leading cause of death, next only to heart disease. Some forms of cancer can be treated successfully, particularly if they are diagnosed early. However, many forms are resistant to even the best current medical techniques. There has been growing effort over the past few decades to learn more about the biology of cancer, with the expectation that this knowledge will provide insights that could lead to the development of new clinical interventions or, perhaps more importantly, finding ways to prevent it. The purpose of this chapter is to inform the reader of our current understanding of cancer biology, with the hope that patients and caregivers alike can reach an informed decision in the treatment and management of the disease.

BASIC CONCEPTS IN CANCER

At the outset, it is important to understand that cancer is not one disease but it includes over 100 different diseases with common features. An underlying feature of all cancers is that they arise from a single cell that undergoes some genetic change and becomes defiant to the normal rules of cell division. Normal cells in a tissue respond continually to external signals that dictate when to divide, differentiate into another cell, or die. Cancer cells, on the other hand, are impervious to such signals and exhibit autonomous growth and proliferation. This results in an outgrowth of tissue, which may appear as a lump, at some site where there was previously no abnormality. This outgrowth, called a tumor, may remain confined to its original location (cancer in-situ), in which case it is considered *benign*. But some tumor cells with additional genetic alterations can invade surrounding tissues including the bloodstream and spread to other parts of the body, where they can settle and grow into secondary tumors that can be fatal. This process of migration of tumor cells from its initial site to other parts of the body is called *metastasis* and such tumors are considered *malignant*. Not all cancers appear as an outgrowth or lump of tissue. The cancers of the blood system are characterized by a great increase in the number of cells, such as white cells and plasma cells, circulating in the blood. Leukemias are such cancers, in which the number of white cells or leukocytes is greatly increased.

Thus, a widely accepted definition of cancer is that it is a new growth, called *neoplasia*. Of course, a new growth can also occur in adult normal tissues. For example, muscle grows with use, the breasts grow as a result of hormonal stimulation, and wound heals when new tissue appears. Such new growth in normal tissues stops when the stimulation is withdrawn, when the hormone level drops, or when the wound heals. By contrast a tumor continues to grow even when there is no known stimulus. Another important difference between normal and cancer cells is in their structural and functional organization. Normal cells in an organ are highly organized and perform specific functions. They may be assembled into channels, ducts, or layers in an organ-specific manner and make specialized products such as hormones or proteins that are exported to other parts of the body to perform specific functions. For example, adrenocorticotropic hormone (ACTH) made in the pituitary gland is released into the bloodstream and exported to adrenal glands, where they bind to cognate receptors and make adrenal glands secret cortisol. The amounts and the rates at which these hormones, i.e., ACTH and cortisol in this example, are made is highly regulated. Excess cortisol production in adrenal glands will lead to Cushing's disease. This can happen when there is tumor growth in either pituitary gland or adrenal gland (Figure 1). Tumor cells in general have abnormal shapes and are scrambled in a random way and the biochemical reactions within the cells are not as well regulated. In some cases, tumor cells may start to make products such as hormones that the normal cells in that organ did not produce. For example, tumor cells in the liver may make ACTH, which is normally made in

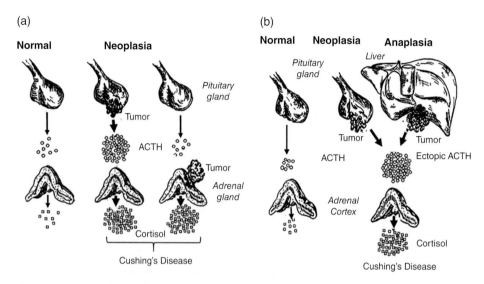

Figure 1. Examples of neoplasia and anaplasia. When cells in a tumor perform a function specific to the organ in an uncontrolled fashion (e.g., excess production of cortisol as a result of a tumor in either the pituitary gland or adrenal gland) is referred to as neoplasia (a). And, when cells in a tumor begin to perform a function that is not typical of the organ (e.g., production of ACTH by tumor cells in the liver) is referred to as anaplasia (b). *Source*: Adapted from Pardee and Reddy [2].

the pituitary gland and not by normal liver cells. This ectopic ACTH made by liver tumor cells can then trigger adrenal glands to secret excess cortisol, resulting in Cushing's disease. This dysregulation of cellular function in cancers is called *anaplasia* (Figure 1).

The cancer can continue to alter its properties and become more neoplastic and more anaplastic as the tumor progresses. Various parts of the same cancer can become different as the tumor grows, and in its later stages a cancer can consist of a variety of cells with different properties. Since an anticancer drug is likely to work against cancer cells with some but not other properties, the presence of a variety of cells with different properties makes the treatment of a late-stage cancer very challenging.

TYPES OF CANCER

There are many types of cancers and they are often named after the tissue or cell type from which they originate. Thus, cancers originating from epithelial cells that cover the surface of our skin and internal organs and derived from the embryonic ectoderm and endoderm are called *carcinomas* (Figure 2); cancers arising from muscle, bone, fat, or connective tissue cells derived from embryonic mesoderm are called *sarcomas*; *leukemia* results from malignant leukocytes (white blood cells);

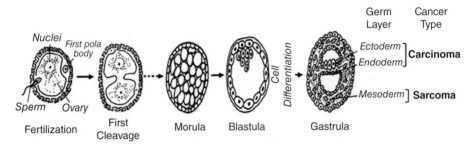

Figure 2. A pictorial representation of the early stages of embryo development and differentiation leading to the formation of three germ layers, viz., ectoderm, endoderm, and mesoderm, at gastrula. Cancers derived from the ectoderm and endoderm are called carcinomas and those derived from the mesoderm are called sarcomas. *Source*: Adapted from Pardee and Reddy [2].

myeloma is also a form of blood cancer that develops in bone marrow and results from modified plasma cells that produce large quantities of abnormal antibodies; and *gliomas* result from altered glial cells (nerve cells). These names are of little value when it comes to cancer treatment. However, since cells with different embryonic origins differentiate in different ways, the growth and biochemistry of these various cancers are expected to be different. Therefore, cancer is not one disease but many diseases that share the common properties of neoplasia, anaplasia, and metastasis.

THE CAUSES OF CANCERS

Cancer is caused largely by agents that are external to body [3]. This is implicit in fact that the incidence of certain cancers in individuals with similar genetic backgrounds differ significantly depending on the country in which they live. For example, in Japan the incidence of prostate cancer is three-times lower than that in the United States. However, when Japanese men migrate to the United States and adapt a western lifestyle, their prostate cancer incidence rates reflect those of Americans [4]. Furthermore, when Japanese-Americans return to Japan their prostate cancer rates decrease to the level in Japan. Similarly, incidence of stomach cancer is six to eight times higher in Japan than in the United States. However, descendants of migrant Japanese settled in the United States exhibit stomach cancer incidence rates like those in the American population [5]. Thus, the diet and lifestyle, but not genetics per se, can be major contributors to cancer susceptibility.

The probability of developing cancer generally increases with age; a 65-year-old person is eight to ten times more likely to get cancer than a 40-year-old [6, 7]. This age-dependent increase suggests that cancer development is a progressive process requiring accumulation of multiple random changes that can be either genetic or epigenetic [6]. Genetic changes are mutations or alterations in DNA nucleotide sequence of genes impacting the structure and function of proteins that they encode. Mutations can be in the form of change in a single DNA base pair (point mutation) of a gene or can involve deletion, insertion, duplication, or translocation of a large segment of chromosome containing multiple genes. Mutations can be either hereditary or acquired. Hereditary (familial) mutations, also called germline mutations, are those that are inherited from parents and are present in virtually every cell in the body. Acquired (or somatic) mutations, on the other hand, are those that occur sometime during person's life and are present in only certain cells.

Epigenetic changes are those that effect gene activity and expression without causing alteration in the DNA nucleotide sequence and therefore are not mutations. Two major epigenetic changes influencing the activity and expression of genes include chemical modification (acetylation, deacetylation, and methylation) of histones, which are special proteins that complex with genomic DNA to form chromatin, and methylation (hypo- and hyper-methylation) of cytosine residues in DNA. Both modifications regulate the structure and function of chromatin, affecting the expression of genes. For example, histone deacetylation or DNA methylation can condense chromatin, making it inaccessible for transcription factors necessary for the expression of genes. In such a scenario, histone hypo-acetylation or DNA hyper-methylation can silence the expression of tumor suppressor genes whose function is to prevent abnormal growth and proliferation of cells, and thereby promote carcinogenesis.

Humans are exposed to a variety of carcinogens daily that can cause genetic or epigenetic changes. These carcinogens can be grouped broadly into three major categories; (i) chemical carcinogens, such as tobacco use, alcohol use, asbestos, and unhealthy diet, (ii) biological carcinogens, such as infections from certain viruses and bacteria, and (iii) physical carcinogens, such as UV and ionizing radiation. Additionally, although less frequently, familial genes can also contribute to cancer development. Besides genetic or epigenetic changes, age-related alterations in immune system and hormonal imbalance may also play a role in carcinogenesis.

CHEMICAL CARCINOGENS

It is estimated that at least 80% of cancers are caused by chemical carcinogens. An earliest report (1775) on chemical carcinogens in humans was the finding that the exposure of chimney sweepers to the coal tar in chimneys was highly correlated with the development of cancer of the scrotum [8]. Since then, many coal tar products such as dibenzanthracene, and benzopyrene have been shown to cause cancer [9]. Hydrocarbons such as benzopyrene are themselves chemically unreactive in our bodies, but are converted by oxidation to highly reactive compounds such as their epoxides. These reactions are catalyzed mainly by enzymes in liver cells whose natural function is to remove toxic chemicals from the blood. The activated carcinogens travel from the liver, and can affect cells in other parts of body.

Smoking is the major cause of many types of cancers [10], including cancer of lung, larynx, mouth, esophagus, throat, bladder, kidney, liver, colon and rectum, and cervix, as well as acute myeloid leukemia. Smokeless tobacco (snuff or chewing tobacco) increases risks of cancer of the mouth, esophagus, and pancreas (NCI) [11]. Tobacco smoke contains over five dozen cancer causing agents, including benzopyrene, acetaldehyde, arsenic, and benzene, that can damage both DNA and protein in cells lining lungs.

Alcohol consumption is also a major cause of many cancers including liver cancer [12]. Many studies have found a link between alcohol use and breast and colorectal cancers. Cancers of mouth, esophagus, larynx, and throat are more common in alcohol users than in nonalcohol users. Alcohol may increase the risk of cancer in multiple ways, including by metabolizing (breaking down) ethanol in alcoholic drinks to acetaldehyde, a potent DNA damaging agent; generating reactive oxygen species that can damage DNA, proteins, and lipids through a process called oxidation; and increasing blood levels of estrogen, a sex hormone linked to the risk of breast cancer. The combination of alcohol use with tobacco smoking can lead to even a higher risk for several of these cancers.

Exposure to asbestos is associated with increased risk of lung cancer and mesothelioma [12], which is a cancer of the thin membrane that lines the chest and abdomen. Asbestos comprises of a set of six naturally occurring fibrous materials, which when lodge deep in the lung can cause irritation leading to fibrosis (scaring) that is inextricably linked to cancer [13, 14]. As a result, asbestos exposure eventually leads to full blown cancer in the lung. This is significant because asbestos fiber particles are released as automobile brake linings and tires wear down. The combined effects on lung tissue of these fibers and smoking greatly increase the incidence of lung cancer.

Dietary factors account for about 30% of all cancers in Western countries [15]. A very high correlation exists between meat consumption and incidence of colorectal, breast, and prostate cancers. Meat contains animal protein, saturated fat, and in some cases carcinogenic compounds such as heterocyclic amines and polycyclic aromatic hydrocarbons (PAH) formed during processing and cooking of meat. In the case of colorectal cancer, bile acids released from the gall bladder chemically modify meat fats in intestine to make them absorbable. Unfortunately, bacteria in intestine turn these bile acids into carcinogen-promoting substances called secondary bile acids. The high fat content of meat and other animal products also increase production of hormones, such as estrogen and testosterone, which increase the risk of hormone-related breast and prostate cancers.

The fact that many chemicals must be changed inside cells to become carcinogenic has several interesting consequences. For example, some drugs and foods alter the enzyme systems that activate carcinogens. The drug phenobarbital stimulates the P450 enzyme system, which activates many carcinogens, such as benzopyrene. Thus, production of cancer can depend on combinations of agents in the diet and in the environment.

BIOLOGICAL CARCINOGENS

It is estimated that viral and bacterial infections contribute to 15–20% of human cancers worldwide [16, 17]. Viruses associated with human cancers are also called tumor- or onco- viruses [18]. There are five DNA tumor viruses, including Epstein-Barr virus (EBV), Hepatitis B virus (HBV), human papillomavirus (HPV), human herpesvirus 8 (HHV8), and Merkel cell polyomavirus (MCV), and two RNA viruses, including hepatitis C virus (HCV) and human T lymphotrophic virus type I (HTLV-I), whose infection is known to cause cancer. These viruses cause cancer by introducing their genetic material into normal host cells. The genetic material is DNA in the case of the DNA viruses, or RNA that is reverse-transcribed into DNA in the case of the RNA viruses. In either case, the portion of DNA involved in causing cancer is localized to a very small part of the virus genome, sufficient to code for one or a few proteins. Generally, virus infection alone is not sufficient for cancer, additional events, such as immunosuppression, somatic mutation, genetic disposition, or exposure to carcinogens, are also needed for infection to manifest as cancer. For example, in the case of acquired immune deficiency syndrome (AIDS), human immunodeficiency virus (HIV) does not directly cause cancer. It severely damages the body's cellular immunity, making the person more susceptible to infection with tumor viruses, such as HHV8, that then cause cancer. Thus, carcinogenesis is a multistep process (see below), only one stage being caused by a virus.

In addition to viruses, bacterial infection can also cause cancer. For example, chronic infection of the wall of stomach with *Helicobacter pylori* is linked to gastric cancer [19] and the infection of *Chlamydia pneumonia* is association with lung cancer [20]. Chronic inflammation resulting from these bacterial infections is implicated in carcinogenesis. Similarly, inflammation caused by parasites in some parts of the world is also linked to cancer. Parasitic flatworm *Schistosomiasis haematobium* infections in developing countries, and *Schistosomiasis japonicum* in the Far East are two examples linked to bladder and colorectal cancers, respectively [21, 22]. There is a functional relationship between inflammation caused by biological infections and cancer [23, 24].

PHYSICAL CARCINOGENS

Melanoma is a common ultraviolet-induced skin cancer [25, 26]. Ultraviolet (UV) radiation emitted by the sun is made up of two types of rays, called UVA and UVB. Although the DNA of skin cells absorbs UVB more readily than UVA, both types of UV rays can cause cancer. UVB absorbed by DNA is known to bond together adjacent thymine base pairs in genetic sequence to form pyrimidine dimers. Normally, these pyrimidine dimers are removed through DNA repair process. Occasionally, when the repair process is defective, these cells survive but are not the same as before irradiation and grow into cancers. UV radiation emitted by sun, sunlamps, and tanning booths can also be carcinogenic.

X-rays and γ-rays emanating from natural sources, such as cosmic rays that hit the earth from outer space and radioactive elements that are present in the earth, or man-made sources, such as nuclear power plants, are potent carcinogens. The γ-rays produced by atomic bombs are also carcinogenic to survivors. Both X-rays and γ-rays are forms of high-frequency ionizing radiation. Ionizing radiation inside a living cell can strike DNA molecule and damage it directly. Alternatively, ionizing radiation can produce free radicals that react with DNA and damage it. If the radiation dose is not too high, human cells can repair the damage and survive. We are exposed constantly to a low background of ionizing radiation emitting from natural sources. Against this uncontrollable background, the dosage of X-rays from occasionally used instruments, such as diagnostic devices in the clinics and full-body scanners at the airports, need to be kept as low as possible [27].

GENETIC PREDISPOSITION

Some genetic conditions are also known to favor the development of cancer, though not to initiate it [28]. For example, the genetic trait familial polyposis is expressed as small growths or polyps in the intestinal tract. Although these are not themselves cancers, they are readily converted into cancers. Genetically defective individuals who repair their damaged DNA poorly have a high incidence of cancer. Hereditary genetic defects in DNA repair system, such as those in Bloom syndrome [29], Werner syndrome [30], and Fanconi anemia [31], are associated with an increased risk of cancer. Similarly, a subset of familial cancers of breast, ovary, prostate, and pancreas can also result from mutations in DNA repair genes, such as BRCA1, BRCA2, Rb, and p53 [32–34]. These genes are also considered as tumor suppressor genes. Ability to repair damaged DNA is, therefore, important for avoiding cancer [35]. However, many cancers are not hereditary; overall, inherited mutations are thought to play a role in 5–10% of all cancers. Inherited genetic mutations that confer increased susceptibility to cancer are currently being investigated for their use in cancer surveillance and prevention.

CANCER DEVELOPS IN MULTIPLE STEPS OF GENETIC AND EPIGENETIC CHANGES

Cancer development in many tissues or organs is a multi-step process involving multiple genetic or epigenetic changes in several genes [36, 37]. Cancer is clonal in the sense that it originates from a single cell. Initial mutations in a cell confer the ability to proliferate spontaneously without relying on external stimuli (growth factors [GFs], cytokines, hormones, etc.) that normal cells depend on for proliferation. This autonomous proliferation predisposes the cells in subsequent generations to acquire additional mutations that progressively alter their behavior.

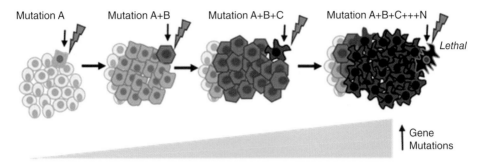

Mutation A   Mutation A+B   Mutation A+B+C   Mutation A+B+C+++N

Lethal

Gene
Mutations

Time to cancer progression (in months and years)

Figure 3. Clonal origin of cancer. Cancer arises from a single abnormal cell that undergoes successive rounds of gene alterations (mutations) and natural selection. With every round of new mutation and selection a tumor cell progresses closer to becoming malignant and lethal. This entire process can take anywhere from just few months to several years.

Selective expansion of surviving cells after successive rounds of mutations will lead to tumor growth and progression (Figure 3). Cancer progression is a stepwise process in which *initiation* events contribute to early stages of neoplastic transition and then *promotion* events constitute frank neoplasia. It can take months and years for a neoplastic transition to become frank neoplasia. Agents that cause initiation are generally mutagens (initiators), such as benzopyrene in tobacco smoke, which cause DNA damage. A cell with damaged DNA must multiply in order to become cancerous, and this is facilitated by a promoter, which often is an irritant, such as asbestos and rare plant products called phorbol esters. Promotors are also known to cause DNA-damaged cells to rearrange their chromosomes as they multiply, thereby causing additional gene alterations. Cancer can develop even after a long time elapse between exposures to an initiator and a promoter.

That a single genetic defect by itself is not sufficient to cause cancer is illustrated in individuals with retinoblastoma (Rb) gene mutation. As mentioned above, familial Rb mutation can *predispose* a person to cancer, but additional genetic changes are needed for a cell with Rb mutation to become cancerous. These additional genetic or epigenetic changes, resulting from constant exposure to environmental carcinogens such as chemicals, viruses, hormones, disrupt the regulatory mechanisms in place to control the ability of normal cell to proliferate, differentiate, and die.

Cells are programmed through a complex network of regulatory circuits to control their proliferation and maintain homeostasis in a tissue or organ of their origin. These regulatory circuits comprise of proteins and enzymes whose expression and/or activity is coupled to mitogenic signals generated by extracellular molecules, such as GFs and cytokines, in the surrounding microenvironment of cells in the tissue. Random mutations in genes that encode these proteins can disrupt their expression or activation and enable the cells to acquire new capabilities that make them cancerous. Although there are many different types of cancers, they all acquire the same six unique and complementary characteristics as they progress through a series of premalignant stages into invasive metastatic cancers [38, 39]. These acquired characteristics or hallmarks include: (i) proliferative autonomy, (ii) resistance to anti-proliferative signals, (iii) evasion of programmed cell death, (iv) unlimited replicative potential, (v) neovascularization of tumor, and (vi) invasion and metastasis.

Proliferative Autonomy

Cell division is an orderly process in which cells progress through four distinct periods or phases (Figure 4). First, cells go through a biochemically active (RNA and protein synthesis) period called G_1 (gap 1) phase when cells grow in preparation for division. This period lasts until the onset of DNA synthesis when cell enters S phase to duplicate genetic material [40]. This is followed by a second "gap" period, G_2 phase, when newly replicated chromatin condenses to form chromosomes. This is followed by mitosis, M phase, when duplicated chromosomes separate and after this comes cell division. However, under suboptimal conditions, such as a lack of proper nutrients or GFs, cells exit the cell cycle and enter a quiescent or resting state called G_0. These cells can reenter the cell cycle when conditions become favorable. The decision to continue cycling, i.e., to enter S phase of DNA synthesis or to go into quiescence (G_0 phase), is made in G_1 phase. Alternatively, post-mitotic cells can also exit the cell cycle and differentiate to perform a variety of their normal functions, such as the conduction of impulses by nerve cells and the contraction of muscle cells.

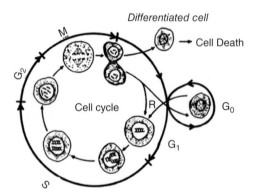

Figure 4. Cell cycle and differentiation. A diagrammatic presentation of the different phases (G_1, S, G_2, and M) of the cell cycle through which cells traverse during the course of each division in order to generate two daughter cells. Normal cells are capable of exiting the cell cycle during the G_1 phase to either differentiate in order to perform functions specific to the organ in which they reside and ultimately die, or enter into a resting or quiescent phase (G_0). Cancer cells, on the other hand, remain in the cycle without differentiating or entering G_0 even under nonphysiological and suboptimal conditions.

Several proteins, including cyclins and cyclin-dependent kinases (CDKs), tightly regulate the events in cell cycle to ensure that cells divide only when necessary and only under permissive circumstances [41]. Normal cells depend on mitogenic signals generated by external GFs, such as platelet-derived growth factor (PDGF), epidermal growth factor (EGF), fibroblast growth factor (FGF) and insulin-like growth factor (IGF), to move from a quiescent (G_0) state into an active proliferative cycle. GFs bind to cognate transmembrane receptors and transmit mitogenic signals necessary for G_0/G_1 phase cells to grow and become committed to DNA synthesis, which is fundamental to cell division. Normal cells cease to divide in the absence of GFs. By contrast, cancer cells continue to divide whether or not these exogenous growth-stimulating factors are present. This proliferative autonomy from external GFs can result from genetic or epigenetic changes that either alter the structure, or cause gross overexpression, of GF receptors. For example, in some breast cancers truncated EGF receptor (EGFR) continues to send growth-stimulating signals even without any EGF binding, and in some other breast cancers the HER2/neu receptor overexpression renders cells hyper-responsive to very low levels of GF that normally are not sufficient to promote cell division [42, 43].

Beside GF receptors, cells contain other transmembrane receptors, called integrins, capable of promoting active cell cycle [44]. Integrin receptors are heterodimer proteins that physically attach the cell cytoskeleton to extracellular matrix (ECM). About 30% of tissue in an organ is made up of ECM, which is a noncellular material composed of a meshwork of proteins, glycoproteins, proteoglycans, and polysaccharides that provide structural and biochemical support to the cells, and play a major role in chemical signaling between cells. There are at least 24 different heterodimer integrin receptors (formed by the combination of 18 α-subunits and 8 β-subunits) with binding specificity to distinct protein moieties on ECM [45]. ECM binding activates integrin receptors to generate signals inside the cells that can influence their behavior in a variety of ways, including proliferation, survival, migration, and invasion. Integrin heterodimers favoring growth and proliferation are highly upregulated in cancer as compared to normal cells. For example, αvβ3, α5β1, and αvβ6, which are usually are at very low levels in adult epithelia, are highly overexpressed in some cancers. Although integrins lack the ability to transform cells, they cooperate with GF receptors and cytokine receptors to enhance tumorigenesis. For example, α6β4 integrins cooperate with ERBB2, a member of EGFR family, to initiate breast cancer in spontaneous mouse models of tumorigenesis [45].

Proliferative autonomy can also come from alterations in components of downstream effector proteins that receive and process the signals from ligand-activated GF receptors and integrin receptors. There are two major pathways, viz., Ras/Raf/MAPK pathway [46] and PI(3)K/Akt/mTOR pathway [47, 48], which link ligand-activated transmembrane signals to the nucleus (Figure 5). In Ras/Raf/MAPK pathway, GFs, such as PDGF, EGF, and FGF, activate (phosphorylate) tyrosine kinase on the cytoplasmic domain of cognate receptors. Phosphorylated tyrosine

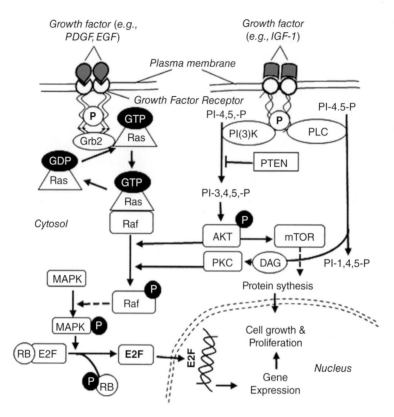

Figure 5. An overview of growth factor stimulated Ras/Raf/MAPK and PI(3)K/Akt,/mTOR signaling pathways involved in the regulation of cell growth and proliferation, i.e., transition of cell from G_1 to S phase. "P" represents phosphate, which communicates the signal. The arrow indicates activation and "T" indicates inhibition.

kinase becomes the binding site for docking proteins such as GRB2, which then recruit and activate downstream effector proteins, including Ras, Raf, and MAPK that transmit mitogenic signals emanating from GF-activated receptors to the nucleus, where proteins and enzymes necessary for cell survival and proliferation are made. Ras activation occurs at multiple points during the progression of cells from G_0/G_1 to S phase. Mutations in Ras and downstream effector genes such as Raf can result in activation of mitogenic signals capable of promoting proliferation of cells independent of ligand-activated GF receptors.

Typically, a single point mutation is sufficient to transform the normal Ras into tumorigenic Ras. There are three major RAS proteins (H-RAS, N-RAS, and K-RAS) in humans with differing binding affinities to their downstream effector RAF proteins. One or the other of these RAS proteins is mutated in over 20% of human tumors; K-Ras mutations are seen in over 50–60% of tumors of colon adenocarcinoma, pancreas ductal adenocarcinoma, and soft tissue angiosarcoma. Similarly, B-Raf, one of the three Raf-isoforms (A-Raf, B-Raf, and C-Raf), is mutated at a very high frequency in several human cancers, particularly melanoma (30–60%), thyroid cancer (30–50%), colorectal cancer (5–20%), and ovarian cancer (~30%). Thus, mutations that activate either Ras or Raf can liberate a cell from its dependence on the signals from GF receptors and integrins for proliferation, and promote tumorigenesis.

In addition to Ras/Raf/MAPK pathways, activation of phasphotidylinositol-3-OH kinase (PI3K) and diacylglycerol (DAG)-dependent forms of protein kinase C (PKC) is essential for the transition of cells from G_1 to S phase, a critical step in the cell cycle when cells commit to DNA replication, a hallmark of proliferation [49]. This is facilitated by GFs, such as IGF-1, EGF, HER2, and VEGF, whose binding to tyrosine kinase linked receptors stimulate membrane bound phospholipase C (PLC), which converts phosphatidylinositol 4,5-bisphosphate (PIP_2) to phosphatidylinositol 1,4,5-trisphosphate (PIP_3) and DAG, and activates PI3K and PKC. PI3K in turn phosphorylates and activates Akt, also known as protein kinase B (PKB), which interacts directly with Ras/Raf/MAPK pathway to regulate the transition of cells from G_1 to S Phase (Figure 5).

Hyperactivation of Akt can promote cell proliferation and survival independent of external stimuli. Akt hyperactivation can result from somatic mutations in PI3K gene, most notably PIK3CA (the gene coding for PI3K-α) that increase PI3K activity, mutation, and/or amplification of the genes (AKT1, AKT2, and AKT3) coding for Akt, or from biallelic loss of a negative regulator of PI3K called phosphatase and tensin homolog (PTEN) [50]. Each of these genetic alterations affecting the PI3K/Akt signaling pathway are prevalent in a diverse variety of cancers, including glioblastoma, endometrial cancer, and prostate cancer [51]. Reduced expression of PTEN is also found in many other tumor types, such as lung and breast, through promotor methylation [52].

An intracellular protein kinase called mTOR (mammalian target of rapamycin) also regulates PI3K/Akt signaling. mTOR acts both as a downstream effector and an upstream negative-feedback inhibitor of PI3K. mTOR plays an important role in the regulation of cell growth, survival, and proliferation by sensing and integrating intracellular and extracellular cues, including GFs, stress, energy level, oxygen, and amino acids, to control ribosomal biogenesis and protein synthesis. One of the mTOR regulated proteins is cyclin D1, which plays a central role in expression of genes necessary for cells to transit from G_1 to S phase. Deregulation of mTOR or loss of mTOR negative-feedback inhibition of PI3K can cause excessive activation of PI3K/Akt/mTOR signaling leading to proliferative independence. Aberrant activation of the mTOR pathway has been implicated in a variety of malignancies including breast cancer, renal cell carcinoma, and lymphoma [52].

Resistance to Anti-proliferative Signals

To avoid an unabated growth, proliferative signals in normal tissues are counterbalanced by multiple anti-proliferative signals that induce cellular quiescence. A fine balance between proliferation and quiescence is essential for tissue homeostasis. Anti-proliferative signals are induced by growth inhibitory molecules that are present on the surface of contacting neighboring cells or embedded within the ECM [53], or by soluble growth inhibitory factors, such as transforming growth factor-β (TGF-β) [54], present in the local tissue microenvironment.

The growth and proliferation of normal cells in a tissue are regulated through an interaction between growth factor signaling, availability of nutrients, and cell density. Normal cells need a minimum amount of space to spread in order to grow. Widely spaced cells grow and divide and eventually cover the entire space available to them. As cell density increases, they come into contact with one another, their movement gets restricted, and they stop growing despite the continuing abundance of nutrients and GFs. This process, known as density dependent or contact inhibition of cell growth and proliferation, is essential to the regulation of proper tissue growth, differentiation, and development. Contact inhibition controls the size that each organ acquires during the development. Its role in control of organ size is also evident during regeneration of an injured tissue. For example, after partial hepatectomy, the hepatocytes in the liver are mobilized to divide rapidly to regenerate the liver. Once the liver reaches its original size, hepatocytes stop dividing to ensure that the regenerating liver does not over grow. Most human cancer cells are refractory to this process of contact inhibition imposed by neighboring cells, and continue to grow and proliferate to become tumorigenic. Significance of contact inhibition in cancer is also evident in a strain of rat, called naked mole (*Heterocephalus glaber*), which shows hypersensitivity to contact inhibition and is highly resistant to cancer [55]. Recently, it has been found that these rats secret a super sugar called high-molecular-mass Hyaluronan (HMM-HA) that prevents cells from overcrowding and forming tumors. When HMM-HA is removed from cells of naked mole rats, they lose sensitivity to contact inhibition and become susceptible to tumors [56].

Besides the physical constraints imposed by cell-cell adhesion, contact inhibition also involves signaling pathways that suppress proliferation. Hippo-Yap is one such signaling pathway that is evolutionarily conserved from fruit flies to mammals to control organ size. Cell-cell contact stimulates transmembrane receptors, such as E-cadherin, CD44, and β1-integrin, which trigger the activation of Hippo pathway comprising of a cascade of protein phosphorylation events involving serine/threonine kinases [57]. In the basal state, YAP (Yes-associated protein) resides primarily within the nucleus of sparsely growing cells and acts as a transcription coactivator to stimulate the expression of growth-promoting and anti-apoptotic genes including Ki67, c-Myc, survivin, and cIAP. When cells come in contact with one another, Hippo kinase phosphorylates (inactivates) YAP, translocating it from the nucleus to cytoplasm. This results in the downregulation of the expression of YAP-target genes and, thereby, the manifestation of contact inhibition. Therefore, either overexpression of YAP or dysregulation of Hippo signaling can abrogate cell contact inhibition and promote tumorigenesis. Currently, inhibitors targeting Yap and upstream regulators of Hippo signaling are being evaluated for the treatment of a variety of cancers, including melanoma, breast cancer, and gastric cancer [58].

Antigrowth signals are also generated by soluble signaling molecules such as TGF-β. TGF-β is a potent inhibitor of epithelial cell proliferation and acts as a tumor suppressor [54, 59]. Antigrowth signals block the proliferation in two distinct ways; one is by forcing the cells to enter transiently into a quiescent (G_0) (metabolically inactive) state from which they can return to proliferative cycle when extracellular conditions become conducive, and the second is by inducing terminal differentiation when the cells permanently exit proliferative cycle and enter a post-mitotic (non-dividing) state as a result of cell development. Most cells *in vivo* are in the post-mitotic state, carrying out their normal functions, such as conduction of impulses by nerve cells and the contraction of muscle cells. TGFβ induces these antigrowth signals by binding to two classes of receptors, the TGFβ type I receptor (TβRI) and the TGFβ type II receptor (TβRII), which are serine/threonine kinase receptors that form heterodimeric complexes upon TGFβ binding (Figure 6). TGFβ bound receptors phosphorylate and activate Smad proteins that induce the expression of two CDK inhibitors, viz., p21^{Cip1} and p15^{Ink4B}, involved in cell cycle arrest, and downregulate the expression of three transcription factors, viz., Myc, Id1, and Id2, involved in proliferation and inhibition of differentiation (Figure 6). Somatic mutations or dysregulation of these components in TGFβ-Smad pathway can result in resistance against

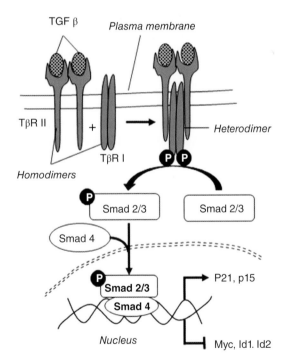

Figure 6. The TGFβ–Smad pathway transmitting antigrowth signals. The binding of TGFβ to its receptor results in activation (phosphorylation) of Smad 2/3, which then complexes with Smad 4 and other proteins (not shown) to enter the nucleus and bind to specific promoter sequences in DNA. DNA bound Smad complexes can induce (e.g., p21, p15) or repress (e.g., Myc, Id1, Id2) the expression of genes to suppress cell proliferation.

antigrowth signals and promote tumorigenesis. For example, mutations yielding truncated or inactive TβRII receptor are seen in colon and gastric cancers, and Smad4 is deleted or mutated in over 50% of the pancreatic carcinomas.

Most of the anti-proliferative signals, including those stemming from TGFβ and cell adhesion molecules, impinge on retinoblastoma protein (pRb) and its family members, p130 and p107, which are tumor suppressor proteins that control the activity of a family of heterodimeric transcription factors called E2Fs, to manifest their anti-proliferative effect [34]. E2F transcription factor, a potent inducer of proliferation, is required for the expression of genes necessary for the progression of cells from G_1 to S phase. Hypophosphorylated pRb represses the expression of these genes by binding to and inactivating E2F (Figure 7). Phosphorylation of pRb liberates E2F to resume its transcriptional activity. Complexes of CDKs and cyclins phosphorylate pRb during G_1 phase to liberate E2F to induce the expression of genes required for the cells to enter S phase. Cdk inhibitors, such as those induced by TGFβ signaling (i.e., p21^{Cip1} and p15^{Ink4B}), can block pRb phosphorylation and, thereby, prevent E2F from becoming free to express the genes required for the entry of cells into the S phase. Thus, pRb plays a central role in controlling E2F transactivation, and the loss of pRb function can result in cells that are resistant to normal anti-proliferative signals. pRb loss-of-function can occur in a variety of ways, including the deletion of genes that encode Cdk/cyclin inhibitors, such as p15^{Ink4B} and p16^{INK4a}; overexpression of cyclins that activate Cdks; mutations in Cdks that confer resistance to inhibitory actions of Cdk-inhibitors; and, most importantly, mutations or deletion of RB gene itself that can liberate E2F to induce proliferation. In case of cervical cancer, pRb loss-of-function also results from its sequestration by E7 oncoprotein of human papilloma virus [60]. Interestingly, the frequency of each of these loss-of-function mutations seems to vary between different types of cancers. For example, mutation or deletion of RB is seen in 80% of small cell lung cancers; loss of Cdk inhibitors in 80% pancreatic cancers, 60%

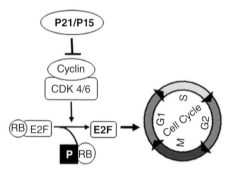

Figure 7. Role of RB in the regulation of E2F transcription factor activity during cell cycle. RB sequesters E2F and blocks E2F-mediated expression of genes necessary for the progression of cells from G_1 to S phase. Whereas cyclin-CDKs phosphorylate RB and restore E2F activity to promote cell proliferation, p21 and p15 inhibit cyclin-CDKs and block E2F activation. The arrow indicates activation and "T" indicates inhibition. (See insert for color representation.)

glioblastomas, 75% T cell acute lymphoblastic leukemias, and 58% non-small cell lung cancers; cyclin D1 overexpression in 90% Mantle cell lymphomas, and >50% breast cancers; and Cdk4 mutation or overexpression in 40% glioblastoma multiforme. Thus, antigrowth signals mediated by pRb are central to the controlled growth and proliferation of normal cells and its loss-of-function, by one mechanism or other, is an important attribute of many cancers.

Anti-proliferative signals can also lead to irreversible terminal differentiation (a permanent nondividing state) of cells. This involves the Myc-Max-Mad transcription factor network [61, 62]. Myc and Mad are two transcription factors with related basic helix-loop-helix leucine zipper (bHLHLZ) domains that dimerize individually with another bHLHLZ containing factor Max and function as cellular antagonists; whereas Myc/Max dimers induce the expression of genes (e.g., p53, ornithine decarboxylase, and α-prothymosin) that promote proliferation, Mad/Max dimers repress a subset of Myc/Max target genes to enforce cell cycle arrest associated with terminal differentiation (Figure 8). Thus, Max is a functional partner of both Myc and Mad, which compete with one another to dimerize with Max for opposing effects. Max is expressed constitutively at high levels in growing, resting, or differentiating cells. In normal cells, Myc expression is tightly controlled by mitogen availability and anti-proliferative signals induce Max dimerization with Mad to promote differentiation. However, Myc overexpression, as it occurs in many cancers, favors Max dimerization with Myc, thereby suppressing differentiation and promoting growth and proliferation. Thus, Myc overexpression or Myc activation mutations result in resistance to anti-proliferative signals [63]. Myc is

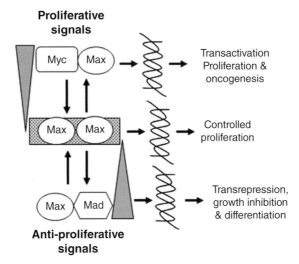

Figure 8. Myc-Max-Mad transcription factor network controlling cell proliferation and differentiation. While Myc and Mad expression changes in response to proliferative and anti-proliferative signals, respectively, the Max expression remains stable in response to these signals. Cones represent relative expression of Myc and Mad, and the vertical bar represents the stable expression of Max.

activated as a consequence of both oncogenic and epigenetic events and is overexpressed and/or activated in more than half of human cancers, including Burkitt lymphomas, and carcinomas of the cervix, lung, breast, stomach, and colon. For these reasons, Myc is considered to be a promising target for the treatment of cancer [64].

Evasion of Programmed Cell Death (Apoptosis)

Programmed cell death, which is referred to as apoptosis, is a genetically regulated form of cell death [65, 66]. It is a naturally occurring process by which dysfunctional or dispensable cells in a tissue commit to self-destruction (cellular suicide). This process of cellular suicide occurs under seemingly normal physiological conditions. In an average adult, more than 10 billion cells go through this process on a daily basis to keep balance with the number of new cells arising from stem cell population in the body. This flux of cells through the body is so immense that the mass of cells equivalent to almost entire body weight can turnover in a year time. Apoptosis is essential for tissue homeostasis, embryogenesis, and induction and maintenance of immune tolerance. This process also plays an important role in preventing unwanted cell proliferation, which otherwise can lead to tumorigenesis. Dysregulation of apoptosis impairs tissue homeostasis leading to either: (i) tissue outgrowth, where cells die much slower than they can divide, as in the case of cancers [65]; or (ii) tissue atrophy, where in cells die faster than they can divide, as in the case of neurodegenerative diseases [67].

Mechanistically, intra cellular cysteine proteases, known as "caspases," play a central role in apoptosis [68, 69]. There are about a dozen or so caspases expressed in human cells as inactive precursors (procaspases) that typically require processing (undergo cleavage at specific sites) in order to be active. Caspases work together in a cascade to cause selective proteolytic breakdown of key components required for normal functioning of cells. These components include cytoskeletal proteins that maintain structural integrity of cells and nuclear proteins essential for the repair of damaged DNA. Caspases can also activate DNases that fragment chromatin. In the process, cells experience membrane blebbing, shrinkage, chromatin condensation, and DNA fragmentation, and ultimately cells are fragmented into small particulate structures called apoptotic bodies containing fragmented DNA and cytosol. These apoptotic bodies are then engulfed by the neighboring cells or macrophages to complete the process of cell death without leaving any trace of dead cells.

Caspases in proteolytic cascades can be divided into two groups; upstream "initiators" (e.g., caspase-8, -9, and -10) and downstream "effectors" or "executioners" (e.g., caspase-3, -6, and -7). While the effectors are the ones that actually execute proteolytic breakdown of target proteins, the initiators sense and process the extracellular signals or intracellular conditions and trigger apoptosis when cell death is warranted (Figure 9). Extracellular signals controlling apoptosis are mediated by cell surface receptors that bind either death factors (e.g., FAS-ligand, TRAIL, and

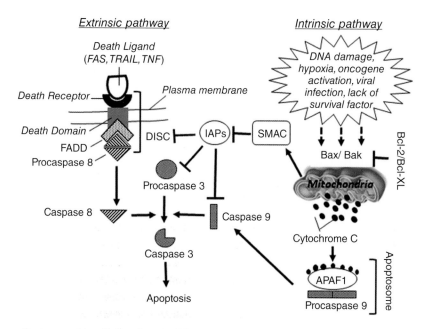

Figure 9. Apoptotic pathways showing proteolytic cascade of caspases. Initiator caspases 8 and 9 activated by extrinsic and intrinsic signals, respectively, converge on caspase 3 to execute apoptosis. The arrow indicates activation and "T" indicates inhibition. (See insert for color representation.)

TNF-α) or survival factors (e.g., IGF-1, IGF-2, IL-1, and IL-2) [70]. Binding of death factors to cognate receptors on cell surface leads to the recruitment of several intracellular proteins, including certain procaspases, to the cytosolic domain of the receptors to form "death-inducing signaling complex" (DISC) that triggers the activation of procaspases. Typically, procaspase-8, and in some cases procaspase-10, are recruited to the DISC for activation. Survival factors on the other hand keep apoptosis in check by preventing the activation of procaspases. These processes of regulation of caspases by extracellular death or survival factors constitute so called "extrinsic" pathway for apoptosis.

Abnormal intracellular conditions, such as DNA damage, hypoxia, growth factor insufficiency, increase in calcium levels, viral infection, and oncogene activation, signal the activation of "intrinsic" pathway for apoptosis [71]. Apoptotic signals of intrinsic pathway are mediated by cytosolic proteins such as members of Bcl-2 (B-cell lymphoma 2) family that act on mitochondria to govern the release of cytochrome-c (cyt-c) from the mitochondrial outer membrane into the cytosol (Figure 9). The Bcl-2 family of proteins can be divided into two groups with either anti-apoptotic (e.g., Bcl-2, Bcl-XL, Bcl-W) or pro-apoptotic (e.g., Bax, Bak, Bid, Bim) function. Anti-apoptotic Bcl-2/Bcl-XL promote cell survival by preserving the integrity of mitochondria, endoplasmic reticulum (ER), and nuclear envelop. Pro-apoptotic Bax/Bak embedded in mitochondria outer membrane, on the other hand, disrupt mitochondrial membrane to make it leaky and release certain factors such as cyt-c and SMACs (second mitochondria-derived activator of caspases) into cytosol to activate initiator procaspase, procaspase-9. Cyt-c orchestrates the assembly of a multiprotein caspase-activating complex, known as apoptosome, in cytosol. Apaf1 (apoptotic protease-activating factor 1), a caspase-activating protein in the complex, then recruits procaspase-9 to apoptosome for activation.

Activated initiator caspases in both the extrinsic (caspase-8 or -10) and intrinsic (caspase-9) pathways converge on to the common downstream effector caspases (e.g., caspase-3 or caspase-7) to execute apoptosis. However, these effector caspases, which are at the core of the apoptotic machinery, can be blocked from executing apoptosis by IAPs (inhibitors of apoptosis proteins). SMACs released from mitochondria by pro-apoptotic Bax/Bak can bind to IAPs and prevent IAPs from inhibiting effector caspases. In addition, anti-apoptotic Bcl-2/Bcl-XL can heterodimerize with pro-apoptotic Bax/Bak and prevent their damaging effect on mitochondria and the release of cyt-c and SMACs to induce apoptosis. Thus, the apoptotic machinery is tightly controlled by anti-apoptotic (survival) and pro-apoptotic (death) signaling pathways to keep the apoptotic machinery in a latent state in almost all normal cells in human body. Any dysregulation of these signaling pathways can adversely affect apoptotic process. For example, an increase in the expression of anti-apoptotic proteins, Bcl-2/Bcl-XL, or a decrease in pro-apoptotic proteins, Bax/Bak, would inactivate the apoptotic machinery and promote cell survival, a key contributing factor in cancer development and progression [72]. A tumor specific nuclear antigen of 53 kilo-Daltons, called p53, is a tumor suppressor protein, whose main function is to sense DNA damage and activate the DNA repair pathway in response to DNA damage. However, when the DNA damage is irreparable, it induces the expression of genes that lead to apoptosis.

p53 induces the expression of pro-apoptotic proteins, such as Bax, in response to a variety of stressful conditions, including DNA damage, hypoxia, and oncogene activation, to promote cell death [73–75]. Therefore, loss of p53 function can result in cellular resistance to apoptosis, allowing unabated growth and proliferation of cancer cells even under extreme conditions. Indeed, gene mutations effecting p53 function are seen in almost every type of human cancer at rates ranging from 10% (e.g., in hematopoietic malignancies) to almost 100% (e.g., in high-grade serous carcinoma of ovary and squamous cell carcinoma) [76].

Apoptosis can also be mitigated through PI(3)K/Akt pathway [77], which is activated by the binding of extracellular GFs, such as IGF-1, IGF-2, and IL-3, to cognate receptors. pTEN, a tumor suppressor, normally suppresses PI(3)K/Akt signaling and, thereby, promotes cell cycle arrest and apoptosis. Therefore, loss of pTEN can result in evasion of apoptosis. pTEN inactivation is a major event that occurs in many cancers including, brain, lung, breast, and prostate cancer. In fact, next to p53, pTEN has the highest frequency of mutations in cancers. Another mechanism of cellular evasion of apoptosis involves expression of a soluble decoy receptor, termed decoy receptor 3 (DcR3), which binds to the death-inducing Fas-ligand and prevents it from binding to its receptor Fas and induce apoptosis [78]. Genomic amplification and/or overexpression of DcR3 is evident in various malignant tumors arising from esophagus, stomach, glioma, lung, colon, rectum, and breast. Thus cellular events that impede apoptosis are major contributing factors in the development and progression of cancer.

Acquisition of Unlimited Replicative Potential

While the acquisition of the three capabilities described above can obviate basic controls that limit the ability of cells to grow, proliferate, and survive, cancer cells still must acquire replicative immortality, in order to proliferate indefinitely, a characteristic of all cancers. In general, normal cells go through a finite number of cell division cycles and then enter an irreversible non-proliferative, but viable, G_0-like state called replicative or cellular senescence (irreversible cell cycle arrest). In order to be tumorigenic, cells must overcome replicative senescence. The major determinant of replicative senescence (replicative potential) are DNA-protein structures at the ends of each chromosome called telomeres [79–81]. Telomeres are critical for protecting the ends of linear chromosomes from fusing with one another and, thereby, for maintaining the genome integrity, stability, and function [80]. Telomeres contain several kilobase pairs of double stranded DNA made of hexanucleotide (TTAGGG) sequence repeats. Owing to a mechanistic limitation in the ability of enzymes of DNA synthesis (DNA polymerases) to fully replicate the ends of linear chromosome DNA duplexes, a condition referred to as the "end-replication problem," the DNA polymerase responsible for duplicating chromosomal DNA leaves 50–200 base pairs in 3'-ends of telomeric DNA strands unreplicated. This results in telomere shortening at the conclusion of each round of DNA replication. Some cells, such as germ cells, early embryonic cells, and some adult stem cells that need to undergo many divisions, express an RNA-dependent DNA polymerase, called telomerase, which adds TTAGGG repeats onto the ends of telomeric DNA strands to avoid telomere shortening [82]. However, most somatic cells do not express telomerase. In these cells, telomeres are shortened progressively with each cell division, and ultimately reach a critically short length when telomeres become dysfunctional. Telomere-dysfunction normally triggers the process of replicative senescence. Cells that fail to enter senescence invariably succumb to p53-mediated apoptotic death. Thus, senescence and apoptosis serve as failsafe mechanisms to prevent proliferation of cells at risk for genome instability, which can lead to neoplastic transformation. Nevertheless, occasionally few (1 in 10 million) cells that survive endure genome instability and acquire certain genome alterations (gene mutations). These rare surviving cells can attain limitless proliferative potential (termed immortalization) through additional genetic or epigenetic changes that cause the expression of the catalytic subunit of telomerase, called telomerase reverse transcriptase (TERT). TERT activates telomerase and, thereby, stabilizes telomeres to sustain continued proliferation. About 90% of cancer cells rely on telomerase for telomere stability. The rest adopt a telomerase-independent recombination-based approach called "alternative lengthening of telomeres" (ALT) to maintain telomeres [83]. Thus, telomere-dysfunction and genome instability in the absence of normal senescence or apoptotic response and subsequent stabilization of telomeres enable cells to acquire limitless proliferative potential, which is fundamental for tumor growth and progression.

Besides telomere shortening due to repeated cell divisions in the absence of telomerase, telomere-dysfunction can also result from changes in telomere-associated proteins, such as TRF1, TRF2, and TIN2 [80, 81]. Moreover, several other hazardous conditions, such as non-telomeric DNA damage caused by radiation or carcinogens, oxidative stress, and activation of certain mitogenic oncogenes, such as mutant RAS, can also elicit senescence response despite telomere function. Cellular senescence requires activation of cell cycle and DNA damage checkpoints controlled by p53 and/or pRb tumor suppressors and expression of their regulators, such as p16^{INK4a}, p21, and ARF [81]. p53 checkpoint-mediated senescence involves several anti-proliferative activities including induction of p21, a cyclin-dependent kinase inhibitor (CDKI). CDKs are critical for the progression of cells through G$_1$, S, and G$_2$/M phases of cell cycle. CDKs phosphorylate and inactivate pRb, which in its activated (hypophosphorylated) state binds to E2F transcription factor and prevents it from transactivation of genes necessary for cell cycle progression. CDKIs, therefore, can keep pRb in an activated state by inhibiting CDKs and block E2F-mediated expression of genes and cause cell cycle arrest. In the absence of pRb checkpoint, an intact p53 checkpoint would activate an apoptotic pathway to kill cells with telomere-dysfunction and genome instability. In the event that neither checkpoint is intact, cells with telomere-dysfunction survive and acquire gross genome rearrangements and tumor-promoting mutations through multiple cell divisions. Subsequent activation of telomerase in these cells then confers unlimited replicative potential. Thus, mutations affecting p53 and pRb enable cancer cells to overcome senescence and apoptotic responses to telomere dysfunction and generate clinically obvious tumors.

Therefore, it is not surprising that p53 and/or pRb mutations, as described above, are seen in almost every type of cancer [73]. However, the timing of the detection of these mutations during tumorigenesis seems to vary from one cancer to another. For example, p53 is generally mutated late during tumorigenesis in most organs, including that in colon, pancreas, prostate, and bladder. By contrast, p53 mutation is found in ductal cell carcinoma in-situ (DISC), a premalignant breast lesion. Similarly, p53 is lost or mutated early in astrocytoma and esophageal adenocarcinoma tumorigenesis. And in liver cancer, both p53 and pRb are thought to be eliminated at early stages of tumorigenesis. Although it is not known whether p53 or pRb mutations are involved in the initiation of neoplastic transformation, germline p53 mutations are associated with the occurrence of several cancers, including carcinomas of breast, adrenal cortex, lung, and gastrointestinal tract, sarcomas of bone and soft tissues, and lymphomas, at much earlier ages than expected in the general population. Likewise, germline pRb mutations are associated with childhood Rb and a predisposition to osteosarcoma. Irrespective of the timing of the detection of these mutations, there seems to be a strong correlation between loss of pRb and lack of functional p53 in several tumors, and the loss of function of both p53 and pRb is required in order for cancer cells to acquire replicative immortality and become tumorigenic.

Tumor Vascularization (Angiogenesis)

Having acquired the cellular and molecular requirements described above, tumor cells, like normal cells, then require an adequate supply of oxygen and nutrients, and an effective means to discharge carbon dioxide and metabolic waste, in order to survive and proliferate for a sustained growth of tumor. Vasculature carrying blood (blood vessels) or lymphatic fluid (lymphatic vessels) are the primary sources for such requirements to the cells in most tissues [84, 85]. However, in pre-malignant stages, tumors are separated from the vascularized peri-tumoral tissue by a basal lamina that prevents blood vessels from infiltrating tumors. In the absence of vasculature, tumors cease to grow beyond the size of about $1-2\,mm^3$ in diameter, and become necrotic or apoptotic. Therefore, tumors must acquire new vasculature (neovascularization) capable of delivering blood to intra-tumoral spaces in order for rapidly proliferating tumor cells to survive, and transit from a premalignant stage to a fully malignant form capable of spreading to other parts of body [85]. Neovascularization is achieved primarily through the process called angiogenesis by which new blood vessels are developed from preexisting vasculature.

Angiogenesis is a multistep process that begins when vascular endothelial cells switch from a quiescent state to proliferate in response to angiogenic stimuli and acquire angiogenic phenotype [86, 87]. In adults, endothelial cells (ECs) that line mature blood vessels are mostly dormant, dividing on average once in about every 1000 days. Initiation of angiogenesis requires stimulation of endothelial cell growth and proliferation. Hypoxic condition or an oncogene activation causes tumor cells in rapidly growing tumors to secrete a variety of proangiogenic factors, such as vascular endothelial growth factor (VEGF) and basic fibroblast growth factor (bFGF), which diffuse through the surrounding tissue and bind to cognate receptor on ECs of preexisting blood vessels in the neighborhood. Besides inducing proliferation and migration of ECs, these factors stimulate ECs to secrete various proteases, including matrix metalloproteases (MMPs), which degrade the basement membrane and peri-vesicular ECM. This allows proliferating ECs to escape from the walls of preexisting blood vessels and migrate toward the source of angiogenic stimulus and form "primary sprouts." Extending sprouts contain two morphologically and functionally distinct forms of ECs, called "tip cells" and "stalk cells." Endothelial tip cells proliferate minimally but facilitate the breakdown of ECM surrounding preexisting blood vessels and lead the growth of new sprouts toward tumor cells that secrete angiogenic factors. Endothelial stalk cells, on the other hand, proliferate actively to extend sprouts and promote the formation of vascular lumen (tube). The ECs then stabilize vasculature by secreting platelet derived growth factor (PDGF), which recruits pericytes or smooth muscle cells to form basement membrane around newly formed vesicles [87, 88]. In the absence of the basement membrane, the nascent vessels are unstable and are prone to regress. Stabilized tube-like structures (blood vessels) thus formed bring blood to tumors and provide sustenance to actively proliferating tumor cells.

Angiogenesis is regulated not only by proangiogenic factors but also by antiangiogenic molecules [85]. Moreover, overexpression of proangiogenic factors is not itself sufficient for angiogenesis. It should be accompanied by downregulation of antiangiogenic molecules that inhibit vascular growth. Although VEGF is the most prevalent proangiogenic factor that stimulates the entire cascade of events required for angiogenesis, depending on the tumor type and tissue context, other GFs such as bFGF, PDGF, angiogenin, TGF-β, tumor necrosis factor-α (TNF-α), can also play important roles in neoplastic vascularization. Expression of several of these angiogenic factors is regulated by some of the same oncogenes that regulate cell proliferation (e.g., Ras and Myc), indicating multitasking capability of oncogenes in neoplastic transformation.

The levels of expression of angiogenic factors and their receptors reflect the aggressiveness with which tumors grow and spread. Therefore, the levels of angiogenic factors and their receptors in tissues can serve as prognostic indicators of cancer progression [85]. For example, there is a correlation between the expression of VEGF family of angiogenic factors and the risk of progression of several cancers, including colon, breast, lung, head and neck, stomach, and uterine (viz; cervix, ovary). Interestingly, there is a great variation in the extent of neovascularization between different tumor types. For example, whereas highly aggressive forms of pancreatic ductal carcinomas remain largely avascular, other tumors including renal and pancreatic neuroendocrine carcinomas are densely vascularized. This variation is due to the possibility that some tumors can develop and progress by exploiting the preexisting vasculature in the absence of angiogenesis [89].

There are several naturally occurring antiangiogenic factors, including, thrombospondin-1 (TSP-1), angiostatin, endostatin, interferon, and inhibitors of MMPs, to counterbalance proangiogenic factors in normal tissues [85]. These factors can induce apoptosis in tumor cells and ECs and block the migration and formation of lumen in sprouting ECs. These endogenous antiangiogenic factors are often downregulated in tumor tissues. This can result from the loss of function of tumor suppressor genes that regulate the expression of these genes. For example, TSP-1 is positively regulated by the p53 tumor suppressor and the loss of p53 function, which is prevalent in most cancers, would result in downregulation of TSP-1. Currently several synthetic inhibitors of angiogenesis, such as sunitinib, sorafenib, and temsirolimus, are being evaluated in combination with chemotherapeutic agents for the treatment of a variety aggressive cancers [90]. Antiangiogenic drugs are reported to enhance the delivery of chemotherapeutic agents to the tumors by normalizing the tumor vasculature. Unlike normal blood vessels, new vasculature in tumors is structurally and functionally abnormal and the blood vessels are immature and leaky [85]. These abnormalities can affect the delivery of chemotherapeutic agents to the tumors. Since abnormalities in tumor vasculature are suggested to result from excessive proangiogenic stimuli, it is possible that antiangiogenic drugs may normalize tumor vasculature by neutralizing some of the angiogenic stimuli. However, there is the concern of side effects with antiangiogenic drugs. Antiangiogenic drugs can interfere with many normal body processes such as wound healing, blood pressure, kidney function, and increased risk of clots in arteries leading to strokes or heart attacks. Although inhibition of angiogenesis offers an attractive strategy for the treatment of a variety of aggressive cancers, newer antiangiogenic drugs with a selective effect on tumor vasculature are needed for an effective treatment of aggressive cancers. In this regard, it is important to note that several integrin heterodimers, such as $\alpha 1\beta 1$, $\alpha 1\beta 2$, $\alpha 5\beta 1$, $\alpha 3\beta 1$, $\alpha v\beta 5$, $\alpha v\beta 8$, on the surface of proangiogenic factor-stimulated endothelial cells play an in important role in angiogenesis by regulating cell growth, survival, and migration [91]. Thus, ongoing efforts to develop inhibitors targeting integrins can also offer a viable opportunity to treat cancers by suppressing tumor vasculature.

Invasion and Metastasis

The spreading of a tumor from its primary site to other sites in body – metastasis – is the deadliest aspect of cancer. Besides invading adjacent tissues, tumor cells use blood or lymphatic circulation to spread through the body under favorable conditions [92]. The choice for tumor cells to enter blood vs. lymphatic circulation is dependent on the relative abundance and accessibility of the respective vessels induced by the tumor. As described above, a network of vasculature formed as a result of tumor angiogenesis makes cancer cells more accessible to blood vessels than lymphatic vessels. Lymphatic vessels inside tumors often do not function as well as those formed outside [93]. Therefore, tumor cells must invade through intervening connective tissue in order to enter functional lymphatic vessels. Thus, the majority of tumor cells use the bloodstream to spread through the body. Even those tumor cells that enter lymphatic vessels eventually gain access to the bloodstream as the lymphatic vessels ultimately drain into the venous blood when they join thoracic duct, a large lymph duct. However, on the way to thoracic duct, tumor cells in lymphatic vessels encounter a series of lymph nodes, which often are the initial sites of metastasis. Tumor cells can also invade local lymph nodes directly by penetrating through the intervening connective tissue. It is for these reasons that the lymph nodes nearest to the primary tumor are examined for the presence of tumor cells as an early indicator of metastasis [94].

The metastatic process involves a number of steps including migration (invasion) of tumor cells at a primary site into the surrounding normal tissue to reach nearby blood or lymphatic vessel, crossing (intravasate) the endothelial barrier to enter circulation, and exit (extravasate) from a distant capillary bed to grow (colonize) at a secondary site (Figure 10).

Invasion and Metastasis

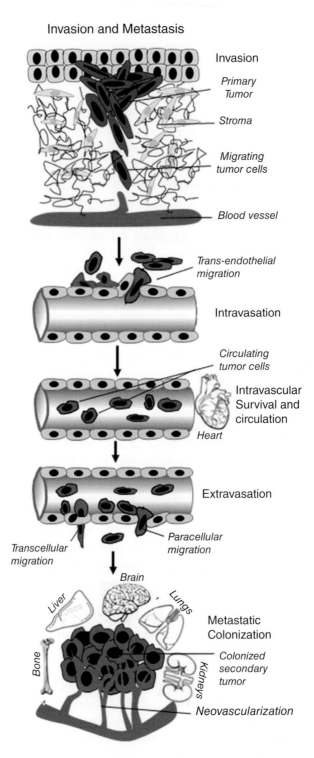

Invasion

Primary Tumor

Stroma

Migrating tumor cells

Blood vessel

Trans-endothelial migration

Intravasation

Circulating tumor cells

Intravascular Survival and circulation

Heart

Extravasation

Transcellular migration

Paracellular migration

Brain

Liver

Lungs

Metastatic Colonization

Bone

Colonized secondary tumor

Kidneys

Neovascularization

Figure 10. Biological processes associated with the spreading of cancer from its primary site to other organs in body. *Source: From invasion to metastasis.* (See insert for color representation.)

Invasion

The first step in metastasis involves migration of tumor cells toward the blood or lymphatic vessels through the process known as "invasion." Invading solid tumor cells, such as those in carcinomas, must first cross the basement membrane (BM) that separates the tumor from the stroma, which is an ECM comprising of collagen and a number of other cell types surrounding the tissue [95]. Metastasizing tumor cells then migrate through ECM to reach blood or lymphatic vessels in order to be carried to distant sites in body. In order to be motile, cancer cells acquire different morphological and migratory characteristics as a result of changes in expression of several oncogenes and tumor suppressor genes, including c-MET, EGFR, RAS, PTEN, TGF-β, and integrins, in response to the environmental conditions. Due to changes in cytoskeletal proteins cancer cells can acquire amoeboid-like morphology or mesenchymal phenotype [96, 97]. While amoeboid-like cells tend to exhibit increased contractility and migrate individually or in stream, cells with mesenchymal phenotype display a lower contractility and migrate collectively in clusters of five or more cells. Cancer cells migrating in clusters, such as those from colorectal and breast tumors, often lose their epithelial characteristics and display epithelial-mesenchymal transition (EMT) markers, such as downregulation of E-cadherin and catenin, and upregulation of N-cadherin, vimentin, and fibronectin. E-cadherin is central to the formation of cell-cell junctions that suppress proliferation and the invasive capability of polarized epithelial cells. In contrast, N-cadherin is a cell-cell adhesion protein that promotes survival and migration as observed in case of migrating neurons and mesenchymal cells during embryogenesis. Therefore, the switch from epithelial (E)- to neuronal (N)-cadherin expression is a key event in EMT and progression of epithelial cancers from adenomas to carcinomas. Moreover, N-cadherin-mediated cell-cell adhesions can facilitate persistent migration of carcinoma cells collectively (in clusters) through ECM [98].

There are several developmentally important genes that code for proteins, such as Snail, Slug, Sip1, and Twist, which bind to the E-cadherin gene promoter and repress its expression and induce EMT [99]. During embryonic development, the expression of these genes is under strict spatial and temporal control. However, in invasive tumor cells one or more of these genes is often overexpressed and their expression is associated with a loss of E-cadherin and the gain of N-cadherin. For example, Twist expression is inversely correlated with E-cadherin expression in invasive lobular carcinomas, and overexpression of Twist, Snail, and Sip1 is associated with downregulation of E-cadherin expression and upregulation of N-cadherin expression in gastric cancer. These observations indicate that tumor cells coopt genes involved in embryonic development to acquire capabilities needed for invasion and metastatic spread. Although, tumor cells acquire EMT-mediated mesenchymal phenotype during invasion, these cells must subsequently revert to epithelial phenotype as the cells in metastasized tumors growing at a secondary site are morphologically and histopathologically similar to those in the primary tumors prior to EMT [97]. This means that the mesenchymal phenotype that tumor cells acquire through EMT is transitory, occurring only during invasion and metastatic spread. In order for invasive mesenchymal tumor cells to reacquire epithelial phenotype, they go through the reverse of EMT process, referred to as "mesenchymal-epithelial transition" (MET) [97]. This reversion from EMT to MET renders metastasized tumor cells noninvasive, permitting their growth and colonization at a secondary site. At present, the regulatory processes controlling the expression of Twist, Snail, Slug, or Sip1 to orchestrate first EMT and then MET of tumor cells at different steps during metastasis are not clear.

Integrin-mediated transmembrane connections between the actin cytoskeleton and the ECM, which are organized in discrete clusters as focal adhesions, also play an important role in the migration of tumor cells through ECM [45, 100]. In migrating cells, the dynamic assembly and disassembly of polarized integrin-mediated adhesions is essential for the cell migration speed and directional persistence. Cancer cells express different integrin $\alpha\beta$ heterodimers on cell surfaces that recognize different ligands, such as collagen (e.g., $\alpha2\beta1$, $\alpha1\beta1$), fibronectin (e.g., $\alpha5\beta1$, $\alpha v\beta3$), and laminin (e.g., $\alpha2\beta1$, $\alpha3\beta1$) in ECM. The composition and concentration of integrins on the surface of migrating tumor cells can change as they encounter different stromal microenvironments. Cell migration speed is determined by the concentration of integrins, the density of ligands, and the affinity between integrins and ligands. For example, breast cancer cells with high level of $\alpha5\beta1$ are three times more invasive than those with low level of $\alpha5\beta1$. Similarly, the level of expression of $\alpha v\beta3$ in melanoma and $\alpha2\beta1$ in rhabdomyosarcoma is correlated with tumor invasion.

As the cancer progresses, collagen type I network in stroma surrounding the tumor also undergo significant changes. In normal stroma, collagen fibers are typically curly. However, in early stages of tumor growth, the amount of collagen in stroma increases and collagen fibers become straightened and line up along the tumor BM. As the tumor turns invasive, collagen fibers get bundled and turn vertical to the BM so as to make a pathway for cancer cells to migrate away from the tumor. Such organization of collagen fibers is associated with poor survival of patients [101].

Intravasation

The entry of cancer cells into the blood or lymph circulation – intravasation – is a rate-limiting step in metastasis that determines the number of tumor cells entering circulation and, thereby, the chances of tumor formation at a secondary site [94]. Intravasation is a naturally occurring process that is vital not only for embryonic development but also for wound healing and for the immune system to fight infections. However, the efficiency with which cancer cells intravasate vasculature is very inefficient as compared to that occurs during embryogenesis, wound healing, or immune response [102]. Tumor cells can intravasate lymphatic vessels with relative ease as the cell-cell junctions of endothelial cells in peripheral lymphatic walls are punctuated with a junction spacing of ~3 μm, which is large enough for cancer cell squeezing. By contrast, endothelial cells that line the blood vessels have a tight cell-cell junction and, moreover, endothelial cells are supported by the basement membrane that is stabilized with pericytes. In order to cross these structural barriers cancer cells undergo cytoskeletal changes and also upregulate integrins and other adhesion molecules that facilitate their attachment to endothelial cells. There seems to be a close spatial relationship between the sites through which macrophages exit from blood vessels and the cancer cells enter the blood vessels. Moreover, macrophages produce EGF that acts as a chemotactic agent to guide the cancer cells toward the perivesicular macrophages. Therefore, it is possible that cancer cells with mesenchymal phenotype, acquired via EMT, may roll along the endothelial cells to enter the blood vessels at the points where macrophages exit. Besides this active entry, cancer cells can enter passively into blood circulation through neovasculature in tumors. Tumor vasculature is typically disorganized and leaky with looser association of vascular BM with endothelial cells and pericytes creating focal holes large ($\sim2.5\,\mu M$) enough for cancer cells to pass through. Space constraints in growing tumors push proliferating tumor cells against the blood vessels forcing their passage through such holes to enter the circulation. Thus, the entry of cancer cells through this passive process would not require major disruption of BM.

Intravascular Survival

As the cancer cells enter bloodstream they encounter high mechanical forces, including shear stress resulting from blood flow, which can be damaging [102, 103]. Shear stress is so intense that most of the cells (>99.9%) die within minutes of entry into circulation. The cytoskeleton of many cancer cells is incapable of withstanding such shear stress and are fragmented into microparticles as they enter blood stream and the microparticles are taken up by the myeloid cells. In order to protect themselves from stress-induced death they shield themselves with platelets and coagulation factors that they activate and assemble into microaggregates of cell-pellets [104]. This aggregation allows cells in the center to survive mechanical forces. In addition, single cells that acquire amoeboid-like morphology tend to survive shear stress as a result of their ability to attach to the substratum and increased contractility. Some cells can also survive lethal mechanical injury by inhibiting apoptosis through a signaling pathway activated by pannexin-1 transmembrane channel protein-mediated release of ATP and activation of P2Y receptor [105]. In general, the few surviving cells within the clusters of circulating tumor cells (CTCs) are the ones that possess a higher metastatic potential than the majority of the single cells that survive circulation.

CTCs must also evade the immune system in order to survive [103]. Cells in solid tumors are protected from the antitumor activity of the immune system by the strong immunosuppressive microenvironment that exists within the tumors. However, once the tumor cells are blood borne they become exposed to an unabated immune system capable of mounting an aggressive antitumor activity and destroying CTCs. Natural killer (NK) cells play a central role in tumor immuno-surveillance and antitumor cell activity. NK cells contain NK group 2d (NKG2D) receptors that recognize MHC class 1 polypeptide-related sequences A and B (MICA and MICB) expressed on the surface of malignant tumor cells and secrete cytolytic granzyme B and porforin that destroy tumor cells. They can also induce apoptosis by activating FasL- and TRAIL-mediated death receptor pathways. To overcome such immunosuppression, CTCs can adopt various methods to encounter antitumor cell activity of NK cells. They include: (i) metalloproteinase-mediated cleavage and shedding of cell surface MICA/B to avoid recognition by NKG2D receptors on NK cells; (ii) secretion of immunomodulatory molecules, such as prostaglandin E_2, TGF-β, and adenosine, that inhibit the antitumor activity of NK cells; and (iii) secretion of lactate dehydrogenase isoform 5 (LDH5) that induces NKG2D ligands on myeloid cells, which cause downregulation of NKG2D on NK cells.

In addition, platelet- and coagulation factor-mediated formation of microaggregate cell-pellet described above can physically shield CTCs from direct contact with NK cells and, thereby, prevent the activation of antitumor cell activity of NK cells. Platelets in the tumor cell microaggregates can also downregulate NKG2D on NK cells to suppress antitumor cell activity. These are some in a growing list of the ways through which a few CTCs are able to endure the harsh immunological and mechanical forces that they encounter in order to survive.

Extravasation

Having survived the hostile intravascular environment, CTCs must then exit from circulation (extravasate) into the surrounding tissue in order to be metastatic [102, 103]. Since tumor cell extravasation is a rare, inefficient, and transient event, it has been difficult to study the mechanism and molecular regulation of tumor cell extravasation directly. However, based on the knowledge of extravasation of white blood cells, which is a normal and frequently occurring event, extravasation of CTCs across blood vessel walls is believed to involve a cascade of steps including: (i) initial capture and rolling of CTC on endothelium under flow conditions before stable adhesion to endothelium (mediated by selectin and cadherins), (ii) arrest, adhesion strengthening and spreading (mediated by receptors such as integrins, CD44, and MUC1), (iii) intravascular crawling (iv) disruption of cellular endothelial cell junctions, and (v) transendothelial migration. Each of these steps in tumor cell extravasation is currently being tested using *in vitro* and *in vivo* models. Due to size constraints, cancer cells are reported to reduce speed and get arrested in small capillaries leading to a stable adhesion of tumor cells to the endothelium. This entrapment of CTCs in capillaries is considered to be the main cause for cancer cell arrest prior to extravasation. However, in the case of tumor cells that cannot enter capillaries because of their relatively large size, adhesion between tumor cells and endothelium in larger blood vessels is shown to be mediated by several ligands and receptors including selectins, cadherins, and integrins. Accordingly, expression of E-selectin is reported to be correlated with higher metastatic potential and poor prognosis [106]. Upon stable attachment, tumor cells are found to perform intravascular crawling before initiating extravasation. This intravascular crawling on endothelium is suggested to allow tumor cells to find an ideal site to extravasate.

Mechanistically, there are two modes of transendothelial migrations; one is paracellular migration (referred to as "diapedesis") in which tumor cells migrate through the cell-cell junction in the vesicular endothelium cell monolayer, and the second is transcellular migration in which tumor cells penetrate the bodies of endothelial cells [103]. Paracellular diapedesis is the predominant form of transmigration that CTCs use to extravasate. There are certain ligand-receptor pairs that have been identified to specifically affect transendothelial migration, without affecting tumor cell adhesion to endothelium. For example, junctional adhesion molecule c (JAM-C)-mediated interaction between tumor and endothelial cells promotes lung metastasis. Using an *in vitro* model, it is shown that disruption of the binding of CXCL12 ligand expressed on endothelial cell surface to the cognate receptor CXCR4 expressed on tumor cells abrogates transendothelial migration of lung tumor cells without affecting their adhesion to endothelium.

Although, these *in vitro* models have provided a molecular explanation for how CTCs attach and extravasate through the endothelium, we are still far away from full understanding of the mechano-molecular mechanisms underlying tumor cell extravasation *in vivo*.

Metastatic Colonization

Colonization of tumor cells at secondary sites is the most inefficient and rate-limiting step in metastatic process. Although primary tumors release millions of tumor cells into circulation, there is a long latent period, sometime lasting years and even decades, before clinical manifestation of metastatic growth, and in some, there may never be an overt metastasis. Most of the tumor cells that survive the hostile intravascular environment and infiltrate distant organs fail to develop overt metastasis. In order for clinical manifestation of metastatic growth, infiltrated tumor cells must find secondary organs conducive to colonize, i.e., survive and proliferate, or possess the intrinsic ability to colonize distant tissues they infiltrate. Despite all these bottlenecks, some cells do survive and metastasize in distant organs and turn lethal.

Some organs, such as liver and bone, seem to be inherently easier for tumor cells to colonize than others, such as lungs and brain. This led to the concept of "seeds and soil," referring to the tumor cells (the "seeds") colonizing favorably in selective distant organs (the "soil") [107]. In this scenario, the correct "seed" must find correct "soil" in order to colonize and grow. This concept is strengthened by the findings that specific organ-dependent conditions enable infiltrated tumor cells to survive. For example, macrophages in a tissue microenvironment can induce signals by binding to VCAM-1 expressed on the surface of infiltrated tumor cells to promote their survival. Effective colonization can also depend on the intrinsic ability of tumor cells to survive the hostile environment in the infiltrated tissues. For example, Src activity in disseminated breast cancer cells is shown to promote their survival and metastasize in bones. In other instances, disseminated tumor cells can transmit signals that stimulate bone marrow progenitors to express receptor tyrosine kinase MET required for bone metastasis. Importantly, infiltrated tumor cells express a subset of specific genes in different organs, such as brain and lung, which can favor their metastatic growth. Thus, the availability of a favorable microenvironment within the host tissue, together with the ability of infiltrating tumor cells to express specific genes necessary for survival and proliferation, are required for metastatic colonization.

Besides the "seeds and soil" concept, in which CTCs extravasate prior to their growth and colonization in surrounding tissue, CTCs lodged in the microvasculature can initiate intravascular tumor growth, forming embolus, which eventually ruptures the blood vessel, or they can extravasate by breaching vascular wall to form a micrometastasis. According to this concept, failure of metastasis in certain organs is suggested to be a result of a failure of CTCs to enter those organs, rather than those organs not being favorable for metastasis [108]. This is consistent with an experimental observation in which the formation of tumors in certain organs is directly correlated with the arrest of CTCs in the small capillaries of those organs [107]. Moreover, due to differences in composition of vascular walls, transendothelial migration of CTCs can be more permissive in some organs (e.g., liver and bone) than in other organs (e.g., lung and brain). Liver and bone marrow contain sinusoidal capillaries made of fenestrated endothelial cells and discontinuous basal lamina, which permit extravasation and, thereby, a high incidence of liver and bone metastasis. Endothelium in the lung, on the other hand, has tight endothelial junctions that are not very conducive for extravasation of CTCs. Similarly, brain capillary walls are additionally reinforced by pericytes and astrocytes that constitute the blood-brain barrier. However certain genes, e.g., angiopoietin-like 4 (ANGPTL4), cyclooxygenase 2 (COX2), MMP1, and osteonectin, have been identified that can mediate endothelial disjunction and vascular permeability to facilitate extravasation of CTCs in tissues like the lung and brain.

Finally, tumor cells that seed distant organs must then rely on the suitability and fertility of the host stromal microenvironment for survival and growth. It has been observed in experimental models that the signals emitted by primary tumors can create "premetastatic niches" in distant organs prior to the arrival of CTCs [109]. For example, tumor derived placental growth factor (PlGF) is shown to act on lung parenchyma to mobilize bone marrow derived VEGFR1+ cells that increase the survival of infiltrating cancer cells. Similarly, macrophage inhibitory factor (MIF) containing exosomes released by pancreatic cancer cells are shown to increase liver metastasis by inducing TGF-β secretion. Tumor cells homing in niches can receive vital support through contact with stromal cells. For example, claudin-2 mediated cell-cell interactions between breast cancer cells and hepatocytes, mediated by c-Met signaling, stimulate metastasis to the liver. While observations such as these are valuable in understanding the ways in which tumor cells adopt themselves to survive and proliferate in distant organs, a lot remains to be learned regarding the symbiotic relationship between the stroma in the host tissue microenvironment and the growing tumors. This can differ significantly between cancer cells derived from tumors not only in different organs but also those in the same organ. A full understanding of these prometastatic interactions in the future may provide effective strategies for the treatment of cancers to prevent metastatic growth.

SUMMARY

Cancer is a new growth of a tissue (neoplasia) arising from genetic changes in a single cell. These genetic changes result in dysregulation of cellular functions (anaplasia) leading to autonomous growth and proliferation, and manifest as a tumor. Unlike normal cells, cancer cells do not respond to external signals that govern cell division and differentiation. As the tumor grows, the cancer cell properties continue to change and eventually a variety of cells arise at later stages of cancer. Although the underlying features are the same, many types of cancers are named after the tissue or the cell of origin – carcinomas arising from ectoderm and endoderm of epithelial cells, sarcomas from mesoderm of muscle, bone, fat, and connective tissues, leukemia from leukocytes, myeloma from plasma cells in bone marrow, and gliomas from glial nerve cells. A solid tissue tumor may be benign when confined to the tissue of origin or may become malignant with more genetic alterations causing the cells to migrate to surrounding tissues and further through the bloodstream to other parts of the body and grow into secondary tumors.

The genetic alterations in cancer are largely caused by carcinogens in the environment and diet. These include chemicals, infectious agents, and physical carcinogens such as UV and ionizing radiation. Less frequently, inherited genes and epigenetics can contribute to cancer development. Further, alterations in immune system and hormonal imbalance during the aging process can play a role in carcinogenesis.

Cancer development is clonal – originating from genetic mutations and epigenetic alterations in a single cell. It is a multistep process involving multiple mutations and alterations of several different genes. The initial genetic changes confer a cell with the ability to proliferate autonomously without responding to the external stimuli such as GFs, cytokines, and hormones. This autonomous proliferation predisposes the cells to acquire additional mutations that further leads to selective survival and expansion of surviving cells in the process of tumor growth and progression.

Cancers of all tissue types acquire the same characteristics that are unique and complementary as they progress through premalignant to invasive metastatic stages. These common characteristics include autonomous proliferation, resistance to anti-proliferative signals, evasion of programmed cell death (apoptosis), unlimited potential to replicate, vascularization of tumors with new blood vessels, and invasion followed by migration and metastasis. In each of the common characteristics, the normal cellular processes are altered following the changes in the genes and proteins involved in each of these processes.

REFERENCES

1. Siegel, R.L., Miller, K.D., and Jemal, A. (2016). Cancer statistics, 2016. *CA Cancer J. Clin.* 66: 7–30.
2. Pardee, A.B. and Reddy, G.P. (1986). Cancer: Fundamental Ideas. In: *Carolina biological readers* (ed. J.J. Head), 32. Burlington, North Carolina: Carolina Biological Supply Company.
3. Anand, P., Kunnumakkara, A.B., Sundaram, C. et al. (2008). Cancer is a preventable disease that requires major lifestyle changes. *Pharm. Res.* 25: 2097–2116.
4. Karr, J.P. (1992). Prostate cancer in the United States and Japan. *Adv. Exp. Med. Biol.* 324: 17–28.
5. Kim, Y., Park, J., Nam, B.H., and Ki, M. (2015). Stomach cancer incidence rates among Americans, Asian Americans and native Asians from 1988 to 2011. *Epidemiol. Health* 37: e2015006.
6. Serrano, M. and Blasco, M.A. (2007). Cancer and ageing: convergent and divergent mechanisms. *Nat. Rev. Mol. Cell Biol.* 8: 715–722.
7. White, M.C., Holman, D.M., Boehm, J.E. et al. (2014). Age and cancer risk: a potentially modifiable relationship. *Am. J. Prev. Med.* 46: S7–S15.
8. Pott, P. (1993). The first description of an occupational cancer in 1777 (scrotal cancer, cancer of chimney sweeps). *Bull. Soc. Liban. Hist. Med.* 4: 98–101.
9. Loeb, L.A. and Harris, C.C. (2008). Advances in chemical carcinogenesis: a historical review and prospective. *Cancer Res.* 68: 6863–6872.
10. Carbone, D. (1992). Smoking and cancer. *Am. J. Med.* 93: 13S–17S.
11. Smokeless tobacco and public health: A global perspective. NIH publication No. 14-7983. Bethesda, MD: U.S. Department of Health and Human Services, Centers for Disease Control and Prevention and National Institutes of Health, National Cancer Institute; 2014. National Cancer Institute and Centers for Disease Control and Prevention; pp. B5–63.
12. Ratna, A. and Mandrekar, P. (2017). Alcohol and cancer: mechanisms and therapies. *Biomolecules* 7: 61.
13. Cox, T.R. and Erler, J.T. (2014). Molecular pathways: connecting fibrosis and solid tumor metastasis. *Clin. Cancer Res.* 20: 3637–3643.
14. Wolff, H., Vehmas, T., Oksa, P. et al. (2015). Asbestos, asbestosis, and cancer, the Helsinki criteria for diagnosis and attribution 2014: recommendations. *Scand. J. Work Environ. Health* 41: 5–15.
15. Key, T.J., Schatzkin, A., Willett, W.C. et al. (2004). Diet, nutrition and the prevention of cancer. *Public Health Nutr.* 7: 187–200.
16. Carrillo-Infante, C., Abbadessa, G., Bagella, L. et al. (2007). Viral infections as a cause of cancer (review). *Int. J. Oncol.* 30: 1521–1528.
17. zur Hausen, H. (1991). Viruses in human cancers. *Science* 254: 1167–1173.
18. Liao, J.B. (2006). Viruses and human cancer. *Yale J. Biol. Med.* 79: 115–122.
19. Ahn, H.J. and Lee, D.S. (2015). Helicobacter pylori in gastric carcinogenesis. *World J. Gastrointest. Oncol.* 7: 455–465.
20. Chaturvedi, A.K., Gaydos, C.A., Agreda, P. et al. (2010). Chlamydia pneumoniae infection and risk for lung cancer. *Cancer Epidemiol. Biomark. Prev.* 19: 1498–1505.

21. Mostafa, M.H., Sheweita, S.A., and O'Connor, P.J. (1999). Relationship between schistosomiasis and bladder cancer. *Clin. Microbiol. Rev.* 12: 97–111.

22. OE, H.S., Hamid, H.K., Mekki, S.O. et al. (2010). Colorectal carcinoma associated with schistosomiasis: a possible causal relationship. *World J. Surg. Oncol.* 8: 68.

23. Maeda, H. and Akaike, T. (1998). Nitric oxide and oxygen radicals in infection, inflammation, and cancer. *Biochemistry (Mosc)* 63: 854–865.

24. Mustaacchi, P. (2000). *Parasites*. Hamilton, Ontario: B.C. Decker.

25. Reichrath, J. and Rass, K. (2014). Ultraviolet damage, DNA repair and vitamin D in nonmelanoma skin cancer and in malignant melanoma: an update. *Adv. Exp. Med. Biol.* 810: 208–233.

26. Runger, T.M. (2016). Mechanisms of melanoma promotion by ultraviolet radiation. *J. Invest. Dermatol.* 136: 1751–1752.

27. Trosko, J.E. (1996). Role of low-level ionizing radiation in multi-step carcinogenic process. *Health Phys.* 70: 812–822.

28. Garber, J.E. and Offit, K. (2005). Hereditary cancer predisposition syndromes. *J. Clin. Oncol.* 23: 276–292.

29. Cunniff, C., Bassetti, J.A., and Ellis, N.A. (2017). Bloom's syndrome: clinical spectrum, molecular pathogenesis, and cancer predisposition. *Mol. Syndromol.* 8: 4–23.

30. Lebel, M. and Jr. Monnat, R.J. (2018). Werner syndrome (WRN) gene variants and their association with altered function and age-associated diseases. *Ageing Res. Rev.* 41: 82–97.

31. Mehta, P.A. and Tolar, J. (1993). Fanconi anemia. In: *GeneReviews((R))* (ed. M.P. Adam, H.H. Ardinger, R.A. Pagon, et al.). Seattle, WA: Available online at: https://www.ncbi.nlm.nih.gov/books/NBK1401.

32. Casaubon, J.T., and Regan, J.P. (2018). BRCA 1 and 2. In StatPearls (Treasure Island [FL]).

33. Lozano, G. (2016). The enigma of p53. *Cold Spring Harb. Symp. Quant. Biol.* 81: 37–40.

34. Sherr, C.J. and McCormick, F. (2002). The RB and p53 pathways in cancer. *Cancer Cell* 2: 103–112.

35. Jeggo, P.A., Pearl, L.H., and Carr, A.M. (2016). DNA repair, genome stability and cancer: a historical perspective. *Nat. Rev. Cancer* 16: 35–42.

36. Farber, E. (1984). The multistep nature of cancer development. *Cancer Res.* 44: 4217–4223.

37. Sadikovic, B., Al-Romaih, K., Squire, J.A. et al. (2008). Cause and consequences of genetic and epigenetic alterations in human cancer. *Curr. Genomics* 9: 394–408.

38. Hanahan, D. and Weinberg, R.A. (2000). The hallmarks of cancer. *Cell* 100: 57–70.

39. Hanahan, D. and Weinberg, R.A. (2011). Hallmarks of cancer: the next generation. *Cell* 144: 646–674.

40. Reddy, G.P. (1994). Cell cycle: regulatory events in G1-->S transition of mammalian cells. *J. Cell. Biochem.* 54: 379–386.

41. Reddy, G.P., Cifuentes, E., Bai, U. et al. (2004). Onset of DNA synthesis and S phase. In: *Cell Cycle and Growth Control: Biological Regulation and Cancer* (ed. G.S. Stein and A.B. Pardee), 149–200. Wiley.

42. Appert-Collin, A., Hubert, P., Cremel, G. et al. (2015). Role of ErbB receptors in cancer cell migration and invasion. *Front. Pharmacol.* 6: 283.

43. Jr. Roskoski, R. (2014). The ErbB/HER family of protein-tyrosine kinases and cancer. *Pharmacol. Res.* 79: 34–74.

44. Moreno-Layseca, P. and Streuli, C.H. (2014). Signalling pathways linking integrins with cell cycle progression. *Matrix Biol.* 34: 144–153.

45. Desgrosellier, J.S. and Cheresh, D.A. (2010). Integrins in cancer: biological implications and therapeutic opportunities. *Nat. Rev. Cancer* 10: 9–22.

46. Molina, J.R. and Adjei, A.A. (2006). The Ras/Raf/MAPK pathway. *J. Thorac. Oncol.* 1: 7–9.

47. Laplante, M. and Sabatini, D.M. (2012). mTOR signaling in growth control and disease. *Cell* 149: 274–293.

48. Porta, C., Paglino, C., and Mosca, A. (2014). Targeting PI3K/Akt/mTOR signaling in cancer. *Front. Oncol.* 4: 64.

49. Khan, A.Q., Kuttikrishnan, S., Siveen, K.S. et al. (2018). RAS-mediated oncogenic signaling pathways in human malignancies. *Semin. Cancer Biol.* 18: 30002–30006.

50. Pulido, R. (2018). PTEN inhibition in human disease therapy. *Molecules* 23: E285.

51. Wise, H.M., Hermida, M.A., and Leslie, N.R. (2017). Prostate cancer, PI3K, PTEN and prognosis. *Clin. Sci. (Lond.)* 131: 197–210.

52. Lim, H.J., Crowe, P., and Yang, J.L. (2015). Current clinical regulation of PI3K/PTEN/Akt/mTOR signalling in treatment of human cancer. *J. Cancer Res. Clin. Oncol.* 141: 671–689.

53. Binder, M.J., McCoombe, S., Williams, E.D. et al. (2017). The extracellular matrix in cancer progression: role of hyalectan proteoglycans and ADAMTS enzymes. *Cancer Lett.* 385: 55–64.

54. Zhao, M., Mishra, L., and Deng, C.X. (2018). The role of TGF-beta/SMAD4 signaling in cancer. *Int. J. Biol. Sci.* 14: 111–123.

55. Seluanov, A., Hine, C., Azpurua, J. et al. (2009). Hypersensitivity to contact inhibition provides a clue to cancer resistance of naked mole-rat. *Proc. Natl. Acad. Sci. U. S. A.* 106: 19352–19357.

56. Tian, X., Azpurua, J., Hine, C. et al. (2013). High-molecular-mass hyaluronan mediates the cancer resistance of the naked mole rat. *Nature* 499: 346–349.

57. Ehmer, U. and Sage, J. (2016). Control of proliferation and cancer growth by the Hippo signaling pathway. *Mol. Cancer Res.* 14: 127–140.

58. Kim, H.B. and Myung, S.J. (2018). Clinical implications of the Hippo-YAP pathway in multiple cancer contexts. *BMB Rep.* 51: 119–125.

59. Heldin, C.H., Landstrom, M., and Moustakas, A. (2009). Mechanism of TGF-beta signaling to growth arrest, apoptosis, and epithelial-mesenchymal transition. *Curr. Opin. Cell Biol.* 21: 166–176.

60. Lee, C. and Cho, Y. (2002). Interactions of SV40 large T antigen and other viral proteins with retinoblastoma tumour suppressor. *Rev. Med. Virol.* 12: 81–92.

61. Amati, B. and Land, H. (1994). Myc-Max-Mad: a transcription factor network controlling cell cycle progression, differentiation and death. *Curr. Opin. Genet. Dev.* 4: 102–108.

62. Zhou, Z.Q. and Hurlin, P.J. (2001). The interplay between Mad and Myc in proliferation and differentiation. *Trends Cell Biol.* 11: S10–S14.

63. Kalkat, M., De Melo, J., Hickman, K.A. et al. (2017). MYC deregulation in primary human cancers. *Genes (Basel)* 8: 2–30.

64. Chen, H., Liu, H., and Qing, G. (2018). Targeting oncogenic Myc as a strategy for cancer treatment. *Signal Transduct. Target. Ther.* 3: 5.

65. Cotter, T.G. (2009). Apoptosis and cancer: the genesis of a research field. *Nat. Rev. Cancer* 9: 501–507.

66. Kerr, J.F., Wyllie, A.H., and Currie, A.R. (1972). Apoptosis: a basic biological phenomenon with wide-ranging implications in tissue kinetics. *Br. J. Cancer* 26: 239–257.

67. Barinaga, M. (1998). Is apoptosis key in Alzheimer's disease? *Science* 281: 1303–1304.

68. Julien, O. and Wells, J.A. (2017). Caspases and their substrates. *Cell Death Differ.* 24: 1380–1389.

69. Salvesen, G.S. and Dixit, V.M. (1997). Caspases: intracellular signaling by proteolysis. *Cell* 91: 443–446.

70. Walczak, H. and Krammer, P.H. (2000). The CD95 (APO-1/Fas) and the TRAIL (APO-2L) apoptosis systems. *Exp. Cell Res.* 256: 58–66.

71. Hengartner, M.O. (2000). The biochemistry of apoptosis. *Nature* 407: 770–776.

72. Reed, J.C. (1999). Mechanisms of apoptosis avoidance in cancer. *Curr. Opin. Oncol.* 11: 68–75.

73. Hickman, E.S., Moroni, M.C., and Helin, K. (2002). The role of p53 and pRB in apoptosis and cancer. *Curr. Opin. Genet. Dev.* 12: 60–66.

74. Lowe, S.W., Cepero, E., and Evan, G. (2004). Intrinsic tumour suppression. *Nature* 432: 307–315.

75. Wu, X. and Deng, Y. (2002). Bax and BH3-domain-only proteins in p53-mediated apoptosis. *Front. Biosci.* 7: d151–d156.

76. Nichols, K.E., Malkin, D., Garber, J.E. et al. (2001). Germ-line p53 mutations predispose to a wide spectrum of early-onset cancers. *Cancer Epidemiol. Biomark. Prev.* 10: 83–87.

77. Stiles, B.L. (2009). PI-3-K and AKT: onto the mitochondria. *Adv. Drug Deliv. Rev.* 61: 1276–1282.

78. Wu, Q., Zheng, Y., Chen, D. et al. (2014). Aberrant expression of decoy receptor 3 in human breast cancer: relevance to lymphangiogenesis. *J. Surg. Res.* 188: 459–465.

79. Kim Sh, S.H., Kaminker, P., and Campisi, J. (2002). Telomeres, aging and cancer: in search of a happy ending. *Oncogene* 21: 503–511.

80. Rodier, F., Kim, S.H., Nijjar, T. et al. (2005). Cancer and aging: the importance of telomeres in genome maintenance. *Int. J. Biochem. Cell Biol.* 37: 977–990.

81. Sharpless, N.E. and DePinho, R.A. (2004). Telomeres, stem cells, senescence, and cancer. *J. Clin. Invest.* 113: 160–168.

82. Shay, J.W. and Wright, W.E. (2011). Role of telomeres and telomerase in cancer. *Semin. Cancer Biol.* 21: 349–353.

83. De Vitis, M., Berardinelli, F., and Sgura, A. (2018). Telomere length maintenance in cancer: at the crossroad between telomerase and alternative lengthening of telomeres (ALT). *Int. J. Mol. Sci.* 19: E606.

84. Folkman, J. (1971). Tumor angiogenesis: therapeutic implications. *N. Engl. J. Med.* 285: 1182–1186.

85. Nishida, N., Yano, H., Nishida, T. et al. (2006). Angiogenesis in cancer. *Vasc. Health Risk Manag.* 2: 213–219.

86. De Palma, M., Biziato, D., and Petrova, T.V. (2017). Microenvironmental regulation of tumour angiogenesis. *Nat. Rev. Cancer* 17: 457–474.

87. Ribatti, D. and Crivellato, E. (2012). "Sprouting angiogenesis", a reappraisal. *Dev. Biol.* 372: 157–165.

88. Darland, D.C. and D'Amore, P.A. (1999). Blood vessel maturation: vascular development comes of age. *J. Clin. Invest.* 103: 157–158.

89. Dome, B., Hendrix, M.J., Paku, S. et al. (2007). Alternative vascularization mechanisms in cancer: pathology and therapeutic implications. *Am. J. Pathol.* 170: 1–15.

90. Rajabi, M. and Mousa, S.A. (2017). The role of angiogenesis in cancer treatment. *Biomedicine* 5: 34.

91. Avraamides, C.J., Garmy-Susini, B., and Varner, J.A. (2008). Integrins in angiogenesis and lymphangiogenesis. *Nat. Rev. Cancer* 8: 604–617.

92. Wong, S.Y. and Hynes, R.O. (2006). Lymphatic or hematogenous dissemination: how does a metastatic tumor cell decide? *Cell Cycle* 5: 812–817.

93. Padera, T.P., Kadambi, A., di Tomaso, E. et al. (2002). Lymphatic metastasis in the absence of functional intratumor lymphatics. *Science* 296: 1883–1886.

94. Chiang, S.P., Cabrera, R.M., and Segall, J.E. (2016). Tumor cell intravasation. *Am. J. Phys. Cell Physiol.* 311: C1–C14.

95. Sahai, E. (2005). Mechanisms of cancer cell invasion. *Curr. Opin. Genet. Dev.* 15: 87–96.

96. Clark, A.G. and Vignjevic, D.M. (2015). Modes of cancer cell invasion and the role of the microenvironment. *Curr. Opin. Cell Biol.* 36: 13–22.

97. Thiery, J.P. (2002). Epithelial-mesenchymal transitions in tumour progression. *Nat. Rev. Cancer* 2: 442–454.

98. Friedl, P. and Alexander, S. (2011). Cancer invasion and the microenvironment: plasticity and reciprocity. *Cell* 147: 992–1009.

99. Kang, Y. and Massague, J. (2004). Epithelial-mesenchymal transitions: twist in development and metastasis. *Cell* 118: 277–279.

100. Huttenlocher, A. and Horwitz, A.R. (2011). Integrins in cell migration. *Cold Spring Harb. Perspect. Biol.* 3: a005074.

101. Conklin, M.W., Eickhoff, J.C., Riching, K.M. et al. (2011). Aligned collagen is a prognostic signature for survival in human breast carcinoma. *Am. J. Pathol.* 178: 1221–1232.

102. Madsen, C.D. and Sahai, E. (2010). Cancer dissemination–lessons from leukocytes. *Dev. Cell* 19: 13–26.

103. Strilic, B. and Offermanns, S. (2017). Intravascular survival and extravasation of tumor cells. *Cancer Cell* 32: 282–293.

104. Stegner, D., Dutting, S., and Nieswandt, B. (2014). Mechanistic explanation for platelet contribution to cancer metastasis. *Thromb. Res.* 133 (Suppl 2): S149–S157.

105. Furlow, P.W., Zhang, S., Soong, T.D. et al. (2015). Mechanosensitive pannexin-1 channels mediate microvascular metastatic cell survival. *Nat. Cell Biol.* 17: 943–952.

106. Laubli, H. and Borsig, L. (2010). Selectins promote tumor metastasis. *Semin. Cancer Biol.* 20: 169–177.

107. Azevedo, A.S., Follain, G., Patthabhiraman, S. et al. (2015). Metastasis of circulating tumor cells: favorable soil or suitable biomechanics, or both? *Cell Adhes. Migr.* 9: 345–356.

108. Coman, D.R., de, L.R., and Mcc, U.M. (1951). Studies on the mechanisms of metastasis; the distribution of tumors in various organs in relation to the distribution of arterial emboli. *Cancer Res.* 11: 648–651.

109. Peinado, H., Zhang, H., Matei, I.R. et al. (2017). Pre-metastatic niches: organ-specific homes for metastases. *Nat. Rev. Cancer* 17: 302–317.

3

CAUSES OF CANCER: GENETIC, EPIGENETIC, VIRAL, MICROENVIRONMENTAL, AND ENVIRONMENTAL CONTRIBUTIONS TO CANCER

Michaela Reagan

Tufts University School of Medicine, Boston, MA, USA
Faculty Scientist I, Maine Medical Center Research Institute,
Scarborough, ME, USA

INTRODUCTION

It is well known that genetic mutation is the commonality linking all cancers across stage, location, and triggering event. The origins of these genetic abnormalities, however, are often diverse, multifactorial, and not well understood between patients, tumor types, and during disease progression. Environmental influences, epigenetic changes, random mutations, and systemic alterations, such as inflammation and metabolic dysfunction, most commonly initiate genetic mutations (known as "acquired" somatic mutations), with only 10% of cancers stemming from inherited genetic mutations (known as "germline" mutations) [1]. The multitude of cancer causes and myriad mutations associated with cancer are what make this disease so intratumorally (within a tumor) and intertumorally (between different tumors) heterogeneous and difficult to treat and cure [2]. In this chapter, the known causes of cancer are outlined and discussed to suggest preventative

Cancer: Prevention, Early Detection, Treatment and Recovery, Second Edition. Edited by Gary S. Stein and Kimberly P. Luebbers.
© 2019 John Wiley & Sons, Inc. Published 2019 by John Wiley & Sons, Inc.

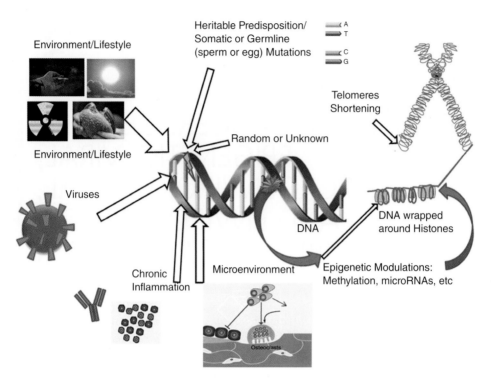

Figure 1. The causes of cancer. Cancerous mutations can be caused by numerous factors. Mutations of the DNA can result from the environment (pollutants, carcinogens, UV exposure, radiation), choices we make and our lifestyles, random or unknown mutations that are not corrected properly, epigenetic alterations, viruses, or even from changes in the local, noncancerous microenvironment of the cell. Mutations can also be inherited from our parents and can be enhanced by inflammatory signals. Shortening of telomeres during aging can also contribute to cancer. (See insert for color representation.)

measures that can be taken by individuals, societies, or governments to decrease cancer prevalence (Figure 1).

The causes of cancer have been studied primarily in mice (often termed "*in vivo*," for "in the body") or in culture conditions, such as glass or plastic petri dishes, using cells removed from humans or other animals (termed "*in vitro*" experiments, from the Latin, "in glass"). These experiments are advantageous in that they give a controlled environment in which to experiment on the exact molecular cause of a mutation with correct controls with everything else held constant except that which is being tested. This of course is not possible with human populations; however, epidemiological (human population) studies give us very strong correlation data that provides relevant insight into real human actions or traits that give rise to cancer. Epidemiological data can demonstrate clear correlations, but assessing causality must be done cautiously; more on this can be found in Chapter 10.

TUMOR GENETICS AND MUTATIONAL LANDSCAPE

Drs. Robert Weinberg and Douglas Hanahan published the seminal descriptive paper summarizing the Hallmarks of Cancer [3]. These Hallmarks of Cancer, described in the journal *Cell* in 2000 [3], comprise of: self-sufficiency in growth signals, insensitivity to anti-growth signals, evading apoptosis, limitless replicative potential, sustained angiogenesis, tissue invasion, and metastasis. In 2011, based on extensive lab research and validation, the authors added evasion of the immune system, inflammation, genomic instability, and abnormal metabolic pathways to the list [4, 5]. More recently, others have added differentiation arrest as a hallmark of certain cancers that appear stuck in an undifferentiated state [6]. Although the line between cause and consequence of many of these hallmarks is often blurred, it is evident that targeting these tumor properties has been successful, to various extents in different cases, and that understanding these hallmarks will help illuminate cancer biology and potential novel treatments.

Most mutations within a cell, such as deletions, substitutions, insertions, or translocations, do not lead to cancer, but instead cause a number of other changes, depending on their location and the type of nucleotide alteration that occurs. Although some mutations are deleterious, many are not harmful at all and will never be known to the mutation-bearer. Some mutations stop the mRNA from being transcribed, or the encoded protein from being made (if, for example, a start codon or promoter region is altered, respectively). This could be either neutral for the cell (due to compensation by other proteins that play a similar role in the cell) or deleterious and cancerous (if the gene is a tumor suppressor, such as p53, discussed later). If the mutated mRNA is transcribed, the mutation may be effectively meaningless due to redundancy of the genetic code. For example, the two codons *G-A-A* and *G-A-G* both specify the amino acid glutamic acid and hence a mutation causing an A to become a G has no effect on downstream function. Lastly, often mutated proteins simply cause cell death. However, when a mutation occurs in a coding region of the genome that is able to cause mutated proteins to emerge, termed an oncogene, this can lead to a signaling cascade within the cell that causes cancer. Importantly, mutations in "noncoding" regions of our genome (regions that do not code for a gene or protein) may also cause cancer if these occur in regions that are functional or required by the cell, such as microRNAs, long noncoding RNAs, enhancer regions, or other regions we are only now exploring.

Cancer-causing mutations result from DNA damage that is not correctly repaired, errors in DNA replication, or from the insertion or deletion of segments of DNA. For example, a chromosomal rearrangement termed the Philadelphia Chromosome, creates a novel Bcr/Abl fusion oncogene and fusion oncoprotein that leads to CML (chronic myeloid leukemia) [7]. This hyperactive protein causes extensive activation of many pathways within the cell that increase the cell's proliferation, analogous to pressing down on the accelerator when driving. Conversely, proteins that are tumor suppressors, sometimes called anti-oncogenes, work as breaks in the cell cycle to assure that cells do not proliferate out of control. These can become mutated, removing the breaking mechanisms, and leading to uncontrolled cell growth and tumor initiation.

Mutations that convert cellular proto-oncogenes (normal genes that become oncogenes when altered by mutations) to oncogenes (genes that are known to cause cancer), are some of the most common causes of cancer. These oncogenes may cause hyperactivation of signaling pathways that cause cells to proliferate out of control, while inactivation of tumor suppressors eliminates critical negative regulators of signaling, essentially removing the breaks on cell proliferation. Importantly, some tumor types are known to be frequently caused by mutations in specific genes (such as Ras in pancreatic cancer), most cancers are driven by more diverse mutations with intermediate frequencies (2–20%) [8]. The most common classes of oncogenes are shown in Table 1. By sequencing 21 different tumor types, mutations related to proliferation, apoptosis, genome stability, chromatin regulation, immune evasion, RNA processing, and protein homeostasis have been discovered recently as the array of cancer causing genes [8]. Interestingly, only 22 genes were found to be significant in three or more tumor types: the well-established cancer genes *TP53, PIK3CA, PTEN, RB1, KRAS, NRAS, BRAF, CDKN2A, FBXW7, ARID1A, MLL2,* and *STAG2,* were significant in four or more tumor types and the ten genes *ATM, CASP8, CTCF, ERBB3, HLA-A, HRAS, IDH1, NF1, NFE2L2* and *PIK3R1* were significant in three tumor types [8]. These are sorted into categories as shown in Table 1.

Besides those genes cited in Table 1, a number of other oncogenes have been identified as mutated in cancer that likely pay a driver role in cancer initiation. These fall into the following categories: anti-proliferative proteins (e.g., *ARHGAP35, DNER, MGA, and IRF6*), cell proliferation proteins (e.g., *SOS1, ELF3, SGK1* and *MYOCD*), pro-apoptotic factors (e.g., *ALPK2, BCLAF1, MAP4K3, ZNF750,* and *TNF*), proteins related to genome stability (e.g., *CEP76, RAD21,TP53BP1, TPX2, ZRANB3,* and *STX2*), chromatin regulation (e.g., *SETDB1, SETD2, MBD1, EZH1, EZH2, CHD8, CHD4, HIST1H4E, HIST1H1E* and *HIST1H3B),* immune evasion, *HLA-B, TAP1,* and *CD1D),* RNA processing and metabolism (e.g., *PCBP1, QKI, RPL5*), and protein homeostasis *(TRIM23)* [8]. Other common proto-oncogenes include genes in the following families: *YAP* and *TAZ* [31], *WNT* [32], *MYC* [33, 34], *ERK* [35], and *TRK* [36], among many others. Importantly, more oncogenes and oncoproteins are being discovered each year, and novel targets and mechanisms to stop tumor growth are being developed worldwide.

The loss of tumor suppressors can be just as dangerous and carcinogenic as the gain of oncogenic activity, and often both of these mechanisms work in concert to drive tumorigenesis. For example, PTEN, a tumor suppressor that inhibits activity of the proto-oncoproteins PI3K and Akt [37] and overexpression of prostate-specific G-protein-coupled receptor have been found to act in concert to promote prostate cancer [38]. Other tumor suppressors include APC, UBC, DAPK, PLAGL1, PYCARD/ASC, RB1, pVHL, RBM5, and YPEL3. CD95 was first found

TABLE 1. Genes significant in three or more cancer types.

Oncogenes	Normal Function	Example	References
GTPases	Signal transduction, regulation of biochemical processes	Ras (mutated in about 20% of human cancers), *RHEB, RHOA. HRAS, KRAS,* and *NRAS* mutated in up to 90% of pancreatic cancers.	[9–11]
Receptor Tyrosine Kinases	Detect and transfer signals between outside and inside of cell to switch on or off signaling pathways	Members of the epidermal growth factor receptor (*HER/EGFR/ERBB*) family, platelet-derived growth factor receptor (*PDGFR*), and vascular endothelial growth factor receptor (*VEGFR*), insulin growth factor receptor (*IGFRs*), ephrin type-B receptor (*EPHB*), fibroblast growth factor receptor (*FGFRs*), *ERBB3*	[12–14]
Transcription factors	Transcriptional regulation, insulator activity, V(D)J recombination, regulators of expression of antioxidant proteins that protect against oxidative damage.	Transcriptional repressor *CTCF*, myc. Nuclear factor (erythroid-derived 2)-like 2 (*Nrf2*)	[8, 15]
Cytoplasmic Serine/threonine kinases and their regulatory subunits	cell cycle regulation, cell proliferation, differentiation, cells survival	*RAF* family (*BRAF*); cyclin-dependent kinases; Ataxia telangiectasia mutated (*ATM*)	[16, 17]
Phosphatidylinositide 3-kinases (PI-3Ks)	Cell growth, proliferation, differentiation, motility, survival and intracellular trafficking		[8, 18]

Tumor Suppressors	Normal Function	Example	Reference
TP53	Encodes p53 proteins, limit cell proliferation, prevents cancer, binds DNA to prevent transcription	Most frequently mutated gene in cancer. Encodes 12 proteins (p53 isoforms)	[8, 19, 20]
Phosphatidylinositide 3-kinases (PI-3Ks)	Cell growth, proliferation, differentiation, motility, survival and intracellular trafficking	*PTEN*, (phosphatase and tensin homolog) dephosphorylates phosphoinositide substrates; *PIK3R1* (phosphoinositide-3-kinase, regulatory subunit 1 alpha)	[8, 18]
Pocket protein family	Signaling. RB1 prevents excessive cell growth by inhibiting cell cycle progression until a cell is ready to divide	*RB1* (retinoblastoma protein) causes retinoblastoma and other cancers	[8, 21]

(continued)

TABLE 1. (*Continued*)

Tumor Suppressors	Normal Function	Example	Reference
Cyclin-dependent kinase Inhibitors	Tumor suppressors	Cyclin-dependent kinase Inhibitor 2A and 2B, (*CDKN2A, CDKN2B*) codes for proteins p16 and p14arf	[8, 22]
F-box proteins	Tumor suppressor ubiquitin ligase that drives protein degradation of oncoproteins	*FBXW7* (F-box/WD repeat-containing protein 7),	[8, 23]
SWI/SNF family	nucleosome remodeling complexes, remodel the way DNA is packaged	ARID1A alters the accessibility of chromatin to a variety of nuclear factors	[8, 24]
Trithorax-group proteins	histone methyltransferases that plays an important role in regulating gene transcription	*KMT2D* (lysine [K]-specific methyltransferase 2D), formerly named MLL2, commonly mutated in Kabuki syndrome	[8, 25]
Cohesin multiprotein complex proteins	Plays a role in essential role in sister chromatin cohesion	*STAG2* (Cohesin subunit SA-2) mutations in glioblastoma, melanoma, Ewing sarcoma, bladder & myeloid neoplasms	[8, 26]
Caspases	Activate cell death; mediated by both receptor-mediated apoptosis and in its absence, necroptosis.	*CASP8* mutations can cause cancer also by causing disruption to maintenance and homeostasis of the adult T-cell population	[8, 27]
Neurofibromin 1 (NF1)	Negative regulation of the RAS pathway	*NF1* mutations often lead to truncation of translated proteins, removing RAS regulation function.	[8, 28]

Immune Evasion Genes	Normal Function	Example	Reference
Major histo-compatibility complexes	Immune system recognition	Human leukocyte antigens (*HLA-A*).	[8, 29]

Metabolism genes	Normal Function	Example	Reference
Isocitrate dehydrogenases	Cytoplasmic and peroxisomal lipid metabolism	*IDH1*, (Isocitrate dehydrogenase 1 [NADP+], soluble)	[8, 30]

to be a tumor suppressor but it has more recently been found that this receptor can transmit non-apoptotic signals, promote inflammation, and contribute to carcinogenesis [39]. Some tumor suppressors, such as ST14, have been shown to be regulated (or independent of regulation from) microRNAs, a topic covered in Epigenetics later [40].

Hormones are also key drivers of many cancers, specifically those that express hormone receptors, and mutations or other changes that cause excess hormonal signaling to cancer cells can drive many tumors. Some examples include the thyroid hormone receptor [41] in thyroid cancer, and androgens such as testosterone in prostate cancer, and estrogen in breast and female reproductive cancers [42, 43].

Certain mutations appear more commonly in certain types of cancers, and the reason behind this remains unknown, making it virtually impossible to target and treat all cancers with the same type of therapy. The mutational landscape across 12 major cancer types was recently reported. It found that mutations in certain genes, such as *TP53*, are common in many different cancers (42% of cancer), while other mutations were tumor-type specific [44]. Inactivating mutations of tumor suppressors, such *TP53* have also been shown in many mouse models to be true cancer drivers, while other mutations have not been verified as drivers and may simply be passenger mutations – common but not participatory in oncogenesis [20, 45]. Importantly, the development of malignant tumors generally requires an accumulation of genetic mutations through a process referred to as "multistep carcinogenesis." Recent evidence from The Cancer Genome Atlas (TCGA) has used the latest sequencing and analysis methods to identify somatic variants across thousands of tumors across 12 common tumor types [44]. They identified 127 significantly mutated genes from well-known pathways, including mitogen-activated protein kinase (MAPK), phosphatidylinositol-3-OH kinase (PI3K), Wnt/β-catenin and receptor tyrosine kinase signaling pathways, and cell cycle control. They also found mutated genes in emerging pathways related to, for example, histones, histone modification, gene splicing, metabolism, and proteolysis. The average number of mutations in these significantly mutated genes varied across tumor types with most tumors having two to six mutations, a relatively small number of driver mutations required for oncogenesis. Tissue specificity was observed in transcriptional factor or regulator mutations, whereas mutations in histone modifiers were often pervasive across several cancer types. They also found mutation patterns that demonstrated clonal and subclonal architecture derived from temporal tumor evolution, one of the most challenging aspects of tumorigenesis [44].

Inherited/Familial Cancers

Certain cancers are known to be "familial," meaning a germline (inherited) mutation either provides a strong predisposition or directly causes a certain cancer. Often cancers are a combination of an inherited predisposition and the right environmental stimuli to cause tumor development, but a strong enough genetic link,

TABLE 2. Familial cancer mutations and cancers.

Cancer Type	Genes with Inherited Mutation	References
Breast Cancer	BRCA1, BRCA2; TP53, STK11, PTEN, ATM and CHEK2	[46]
Gynecological Cancers	BRCA1, BRCA2; Variants at 1p36 (nearest gene, WNT4), SYNPO2, ABO, ATAD5, RSPO1, GPX6	[47, 48]
Acute Myeloid Leukemia (AML)	DDX41	[49]
Renal Cell Carcinoma	PBRM1	[50]
Ovarian Small Cell Carcinoma	SMARCA4	[51]
Small Intestinal Carcinoid	Inositol polyphosphate multikinase (IPMK)	[52]
Diffuse Gastric Cancer	CDH1 (E-cadherin) and CTNNA1 (α-E-catenin), p53	[53, 54]
Colorectal Cancer	APC (adenomatous polyposis coli), MUTYH	[55, 56]
Pancreatic cancer	PRSS1 (protease, serine, 1 [trypsin 1])	[57, 58]
Melanoma	CDKN2A	[59]
Meningioma	SHH-GLI1 signaling pathway gene, SUFU	[60]
Various (Sarcoma, breast, brain, retinoblastoma, adrenocortical carcinoma)	P53, RB1	[21, 61]
Paraganglioma	MDH2	[62]

with affected family members and significant increases in cancer risk factors, signifies that these cancers are familial.

Many of the genes related to predispositions for certain cancers are shown in Table 2 and also described in more detail elsewhere [57, 63]. Hereditary breast cancer is a common outcome from a hereditary predisposition: women who inherit pathogenic mutations in the BRCA1 or BRCA2 genes have up to an 85% risk of developing breast cancer in their lifetimes, and more similar genes have been identified with GWAS (genome wide association studies) [46]. GWAS and familial cancer study results are implicating more germline mutations as causing predispositions for cancer every year.

EPIGENETIC DRIVERS OF CANCER

As opposed to genetic alterations, which directly change the way cells express genes due to direct mutations within the gene (or its promoter/enhancer region), epigenetic alterations modulate the way a gene is expressed through other mechanisms. Although these two alterations have long been thought to be mechanistically distinct in carcinogenesis, it is becoming increasingly clear that these are closely linked. Building on our increased understanding of the epigenetic regulation and signaling in physiological and pathological states, we now know that epigenetic and genetic drivers of cancer are not independent, but rather are tightly intertwined. Epigenetic changes that are commonly associated with cancer include DNA methylation, histone modification, nucleosome positioning, and noncoding

RNA (e.g., microRNA) expression; all these changes alter gene transcription, splicing, expression or translation of proteins. Recent evidence from whole exome sequencing demonstrates that many inactivating mutations in human genes actually control epigenomic aspects of the cell, as discussed in Table 1. Epigenome changes and mutations also have the potential to cause point mutations and disable DNA repair functions. This crosstalk between the genome and the epigenome offers new possibilities for therapy [64].

MicroRNAs (or miRs) are short, noncoding RNA molecules that regulate the expression of one of many target genes. OncomiRs are microRNAs that play a role in the onset or development of tumors and are overexpressed in certain cancers [65]. Inhibition of these miRs using various strategies is a possibility being explored in the field now, which may be an effective therapy for tumors that are considered miR-addicted, which require the over- or under-expression of a certain miR to proliferate or survive. miRs may also play a role in initial carcinogenesis of tumor cells, but determining this requires more *in vivo* investigations [66]. Lastly, epigenetic alterations are also seen within noncancerous cells surrounding tumor cells, and represent one way in which the local microenvironment may be modulated by tumor cells to support their development (see Microenvironment later) [67, 68].

HDACs (Histone deacetylases) are proteins that modify histones to lead to increased binding of histones to DNA, and subsequent decreased transcription. HDAC-inhibitors are being explored and used as anticancer therapies; they cause increased expression of tumor-suppressor genes, which have been shut-off via this epigenetic mechanism, and are a novel class of drugs that hold great potential in stopping tumor growth by targeting epigenetic rather than genetic dysfunction. Of course, why these HDACs may be over-expressed or overactive may lead back to a genetic mutation, again highlighting the interconnectedness of genetic and epigenetic changes contributing to tumor growth [69, 70]. Aberrant DNA methylation is another key tenet of epigenetic-driven oncogenesis, described by Tenen et al. [71].

VIRAL

Certain viruses are also known to cause cancers, with six distinct human viruses currently termed oncogenic [72]. These are summarized in Table 3. Some of these viruses are now the most preventable causes of cancer with the invention of safe and effective anti-HPV vaccines, which provide long-term immunization from these viruses [81].

ENVIRONMENTAL STIMULI

External influences such as toxins, pollutants, carcinogens, and radiation, are one of the largest contributors to cancer. Chronic inflammation is arguably the largest risk factor for cancer, with 20% of cancers being linked to chronic infection and 30%

TABLE 3. Viral-associated cancers.

Virus	Resulting Cancer
Human papillomaviruses (HPVs)	~100% of cervix/uterine tumors, significant proportion of anal, penile, and oral carcinomas [72], some head and neck squamous cell carcinomas (HNSCC) [73–75].
Hepatitis B Virus (HBV)	Hepatocellular (liver) cancer [76],
Hepatitis C virus (HCV)	Hepatocellular (liver) cancer [77]
Epstein-Barr virus (EBV)	Burkitt lymphoma, some Hodgkin lymphomas, nasopharyngeal carcinoma, and some gastric cancers
Kaposi sarcoma virus (KSHV), a herpes virus	Kaposi sarcoma [78]
human T-cell leukemia/lymphoma virus type 1 (HTLV-1)	Adult T leukemia/lymphoma [79]
Merkel cell polyomavirus	Merkel cell carcinoma [80]
HIV	Not directly oncogenic, but can be associated with cancer through immunosuppression (AIDS) [72]

of cancers being linked to chronic inhalation of pollutants such as tobacco smoke and silica [82]. Up to 35% of cancers can be attributed to dietary factors, including 20% of cancers being linked with obesity [82]. The link between inflammation and cancer has become stronger as more epidemiological and *in vivo* evidence has accumulated [82]. Gut and skin microbiomes, influenced by diet, genetics, initial colonization of microbes, antibiotics, and other variables, are also gathering potential as new mediators of many diseases, including cancer [83, 84]. Some of the most well-known and strongly supported environmental causes of cancers, with both epidemiological and *in vivo/in vitro* data are shown in Table 4. A more in depth review of this area can also be found in Sankpal et al. [120].

Other epidemiological data correlates risk with certain variables, but the direct causative links may be less biologically understood or more complicated. For example, exercise decreases the risks of many cancers, likely through whole-systemic effects such as decreased inflammation, decreased risk of obesity, decreased metabolic syndrome, and better overall immune function, but the pathways governing these changes are highly interconnected. Similarly, deceased "sitting time" is significantly associated with decreased cancer risk [130], but "sitting time" is a complex variable containing multiple other variables and no direct studies have proven that "increased sitting time" itself directly causes cancer (unlike, say, cigarette smoke). Other data tell us increased risks for cancer without a known or hypothesized mechanism; for example, African Americans (vs Caucasians) and males (vs females) have an increased risk of multiple myeloma (MM) but no theories or biological scientific studies can yet explain this. This may suggest that there are underlying mechanisms by which MM progresses that we do not fully understand, or, conversely, that there are simply correlations that may not suggest causation. These are important to differentiate when determining best treatments, prevention, and risk stratification categories.

TABLE 4. Common environmental causes of cancer.

Agent	Cancer	Notes	References
Cigarette Smoke	Lung, Esophageal, Larynx, Oral		[85–88]
Asbestos	Mesothelioma, Stomach, Lung, others		[89–93]
Pesticides (e.g., Dichlorodiphenyl-trichloroethane [DDT], etc.)	All. Notably blood, skin cancer, colorectal and breast cancer	Most up-to-date and comprehensive review can be found in Guyton et al. [94].	[95–102]
"Western Diet" (High Fat/carb/red meat, Low Fiber, Low vegetable)	Colorectal, Liver. Associations with other cancers	Restricted calorie diet can also reduce cancer prevalence [103, 104].	[84, 105–109]
Alcohol (Chronic/high doses)	Liver, mouth, pharynx, larynx, esophagus, colon, rectum, Upper aerodigestive tract	No increased risk of breast cancer [110], leukemia [111] or multiple myeloma [112]	[113–115]
Ionizing Radiation (UV sunlight/ tanning beds)	Nonmelanoma and melanoma skin cancers	Optimal dosages should maximize Vitamin D and minimize cancer risks.	[116–118]
Obesity (often interdependent with Metabolic Syndrome/Diabetes)	Liver, endometrial, pancreatic, colon, potentially others.	Confounded with microbiome [119], adipokines, inflammation, metabolism changes.	[120–124]
Low Physical Activity	Breast cancer, colon cancer, skin cancer	Many confounding effects, depending on type of activity, type of organ, etc.	[125–127]
Pollution, certain Chemicals, and Carcinogens	Lung cancer (Air); Liver, lung, bladder, and kidney cancer (Water)	Arsenic, by-products of chlorination, and other pollutants in water.	[128, 129]

MICROENVIRONMENT-DRIVEN ONCOGENESIS

Interestingly, it has recently surfaced that microenvironmental stimuli, meaning changes in the local microenvironment surrounding a cell within a certain niche or organ within the body, can also directly cause cancerous mutations and tumor development in neighboring cells [131]. For example, new mouse models have described ways in which modulations of the local bone microenvironment (deletion of the gene Dicer1 specifically in mouse osteoprogenitors), can give rise to mutations in adjacent cells; tumor of local blood cells, specifically leukemia, were instigated through interaction with this modulated, dysfunctional local environment [132]. Many of the mechanisms behind this process remain to be elucidated, but may be related to changes in metabolism in susceptible cells and increases in oxidative stress leading to DNA damage [133]. This data defy the dogma of cell-autonomous events that lead to initiation of cancer, suggesting that leukemias may develop through non-cell autonomous pathways, which may be true for other hematological malignancies such as myeloma, or solid tumors as well. Hence, aberrant microenvironments can cause cancer by creating a niche, or tumor-enabling milieu, with an ability to prime or directly initiate tumorigenesis. The possibility that disruption of the normal cells surrounding the tumor may be not just a consequence, but potentially an initiator of cancer, deserves further interrogation. Moreover, mechanisms by which the abnormal local environment signals and causes neighboring cells to become cancer cells must also be explored more fully, as this may involve cell fusion or cell–cell communication via exosome, microvesicles, microparticles, excreted proteins, RNA, or even DNA. Research into microenvironment-directed carcinogenesis may have profound implications for anticancer therapies of the future.

TUMOR HETEROGENEITY AND CLONALITY

Tumors are often spatially and temporally heterogeneous within a patient, causing a huge challenge for doctors and researchers in finding the "cause" of a certain cancer. Tumors are also often heterogeneous across patients and develop, spread, and respond to treatments differently. Moreover, anticancer treatments that do not kill 100% of tumor cells (which is most treatments), often allow a certain resistant clone to grow from the rainbow of pretreatment clones or evolve in response to the treatment, leading to the development of a whole new type of drug-resistant tumor with new challenges and traits. It is also common (found in 2–6% of all cancers) that the first, or primary, tumor cannot be identified or located if the tumor has spread (metastasized) or remutated (picked up a myriad of other genetic mutations). This type of tumor, known as "carcinoma of unknown primary origin" (CUP) [134], is detected when it has already begun to spread or disseminate throughout the body, undergoing EMT (Epithelial to Mesenchymal Transition, essential for carcinoma metastasis) that leads to a change in cell marker expression and makes it harder to detect the primary location or type of cell that became mutated. Moreover, even when healthcare providers can diagnose a type of cancer, it can often be difficult to identify the cause because there may be so many different types of mutations within the cancer – each cell may represent a different clone – and mathematical models to describe which clone likely stemmed from which clone can be incredibly challenging in humans, as well as mice. For example, multiple myeloma (MM), a type of blood cancer that grows and spreads in the bone marrow (BM), is composed of many different clones of MM cells, and certain clones emerge after different rounds of chemotherapy and treatment, making the identification of the driver clone or mutation challenging [135].

CAUSES OF DISEASE PROGRESSION

When searching for the cause of cancer, it is important to recognize that an initial mutation causing an abnormal growth is not necessarily a cancer (it may be simply a benign growth or precancerous carcinoma *in situ*). In this way, the causes of cancer go beyond the initial mutations; cancer can be stopped if invasion is prohibited, even if all mutation-containing cells are not eradicated. For these mutated cells to develop beyond an isolated, mutated cell and become overt, symptom-causing cancer, the cells must adapt, mutate, or otherwise change their local environment (Figure 2). Most cancers, though not all (e.g., brain tumors) must invade the surrounding tissue and metastasize to wreak havoc. To do this, solid tumors undergo an EMT (epithelial-to-mesenchymal transition) that allows them to migrate to distant locations. Breast cancers must invade and migrate before to make the transition from benign (precancerous carcinoma *in situ*) to invasive carcinoma stages. In this way, there are actually many

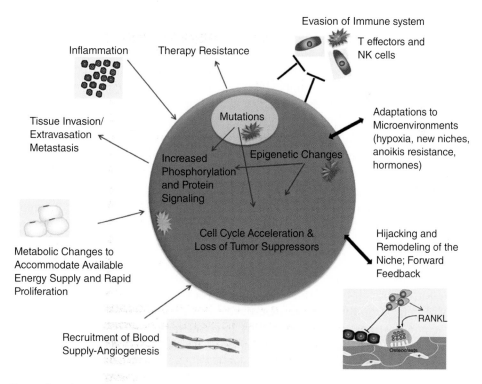

Figure 2. Drivers of cancer. Genetic and epigenetic mutations or alterations typically lead to altered signaling within a cell that alter cell proliferation, cell cycle progression, apoptosis, or cell death, such that the cell becomes immortal. Inflammation, metabolic changes, immune evasion, and signals from the local microenvironment can also contribute to tumorigenesis and disease progression. These mutations then lead to therapy resistance, increased survival ability, blood vessel recruitment (angiogenesis), invasion and metastasis, and modulation of the local noncancerous environment to further support the tumor through positive feedback loops. RANKL, Receptor activator of nuclear factor kappa-B ligand. (See insert for color representation.)

other causes of cancer, including all the steps that support this invasive and migratory capacity of tumor cells. Depending on the tumor, this may include many stages of metastasis including local migration through the basement extracellular matrix membrane, angiogenesis (recruitment of blood vessels to the tumor), evasion of the immune system, transendothelial migration into (intravasation) and out of (extravasation) the blood vessels, survival while traveling through the circulatory or lymphatic systems, homing to distant organs, reinitiation of proliferation at the distant site, metabolic changes and adaptations to the new microenvironment, and hijacking of the new environment to support tumor growth and drug/therapy resistance.

MICROENVIRONMENT HIJACKING FOR CANCER PROGRESSION

As mentioned above, the microenvironment often acts as a tumor-supportive niche essential for tumor cell growth. Many tumor cells are completely reliant on the correct local microenvironment: metastatic breast cancer cells commonly grow in the BM niche but do not grow in other organs, such as the pancreas [136]. Organotropism, or the specific homing of cancer cells for certain types of distant organs, often differs across tumor-types [137, 138]. This has been seen in humans and in mouse models and demonstrates that, although not always crucial in the cause of cancer, the microenvironment may be key in determining if the tumor is able to disseminate and colonize other organs within the patient [139–141]. As tumor cells often hijack and modulate the microenvironment that they spread or metastasize into, understanding the bi-directional communication between the microenvironment or host cells (such as immune cells [142, 143], BM stromal cells [67, 144], osteoblasts [145], cancer-associated fibroblasts [146], or tumor-associated macrophages [147]) and the resident tumor cell may be of even greater importance in designing ways to stop tumor growth after the initial mutation has occurred.

For example, the unique properties of the BM niche make it exceptionally conducive for tumor cell colonization. Breast cancer (BrCa), prostate cancer (PrCa), and MM cells strongly prefer to metastasize and grow within BM, compared to other anatomical locations, due not only to characteristic properties of the bone described above, but also due to positive feedback loops initiated by tumor cells within the niche. By causing osteolytic (bone-destructive, common in BrCa and MM) or osteoblastic (bone forming, common in PrCa) lesions in bone, cellular crosstalk is initiated that supports tumor growth and uncoupling of bone remodeling. Osteolytic cancers induce a forward feedback loop termed the "vicious cycle," where bone-embedded growth factors, ECM (extracellular matrix) proteins, and calcium are released as bone is resorbed, which then signal to tumor cells to support their growth [148, 149]. As tumor cells (seeds) take hold and begin to grow in the fertile soil of the BM, dynamic interactions occur where the seeds modulate and usurp the BM to support their own growth, at the expense of normal bone homeostasis, leading to increased fractures, hypercalcemia, spinal cord compression, and immune cell dysfunction [150]. This fertile soil has been shown to be useful to cancer cells in numerous ways, including providing protection from anticancer therapies through cell adhesion mediated drug resistance (CAM-DR). This occurs, in part, due to the quiescent state induced in cells in the BM niche, allowing for long-term survival of malignant cells. Thus, cancer stemness is not a fixed entity, but can be instilled and nurtured by the niche microenvironment. The soil also provides numerous other growth factors that facilitate tumor quiescence or growth with the marrow; a clearer understanding of what dictates if a tumor cell will remain dormant for years or proliferate and initiate the vicious cycle is still needed. Importantly, the plasticity of differentiation of putative cancer stem cells (CSCs) that may be driven by BM interactions suggests that eradicating CSCs will not stop tumor growth, as more CSCs may be reinitiated upon association with the HSC niche. Although evidence for this is clear with leukemia initiating stem cells, the presence of stem cells or tumor initiating cells remains controversial in other cancers, and hence the role of the BM is likely be dependent on tumor type and clone properties.

CONCLUSIONS

This chapter discussed the causes of cancer in the following order: Genetic (somatic and inherited), Epigenetic, Viral, and Environmental. We then summarized the concept of Microenvironment-Driven Oncogenesis (a novel paradigm), and tumor heterogeneity and clonality, important tumor properties that make these exceedingly evasive. We then explored why it is that cancer, as a disease, is more than just a result of mutations and uncontrolled cell growth, but also typically progresses through stages of invasion and metastasis to other organs to carry out a full attack. As part of this, we finally examined how tumors hijack and modulate their local environment to support their own growth; without this capacity, cancer would not be the terrible disease that it is.

REFERENCES

1. Grivennikov, S.I., Greten, F.R., and Karin, M. (2010). Immunity, inflammation, and cancer. *Cell* 140: 883–899.
2. McGranahan, N. and Swanton, C. (2015). Biological and therapeutic impact of intratumor heterogeneity in cancer evolution. *Cancer Cell* 27: 15–26.
3. Hanahan, D. and Weinberg, R.A. (2000). The hallmarks of cancer. *Cell* 100: 57–70.
4. Hanahan, D. and Weinberg, R.A. (2011). Hallmarks of cancer: the next generation. *Cell* 144: 646–674.
5. Parker, S.J. and Metallo, C.M. (2015). Metabolic consequences of oncogenic IDH mutations. *Pharmacology & Therapeutics* 152: 54–62. https://doi.org/10.1016/j.pharmthera.2015.05.003.

6. Koschmieder, S., Halmos, B., Levantini, E. et al. (2009). Dysregulation of the C/EBPalpha differentiation pathway in human cancer. *Journal of Clinical Oncology* 27: 619–628.

7. Ben-Neriah, Y., Daley, G.Q., Mes-Masson, A.M. et al. (1986). The chronic myelogenous leukemia-specific P210 protein is the product of the bcr/abl hybrid gene. *Science* 233: 212–214.

8. Lawrence, M.S., Stojanov, P., Mermel, C.H. et al. (2014). Discovery and saturation analysis of cancer genes across 21 tumour types. *Nature* 505: 495–501.

9. Downward, J. (2003). Targeting RAS signalling pathways in cancer therapy. *Nature Reviews. Cancer* 3: 11–22.

10. Parker, J.A. and Mattos, C. (2015). The Ras-membrane interface: isoform-specific differences in the catalytic domain. *Molecular Cancer Research* 13: 595–603.

11. Bryant, K.L., Mancias, J.D., Kimmelman, A.C. et al. (2014). KRAS: feeding pancreatic cancer proliferation. *Trends in Biochemical Sciences* 39: 91–100.

12. Gschwind, A., Fischer, O.M., and Ullrich, A. (2004). The discovery of receptor tyrosine kinases: targets for cancer therapy. *Nature Reviews. Cancer* 4: 361–370.

13. Hojjat-Farsangi, M. (2014). Small-molecule inhibitors of the receptor tyrosine kinases: promising tools for targeted cancer therapies. *International Journal of Molecular Sciences* 15: 13768–13801.

14. Montemurro, F., Di Cosimo, S., and Arpino, G. (2013). Human epidermal growth factor receptor 2 (HER2)-positive and hormone receptor-positive breast cancer: new insights into molecular interactions and clinical implications. *Annals of Oncology* 24: 2715–2724.

15. Qi, C.-F., Kim, Y.S., Xiang, S. et al. (2012). Characterization of ARF-BP1/HUWE1 interactions with CTCF, MYC, ARF and p53 in MYC-driven B cell neoplasms. *International Journal of Molecular Sciences* 13: 6204–6219.

16. Xu, W. and Ji, J.-Y. (2011). Dysregulation of CDK8 and cyclin C in tumorigenesis. *Journal of Genetics and Genomics* 38: 439–452.

17. Akli, S. and Keyomarsi, K. Cyclin E and its low molecular weight forms in human cancer and as targets for cancer therapy. *Cancer Biology & Therapy* 2: S38–S47.

18. Mukohara, T. (2015). PI3K mutations in breast cancer: prognostic and therapeutic implications. *Breast Cancer (Dove Med. Press.)* 7: 111–123.

19. Duffy, M.J., Synnott, N.C., PM, M.G. et al. (2014). p53 as a target for the treatment of cancer. *Cancer Treatment Reviews* 40: 1153–1160.

20. Bradner, J.E. (2015). Cancer: an essential passenger with p53. *Nature* 520: 626–627.

21. De Jong, M.C., Kors, W.A., de Graaf, P. et al. (2014). Trilateral retinoblastoma: a systematic review and meta-analysis. *The Lancet Oncology* 15: 1157–1167.

22. Walker, G.J., Flores, J.F., Glendening, J.M. et al. (1998). Virtually 100% of melanoma cell lines harbor alterations at the DNA level within CDKN2A, CDKN2B, or one of their downstream targets. *Genes, Chromosomes & Cancer* 22: 157–163.

23. Davis, R.J., Welcker, M., and Clurman, B.E. (2014). Tumor suppression by the Fbw7 ubiquitin ligase: mechanisms and opportunities. *Cancer Cell* 26: 455–464.

24. Wu, R.-C., Wang, T.-L., and Shih, I.-M. (2014). The emerging roles of ARID1A in tumor suppression. *Cancer Biology & Therapy* 15: 655–664.

25. Ng, S.B., Bigham, A.W., Buckingham, K.J. et al. (2010). Exome sequencing identifies MLL2 mutations as a cause of Kabuki syndrome. *Nature Genetics* 42: 790–793.

26. Tirode, F., Surdez, D., Ma, X. et al. (2014). Genomic landscape of Ewing sarcoma defines an aggressive subtype with co-association of STAG2 and TP53 mutations. *Cancer Discovery* 4: 1342–1353.

27. Liu, B., Zhang, Y., Jin, M. et al. (2010). Association of selected polymorphisms of CCND1, p21, and caspase8 with colorectal cancer risk. *Molecular Carcinogenesis* 49: 75–84.

28. Abramowicz, A. and Gos, M. Neurofibromin in neurofibromatosis type 1 - mutations in NF1 gene as a cause of disease. *Developmental Period Medicine* 18: 297–306.

29. Smith, A.G., Fan, W., Regen, L. et al. (2012). Somatic mutations in the HLA genes of patients with hematological malignancy. *Tissue Antigens* 79: 359–366.

30. Bogdanovic, E. (2015). IDH1, lipid metabolism and cancer: shedding new light on old ideas. *Biochimica et Biophysica Acta* 1850: 1781–1785.

31. Moroishi, T., Hansen, C.G., and Guan, K.-L. (2015). The emerging roles of YAP and TAZ in cancer. *Nature Reviews. Cancer* 15: 73–79.

32. Webster, M.R. and Weeraratna, A.T. (2013). A Wnt-er migration: the confusing role of β-catenin in melanoma metastasis. *Science Signaling* 6: pe11.

33. Chesi, M., Robbiani, D.F., Sebag, M. et al. (2008). AID-dependent activation of a MYC transgene induces multiple myeloma in a conditional mouse model of post-germinal center malignancies. *Cancer Cell* 13: 167–180.

34. Petrich, A.M., Nabhan, C., and Smith, S.M. (2014). MYC-associated and double-hit lymphomas: a review of pathobiology, prognosis, and therapeutic approaches. *Cancer* 120: 3884–3895.

35. Sever, R. and Brugge, J.S. Signal transduction in cancer. *Cold Spring Harbor Perspectives in Medicine* 5: a006098.

36. Brzeziańska, E., Pastuszak-Lewandoska, D., and Lewiński, A. (2007). Rearrangements of NTRK1 oncogene in papillary thyroid carcinoma. *Neuro Endocrinology Letters* 28: 221–229.

37. Sansal, I. and Sellers, W.R. (2004). The biology and clinical relevance of the PTEN tumor suppressor pathway. *Journal of Clinical Oncology* 22: 2954–2963.

38. Rodriguez, M., Siwko, S., Zeng, L. et al. (2015). Prostate-specific G-protein-coupled receptor collaborates with loss of PTEN to promote prostate cancer progression. *Oncogene* 35 (9): 1153–1162. https://doi.org/10.1038/onc.2015.170.

39. Fouqué, A., Debure, L., and Legembre, P. (2014). The CD95/CD95L signaling pathway: a role in carcinogenesis. *Biochemica et Biophysica Acta* 1846: 130–141.

40. Wang, Y., Rathinam, R., Walch, A. et al. (2009). ST14 (suppression of tumorigenicity 14) gene is a target for miR-27b, and the inhibitory effect of ST14 on cell growth is independent of miR-27b regulation. *The Journal of Biological Chemistry* 284: 23094–23106.

41. Park, J.W., Zhao, L., Willingham, M. et al. (2015). Oncogenic mutations of thyroid hormone receptor β. *Oncotarget* 6: 8115–8131.

42. Gleave, M., Hsieh, J.T., Gao, C.A. et al. (1991). Acceleration of human prostate cancer growth in vivo by factors produced by prostate and bone fibroblasts. *Cancer Research* 51: 3753–3761.

43. Santen, R.J. (2011). Clinical review: effect of endocrine therapies on bone in breast cancer patients. *The Journal of Clinical Endocrinology and Metabolism* 96 (2): 308–319. https://doi.org/10.1210/jc.2010-1679.

44. Kandoth, C., MD, M.L., Vandin, F. et al. (2013). Mutational landscape and significance across 12 major cancer types. *Nature* 502: 333–339.

45. Liu, Y., Zhang, X., Han, C. et al. (2015). TP53 loss creates therapeutic vulnerability in colorectal cancer. *Nature* 520: 697–701.

46. Aloraifi, F., Boland, M.R., Green, A.J. et al. (2015). Gene analysis techniques and susceptibility gene discovery in non-BRCA1/BRCA2 familial breast cancer. *Surgical Oncology* https://doi.org/10.1016/j.suronc.2015.04.003.

47. Kuchenbaecker, K.B., Ramus, S.J., Tyrer, J. et al. (2015). Identification of six new susceptibility loci for invasive epithelial ovarian cancer. *Nature Genetics* 47: 164–171.

48. Hirasawa, A., Masuda, K., Akahane, T. et al. (2014). Family history and BRCA1/BRCA2 status among Japanese ovarian cancer patients and occult cancer in a BRCA1 mutant case. *Japanese Journal of Clinical Oncology* 44: 49–56.

49. Polprasert, C., Schulze, L., Sekeres, M.A. et al. (2015). Inherited and somatic defects in DDX41 in myeloid neoplasms. *Cancer Cell* https://doi.org/10.1016/j.ccell.2015.03.017.

50. Benusiglio, P.R., Couvé, S., Gilbert-Dussardier, B. et al. (2015). A germline mutation in PBRM1 predisposes to renal cell carcinoma. *Journal of Medical Genetics* https://doi.org/10.1136/jmedgenet-2014-102912.

51. Moes-Sosnowska, J., Szafron, L., Nowakowska, D. et al. (2015). Germline SMARCA4 mutations in patients with ovarian small cell carcinoma of hypercalcemic type. *Orphanet Journal of Rare Diseases* 10: 32.

52. Sei, Y., Zhao, X., Forbes, J. et al. (2015). A hereditary form of small intestinal carcinoid associated with a germline mutation in inositol polyphosphate multikinase. *Gastroenterology* https://doi.org/10.1053/j.gastro.2015.04.008.

53. Pinheiro, H., Oliveira, C., Seruca, R. et al. (2014). Hereditary diffuse gastric cancer - pathophysiology and clinical management. *Best Practice & Research. Clinical Gastroenterology* 28: 1055–1068.

54. Pinheiro, H., Oliveira, P., and Oliveira, C. (2015). Hereditary cancer risk assessment: challenges for the next-gen sequencing era. *Frontiers in Oncology* 5: 62.

55. Mazzoni, S.M. and Fearon, E.R. (2014). AXIN1 and AXIN2 variants in gastrointestinal cancers. *Cancer Letters* 355: 1–8.

56. Balaguer, F., Castellvi-Bel, S., Castells, A. et al. (2007). Identification of MYH mutation carriers in colorectal cancer: a multicenter, case-control, population-based study. *Clinical Gastroenterology and Hepatology* 5: 379–387.

57. Stoffel, E.M. (2015). Screening in GI cancers: the role of genetics. *Journal of Clinical Oncology* 33 (16): 1721–1728. https://doi.org/10.1200/JCO.2014.60.6764.

58. Mastoraki, A., Chatzimavridou-Grigoriadou, V., Chatzipetrou, V. et al. (2014). Familial pancreatic cancer: challenging diagnostic approach and therapeutic management. *Journal of Gastrointestinal Cancer* 45: 256–261.

59. Hansen, C.B., Wadge, L.M., Lowstuter, K. et al. (2004). Clinical germline genetic testing for melanoma. *The Lancet Oncology* 5: 314–319.

60. Smith, M.J. (2015). Germline and somatic mutations in meningiomas. *Cancer Genetics* https://doi.org/10.1016/j.cancergen.2015.02.003.

61. Gonzalez, K.D., Noltner, K.A., Buzin, C.H. et al. (2009). Beyond Li Fraumeni syndrome: clinical characteristics of families with p53 germline mutations. *Journal of Clinical Oncology* 27: 1250–1256.

62. Cascón, A., Comino-Méndez, I., Currás-Freixes, M. et al. (2015). Whole-exome sequencing identifies MDH2 as a new familial paraganglioma gene. *Journal of the National Cancer Institute* 107: djv053.

63. Merino, D. and Malkin, D. (2014). p53 and hereditary cancer. *Sub-Cellular Biochemistry* 85: 1–16.

64. You, J.S. and Jones, P.A. (2012). Cancer genetics and epigenetics: two sides of the same coin? *Cancer Cell* 22: 9–20.

65. Roccaro, A.M., Sacco, A., Thompson, B. et al. (2009). MicroRNAs 15a and 16 regulate tumor proliferation in multiple myeloma. *Blood* 113: 6669–6680.

66. Cheng, C.J., Bahak, R., Babar, I.A. et al. (2014). MicroRNA silencing for cancer therapy targeted to the tumour microenvironment. *Nature* 518: 107–110.

67. Reagan, M.R. and Ghobrial, I.M. (2012). Multiple myeloma-mesenchymal stem cells: characterization, origin, and tumor-promoting effects. *Clinical Cancer Research* 18: 342–349.

68. Reagan, M.R., Mishima, Y., Glavey, S.V. et al. (2014). Investigating osteogenic differentiation in multiple myeloma using a novel 3D bone marrow niche model. *Blood* 124: 3250–3259.

69. Fiskus, W., Sharma, S., Qi, J. et al. (2014). Highly active combination of BRD4 antagonist and histone deacetylase inhibitor against human acute myelogenous leukemia cells. *Molecular Cancer Therapeutics* 13: 1142–1154.

70. Stubbs, M.C., Kim, W., Bariteasu, M. et al. (2015). Selective inhibition of HDAC1 and HDAC2 as a potential therapeutic option for B-ALL. *Clinical Cancer Research* 21 (10): 2348–2358. https://doi.org/10.1158/1078-0432.CCR-14-1290.

71. Amabile, G., Di Ruscio, A., Müller, F. et al. (2015). Dissecting the role of aberrant DNA methylation in human leukaemia. *Nature Communications* 6: 7091.

72. Hibner, U. and Grégoire, D. (2015). Viruses in cancer cell plasticity: the role of hepatitis C virus in hepatocellular carcinoma. *Contemporary Oncology (Pozn)* 19: A62–A67.

73. Stier, E.A., Sebring, M.C., Mendez, A.E. et al. (2015). Prevalence of anal human papillomavirus infection and anal HPV-related disorders in women: a systematic review. *American Journal of Obstetrics and Gynecology* https://doi.org/10.1016/j.ajog.2015.03.034.

74. Rusan, M., Li, Y.Y., and Hammerman, P.S. (2015). Genomic landscape of human papillomavirus-associated cancers. *Clinical Cancer Research* 21: 2009–2019.

75. Smola, S. (2014). Human papillomaviruses and skin cancer. *Advanced Experimental Medicine and Biology* 810: 192–207.

76. MacLachlan, J.H. and Cowie, B.C. Hepatitis B virus epidemiology. *Cold Spring Harbor Perspectives in Medicine* 5: a021410.

77. Hsu, Y.-C., Chun-Ying, W., and Lin, J.-T. (2014). Hepatitis C virus infection, antiviral therapy, and risk of hepatocellular carcinoma. *Seminars in Oncology* 42: 329–338.

78. Bhutani, M., Polizzotto, M.N., Uldrick, T.S. et al. (2015). Kaposi sarcoma-associated herpesvirus-associated malignancies: epidemiology, pathogenesis, and advances in treatment. *Seminars in Oncology* 42: 223–246.

79. De Martel, C., Ferlay, J., Vranceschi, S. et al. (2012). Global burden of cancers attributable to infections in 2008: a review and synthetic analysis. *The Lancet Oncology* 13: 607–615.

80. Batinica, M., Akgül, B., Silling, S. et al. (2015). Correlation of Merkel cell polyomavirus positivity with PDGFRα mutations and survivin expression in Merkel cell carcinoma. *Journal of Dermatological Science* 79 (1): 43–49. https://doi.org/10.1016/j.jdermsci.2015.04.002.

81. Ferris, D., Samakoses, R., Block, S.L. et al. (2014). Long-term study of a quadrivalent human papillomavirus vaccine. *Pediatrics* 134: e657–e665.

82. Aggarwal, B.B., Vijayalekshmi, R.V., and Sung, B. (2009). Targeting inflammatory pathways for prevention and therapy of cancer: short-term friend, long-term foe. *Clinical Cancer Research* 15: 425–430.

83. Yu, Y., Champer, J., Beynet, D. et al. (2015). The role of the cutaneous microbiome in skin cancer: lessons learned from the gut. *Journal of Drugs in Dermatology* 14: 461–465.

84. O'Keefe, S.J.D., Li, J.V., Lahti, L. et al. (2015). Fat, fibre and cancer risk in African Americans and rural Africans. *Nature Communications* 6: 6342.

85. Hackshaw, A.K., Law, M.R., and Wald, N.J. (1997). The accumulated evidence on lung cancer and environmental tobacco smoke. *British Medical Journal* 315: 980–988.

86. Stinn, W., Berges, A., Meurrens, K. et al. (2013). Towards the validation of a lung tumorigenesis model with mainstream cigarette smoke inhalation using the A/J mouse. *Toxicology* 305: 49–64.

87. Witschi, H., Espiritu, I., Dance, S.T. et al. (2002). A mouse lung tumor model of tobacco smoke carcinogenesis. *Toxicological Sciences* 68: 322–330.

88. Keast, D., Ayre, D.J., and Papadimitriou, J.M. (1981). A survey of pathological changes associated with long-term high tar tobacco smoke exposure in a murine model. *The Journal of Pathology* 135: 249–257.

89. Fortunato, L. and Rushton, L. (2015). Stomach cancer and occupational exposure to asbestos: a meta-analysis of occupational cohort studies. *British Journal of Cancer* https://doi.org/10.1038/bjc.2014.599.

90. Baur, X., Soskolne, C.L., Lemen, R.A. et al. (2015). How conflicted authors undermine the World Health Organization (WHO) campaign to stop all use of asbestos: spotlight on studies showing that chrysotile is carcinogenic and facilitates other non-cancer asbestos-related diseases. *International Journal of Occupational and Environmental Health* 37: 176–179.

91. Lenters, V., Vermeulen, R., Dogger, S. et al. (2011). A meta-analysis of asbestos and lung cancer: is better quality exposure assessment associated with steeper slopes of the exposure-response relationships? *Environmental Health Perspectives* 119: 1547–1555.

92. Robinson, C., Alfonso, H., Woo, S. et al. (2014). Effect of NSAIDS and COX-2 inhibitors on the incidence and severity of asbestos-induced malignant mesothelioma: evidence from an animal model and a human cohort. *Lung Cancer* 86: 29–34.

93. Robinson, C., Alfonso, H., Woo, S. et al. (2014). Statins do not alter the incidence of mesothelioma in asbestos exposed mice or humans. *PLoS One* 9: e103025.

94. Guyton, K.Z., Loomis, D., Grosse, Y. et al. (2015). Carcinogenicity of tetrachlorvinphos, parathion, malathion, diazinon, and glyphosate. *The Lancet Oncology* 16: 490–491.

95. Salerno, C., Carcagni, A., Sacco, S. et al. (2015). An Italian population-based case-control study on the association between farming and cancer: are pesticides a plausible risk factor? *Archives of Environmental & Occupational Health* https://doi.org/10.1080/19338244.2015.1027808.

96. Schinasi, L. and Leon, M.E. (2014). Non-Hodgkin lymphoma and occupational exposure to agricultural pesticide chemical groups and active ingredients: a systematic review and meta-analysis. *International Journal of Environmental Research and Public Health* 11: 4449–4527.

97. Zendehdel, R., Tayefeh-Rahimian, R., and Kabir, A. (2014). Chronic exposure to chlorophenol related compounds in the pesticide production workplace and lung cancer: a meta-analysis. *Asian Pacific Journal of Cancer Prevention* 15: 5149–5153.

98. George, J. and Shukla, Y. (2013). Early changes in proteome levels upon acute deltamethrin exposure in mammalian skin system associated with its neoplastic transformation potential. *The Journal of Toxicological Sciences* 38: 629–642.

99. National Toxicology Program 2011. Toxicology and carcinogenesis studies of diethylamine (CAS No. 109–89-7) in F344/N rats and B6C3F1 mice (inhalation studies). *Natl. Toxicol. Program Tech. Rep. Ser.* 1–174. at https://www.ncbi.nlm.nih.gov/pubmed/22127322.

100. Song, L. et al. (2014). The organochlorine p,p'-dichlorodiphenyltrichloroethane induces colorectal cancer growth through Wnt/β-catenin signaling. *Toxicology Letters* 229: 284–291.

101. Song, L., Zhao, J., Xiaoting, J. et al. (2014). p, p'-Dichlorodiphenyldichloroethylene induces colorectal adenocarcinoma cell proliferation through oxidative stress. *PLoS One* 9: e112700.

102. Jin, J., Yu, M., Hu, C. et al. (2014). Pesticide exposure as a risk factor for myelodysplastic syndromes: a meta-analysis based on 1,942 cases and 5,359 controls. *PLoS One* 9: e110850.

103. Astagimath, M.N. and Rao, S.B. (2004). Dietary restriction (DR) and its advantages. *Indian Journal of Clinical Biochemistry* 19: 1–5.

104. Kritchevsky, D. Caloric restriction and experimental mammary carcinogenesis. *Breast Cancer Research and Treatment* 46: 161–167.

105. Lwin, S.T., Olechnowicz, S.W.Z., Fowler, J.A. et al. (2015). Diet-induced obesity promotes a myeloma-like condition in vivo. *Leukemia* 29: 507–510.

106. Yang, Y., Zhang, D., Feng, N. et al. (2014). Increased intake of vegetables, but not fruit, reduces risk for hepatocellular carcinoma: a meta-analysis. *Gastroenterology* 147: 1031–1042.

107. Tárraga López, P.J., Albero, J.S., and Rodríguez-Montes, J.A. (2014). Primary and secondary prevention of colorectal cancer. *Clinical Medicine Insights. Gastroenterology* 7: 33–46.

108. Takahashi, H., Hosono, K., Endo, H. et al. (2013). Colon epithelial proliferation and carcinogenesis in diet-induced obesity. *Journal of Gastroenterology and Hepatology* 28 (Suppl 4): 41–47.

109. Turner, D.P. (2015). Advanced glycation end-products: a biological consequence of lifestyle contributing to cancer disparity. *Cancer Research* 75 (10): 1925–1929. https://doi.org/10.1158/0008-5472.CAN-15-0169.

110. Zakhari, S. and Hoek, J.B. (2015). Alcohol and breast cancer: reconciling epidemiological and molecular data. *Advances in Experimental Medicine and Biology* 815: 7–39.

111. Rota, M., Porta, L., Pelucchi, C. et al. (2014). Alcohol drinking and risk of leukemia-a systematic review and meta-analysis of the dose-risk relation. *Cancer Epidemiology* 38: 339–345.

112. Rota, M., Porta, L., Pelucchi, C. et al. (2014). Alcohol drinking and multiple myeloma risk–a systematic review and meta-analysis of the dose-risk relationship. *European Journal of Cancer Prevention* 23: 113–121.

113. Gonzales, J.F., Barnard, N.D., Jenkins, D.J. et al. (2014). Applying the precautionary principle to nutrition and cancer. *Journal of the American College of Nutrition* 33: 239–246.

114. Li, Y., Mao, Y., Zhang, Y. et al. (2014). Alcohol drinking and upper aerodigestive tract cancer mortality: a systematic review and meta-analysis. *Oral Oncology* 50: 269–275.

115. Mercer, K.E., Hennings, L., and Ronis, M.J.J. (2015). Alcohol consumption, Wnt/β-catenin signaling, and hepatocarcinogenesis. *Advances in Experimental Medicine and Biology* 815: 185–195.

116. Mancebo, S.E., Hu, J.Y., and Wang, S.Q. (2014). Sunscreens: a review of health benefits, regulations, and controversies. *Dermatologic Clinics* 32: 427–438.

117. Reichrath, J. and Reichrath, S. (2014). Sunlight, vitamin D and malignant melanoma: an update. *Advances in Experimental Medicine and Biology* 810: 390–405.

118. Mancebo, S.E. and Wang, S.Q. (2014). Skin cancer: role of ultraviolet radiation in carcinogenesis. *Reviews on Environmental Health* 29: 265–273.

119. Ellekilde, M., Selfjord, E., Larsen, C.S. et al. (2014). Transfer of gut microbiota from lean and obese mice to antibiotic-treated mice. *Scientific Reports* 4: 5922.

120. Sankpal, U.T., Pius, H., Khan, M. et al. (2012). Environmental factors in causing human cancers: emphasis on tumorigenesis. *Tumour Biology* 33: 1265–1274.

121. McGlynn, K.A., Petrick, J.L., and London, W.T. (2015). Global epidemiology. *Clinics in Liver Disease* 19: 223–238.

122. Durko, L. and Malecka-Panas, E. (2014). Lifestyle modifications and colorectal cancer. *Current Colorectal Cancer Reports* 10: 45–54.

123. Nagaraju, G.P., Aliya, S., and Alese, O.B. (2015). Role of adiponectin in obesity related gastrointestinal carcinogenesis. *Cytokine & Growth Factor Reviews* 26: 83–93.

124. Inoue, M. and Tsugane, S. (2012). Insulin resistance and cancer: epidemiological evidence. *Endocrine-Related Cancer* 19: F1–F8.

125. Gao, Y., Huang, Y.B., Liu, X.O. et al. (2013). Tea consumption, alcohol drinking and physical activity associations with breast cancer risk among Chinese females: a systematic review and meta-analysis. *Asian Pacific Journal of Cancer Prevention* 14: 7543–7550.

126. Kruk, J. and Duchnik, E. (2014). Oxidative stress and skin diseases: possible role of physical activity. *Asian Pacific Journal of Cancer Prevention* 15: 561–568.

127. Filaire, E., Dupuis, C., Galvaing, G. et al. (2013). Lung cancer: what are the links with oxidative stress, physical activity and nutrition. *Lung Cancer* 82: 383–389.

128. Claxton, L.D. (2014). The history, genotoxicity, and carcinogenicity of carbon-based fuels and their emissions: part 5. Summary, comparisons, and conclusions. *Mutation Research/Reviews in Mutation Research* 763: 103–147.

129. Morris, R.D. (1995). Drinking water and cancer. *Environmental Health Perspectives* 103 (Suppl): 225–231.

130. Lee, J., Kuk, J.L., and Ardern, C.I. (2015). The relationship between changes in sitting time and mortality in post-menopausal US women. *Journal of Public Health (Oxford, England)* https://doi.org/10.1093/pubmed/fdv055.

131. Kang, Y. and Pantel, K. (2013). Tumor cell dissemination: emerging biological insights from animal models and cancer patients. *Cancer Cell* 23: 573–581.

132. Raaijmakers, M.H.G.P., Mukherjee, S., Guo, S. et al. (2010). Bone progenitor dysfunction induces myelodysplasia and secondary leukaemia. *Nature* 464: 852–857.

133. Raaijmakers, M.H.G.P. (2012). Myelodysplastic syndromes: revisiting the role of the bone marrow microenvironment in disease pathogenesis. *International Journal of Hematology* 95: 17–25.

134. Varadhachary, G.R. (2011). Carcinoma of unknown primary: focused evaluation. *Journal of the National Comprehensive Cancer Network* 9: 1406–1412.

135. Narayanan, N.K., Duan, B., Butcher, J.T. et al. (2014). Characterization of multiple myeloma clonal cell expansion and stromal Wnt/β-catenin signaling in hyaluronic acid-based 3D hydrogel. *In Vivo* 28: 67–73.

136. Moreau, J.E., Anderson, K., Mauney, J.R. et al. (2007). Tissue-engineered bone serves as a target for metastasis of human breast cancer in a mouse model. *Cancer Research* 67: 10304.

137. Lu, X. and Kang, Y. (2007). Organotropism of breast cancer metastasis. *Journal of Mammary Gland Biology and Neoplasia* 12: 153–162.

138. Lu, X. and Kang, Y. (2009). Efficient acquisition of dual metastasis organotropism to bone and lung through stable spontaneous fusion between MDA-MB-231 variants. *Proceedings of the National Academy of Sciences of the United States of America* 106: 9385–9390.

139. Swami, A., Reagan, M.R., Basto, P. et al. (2014). Engineered nanomedicine for myeloma and bone microenvironment targeting. *Proceedings of the National Academy of Sciences of the United States of America* 111: 10287–10292.

140. Maiso, P., Huynh, D., Maschetta, M. et al. (2015). Metabolic signature identifies novel targets for drug resistance in multiple myeloma. *Cancer Research* https://doi.org/10.1158/0008-5472. CAN-14-3400.

141. Azab, A.K., Hu, J., Quang, P. et al. (2012). Hypoxia promotes dissemination of multiple myeloma through acquisition of epithelial to mesenchymal transition-like features. *Blood* 119: 5782–5794.

142. Topalian, S.L., Drake, C.G., and Pardoll, D.M. (2015). Immune checkpoint blockade: a common denominator approach to cancer therapy. *Cancer Cell* 27: 450–461.

143. Kawano, Y., Moschetta, M., Manier, S. et al. (2015). Targeting the bone marrow microenvironment in multiple myeloma. *Immunological Reviews* 263: 160–172.

144. Glavey, S.V., Manier, S., Natoni, A. et al. (2014). The sialyltransferase ST3GAL6 influences homing and survival in multiple myeloma. *Blood* https://doi.org/10.1182/blood-2014-03-560862.

145. Reagan, M.R., Liaw, L., Rosen, C.J. et al. (2015). Dynamic interplay between bone and multiple myeloma: emerging roles of the osteoblast. *Bone* 75: 161–169.

146. Olechnowicz, S.W.Z. and Edwards, C.M. (2014). Contributions of the host microenvironment to cancer-induced bone disease. *Cancer Research* 74: 1625–1631.

147. Ribatti, D., Moschetta, M., and Vacca, A. (2014). Macrophages in multiple myeloma. *Immunology Letters* 161: 241–244.

148. Guise, T.A., Mohammad, K.S., Clines, G. et al. (2006). Basic mechanisms responsible for osteolytic and osteoblastic bone metastases. *Clinical Cancer Research* 12: 6213s–6216s.

149. Weilbaecher, K.N., Guise, T.A., and McCauley, L.K. (2011). Cancer to bone: a fatal attraction. *Nature Reviews. Cancer* 11: 411–425.

150. Roodman, G.D. (2012). Genes associate with abnormal bone cell activity in bone metastasis. *Cancer Metastasis Reviews* 31: 569–578.

III

A CANCER DIAGNOSIS

4

EARLY WARNING SIGNS

Solomon A. Graf[1,2,3] and F. Marc Stewart[1,2]

[1] *Clinical Research Division, Fred Hutchinson Cancer Research Center,*
Seattle, WA, USA
[2] *Department of Medicine, University of Washington, Seattle, WA, USA*
[3] *Veterans Affairs Puget Sound Health Care System, Seattle, WA, USA*

INTRODUCTION

Cancer encompasses a broad array of diseases and states of health. It can affect any tissue in the body and consequently can cause virtually any symptom. The evolution and proliferation of neoplastic, or abnormal, cells occurs stepwise toward an increasingly unregulated behavior that ceases to work in harmony with the host, or body, and instead sets its own course. This course defines the signs and symptoms of each cancer and, ultimately, its impact on a patient's life. As a rule, the earlier in a cancer's course that it is detected, the greater the likelihood that it can be effectively managed. Cancer screening, a preventative health measure instituted on a public health scale, has saved countless lives since its inception several decades ago. Despite this, the vast majority of cancers are not identified during scheduled screening but rather by patients themselves. This chapter will explore the nature of the initial symptoms of cancer and the challenges inherent in cancer diagnosis.

Cancer: Prevention, Early Detection, Treatment and Recovery, Second Edition. Edited by Gary S. Stein and Kimberly P. Luebbers.
© 2019 John Wiley & Sons, Inc. Published 2019 by John Wiley & Sons, Inc.

EARLY DETECTION: TECHNIQUES

The 2013 film "Elysium" features a futuristic scene in which a machine scans a human body from head to toe, declares the detection of cancer cells, and then eradicates them as it might viruses on a personal computer. This fantasy captures a core ideal of *oncology*, or cancer medicine: namely, every cancer originates in a single cell, and the ability to identify and destroy these cells at their inception would render the disease obsolete. In practice, we are limited in our ability to detect cancerous cells by available technology and its cost.

In 1895 Wilhelm Roentgen used X-rays to image internal structures of the human body. Within months of its report the technology was being employed on the battlefield to locate bullets in wounded soldiers and, soon thereafter, in identifying other abnormalities, including tumors. Today, a variety of imaging techniques are used to detect cancer at early stages. *Mammograms* are X-ray images of breast tissue and since the 1950s have been detecting tumors in the breast before they can be identified by the patient or physician. *Computed tomography* (CT), or CT-scanning, compiles and processes many rapidly acquired X-ray images to render cross-sectional imaging of sections of the body. With the interpretation of a radiologist, these images reliably show abnormalities as small as a centimeter across, or less.

Other techniques for cancer screening do not require radiology. In 1923 George Papanicolaou described his technique of brushing the cervix of a woman to collect cells that could then be stained and examined under a microscope. Though it was not widely adopted until the 1960s, this technique is largely responsible for the approximately 70% reduction in cervical cancer death in the United States in the last 30 years [1]. Cancer of the colon and rectum, the second most deadly cancer in this country (behind lung cancer) is effectively screened with a colonoscopy, a procedure in which a thin, flexible tube is passed through the colon to visualize the *epithelium*, or lining, of the tract, and even allow for minimally invasive surgical removal of abnormal growths. Indeed, it is the one cancer screening technique that can double as a treatment procedure.

While screening for cancers of the breast, cervix, and colon are of proven benefit and universally recommended depending on a patient's age, certain situations constitute high enough risk to warrant additional screening measures. Patients with liver disease and viral hepatitis are at increased risk for liver cancer and undergo periodic examinations with blood tests and ultrasound exams. The blood tests include analyzing for the marker *alpha-fetoprotein*, a glycoprotein that while normally absent in adults can be secreted by cancerous liver cells and lead to elevated levels in the blood plasma. Additionally, patients with known genetic lesions that predispose them to develop cancer in particular organs at an early age, such as mutations in the *BRCA* gene or the *PTEN* gene, may be recommended to undergo ultrasound screening of the ovaries, thyroid, or kidneys. A medical geneticist, a physician specially trained in uncommon genetic syndromes, often assists in guiding screening programs for such individuals.

GENERAL SIGNS AND SYMPTOMS

Though cancer screening methods are increasingly effective and efficient, most cancers are found because they have either caused a physical *sign* observed by someone else (such as a spouse or a physician) or a *symptom* experienced by the patient. The signs and symptoms of cancer are diverse, and can inform both the origin and stage of a cancer. Essentially any cancer type can cause fatigue, fever, reduced appetite, weight loss, and fevers. These symptoms reflect the cancer's production of inflammatory chemicals known as *cytokines* that circulate through the bloodstream. In more advanced cases, these symptoms result in cancer cachexia, a *hypercatabolic state* (condition of excess metabolic activity) in which normal tissues, including lean body mass, waste away, even if a patient is consuming a relatively normal amount of food. Cancer cachexia is a complication of late-stage cancer, and can be profoundly debilitating. Its reversal depends on successful control of the underlying malignancy which itself can be compromised by these symptoms. Potentially effective treatments for the cancer such as surgery and chemotherapy may need to be minimized or foregone altogether if patients are so weakened by cancer cachexia that the treatment would risk too much harm. Designation of patient's *performance status* according to how active and independent he or she is summarizes these symptoms (and other aspects of a patient's condition such as age and other medical problems) and is used to predict how well a patient will tolerate the toxicities of therapy.

Certain cancers are more prone to cause such systemic signs and symptoms. Hematologic cancers, or cancers of the blood and blood-forming organs, frequently cause fevers, drenching night sweats, and abnormal weight loss, which in such a situation is defined as an unintentional loss of at least 10% of the normal body weight over six months or less. Indeed, these symptoms are collectively referred to as *B symptoms* and are associated with lymphomas, in particular. A standard classification scheme of lymphomas, the *Ann Arbor staging* system, factors in the presence of B symptoms in assigning disease stage, as their presence is associated with a more aggressive and difficult-to-treat disease.

These general signs and symptoms of cancer are not unique to cancer; they are present in many other systemic illnesses, as well. The influenza virus, for example, classically causes many of these aforementioned B symptoms but typically is self-limited, with improvement after several days or a week. The persistence of these signs or symptoms over weeks or months must raise concern for the existence of underlying cancer, but can also occur in cases of uncontrolled chronic infections such as tuberculosis or endocarditis and autoimmune diseases such as rheumatoid arthritis and systemic lupus erythematous. As a result, the cause of these signs and symptoms is often not immediately apparent to healthcare providers presented with such complaints and no additional clues. The level of suspicion for an underlying cancer, which itself is influenced by the patient's age, family history, risk factors such as cigarette smoking history, and overall appearance and physical exam, dictates the breadth and depth of medical evaluation. Such an evaluation typically is performed in a tiered fashion, initially involving a comprehensive exam and basic blood work (investigating the numbers of red blood cells, white blood cells, kidney function, and liver function) and graduating to more involved testing, such as with specialized blood tests and a CT or ultrasound scan either to hone in on a suspected diagnosis or in the uncommon instances in which the initial evaluation is unrevealing. Nevertheless, general signs and symptoms of cancer often require multiple discussions and the persistence of the patient and the patient's doctor to establish the diagnosis.

One example of the potential complexity of such a situation is a *venous thrombus*, or a blood clot found in the veins. Venous thromboses can cause local symptoms in an extremity, for instance swelling and pain in a leg, and on occasion break off and travel, or *embolize*, to the lungs, putting a patient at risk for shortness of breath, chest pain, and even death. Venous thromboses are quite uncommon in the general population – approximately three such events occur for every 10000 years of life in people who do not possess any risk factors, but the risk increases 10 or even 100 fold in individuals who are predisposed to venous thromboses, such as in the setting of a recent orthopedic surgery or a strong family history of venous thromboses. Cancer, probably because of the circulating inflammatory factors it is associated with, also substantially increases the risk for developing a venous thrombus [2]. Interestingly, certain cancers such as pancreatic cancer appear to confer a much higher risk for venous thrombosis than other cancers, such as the prostate. Because of the clear association between venous thrombosis and malignancy, some experts advocate for undertaking a search for an underlying cancer in patients in whom a venous thrombosis is diagnosed with no alternative risk factor. There is no evidence, however, to suggest that identifying a cancer through such an investigation leads to better treatment and control of the cancer, presumably since such a diagnosis is often of advanced-stage disease for which limited treatments are available.

Cancer that spreads, or *metastasizes*, far from its site of origin tends to do so either through the lymph system (*lymphatic* metastasis) or blood (*hematogenous* metastasis). Lymph node metastases represent deposits of cancer and cause enlargement of the nodes. Depending on their proximity to the skin, these lymph nodes may appear as lumps or masses that are usually initially painless. Metastases spread hematogenously and are liable to grow anywhere in the body but, due to the architecture of the vascular system, find residence first in the liver or the lungs.

ORGAN-SPECIFIC SIGNS AND SYMPTOMS

Tumors by their nature are unregulated in their proliferation and spread. They ignore boundaries of normal tissues, invade or compress anatomic structures, and in doing so can damage healthy organ function.

Central Nervous System

The most dramatic example of symptoms resulting from compression by tumor is in the case of cancers that grow in or against the central nervous system, which is comprised of the brain and spinal cord. Essentially any kind of tumor can do this, though lung cancer, breast cancer, and prostate cancer are most commonly responsible. Tumors that compress the *thecal sac* surrounding the spinal cord (or in the low back, the *cauda equina*, which are nerve roots stemming from the bottom of the spinal cord and extending down through the spinal canal to the level of the pelvis) can cause severe pain and damage to the nerves resulting in loss of sensation, loss of strength, and incontinence of bowel and bladder. These neurological deficits can become permanent within a matter of hours of their onset. Spinal cord compression by cancer, therefore, which first manifests usually in the form of back pain with or without extension of pain down a limb, is therefore considered a medical emergency and requires immediate radiographic diagnosis and treatment with medications (steroids relieve some of the inflammation associated with the tumors) and, often, surgical and radiation therapy.

Tumors in the brain may arise from brain cells (such as *glioblastoma multiforme*) in which case they are referred to as primary brain tumors, or may have spread, or *metastasized*, from other sites of disease. They can cause headaches, seizures, nausea, and visual disturbances due to increased pressure inside the skull; depending on the exact location of lesions within the brain, symptoms similar to a stroke can also manifest, such as deficits in language, thinking, or balance.

Lung Cancer

Most commonly, the first signs and symptoms of cancer correspond to the organ that gave rise to the malignancy. In the lungs, tumors replace functional *alveoli*, the structure that exchanges carbon dioxide for oxygen, and can lead to shortness of breath, particularly with exertion. Approximately 50% of patients newly diagnosed with lung cancer describe this complaint at the time of their initial presentation. Lung tumors also cause localized irritation and inflammation triggering a cough, particularly if the tumor is located centrally in the chest, in the vicinity of larger airways; if the tumors invade blood vessels, patients may cough up sputum tinged or streaked with blood. Since lung cancer is strongly associated with smoking, it can be challenging to determine whether such symptoms are a result of lung cancer or other diseases of smoking, such as emphysema or bronchitis. Moreover, all of these conditions predispose to infection, which in the setting of lung cancer can result from a tumor blocking the expectoration of bacteria-filled sputum, a condition termed *post-obstructive pneumonia*. As with any cancer, lung cancer can also grow against and into nerves and bones, resulting in significant, often unremitting, pain. Up to 20% of patients are found to have involvement of the skeleton by the lung cancer at the time of their diagnosis.

On occasion, tumors in the right lung can push on the *superior vena cava*, the blood vessel that returns blood from much of the upper half of the body to the heart. This compression, termed *superior vena cava syndrome*, causes congestion of venous return and swelling in the head, neck, and upper torso. A *Pancoast tumor*, named after the radiologist Henry Pancoast who described the phenomenon in the 1920s, describes a tumor found at the very top of the lungs. This area is an important thoroughfare for key nervous pathways and so a variety of neurological deficits can ensue. These include *Horner's syndrome*, characterized by one-sided pupillary constriction, eyelid drooping, *enopthalmos* (sunken eyeball), and *anhidrosis* (lack of sweating); *Pancoast syndrome*, characterized by compression of the nerves supplying the arm, resulting in pain and weakness in the limb; and hoarseness due to damage of the *recurrent laryngeal nerve*, which controls the larynx. Physicians who are presented with these symptoms, particularly in patients older in age and with a history of smoking, should be on the look-out for lung cancer and have a low threshold to perform diagnostic testing with a chest X-ray or CT scan. Indeed, the link between cigarette smoking and lung cancer is so strong that even in the absence of any concerning symptoms, patients with a significant smoking history may benefit from *screening* CT scans performed in asymptomatic individuals to identify early and potentially curable lung cancers [3].

Colorectal Cancer

Cancer of the colon or rectum represents the classic example of stepwise development of cancer from normal cells: typically over the course of years normal *epithelium*, the cells that line the surface of an organ, acquire alterations in genes that lead initially to *hyperplasia* (excess number of microscopically normal-appearing cells), then the development of an *adenoma* (an abnormal, but benign, tumor), and, ultimately, *carcinoma* (a malignant tumor derived from epithelium). As such, cancer of the colon or rectum can frequently be prevented with a regular screening *endoscopy*, a minor medical procedure in which a camera is passed through the rectum and colon to visualize the epithelium and allow for biopsies and removal of abnormal appearing lesions before they become cancerous.

Colorectal cancers that develop despite screening measures or in patients that do not engage in routine screening can cause symptoms such as blood in the stool, abdominal pain, and changes in bowel function. Constipation associated with nausea and vomiting occurs in a small proportion of patients that develop *bowel obstruction* due to their colorectal cancer, a severe complication that requires prompt surgical attention. Most commonly, the first sign of colorectal cancer is an iron-deficiency anemia, often accompanied by fatigue, due to occult and intermittent loss of small amounts of blood in the stool. A diagnosis of iron-deficiency anemia is common in the population at large and might reflect blood loss due to menstruation or iron deficiency due to limitations in intestinal absorption or dietary restrictions. In the absence of such explanations, therefore, endoscopic evaluation may be indicated to look for a cancer of the gastrointestinal tract. Similar to other malignancies, colorectal cancer in the most advanced stages metastasizes, first to regional lymph nodes and eventually to distant sites such as the liver, the lungs, and the *peritoneum*, or the membrane lining of the inside of the abdomen. Spread to these organs can cause, respectively, pain in the upper-right quadrant of the abdomen with or without jaundice, a cough or shortness of breath, and symptoms of abdominal distension resulting from an accumulation of *ascites*, or fluid in the abdomen. Approximately one in five patients with colorectal cancer in the United States have distant, metastatic disease at the time of diagnosis.

Other Cancers of the Gastrointestinal System

In addition to the colon and rectum, the pancreas, liver, stomach, and esophagus can develop cancer with symptoms specific to the site of origin. Cancer of the pancreas is notorious for causing pain, even from small tumors. Pancreatic cancer pain is typically located in the middle of the abdomen and often *radiates*, or gives the sensation of traveling, to the sides or back. Jaundice due to compression of a bile duct from the pancreatic tumor, weight loss, and fatigue commonly accompany pancreatic cancer, as well. As noted above, blood clots are a frequent complication of advanced pancreatic cancer; *Trousseau's Syndrome*, the phenomenon of multiple episodes of blood clots associated with blood vessel inflammation and tenderness is classically associated with pancreatic cancer.

Hepatocellular carcinoma, or cancer of the liver, is associated with chronic liver diseases including cirrhosis and viral hepatitis; as noted above, patients with these underlying conditions should receive periodic screening for liver cancer with imaging and blood tests for alpha fetoprotein. Hepatocellular carcinoma can tip a marginally functioning liver into failure. Symptoms of liver failure including confusion, ascites, and jaundice in the setting of previously stable liver disease may warrant evaluation for the development of hepatocellular carcinoma.

Cancers of the upper gastrointestinal tract including the stomach and esophagus can cause iron-deficiency anemia from blood loss similar to colorectal cancer. Owing to blockage of the swallowing mechanism, tumors in the esophagus and first portion of the stomach sometimes cause *dysphagia*, or difficulty swallowing, and *odynophagia*, or pain with swallowing, and so lead to significant weight loss and malnourishment. Nausea and *early satiety*, or the sensation of being full after ingesting a small meal, may arise from tumors that inhibit the emptying of the stomach into the small intestine, termed *gastric outlet obstruction*. *Virchow's node*, a lymph node found in the left *supraclavicular fossa* (the area above the collarbone on the left), receives lymph from the upper gastrointestinal tract. The node is named after Rudolf Virchow, a German pathologist, who observed that cancers in the abdomen, especially the stomach, present with enlargement of this node due to the lymph drainage pattern.

Breast Cancer

Comprising more than 20% of all invasive cancers in women, breast cancer is the single most common such malignancy. As with colon cancer, screening identifies many cases before they cause any physical signs or symptoms and, by identifying the disease early on, has resulted in significant improvements in mortality due to this disease. Nevertheless, routine mammography can miss a substantial minority of palpable breast cancers (up to 15%) due to the radiographic characteristics of the tumor and surrounding breast tissue, and another 30% of breast cancer is diagnosed between scheduled mammographic exams. Furthermore, screening mammography can reveal abnormalities that on subsequent biopsy prove to be benign; these *false positives* result in unnecessary anxiety on the part of the patient, medical expense, and, rarely, procedural complications. In addition, the effectiveness of screening, whether with regular mammography or other imaging modalities and, in certain cases, combined with patient self-examinations, is dictated to a significant degree by the likelihood of the individual patient developing a breast cancer; this likelihood is influenced by patient age, gynecologic history, and other factors. Guidelines for screening are consequently highly controversial, though as a rule patients should be counseled on the risks and benefits of screening based on their personal risk of developing breast cancer, which is calculated according to published models such as the *Breast Cancer Risk Assessment Tool* [4].

In those instances in which breast cancer is not identified by screening, the presenting symptom is typically a firm, immobile mass within the breast tissue; associated skin changes in the area of the tumor can include dimpling and inflammation, and the nipple can appear pulled in, or inverted, and/or discharge clear or bloody fluid. Breast cancer that metastasizes often does so first to the lymph nodes in the armpit which is the first stop in the lymph drainage of the breast tissue. Therefore a mass in the armpit may accompany a newly diagnosed breast tumor and, indeed, be the initial finding in patients in whom the primary tumor is deep in the breast or relatively small. Beyond the lymph nodes in the armpit, breast cancer is prone to subsequently spread to the bones, which can cause pain or even fracture, and the liver and lungs. Cancer metastases in the bones, regardless of their original site, can cause sufficient breakdown of the skeleton to cause *hypercalcemia*, an increase in the level of calcium in the bloodstream. Hypercalcemia itself can cause a variety of signs symptoms, such as profound fatigue, constipation, nausea, and kidney damage. In addition to treating the underlying cancer, the treatment of hypercalcemia requires measures aimed at flushing the mineral out of the bloodstream and into the urine and inhibiting the turnover of calcium within bones.

As with any metastatic disease, the dysregulation of a breast cancer cell's suicide pathways and homing mechanisms render it capable of seeding and growing in any distant organ, resulting in potentially myriad signs and symptoms.

Prostate Cancer

The prostate gland is the organ found in males positioned between the urinary bladder and rectum that secretes fluids into the ejaculate supporting the survival and motility of sperm. Cancer of the prostate is an extremely common malignancy, and is estimated to affect approximately one in six men; the rate of mortality from prostate cancer is approximately 1 in in 34 men. *Prostate specific antigen* (PSA), is a protein secreted by and unique to prostate cells. The PSA level in the bloodstream is easily measured, and can be elevated due to prostate cancer or other processes such as benign prostatic hypertrophy or an infectious prostatitis. For many of the reasons outlined above regarding the specifics of breast cancer screening, routine prostate cancer screening by PSA measurement is controversial. Moreover, while some prostate cancers are potentially lethal and require comprehensive treatment, others are slow-growing and pose no danger over the lifetime of the patient, particularly since they often appear in a man's older age; indeed, autopsy series have shown that 80% of men who die in their 70s or later have prostate cancer. Diagnosing these cancers that through a natural course would cause no ill-effects during a patient's lifetime is termed *overdiagnosis*. Since the relatively indolent course of such prostate cancers cannot be reliably predicted at the time of the diagnosis, however, the diagnosis of prostate cancer, which itself can result in complications stemming from the biopsy of the gland, leads to treatments with inherent potential side effects including urinary incontinence, sexual dysfunction, and bowel symptoms. The role of PSA screening, therefore, is complex and requires in-depth discussion between a patient and his doctor.

More than three out of four diagnoses of prostate cancer are made due to evaluation of an abnormal PSA test; a subset of these abnormal PSA values were checked to investigate an abnormality detected on *digital rectal examination*, a technique in which a clinician palpates the surface of the prostate gland using a finger inserted into the rectum. Men with early prostate cancer limited to the gland itself are generally without symptoms. An abnormal flow of urine, including hesitancy and frequency, results from enlargement of the prostate which can be due to cancer or, more often, benign prostatic hypertrophy. Blood in the urine or sperm are more ominous symptoms that necessitate an evaluation for cancer. Prostate cancer is notorious for spreading to the bones as a first site of metastasis, and pain and neurological deficits are common complications of this.

Head and Neck Cancer

The anatomy of the head and neck is intricate, and involves anatomic structures that are critical for many different bodily functions. The first signs and symptoms of cancer in this region of the body therefore encompass a wide spectrum. Pain localizing to the site of the primary tumor is a common complaint, but pain may be referred as well. Generally, tumors of the tongue, lip, floor of the mouth, and buccal mucosa are identified earlier in their course than those involving deeper structures in the throat, such as the *hypopharynx* and tonsils, which are relatively less well enervated. *Otalgia*, or pain experienced in the ear, can be referred from disruption of various cranial and cervical nerves that supply the sinuses and oral cavity down through the larynx. Lesions in the oral cavity, masses in the neck, dysphagia, odynophagia, abnormalities of speech, and bleeding are routinely observed in patients with cancers affecting the oral cavity and pharynx. Hoarseness, cough, and difficulty breathing accompany tumors of the larynx. Since tobacco or alcohol use are the chief risk factors for the development of head and neck cancer, patients with such a history should be mindful to report new symptoms and receive regular medical and dental follow-up.

Genitourinary Cancers

Historically, ovarian cancer was labeled a "silent killer" because of the impression that signs and symptoms did not appear until the most advanced stages. More recent studies have shown, however, that symptoms of bloating, pain in the abdomen or pelvis, early satiety, and urinary symptoms manifest even in the early stages of ovarian cancer and that evaluation for this possibility be considered in the appropriate circumstances. A mass in the *adnexa of the uterus* found during routine physical examination or imaging of the pelvis performed for a separate reason may be the initial sign of ovarian cancer in asymptomatic women. A *Sister Mary Joseph Nodule* describes a mass in the navel that represents metastatic cancer. While nonspecific for a site of origin, cancers of the ovary, in addition to organs of the gastrointestinal tract, are most typically implicated. Postmenopausal bleeding, while observed infrequently as a result of ovarian cancer, is more indicative of abnormalities, including cancer, of the uterus or cervix.

Though rare, testicular cancer is the most common solid tumor in young men; additionally, testicular cancer is highly treatable, with cure rates exceeding 95%. As a result, the successful treatment of testicular cancer can result in decades of essentially normal life for an individual who otherwise would have succumbed to disease at an early age. For this reason, it is critical that a diagnosis of testicular cancer is not missed. A painless mass or unilateral enlargement of the testicle is the most common presentation; pain or discomfort in the lower abdominal or scrotal regions can nevertheless occur. Such symptoms require prompt medical attention and urological evaluation. Since testicular tumors can secrete functional hormones, *gynecomastia*, or enlargement of the breast, and symptoms of hyperthyroidism may result. Testicular cancer metastasizes most often to the lymph nodes in the pelvis and abdomen, the lungs, and the bones;

metastatic disease is found in about one in ten patients newly diagnosed with testicular cancer.

Cancers of the kidney are associated with a "classic triad" of symptoms, including flank pain, blood in the urine, and a mass in the abdomen. Despite the nomenclature, all three symptoms are found in less than 10% of patients with kidney cancer. Increasingly, as CT and ultrasound scans are being performed for other indications, kidney cancers are diagnosed *incidentally*, that is, as a chance finding during evaluation of a separate issue. Kidney cancers are relatively commonly implicated in *paraneoplastic syndromes*, the phenomenon in which a cancer secretes a hormone or protein into the bloodstream that causes effects seemingly unrelated to the site of the tumor. Kidney cancers can, for example secrete the hormone *erythropoietin*, resulting in increased red blood cells, *renin*, causing hypertension, and *insulin*, causing hypoglycemia. Treatment of paraneoplastic syndromes fundamentally requires effective treatment of the tumor itself.

Melanoma

Melanoma, a potentially fatal cancer of the skin, epitomizes the value of recognizing and attending to the early warning signs and symptoms in cancer. Melanomas are curable when found early enough and removed surgically. Differentiating *benign nevi*, or moles, from potentially malignant melanoma requires the expertise of a dermatologist and, for any questionable lesion, pathologic examination of a biopsy. Yet, there are several features of pigmented skin lesions that can be reliably observed by the lay public and warrant medical attention: these include a change in size, color, or shape of the lesion, and any associated inflammation, bleeding, crusting, change in sensation, and size exceeding 6 mm. Similar to the theme described above in many types of malignancy, familiarity with what constitutes concerning signs and symptoms of disease combined with an awareness of unexpected physical changes are important for the earliest possible detection of cancer.

SUMMARY

The signs and symptoms of cancer are many and diverse. They overlap with diseases of other etiologies, particularly infectious and autoimmune, and they cover a range of intensities. The majority of cancers give no signs or symptoms, at least in the early stages. Timely diagnosis of cancer requires vigilance of the individual and the healthcare team. The recognition of abnormal signs and symptoms and an appropriate, measured pursuit of an underlying diagnosis yields the best chance of finding a cancer early and successfully managing it.

REFERENCES

1. Cannistra, S.A. and Niloff, J.M. (1996). Cancer of the uterine cervix. *New England Journal of Medicine* 334 (16): 1030–1037.
2. Timp, J.F., Braekkan, S.K., Versteeg, H.H. et al. (2013). Epidemiology of cancer-associated venous thrombosis. *Blood* 122 (10): 1712–1723.
3. Aberle, D.R., Adams, A.M., Berg, C.D. et al. (2011). Reduced lung-cancer mortality with low-dose computed tomographic screening. *New England Journal of Medicine* 365 (5): 395–409.
4. http://www.cancer.gov/bcrisktool

5

IMPORTANCE OF IMAGING FOR DIAGNOSIS AND TREATMENT OF CANCER

Kristin DeStigter, Sally Herschorn, Anant Bhave,
Benjamin Lange, and Sean Reynolds

Larner College of Medicine, University of Vermont, Burlington, VT, USA

Radiological imaging tests are an important part of the process of cancer diagnosis and treatment. Radiologic imaging encompasses techniques that use X-rays, sound waves (ultrasound), magnetic fields, and radioactive particles to look inside the body (Table 1). These tests in the context of good clinical information can play an important role in all stages of detection and treatment. Imaging can be used to find cancer in its earliest stages before there are any signs or symptoms. This type of a "screening" exam is exemplified by screening mammography, a familiar test routinely performed to detect early breast cancer. Recently lung cancer screening Computed Tomography (CT) has gained favor to detect early disease in high-risk patients. Radiology tests can also be used to search for suspected cancer when signs and symptoms are suspicious. Cross-sectional imaging tests like CT and Magnetic Resonance Imaging (MRI), sometimes combined with Positron Emission Tomography (PET), can be useful in both localizing a tumor and to help classify stage disease spread. In addition, these techniques can assist with biopsy planning as well as subsequent treatment planning tailored to provide the best optimal outcome. For example CT or MRI can show if a pancreatic cancer has invaded important blood vessels or other organs and this will direct subsequent management. Baseline radiological imaging studies help to follow response to therapy and look for recurrence.

Cancer: Prevention, Early Detection, Treatment and Recovery, Second Edition. Edited by Gary S. Stein and Kimberly P. Luebbers.
© 2019 John Wiley & Sons, Inc. Published 2019 by John Wiley & Sons, Inc.

TABLE 1. Comparison of radiological modalities commonly used for imaging cancer.

Modality	Radiography	Ultrasound (US)	Computed Tomography (CT)	Magnetic Resonance Imaging (MRI)	Positron Emission Tomography (PET)	SPECT	Mammography/ Tomosynthesis
How Images are Produced	X-rays (ionizing radiation) passed through the body with differences in attenuation of X-ray beam producing the images	High frequency sound (no ionizing radiation) with computer generating cross-sectional images based on time of sound return to the machine combined with attenuation of sound waves	Scans through the body using X-rays with computer processing to construct cross-sectional images	Magnetic field (no ionizing radiation) used to evaluate differences in the hydrogen content of the tissues. Computer processing produces cross-sectional images.	Injection of positron emitting radiotracer (ionizing radiation). Computer processing of data on radiotracer location generates cross-sectional images.	Injection of gamma-ray or X-ray emitting radiotracer (ionizing radiation). Computer processing of data on radiotracer location generates cross-sectional images.	X-rays (ionizing radiation) as with radiography passed through the breast with differences in X-ray attenuation producing images. Optimized for high spatial resolution and high soft-tissue contrast. Tomosynthesis adds computer postprocessing to produce partially cross-sectional images.
Advantages	Fast Portable High-Spatial Resolution	Fast Portable Enables dynamic (Real-Time) evaluation	Fast (especially with helical acquisition) Excellent soft tissue contrast Limited dynamic evaluation	Outstanding soft tissue contrast Some sequences enable direct diagnoses (e.g., DWI for stroke) Inherently multiplanar	Allows evaluation of physiologic processes More sensitive than CT or MR for metastatic disease in the proper setting Higher resolution than SPECT Inherently cross-sectional	Cross-sectional evaluation of gamma emitting radiotracer localization – improves sensitivity and specificity over planar images	Fast Excellent soft tissue contrast High spatial resolution – aids in detecting and characterizing calcifications

Disadvantages	Poor soft tissue contrast	Highly operator dependent Low soft tissue contrast (compared to CT/MR)	High radiation doses, particularly for dynamic evaluation Not portable	Slow Interior of scanner uncomfortable for some patients Requires extensive shielding Not portable	Slow Low spatial resolution relative to other modalities Patient must hold position for a long period of time Relatively high radiation exposure	Slow Low resolution Relatively high radiation exposure	Nonspecific – exams often require follow-up ultrasound and/or biopsy Compression uncomfortable for patients
Applications							
Relative cost	Cheap	Cheap	Expensive	Most Expensive	Most Expensive	Expensive	Cheap

INTERVENTIONAL RADIOLOGY

Procedures performed by interventional radiologists are becoming increasingly important in the management of patients with cancer (Table 2). Interventional radiology procedures are now widely used as the best minimally invasive option to obtain biopsy samples from solid tumors for cytology and pathology testing. Image-guided procedures are also useful to sample fluid collections that may contain tumor cells. Transcatheter embolization is an interventional radiology procedure that delivers cancer-treating agents (including chemotherapy and radiation therapy) directly through the blood vessels supplying the tumor. Interventional therapies also include tumor ablation, procedures that kill cancer cells directly using a variety of techniques including heating, freezing, and electrical stimulation. Image-guidance is useful to gain access to veins to insert catheters that deliver treatment to cancer patients. Some of these catheters may also be used instead of a traditional IV during CT and MR scans performed for cancer monitoring. Lastly, complications from cancer, like infection and bleeding, can be treated with interventional techniques. Compared to traditional surgery, interventional radiology techniques can often be performed at lower cost with fewer side effects, often as outpatient procedures.

IMAGING SCREENING, DIAGNOSIS AND STAGING, AND FOLLOW-UP

Before approaching the topic of the integration of imaging into modern oncologic care, a few definitions and a discussion on the distinct imaging modalities utilized in providing care to cancer patients is necessary. All the imaging modalities utilized in the modern practice of radiology have been used at some point for cancer care. The modern paradigm of cancer care consists of four distinct settings in which imaging tests play a role: large-scale screening of populations without known disease, cancer diagnosis, cancer staging, and follow-up to evaluate for treatment efficacy and cancer recurrence.

Large-scale mammographic screening of the female population has been performed for many years. As noted above, with the publication of the National Lung Cancer Screening Trial in 2011, many centers have begun programs using low-dose CT to screen for Lung Cancer in smokers who meet specific criteria.

BREAST CANCER SCREENING, DIAGNOSIS AND STAGING

Breast cancer is a useful model for understanding the more general integration of imaging into the diagnosis and treatment of oncologic disease. Breast cancer has a well-established imaging-based screening regime. The success of breast cancer

TABLE 2. Commonly performed interventional radiology procedures.

IR Procedures	Description	Imaging Guidance	Examples
Biopsy and Drainage	– A wide variety of tumors may be sampled with small needles under radiology guidance – Abnormal fluid collections within the body may be drained to help diagnose cancers or to improve cancer symptoms	– X-ray – Ultrasound – CT – MRI	– Sample tumors and lymph nodes throughout the body – Drain fluid collections throughout the body
Venous Access	– Temporary or semi-permanent catheters can be placed in veins using minimally invasive techniques – Uses include chemotherapy, antibiotics, nutrition, and IV contrast for CT or MR scans	– X-ray – Ultrasound	Access devices include: – PICC lines – Groshong or Hickman catheters – Chest ports
Dialysis Access	– Temporary or semi-permanent dialysis catheters can be placed in veins or the abdomen (peritoneal space) for patients with kidney failure due to cancer	– X-ray – Ultrasound	Access devices include: – Tunneled venous dialysis catheters – Peritoneal dialysis catheters
Tumor Ablation	– Many cancers may be treated using needles placed directly into the tumor – Tumor cells are killed using a variety of techniques including heating, freezing, and electrical stimulation	– X-ray – Ultrasound – CT – MRI	Types of Ablation: – Radiofrequency – Cryoablation – Microwave – Electroporation Frequently treated cancers: – Lung cancer – Liver cancer – Kidney cancer
Transcatheter Embolization	– Some cancers may be treated using catheters to deliver potent medications directly into the tumor – Tiny particles containing chemotherapy or radiation therapy are administered directly into the tumor through feeding blood vessels	– X-ray	Frequently treated cancers: – Liver cancer – Metastatic colon cancer – Cholangiocarcinoma (bile duct cancer) – Metastatic neuroendocrine cancer

screening has resulted in the majority of breast cancers being discovered during screening. Screening allows early detection of breast cancer, which improves survival and often allows less invasive and less costly treatment. There are many studies on the efficacy of screening and they have been interpreted in different ways in the literature and media, leading to controversy. The potential benefits of screening (lower chances of dying from breast cancer if diagnosed) must be weighed against the potential harms (call backs for additional imaging when there is no cancer, biopsies of benign findings, and treatment of cancers that would not have been responsible for the patient's death if left untreated).

Screening is not intended to determine whether or not cancer is present. Rather screening aims to separate patients without cancer from those who *might potentially* have cancer (these women get called back for additional evaluation). The greatest mortality benefit to the overall population is gained from annual screening mammography for all women ages 40 and up. However, some women (those with genetic mutations such as BRCA, prior radiation treatment to the chest in their teens and 20s, or a strong family history of breast cancer) are at increased risk for developing breast cancer. This is defined as a greater than 20% lifetime risk of breast cancer. For comparison, the average lifetime risk of breast cancer in women in the general population is 13%. In this select population, high-risk screening regimens are used (more information can be found at http://www.cancer.org/cancer/breastcancer/moreinformation/breastcancerearlydetection/breast-cancer-early-detection-acs-recs).

Mammographic screening is conducted with two standard mammographic views, a top to bottom image (craniocaudal [CC]) and an oblique image angled along the chest wall (mediolateral oblique [MLO]) of each breast (Figure 1a shows these two views). 3D mammography, also called tomosynthesis, is an augmentation of the standard two-view mammogram also now available for both screening and diagnosis. 3D mammography has been shown to reduce the number of patients called back when there is actually no cancer as well as improving cancer detection, especially for women with dense breasts. This is important because mammography is less sensitive for detecting breast cancer in women with dense breasts (approximately 40–50% of the population) and because women with dense breasts are at slightly increased risk of developing breast cancer. These women may choose to have supplementary screening in addition to mammography with ultrasound or MRI. MRI is typically reserved for those at very high risk (over 20% lifetime risk) while ultrasound is usually used for those at moderate risk (15–20%).

Diagnostic breast imaging is used for women (and sometimes men) with symptoms (lump, pain, nipple discharge) or women whose screening mammogram showed an abnormality. In the diagnostic setting, the focus is on the problem that caused the patient to come for imaging. The goal is to determine whether or not a given patient has or does not have cancer. Ultrasound is an essential part of breast diagnosis. Not all cancers are detectable on mammography, therefore, women with lumps must be evaluated with ultrasound to make sure that an abnormality causing the lump is detected (Figure 1b shows an example of the ultrasound appearance of breast cancer). If there is an abnormality, ultrasound is helpful to determine whether a finding is suspicious for cancer or is a definitely benign finding, such as a cyst. Sometimes, a biopsy is needed to establish whether or not cancer is present. Breast imagers perform needle biopsies using ultrasound,

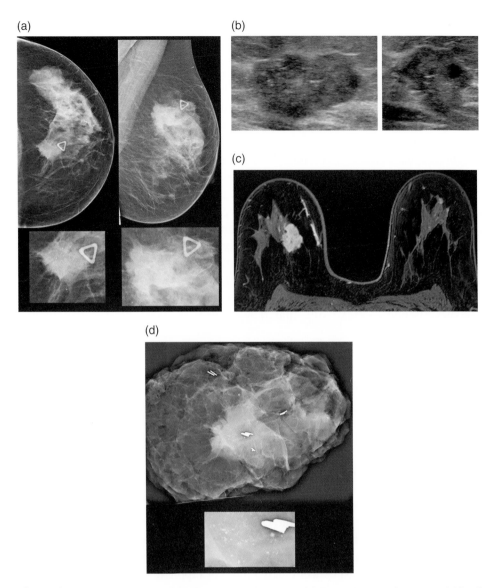

Figure 1. Breast Cancer, screening diagnosis and staging. (a) Top to bottom (craniocaudal [CC]) (left) and midline to lateral (mediolateral oblique [MLO]) (right) images of the right breast obtained in a patient with a palpable lump shown by the triangular palpable lump marker. The magnified images centered on the palpable lump marker show a spiculated mass in the upper inner portion of the breast. (b) As part of the workup of the mass seen on the mammogram ultrasound was performed. The mass is seen centered on the ultrasound images. It is decreased in ultrasound attenuation as compared to the surrounding fat and breast tissue. The tiny bright dots within the mass are concerning for micro-calcifications, raising the risk that this is a malignant mass. (c) Breast MRI obtained to evaluate the extent of the mass. The mass is seen on the left as the bright region, very different in its appearance from the remaining glandular tissue of the breast. (d) Images from a so-called specimen mammogram obtained after a breast surgeon removed the mass seen on (a)–(c). These are obtained at higher dose than the normal images, as dose is no longer a concern. The magnified imaged below centered on the surgical clip shows to better advantage the micro-calcifications also seen at ultrasound. This cancer proved to be an invasive ductal carcinoma, the most common type of breast cancer.

mammography, or MRI for guidance. This ensures that the biopsied lesion was truly the abnormality detected at imaging. These procedures are done using local anesthetic and are typically well tolerated with minimal pain. Approximately 70% of biopsies are benign (no cancer is present). When a cancer is diagnosed, the patient can be referred for treatment to a surgeon (Figure 1d shows a mammogram of a breast mass after it was removed by a breast surgeon). Sometimes additional testing is also necessary. MRI can sometimes be helpful in assessing the size of a cancer, how much of the breast contains cancer cells, and whether or not there are additional sites in the same breast or opposite breast (Figure 1c). This information may be helpful to the surgeon in planning the approach to a given cancer. Breast imaging radiologists work very closely with surgeons, medical oncologists, and radiation oncologists to review each patient's individual case and come to a consensus on optimum treatment. The recurrence rate for breast cancer after treatment is low however, patients usually visit the breast imaging department annually for surveillance to make sure that if a recurrence develops it is caught and treated early.

DIAGNOSIS, STAGING, AND FOLLOW-UP

The process of oncologic imaging in most cancer patients is similar to that described for breast cancer. The crucial difference is that, for the majority of cancers, screening is not used for diagnosis. In the absence of a defined screening regime, the process of diagnosing cancer and staging it is not as clearly defined as it is in the case of breast cancer. Most cancer patients present for initial imaging in a scenario similar to that described above in the Diagnostic setting. However, the focus is not necessarily on answering a question as clear-cut as "is there cancer here or not?" The vast majority of patients who present for Diagnostic Mammography either have a symptom of breast cancer (such a lump or nipple discharge) or have had an abnormality on their screening mammogram that might potentially be cancer. This is not true of patients who present for Diagnostic Imaging as a whole. Imaging plays a large and vital role in all facets of modern medical care, which changes the scope of the clinical question to be answered. Patients who will ultimately be diagnosed with a cancer therefore have two common presentations. The first group presents to a physician with symptoms suspicious for cancer. The physician will then order an imaging test to confirm his or her suspicion that cancer is present. Such presentations are analogous to the diagnostic setting of breast cancer imaging – cancer is suspected and imaging is used to confirm this diagnosis.

However, this is not the only, or even necessarily the most common, way in which imaging plays a role in diagnosis of cancer. The other common presentation involves people imaged for symptoms that are either nonspecific or unrelated to cancer. The clinician ordering the test has no reason to suspect a cancer. Not all cancers produce symptoms. Indeed, cancers capable of producing symptoms are

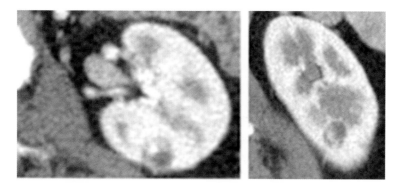

Figure 2. Axial and Coronal CT images of the left kidney show a 1 cm mass. It is predominantly fluid however, a small enhancing nodule is seen in the wall. This appearance is concerning for a cystic renal cell carcinoma, which this lesion proved to be at the time of ablation.

generally relatively advanced – they have been in the body for a long time before they actually cause symptoms. Imaging is quite good at detecting these unsuspected cancers. A small enhancing renal mass seen during a CT for appendicitis is a good example of such a cancer (Figure 2).

The detection of a cancer, whether suspected or not prior to imaging, is the first of many tasks that a radiologist faces when interpreting an imaging study. The second step is to provide information to the clinician that will be helpful in treatment. This can be thought of as determining how much cancer is present and where in the body that cancer is seen. Ultimately, this information is used for a process called staging. Staging is a method for subcategorizing patients into groups with similar amounts and/or types of disease. It is typically specific to the type of cancer encountered, for example, lymphoma is staged differently to breast cancer. By devising specific treatments for groups with similar disease, it is possible to improve overall outcomes and tailor care to a particular patient. The distinction between staging and diagnosis is important both for radiologists interpreting images and for oncologists and surgeons who use the information provided to treat patients. Imaging tests have specific strengths and weaknesses. In particular, some tests are very good for making diagnoses while others are very good at staging. The choice of what test to use varies not only by the setting, that is, diagnostic versus staging, but also based on the type of cancer being imaged.

Finally, once a patient has been diagnosed and staged he or she will begin treatment. Imaging is used on patients undergoing treatment to determine whether or not the treatment is effective. This process is normally termed follow-up. It is similar to staging in that the amount and extent of disease is being evaluated. The duration of this follow-up period varies based on the type of cancer, the extent of disease, the treatment administered, and even the discretion of the physician or physicians treating a given patient. As with breast cancer, most cancer patients are imaged at set intervals for a while after treatment has been completed to determine whether or not the cancer has returned in order to allow prompt treatment of such a recurrence.

An easy way to think of all diagnostic tests, imaging included, is that they are ways of discriminating between normal and abnormal (disease) states. Imaging tests accomplish this in a variety of ways. X-rays, CT, and mammography all depend on the differential attenuation of ionizing radiation (essentially high energy light) by the tissues of the body. Ultrasound operates on the same principle, but with high-frequency sound waves rather than high-energy light. MRI works by evaluating the differences in the concentration and milieu of hydrogen atoms within a volume. Ultimately, these exams are all methods of noninvasively producing images of the anatomy. They are therefore reliant on changes in patient anatomy for diagnosis. This is very useful for oncologic imaging because the sensitivity of these tests for even small cancers is high. Most cancers present as masses and therefore distort the anatomy in ways easily detected at imaging.

This is not necessarily true in staging a tumor. It is very important to detect even tiny metastases because they change the imaging assessment of the amount and extent of disease in a given patient. This in turn changes the patient's stage and therefore the patient's treatment. If one defines a metastasis as a group of cells related to the original cancer but outside of the original mass, one begins to see the problem. A small group of cells that should not be in a particular location is difficult to detect. For example, it takes a large number of cancer cells to enlarge a lymph node until it becomes clearly abnormal by Ultrasound, CT, or MRI. The presence or absence of lymph node metastases is very important for staging in the most commonly used staging scheme (the Tumor, Node, Metastases [TNM] system). It is in this setting that the "other" frequently used imaging modality, nuclear medicine, is most useful.

ONCOLOGIC APPLICATIONS OF NUCLEAR MEDICINE

Nuclear medicine imaging tests are widely used in oncologic imaging, primarily for the staging of disease. They are useful because they provide information not obtainable from anatomic imaging. Anatomic imaging is the primary mode in which medical imaging is used, as described previously. However, the effect of cancer cells on the local anatomy is not the only thing differentiating cancer cells from normal cells. Cancer cells, after all, are very different from normal cells in physiologic terms. Their DNA is similar to the cells they arise from but it isn't the same. They typically have a suite of mutations allowing them to escape the physiologic constraints placed on normal cells. They often have mutations disabling the upper limit on cell divisions and bypassing the normal checkpoints for DNA replication for example. In essence, the suite of mutations present in a cancer cells make them different from normal cells, particularly in their physiology, specifically in the ways in which they acquire and use energy. Nuclear medicine exams used for the purposes of oncologic imaging exploit this aberrant physiology to render even anatomically insignificant numbers of cancer cells visible.

The fundamental difference between nuclear medicine exams and other imaging exams is that the patient is the source of the energy used to generate the image. In X-ray, CT, MRI, and ultrasound, a source external to the patient produces the energy that will ultimately be used to form the image. The patient is exposed to energy that can pass through the patient and then is used to form an image of the inside of the patient. In nuclear medicine, the patient is given one of a vast array of so-called radiopharmaceuticals or radiotracers. These tracers are labeled with radioactive compounds. When they undergo radioactive decay, they emit gamma or X-ray radiation. For context, the most commonly used radionuclide in nuclear medicine, Technetium 99 m, produces gamma rays with about twice the energy of the average energy of the X-ray source used in CT. Different nuclides produce radiation with different energies.

Nuclides are chemically active and therefore may play a role in the various chemical reactions that occur within the body to make life possible. This makes nuclides very useful in that their distribution can be representative of a patient's underlying physiology. Some nuclides are used in their elemental form (Gallium-67 or Iodine-123 for example) while others are used to label an array of compounds that follow a specific distribution in the body. The combination of the two (a radionuclide and its distribution altering compound) is known as a radiopharmaceutical. For example, the aforementioned ^{99}Tc can be complexed with methyl diphosphate (MDP) to produce the commonly used bone localizing radiopharmaceutical ^{99}Tc-MDP. This compound is incorporated into the matrix that forms the bones and therefore can show areas where the bone is more active such as fractures or points where it is actively remodeling as it does around a metastasis (Figure 3a and b).

There are limitations to this approach. The biggest is that amount of radiation available to create the image is much smaller than that available with X-ray or CT. This is because the "source" (the radiotracer) is inside the patient. This means that the patient will be exposed to ionizing radiation from the radionuclide until it has completely decayed and/or been excreted. Image quality (both spatial resolution and contrast) improves as the amount of radiation used to produce the image increases. In practice, this means that acquisition of nuclear medicine images takes a longer time and requires patients to remain still in order to obtain sufficient diagnostic quality. Because only a small amount of energy is used to produce a given image, the spatial resolution and the anatomic information of the image are both poor. Despite this, because many of the processes evaluated with nuclear medicine are physiologically very different in the cancer cell as compared to a normal cell, the images carry information not obtainable with anatomical imaging such as CT. Hybrid techniques allowing fusion of functional processes detected with nuclear medicine and anatomical information from CT have been found to be even more beneficial than either technique alone.

The most commonly used radiopharmaceutical in oncologic imaging is 2-deoxy-2-(^{18}F)fluoro-D-glucose, normally abbreviated as ^{18}F-FDG or simply as

(a)

(b)

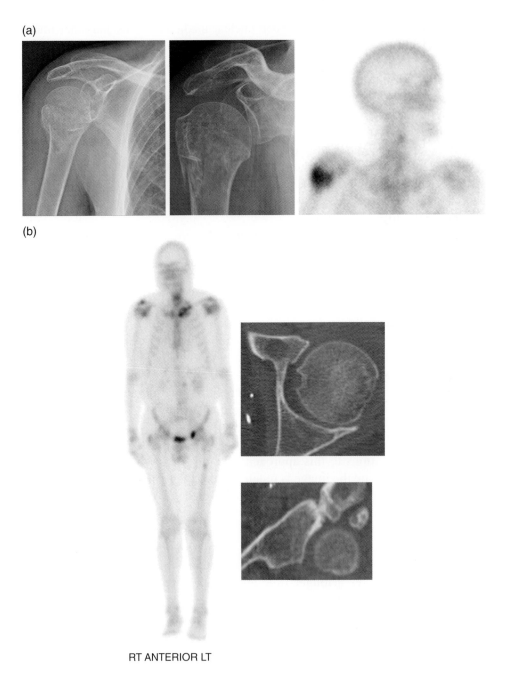

RT ANTERIOR LT

Figure 3. (a) Two views of the right shoulder as well as an image from ^{99m}Tc-MDP bone scan show a fracture of the surgical neck of the humerus. Increased MDP uptake is seen at the same point on the bone scan due to increased remodeling of the bone. (b) Anterior projection from a ^{99m}Tc-MDP whole body bone scan as well as axial and coronal images from a CT obtained within one week of the bone scan show increased radiopharmaceutical uptake at the left puboacetabular junction corresponding to a lytic metastasis seen on CT. There are additional metastases present in the spine and proximal left femoral metadiaphysis.

FDG. The radionuclide is Fluorine-18. It works slightly differently to most nuclides in that, though it is radioactive, it does not emit light directly when it decays. Instead, it produces a positron (the antimatter electron and the "P" in PET scan). When this particle encounters an electron both particles are annihilated, again producing light of ionizing intensity (ionizing radiation). Rather than a single beam (as with gamma rays), two beams are produced traveling at 180° from each other. These beams are used to localize the annihilation event and therefore the location where the Flurorine-18 atom was when it emitted the positron.

FDG is a glucose-analogue. All cells in the body absorb this analogue just as they absorb glucose but they cannot actually break the analogue down to produce energy (it lacks the 2′ hydroxyl group necessary to complete glycolysis). Therefore it shows areas of the body where there is high demand for glucose. Cancer cells, particularly aggressive cancer cells, often have a very high demand for energy with a correspondingly high demand for substances that can be used to produce energy such as glucose. The demand for the glucose further increases when the tumor cells switch to nonoxidative metabolism. This metabolism requires around 20 times more glucose molecules to produce same amount of energy. This effect described by Warburg, in turn, means that on FDG PET scans they have a disproportionate demand for FDG – they are "FDG avid (Figure 4)."

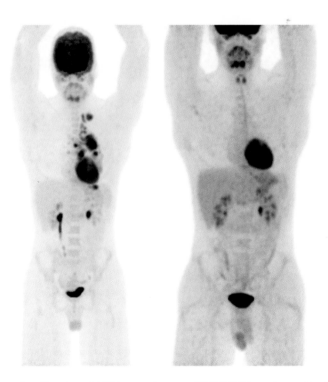

Figure 4. Pre- and posttreatment MIP images from an ^{18}F-FDG PET scan in a patient with Hodgkin lymphoma. There is abnormal uptake seen throughout the chest. This resolves on the posttreatment study with the normal, physiologic distribution of the radiopharmaceutical shown.

From the standpoint of oncologic imaging this technique is very useful because it can indicate the presence of cancer cells even in anatomically normal tissues (Figure 5). This makes FDG PET more sensitive for metastatic disease than anatomic scans and in turn allows more accurate and much less invasive staging.

FDG has become the workhorse molecule for staging the majority of cancers because it can show a physiologic change common to the majority of cancers, regardless of the underlying subtype or the organ of origin. There are a number of other radiopharmaceuticals that are also used in oncologic practice. These radiopharmaceuticals depend on the clonal nature of cancers – they normally are very similar genetically, arising from either a single deranged cell or a very small group of cells. As a result, the other cells in the clone often have similar genetic profiles to the progenitor cells. These radiopharmaceuticals

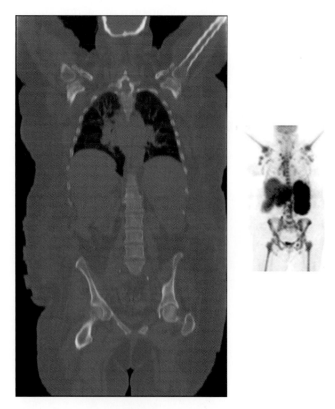

Figure 5. ^{18}F-FDG PET as well as coronal images from the accompanying attenuation correction CT shows a normal CT appearance of the bones. However, bone marrow uptake is diffusely increased in this patient with non-Hodgkin lymphoma. This finding is diagnostic of bone marrow involvement, which modifies the patient's staging. Also note splenomegaly and nodal involvement both below and above the diaphragm.

tend to be useful only for certain specific types of tumor because the cell must have the ability to take them in order for the radiopharmaceuticals to be useful for imaging.

The most commonly used of these radiopharmaceuticals are: radioiodine (2 different commonly used radionuclides Iodine-123 and Iodine-131), Indium-111 Pentetreotide, Iodine-123 or -131 Metaiodobenzylguanidine (MIBG), and various radiolabeled monoclonal antibodies.

Radioiodine is of great use in the staging and treatment of thyroid neoplasms. It comes in two variants. Iodine-123 (I-123) is the variant more commonly used for imaging for a number of reasons. Ionizing radiation produced from its decay is less intense than that produced by Iodine-131 (I-131). Also, its shorter half-life allows relatively more of the radiopharmaceutical to be administered, which produces better images. Finally, it does not emit charged particles as well as light (as I-131 does) which means that the dose (the cumulative sum of all radiation) a patient receives from I-123 is substantially smaller than the dose from I-131 which undergoes the same number of decays. The latter property of I-131 (emission of charged particles which results in a high-radiation dose) can be useful in the proper context. In modern practice, it is used for the treatment of thyroid cancer. Once the thyroid has been excised surgically, I-131 can be used to administer a dose of radiation to the residual thyroid sufficiently high to destroy residual thyroid tissue and also micrometastases if present (Figure 6).

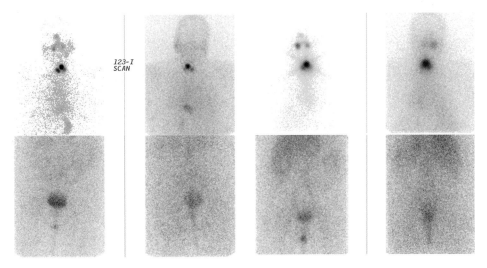

123-I
SCAN

Figure 6. Normal planar images from I-123 (left) and I-131 (right) scans. Uptake in the neck relates to residual thyroid tissue – the patient was status post thyroidectomy for thyroid cancer. The I-131 will ablate this tissue. Note the uptake seen in the liver on the I-131 study not seen on the I-123 study. This is physiologic, it occurs due the metabolization of I-131 containing thyroid hormone by the liver. The posttreatment study is generally obtained one week after the administration of I-131.

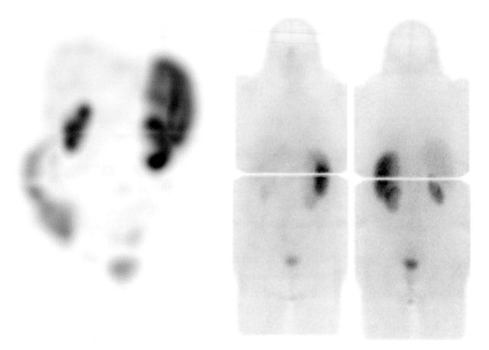

Figure 7. MIP from SPECT and planar images from a normal Indium-111 Pentreotide (Octreoscan). This is the normal uptake of this radiopharmaceutical. Notice the low uptake in the lungs, pancreatic fossa, liver, abdomen, and pelvis. This is helpful as these are the areas in which the neuroendocrine tumors, which this radiopharmaceutical is used for imaging, are most frequently seen.

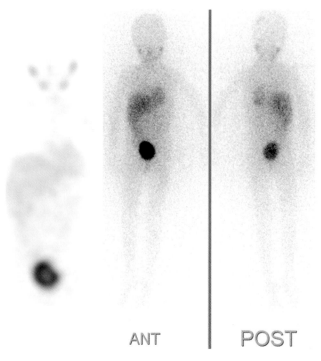

ANT | POST

Figure 8. MIP and whole body planar Images from an I-123 MIBG scan show the normal distribution of this radiopharmaceutical. This study, as is the case for most MIBG studies, was performed to evaluate for neuroblastoma.

Indium-111 Pentetreotide is an analog of somatostatin, a naturally occurring neuropeptide that is synthesized and released by some cells in the nervous system and the endocrine system. Some neuroendocrine tumors express receptors for somatostatin in large numbers. This makes Pentetreotide very useful both for diagnosis of primary tumors too small to be easily detected on CT or MR as well as for staging of these tumors. Tumors where this radiopharmaceutical can be useful include neuroblastoma, pheochromocytoma, carcinoid tumor, pituitary adenoma, medullary thyroid cancer, pancreatic islet cell tumor, paragangliomas, and small cell lung cancers (Figure 7)

Iodine-123 or Iodine-131 MIBG is an analog of guanethidine, similar to norepinephrine. It is therefore taken up by the cells of the adrenal medulla and by tumors derived from these cells, specifically neuroblastomas and pheochromocytomas. For these tumor subtypes it is slightly more sensitive and specific than Pentetreotide, however, some tumors derived from the adrenal medulla take up Pentetreotide though they do take up MIBG and vice versa (Figure 8).

6

CANCER BIOMARKERS

Robert H. Getzenberg

Dr. Kiran C. Patel College of Allopathic Medicine, Nova Southeastern University, Fort Lauderdale, FL, USA

WHAT IS A CANCER BIOMARKER?

Although this may seem like a semantic point, it is important to start this chapter with a definition of what a cancer biomarker is. This is relevant for the obvious reason that there should be a clear understanding of what is being discussed here. Further, it raises the point that how we define a cancer biomarker is evolving as we develop more and better tools by which to identify such markers as well as optimize our approaches to validating and implementing these markers along with optimizing the clinical questions that they address. For the purpose of this chapter, the focus will be on blood and tissue-based biomarkers. Using the definition of a biomarker found on dictionary.com, a biomarker is, "a distinct biochemical, genetic, or molecular characteristic or substance that is an indicator of a particular biological condition or process." The descriptor of biomarker includes both individual and combinations of markers. Although imaging can be considered as one of the earliest types of cancer biomarkers that has been widely adapted across many cancer types, imaging approaches will not be discussed here. There are many types of biomarkers that are utilized in cancer and the attention of this chapter will be on the serum and tissue type markers in solid tumors with no

Cancer: Prevention, Early Detection, Treatment and Recovery, Second Edition. Edited by Gary S. Stein and Kimberly P. Luebbers.
© 2019 John Wiley & Sons, Inc. Published 2019 by John Wiley & Sons, Inc.

discussion of their utility in hematologic cancers. Finally, examples of cancer biomarkers will be provided below.

HOW ARE CANCER BIOMARKERS BEING USED?

In the era of what is being described as the personalized approach to cancer, biomarkers are front and center. They are being developed or in current use in the entire spectrum of cancer from screening to diagnosis to prognosis and therapeutic selection/optimization. There are a number of factors that enter into the equation of the utility of a cancer biomarker. Obviously, these are based primarily on the clinical and performance characteristics of the tool. Financial factors are considered as well including the cost of the test, the clinical value that the biomarker brings to the individual patient, and the more global cost to the healthcare system that is expensed as part of the biomarker utilization.

TYPES OF CANCER BIOMARKERS

As stated above, for the purpose of this chapter, a biomarker will be defined as, "a distinct biochemical, genetic, or molecular characteristic or substance that is an indicator of a particular biological condition or process" and the focus will be upon those related to cancer. This remains a very broad definition with many entities fulfilling it. There are many types of cancer biomarkers and they are defined by the type of biomarker that they are, i.e., protein, DNA, RNA, exosomes, Circulating Tumor Cell (CTC) etc., as well as the source of the biomarker, i.e., body fluids such as blood (serum, plasma, whole blood), urine, semen, and saliva or tissue such as from biopsies or resections. From the tissue perspective, morphological changes are the oldest cancer biomarkers since Virchow originally identified the cellular changes associated with cancer [1]. Immunohistochemical (IHC) analyses are used to stain for proteins that are associated with cancer. These stains are often used in conjunction with the morphological analysis performed by a pathologist in examining a biopsy or tissue sample. Fluorescence in-situ hybridization (FISH) is a complementary approach to IHC analysis with the exception of its use in the detection of nucleic acid (DNA or RNA) probes as opposed to proteins. Proteomic approaches have revealed individual markers and panels of proteins both in tissue and in blood/urine that have apparent clinical utility. These include analyses of serum proteins as well as proteins within the cancer cell along with the microenvironment in which the cancer cell exists. Whole genome/epigenome-based approaches are becoming more commonplace and the signatures obtained from these comprehensive analyses of the tumors are just being implemented into the clinical care paradigms for cancer patients. Despite the increased use of genomics as a tool for selecting the most appropriate course of cancer treatment, there are still questions about the current state of its utility with agreement that there is still a great deal to do to make the genomic analysis of tumors of commonplace [2].

There are a variety of sources for cancer biomarkers. Tumor specific markers are those that appear to be relatively uniquely expressed within the cancer that exists. Cell cycle markers are associated with the cellular progression through the cell cycle and with the faster mitotic rates typically associated with cancers and are often used as indicators of tumor growth most often at the tissue level. Similarly, metabolic changes are hallmarks of the cancer process and cancer associated changes can be utilized as biomarkers [3]. Changes in energy metabolism were an early recognized characteristic of cancer cells [4] and modern approaches to deciphering the metabolic changes associated with cancer have revealed cancer biomarkers. Some of the biomarkers that have been used the longest are those that depict the genetic basis of cancers that can be somatic or germline. Although the existence of epigenetic changes associated with cancer have been identified more recently than their genetic counterparts, a significant amount of effort has been dedicated to the identification and validation of these genetic changes along with their potential use as biomarkers. Changes that are associated with the metastatic process have also been used as biomarkers of cancer. Finally, systemic responses to the presence of the cancer(s) can also be utilized as biomarkers. As an example, indicators of an immune response to the cancer can be used as a sentinel of the presence of cancer as well as potential predictors of therapeutic response [5].

CHALLENGES IN THE IMPLEMENTATION OF CANCER BIOMARKERS

Although there has been a relative explosion of cancer biomarkers over the past decade, with the advances that have been made in our understanding of cancer, it would seem that there should be more. There are complexities at every level of cancer biomarker discovery and development that make clinically useful biomarkers difficult to obtain. These complexities are related to issues such as heterogeneity, the lack of translatability from the lab to the clinic and the apparent lack of many truly cancer specific changes (i.e., fingerprints) that can serve as highly specific biomarkers.

Tumor heterogeneity is among the most perplexing issues related to the implementation of cancer biomarkers and similarly to therapeutic efficacy [6]. Simply stated, tumors rarely uniformly express a single biomarker. For example, in what is considered to be estrogen receptor (ER) or progesterone receptor (PR) positive breast cancer, not all of the cells express the receptors. In fact, the tumor is considered to be ER and/or PR positive if at least 1% of the cells examined have estrogen and/or progesterone receptors [7]. Prostate specific antigen (PSA), one of the most commonly used clinical biomarkers in cancer as well as the androgen receptor, a hormone driver of its expression, are typically not expressed in every cell within a tumor. Similarly genetic and epigenetic markers are typically not uniformly expressed within a tumor. To make things more complex, the surrounding pathologically normal tissue may demonstrate some heterogeneity and represent a type of field effect with some of the signatures that are found in the associated tumor.

Heterogeneity may also exist within an individual. For example, if multiple foci of disease exist, these may differ in the expression of biomarkers [8]. While the biology of these differences is beyond the scope of this chapter, these are considered to be either as a result of tumors that have arisen independently or offshoots of a single lesion. This makes the identification and validation of cancer biomarkers very difficult since it is not always clear what the specific signature of a tumor is and therefore the biomarkers that define it may be difficult to discern. Further, the primary lesion(s) may differ in expression of biomarkers from the resulting metastatic lesions and even these metastatic lesions may be heterogeneous. Therefore this biologically derived heterogeneity makes the description of true cancer-specific biomarkers difficult, if not impossible.

Another level of heterogeneity exists between individuals. This occurs even within normal individuals. For example, for normal laboratory tests, there are ranges of what is considered to be normal. We are all unique. This problem is quite evident in prostate cancer where, as will be described in more detail below, PSA is used as a blood-based biomarker. PSA is a kallikrein protein that is expressed normally within the prostatic epithelia but is often aberrantly secreted into the bloodstream in men with prostate cancer. The blood levels that are considered to be "normal" are a range and can be influenced by noncancerous conditions including inflammation, benign growth, age, and the size of the prostate. Therefore normal variation can muddy the value of cancer biomarkers.

Finally, as mentioned below, one of the challenges in the clinical implementation of cancer biomarkers has been assay heterogeneity. In other words, even when a biomarker is identified, the development of standardized and reproducible assays with which to detect the marker(s) can be problematic [9]. The same result needs to be obtained every time an assay is run regardless of where it is run to increase the utility of such biomarkers.

Cancer Specificity?

The complexity of cancer, as described briefly above, makes the existence and discovery of cancer-specific biomarkers difficult. As described earlier, many of the identified cancer biomarkers are not truly cancer specific and result from altered or de-differentiated expression changes. As a result, the struggle with the specificity of cancer biomarkers starts at the biology and continues on to the development of clinical tests. The clinical tests can only be as good as the biomarkers that they are assaying and this limitation has certainly been at the center of the difficult path toward clinically useful cancer biomarkers.

Translation From Laboratory to Clinic

The journey from the discovery of a cancer biomarker to its clinical utility is an arduous one that few cancer biomarkers have survived. There are many reasons to explain this. As described above, tumor heterogeneity and the issues related to cancer specificity make the actual discovery process, difficult. Margaret Pepe and colleagues developed a concept of the five phases of cancer biomarker development [10]. These include the preclinical exploratory studies that result in the identification of the biomarker(s) of interest (Phase 1); the development of a clinical assay for the detection of the biomarker(s) of interest (Phase 2); the utilization of the clinical assay to probe a retrospectively collected longitudinal repository, the most common type of sample collections that exist today (Phase 3); the evaluation of the clinical assay in a prospective study, i.e., how does it work in a real world scenario (Phase 4); and, if appropriate, an evaluation of the clinical assay in a screening type environment, the holy grail of many cancer biomarkers (Phase 5). These stages of biomarker development and evaluation are logical and follow, in a parallel fashion, the phase of trials required for drug development. Similar to drug development, few cancer biomarkers make their way through this process. Unlike the development of therapeutics, until recently, few cancer biomarkers actually went through this logical series of studies and therefore the waters have been muddied by small, improperly designed and executed studies. Despite this, efforts such as that being conducted by the NCI's Early Detection Research Network (http://edrn.nci.nih.gov), have brought this type of scientific rigor to the process of cancer biomarker development and are producing the types of data that are necessary for clinical adaptation of cancer biomarkers.

EXAMPLES OF MAJOR CANCER BIOMARKERS BY TUMOR TYPE

This chapter will focus on selected cancer types along with markers that have demonstrated clinical utility in order to provide a perspective on the types of cancer biomarkers and how they may be utilized today.

Breast Cancer

Breast cancer is an example of where biomarkers have greatly impacted the clinical course of the disease at different levels. Biomarkers are utilized in aiding the determination of an individual's risk for developing the disease and therefore helping to guide the course of screening and monitoring. Genetic alterations such as those found in the *BRCA1* and *BRCA2* genes [11] have been used to identify women at increased risk for breast and ovarian cancer [12]. Although these mutations are only associated with a relatively small percentage of either breast or ovarian cancers, they have provided guidance to women and may result in enhancing their screening approaches and/or chemoprevention [13]. These are among the growing number of germline mutations that are found in DNA typically isolated from a blood or saliva sample.

Once a woman is diagnosed with breast cancer, based upon pathologic examination of biopsies of their breast tissue, biomarkers are again used to aid in the characterization of the type of breast cancer and therefore provide both prognostic information as well as guidance related to the most effective therapeutic interventions. In addition to the pathologic subtype of breast cancer that is determined by its morphology, IHC staining is performed in order to determine the expression of hormone receptors (estrogen and progesterone) as well as to determine the Her2 status of the patient [7]. Her2 status can also be determined by examining the expression of the encoding RNA using FISH and androgen receptor status is now being more commonly analyzed by IHC as with the other hormone receptors. The expression of the androgen receptor, particularly when combined with the other steroid hormones described above, can also serve as a prognostic factor as well as guide treatment planning [14]. The resulting expression patterns provide clinically useful information and are being utilized to guide therapeutic strategies. In addition to the information provided from these biomarkers, molecular classifications of breast cancer samples has provided for the identification of subgroups that may indeed have differential therapeutic responses [15, 16]. To increase the predictive ability of an individual gene or multiple genes, a 21-gene test has been developed to predict recurrence in postmenopausal women treated with anastrozole or tamoxifen. In this study of 1231 evaluable patients, based upon low, intermediate, or high score using the 21-gene set, this set of genes was shown to be an independent predictor of recurrence [17]. A recent proteomic analysis revealed that the tissue expression of ubiquitin and truncated S100P identified the presence of breast cancer with a high level of sensitivity and specificity ([18]).

Examination and/or enumeration of circulating cells, which are typically described as circulating tumor cells (CTCs), has been applied to many cancer types but none as much as they have to breast cancer. The clinically utilized tool of CTC enumeration has been FDA (Food and Drug Administration) approved utilizing the CellSearch system (Veridex Inc.). A number of studies have validated that less than five CTCs per 7.5 ml sample is correlative with a better outcome [19, 20]. Despite this, a recent Phase 3 study found that simple enumeration of CTCs with the CellSearch system to support treatment decisions was not associated with improved outcomes [21]. An alternative approach to quantifying CTCs associated with breast cancer has also recently been described [22]. It is apparent from these studies that while counting the number of CTCs may have value to truly impact clinical decisions we need different approaches. These approaches are focusing on two major areas. The first is in delineating and characterizing the cells that are found circulating in the bloodstream of cancer patients. Historically, CTCs are identified as cells possessing epithelial cell adhesion molecule, EpCAM, on their surface. Utilizing anti-EpCAM antibodies, these cells are then separated from the other cell types found in the blood and then further characterized. Recent studies have demonstrated that many important cells involved in the cancer process may not possess EpCAM on their surface and therefore go undetected by these approaches. This will be discussed in more detail in the section on prostate cancer biomarkers. Further, even in the EpCAM containing CTCs, enumeration may not be sufficient and molecular characterization of the isolated cells may be critical to defining the essential clinical information [23–25]. Despite the complexity of cell types found within the blood, we are clearly just beginning to decipher the information contained within along with its correlation with the clinical value of these signatures.

Prostate Cancer

As we look at the utility of body-fluid-based biomarkers for the detection of cancer, the one that has been in clinical use the longest is a biomarker of prostate cancer, PSA. PSA was approved by the US FDA in 1986 along with DRE (digital rectal examination) for the early detection of prostate cancer [26]. In order to determine the potential clinical utility of PSA, it is important to understand some of the biology that underlies this biomarker. PSA is distinct from many of the types of cancer-related biomarkers described above in that its expression is not actually cancer specific. PSA is a member of the kallikrein-related peptidase family and is a glycoprotein enzyme encoded in humans by the KLK3 gene. PSA is secreted by the epithelial cells of the prostate gland into the prostatic acini where it becomes a component of the ejaculate and contributes to the liquefaction of semen and may also be involved in degrading cervical mucus in the female. It is found primarily in the prostate but has also been shown to be expressed in the salivary glands and in breast tissue among others. While the expression of PSA is not cancer specific, as a secretory epithelial protein, it is normally not found in the bloodstream at significant levels. In cancer, as well as in some benign states like prostatic inflammation and BPH, the prostatic architecture can be disrupted and instead of PSA being secreted into the ejaculate, it ends up in the blood as well. The detection of PSA in the bloodstream has been utilized as a tool with which to identify those men at risk for having prostate cancer as well as a tool to follow the extent of prostate cancer in men already diagnosed with the disease [26].

Therefore, PSA is not a classic cancer biomarker but as a single blood-based cancer biomarker has had more impact than any other. Since its widespread clinical adoption, the number of men that present for the first time with metastatic prostate cancer has decreased significantly. The European Randomized Study of Screening for Prostate Cancer did find that screening with PSA does result in a significant reduction of deaths from prostate cancer [27]. Although beyond the scope of this chapter, controversy regarding its use rests primarily with what are considered to be elevated levels of PSA and how the findings are handled clinically. Furthermore, despite evidence of limited efficacy in older men, the number of men 75 years of age and older being screened for prostate cancer with PSA remains high [28]. The US Preventive Services Task Force has recently updated its recommendation to indicate that men aged 55–69 should make the personal decision, together with their clinician and considering the potential harms and benefits, on whether to undergo periodic PSA-based screening for prostate cancer [29].

Several approaches have been used to improve upon the clinical utility of PSA. The prostate health index (phi) takes into consideration the precursor form of PSA, proPSA, as well as the unbound or free PSA, fPSA in a formula (phi = [−2] proPSA/ fPSA × PSA1/2). Phi has been shown to have increased accuracy at predicting the presence of prostate cancer on biopsy than PSA alone [30]. A panel of four kallikreins, the 4Kscore (OPKO Health, Inc.) detected within the blood has been demonstrated to predict the risk of having prostate cancer as well as its potential aggressiveness. In a recent study of 4765 men, the 4Kscore was able to predict the detection of prostate cancer by 10-core biopsy as well as detect high-grade disease [31]. This test is now being rapidly incorporated into clinical practice with large studies supporting its utility [32].

A proteomic approach has identified an eight biomarker assay based upon proteins that were qualified from a candidate list of 11 such biomarkers. In a clinical validation of these eight protein biomarkers, 318 patients that were biopsied along with matching prostatectomy specimens demonstrated that the this assay could separate those with favorable and nonfavorable pathology (AUC 0.68, OR 20.9) as well as Gleason Score 6 versus non-Gleason Score 6 disease (AUC 0.65, OR 12.95) [33]. This assay being commercialized by Metamark Genetics Inc., appears to provide information that is additive to the currently available clinical tools.

A unique aspect of urologic cancers, including prostate cancer, is their exposure to the urinary system and therefore the ability to detect biomarkers within the urine. With the compartmentalized exposure of these organs, this media provides an added degree of specificity to the detection of cancer related biomarkers. One such biomarker is PCA3, an mRNA that does not appear to encode a protein but which appears within the urine of men with prostate cancer [34]. Urine-based PCA3 is approved for the detection of prostate cancer in men who have previously had a negative biopsy. In a recent meta-analysis, using a cutoff of 20, investigators reported a sensitivity of 72%, a specificity of 53%, and a diagnostic odds ratio of 3.18 [35]. Another source of urine-based biomarker of prostate cancer is exosomes that are isolated from the urine and are then used to decipher an RNA signature that has focused upon ERG and PCA3. In a recent study, the negative predictive value (NPV) and positive predictive value (PDV) were 98.6% and 34.7% respectively for the detection of more aggressive prostate cancer [36]. In another example of using urine as a media in which to noninvasively study what is going on within the prostate, a three-gene panel of urinary markers was recently identified and in preliminary studies was shown to be correlative with higher Gleason score ≥ 7, i.e., more aggressive prostate cancer [37].

There have been significant advances in the development and validation of genomic and epigenetic tools that appear to provide clinically useful diagnostic and prognostic information for individuals with prostate cancer. The Oncotype DX test (Genomic Health, Inc.) analyzes a genomic prostate score (GPS) that combines the expression of 17 genes originally identified from analyzing prostatectomy samples to identify patients with aggressive prostate cancer [38, 39]. The 17 gene set, which contains genes from a number of cancer associated pathways, has been validated using biopsy samples obtained at the time of diagnosis and is predictive of the presence of high grade and high stage disease (OR = 2.1) [40].

Prolaris (Myriad Genetics, Inc.) is a test that utilizes a set of 46 genes that are focused upon cell cycle progression (CCP). Although the mitotic index of prostate cancer is not considered to be high, cancer cells in general, proceed through the cell cycle at higher rates than their normal counterparts and therefore biomarkers of this proliferative state have been of clinical value. In a study of biopsy samples from 761 men that then went on to have radical prostatectomies, the Prolaris test provided significant predictive information regarding the risk of progression of these men to either biochemical recurrence and/or the development of metastases using either univariate (hazard ration = 2.08) or multivariate analysis (hazard ratio = 1.76) [41].

The third example of the genetic tests being utilized in prostate cancer is the Decipher Prostate Cancer Classifier (GenomeDx Biosciences). This test is designed to provide prognostic information post-radical prostatectomy. In a study of 260 men, 99 of which went on to develop metastases, the Decipher test was significantly correlated with biochemical recurrence, metastasis as well as prostate cancer specific mortality. In individuals with high and low Decipher scores, the cumulative incidence of metastasis was 47 and 12%, respectively [42]. Other studies have supported these findings.

The ConfirmMDx is an epigenetic assay that uses methylation specific PCR to examine the hypermethylation of selected genes including GSTP1, APC, and RASSFI in a prostatic biopsy core. This test was developed to help to rule out false-negative biopsies and provides strong predictive information regarding the potential to detect prostate cancer on a repeated prostatic biopsy (Odds Ratio of 3.01) [43].

MicroRNAs signatures have been developed and are being applied. At this point they still need clinical validation but provide yet additional genomic tools of potential value [44].

Although not as clinically advanced as their use has been in breast cancer, CTCs have also begun to take their place in the prostate cancer paradigm. One limitation to the enumeration of CTCs using the CellSearch Instrument (Veridex, Inc.) is that approximately 50% of the men with prostate cancer have undetectable CTCs [45]. On the contrary, the detection of CellSearch defined CTCs has been rare in men with clinically localized disease [46]. The baseline number of CTCs has been shown to be predictive of prostate cancer specific survival with those with five or more cells per 7.5 ml having a poorer outcome [47]. Combination of CTC enumeration and the blood marker, lactate dehydrogenase (LDH) has been shown to be a surrogate of survival in men on a study of abiraterone plus prednisone versus prednisone alone [48]. Combining other markers like LDH with the CTC enumeration may provide additional value to these measurements. Since the original enumeration studies, a number of groups have been developing approaches to characterizing the cells that are found within the blood. An example of the type of relevance of this approach is the exciting work from Jun Luo, Emmanuel Antonarakis, and colleagues in which they have identified the existence of a variant of the androgen receptor (AR-V7) within the circulating cells of men with prostate cancer and have shown the expression of this variant in the circulating cells is predictive of response to commonly used therapies for the disease [49, 50].

Work from Epic Sciences is going beyond the EpCAM selection of the CTCs and characterizing all of the cells found within the blood using morphologic and genetic markers of the isolated cells [51, 52]. We are just entering the era in which circulating cells can be utilized as liquid biopsies and many approaches are being investigated with direct clinical correlation.

In addition to the markers/tests described above, there are many novel biomarkers that are being evaluated for their clinical utility. As with breast cancer, in addition to the classical utility of cancer biomarkers, prostate cancer biomarkers are being developed to determine those most appropriate or inappropriate for active surveillance as well as men that may respond to individual therapeutic approaches. In addition to individual biomarkers, panels of markers are being developed. Among these panels of markers being explored are cancer/testis antigens, which appear to be indicative of more advanced prostate cancer [53] and can be detected utilizing a nanowire-based approach [54]. Another panel of markers includes three serum phospholipids that may be indicative of the presence of the disease [55].

Bladder Cancer

As described briefly above related to prostate cancer, bladder cancer is relatively unique in its access to urine as a source of biomarkers. Examination of cells found within the urine, cytology, has been a mainstay of bladder cancer diagnosis. While cytology has a high degree of specificity, if you see cancer cells in the urine, they must have come from somewhere, sensitivity is considered to be limited [56]. As an additive tool to cytology, FISH, which can detect aneuploidy in chromosomes 3, 7, and 17 as well as loss of the 9p21 locus (UroVysion, Abbott Molecular Inc.), has been added. While the bladder is seemingly an ideal cancer type to develop specific biomarkers, based upon its constant bathing in urine, this media has been a difficult one with which to develop biomarkers. Markers, including NMP-22, AccuDx, a fibrin-fibrinogen degradation product and BTA have been FDA approved but are not widely utilized. Prognostic biomarkers for bladder cancer have been reviewed [57]. BLCA-4 is a urine-based biomarker for bladder cancer that has now been validated in a number of studies by different groups [58–61]. Despite its potential promise, the absence of commercial development of this biomarker has limited its further validation. Genomic Health has recently developed a liquid biopsy test in which they examine multiple biomarkers within the urine. In a recent report from their proof-of-concept study, they describe a negative predictive value for high-risk recurrence of 95% [62]. Obviously, further validation is needed but this approach appears promising.

Lung Cancer

Although lung cancer is the leading cause of cancer deaths in the United States, there are few biomarkers for the disease and the disease is often detected either through incidental imaging tests or by presenting symptoms but typically not until it is already advanced. Among the blood-based biomarkers are those that are found to be elevated in several cancer types, in particular aggressive variants. These biomarkers include carcinoembryonic antigen (CEA), cytokeratin 19, carbohydrate antigen-125 (CA–125), and neuron-specific enolase (NSE) [63]. These are considered to be relatively nonspecific markers that represent either a more embryonic state, mesenchymal transition, and/or de-differentiation of the epithelial cell and, as such, are also utilized as biomarkers for a number of cancer types. There are numerous new markers currently under investigation. Included among these are proteomic markers, miRNAs, genomic changes, metabolites as well as circulating DNA.

Melanoma

While the visualization of melanoma lesions on the skin provides for an easier risk stratification and biopsy for diagnosis, the prognosis of individuals with such lesions covers a spectrum. A number of serum biomarkers have been explored to determine their utility as prognostic markers. These include, LDH, tyrosinase, and vascular endothelial growth factor [64]. At least some melanomas have been shown to induce an immune response and indicators of such a response have likewise been developed into biomarkers of the disease as well as potential predictors of therapeutic response.

Colon Cancer

The landscape of colon cancer has changed significantly with the incorporation of screening colonoscopies into clinical practice. This has resulted in most individuals with colorectal tumors presenting with localized disease [64]. Based upon the typical availability of tissue resulting from biopsies and resections, genetic biomarkers have been extensively examined. Mutations in the *APC* gene are very common and mutations in the *BRAF* gene along with microsatellite instability as a result of inactivation of the DNA mismatch repair system are more typically found in Lynch syndrome [65]. Oncotype Dx (Genomic Health Inc.) has developed a multigene colon cancer assay, which may add to the prediction of recurrence in patients with stage 2 and 3 colon cancers [66]. As with the other cancer types described above, there are a number of novel biomarkers being developed and validated.

CONCLUSIONS

This is an exciting time in the development and application of novel cancer biomarkers. The latest technologies are being applied to identify not just the typical biomarkers associated with the diagnosis or prognosis of the disease but biomarkers that can aid in the prediction of response to individual therapies. These markers are guiding the personalization of cancer therapy and helping to direct whether surgery, chemotherapy, hormonal therapy, radiation therapy, and/or other approaches have the best chance for success. The next several years will provide cancer patients with the realization of these efforts.

ACKNOWLEDGMENTS

I would like to thank Ms. Suzanne Hendrix for her valuable editorial assistance.

REFERENCES

1. Mukherjee, S. (2010). *The Emperor of All Maladies: A Biography of Cancer*. New York: Scribner.
2. Kaiser, J. (2018). Is genome-guided cancer treatment hyped? *Science* 360 (6387): 365.
3. Vermeersch, K.A. and Styczynski, M.P. (2013). Applications of metabolomics in cancer research. *J. Carcinog.* 12: 9.
4. Warburg, O. (1925). The metabolism of carcinoma Cells. *J. Cancer Res.* 9 (1): 148–163.
5. Bigbee, W.H.R. (2003). Tumor markers and immunodiagnosis. In: *Holland-Frei Cancer Medicine*, 6e (ed. P.R. Kufe, R.E. Pollock, R.R. Weichselbaum, et al.). Hamilton, Ontario, Canada: BC Decker Inc.
6. Pribluda, A., de la Cruz, C.C., and Jackson, E.L. (2015). Intratumoral heterogeneity: from diversity comes resistance. *Clin. Cancer Res.* 21 (13): 2916–2923.
7. Hammond, M.E., Hayes, D.F., Dowsett, M. et al. (2010). American society of clinical oncology/college of American pathologists guideline recommendations for immunohistochemical testing of estrogen and progesterone receptors in breast cancer. *J. Clin. Oncol.* 28 (16): 2784–2795.
8. Aryee, M.J., Liu, W., Engelmann, J.C. et al. (2013). DNA methylation alterations exhibit intraindividual stability and interindividual heterogeneity in prostate cancer metastases. *Sci. Transl. Med.* 5 (169): 169ra110.
9. de Gramont, A., Watson, S., Ellis, L.M. et al. (2015). Pragmatic issues in biomarker evaluation for targeted therapies in cancer. *Nat. Rev. Clin. Oncol.* 12 (4): 197–212.
10. Pepe, M.S., Etzioni, R., Feng, Z. et al. (2001). Phases of biomarker development for early detection of cancer. *J. Natl. Cancer Inst.* 93 (14): 1054–1061.
11. Welcsh, P.L. and King, M.C. (2001). BRCA1 and BRCA2 and the genetics of breast and ovarian cancer. *Hum. Mol. Genet.* 10 (7): 705–713.
12. Easton, D.F. (1999). How many more breast cancer predisposition genes are there? *Breast Cancer Res.* 1 (1): 14–17.
13. King, M.C., Wieand, S., Hale, K. et al. (2001). Tamoxifen and breast cancer incidence among women with inherited mutations in BRCA1 and BRCA2: national surgical adjuvant breast and bowel project (NSABP-P1) breast cancer prevention trial. *JAMA* 286 (18): 2251–2256.
14. Zhang, L., Fang, C., Xu, X. et al. (2015). Androgen receptor, EGFR, and BRCA1 as biomarkers in triple-negative breast cancer: a meta-analysis. *Biomed. Res. Int.* 2015: 357485.
15. Curtis, C., Shah, S.P., Chin, S.F. et al. (2012). The genomic and transcriptomic architecture of 2,000 breast tumours reveals novel subgroups. *Nature* 486 (7403): 346–352.
16. Lehmann, B.D., Bauer, J.A., Chen, X. et al. (2011). Identification of human triple-negative breast cancer subtypes and preclinical models for selection of targeted therapies. *J. Clin. Invest.* 121 (7): 2750–2767.
17. Dowsett, M., Cuzick, J., Wale, C. et al. (2010). Prediction of risk of distant recurrence using the 21-gene recurrence score in node-negative and node-positive postmenopausal patients with

breast cancer treated with anastrozole or tamoxifen: a TransATAC study. *J. Clin. Oncol.* 28 (11): 1829–1834.

18. Chung, L., Shibli, S., Moore, K. et al. (2013). Tissue biomarkers of breast cancer and their association with conventional pathologic features. *Br. J. Cancer* 108 (2): 351–360.

19. Cristofanilli, M., Hayes, D.F., Budd, G.T. et al. (2005). Circulating tumor cells: a novel prognostic factor for newly diagnosed metastatic breast cancer. *J. Clin. Oncol.* 23 (7): 1420–1430.

20. Pierga, J.Y., Hajage, D., Bachelot, T. et al. (2012). High independent prognostic and predictive value of circulating tumor cells compared with serum tumor markers in a large prospective trial in first-line chemotherapy for metastatic breast cancer patients. *Ann. Oncol.* 23 (3): 618–624.

21. Smerage, J.B., Barlow, W.E., Hortobagyi, G.N. et al. (2014). Circulating tumor cells and response to chemotherapy in metastatic breast cancer: SWOG S0500. *J. Clin. Oncol.* 32 (31): 3483–3489.

22. Magbanua, M.J., Carey, L.A., DeLuca, A. et al. (2015). Circulating tumor cell analysis in metastatic triple-negative breast cancers. *Clin. Cancer Res.* 21 (5): 1098–1105.

23. Jiang, Z.F., Cristofanilli, M., Shao, Z.M. et al. (2013). Circulating tumor cells predict progression-free and overall survival in Chinese patients with metastatic breast cancer, HER2-positive or triple-negative (CBCSG004): a multicenter, double-blind, prospective trial. *Ann. Oncol.* 24 (11): 2766–2772.

24. Magbanua, M.J. and Park, J.W. (2014). Advances in genomic characterization of circulating tumor cells. *Cancer Metastasis Rev.* 33 (2–3): 757–769.

25. Smirnov, D.A., Zweitig, D.R., Foulk, B.W. et al. (2005). Global gene expression profiling of circulating tumor cells. *Cancer Res.* 65 (12): 4993–4997.

26. De Angelis, G., Rittenhouse, H.G., Mikolajczyk, S.D. et al. (2007). Twenty years of PSA: from prostate antigen to tumor marker. *Rev. Urol.* 9 (3): 113–123.

27. Schroder, F.H., Hugosson, J., Roobol, M.J. et al. (2012). Prostate-cancer mortality at 11 years of follow-up. *N. Engl. J. Med.* 366 (11): 981–990.

28. Drazer, M.W., Huo, D., and Eggener, S.E. (2015). National prostate cancer screening rates after the 2012 US preventive services task force recommendation discouraging prostate-specific antigen-based screening. *J. Clin. Oncol.* 33 (22): 2416–2423.

29. Grossman, D.C., Curry, S.J., Owens, D.K. et al. (2018). Screening for prostate cancer: US preventive services task force recommendation statement. *JAMA* 319 (18): 1901–1903.

30. Stephan, C., Vincendeau, S., Houlgatte, A. et al. (2013). Multicenter evaluation of [-2] proprostate-specific antigen and the prostate health index for detecting prostate cancer. *Clin. Chem.* 59 (1): 306–314.

31. Bryant, R.J., Sjoberg, D.D., Vickers, A.J. et al. (2015). Predicting high-grade cancer at ten-core prostate biopsy using four kallikrein markers measured in blood in the protecT study. *J. Natl. Cancer Inst.* 107 (7): Available online at: https://academic.oup.com/jnci/article/107/7/djv095/913174.

32. Punnen, S., Pavan, N., and Parekh, D.J. (2015). Finding the wolf in sheep's clothing: the 4Kscore is a novel blood test that can accurately identify the risk of aggressive prostate cancer. *Rev. Urol.* 17 (1): 3–13.

33. Blume-Jensen, P., Berman, D.M., Rimm, D.L. et al. (2015). Development and clinical validation of an in situ biopsy-based multimarker assay for risk stratification in prostate cancer. *Clin. Cancer Res.* 21 (11): 2591–2600.

34. Bussemakers, M.J., van Bokhoven, A., Verhaegh, G.W. et al. (1999). DD3: a new prostate-specific gene, highly overexpressed in prostate cancer. *Cancer Res.* 59 (23): 5975–5979.

35. Luo, Y., Gou, X., Huang, P. et al. (2014). The PCA3 test for guiding repeat biopsy of prostate cancer and its cut-off score: a systematic review and meta-analysis. *Asian J. Androl.* 16 (3): 487–492.

36. Donovan, M. J., Noerholm, M., Bentink, S., et al.(2015). "A first catch, non-DRE urine exosome gene signature to predict Gleason 7 prostate cancer on an initial prostate needle biopsy." 2015 Genitourinary Cancers Symposium (ASCO GU).

37. Leyten, G.H., Hessels, D., Smit, F.P. et al. (2015). Identification of a candidate gene panel for the early diagnosis of prostate cancer. *Clin. Cancer Res.* 21 (13): 3061–3070.

38. Badani, K.K., Kemeter, M.J., Febbo, P.G. et al. (2015). The impact of a biopsy based 17-gene genomic prostate score on treatment recommendations in men with newly diagnosed clinically prostate cancer who are candidates for active surveillance. *Urol. Pract.* 2 (4): 181–189.

39. Cullen, J., Rosner, I.L., Brand, T.C. et al. (2015). A Biopsy-based 17-gene genomic prostate score predicts recurrence after radical prostatectomy and adverse surgical pathology in a racially diverse population of men with clinically low- and intermediate-risk prostate cancer. *Eur. Urol.* 68 (1): 123–131.

40. Klein, E.A., Cooperberg, M.R., Magi-Galluzzi, C. et al. (2014). A 17-gene assay to predict prostate cancer aggressiveness in the context of Gleason grade heterogeneity, tumor multifocality, and biopsy undersampling. *Eur. Urol.* 66 (3): 550–560.

41. Cuzick, J., Stone, S., Fisher, G. et al. (2015). Validation of an RNA cell cycle progression score for predicting death from prostate cancer in a conservatively managed needle biopsy cohort. *Br. J. Cancer* 113 (3): 382–389.

42. Ross, A.E., Johnson, M.H., Yousefi, K. et al. (2015). Tissue-based genomics augments post-prostatectomy risk stratification in a natural history cohort of intermediate- and high-risk men. *Eur. Urol.* 69 (1): 157–165.

43. Partin, A.W., Van Neste, L., Klein, E.A. et al. (2014). Clinical validation of an epigenetic assay to predict negative histopathological results in repeat prostate biopsies. *J. Urol.* 192 (4): 1081–1087.

44. Wen, X., Deng, F.M., and Wang, J. (2014). MicroRNAs as predictive biomarkers and therapeutic targets in prostate cancer. *Am. J. Clin. Exp. Urol.* 2 (3): 219–230.

45. Halabi, S., Small, E.J., Hayes, D.F. et al. (2003). Prognostic significance of reverse transcriptase polymerase chain reaction for prostate-specific antigen in metastatic prostate cancer: a nested study within CALGB 9583. *J. Clin. Oncol.* 21 (3): 490–495.

46. Khurana, K.K., Grane, R., Borden, E.C. et al. (2013). Prevalence of circulating tumor cells in localized prostate cancer. *Curr. Urol.* 7 (2): 65–69.

47. Danila, D.C., Heller, G., Gignac, G.A. et al. (2007). Circulating tumor cell number and prognosis in progressive castration-resistant prostate cancer. *Clin Cancer Res.* 13 (23): 7053–7058.

48. Scher, H.I., Heller, G., Molina, A. et al. (2015). Circulating tumor cell biomarker panel as an individual-level surrogate for survival in metastatic castration-resistant prostate cancer. *J. Clin. Oncol.* 33 (12): 1348–1355.

49. Antonarakis, E.S., Lu, C., Wang, H. et al. (2014). AR-V7 and resistance to enzalutamide and abiraterone in prostate cancer. *N. Engl. J. Med.* 371 (11): 1028–1038.

50. Nakazawa, M., Lu, C., Chen, Y. et al. (2015). Serial blood-based analysis of AR-V7 in men with advanced prostate cancer. *Ann Oncol.* 6 (9): 1859–1865.

51. Beltran, B., Jendrisak, A., Landers, M. et al. (2015). Phenotypic characterization of circulating tumor cells (CTCs) from neuroendocrine prostate cancer (NEPC) and metastatic castration-resistant prostate cancer (mCRPC) patients to identify a novel diagnostic algorithm for the presence of NEPC. 2015 Genitourinary Cancers Symposium. *J. Clin. Oncol.* 33 (suppl 7): abstr 197.

52. Scher, H., A. Jendrisak, J. Louw, et al. (2015). Predictive biomarkers of sensitivity to androgen receptor signaling (ARS) and taxane-based chemotherapy in circulating tumor cells (CTCs) of patients (pts) with metastatic castration resistant prostate cancer (mCRPC). 2015 Genitourinary Cancers Symposium, J Clin Oncol 33, (suppl 7; abstr 147).

53. Kulkarni, P., Shiraishi, T., Rajagopalan, K. et al. (2012). Cancer/testis antigens and urological malignancies. *Nat. Rev. Urol.* 9 (7): 386–396.

54. Takahashi, S., Shiraishi, T., Miles, N. et al. (2015). Nanowire analysis of cancer-testis antigens as biomarkers of aggressive prostate cancer. *Urology* 85 (3): 704 e701–704 e707.

55. Patel, N., Vogel, R., Chandra-Kuntal, K. et al. (2014). A novel three serum phospholipid panel differentiates normal individuals from those with prostate cancer. *PLoS One* 9 (3): e88841.

56. Ye, F., Wang, L., Castillo-Martin, M. et al. (2014). Biomarkers for bladder cancer management: present and future. *Am. J. Clin. Exp. Urol.* 2 (1): 1–14.

57. Habuchi, T., Marberger, M., Droller, M.J. et al. (2005). Prognostic markers for bladder cancer: international consensus panel on bladder tumor markers. *Urology* 66 (6 Suppl 1): 64–74.

58. Guo, B., Che, T., Shi, B. et al. (2015). Interaction network analysis of differentially expressed genes and screening of cancer marker in the urine of patients with invasive bladder cancer. *Int. J. Clin. Exp. Med.* 8 (3): 3619–3628.

59. Konety, B.R., Nguyen, T.S., Brenes, G. et al. (2000). Clinical usefulness of the novel marker BLCA-4 for the detection of bladder cancer. *J. Urol.* 164 (3 Pt 1): 634–639.

60. Santoni, M., Catanzariti, F., Minardi, D. et al. (2012). Pathogenic and diagnostic potential of BLCA-1 and BLCA-4 nuclear proteins in urothelial cell carcinoma of human bladder. *Adv. Urol.* 2012: 397412.

61. Wang, Z-Y., Li, H-Y., Wang, H. et al. (2018). Bladder cancer-specific nuclear matrix proteins-4 may be a potential biomarker for non-muscle-invasive bladder cancer detection. *Dis Markers*, article ID 5609395.

62. Genomic Health, I. (2015). Genomic Health Presents Positive Proof-of-Concept Data for Bladder Cancer Liquid Biopsy Test REDWOOD CITY, Calif., PRNewswire.

63. Li, X., Asmitananda, T., Gao, L. et al. (2012). Biomarkers in the lung cancer diagnosis: a clinical perspective. *Neoplasma* 59 (5): 500–507.

64. Karagiannis, P., Fittall, M., and Karagiannis, S.N. (2015). Evaluating biomarkers in melanoma. *Front. Oncol.* 4: 383.

65. Bartley, A.N. and Hamilton, S.R. (2015). Select biomarkers for tumors of the gastrointestinal tract: present and future. *Arch. Pathol. Lab. Med.* 139 (4): 457–468.

66. You, Y.N., Rustin, R.B., and Sullivan, J.D. (2015). Oncotype DX colon cancer assay for prediction of recurrence risk in patients with stage II and III colon cancer: a review of the evidence. *Surg. Oncol.* 24 (2): 61–66.

7

MOLECULAR DIAGNOSIS

Mark F. Evans

*Department of Pathology & Laboratory Medicine,
University of Vermont, Burlington, VT, USA*

Carcinogenesis is characterized by the dysregulation of mechanisms that maintain cell, tissue, and body integrity. Molecular signatures of these processes are increasingly used as biomarkers for improved diagnostics and patient care. This chapter reviews principles of clinical laboratory testing, molecular techniques, and their diagnostic applications.

CLINICAL DIAGNOSTICS

Modern clinical diagnostics is founded on the pioneering studies of Giovanni Battista Morgagni of Padua (1682–1771), the "father of anatomic pathology" and Rudolf Virchow of Berlin (1821–1902), the "father of cellular pathology." Dr. Morgagni documented the relationship between diseases and the gross changes observed in autopsy specimens; Dr. Virchow established the correlation of cellular changes with disease having formulated the cellular theory of life, *omnis cellula e cellula* (all cells come from cells). Molecular diagnostic techniques evaluate disease at the genetic and protein expression levels and are used to refine, define, and extend the diagnostic capabilities of twenty-first century medicine.

Cancer: Prevention, Early Detection, Treatment and Recovery, Second Edition. Edited by Gary S. Stein and Kimberly P. Luebbers.
© 2019 John Wiley & Sons, Inc. Published 2019 by John Wiley & Sons, Inc.

Laboratory Testing

Clinical laboratory testing is highly regulated. In the United States, all diagnostic tests are performed in Clinical Laboratory Improvement Amendments (CLIA) certificated laboratories. The CLIA Program is overseen by the Centers for Medicare and Medicaid Services (CMS) and enforces national standards for test performance. More than 250000 US labs are CLIA regulated. Additionally, the College of American Pathologists (CAP) provides a CMS recognized CAP Laboratory Accreditation Program that involves periodic inspections of clinical testing laboratories for compliance with an extensive checklist of testing criteria. Around 6900 US labs are CAP accredited.

Many diagnostic tests have undergone US Federal Drug Administration (FDA) validation for laboratory implementation. The FDA reviews proprietary *in vitro* diagnostic tests for a variety of criteria including test repeatability, reproducibility, comparability to or improvement over current standard of care procedures, and dependent on the specific purpose of the test, test performance on different populations (e.g., by gender, age group, smoking status, concurrent health issues). Tests may be "approved" (Premarket Approval [PMA] for devices that have undergone clinical trials to demonstrate safety and effectiveness) or "cleared" (Premarket Notification [510(k)] for devices and tests that are substantially equivalent to existing PMA certified diagnostic assays).

An extensive menu of molecular diagnostic tests is now offered in CLIA/CAP certified labs. Those tests that are also FDA approved or cleared are highlighted in this chapter to illustrate the scope of current molecular diagnostic tests.

MOLECULAR DIAGNOSTIC TESTS

Broadly, molecular diagnostic tests have four closely interrelated aims.

1. *Denotative*. The results of these tests have significant clinical implications as they help support the formal naming and classification of a malignant condition.
2. *Prognostic*. Favorable or unfavorable cancer prognoses are identified by these assays.
3. *Theranostic*. Specific pharmacologic interventions are recommended on the basis of these tests.
4. *Patient Management*. These procedures are for deciding the course of patient monitoring, additional lab tests or clinical interventions.

Molecular diagnostics is concerned with the identification of "actionable" biomarkers that have significant diagnostic, prognostic, therapeutic, or patient management implications among cancer patient subgroups. Test results are incorporated into decision trees so that patients are assigned to the most appropriate and timely healthcare support.

Guidelines for Test Selection

Guidance to physicians about which test(s) to choose for a particular cancer is available from a variety of sources. These include recommendations by organizations such as the American Society of Clinical Oncology (ASCO) or the National Comprehensive Cancer Network (NCCN). Additionally, medical subspecialty governing bodies such as CAP [1] and national clinical centers such as the Mayo Clinic provide guidelines. Diagnostic precision is further advanced by way of clinical research publications, conferences, and experimental protocols.

MOLECULAR TECHNIQUES AND THEIR APPLICATIONS

The range of molecular diagnostic tests with FDA-listed applications includes: immunohistochemistry (IHC), in situ hybridization (ISH), polymerase chain reaction (PCR), sequencing technologies, microarray assay, transcription mediated amplification (TMA), Invader chemistry, NanoString technology, hybrid capture (HC), circulating tumor cell (CTC) detection, and proteomic assays.

IHC and ISH are applied to tissue sections and the data is reviewed by microscopy. These tests have the advantage of allowing protein and nucleic acids molecular expression to be interpreted directly in relationship to tumor morphology. Other tests analyze nucleic acids or proteins extracted from ground-up tissues but allow the investigation of submicroscopic levels of molecular detail such as gene sequence mutations.

Patient Specimens

Molecular diagnostic tests are applied to a wide range of patient samples including freshly excised or flash-frozen surgical excisions, formalin-fixed, paraffin-embedded (FFPE) tissues, fine needle aspirations (FNAs), cell scrapes, urine or other cytological specimens, blood serum or lymphocytes, and stool samples.

Fresh tissues specimens or tissue frozen under liquid nitrogen shortly after surgery retain the greatest preservation of in situ biological features, especially DNA and RNA integrity. Other specimens require some form of fixation to preserve tissues from cell autolysis or putrefaction due to bacterial or fungal infections. Surgical/biopsy specimens are typically fixed in 10% neutral buffered formalin, which induce covalent cross-links between proteins. The fixed tissues are then embedded in paraffin to enable tissue storage and sectioning.

Immunohistochemistry (IHC)

IHC is used to screen for abnormal patterns of protein expression in a tissue by microscopy [2]. The technique allows the localization and identification of cellular or tissues constituents (antigens) by means of antigen–antibody interactions. The assay was first conceived by Albert Coons in the early 1940s and developed for use on FFPE specimens by Taylor and Burns in the mid-1970s using enzyme-conjugated antibodies. IHC has been a routine diagnostic test since the 1990s following the development of improved antigen specific antibody generating technologies and sensitive detection methodologies especially for IHC of FFPE specimens.

An antigen (<u>anti</u>body <u>gen</u>erator) is an entity recognized as foreign by the host immune system and may be a protein or other biochemical construct. A given antigen may display a variety of features that can evoke an antibody response; these features are referred as epitopes (antigenic determinants). Lab test antibodies may be monoclonal or polyclonal. Monoclonal antibodies are mostly produced from mice or rabbits and are in the form of epitope-specific antibody clones. Polyclonal antibodies detect multiple epitopes on a given antigen.

The IHC methodology involves pretreating slide mounted tissue sections to make accessible antigens/epitopes that may have become masked during fixation. A primary antibody is then applied to bind to the target antigen; the primary antibody is then detected using a secondary antibody conjugated with an enzyme (typically horseradish peroxidase [HRP] or alkaline phosphatase [AP]). The enzyme is demonstrated by application of an appropriate substrate that results in a colored stain. In immunofluorescence (IF) the end point is a fluorescent dye.

IHC is assessed according to cellular compartment stained (nuclear, cytoplasmic, or membranous); the proportion of cells showing staining; and, staining intensity (typically rated negative, weak, moderate, or strong). For many clinical applications, histopathologists are trained specifically for the evaluation of a given IHC marker to ensure that there is interpretative consensus between pathologists and across institutes.

Current FDA listed IHC applications are shown in Table 1. There are numerous other IHC cancer diagnostic tests performed in CLIA certified/CAP accredited labs. Our own institute has a battery of more than 200 diagnostic antibodies; some are used on a daily basis others only very occasionally. Many clinical applications utilize multimarker panels. For example, the IHC4 breast cancer test can be used to assess residual risk of cancer recurrence for patients with estrogen receptor (ER) positive or progesterone receptor (PgR) positive tumors treated with hormone therapies such as tamoxifen. Tumor specimens are screened with antibodies against four antibodies – standard of care markers: ER, PgR, human epidermal growth factor receptor 2 (HER2), and a cell proliferation marker (Ki-67). The stains are scored semi-quantitatively with reference to percentage of cells stained and strength of staining; the scores are then interrelated in an algorithm to assess risk of tumor recurrence [3]. Figure 1 includes examples of IHC used in the diagnosis of breast cancer.

Cytogenetics and In Situ Hybridization (ISH)

Clinical cytogenetics involves the short-term culture of fresh tissues; metaphase chromosome spreads are then prepared onto glass slides from dividing culture cells. Staining procedures show up chromosome specific banding patterns (karyotyping). The detection of abnormal karyotypes (structural or numerical) is utilized in the diagnosis of a range of cancers including leukemias, lymphomas, renal cell carcinoma, and soft tissue neoplasms. ISH is widely used to supplement traditional karyotyping.

TABLE 1. FDA approved immunohistochemical tests for cancer diagnostics.

Cancer Type	Test Name	FDA PMA Dates
Breast	[a]**HercepTest**[1] **(also approved for**	**09/25/1998**
	metastatic gastric or	**10/20/2010***
	gastroesophageal junction	**06/08/2012****
	adenocarcinomas*)	**02/22/2013*****
	[b]**PATHWAY anti-HER-2/neu (4B5)**[2]	**11/28/2000**
	[c]**InSite HER-2/neu KIT (CB11)**[1]	**12/22/2004**
	[d]**Bond Oracle Her2 IHC System (CB11)**[1]	**04/25/2012**

Sample: FFPE specimens
Methods: Antibody: [1]Rabbit polyclonal, [2]Rabbit monoclonal
Theranostic Test: HER2 positive patients are candidates for treatment with Herceptin®
   (transtuzumab). The HercepTest is also approved for identifying breast cancer patients
   eligible for Perjeta (pertuzumab),** and for Kadcyla (ado-trastuzumabemtansine).***

Colorectal	[a]**DAKO EGFR PharmDx Kit**	**02/12/2004***
		09/27/2006**

Sample: FFPE specimens
Method: Antibody: Mouse monoclonal
Theranostic Test: patients overexpressing epidermal growth factor receptor (EGFR) are
   candidates for treatment with Erbitux (cetuximab)* or Vectibix (panitumumab).**

Colorectal	[b]**Ventana MMR IHC Panel**	**10/27/17**

Sample: FFPE specimens
Method: Five antibody immunohistochemistry (IHC) panel: Four mouse monoclonal: anti-
   MLH1, anti-PMS2, anti-MSH2, anti-BRAF V600E; one rabbit monoclonal: anti-MSH6
Denotative Test: The panel detects mismatch repair (MMR) proteins deficiency in patients with
   colorectal cancer for the identification of probable Lynch syndrome and to detect BRAF V600E
   protein to differentiate between sporadic colorectal cancer and probable Lynch syndrome.

Gastric	[a]**DAKO C-Kit PharmDx kit**	**06/27/2005**

Sample: FFPE Gastro-intestinal stromal tumors (GISTs)
Method: Antibody: Rabbit polyclonal
Denotative and Theranostic Test: c-Kit (CD117) expression aids the differential diagnosis of
   GISTs from tumors of similar histologic appearance. GIST patients may be candidates for
   treatment with Gleevec (imatinib mesylate).

Suppliers:
[a]Dako North America, Inc., Carpinteria, CA;
[b]Ventana Medical Systems, Inc., Tucson, AZ;
[c]Biogenex Laboratories, Inc., Freemont, CA;
[d]Leica Biosystems Inc., Buffalo Grove, IL.
Notes: FFPE: formalin-fixed, paraffin-embedded. * Asterisks per individual test link a test element
to date of approval. [1,2] Superscript numbers per individual test link an antibody to its clonal status.

Additionally, ISH enables the detection of finer level chromosome abnormalities as
well as the visualization of DNA or RNA expression patterns in relation to cytologic
or histologic features, and is applicable to interphase (nondividing) cells [4].

   The ISH assay was independently described in 1969 by Gall and Pardue [5],
and by John et al. [6]. ISH involves hybridizing a labeled nucleic acid probe
complementary to target sequences in slide-mounted metaphase chromosome

spreads, cell preps, or tissue sections (fresh/frozen or FFPE). Originally, the technique was performed using isotopic labels such as ^{32}P, ^{35}S, or ^{3}H. In the 1980s and 1990s, hapten labels (e.g., biotin, digoxigenin, dinitrophenol [DNP]) and fluorescent dyes were developed along with sensitive detection modalities.

Broadly, there are two categories of non-isotopic ISH: chromogenic ISH (CISH) and fluorescence ISH (FISH). In CISH, following hybridization, a hapten labeled probe is detected by application of AP or HRP enzyme linked reagents (such as antibodies) using colorimetric substrates. FISH uses probes directly labeled with a fluorescent dye or uses fluorescently labeled detection reagents for the detection of a hapten label. The main advantage of CISH is the ability to counterstain specimens with histochemical stains and review the data by bright-field microscopy. CISH stained slides are easily archived. FISH signal clinical data interpretation requires fluorescence microscopy and digital imaging technology. Multiplex FISH allows the detection of two or more targets using different fluorescent labels; the multiplex limit with CISH is two probes.

As with IHC, diagnostic ISH assays frequently have detailed guidelines to ensure standardized interpretations between observers and institutions. These rules may include counting ISH signals in a minimum numbers of cells from different areas within a tumor as well as counting ISH signals among normal adjacent tissues.

For clinical testing purposes, ISH probes can be classed into four categories: chromosome enumeration probes (CEN) (that target centromeric sequences located between the short (p) and long arms (q) of a chromosome); locus specific identifier (LSI) probes (that localize to gene regions on the p or q arms); dual fusion LSI probes; and break-apart LSI probes. CEN probes allow determination of changes in chromosome copy number. LSI probes provide data about gene amplification events and may be used in conjunction with CEN probes to allow a comparison of LSI gene signal counts to chromosome copy number (Figure 1). Dual fusion and break-apart probes are used to investigate translocation events. FDA-listed ISH tests are detailed in Table 2. As with IHC, there are many other CLIA/CAP established ISH tests especially in the field of clinical cytogenetics.

Figure 1. Breast cancer case study molecular diagnostics: immunohistochemistry (IHC) and chromogenic in situ hybridization (CISH). All assays were applied to the same tumor. (a) Hematoxylin and eosin (H&E) stained section of the breast tumor. (b) IHC for breast cancer standard of care marker estrogen receptor (ER): >90% of tumor cell nuclei show are stained; strong intensity on average. (c) IHC for standard of care marker progesterone receptor (PgR): 31–40% of tumor cell nuclei are stained; moderate intensity on average. (d) IHC for HER2 using an FDA approved IHC assay (antibody clone 4B5 [Table 1]) ~90% tumors cells show uniform intense complete membrane staining demonstrating HER2 overexpression. (e) ISH for *HER2* DNA (INFORM dual ISH probe [Table 2]). Red CISH signals indicate signals from the centromere of chromosome 17; the black signals represent the *HER2* DNA locus and are generated by the enzymatic deposition of silver atoms (silver-enhanced ISH [SISH]). The excess of black signals relative to red indicates *HER2* DNA amplification. (f) An experimental ISH assay for *HER2* RNA (RNAscope®, Advanced Cell Diagnostics, Newark, CA) showing abundant *HER2* RNA expression (dot signals). (g) IHC for Ki67 (mouse monoclonal antibody MIB-1). Ki67 is a cell proliferation marker often used in diagnostic assessments; however, because of a lack of consensus about scoring criteria Ki67 is not recommended for routine breast IHC. (h) RNAscope ISH for *MKI67* for comparison with the protein stain. Scale bar: 20 μm. (See insert for color representation.)

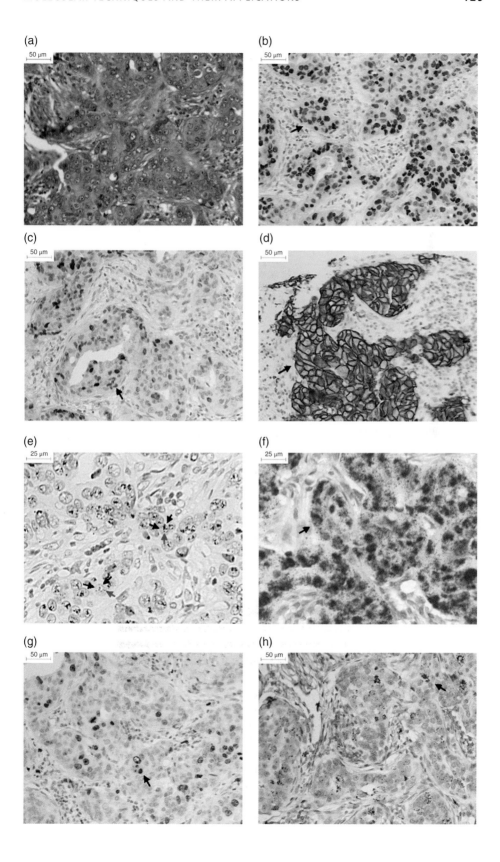

TABLE 2. FDA approved or cleared in situ hybridization tests for cancer diagnostics.

Cancer Type	Test Name	FDA Dates
Acute Myeloid Leukemia (AML)/ Myelodysplastic Syndrome (MDS)	[a]**Vysis D7S486/CEP 7 FISH Probe Kit**	[1]**09/13/2013**

Sample Type: Bone marrow or peripheral blood lymphocytes
Method: Dual-color FISH (LSI 7q31 and CEP 7)
Denotative Test: Deletion of chromosome 7q and loss of a complete chromosome 7 (monosomy 7) aids AML and MDL diagnosis.

Acute Myeloid Leukemia (AML)	[a]**Vysis EGR1(Early growth response protein-1) FISH Probe Kit**	[1]**08/29/2011**

Sample Type: Bone marrow lymphocytes
Method: Dual-color FISH (LSI 5q31 and CEP 5p15.2)
Prognostic Test: Deletion of chromosome 5q is associated with an unfavorable prognosis in AML patients.

Chronic Lymphocytic Leukemia (CLL)	[a]**Vysis CLL FISH Probe Kit**	[1]**08/09/2011**

Sample Type: Peripheral blood lymphocytes
Method: Multicolor FISH (LSI TP53, ATM, 13q34 and CEP 12)
Denotative Test: Screens for p53, ATM, 11q deletions, and chromosome 12 trisomy enabling CLL sub-classification.

Chronic Lymphocytic Leukemia (CLL)	[a]**CEP 12 SpectrumOrange**	[1]**01/13/1997**

Sample Type: Peripheral blood lymphocytes
Method: Mono-color FISH (CEP 12)
Denotative Test: Chr. 12 enumeration aids CLL diagnosis.

Bladder	[a]**Vysis UroVysion Bladder Cancer Recurrence Kit**	[1]**08/03/2001**

Sample Type: Urine specimens from persons with hematuria suspected of having bladder cancer.
Method: Multicolor FISH (LSI 9p21 and CEP 3, 7, 17)
Denotative and Prognostic Test: The test aids initial diagnosis of bladder cancer and the monitoring of tumor recurrence for previously diagnosed bladder cancer.

Breast	[b]**SPOT-LIGHT HER2 CISH Kit**	[2]**07/01/2008**
	[c]**INFORM HER2 Dual CISH/ SISH DNA Probe**	[2]**06/14/2011**
	[d]*HER2* **CISH PharmDxTM Kit**	[2]**11/30/2011**
	[a]**PATHVYSION HER-2 DNA Probe FISH Kit**	[2]**12/11/1998**
	[d]**HER2 FISH PharmDx Kit**	[2]**05/03/2005** [2]**10/20/2010***

Sample Type: FFPE specimens
Method: "Spot-Light" mono-color CISH; Others dual color FISH of CISH (LSI *HER2* [17p12] and CEP 17)
Theranostic Test: HER2 amplification indicates patient eligibility for treatment with Herpceptin (Trastuzumab). *Also approved for use on metastatic gastric and gastro-esophageal junction adenocarcinomas.

TABLE 2. (*Continued*)

Cancer Type	Test Name	FDA Dates
Breast	**[d]Dako *TOP2A* FISH PharmDx Kit**	**[2]01/11/2008**

Sample Type: FFPE surgical specimens
Method: Dual-color FISH (LSI *TOP2A* [17q21.2] and CEP 17)
Prognostic and Theranostic Test: Topoisomerase 2 alpha (*TOP2A*) amplification is associated with elevated risk of recurrence or decreased long term survival; patients may be treated by TOP2A inhibitor anthracycline therapies (Doxorubicin, Epirubicin).

Myeloid Disorders	**[a]CEP 8 SpectrumOrange DNA FISH Probe Kit**	**[1]11/29/1996**

Sample Type: Bone marrow specimens
Method: Monocolor FISH (CEP 8)
Denotative Test: Trisomy 8 is common in: chronic myelogenous leukemia (CML), acute myeloid leukemia (AML), Myeloproliferative disorder (MPD), myelodysplastic syndrome (MDS), hematological disorders not otherwise specified (HDNOS).

Non-small cell lung cancer (NSCLC)	**[a]Vysis ALK Break Apart FISH Probe Kit**	**[2]08/26/2011**

Sample Type: FFPE tissue
Method: Dual–color FISH (LSI 2p23, break apart probes)
Theranostic Test: Utilizes two LSI probes to screen for *ALK* translocation events. Patients with the rearrangement may be eligible for treatment with Xalkori (Crizotinib).

Suppliers:
[a]Abbott Molecular, Inc., Des Plaines, IL;
[b]Life Technologies, Inc., Grand Island, NY;
[c]Ventana Medical Systems, Inc., Tucson, AZ;
[d]Dako North America, Inc., Carpinteria, CA.
Notes: FFPE: formalin-fixed, paraffin-embedded. FDA: [1]510(k); [2]PMA. LSI: locus specific identifier; CEP: chromosome enumeration probe.

Polymerase Chain Reaction (PCR)

PCR is an *in vitro* method for the amplification of DNA [7]. Kary Mullis was awarded the Noble Prize for Chemistry in 1993 for developing the technique in the early 1980s. The procedure involves heat denaturation of specimen DNA into single strand fragments. Oligonucleotide DNA primers designed to be complementary to sequences upstream and downstream of a genomic region of interest anneal to the denatured DNA forming double stranded hybrids; this initiates DNA synthesis (extension) by the enzyme DNA polymerase. The assay uses a thermostable DNA polymerase that maintains activity despite repeated heating to >95 °C during denaturation. These enzymes are extractable from microorganisms that live in thermal springs or hydrothermal vents in the ocean bed. The most commonly used enzyme is *Taq* DNA polymerase extracted from the bacterium *Thermus aquaticus*. Denaturation, annealing and extension steps constitute one cycle of PCR. Automated PCR cycles result in the targeted exponential amplification of the

DNA of interest. Theoretically, after 30 cycles more than one billion amplicon copies (2^{30}) of a target sequence will be generated accounting for PCR sensitivity.

End-point PCR (first generation PCR) consists in performing 30–50 cycles of amplification followed by analytical tests on the PCR product. In real-time quantitative PCR (QPCR [second generation PCR]), the accumulation of fluorescently labeled PCR amplicons is continuously monitored at each cycle of PCR. Different colored fluorescent dyes may be used to enable multiplex assays that simultaneously assess different targets within the same PCR. QPCR allows the quantitation of a sample target by reference to a standard curve generated from known quantities of the target. The simplest form of QPCR utilizes a fluorescent dye (e.g., SYBR® Green) that binds to double stranded DNA. Other approaches use amplicon-specific probes labeled with a fluorescent marker. There are several varieties of probes including TaqMan, Molecular Beacons, and Scorpions.

Digital PCR (dPCR) (third generation PCR) assays a single tube 20 µl reaction partitioned by emulsification into 20 000 1 nl droplets; each represents a unique PCR and involves a fluorescent TaqMan probe. After ~40 cycles of PCR, droplets are analyzed one-by-one for the detection of amplicon-generated fluorescence. The test has the highest sensitivity of any PCR assay. dPCR allows direct target quantification by Poisson distribution analyses and does not require a standard reference curve.

Reverse transcription PCR (RT-PCR) allows the investigation of RNA expression by first, second, or third generation PCR. RNA extract is converted to complementary DNA (cDNA) by reverse transcriptase (RT) and the DNA can then be used for PCR procedures.

Amplification-refractory mutation system (ARMS) PCR, also known as allele-specific PCR (ASPCR) is used for detecting single base changes/mutations in a target sequence. PCR will only proceed when the 3′-terminal base in an primer anneals with its target (i.e., A:T or C:G). By the judicious use of primers to detect wild type or mutated sequences the presence of a mutation or other sequence of specific interest may be inferred.

Epigenetic changes can also be investigated by PCR. Pretreating DNA extract with bisulfite converts cytosine residues into uracil whereas methylated cytosine is unconverted. When the DNA is subjected to PCR, uracil residues are converted to thymidine ones; i.e., the net result is a C to T conversion in the PCR generated amplicons. This DNA base change can be detected by sequencing or restriction enzyme analyses. Altered methylation state is associated with gene expression deregulation and is used as a diagnostic maker (Table 4).

Immuno-PCR utilizes an antibody to detect a protein target in solution. The antibody is directly labeled with an oligonucleotide sequence. Real time PCR is used to detect PCR amplicons amplified from the oligonucleotide sequences. The test was developed as a more sensitive alternative to the *in vitro* enzyme-linked immunosorbent assay (ELISA). In ELISA, enzyme conjugated antibody is used to screen for its target immobilized in a microtitre plate; the enzyme is detected using a colorimetric or chemiluminescent substrate.

There are far more variations and clinical applications of the PCR technique than can be accounted for in this chapter. There are also varieties of "do-it-all" automated platforms that perform nucleic acids extractions and incorporate PCR techniques (e.g., ARMS) for high throughput multi-analytic research or diagnostic tests. These include the AutoGenomics INFINITY® System (Carlsbad, CA) and the Luminex xMAP® Technology (Austin, TX) that can screen up to 240 or 500 different biomarkers per specimen respectively.

Whereas IHC and ISH frequently rely upon qualitative judgments by expert observers, PCR data is more quantitative. Nevertheless there are detailed recommendations for the standardization of PCR, such as the MIQE guidelines [8].

FDA listed PCR-based tests are shown in Table 3. Although not FDA listed, the Oncotype DX® test (Genomic Health® Inc., Redwood City, CA) is approved by the NCCN and by ASCO and as a multigene assay typifies contemporary molecular diagnostic approaches. The test is a RT-QPCR assay for 21 genes (16 cancer and 5 reference) and is recommended for node-negative, ER-positive, HER2-negative invasive breast tumors <1.0 cm in size. Based on gene expression levels, a breast cancer recurrence score is calculated to stratify patients into those requiring hormone therapy versus those who need more aggressive chemotherapy. Other ACSCO-approved breast cancer QRT-PCR tests include the EndoPredict® test (Sividon Diagnostics GmbH, Cologne, Germany) and the Breast Cancer Index test (Biotheranostics, Inc., San Diego, CA) that both screen for 12 genes and have similar applications as the Oncotype DX test.

Sequencing Technologies

Sanger Sequencing: Nucleic acid sequencing was developed in the mid-1970s by Frederic Sanger who was awarded the Noble Prize for Chemistry for his discovery in 1980. A basic reaction involves a DNA sample, a sequence specific DNA oligonucleotide primer, DNA polymerase enzyme, the four DNA deoxynucleotide triphosphates (dNTPs) building blocks (dNTPs, i.e., deoxyadenosine triphosphate [dATP], deoxythymidine triphosphate [dTTP], deoxycytidine triphosphate [dCTP], deoxyguanosine triphosphate [dGTP]) and one of four corresponding dideoxynucleotides (ddNTPs). The incorporation of a ddNTP into a DNA molecule prevents further extension of the sequence by the DNA polymerase. The assay originally used isotopically labeled ddNTPs and required four separate individual reactions run in parallel (i.e., each with a different labeled ddNTP). Sequencing reaction data were assessed by autoradiography following polyacrylamide gel electrophoresis.

Dye Terminator Sequencing: The Sanger method was adapted so that each ddNTP is labeled with a different fluorescent dye allowing a one tube reaction. Additionally, PCR was incorporated to improve reaction efficiency. The generated fragments are sieved by capillary gel electrophoresis. DdNTP fluorophore emissions are detected as the fragments pass through the gel revealing the color-coded base sequence order and are assembled into a linear sequence by dedicated software data analyses.

Next Generation Sequencing (NGS) also called *Massively Parallel Sequencing* (MPS) refers to non-Sanger sequencing modalities conceived and developed in the 1990s/early 2000s that allow high-throughput sequencing analyses of millions of

T A B L E 3. FDA approved or cleared PCR-based tests for cancer diagnostics.

Cancer Type	Test Name	FDA Dates
Breast	[a]**GeneSearch Breast Lymph Node (BLN) Test Kit**	**07/16/2007**

Sample Type: Lymph node biopsy excision, fresh tissue
Method: Real-Time Reverse Transcription PCR
Denotative and Patient Management Test: Rapid detection test for >0. 2 mm metastases in sentinel lymph node biopsies. Test results assist the intra- or postoperative decision to remove additional tissue. The expression of mammaglobin (MG) and cytokeratin 19 (CK19) is measured. Marker detection at or below a clinical sensitivity threshold indicates metastases.

Cervical Cytology	[b]**COBAS Human Papillomavirus (HPV) Test**	**04/19/2011**

Sample Type: Cervical cytology DNA
Method: Real Time PCR
Patient Management Test: Screens for high-risk HPV (HR-HPV) infections in low-grade abnormal cervical cytology. HR HPV negative patients may be excluded from biopsy test referral.

Colorectal (CRC)	[c]***therascreen* KRAS RGQ PCR Kit**	**07/06/2012**

Sample Type: FFPE CRC tissue
Method: Amplification Refractory Mutation System/Scorpions Real-Time PCR (ARMS/S PCR)
Theranostic Test: Detects seven somatic mutations in the *KRAS* oncogene. CRC patients without mutations are candidates for treatment with Erbitux (cetuximab) or Vectibix (panitumumab).

Colorectal (CRC)	[d]***Cologuard***	**08/11/2014**

Sample Type: Stool sample
Methods: QPCR, bisulfite conversion, Enzyme-Linked Immunosorbent Assay (ELISA)
Patient Management Test: For adults >50 years; screens for 7 *KRAS* mutations, methylation of *NDRG4 and BMP3*, hemoglobin. A composite score is generated from the data. A positive result suggests CRC and referral for diagnostic colonoscopy.

Melanoma	[e]**THxID™ BRAF Kit**	**05/29/2013**

Sample Type: FFPE melanoma tissue
Method: Real-Time PCR for *BRAF* V600E / V600K somatic mutations
Theranostic Test: Patients with *BRAF* V600E mutation may be treated with Dabrafenib [tafinlar]; patients with *BRAF* V600E or V600K with Trametinib [mekinist].

Melanoma	[b]**COBAS 4800 BRAF V600 Mutation Test**	**8/17/2011**

Sample Type: FFPE melanoma tissue
Method: Real-Time PCR
Theranostic Test: Aids identification of V600E mutation positive patients for treatment with Vemurafenib.

Non-small cell lung cancer (NSCLC)	[c]***therascreen* EGFR RGQ PCR Kit**	**07/12/2013**

Sample Type: FFPE NSCLC tissue
Method: Amplification Refractory Mutation System/Scorpions Real-Time PCR (ARMS/S PCR)

TABLE 3. (*Continued*)

Cancer Type	Test Name	FDA Dates

Theranostic Test: Detects exon 19 deletions and exon 21 substitution mutations of the epidermal growth factor receptor gene to select patients for Gilotrif (afatinib) treatment.

Non-small cell lung cancer (NSCLC) **[b]COBAS EGFR Mutation Test** **05/14/2013**

Sample Type: FFPE NSCLC tissue

Method: Real-Time PCR

Theranostic Test: Detects of *EGFR* gene exon 19 deletions and exon 21 (I858r) substitution mutations to select patients with metastatic NSCLC for whom Tarceva (erlotinib) is indicated.

Ovarian **[f]BRACAnalysis CDx** **12/19/2014**

Sample Type: Whole blood specimen genomic DNA extract

Method: PCR, Multiplex PCR, Sanger Sequencing

Theranostic Test: Screens for variants in the protein coding regions and intron/exon boundaries of the *BRCA1* and *BRCA2* genes; also detects single nucleotide variants and small insertions and deletions (indels) and large deletions and duplications. Test results identifying patients with germline *BRCA* variants eligible for treatment with Lynparza (olaparib).

Prostate **[g]NADiA ProsVue** **[1]09/20/2011**

Sample Type: Blood serum DNA

Method: Quantitative, Immuno-PCR

Prognostic Test: Measures the rate of change of serum total prostate specific antigen (PSA) over a period of time to identify patients at reduced prostate cancer recurrence risk for an eight year period following prostatectomy.

Suppliers:
[a]Janssen Diagnostics, Inc., Raritan, NJ;
[b]Roche Molecular Systems, Inc., Pleasanton, CA;
[c]Qiagen Inc., Valencia, CA;
[d]Exact Sciencses Corporation, Madison, WI;
[e]bioMérieux Inc., Durham, NC;
[f]Myriad Genetic Laboratories, Inc., Salt Lake City, UT;
[g]Iris Molecular Diagnostics, Chatsworth, CA.
Note: All assays have FDA PMA status, except [1]510(k). FFPE: formalin-fixed, paraffin-embedded.

sequences. These proprietary automated technologies include Solexa sequencing (Illumina), pyrosequencing (Roche 454, Qiagen), Ion torrent (Life Technologies), and SOLiD sequencing (Applied Biosystems). Dedicated software data analyses are required.

Sanger/dye terminator sequencing has been widely used in conjunction with PCR/RT-PCR to identify the presence of a mutation, single nucleotide polymorphism (SNP), translocation/chimeric transcript events, or other genomic rearrangements (Table 3). NGS has been developed for numerous cancer research applications including genotyping (for structural changes through to SNP changes), aneuploidy and copy number variation (CNV) analysis, transcriptome sequencing (RNA-Seq), small noncoding RNA and microRNA sequencing, viral and microbial sequencing, through to the entire genome or exome sequencing. Presently, diagnostic NGS applications are largely theranostic and for targeted panels of genes associated with actionable pharmacologic agents [9] (Table 4).

T A B L E 4. Other molecular tests FDA approved or cleared for cancer diagnostics.

Cancer Type	Test Name	FDA Date
Breast	[a]**Agendia MammaPrint**	[1]**06/22/2007**[*] [1]**02/10/2015**[**]

Sample Type: RNA extract from fresh-frozen* or FFPE tissues**
Method: Microarray
Prognostic Test: Expression profiling of 70 genes to assess risk for distant metastasis (up to 10 years for patients <61 years old, up to 5 years for ≥61 years). The test is for Stage I or II breast tumors ≤5.0 cm size and lymph node negative.

Metastatic Cancer of Unknown Origin	[b]**Tissue of Origin Test**	[1]**07/30/2008**[*] [1]**06/08/2010**[**]

Sample Type: RNA extract from fresh-frozen* or FFPE** tissues
Method: Microarray
Denotative Test: Quantifies the similarity of tumor specimens of unknown primary to 15 cancer types representing 60 morphologies by measuring the degree of similarity between the RNA expression pattern in a patient's tumor and the RNA expression patterns in a database of tumor samples diagnosed according to current clinical and pathological practice. This microarray based test assesses the expression of ~2000 genes.

Breast	[c]**Prosigna™ Breast Cancer Prognostic Gene Signature Assay**	09/06/2013

Sample Type: RNA extracted from FFPE specimens
Method: NanoString technology
Prognostic Test: Predicts distant recurrence-free survival at 10 years in postmenopausal, hormone receptor positive, Stage I or II breast cancer, 0–3 positive lymph nodes and treated with adjuvant endocrine therapy alone. RNA expression levels of 50 genes generate a risk of recurrence score.

Breast, Ovarian, Prostate	[c]**23andMe Personal Genome Service Genetic Health Risk Report for BRCA1/BRCA2 (Selected Variants)**	[2]**03/06/2018**

Sample Type: Patient saliva
Method: Illumina BeadArray microarray technology assay of 185delAg and 5382insC variants in the *BRCA1* gene and 6174delT variant in the *BRCA2* gene.
Personal Genome Information Test: The three gene variants detected are most common in people of Ashkenazi Jewish descent and do not represent the majority of BRCA1/BRCA2 variants in the general population. Consumers consult with a qualified healthcare provider subsequent to test results.

Cervical Cytology Abnormalities	[d]*digene* **HC2 HR HPV DNA Test**	[2]**03/31/2003**
	[e]**Cervista HR HPV and 16/18 Tests**	[2]**03/12/2009**
	[e]**Aptima *HR HPV RNA and **HPV 16 18/45 RNA Genotype Assay**	[2]**10/28/2011**[*] [2]**10/12/2012**[**]

Sample Type: Cervical cytology RNA (Aptima) or DNA (Cervista, HC2)
Methods: HC2 – Hybrid Capture; Cervista – Invader chemistry; Aptima – Transcription Mediated Amplification (TMA)
Patient Management Tests: These assays screen for high-risk human papillomavirus (HR-HPV) infections in low-grade abnormal cervical cytology. HR HPV negative patients may be excluded from biopsy test referral.

Non-Small Cell Lung Cancer	[f]**Oncomine Dx Target Test**	[3]**06/23/2017**

Sample Type: DNA and RNA isolated from FFPE tumor samples
Methods: High throughput, parallel-sequencing technology (next generation sequencing) for qualitative detection of single nucleotide variants and deletions in 23 genes from DNA, and fusions in *ROS1* from RNA.

136

TABLE 4. (*Continued*)

Cancer Type	Test Name	FDA Date

Theranostic Test: The test aids in selecting non-small cell lung cancer patients for treatment with targeted therapies according to test data.

Ovarian Cancer [g]**FoundationFocus CDxBRCA** [3]**12/19/2016**

Sample Type: FFPE ovarian tumor tissue

Methods: Next generation sequencing based *in vitro* diagnostic device for qualitative detection of *BRCA1* and *BRCA2* gene sequence alterations in ovarian tumor tissue.

Theranostic Test: the assay identifies patients who may be eligible for treatment with Rubraca (rucaparib).

Prostate Cancer [e]**Progensa™ PCA3 Assay** [2]**02/13/2012**

Sample Type: RNA extract from urine sample cells

Method: TMA

Patient Management Test: The test is for men ≥50 years who have had one or more previous negative prostate biopsies and for whom a further biopsy would be recommended based on current standard of care. The assay measures the concentration of prostate cancer gene 3 (PCA3) and prostate-specific antigen (PSA) RNA molecules and calculates the ratio (PCA3 score). A low score is associated with a decreased likelihood of a positive biopsy obviating the need for another biopsy.

Metastatic breast, colorectal [f]**CELLSEARCH® Circulating Tumor** [1]**11/20/2007**
or prostate **Cell (CTC) Test**

Sample Type: Blood from metastatic cancer patients

Method: Circulating ell capture

Prognostic Test: Circulating cells expressing EpCAM (epithelial cell adhesion molecule) are captured by antibody coated magnetic "nanoparticles." This epithelial cell enriched sample is then incubated with a fluorescent cell nuclear dye, and fluorescently labeled monoclonal antibodies for epithelial cell specific cytokeratins (CK 8, 18, & 19) and with CD45 that is specific for leukocytes. CTCs are defined as nucleated cells lacking CD45 and expressing cytokeratin. CTC detection is associated with decreased progression free survival and decreased overall survival for metastatic cancer patients.

Tumor Profiling [j]**MSK-IMPACT (Integrated** [2]**11/15/2017**
Mutation Profiling Of Actionable
Cancer Targets):A Hybridization-
Capture Based Next Generation
Sequencing Assay

Sample Type: FFPE tumor and normal tissue samples

Method: Next generation sequencing; detects somatic alterations (point mutations, small insertions, deletions, microsatellite instability) in a wide range of tumor specimens using a 468 gene panel.

Multipurpose Test: Mutations are reported as "cancer mutations with evidence of clinical significance" or "cancer mutations with potential clinical significance"; particular findings are used to guide diagnosis and treatment plans.

Suppliers:

[a]Agendia Inc., USA, Irvine, CA;

[b]Cancer Genetics, Inc., Rutherford, NJ;

[c]23andMe, Inc., Mountain View, CA;

[d]Qiagen Inc. Valencia, CA;

[e]Hologic Inc., Bedford, MA;

[f]Life Technologies Corporation, Carlsbad, CA;

[g]Foundation Medicine, Inc., Cambridge, MA;

[h]Menarini Silicon Biosystems, Inc., Huntington Valley, PA; [i]Cancer Genetics, Inc., Rutherford, NJ;

[j]Memorial Sloan-Kettering Cancer Center, New York, NY.

Notes FDA: [1]510(k); [2]513(f)(2)de novo; [3]PMA. *Asterisks per individual test link a test element to date of approval. FFPE: formalin-fixed, paraffin-embedded.

Microarray Assay

Microarray technology was developed in the early 1990s. A microarray chip consists of a silicon-based solid support spotted with covalently bound single-strand DNA oligonucleotide probes complementary to gene or other genomic sequences of interest. Current technology enables the investigation of the entire human exome by way of hundreds of thousands of probes imprinted onto a microarray chip. Microarrays for cancer diagnostics consist of limited panels of markers addressing specific clinical issues [10].

Microarrays are most widely utilized for gene expression profiling. The assay involves extracting RNA from a tumor sample and normal companion tissue and converting these preparations to cDNA using RT. Fresh-frozen tissue specimens were originally preferred but assays have now been adapted to work well FFPE specimens' RNA extract despite the degraded quality of the RNA recovered. If the amount of cDNA generated is limiting, PCR amplification of the cDNA may be performed prior to the microarray assay. The two cDNA preparations are each labeled with a different fluorescent dye. These are then hybridized to the microarray. Analysis of the fluorescence data shows which mRNAs are up- or down-regulated in the tumor sample relative to the normal: oligonucleotide probe spots that fluoresce with the color of the tumor cDNA indicate overexpression in the tumor and vice versa; coequal tumor-normal expression levels show a blended coloration. Automated generation of the fluorescent intensities is the basis of data interpretation.

There are three FDA-listed microarray based assays (Table 4). The Agendia MammaPrint® (Agendia Inc. USA, Irvine, CA) interrogates the expression levels of 70 genes to calculate risk of recurrence (ROR) and thereby the appropriateness of hormone or chemotherapy for a patient. The Tissue of Origin Test (Cancer Genetics Inc., Rutherford, NJ) compares the mRNA expression levels of a metastatic tumor of uncertain anatomic origin against the expression levels of known mRNA expression signatures for 15 different organs by assessing 2000 different genes. More than 85 000 newly diagnosed metastatic cancers in the United States have an unclear diagnosis. Knowledge of the primary tumor anatomic location is significant for treatment selection for metastatic cancer patients. Tissue of origin can also be assessed by IHC using antibody marker panels. A personal genome information test developed by 23andMe (Mountain View, CA) provides limited information about BRCA1/BRCA2 mutations that may be associated with an increased risk for breast, ovarian or prostate cancer; consultation with a healthcare provider is required for test interpretation.

Transcription Mediated Amplification (TMA)

Developed in the 1990s, TMA (HOLOGIC®, Bedford, MA) is a method that like PCR amplifies target sequences; unlike PCR the TMA reaction is isothermal and generates RNA rather than DNA amplicons [11]. Clinical TMA assays utilize

RNA extracted from a patient specimen. The TMA amplification reaction includes two enzymes: a RT enzyme that also has RNase H activity, and RNA polymerase. Similar to PCR the assay utilizes two oligonucleotide primers; however, in TMA one primer incorporates an RNA polymerase promoter site into the amplicons; the promoter causes multiple RNA copies to be generated. Exponential amplification ensues generating a billion-fold amplification within a ~30 minute reaction. Finally, RNA amplicons are detected by hybridization to a labeled DNA probe for either a chemiluminescence or fluorescence detection assay.

Table 4 shows TMA diagnostic applications. These include the detection of high-risk type human papillomavirus (HPV) *E6/E7* RNA in cervical cytology samples for the identification of patients requiring a biopsy referral, and a urine cytology test for *PCA3* for selecting patients who may need a prostate biopsy.

Invader Chemistry

Invader® chemistry (HOLOGIC, Bedford, MA) also developed in the 1990s is a proprietary technique for the detection of single-base changes, insertions, deletions, and changes in gene and chromosome number. Unlike PCR or TMA the assay does not generate multiple copies of the target, instead, Invader chemistry works by generating an amplified signal cascade from a detected target [12]. The method involves two simultaneous isothermal reactions; one reaction detects the DNA target of interest and a second reaction generates an amplified signal. A standard 4-hour reaction produces more than 10 million-fold signals per target sequence. The technology can be adapted for combined use with pre-PCR (InvaderPLUS) for even greater detection sensitivity.

The Cervista HPV assay (Table 4) is used in the management of patients with cervical cytologic abnormalities. There is also a theranostic Invader assay for metastatic colorectal cancer patients. These patients are candidates for treatment with the chemotherapeutic drug irinotecan. Uridine diphosphate glucoronosyltransferase 1A1 (UGT1A1) is an enzyme found in the liver that normally inactivates irinotecan limiting its toxic effects upon the patient. However, patients with certain variant sequences of UGT1A1 have lowered enzyme activity putting them at risk of severe toxic side effects; these patients need to be treated at lower doses of irinotecan. The Invader UGT1A1 Molecular Assay identifies patients with the variant sequences that are contraindications for standard irinotecan treatment.

NanoString Technology

NanoString technology (NanoString® Technologies, Seattle, WA) is an adaptation of microarray principles and is used for RNA expression analyses. The technique was established in the 2000s. Total RNA (extracted from >10000 cells) is mixed with two DNA oligonucleotide probes: a 3'-biotin end labeled target-specific capture probe, and a target-specific reporter probe labeled at the 5' - end with a series of differently colored fluorescent tags; the tag color order acts as a "bar code" target-specific identifier of the sequence detected by the probe. If the target sequence is present in the RNA extract, both probes will hybridize. The tripartite probes/target complex is purified and then passed across a NanoString cartridge coated with streptavidin. Streptavidin has a high affinity for biotin and immobilizes hybrid targets. A voltage is applied across the cartridge and causes the negatively charged nucleic acid hybrids to linearize over the cartridge surface. The detected fluorescent tag codes allow quantification of gene target expression in the sample. The assay is highly multiplexable and can screen for the detection of up to 800 targets. Unlike PCR, TMA, or Invader, the assay includes no target or signal amplification steps yet is sensitive enough to detect single copy gene expression. However, if specimen is limiting (<10000 cells) the assay can be combined with pre-PCR steps.

The Prosigna™ NanoString assay is FDA-cleared (Table 4) for breast cancer 10 year risk of distant recurrence (ROR) testing. The test screens for the PAM50 (Prediction Analysis of Microarray 50) gene set that assesses patient specimens on the basis of the expression levels of 50 different genes into the four main molecular subtypes of breast cancer (Luminal A, Luminal B, HER2-enriched, and Basal-like). On the basis of PAM50 data a 10-year ROR score (0–100) is generated [13].

Hybrid Capture

In this test developed in the 1990s, patient DNA extract is screened for a target by hybridization assay with a sequence-specific RNA probe. RNA/DNA hybrids are "captured" in microwell plates coated with antibodies specific for RNA/DNA hybrids. Following well plate immobilization, AP labeled anti-RNA/DNA antibodies are added and enzyme activity is demonstrated using a chemiluminescnt substrate. The assay is FDA-approved for cervical cytologic HPV testing [14] (Table 4).

Flow Cytometry and Circulating Tumor Cell (CTC) Detection

Flow cytometry science dates back to the 1950s. The technique involves the assay of fluid suspensions of single cells simultaneously for their intrinsic light scattering properties and for one or more extrinsic properties using fluorescent probes; for example, antibody probes for the detection of specific proteins [15]. Clinically, flow cytometry is used in the diagnosis of hematological malignancies, such as for distinguishing acute myeloid leukemia (AML) from acute lymphoblastic leukemia (ALL). ALL and AML have similar features but require different treatments. The different immunophenotypes of these conditions can be discerned by flow cytometric cell sorting using antibody panels. Non-Hodgkin lymphoma and other lymphoproliferative disorders can also be diagnosed by immunophenotyping. Flow cytometry is also used to measure minimal residual disease in leukemia patients in remission [15].

Tumor signatures are detectable in the blood by way of cells shed into the vascular from primary or metastatic cancers. The FDA cleared CellSearch® system (Menarini Silicon Biosystems, Inc., Huntington Valley, PA) developed in the 2000s (Table 4) is an adaptation of flow cytometry for the detection of rare CTCs. Circulating cells expressing EpCAM (epithelial cell adhesion molecule) are captured by antibody coated magnetic "nanoparticles." This epithelial cell enriched sample is then incubated with three differently colored fluorescent markers: a cell nuclear dye; monoclonal antibodies for three cytokeratins (CK 8, 18, and 19 that are epithelial specific); and CD45 (specific for leukocytes). CTCs are defined as nucleated cells lacking CD45 and expressing cytokeratin. The detection of CTCs is associated with decreased progression free survival and decreased overall survival for patients with metastatic cancer [16].

Clinical Proteomics and Genomics by Mass Spectrometry

The entire complement of proteins expressed by a cellular system is known as the proteome. The human genome is estimated to comprise up to ~25000 protein encoding genes; these may generate more than 250000 (even as many as 1 million) different protein forms [17]. Clinical proteomics is concerned with the investigation of pathognomonic proteomic signatures detectable from patient tissues or body fluid. Currently, the main technologies available for clinical proteomic applications are protein microarrays and mass spectrometry (MS) [17].

Protein microarrays were first envisaged in the 1980s. Whereas DNA microarrays are imprinted with oligonucleotide sequences, the capture features of protein microarrays may be antibodies, antigens, aptamers (engineered nucleic acid oligonucleotides or peptides that will bind protein targets), or affibodies (engineered proteins that act like antibodies). Patient protein lysate is reacted with the capture features (typically 1000 to >10000 per array). Captured proteins are detected by the application of secondary elements labeled with fluorescent dyes. Protein microarrays are currently primarily used in research applications to validate candidate protein biomarker relationship to a disease.

MS allows the sensitive and specific detection and quantification of proteins by measuring the mass-to-charge ratio of individual charged molecules following enzymatic fragmentation of a protein extract into peptide units. Clinically, the most widely used MA approach is Matrix Assisted Laser Desorption-Time of Flight (MALDI-TOF) that was developed in the late 1980s. The peptide sample is immobilized within a UV-light absorbing matrix compound and ionized by a UV laser pulse (337 nm) inside a MS instrument chamber; the charged molecules are then mobilized by an electric field and the "time-of-flight" to reach a detector is measured. These data allow "peptide mass fingerprinting" of a protein; this fingerprint is dependent on the original protein sequence. The fingerprint generated by a MALDI-TOF assay is evaluated by a software program that performs a virtual enzymatic digestion of all proteins in a database and calculates what a mass spectrum may look like for those proteins. The software then performs a "best fit" assessment comparison of the test sample against the virtual data to identify the protein under investigation.

The VITEK MS system (bioMériuex, Inc., Durham, NC) is an automated MALDI-TOF instrument that is FDA 510(k) cleared for the rapid accurate identification of up to 193 microorganisms (bacteria and fungi including mycobacteria, *Nocardia* and molds) cultured from a patient sample. Patients who are immunosuppressed for reasons such as organ or bone marrow transplant, or cancer therapies are at an especially increased risk for life-threatening infections requiring immediate treatment.

The VeriStrat® test (Biodesix, Boulder, CO) is a MALDI-TOF assay for estimating survival outcomes for patients with advanced non-small cell lung carcinoma (NSCLC); more specifically, for patients who have progressed after or are ineligible for platinum-based therapy and who have wild type (or unknown) epidermal growth factor receptor (EGFR) gene status. The assay is applied to serum samples and detects isoforms of inflammatory proteins that correlate with survival. Patient outcomes are rated as "Good" (candidates for treatment by platinum doublet or single-agent therapies [EGFR tyrosine kinase inhibitors]) or "Poor" (recommended for single agent chemotherapies).

The MassARRAY® System (Agena Bioscience, Inc., San Diego CA) is an adaptation of MALDI-TOF MS for the detection of nucleic acids. The system has been developed as a high-throughput alternative to NGS for screening targeted gene panels for mutation, methylation, gene transcript, or copy number status. Experiments are designed and performed using dedicated software and reagents. DNA or RNA is extracted from a patient sample and amplified by PCR or RT-PCR respectively. The amplicons are extended with Agena system-specific primers that confer mass identification. The final extension product is mixed with an absorption matrix and loaded onto a multi-sample chip platform and placed within a MS chamber. Ionized product is detected and identified by software comparison evaluation. For cancer diagnostics, oncology panels for multiple actionable (theranostic) cancer genes are used for research and in CAP/CLIA-approved diagnostic labs.

PRECISION DIAGNOSTICS AND OVERDIAGNOSIS

Precision medicine (also referred to as personalized medicine or individualized medicine) has been defined by the National Cancer Institute as a form of medicine that uses information about a person's genes, proteins, and environment to prevent, diagnose, and treat disease and alleviate symptoms [18]. Precision medicine is a standard of care goal for modern medical practice with the aims of ensuring that the right therapy is provided to the right patient at the right time and in the right measure as well as leading to the discovery of new targets for cancer treatment and prevention [18, 19]. Included in these aims is the prevention of overdiagnosis, which is defined as diagnosis of a disease that will cause neither symptoms nor death during the lifetime of an individual, or, as detection of disease by screening that in the absence of screening would not have been diagnosed within the lifetime of the patient [20].

The possibility of precision diagnostics has arisen largely because of the improved understanding of cancer and other diseases that has emerged from the human genome project, together with the technological advances in molecular techniques that have allowed standardized, high-throughput, and affordable clinical laboratory testing. These techniques are increasingly supporting improved patient care (Tables 1–4). A greatly increased expansion in the range and scope and inter-combination of these modalities is expected further to continuing genomic and proteomic research as well as investigations into the human epigenome [21] and human microbiome [22].

REFERENCES

1. College of American Pathologists. Cancer Protocols. https://www.cap.org/protocols-and-guidelines (Accessed February 14, 2019).
2. Taylor, C.R. and Shi, S.-R. (2014). Techniques of immunohistochemistry: principles, pitfalls, and standardization. In: *Diagnostic Immunohistochemistry: Theranostic and Genomic Applications*, 4e (ed. D.J. Dabs), 1–38. Philadelphia, PA: Elsevier Saunders.
3. Cuzick, J., Dowsett, M., Pineda, S. et al. (2011). Prognostic value of a combined estrogen receptor, progesterone receptor, Ki-67, and human epidermal growth factor receptor 2 immunohistochemical score and comparison with genomic health recurrence score in early breast cancer. *J. Clin. Oncol.* 29 (32): 4273–4278.
4. Lloyd, R.V., Qian, X., and Jin, L. (2009). Bright-field *In Situ* hybridization. In: *Cell and Tissue Based Molecular Pathology*, 1e (ed. R.R. Tubbs and M.H. Stoler), 114–126. Philadelphia, PA: Churchill Livingstone Elsevier.
5. Gall, J.G. and Pardue, M.L. (1969). Formation and detection of RNA-DNA hybrid molecules in cytological preparations. *Proc. Natl. Acad. SCi. USA.* 63 (2): 378–383.
6. John, H.A., Birnstiel, M.L., and Jones, K.W. (1969). RNA-DNA hybrids at the cytological level. *Nature.* 223 (5206): 582–587.
7. McPherson, M.J. and Møller, S.G. (2006). Understanding PCR. In: *PCR (The Basics)*, 2e (ed. M.J. McPherson and S.G. Møller), 9–22. NY: Taylor & Francis Group. New York.
8. Bustin, S.A., Benes, V., Garson, J.A. et al. (2009). The MIQE guidelines: minimum information for publication of quantitative real-time PCR experiments. *Clin Chem.* 55 (4): 611–622.
9. Chang, F. and Marilyn, M.L. (2013). Clinical application of amplicon-based next generation sequencing in cancer. *Cancer Genet.* 206 (12): 413–419.
10. Ross, J.S., Hatzis, C., Symmans, W.F. et al. (2008). Commercialized multigene predictors of clinical outcome for breast cancer. *Oncologist* 13 (5): 477–493.
11. Hessels, D. and Schalken, J.A. (2008). The use of PCA3 in the diagnosis of prostate cancer. *Nat. Rev. Urol.* 6 (5): 255–261.
12. Wong, A.K., Chan, R.C.-K., Stephen Nichols, W. et al. (2008). Human papillomavirus (HPV) in atypical squamous cervical cytology: the Invader HPV test as a new screening assay. *J. Clin. Micro.* 46 (3): 869–875.
13. Dowsett, M., Sestak, I., Lopez-Knowles, E. et al. (2013). Comparison of PAM50 risk of recurrence score with oncotype DX and IHC4 for predicting risk of distant recurrence after endocrine therapy. *J. Clin. Oncol.* 31 (22): 2783–2790.

14. Moor, K.N. and Walker, J.L. (2004). High risk human papillomavirus testing: guidelines for use in screening, triage, and follow-up for the prevention an early detection of cervical cancer. *J. Natl. Compr. Canc. Netw.* 2 (6): 589–596.

15. DiGiuseppe, J.A. (2006). Flow cytometry. In: *Molecular Diagnostics For the Clinical Laboratorian*, 2e (ed. W.B. Coleman and G.J. Tsongalis), 163–172. Totowa, NJ: Humana Press.

16. Beije, N., Jager, A., and Sleijfer, S. (2015). Circulating tumor cell enumeration by the CellSearch system: the clinician's guide to breast cancer treatment? *Cancer Treat. Rev.* 41 (2): 144–150.

17. Office of Cancer Clinical Proteomics Research. Cancer proteomics and proteogenomics. http://proteomics.cancer.gov (Accessed February 14, 2019)

18. National Cancer Institute. Precision medicine initiative and cancer research. http://www.cancer.gov/news-events/cancer-currents-blog/2015/precision-medicine-initiative-2016 (Accessed February 14, 2019).

19. Mayo Clinic. Center for Individualized Medicine. http://mayoresearch.mayo.edu/center-for-individualized-medicine/ (Accessed February 14, 2019).

20. Srivastava, S., Reed, B.J., Ghosh, S. et al. (2016). Research needs for understanding the biology of overdiagnosis in cancer screening. *J. Cell. Physiol.* 231 (9): 1870–1875.

21. National Institutes of Health Roadmap Epigenomics Mapping Consortium. http://www.roadmapepigenomics.org/ (Accessed February 14, 2019).

22. National Institutes of Health Human Microbiome Project. http://commonfund.nih.gov/hmp/index (Accessed February 14, 2019).

IV

NAVIGATING THE CANCER EXPERIENCE

8

NAVIGATING THE CANCER EXPERIENCE

David C. Nalepinski

Duke Cancer Institute, Durham, NC, USA

INTRODUCTION

The concept of cancer patient navigation has varying definitions, depending on the specifics of the disease, the patient population, and the care setting and resources. Cancer is a complex disease that usually requires the expertise of many healthcare specialties and resources that need to be integrated and coordinated to provide the best care for cancer patients. Patient navigators, and the navigation they provide, vary depending on the specific setting. For example, patient navigation may refer to: increasing access to screening for underserved populations; helping patients overcome barriers they encounter within the healthcare system during their treatment; ensuring patients have access to supportive care during treatment; ensuring patients have a smooth transition into survivorship (life beyond treatment); increasing access to clinical trials; or ensuring care is well planned and coordinated between the multiple disciplines involved in treatment. All of these examples are considered patient navigation. As such, patient navigators may have differing expertise and expectations placed upon them. For example, a navigator may refer to: a trained volunteer familiar with the culture of an underserved population; a nurse specialist that is familiar with the trajectory of a specific disease and the resources required to treat it; a research coordinator familiar with a specific research

Cancer: Prevention, Early Detection, Treatment and Recovery, Second Edition. Edited by Gary S. Stein and Kimberly P. Luebbers.
© 2019 John Wiley & Sons, Inc. Published 2019 by John Wiley & Sons, Inc.

trial; or an administrative coordinator that increases the coordination of care. These are only a few examples of the types of patient navigators and their activities. However, what is common between these examples is the increased access to needed care, ensuring care is comprehensive, and ensuring the care is well coordinated. This chapter will explain the need for patient navigation and why the application of this navigation has wide variation. Some of the explanation centers on a deeper understanding of the healthcare delivery system, the payment structure, and the complexity of cancer. Often, the sum of these variables is what necessitates the need for patient navigation and patient navigators.

A person who has just been informed they have cancer is likely to be in crisis and has many questions about their diagnosis. What does this mean? What happens now? How will I get the care I need? How will I pay for this? Will I survive? While the severity of this crisis may differ based on a person's current access to family and community support, healthcare resources and the type of disease they were diagnosed with, one thing is almost certain: Cancer can be an overwhelming and confusing disease to navigate. A newly diagnosed person's anxiety can be further exacerbated by the complexity of the healthcare system itself. Is the patient to be seen within one comprehensive center that offers all treatment and supportive services, or will they be seen by multiple centers within a given region with each only providing a specific part of the care? Are all needed services even available to the patient? Patient navigation may actually begin before a patient even knows they have cancer, which presents broader questions. Does every segment of our population have access to primary care and screening? Are some patients left undiagnosed until the disease has progressed to more severe stages?

While the concept of patient navigation and patient navigators may not provide the answers to all of these questions, navigation may greatly improve the patient experience and may help ensure patients are accessing the healthcare resources they need, with a high degree of coordination. What roles have navigation played for patients? What might cancer navigation look like in a specific patient's local healthcare environment?

THE COMPLEXITY OF CANCER CARE

Introduction

Before we further explore how patient navigation is used in cancer care, it is important to understand why patient navigation is needed in the first place. There are many important variables to understand that contribute to the challenges a cancer patient may experience, including: the various types and severity of cancer a patient may have, the local healthcare system in which a patient is treated, and the United States healthcare delivery system model and its payment structure. Having a deeper understanding of these variables will provide needed context for exploring the various ways the concept of patient navigation has developed and the various ways patient navigators are deployed.

Treating Cancer

Cancer has been a recognized human health condition throughout recorded history. However, only during the past 100 years have we begun to understand the disease, treat it, and sought to prevent and cure it. During the nineteenth century, surgery and anesthesia advanced as the primary treatment to remove tumors [1]. Diagnosis and the success of treatment was soon aided by the evolution of cellular pathology and the use of the microscope to better understand tumors and margins of healthy tissue [1]. In the twentieth century, cancer screening, early detection, prevention, and epidemiology became part of the cancer paradigm. While surgery continued to advance as a primary treatment option, the second half of the twentieth century witnessed the evolution of chemotherapy as an option to eradicate tumor cells [1]. The last quarter of the twentieth century saw advances in radiation physics and computer technology, adding various forms of radiation therapy as viable treatment options [1]. Immunotherapy and targeted therapy have also been added to the menu of treatment options [1]. Our knowledge about cancer and cancer treatment has advanced more in the past two decades than it has in the past few centuries [1]. We are advancing our understanding of cancer, its cause, and its treatment at rapid rates through expansive clinical trials, advances in molecular biology, completion of the human genome and the growth of cancer genetics, and gene expression profiling. The numerous types of cancer and the complexities inherent in its prevention, detection, progression, and treatment defined cancer as one of the most complex groups of diseases. Further, effectively managing cancer goes far beyond the treatment of tumors and cancer cells. Treatment needs to be patient centric to succeed.

Treating the Cancer Patient

Often, patient navigation plays critical roles in ensuring patients have access to all levels of care required for successful cancer treatment. While it is certainly important to understand the complexity of treating cancer tumors and cells, it is also important to understand the additional needs of cancer patients. Cancer patients often have many side effects from treatment and comorbidities that must be managed. Even beyond cancer treatment, cancer survivors may continue to have increased needs related to their specific cancer disease. In 2005, the Institute of Medicine released their landmark report *From Cancer Patient to Cancer Survivor: Lost in Transition* which highlighted the disconnect between a successful cancer treatment and life after cancer [2]. As survivor numbers continue to increase, more and more patients recognize that their primary healthcare providers are ill prepared for the unique health issues experienced by growing numbers of survivors.

Survivorship care planning highlights the monitoring, comorbidities, and second-ary issues that a cancer patient and their care team must focus on, beyond treating the disease itself. Survivorship programs and their associated supportive services focus on treating *the whole patient* through services that address issues such as: treatment tracking and summary, cardiomyopathy, nerve pain, impaired mobility, osteoporosis, anxiety, depression, cognitive dysfunction, spiritual distress, advanced care planning, advanced directives, nutrition, financial conciliation, sup-port groups, massage, art therapy, and coping skills. Successful cancer treatment combines the various treatment modalities to eliminate the tumor *and* the support-ive services to successfully treat the whole patient. However, as we have increased our ability to successfully treat the whole cancer patient, we have also increased the complexity of the treatment and the number of variables that need to be managed. Often, patient navigation is the safety net that ensures a patient has access to all of the services required in the complex healthcare system.

The Complex System

The US Healthcare Landscape

The US healthcare delivery system, its allocation of resources, the policies that govern it, and its financial drivers are incredibly complex. Patient navigation can have a tremendous impact on a cancer patient with a progressive disease that uses many disparate resources within the healthcare system. While national healthcare delivery systems vary somewhat by the level of economic and industrial develop-ment, we find further variance even among affluent and industrialized nations [3]. Countries like the Soviet Union and Czech Republic have centrally planned health-care services, some European countries like Great Britain and Norway may have universal and comprehensive healthcare services, and countries like Canada and Japan have universal basic coverage [3]. A country's government structure and associated healthcare policies often dictate its healthcare infrastructure. Although the US system is considered primarily "entrepreneurial and permissive," [3] it includes many aspects of other systems, further complicating the delivery of health-care services. The United States combines centrally planned services like the Veterans Administration, welfare-oriented services like Medicare and Medicaid, and private sector services. All services are delivered through a dispersed model of care, which is characterized by less structure, high competition, minimal central planning, entry at all levels of care, and a focus on secondary and tertiary levels of care (see box) [4]. Many other affluent industrialized nations have adopted a region-alized model of care that has central planning of resources, dictating the levels of primary, secondary, and tertiary care resources. The result of central planning is systems with roughly two-thirds primary care and one-third specialized care – the inverse of the United States' system of care [4]. While we will not discuss all the pros and cons of each system, it is worth mentioning that navigating the United

---

Primary, Secondary, and Tertiary Care [4]

Most healthcare systems in affluent industrialized nations include the following levels of care:

Primary Care – Performs basic ongoing health maintenance and preventative services for a population. May address common health problems, perform screening services, and vaccinations. Primary care may also act as a "gatekeeper" to higher levels of care.

Secondary Care – Addresses problems that require a higher level of care and may include ambulatory and inpatient services of a regional hospital. Secondary care may also play a role in regular treatment of chronic health issues, such as diabetes.

Tertiary Care – Provides the highest level of subspecialized care for complex and uncommon health conditions.

Quaternary Care – While not as common as other care resources, it is important to mention the important role of quaternary care, which includes research and clinical trials, usually at Academic Medical Centers.

Most cancer patients will use each level of care at various times during the trajectory of their disease.

States' dispersed healthcare landscape is considerably complicated. Further, in a dispersed model there is much less focus on the coordination of care and preventative and primary care, which may add to the difficulty for a cancer patient trying to navigate their treatment journey alone, without the assistance of a patient navigator.

Accountable Care Organizations

In recent years, the concept of Accountable Care Organizations (ACO) has emerged as a possible solution to reduce the complexity of our dispersed model of care. In this model of healthcare resources existing within a geographic region are incentivized to work together to manage patient care across the continuum of care. They are collectively accountable for the capacity of the system, the costs of care, and the quality of the care delivered. Incentives are through shared savings from participating payers. While this model holds promise for increased coordination of care, increased quality of care, and reduced costs of care delivery, it is fairly complicated to implement in the existing entrepreneurial and dispersed system. As such, it has not yet been widely adopted and has not yet lessened the need for patient navigation.

Influence of Payment Structure [4]

The payment structure in the dispersed model of care in the United States increases the challenges for cancer patients. Access to services, coordination of care, access to supportive services, and the overall quality of care received may all have increased complexity based on payment structure and incentives. Our goal here is not to fully define all aspects of the payment structure, but to highlight the variation in the payment structure.

- Modes of Financing – Includes: out-of-pocket, individual private insurance, employment-based insurance, government sponsored
- Types of Health Plans – Includes: Indemnity, Health Maintenance Organizations, Preferred Provider Organizations
- Reimbursement Methods – Includes: Fee for Service, Per Diem, Per Episode, Capitation, Global
- This diverse payment structure may often dictate when and where a cancer patient can be seen, possibly usurping well-coordinated care.

Many different combinations of these modes of financing, types of health plans, and reimbursement models may coexist in any given geographic location within our system of healthcare delivery. Further, each combination may exert differing levels of incentives, access, and treatment options. The influence of this diverse payment structure has mostly served to exacerbate the dispersed nature of the US healthcare delivery system and dilute any efforts toward regional planning and coordination of care.

A Complex Disease: Components of Cancer Care

While it is true that the US healthcare system is complex overall, it is particularly complex and resource intensive for the treatment of cancer. Figure 1 highlights common components of the healthcare delivery system for cancer patients. The cancer care paradigm starts with effective cancer control at the population level and may include education, promotion of healthy lifestyles, preventative care, risk reduction, and access to primary care – all aimed at reducing the burden of cancer [5]. Through effective screening, cancers are diagnosed and patients are referred into treatment. Treatment usually consists of combinations of cancer treatment modalities such as surgery, chemotherapy, and radiation. Following treatment, patients may resume relationships with their primary care provider who may continue to manage comorbidities and cancer treatment related issues, as well as provide surveillance for recurrence. In the best case scenario, treatment would be well coordinated between disciplines and include access to clinical trials, navigation, and supportive services.

Generally, this healthcare delivery is within the context of the overarching system of care described above, and, as such, can exist within several different configurations that vary by geographic location. The most comprehensive cancer centers are those designated as such by the National Cancer Institute. These centers usually include robust research, multidisciplinary state of the art treatment, access to treatment for rare cancers, education, outreach, and cancer control programs [6].

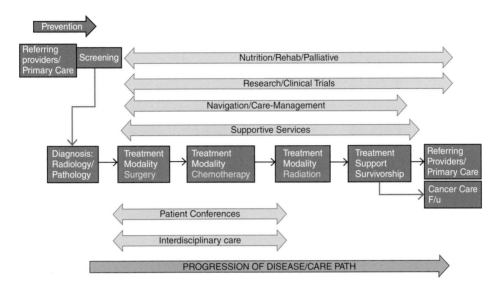

Figure 1. Common components of the healthcare delivery system for cancer patients.

Further, the comprehensive services, such as those highlighted in Figure 1, are usually available within one center, increasing the opportunity and likelihood that services are comprehensive and well coordinated. However, there are fewer than 50 National Cancer Institute (NCI) designated cancer centers in the United States. Often, cancer patients receive different aspects of their treatments from multiple centers that each feature different specialized treatment modalities located within a given geographic location. In the best case scenario, disparate resources located in a geographic location would have relationships and systems that foster well-coordinated care for cancer patients requiring multidisciplinary services. Often, this is not the case and patients are required to be their own advocate to navigate the system. Further, they may have limited or no access to supportive services.

Even if a patient is fortunate enough to be treated within a region that provides comprehensive well-coordinated care, either by one comprehensive center or by a combination of centers with optimal relationships and delivery systems, the trajectory of the disease and the sheer number of different resources needed may present immense challenges for a patient trying to navigate the system. Figure 2a below is a sample care path for a suspect breast cancer patient entering the complex system of care. In this example, steps are taken to diagnose and/or to confirm a diagnosis, to triage for the type of care the patient may require, and to align the necessary resources the patient will require for treatment. As the sample demonstrates, there are multiple decision points in the care path, as well as multiple clinical and administrative resources required *just to get the patient to the treatment they require.* Figure 2b shows the complex care path for a patient receiving treatment, after the steps in Figure 2a are complete. Again, there are multiple decision points, as well as multiple clinical specialties, each with their own modality specific aligned clinical and administrative resources.

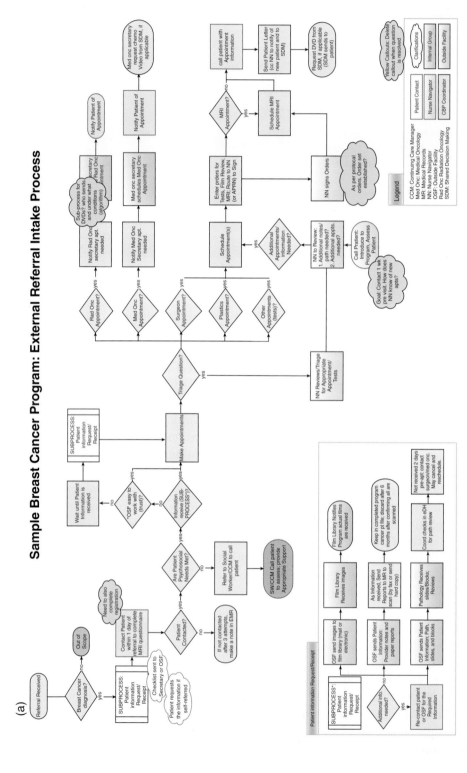

Figure 2. (a) Patient care path for suspect breast cancer patient entering a cancer care delivery. (b) Patient care path for a breast cancer patient starting out.

154

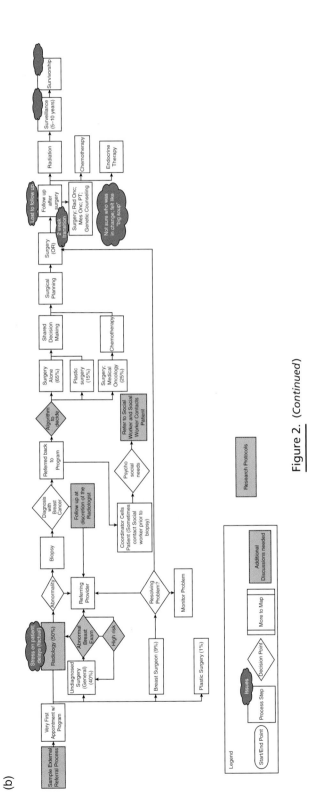

Figure 2. (*Continued*)

155

Recall that the resources in this example may exist within a single comprehensive center or a combination of centers and regional resources within a specific geographic location. Additionally, this is an example for one type of cancer. The trajectory of care may vary based on the type of cancer and the resources required for that cancer in a specific geographic location. To this end, there would be many variations by geographic location, cancer type, and patient insurance type and status.

In Figure 2a and b, we are viewing a sample system for a breast cancer patient *from the system perspective*. Figure 3 demonstrates what the healthcare delivery system may look like *from the patient perspective*. In this example, a patient was seen at a comprehensive center with a high degree of care coordination. In over six months, the patient had 55 encounters with the healthcare delivery system, demonstrating the complexity of treatment even in an optimal treatment delivery environment.

Summary

There are many components that contribute to the complexity of cancer care in the US healthcare system. Generally, cancer treatment has evolved to include the expertise of multiple treatment modalities. Most cancer patients also benefit from the addition of supportive services that meet a patient's additional needs related to psychosocial issues, comorbidities, and the common side-effects of treatment. Complexity is further increased by the dispersed model of healthcare delivery and its complicated payer system in the United States. We have to ensure that, regardless of the specific disease or geographic location, a patient is gaining access to the resources they need for proper preventative and primary care, screening, and treatment should they present with disease. There is no question that, on many levels, patients may benefit from resources that help them navigate the complex cancer care environment. The next section highlights the various ways navigation is used to overcome the complexities of the healthcare environment.

PATIENT NAVIGATION IN CANCER CARE

Evolution of Patient Navigation

The concept of patient navigation in cancer care was not originally predicated on navigating the complexity of the disease and the system of care in which it is treated. The concept originated to address disparities in access to cancer care in the poor populations of the nation [8]. In 1988 Dr. Harold P. Freeman, then President of the American Cancer Society (ACS), presented findings at an ACS Quality Assurance workshop that highlighted these disparities. At the time, advances in treatment and diagnosis of cancer were making significant strides in increasing

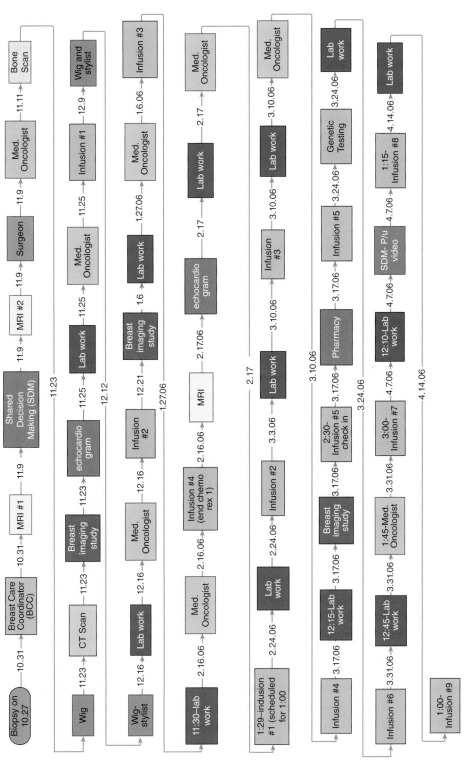

Figure 3. Breast cancer healthcare delivery system from the perspective of a patient. *Source:* Courtesy of Laam [7].

survival rates and the quality of life for cancer patients. However, these improvements were not having equal impact on groups of lower socioeconomic status [9]. The report highlighted gaps in these groups' ability to pay for and obtain care. Moreover, programs aimed at educating the public about cancer prevention and care were not reaching these groups, leaving them with gaps in care and a sense of hopelessness related to the disease [8]. Subsequently, In 1990 Dr. Freeman started what is widely considered to be the first patient navigation program at a Harlem New York public hospital aimed at identifying breast cancer earlier in poor African American woman [8]. While initially aimed at reducing financial, communication, system, and emotional barriers to screening and diagnosis, efforts soon expanded to guide patients across the cancer care continuum [8]. Retrospective studies of these efforts demonstrated significant improvements in early detection and survival rates for this population [10]. The concept of patient navigation in cancer care has continued to evolve since Dr. Freeman's ground-breaking efforts.

While there are many variations in how patient navigation meets the needs of cancer patients, most are based on a few common themes. Cancer care is a complex disease that is treated in a complex dispersed healthcare system. There are underserved populations that need access to screening and care. Once diagnosed, patients need access to timely treatment that is comprehensive and well coordinated. Additionally, patients benefit from supportive services that meet all of a patient's needs. In the following sections, we will review the ways patient navigation has been implemented, how benefits have been measured, and how a changing paradigm in healthcare may or may not warrant an expansion of navigation in cancer care.

Uses of Patient Navigation in Cancer Care

Access to Screening and Care for Underserved Populations

Since Dr. Freeman's use of patient navigation in 1990 Harlem, NY to correct disparities in timely access to the healthcare system for poor African American women with breast cancer, access to cancer care for underserved populations has been a key aim for the use of patient navigators [10]. For example, the NCI has long attributed higher cancer mortality rates for Native Americans to their presenting with more advanced stages of cancer [11]. Under NCI's Cancer Disparities Research Partnership Program, many Native Americans have been targeted for education programs encouraging screening [11]. Additional community-based research, such as the Native Navigators and the Cancer Continuum, have worked to understand the unique and complex cultural barriers that exist to thwart Native American access to screening and early detection. In this case, language, customs, and attitudes are significant barriers to members of tribes seeking early interventions for disease (Even using the word "Cancer" is believed to invite the disease into the host) [12]. For this population, it is important for the navigator to be imbedded in the tribal community with a full understanding of language and customs. Cancer education at the community level is beyond the common scope of cancer navigators, but in this case has proven to be very successful at increasing knowledge, referrals, and access to care [13, 14]. In these settings, communication, trust, and relationships seem to be the most important attributes for patient navigators.

Similar models of community-embedded lay patient navigators have proven successful for other underserved populations as well [15]. Many of these navigation efforts have focused on increasing colorectal cancer screening for minority populations in an urban setting. Similar to the efforts with Native Americans, these efforts used lay and peer patient navigators that were culturally similar to the underserved populations. These efforts resulted in statistically significant increases in screening for these populations [16–20]. Further, these interventions have proven to be cost effective for the healthcare system [21–23].

Various patient navigation interventions for underserved populations have similar criteria for success. There must be a recognized health disparity for an underserved population; such as access to screening and its associated early detection of cancer or the lack of adequate preventative and primary care [24]. The cause of these disparities must be identified; such as language, socioeconomic, or cultural barriers [24]. Then, the role of a patient navigator can be developed to reduce these health disparities. In these instances, the navigator is often a trained lay or peer navigator that understands the unique issues for an underserved population, with the ability to guide patients to the resources they need, removing barriers along the way. While the use of patient navigation at the front end of the cancer care continuum does not seem to have a broad application for all populations, it has proven immensely successful in reducing screening and access disparities for underserved populations.

Community/PCP-Based Navigation

In considering the common components of the healthcare delivery system for cancer care (Figure 1), many think of the role of primary care as having responsibility for prevention and screening, as a referral source for treatment subspecialties, and as the medical home for a patient *before* and *after* treatment. Most patient navigators in cancer care are part of the cancer care system and partner with the patient when they enter the diagnostic and treatment phases of care. However, in many rural settings where the components of treatment are dispersed, patient navigation that is based within the primary care medical home may be the preferred model for success. In these settings, community or primary care provider (PCP)-based navigators can help diagnosed patients identify and overcome barriers to care at various treatment facilities [25]. Patients that have a navigation resource to provide logistical and psychosocial support have a high rate of treatment adherence in a dispersed system [25]. In this scenario, we may imagine the patient navigator as an added resource. However, we should consider the role the PCP may play as the patient advocate in cancer care navigation. In a dispersed system, the patient's primary care provider may be the best resource to help the patients and their families understand the diagnosis, research and choose treatment teams, obtain support resources, assist with rehabilitation, monitor for reoccurrence, provide access to palliation, address psychosocial issues, and manage the patient's comorbidities [26]. Often, however, the PCP may simply refer the patient to an oncologist and expect the medical home to transfer to the oncologist during treatment, and hope to resume their relationship with the patient after treatment [26]. This may be acceptable if the PCP is referring the patient into a well-coordinated and comprehensive system of cancer care. However, if the system of cancer care is disseminated to various resources in a region, the patient may be best served with their PCP remaining as the medical home and patient advocate to navigate the patient through their treatment journey.

Overcoming Barriers to Care

As we have established, the idea of patient navigation originated to help underserved populations gain access to screening and needed care. The concept was expanded through all phases of care to assist patients in dealing with the complexity of the disease of cancer and the system in which it is treated. Often this complexity manifests as barriers to care. In fact, barriers exist through the treatment journey, from access to screening through survivorship, and are present in various forms. Barriers may include: access to screening, timely access to care, language and communication challenges, scheduling difficulties, lack of education about treatment options, fear, employment challenges, insurance and financial issues, transportation, social issues, lack of transitional care, and lack of practical support. Patient navigation has been shown to help patients overcome these barriers and have helped make care assessable, affordable, and appropriate [27]. More importantly, navigators serve as a point of contact across the system of care that can continually assess barriers as they arise and provide the patient with efficient access to resources that will help them overcome these barriers [28]. Often the barriers and the combination of barriers are as unique as the cancer itself and the system of care in which it is treated. As such, the concept of patient navigation is deployed in a wide variety of ways.

Patient Navigation Across the Cancer Care Continuum

A patient may encounter many types of patient navigation across the cancer care continuum. In primary care and prevention the navigator may be a trained lay person, with a unique understanding of the population being served, who can guide them to screening if warranted. In disease diagnosis, the navigator may be a social worker that is able to help the patient cope with the new reality they are facing. In treatment, the navigator may be a clinically trained nurse navigator that provides education and a thorough understanding of the various phases of treatment and the resources needed [29]. In survivorship, the navigator may be a nurse practitioner that can help develop a long-term posttreatment plan [29]. Assuming a cancer patient has overcome barriers to screening and diagnosis and has entered a system of cancer care, the need for navigating the complexity of the disease and its complex treatment remains. As Figure 2a and b demonstrate, a patient going through cancer treatment, even in a comprehensive well-coordinated setting, will encounter many touch points, decision points, and potential trajectories. While those involved in care coordination and delivery (doctors, nurses, scheduling secretaries, supportive services, etc.) may view the system in which they work as fairly straightforward, from the patient's point of reference there is ample opportunity to feel lost within the system and disconnected from their treatment plan. Increasingly, an Oncology Nurse Navigator is used to partner with the patient and serve as a common touchpoint through the treatment phase of cancer care.

THE ONCOLOGY NURSE NAVIGATOR

Though Dr. Freeman first introduced the concept of cancer navigation long ago, the role and function of the Oncology Nurse Navigator (ONN) has not been formally defined. A contributing factor may be the variability of the different cancer care delivery systems. The concept of patient navigation is often developed to meet the unique needs of a given system, thereby increasing the variability in the navigation role itself. In recent years, the Oncology Nursing Society (ONS) has worked to delineate the role of an ONN and provide a better foundation for the role [30]. In their 2012 report, "Oncology Nurse Navigator Role Delineation Study: An Oncology Nursing Society Report," they identified the tasks, knowledge areas, and skills for ONNs [30].

As the role delineation above demonstrates, the tasks of an ONN goes far beyond treating the disease, with an increased focus on the treating the whole patient and improving the patient experience [31]. Not only does the role of an ONN improve the experience and outcomes for the patient, it improves the experience for those working in the system of care by increasing communication and collaboration [31]. While other members of the care delivery team may have similar knowledge and perform similar tasks to that of the ONN, having this role dedicated to these activities may help insure they actually happen, and they provide a singular contact point for the patient through their journey. While this role may be additive to the system, current data suggests that there is a significant return on investment. It is important, however, that the role of the ONN is clearly defined as it relates to the patient care activities of other providers in the system. Any duplication of efforts between the ONN and the rest of the care team may actually add complexity and confusion to the system. In the best-case scenario, a clear care path has been developed that defines the possible trajectories of care through the system from the patient's perspective. In this way, the roles of the care team and how they interrelate can be clearly defined and it provides a plan that the navigator and patient can follow. One of the benefits of an ONN resource that is developed from within the care delivery system is that the role can be clearly defined as it relates to other parts of a specific system, providing clarity for all involved. However, this is not the only navigation model that is successful.

CURRENT ROLE OF THE AMERICAN CANCER SOCIETY IN PATIENT NAVIGATION

The ACS was involved in cancer patient navigation from the very beginning (with Dr. Freeman's work) and they still play a very active role. While some cancer care delivery systems have the resources to develop internally sourced ONNs, not all cancer systems have been able to do so. Regardless of a specific cancer care delivery system's ability to develop navigation resources from within, the ACS has made it part of their mission to train nonmedical ACS navigators, identify barriers, and adapt to any system to assist patients and care providers with navigating the complex system of cancer care [32]. Like the ONN, the ACS navigators focus on the care provided after a definitive diagnosis has been made. There are currently 119 ACS patient navigation sites in the United States, with navigators of varied background, but usually with some experience related to healthcare [32]. In the ACS model, the patient navigator becomes familiar with all parts of the care delivery system and becomes integrated with the various parts of the care team. While this resource is sourced externally from the care delivery system, it is still a significantly effective navigation model. Unlike an internally sourced ONN, the externally sourced ACS patient navigator is less likely to have specific program bias and is able to more freely engage all system resources on behalf of the patient. This may be especially true when care delivery system consists of many dispersed regionally components (as opposed to a comprehensive center). However, like the ONN model, the ACS navigation model does require additional resources and infrastructure, which may not be an option in systems where resources are more constrained.

THE ROLE OF THE LAY PERSON IN PATIENT NAVIGATION

In some healthcare systems, there has been some success developing patient navigation programs from limited resources. In Canada, some successful programs have been developed through the use of volunteers [33]. The Canadian Partnership Against Cancer's Cancer Journey Action Group defines cancer navigation as, "a proactive, intentional process of collaborating with a person and his or her family to provide guidance as they negotiate the maze of treatments, services and potential barriers through the cancer journey" [33]. In several provinces they have developed programs that train, screen, and provide support to volunteer navigators. They carefully evaluate what role the navigator will play in a particular system of care, clearly describe the scope of influence of the navigator, and imbed the navigator in a particular program [33]. There are attributes of this program that contribute to its success that may serve as lessons for all patient navigation programs. Since one aim of the program is to have a high impact with few resources, considerable effort is put into understanding the system and developing the navigator's role. While the navigators are "lay persons," the selection, training, support, and evaluation of navigators is fairly rigorous, as is the specificity in how the resource is used in the system. As a result of this effort, the programs have had high retention of volunteers and a high demand for services [33].

The Clinical Nurse Specialist

While similar to the role of an ONN, the role of the Clinical Nurse Specialist (CNS) has been developed in many countries with a regionalized healthcare delivery system (as opposed to the United States' dispersed system of care). In many European countries, the number of CNS's has continued to increase as medicine and surgical fields have continued to subspecialization [34]. Increasingly, the CNS has been seen as serving an important role in providing continuity of care for patients in the healthcare system. Unlike the ONN, a CNS is used for many applications in the healthcare delivery system, not just for patients with cancer. However, for a CNS that is used within a cancer care delivery system, the role is remarkably similar to an ONN [35–40].

Meeting Additional Patient Needs

Patient navigation in cancer care can increase access to needed care, ensure care is comprehensive, and ensure the care is well coordinated. Often comprehensive care goes beyond treating the disease itself and addresses issues that enable a patient to cope with the side effects and management of their disease and treatment. Two important elements of comprehensive care are patient distress and access to clinical trials. Patient Navigators may play an important role related to these elements.

A diagnosis of cancer and its subsequent treatment usually causes distress for a cancer patient. High levels of distress can impact a patient's ability to comply with treatment and can detract from the patient experience [41]. Distress can be effectively managed if detected [42]. The benefits of a patient navigator that partners with a patient through their treatment journey are twofold. First, the navigator can play a role in screening and identifying patients that have high levels of distress and ensure they get access to needed specialists that can help deal with more specific issues. Second, the overall involvement of a navigator may actually decrease the levels of distress for a patient through increased satisfaction, increase education, and increased adherence to treatment [41].

One of the defining attributes of comprehensive cancer care is access to the best treatments available, which often includes clinical trials [6]. Access and accrual to a clinical trial is mutually beneficial for the patient and the facility conducting the research. Patients get access to the latest treatments and research programs meet accrual goals that increase the power of their research. Often, a patient navigator can help patients overcome barriers to clinical trial enrolment through initial patient education and initial screening for possible eligibility [43–45].

THE CURRENT NEED FOR PATIENT NAVIGATION (CONCLUSION)

As with healthcare, patient navigation is evolving. It is important to recognize that navigation has varying definitions, depending on the specifics of the disease, the patient population, and the care setting and resources. While it is true that cancer is a complex disease that usually requires the expertise of many healthcare specialties and resources, it is the complexity the US healthcare delivery system, its allocation of resources, the policies that govern it, and its financial drivers that drive most of the need for navigation. In many instances, patient navigation is an added resource for a complex system that is already considered very expensive, with diminishing returns for the increased expense. Most of the measures of success for patient navigation relate to addressing the system's shortcomings and barriers, including: increased screening, increased access to care, timely care, patient satisfaction, adherence to treatment, decreased distress, and increased clinical trial enrollment. While some models discussed have used existing resources as the foundation for the development of navigation, navigation is often supplementary and perhaps superfluous.

The continuing evolution of patient navigation in cancer care should focus on further differentiating those needs that are *only* met through the use of a navigator and needs that are met by a navigator *to compensate for system deficiencies*. Perhaps, in the long term, the greatest value derived from the current mottled patient navigation model(s) will be its ability to reveal the inadequacies of the current system of care, that they can be addressed through other system improvements [46].

REFERENCES

1. 2015. (Accessed 10 May 2015, at http://www.cancer.org/cancer/cancerbasics/thehistoryofcancer/index?sitearea.)
2. Stovall, E., Greenfield, S., and Hewitt, M. (2005). *From Cancer Patient to Cancer Survivor: Lost in Transition*. Washington, DC: National Academies Press.
3. Barton, P.L. (1999). *Understanding the US Health Services System*. Chicago: Health Administration Press.
4. Bodenheimer, T.S. and Grumbach, K. (2004). *Understanding Health Policy: A Clinical Approach*, 4e. New York, NY: McGraw-Hill Companies, Inc.
5. National Comprehensive Cancer Control Program (NCCCP). 2015. (Accessed August 18 2015, 2015, at http://www.cdc.gov/cancer/ncccp/what_is_cccp.htm.)
6. 2015. (Accessed 14 May 2015, at https://www.cancer.gov/research/nci-role/cancer-centers.)
7. Laam, L.A. (April 2006). *CBP Improvement Task Force Update. Dartmouth-Hitchcock Medical Center CBP Tumor Board*. Lebanon, NH.
8. Freeman, H.P. (2012). The origin, evolution, and principles of patient navigation. *Cancer Epidemiology, Biomarkers & Prevention* 21: 1614–1617.
9. Freeman, H.P. (1989). Cancer in the socioeconomically disadvantaged. *CA: A Cancer Journal for Clinicians* 39: 266–288.
10. Freeman, H.P. and Wasfie, T.J. (1989). Cancer of the breast in poor black women. *Cancer* 63: 2562–2569.
11. Petereit, D.G., Guadagnolo, B.A., Wong, R., and Coleman, C.N. (2011). Addressing cancer disparities among American Indians through innovative technologies and patient navigation: the walking forward experience. *Frontiers in Oncology* 1: 11.
12. Harjo, L.D., Burhansstipanov, L., and Lindstrom, D. (2014). Rationale for "cultural" native patient navigators in Indian country. *Journal of Cancer Education: The Official Journal of the American Association for Cancer Education* 29: 414–419.

13. Burhansstipanov, L., Krebs, L.U., Harjo, L. et al. (2014). Providing community education: lessons learned from native patient navigators. *Journal of Cancer Education: The Official Journal of the American Association for Cancer Education* 29: 596–606.

14. Burhansstipanov, L., Krebs, L.U., Dignan, M.B. et al. (2014). Findings from the native navigators and the cancer continuum (NNACC) study. *Journal of Cancer Education: The Official Journal of the American Association for Cancer Education* 29: 420–427.

15. Meade, C.D., Wells, K.J., Arevalo, M. et al. (2014). Lay navigator model for impacting cancer health disparities. *Journal of Cancer Education: The Official Journal of the American Association for Cancer Education* 29: 449–457.

16. Jandorf, L., Cooperman, J.L., Stossel, L.M. et al. (2013). Implementation of culturally targeted patient navigation system for screening colonoscopy in a direct referral system. *Health Education Research* 28: 803–815.

17. Horne, H.N., Phelan-Emrick, D.F., Pollack, C.E. et al. (2015). Effect of patient navigation on colorectal cancer screening in a community-based randomized controlled trial of urban African American adults. *Cancer Causes and Control: CCC* 26: 239–246.

18. Braschi, C.D., Sly, J.R., Singh, S. et al. (2014). Increasing colonoscopy screening for Latino Americans through a patient navigation model: a randomized clinical trial. *Journal of Immigrant and Minority Health/Center for Minority Public Health* 16: 934–940.

19. Sly, J.R., Edwards, T., Shelton, R.C., and Jandorf, L. (2013). Identifying barriers to colonoscopy screening for nonadherent African American participants in a patient navigation intervention. *Health Education & Behavior: The Official Publication of the Society for Public Health Education* 40: 449–457.

20. Jandorf, L., Braschi, C., Ernstoff, E. et al. (2013). Culturally targeted patient navigation for increasing African Americans' adherence to screening colonoscopy: a randomized clinical trial. *Cancer Epidemiology, Biomarkers & Prevention: A Publication of the American Association for Cancer Research, Cosponsored by the American Society of Preventive Oncology* 22: 1577–1587.

21. Ladabaum, U., Mannalithara, A., Jandorf, L. et al. (2015). Cost-effectiveness of patient navigation to increase adherence with screening colonoscopy among minority individuals. *Cancer* 121: 1088–1097.

22. Wilson, F.A., Villarreal, R., Stimpson, J.P. et al. (2014). Cost-effectiveness analysis of a colonoscopy screening navigator program designed for Hispanic men. *Journal of Cancer Education: the Official Journal of the American Association for Cancer Education* 30 (2): 260–267.

23. Jandorf, L., Stossel, L.M., Cooperman, J.L. et al. (2013). Cost analysis of a patient navigation system to increase screening colonoscopy adherence among urban minorities. *Cancer* 119: 612–620.

24. Natale-Pereira, A., Enard, K.R., Nevarez, L. et al. (2011). The role of patient navigators in eliminating health disparities. *Cancer* 117: 3543–3552.

25. Fouad, M., Wynn, T., Martin, M. et al. (2010). Patient navigation pilot project: results from the community health advisors in action program (CHAAP). *Ethnicity & Disease* 20: 155–161.

26. Bernay, T. (2001). Becoming a professional cancer patient advocate: a new niche market practice for primary care physicians. *The Western Journal of Medicine* 175: 342–343.

27. Braun, K.L., Kagawa-Singer, M., Holden, A.E. et al. (2012). Cancer patient navigator tasks across the cancer care continuum. *Journal of Health Care for the Poor and Underserved* 23: 398–413.

28. Ullman, K. (2014). Navigating cancer treatment. *Journal of the National Cancer Institute* 106: dju031.

29. Hopkins, J. and Mumber, M.P. (2009). Patient navigation through the cancer care continuum: an overview. *Journal of Oncology Practice/ American Society of Clinical Oncology* 5: 150–152.

30. Brown, C.G., Cantril, C., McMullen, L. et al. (2012). Oncology nurse navigator role delineation study: an oncology nursing society report. *Clinical Journal of Oncology Nursing* 16: 581–585.

31. McMullen, L. (2013). Oncology nurse navigators and the continuum of cancer care. *Seminars in Oncology Nursing* 29: 105–117.

32. Esparza, A. (2013). Patient navigation and the American Cancer Society. *Seminars in Oncology Nursing* 29: 91–96.

33. Lorhan, S., Cleghorn, L., Fitch, M. et al. (2013). Moving the agenda forward for cancer patient navigation: understanding volunteer and peer navigation approaches. *Journal of Cancer Education: The Official Journal of the American Association for Cancer Education* 28: 84–91.

34. Pollard, C.A., Garcea, G., Pattenden, C.J. et al. (2010). Justifying the expense of the cancer Clinical Nurse Specialist. *European Journal of Cancer Care* 19: 72–79.

35. McPhillips, D., Evans, R., Ryan, D. et al. (2015). The role of a nurse specialist in a modern lung-cancer service. *British Journal of Nursing (Mark Allen Publishing)* 24 (Suppl 4): S21–S27.

36. (2013). Metastatic breast cancer: the clinical nurse specialist. *British Journal of Nursing (Mark Allen Publishing)* 22: S10. https://doi.org/10.12968/bjon.2013.22.Sup2.S10.

37. Admi, H., Zohar, H., and Rudner, Y. (2011). "Lighthouse in the dark": a qualitative study of the role of breast care nurse specialists in Israel. *Nursing & Health Sciences* 13: 507–513.

38. Willard, C. and Luker, K. (2007). Working with the team: strategies employed by hospital cancer nurse specialists to implement their role. *Journal of Clinical Nursing* 16: 716–724.

39. Moore, S. (2004). Guidelines on the role of the specialist nurse in supporting patients with lung cancer. *European Journal of Cancer Care* 13: 344–348.

40. Semple, C. (2001). The role of the CNS in head and neck oncology. *Nursing Standard (Royal College of Nursing [Great Britain]: 1987)* 15: 39–42.

41. Swanson, J. and Koch, L. (2010). The role of the oncology nurse navigator in distress management of adult inpatients with cancer: a retrospective study. *Oncology Nursing Forum* 37: 69–76.

42. Sellick, S.M. and Edwardson, A.D. (2007). Screening new cancer patients for psychological distress using the hospital anxiety and depression scale. *Psycho-Oncology* 16: 534–542.

43. Clair McClung, E., Davis, S.W., Jeffrey, S.S. et al. (2013). Impact of navigation on knowledge and attitudes about clinical trials among Chinese patients undergoing treatment for breast and gynecologic cancers. *Journal of Immigrant and Minority Health/Center for Minority Public Health* 17 (3): 976–979.

44. Ghebre, R.G., Jones, L.A., Wenzel, J.A. et al. (2014). State-of-the-science of patient navigation as a strategy for enhancing minority clinical trial accrual. *Cancer* 120 (Suppl 7): 1122–1130.

45. Steinberg, M.L., Fremont, A., Khan, D.C. et al. (2006). Lay patient navigator program implementation for equal access to cancer care and clinical trials: essential steps and initial challenges. *Cancer* 107: 2669–2677.

46. Clark, J.A., Parker, V.A., Battaglia, T.A. et al. (2014). Patterns of task and network actions performed by navigators to facilitate cancer care. *Health Care Management Review* 39: 90–101.

V

CLINICAL DIMENSIONS TO CANCER

9

CLASSIFICATION OF TUMORS, THEIR FREQUENCY AND PROGRESSION

Richard C. Zieren[1,2], Liang Dong[1,3],
Sarah R. Amend[1], and Kenneth J. Pienta[1]

[1]*The James Buchanan Brady Urological Institute, Johns Hopkins University School of Medicine, Baltimore, MD, USA*
[2]*Department of Urology, Amsterdam UMC, University of Amsterdam, Amsterdam, The Netherlands*
[3]*Department of Urology, Renji Hospital, Shanghai Jiao Tong University, School of Medicine, Shanghai, China*

In this chapter, we will discuss cancer: how tumors are classified, their frequency (or how common they are), and their prognosis (or how they behave over time).

TUMOR CLASSIFICATION

Tumors are classified by the type of tissue in which they originated, or histologic type. To understand the tumor classification system, it is helpful to step back briefly and look at how tissues were formed.

The tissues and organs of the body are formed during **embryonic development**. The cells of the early embryo differentiate, or begin to specialize, and then arrange themselves into three distinct layers, known as embryonic germ cell layers. An inner

Cancer: Prevention, Early Detection, Treatment and Recovery, Second Edition. Edited by Gary S. Stein and Kimberly P. Luebbers.
© 2019 John Wiley & Sons, Inc. Published 2019 by John Wiley & Sons, Inc.

layer is the **endoderm**, the outer layer is the **ectoderm**, and the middle layer is the **mesoderm**.

The three embryonic germ cell layers form the four types of tissues in the body: nervous tissue (which forms *neurons*), **connective tissue** (which forms *loose, dense, adipose, blood, cartilage, bone*), **epithelial tissue** (which forms *ciliated columnar epithelium, columnar epithelium, cuboidal epithelium, squamous epithelium* – all of which are internal and external lining in the body), and **muscular tissue** (which forms *cardiac muscle, smooth muscle, skeletal muscle, myoepithelium*). In this way: the **endoderm**-inner layer becomes the digestive system, liver, pancreas, and lungs (inner layers); the **ectoderm**-external cell layer becomes hair, nails, skin, and the nervous system; and the **mesoderm**-middle layer becomes the circulatory system, lungs (epithelial layers), skeletal system, and muscular system.

Cancers, or malignant tumors, are classified according to the tissue where they originated, also called the primary tumor site. There are **six** main histologic groups: carcinoma, sarcoma, myeloma, leukemia, lymphoma, and mixed types. Carcinoma is the cancer from **epithelial tissue** that is found in the internal and external lining of the body. There are two types of carcinomas: adenocarcinoma, which develops in an organ or gland, and squamous cell carcinoma, which develops in the *squamous epithelium* or skin. Sarcoma is a cancer from **connective tissue** that is found in bones, tendons, cartilage, muscle, and fat. Myeloma is a cancer from **connective tissue** that is found in plasma cells of bone marrow. Leukemias are cancers that originate in bone marrow, which derived from **connective tissue**. Lymphomas are cancers that originate in the lymphatic system, which derived from **connective tissue**. Mixed types consist of more than one cancer type from either one histologic group, or from different groups [1].

METASTATIC CANCER

Much of cancer prognosis is dependent on whether the cancer has metastasized. Tumor metastasis is a cancerous tumor that has spread from the place where it started (primary tumor site) to another place in the body. Even though the tumor has spread to a new area, it retains the same type of cancer cells as the original tumor and it retains the same name. For example, when a breast cancer spreads to the bone and forms a new tumor, the new tumor is a metastatic breast cancer, not a bone cancer. The most common sites of tumor metastasis are the lung, liver, brain, and bone. It is not easy for a tumor to metastasize. The following steps are required for metastasis.

The cancer cells:

- leave the primary tumor and spread into nearby normal tissue (*local invasion*);
- spread from nearby normal tissue into the walls of nearby blood vessels and lymph vessels (*intravasation*);

- travel through the bloodstream and the lymphatic system to other parts of the body (*circulation*);
- stop in small blood vessels (capillaries) in another, distant part of the body. Spread through the walls of the distant capillaries and into the surrounding tissue (*arrest* **and** *extravasation*);
- multiply in the new, distant tissue, to form new tumors known as micrometastases (*proliferation*); and
- then, in the new location, the micrometastases form new blood vessels in order to grow (*angiogenesis*) [2].

TUMOR FREQUENCY

Next, we will look at how often the tumors occur, or their frequency. The frequency of a certain tumor being diagnosed depends on geographic location; in this chapter we will focus on tumor frequency within the United States. Specific tumor frequency is often influenced by a person's characteristics and behaviors. When applicable, we will include the gender, age, and race of those most likely to have the tumor, and we will include well-known risk factors, such as inherited syndromes, occupational exposures to chemicals, or a history of smoking or alcohol consumption.

We will also discuss how common the cancer is in relation to other cancers, based on incidence (total number of new cases each year) and prevalence (total number of cases). We will graphically show the incidence trend of cancers since 1975 until present, which will tell us if the tumor is becoming more frequent, less, or has remained stable over time.

TUMOR PROGNOSIS

Statistics that help to show the prognosis of cancer are the overall five-year survival rate (percentage of cancer patients that remained alive at least five years), the incidence rate (new cases per 100000 people per year), and the mortality rate (number of cancer caused deaths per 100000 people per year).

While the overall survival is helpful to know, this is heavily influenced by the stage of the tumor at diagnosis – generally a tumor found in an earlier stage will have a better prognosis. Therefore, we include the five-year relative survival based on the stage of the tumor. Another interesting prognostic statistic is the comparison of incidence and mortality trends. If the incidence and mortality lines are far apart, not many people are dying from the tumor as compared to those who have been diagnosed with it, suggesting early diagnosis or effective treatment. In contrast, if the two lines are close, patients are diagnosed with later stages and treatments are less effective. Also, we include the five-year survival trend, which demonstrates if there have been significant improvements per decade, indicating potential progress with new and better therapies.

COMMON CANCER TYPES

Breast Cancer

Breast cancers are classified by their location and by their histologic (tissue or cellular) subtype. Of primary breast cancers 70–80% are infiltrating ductal cancers, followed by lobular and nipple cancers. Atypical subtypes include phyllodes tumor, angiosarcoma, and primary lymphoma [3].

Breast cancer is the second most frequent in women after skin cancer, and is the second leading cause of cancer death in women, following lung cancer. Men account for less than 1% of breast cancer cases [4]. Risk factors for breast cancer include increasing age, a positive family history, and inheritance of specific genes (for example, BRCA 1 and BRCA 2), and an extended exposure to estrogen. Approximately 12.4% of women will be diagnosed with breast cancer during their lifetime; the median age at diagnosis is 62 years.

The incidence (new cases) and prevalence (all cases) of breast cancer, based on the 2011–2015 SEER data (a data collection program of the National Cancer Institute), is 126.0 per 100 000 women per year, and 3 418 124 women, respectively. In 2018, the estimated total number of new cases and mortality (cancer deaths) of breast cancer is 266 120 and 40 920, respectively.

The overall five-year survival rate of all breast cancer cases combined is 89.7%. The distribution of cases over the disease stages and corresponding relative survival is shown in Table 1. Over the past 10 years, the incidence of breast cancer has been increasing, on average 0.3% per year. However, in the same timeframe, the mortality of breast cancer has been declining, on average 1.8% per year (Figure 1) [5].

TABLE 1. **Breast cancer stages:** distribution and five-year relative survival.

Stage of cancer	Patients diagnosed in stage (%)	5-year survival (%)
Local	62	98.7
Regional	31	85.3
Distant	6	27.0
Un-staged	2	54.5

Local – cancer is confined to the breast.
Regional – cancer is spread outside of but remains close to the breast.
Distant – cancer is outside of and distant from the breast.
Un-staged – cancer stage is unknown.
*Stage distribution percentages do not sum to 100 due to rounding.

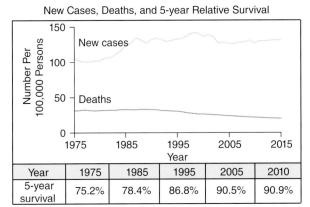

New Cases, Deaths, and 5-year Relative Survival

Year	1975	1985	1995	2005	2010
5-year survival	75.2%	78.4%	86.8%	90.5%	90.9%

SEER 9 Incidence & U.S. Mortality 1975–2015, All Races, Females. Rates are Age-Adjusted

Figure 1. Graph shows 40-year trends of new cases (light gray) and deaths (dark gray) of breast cancer. Table shows five-year survival by year. Data from *SEER 9 Incidence & 5-year Relative Survival Percent from 1975–2010, and U.S. Mortality 1975–2015*. All Races, females. Rates are age-adjusted. *Source*: https://www.cancer.gov

Skin Cancer

Skin cancer originates in the epidermis, or outer layer of the skin. The most common skin cancers are the non-melanoma skin cancers: basal cell carcinoma, the more common of the two, and squamous cell carcinoma. Actinic keratosis is a skin condition that occasionally is a precursor of squamous cell carcinoma. Melanoma is an uncommon form of skin cancer that originates in the lowest part of the epidermis and it is the most likely skin cancer to metastasize. The remaining, less common, primary skin cancers include mycosis fungoides, the Sezary syndrome, Kaposi sarcoma, and Merkle cell carcinoma.

Melanoma is a malignant tumor of melanocytes, which are cells that make the pigment melanin. Cutaneous melanoma may originate in the skin or mucosal surfaces, while uveal melanoma, which behaves much differently than cutaneous melanoma, originates in the uveal tract (the pigmented, center tract of three layers of tissue that make up the eye). The remainder of this section will focus on cutaneous, or skin, melanoma [6, 7].

Melanoma is the fifth most commonly diagnosed cancer, and it occurs more frequently in men than in women. Risk factors for melanoma include a fair complexion, Caucasian race, and a history of extensive exposure to natural or artificial sunlight. Approximately 2.3% of men and women will be diagnosed with melanoma during their lifetime; the median age at diagnosis is 64 years.

The incidence (new cases) and prevalence (all cases) of melanoma based on the 2011–2015 SEER data, is 22.8 per 100 000 men and women per year, and 1 222 023 people, respectively. In 2018, the estimated total number of new cases and mortality (cancer deaths) of melanoma of the skin is 91 270 and 9320 respectively.

The overall five-year survival rate of all melanoma cases combined is 91.8%. The distribution of cases over the disease stages and corresponding relative survival is shown in Table 2. Over the past 10 years, the incidence of melanoma has been increasing, on average 1.5% per year. However, in the same time-frame, the mortality of melanoma has been decreasing, on average 1.2% (Figure 2) [8].

TABLE 2. **Melanoma stages:** distribution and five-year relative survival.

Stage of cancer	Patients diagnosed in stage (%)	5-year survival (%)
Local	84	98.4
Regional	9	63.6
Distant	4	22.5
Un-staged	4	83.7

Local – cancer is confined to the primary site.
Regional – cancer is spread outside of but remains close to the primary site.
Distant – cancer is outside of and distant from the primary site.
Un-staged – cancer stage is unknown.
*Stage distribution percentages do not sum to 100 due to rounding.

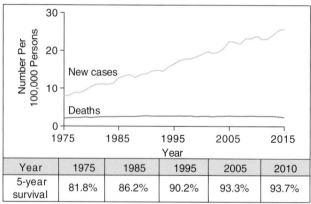

New Cases, Deaths, and 5-year Relative Survival

Year	1975	1985	1995	2005	2010
5-year survival	81.8%	86.2%	90.2%	93.3%	93.7%

SEER 9 Incidence & U.S. Mortality 1975–2015, All Races, Both Sexes. Rates are Age-adjusted

Figure 2. Graph shows 40-year trends of new cases (light gray) and deaths (dark gray) of melanoma. Table shows five-year survival by year. Data from *SEER 9 Incidence & 5-year Relative Survival Percent from 1975-2010, and U.S. Mortality 1975–2015.* All Races, females. Rates are age-adjusted. *Source:* https://www.cancer.gov

Brain and Other Central Nervous System Cancer

The brain and spinal cord comprise the central nervous system (CNS). Brain tumors account for 85–90% of all primary CNS cancers. The CNS tumors are classified based by the type of cells they originate in, and by their location. Astrocytic tumors begin in astrocytes (a type of brain cell, a glial cell) and account for 38% of primary brain tumors; oligodendroglial tumors begin in oligodendrocytes, another type of glial cell, and account for 27% of primary brain tumors; mixed gliomas, a tumor with two types of cells (oligodendrocytes and astrocytes) called an oligoastrocytoma; ependymal tumors, begin in cells that line the fluid filled spaces in the CNS; medulloblastoma is a type of embryonal tumor, most common in children and young adults; pineal parenchymal tumors from in parenchymal cells that make up the pineal gland (produces melatonin, a hormone that controls the daily circadian cycle); meningiomas form in the layers of tissue that cover the brain and spinal cord and account for 79% of primary spinal tumors; germ cell tumors form in germ cells (sperm or ova); and craniopharyngioma, a rare tumor that usually forms just above the pituitary gland. Brain tumors rarely metastasize to distant areas outside the CNS [9, 10].

Brain and other CNS cancer is relatively rare, and is the 10th leading cause of cancer death. This cancer type is slightly more common in men. There are few known risk factors for brain and CNS cancer, however, exposure to vinyl chloride, Epstein–Barr virus infection, and acquired immune deficient syndrome following organ transplants are associated with brain and other CNS cancer. Approximately 0.6% of men and women will be diagnosed with brain and other CNS cancer during their lifetime; the median age at diagnosis is 58 years.

The incidence (new cases) and prevalence (all cases) of brain and other nervous system cancer is 6.4 per 100 000 men and women per year, and 116 039 people, respectively. In 2018, the estimated total number of new cases and mortality (cancer deaths) of brain and other nervous system cancer is 23 880 and 16 830, respectively.

The overall five-year survival rate of all brain and other CNS cancer cases combined is 33.2%. The distribution of cases over the disease stages and corresponding relative survival is shown in Table 3. Over the past 10 years, the incidence of brain and other nervous system cancer has been declining, on average 0.2% per year. However, in the same time frame, the mortality of brain and other nervous system cancer has been increasing, on average 0.4% per year (Figure 3) [11].

Thyroid Cancer

The most common primary thyroid cancers are papillary tumors, followed by follicular tumors. Both tumors, being well differentiated, are highly treatable and usually curable. The remaining, less common, thyroid cancers are medullary and anaplastic tumors, which are poorly differentiated and aggressive [12].

TABLE 3. Brain and nervous system cancer stages: distribution and five-year relative survival.

Stage of cancer	Patients diagnosed in stage (%)	5-year survival (%)
Local	77	36.2
Regional	15	20.4
Distant	2	32.7
Un-staged	5	27.3

Local – cancer is confined to the brain/nervous system.

Regional -cancer is spread outside of but remains close to the brain/nervous system.

Distant – cancer is outside of and distant from the brain/nervous system.

Un-staged – cancer stage is unknown.

*Stage distribution percentages do not sum to 100 due to rounding.

New Cases, Deaths, and 5-year Relative Survival

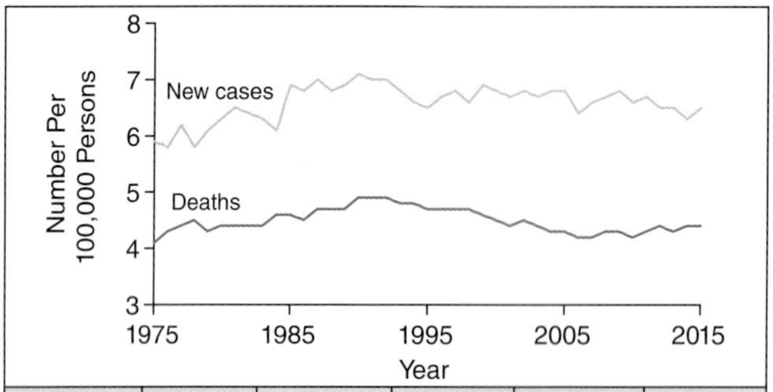

Year	1975	1985	1995	2005	2010
5-year survival	22.8%	24.7%	33.3%	34.8%	34.0%

SEER 9 Incidence & U.S. Mortality 1975–2015, All Races, Both Sexes. Rates are Age-adjusted

Figure 3. Graph shows 40-year trends of new cases (light gray) and deaths (dark gray) of brain and nervous system cancer. Table shows five-year survival by year. Data from *SEER 9 Incidence & 5-year Relative Survival Percent from 1975–2010, and U.S. Mortality 1975–2015.* All races, both sexes. Rates are age-adjusted. *Source*: https://www.cancer.gov

Thyroid cancer is fairly common, representing almost 3.1% of all new cancers. Risk factors for thyroid cancer include excessive exposure to radiation, female gender, and a positive family history. Approximately 1.2% of men and women will be diagnosed with thyroid cancer during their lifetime; the median age at diagnosis is 51 years.

The incidence (new cases) and prevalence (all cases) of thyroid cancer, based on the 2011–2015 SEER data, is 14.5 per 100 000 men and women per year, and 765 547 people, respectively. In 2018, the estimated total number of new cases and mortality (cancer deaths) of thyroid cancer is 53 990 and 2060 respectively.

The overall five-year survival rate of all thyroid cancer cases combined is 98.1%. The distribution of cases over the disease stages and corresponding relative survival is shown in Table 4. Over the past 10 years, the incidence and mortality of thyroid cancer has been increasing, on average 3.1% and 0.7% per year, respectively (Figure 4) [13].

Bone and Joint Cancer

The most common type of primary bone and joint cancer is osteosarcoma, which originates in the osteoid and forms hard, bone tissue, followed by chondrosarcoma, which originates in the cartilaginous tissue, and Ewing sarcoma, which can

TABLE 4. Thyroid cancer stages: distribution and five-year relative survival.

Stage of cancer	Patients diagnosed in stage (%)	5-year survival (%)
Local	67	99.9
Regional	27	98.0
Distant	4	55.5
Un-staged	2	89.1

Local – cancer is confined to the thyroid gland.
Regional – cancer is spread outside of but remains close to the thyroid gland.
Distant – cancer is outside of and distant from the thyroid gland.
Un-staged – cancer stage is unknown.

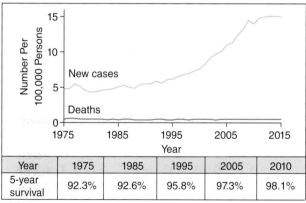

New Cases, Deaths, and 5-year Relative Survival

Year	1975	1985	1995	2005	2010
5-year survival	92.3%	92.6%	95.8%	97.3%	98.1%

SEER 9 Incidence & U.S. Mortality 1975–2015, All Races, Both Sexes. Rates are Age-adjusted

Figure 4. Graph shows 40-year trends of new cases (light gray) and deaths (dark gray) of thyroid cancer. Table shows five-year survival by year. Data from *SEER 9 Incidence & 5-year Relative Survival Percent from 1975–2010, and U.S. Mortality 1975–2015*. All races, both sexes. Rates are age-adjusted. *Source*: https://www.cancer.gov

originate in both osteoid and cartilaginous tissue. Uncommon types include <u>chor-domas</u> and malignant fibrous histiocytomas [14, 15].

Bone and joint cancer is rare, as it represents 0.2% of all new cancer cases in the United States. Osteosarcoma is most common in teenagers. Ewing sarcoma is most common in teenagers and young adults.

Bone cancers are more common in males and in the Caucasian race. Risk factors for bone cancer include certain inherited genetic syndromes, Paget's disease of the bone (a precancerous condition), and a history of radiation therapy. Approximately 0.1% of men and women will be diagnosed with bone and joint cancer during their lifetime; the median age at diagnosis is 42 years.

The incidence (new cases) of bone cancer is 0.9 per 100 000 men and women per year. In 2018, the estimated total number of new cases and mortality (cancer deaths) of bone and joint cancer is 3450 and 1590, respectively.

The five-year overall survival of bone and joint cancer is 66.9%. Over the past 10 years, the incidence of bone and joint cancer has been increasing, on average 0.4% per year. However, in the same timeframe, the mortality of bone and joint cancer has been declining, on average 0.3% per year (Figure 5) [16].

Lung and Bronchus Cancer

Lung and bronchus cancer has two main categories: non-small cell cancer and small cell cancer. Non-small cell lung cancer (NSCLC) is any type of epithelial lung cancer other than small cell lung cancer (SCLC). The most common types of

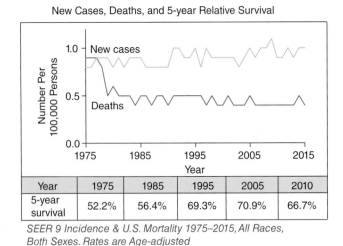

New Cases, Deaths, and 5-year Relative Survival

Year	1975	1985	1995	2005	2010
5-year survival	52.2%	56.4%	69.3%	70.9%	66.7%

SEER 9 Incidence & U.S. Mortality 1975–2015, All Races, Both Sexes. Rates are Age-adjusted

Figure 5. Graph shows 40-year trends of new cases (light gray) and deaths (dark gray) of bone and joint cancer. Table shows five-year survival by year. Data from *SEER 9 Incidence & 5-year Relative Survival Percent from 1975–2010, and U.S. Mortality 1975–2015*. All races, both sexes. Rates are age-adjusted. *Source*: https://www.cancer.gov

NSCLC are squamous cell carcinoma (25%), large cell carcinoma (10%), and adenocarcinoma (40%). SCLC accounts for approximately 15% of all lung and bronchus cancers and is classified into small cell carcinoma, combined small cell carcinoma, and neuroendocrine tumors [17, 18].

Lung and bronchus cancer is the second most commonly diagnosed cancer, and it is the leading cause of cancer death among men and women. Lung cancer is more common in men and particularly in African American men. Risk factors for lung cancer include smoking, exposure to several environmental pollutants, and a history of radiation therapy to the breast or chest. Approximately 6.2% of men and women will be diagnosed with lung and bronchus cancer during their lifetime; the median age at diagnosis is 70 years.

The incidence (new cases) and prevalence (all cases) of lung and bronchus cancer is 54.6 per 100 000 men and women per year, and 541 035 people, respectively. In 2018, the estimated total number of new cases and mortality (cancer deaths) of lung and bronchus cancer is 234 030 and 154 050, respectively.

The overall five-year survival rate of all lung and bronchus cancer cases combined is 18.6%. The distribution of cases over the disease stages and corresponding relative survival is shown in Table 5. Over the past 10 years, the incidence and mortality of lung and bronchus cancer has been declining, on average 2.1 and 2.7% per year, respectively (Figure 6) [19].

Head and Neck Cancer

Oral Cavity and Pharynx Cancer

The pharynx is the part of the throat that is behind the nasal and oral cavities and above the esophagus and voice box. The oropharynx is the part of the pharynx on the level of the oral cavity. Most oral cavity and pharynx cancers start in the squamous cells that line the lips and oral cavity. Leukoplakia (a white patch that does

TABLE 5. **Lung and bronchus cancer stages:** distribution and five-year relative survival.

Stage of cancer	Patients diagnosed in stage (%)	5-year survival (%)
Local	16	56.3
Regional	22	29.7
Distant	57	4.7
Un-staged	5	7.8

Local – cancer is confined to the lung/bronchus.
Regional – cancer is spread outside of but remains close to the lung/bronchus.
Distant – cancer is outside of and distant from the lung/bronchus.
Un-staged – cancer stage is unknown.

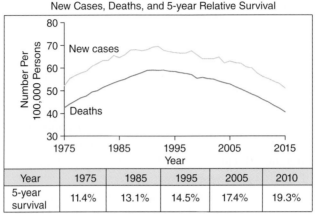

Year	1975	1985	1995	2005	2010
5-year survival	11.4%	13.1%	14.5%	17.4%	19.3%

SEER 9 Incidence & U.S. Mortality 1975–2015, All Races, Both Sexes. Rates are Age-adjusted

Figure 6. Graph shows 40-year trends of new cases (light gray) and deaths (dark gray) of lung and bronchus cancer. Table shows five-year survival by year. Data from *SEER 9 Incidence & 5-year Relative Survival Percent from 1975–2010, and U.S. Mortality 1975–2015*. All races, both sexes. Rates are age-adjusted. *Source*: https://www.cancer.gov

not rub off), erythroplakia (a smooth, red patch), and the mixed forms, erythroleukoplakia, are precursors to oral cancer. Most oropharyngeal cancers are squamous cell carcinomas. Other oropharyngeal cancers are minor salivary gland tumors, lymphomas, and lymphoepitheliomas [20, 21].

Oral cancer is relatively rare. It is more common in men. Risk factors include a history of heavy tobacco and alcohol use and infection with the human papillomavirus (HPV). Approximately 1.2% of men and women will be diagnosed with oral cavity and pharynx cancer during their lifetime; the median age at diagnosis is 63 years.

The incidence (new cases) and prevalence (all cases) of oral cavity and pharynx cancer is 11.3 per 100 000 men and women per year, and 359 718 people, respectively. In 2018, the estimated total number of new cases and mortality (cancer deaths) of oral cavity and pharynx cancer is 51 540 and 10 030, respectively.

The overall five-year survival rate of all oral cavity and pharyngeal cancer cases combined is 64.8%. The distribution of cases over the disease stages and corresponding relative survival is shown in Table 6. Over the past 10 years, the incidence of oral cavity and pharynx cancer has been increasing, on average 0.7% per year. However, in the same timeframe, the mortality of oral cavity and pharynx cancer has been stable (Figure 7) [22].

Larynx Cancer

Most laryngeal cancers originate from squamous cells, which are the cells that line the inside of the larynx. The larynx is the respiratory part of the throat, also called voice box [23].

T A B L E 6. **Oral cavity and pharyngeal cancer stages:** distribution and five-year relative survival.

Stage of cancer	Patients diagnosed in stage (%)	5-year survival (%)
Local	29	83.7
Regional	47	65.0
Distant	20	39.1
Un-staged	4	49.2

Local – cancer is confined to the oral cavity and pharynx.

Regional – cancer is spread outside of but remains close to the oral cavity and pharynx.

Distant – cancer is outside of and distant from the oral cavity and pharynx.

Un-staged – cancer stage is unknown.

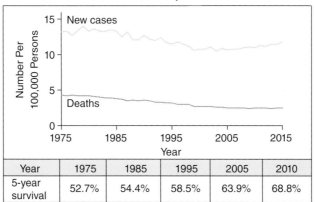

New Cases, Deaths, and 5-year Relative Survival

Year	1975	1985	1995	2005	2010
5-year survival	52.7%	54.4%	58.5%	63.9%	68.8%

SEER 9 Incidence & U.S. Mortality 1975–2015, All Races, Both Sexes. Rates are Age-adjusted

Figure 7. Graph shows 40-year trends of new cases (light gray) and deaths (dark gray) of oral cavity and pharyngeal cancer. Table shows five-year survival by year. Data from *SEER 9 Incidence & 5-year Relative Survival Percent from 1975–2010, and U.S. Mortality 1975–2015.* All races, both sexes. Rates are age-adjusted. *Source:* https://www.cancer.gov

Larynx cancer is relatively rare. It is more common with older age and in men. Risk factors for laryngeal cancer include smoking, alcohol consumption, and development of squamous cell cancers in the upper airway. Approximately 0.3% of men and women will be diagnosed with larynx cancer during their lifetime; the median age at diagnosis is 65 years.

The incidence (new cases) and prevalence (all cases) of larynx cancer is 3.0 per 100 000 men and women per year, and 99 756 people, respectively. In 2018, the estimated total number of new cases and mortality (cancer deaths) of larynx cancer is 13 150, and 3710, respectively.

The overall five-year survival rate of all laryngeal cancer cases combined is 60.9%. The distribution of cases over the disease stages and corresponding relative survival is shown in Table 7. Over the past 10 years, the incidence and mortality of larynx cancer has been declining, on average 2.4% and 2.3% per year, respectively (Figure 8) [24].

Gastrointestinal Cancer

Colorectal Cancer

Most primary colon cancers are adenocarcinomas (mucinous or Signet ring), meaning that they originate in the epithelial tissue that lines the cavity of the colon. Most

TABLE 7. Laryngeal cancer stages: distribution and five-year relative survival.

Stage of cancer	Patients diagnosed in stage (%)	5-year survival (%)
Local	55	77.5
Regional	23	45.6
Distant	19	33.5
Un-staged	3	54.6

Local – cancer is confined to the larynx.
Regional – cancer is spread outside of but remains close to the larynx.
Distant – cancer is outside of and distant from the larynx.
Un-staged – cancer stage is unknown.

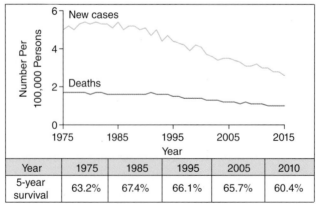

Year	1975	1985	1995	2005	2010
5-year survival	63.2%	67.4%	66.1%	65.7%	60.4%

SEER 9 Incidence & U.S. Mortality 1975–2015, All Races, Both sexes. Rates are Age-Adjusted

Figure 8. Graph shows 40-year trends of new cases (light gray) and deaths (dark gray) of laryngeal cancer. Table shows five-year survival by year. Data from *SEER 9 Incidence & 5-year Relative Survival Percent from 1975–2010, and U.S. Mortality 1975–2015*. All races, both sexes. Rates are age-adjusted. *Source:* https://www.cancer.gov

rectal cancers are adenocarcinomas as well. The remaining 2–5% of colorectal cancers are carcinoid tumors, lymphomas, neuroendocrine, and other tumors [25, 26].

Colorectal cancer is the fourth most commonly diagnosed cancer, and the second leading cause of cancer death. It is more common in men and in those of African American race. Strong risk factors for colorectal cancer include hereditary conditions, a positive family history, and a history of a gynecologic cancer. Approximately 4.2% of men and women will be diagnosed with colorectal cancer during their lifetime; the median age at diagnosis is 67 years.

The incidence (new cases) and prevalence (all cases) of colorectal cancer is 39.4 per 100 000 men and women per year, and 1 332 085 people, respectively. In 2018, the estimated total number of new cases and mortality (cancer deaths) of colorectal cancer is 140 250, and 50 630, respectively.

The overall five-year survival rate of all colorectal cancer cases combined is 64.5%. The distribution of cases over the disease stages and corresponding relative survival is shown in Table 8. Over the past 10 years, the incidence and mortality of colorectal cancer have been declining, on average 2.6 and 2.4% per year, respectively (Figure 9) [27].

Pancreatic Cancer

More than 90% of primary pancreatic cancers are adenocarcinomas, arising from the exocrine portion of the pancreas, where the digestive enzymes are produced. Other types include acinar cell carcinoma, adenosquamous carcinoma, cystadenocarcinoma, and including several other types [28].

Pancreatic cancer is rare, however it is the third leading cause of cancer deaths. It is more common with advancing age, and slightly more common in men. Risk factors for pancreatic cancer include a positive family history, cigarette smoking, obesity, and chronic pancreatitis (inflammation of pancreas). Approximately 1.6% of men and women will be diagnosed with pancreatic cancer during their lifetime; the median age at diagnosis is 70 years.

T A B L E 8. **Colorectal cancer stages:** distribution and five-year relative survival.

Stage of cancer	Patients diagnosed in stage (%)	5-year survival (%)
Local	39	89.8
Regional	35	71.1
Distant	21	13.8
Un-staged	4	35

Local – cancer is confined to the colon/rectum.
Regional – cancer is spread outside of but remains close to the colon/rectum.
Distant – cancer is outside of and distant from the colon/rectum.
Un-staged – cancer stage is unknown.
* Stage distribution percentages do not sum to 100 due to rounding.

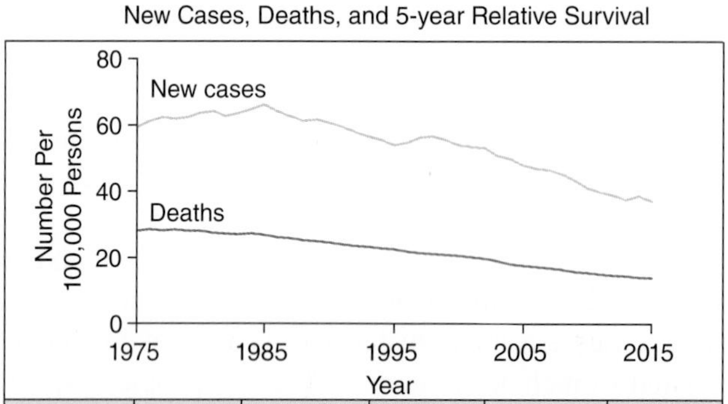

New Cases, Deaths, and 5-year Relative Survival

Year	1975	1985	1995	2005	2010
5-year survival	48.6%	58.1%	59.7%	66.2%	66.2%

SEER 9 Incidence & U.S. Mortality 1975–2015, All Races, Both sexes. Rates are Age-Adjusted

Figure 9. Graph shows 40-year trends of new cases (light gray) and deaths (dark gray) of colorectal cancer. Table shows five-year survival by year. Data from *SEER 9 Incidence & 5-year Relative Survival Percent from 1975–2010, and U.S. Mortality 1975–2015*. All races, both sexes. Rates are age-adjusted. *Source*: https://www.cancer.gov

TABLE 9. Pancreatic cancer stages: distribution and five-year relative survival.

Stage of cancer	Patients diagnosed in stage (%)	5-year survival (%)
Local	10	34.3
Regional	29	11.5
Distant	52	2.7
Un-staged	8	5.5

Local – cancer is confined to the pancreas.
Regional – cancer is spread outside of but remains close to the pancreas.
Distant – cancer is outside of and distant from the pancreas.
Un-staged – cancer stage is unknown.
* Stage distribution percentages do not sum to 100 due to rounding.

The incidence (new cases) and prevalence (all cases) of pancreatic cancer is 12.6 per 100 000 men and women per year, and 68 615 people, respectively. In 2018, the estimated total number of new cases and mortality (cancer deaths) of pancreatic cancer is 55 440 and 44 330, respectively.

The overall five-year survival rate of all pancreatic cancer cases combined is 8.5%. The distribution of cases over the disease stages and corresponding relative survival is shown in Table 9. Over the past 10 years, both the incidence and mortality of pancreatic cancer have been stable (Figure 10) [29].

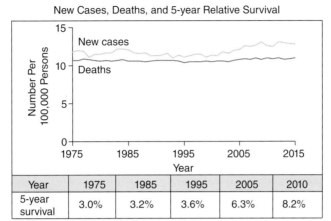

New Cases, Deaths, and 5-year Relative Survival

Year	1975	1985	1995	2005	2010
5-year survival	3.0%	3.2%	3.6%	6.3%	8.2%

SEER 9 Incidence & U.S. Mortality 1975–2015, All Races, Both sexes. Rates are Age-Adjusted

Figure 10. Graph shows 40-year trends of new cases (light gray) and deaths (dark gray) of pancreatic cancer. Table shows five-year survival by year. Data from *SEER 9 Incidence & 5-year Relative Survival Percent from 1975–2010, and U.S. Mortality 1975–2015*. All races, both sexes. Rates are age-adjusted. *Source*: https://www.cancer.gov

Liver and Intrahepatic Bile Duct Cancer

About 90% of primary liver cancers are hepatocellular carcinomas. The remaining tumor types include intrahepatic cholangiocarcinoma, hepatocellular carcinoma with fibrolamellar variant, and mixed hepatocellular carcinoma [30, 31].

Liver and intrahepatic bile duct cancer is relatively rare. However, it is the fifth leading cause of cancer death. It is more common in men and in Asian/Pacific Islander and American Indian/Alaska Native populations. Risk factors for liver and intrahepatic bile duct cancer include any chronic liver disease, such as cirrhosis, hepatitis B or C virus, alcoholic cirrhosis, fatty liver disease, and hemochromatosis. Approximately 1.0% of men and women will be diagnosed with liver and intrahepatic bile duct cancer during their lifetime; the median age at diagnosis is 64 years.

The incidence (new cases) and prevalence (all cases) of liver and intrahepatic bile duct cancer is 8.8 per 100 000 men and women per year, and 71 990 people, respectively. In 2018, the estimated total number of new cases and mortality (cancer deaths) of liver and intrahepatic bile duct cancer is 42 220 and 30 200, respectively.

The overall five-year survival rate of all liver and intrahepatic bile duct cancer cases combined is 17.7%. The distribution of cases over the disease stages and corresponding relative survival is shown in Table 10. Over the past 10 years, the incidence and mortality of liver and intrahepatic bile duct cancer has been increasing, on average 2.6 and 2.5% per year, respectively (Figure 11) [32].

TABLE 10. **Liver and intrahepatic bile duct cancer stages:** distribution and five-year relative survival.

Stage of cancer	Patients diagnosed in stage (%)	5-year survival (%)
Local	44	31.3
Regional	27	10.6
Distant	18	2.4
Un-staged	12	6.3

Local – cancer is confined to the liver/intrahepatic bile duct.

Regional – cancer is spread outside of but remains close to the liver/ intrahepatic bile duct.

Distant – cancer is outside of and distant from the liver/intrahepatic bile duct.

Un-staged – cancer stage is unknown.

* Stage distribution percentages do not sum to 100 due to rounding.

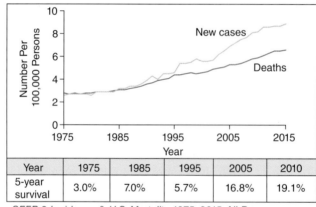

New Cases, Deaths, and 5-year Relative Survival

Year	1975	1985	1995	2005	2010
5-year survival	3.0%	7.0%	5.7%	16.8%	19.1%

SEER 9 Incidence & U.S. Mortality 1975–2015, All Races, Both sexes. Rates are Age-Adjusted

Figure 11. Graph shows 40-year trends of new cases (light gray) and deaths (dark gray) of liver and intrahepatic duct cancer. Table shows five-year survival by year. Data from *SEER 9 Incidence & 5-year Relative Survival Percent from 1975–2010, and U.S. Mortality 1975–2015*. All races, both sexes. Rates are age-adjusted. *Source:* https://www.cancer.gov

Stomach Cancer

More than 90% of primary stomach cancers are adenocarcinomas, meaning that they originate in the epithelial tissue that lines the cavity of the stomach. There are two major types of gastric adenocarcinoma, intestinal, and diffuse. Intestinal adenocarcinomas are well differentiated (more uniform and organized), whereas diffuse adenocarcinomas are poorly differentiated (disordered and poorly organized) [33].

Stomach cancer is relatively rare, and the 15th leading cause of cancer death. It is more common in men and in non-Caucasian races. Risk factors for stomach cancer include *helicobacter pylori* (a bacteria found in the stomach) gastric infection, advanced age, and a positive family history. Approximately 0.8% of men and women will be diagnosed with stomach cancer during their lifetime; the median age at diagnosis is 68 years.

The incidence (new cases) and prevalence (all cases) of stomach cancer is 7.2 per 100 000 men and women per year, and 97 915 people, respectively. In 2018, the estimated total number of new cases and mortality (cancer deaths) of stomach cancer is 26 240 and 10 800, respectively.

The overall five-year survival rate of all stomach cancer cases combined is 31.0%. The distribution of cases over the disease stages and corresponding relative survival is shown in Table 11. Over the past 10 years, the incidence and mortality of stomach cancer have been declining, on average 1.5 and 2.3% per year, respectively (Figure 12) [34].

TABLE 11. **Stomach cancer stages:** distribution and five-year relative survival.

Stage of cancer	Patients diagnosed in stage (%)	5-year survival (%)
Local	28	68.1
Regional	27	30.6
Distant	35	5.2
Un-staged	10	22.7

Local – cancer is confined to the stomach.
Regional – cancer is spread outside of but remains close to the stomach.
Distant – cancer is outside of and distant from the stomach.
Un-staged – cancer stage is unknown.

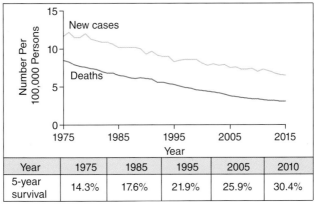

New Cases, Deaths, and 5-year Relative Survival

Year	1975	1985	1995	2005	2010
5-year survival	14.3%	17.6%	21.9%	25.9%	30.4%

SEER 9 Incidence & U.S. Mortality 1975–2015, All Races, Both sexes. Rates are Age-Adjusted

Figure 12. Graph shows 40-year trends of new cases (light gray) and deaths (dark gray) of stomach cancer. Table shows five-year survival by year. Data from *SEER 9 Incidence & 5-year Relative Survival Percent from 1975–2010, and U.S. Mortality 1975–2015.* All races, both sexes. Rates are age-adjusted. *Source:* https://www.cancer.gov

Esophagus Cancer

More than 50% of primary esophageal cancers are adenocarcinomas, meaning that they originate in the epithelial tissue that lines the esophagus. Esophageal adenocarcinomas are associated with Barrett's esophagus, a premalignant condition characterized by replacement of the epithelial lining, from one cell type to another (stratified squamous to simple columnar), located in the lower esophagus. The remaining esophageal cancers are squamous cell carcinomas, usually located in the upper esophagus [35].

Esophageal cancer is relatively rare. It is more common in men, and in advanced age. Risk factors include heavy alcohol use and tobacco use. Approximately 0.5% of men and women will be diagnosed with esophageal cancer during their lifetime; the median age at diagnosis is 68 years.

The incidence (new cases) and prevalence (all cases) of esophageal cancer is 4.0 per 100 000 men and women per year, and 47 284 people, respectively. In 2018, the estimated total number of new cases and mortality (cancer deaths) of esophageal cancer is 17 290 and 15 850, respectively.

The overall five-year survival rate of all esophagus cancer cases combined is 19.2%. The distribution of cases over the disease stages and corresponding relative survival is shown in Table 12. Over the past 10 years, the incidence and mortality of esophageal cancer have been declining, on average 1.2 and 0.9% per year, respectively (Figure 13) [36].

Small Intestine Cancer

Of the primary cancers 25–50% in the small intestine are adenocarcinomas, meaning that they originate in the epithelial tissue that lines the cavity of the small intestine. Other types of small intestine cancer include sarcoma, carcinoid tumor, gastrointestinal stromal tumor, and lymphoma. Combined these tumors account for 1–2% of all gastrointestinal cancers [37].

TABLE 12. **Esophagus cancer stages:** distribution and five-year relative survival.

Stage of cancer	Patients diagnosed in stage (%)	5-year survival (%)
Local	19	45.2
Regional	32	23.6
Distant	39	4.8
Un-staged	10	12.0

Local – cancer is confined to the esophagus.
Regional – cancer is spread outside of but remains close to the esophagus.
Distant – cancer is outside of and distant from the esophagus.
Un-staged – cancer stage is unknown.

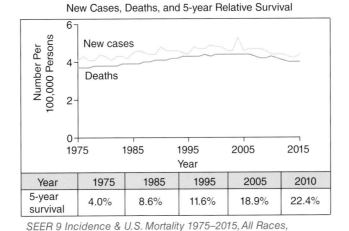

New Cases, Deaths, and 5-year Relative Survival

Year	1975	1985	1995	2005	2010
5-year survival	4.0%	8.6%	11.6%	18.9%	22.4%

SEER 9 Incidence & U.S. Mortality 1975–2015, All Races, Both sexes. Rates are Age-Adjusted

Figure 13. Graph shows 40-year trends of new cases (light gray) and deaths (dark gray) of esophageal cancer. Table shows five-year survival by year. Data from *SEER 9 Incidence & 5-year Relative Survival Percent from 1975–2010, and U.S. Mortality 1975–2015.* All races, both sexes. Rates are age-adjusted. *Source:* https://www.cancer.gov

Compared to other cancers, small intestine cancer is relatively rare. It is slightly more common among men than women. Risk factors include high fat diet and having a health history of Crohn's disease, celiac disease, or familial adenomatous polyposis (FAP). Approximately 0.3% of men and women will be diagnosed with small intestine cancer during their lifetime; the median age at diagnosis is 66 years.

The incidence (new cases) of small intestinal cancer is 2.3 per 100 000 men and women per year. In 2018, the estimated total number of new cases and mortality (cancer deaths) of small intestine cancer is 10 470 and 1450, respectively.

The overall five-year survival rate of all small intestinal cancer cases combined is 67.6%. The distribution of cases over the disease stages and corresponding relative survival is shown in Table 13. Over the past 10 years, the incidence of small intestine cancer has been increasing, on average 2.3% per year. However, in the same timeframe, the mortality of oral cavity and pharynx cancer has been stable (Figure 14) [38].

Anus Cancer

The majority of anal cancers are HPV-associated squamous cell carcinomas, which originate from in epithelial tissue that lines the anal canal. Basaloid transitional cell tumors account for most of the remainder, which are located internally and close to the rectum. Adenocarcinomas originating from anal glands and anal fistulas are rare [39].

Compared to other cancers anal cancer is rare. It is slightly more common in women. A strong risk factor for anal cancer is HPV infection. Other risk factors include smoking

TABLE 13. **Small Intestine cancer stages:** distribution and five-year relative survival.

Stage of cancer	Patients diagnosed in stage (%)	5-year survival (%)
Local	32	85.0
Regional	35	74.6
Distant	27	42.1
Un-staged	7	51.2

Local – cancer is confined to the small intestine.
Regional – cancer is spread outside of but remains close to the small intestine.
Distant – cancer is outside of and distant from the small intestine.
Un-staged – cancer stage is unknown.
* Stage distribution percentages do not sum to 100 due to rounding.

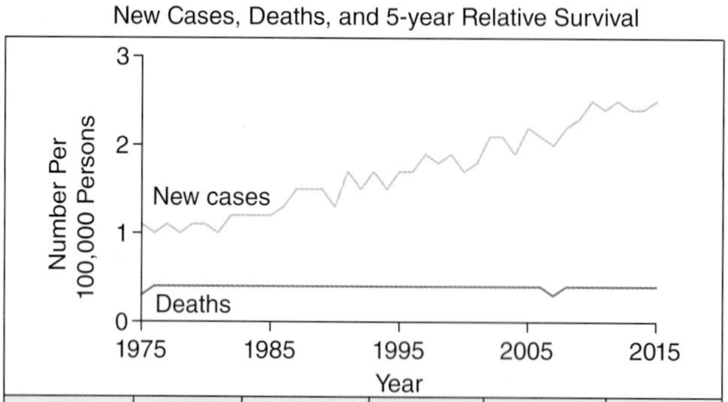

New Cases, Deaths, and 5-year Relative Survival

Year	1975	1985	1995	2005	2010
5-year survival	32.8%	39.0%	53.5%	66.6%	68.9%

SEER 9 Incidence & U.S. Mortality 1975–2015, All Races,
Both sexes. Rates are Age-Adjusted

Figure 14. Graph shows 40-year trends of new cases (light gray) and deaths (dark gray) of small intestinal cancer. Table shows five-year survival by year. Data from *SEER 9 Incidence & 5-year Relative Survival Percent from 1975–2010, and U.S. Mortality 1975–2015*. All races, both sexes. Rates are age-adjusted. *Source:* https://www.cancer.gov

and aging. Approximately 0.2% of men and women will be diagnosed with stomach cancer during their lifetime; the median age at diagnosis is 62 years.

The incidence (new cases) of anal cancer is 1.8 per 100 000 men and women per year. In 2018, the estimated total number of new cases and mortality (cancer deaths) of anus cancer is 8580 and 1160, respectively.

The overall five-year survival rate of all anal cancer cases combined is 67.4%. The distribution of cases over the disease stages and corresponding relative survival is shown in Table 14. Over the past 10 years, the incidence and mortality of anus cancer have been increasing, on average 2.2 and 2.9% per year, respectively (Figure 15) [40].

TABLE 14. **Anal cancer stages:** distribution and five-year relative survival.

Stage of cancer	Patients diagnosed in stage (%)	5-year survival (%)
Local	48	81.5
Regional	32	63.9
Distant	13	29.8
Un-staged	7	57.4

Local – cancer is confined to the anus.
Regional – cancer is spread outside of but remains close to the anus.
Distant – cancer is outside of and distant from the anus.
Un-staged – cancer stage is unknown.

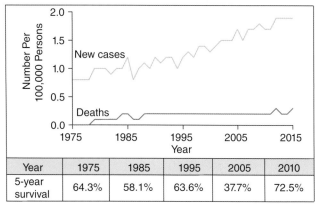

New Cases, Deaths, and 5-year Relative Survival

Year	1975	1985	1995	2005	2010
5-year survival	64.3%	58.1%	63.6%	37.7%	72.5%

SEER 9 Incidence & U.S. Mortality 1975–2015, All Races, Both sexes. Rates are Age-Adjusted

Figure 15. Graph shows 40-year trends of new cases (light gray) and deaths (dark gray) of anal cancer. Table shows five-year survival by year. Data from *SEER 9 Incidence & 5-year Relative Survival Percent from 1975–2010, and U.S. Mortality 1975–2015.* All races, both sexes. Rates are age-adjusted. *Source:* https://www.cancer.gov

Urologic Cancers

Prostate Cancer

More than 95% of primary prostate cancers are adenocarcinomas, meaning that they originate in the epithelial tissue that lines the cavity of the prostate gland. Prostatic intraepithelial neoplasia (PIN) (collections of irregular cells within the epithelial tissue of the prostate gland) is often present with adenocarcinoma and in its advanced stage may be a precursor of the cancer. The remaining 5% of primary prostate cancers consist of small-cell tumors, intralobular acinar carcinomas, ductal carcinomas, clear cell carcinomas, and mucinous carcinomas [41].

TABLE 15. **Prostate cancer stages:** distribution and five-year relative survival.

Stage of cancer	Patients diagnosed in stage (%)	5-year survival (%)
Local	78	100
Regional	12	100
Distant	5	30.0
Un-staged	4	80.9

Local – cancer is confined to the prostate.
Regional – cancer is spread outside of but remains close to the prostate.
Distant – cancer is outside of and distant from the prostate.
Un-staged – cancer stage is unknown.
* Stage distribution percentages do not sum to 100 due to rounding.

Prostate cancer is the most frequent non-dermatologic cancer in men, and is the second leading cause of cancer death in men, following lung cancer. Strong risk factors for prostate cancer include a positive family history and African American race. Approximately 11.2% of men will be diagnosed with prostate cancer during their lifetime; the median age at diagnosis is 66 years.

The incidence (new cases) and prevalence (all cases) of prostate cancer is 112.6 per 100 000 men per year, and 3 120 176 men, respectively. In 2018, the estimated total number of new cases and mortality (cancer deaths) of prostate cancer is 164 690, and 29 430, respectively.

The overall five-year survival rate of all prostate cancer cases combined is 98.2%. The distribution of cases over the disease stages and corresponding relative survival is shown in Table 15. Over the past 10 years, the incidence and mortality of prostate cancer has been declining, on average 5.7 and 2.9% per year, respectively (Figure 16) [42].

Bladder Cancer

More than 90% of primary bladder cancers are transitional cell carcinomas, meaning they originate in the transitional cells (cells that can contract and expand) in the lining of the bladder. Other primary bladder cancers include about 2–7% squamous cell carcinomas, meaning they originate from squamous cells found in the lining of the bladder, and 2% adenocarcinomas that originate from glandular (secretory) cells in the lining of the bladder [43].

Bladder cancer is the sixth most commonly diagnosed cancer, and it is the ninth leading cause of cancer death. Risk factors for bladder cancer include male gender, Caucasian race, and increasing age. Approximately 2.3% of men and women will be diagnosed with bladder cancer during their lifetime; the median age at diagnosis is 72 years.

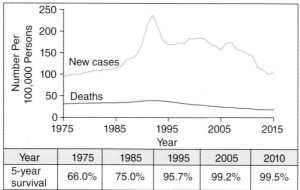

SEER 9 Incidence & U.S. Mortality 1975–2015, All Races,
Males. Rates are Age-Adjusted

Figure 16. Graph shows 40-year trends of new cases (light gray) and deaths (dark gray) of prostate cancer. Table shows five-year survival by year. Data from *SEER 9 Incidence & 5-year Relative Survival Percent from 1975–2010, and U.S. Mortality 1975–2015*. All Races, males. Rates are age-adjusted. *Source:* https://www.cancer.gov

The incidence (new cases) and prevalence (all cases) of bladder cancer, is 19.5 per 100 000 men and women per year, and 708 444 people, respectively. In 2018, the estimated total number of new cases and mortality (cancer deaths) of bladder cancer is 81 190 and 17 240 respectively.

The overall five-year survival rate of all bladder cancer cases combined is 76.8%. The distribution of cases over the disease stages and corresponding relative survival is shown in Table 16. Over the past 10 years, the incidence of bladder cancer has been decreasing, on average 1.0% per year. However, in the same timeframe, the mortality of bladder cancer has been stable (Figure 17) [44].

T A B L E 16. **Bladder cancer stages:** distribution and five-year relative survival.

Stage of cancer	Patients diagnosed in stage (%)	5-year survival (%)
In situ	51	95.4
Local	34	69.4
Regional	7	34.9
Distant	4	4.8
Un-staged	3	46.0

In situ – abnormal, non-invasive cells: pre-cancer.
Local – cancer is confined to the bladder.
Regional – cancer is spread outside of but remains close to the bladder.
Distant – cancer is outside of and distant from the bladder.
Un-staged – cancer stage is unknown.
* Stage distribution percentages do not sum to 100 due to rounding.

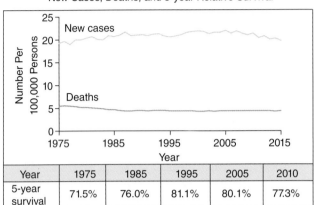

New Cases, Deaths, and 5-year Relative Survival

Year	1975	1985	1995	2005	2010
5-year survival	71.5%	76.0%	81.1%	80.1%	77.3%

SEER 9 Incidence & U.S. Mortality 1975–2015, All Races, Both sexes. Rates are Age-Adjusted

Figure 17. Graph shows 40-year trends of new cases (light gray) and deaths (dark gray) of bladder cancer. Table shows five-year survival by year. Data from *SEER 9 Incidence & 5-year Relative Survival Percent from 1975–2010, and U.S. Mortality 1975–2015*. All races, both sexes. Rates are age-adjusted. *Source*: https://www.cancer.gov

Kidney Cancer

Up to 85% of primary kidney cancers are renal cell carcinomas, or renal adenocarcinomas, which originate in the renal tubules where the blood is filtered and waste is removed – creating urine. About 7% of primary kidney cancers are transitional cell carcinomas of the renal pelvis, where the urine is collected. The remaining primary kidney cancers consist of several uncommon tumors, including the Wilms tumor, or nephroblastoma, cancers that typically occur in childhood [45, 46].

Kidney cancer is the eighth most commonly diagnosed cancer, it represents 3.8% of all new cancers diagnosed. Risk factors for kidney cancer include female gender and race; it is more common in African Americans, American Indian, and Alaska Native populations. Approximately 1.7% of men and women will be diagnosed with kidney cancer during their lifetime; the median age at diagnosis is 64 years.

The incidence (new cases) and prevalence (all cases) of renal cancer, based on the 2011–2015 SEER data, is 15.9 per 100 000 men and women per year, and 505 380 people, respectively. In 2018, the estimated total number of new cases and mortality (cancer deaths) of kidney cancer is 65 340 and 14 970, respectively.

The overall five-year survival rate of all kidney cancer cases combined is 74.5%. The distribution of cases over the disease stages and corresponding relative survival is shown in Table 17. Over the past 10 years, the incidence of kidney cancer has been increasing, on average 0.6% per year. However, in the same timeframe, the mortality of kidney cancer has been decreasing, on average 0.7% per year (Figure 18) [47].

TABLE 17. **Kidney cancer stages:** distribution and five-year relative survival.

Stage of cancer	Patients diagnosed in stage (%)	5-year survival (%)
Local	65	92.6
Regional	16	68.7
Distant	16	11.6
Un-staged	3	38.0

Local – cancer is confined to the kidney.
Regional – cancer is spread outside of but remains close to the kidney.
Distant – cancer is outside of and distant from the kidney.
Un-staged – cancer stage is unknown.

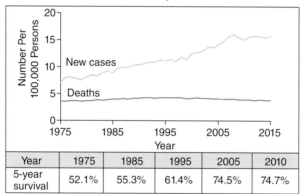

New Cases, Deaths, and 5-year Relative Survival

Year	1975	1985	1995	2005	2010
5-year survival	52.1%	55.3%	61.4%	74.5%	74.7%

SEER 9 Incidence & U.S. Mortality 1975–2015, All Races, Both sexes. Rates are Age-Adjusted

Figure 18. Graph shows 40-year trends of new cases (light gray) and deaths (dark gray) of kidney and renal pelvis cancer. Table shows five-year survival by year. Data from *SEER 9 Incidence & 5-year Relative Survival Percent from 1975–2010, and U.S. Mortality 1975–2015*. All races, both sexes. Rates are age-adjusted. *Source*: https://www.cancer.gov

Testicular Cancer

Most testicular cancers are germ cell tumors, meaning that they originate in the cells that make sperm. Germ cell tumors may be either seminomas or nonseminomas, with nonseminomas further subdivided into embryonal carcinomas, yolk sac tumors, choriocarcinomas, teratomas, and mixed germ cell tumors.

Both seminomas and nonseminomas are highly treatable; the cure rate for seminomas is >90%, while for nonseminomas it approaches 100% [48].

Testicular cancer is most common in young or middle-aged men. Testicular cancer is more common in Caucasian men, followed by American Indian/Alaska Native, and then Hispanic men. Risk factors for testicular cancer include cryptorchidism (an undescended testis) or a positive family history. Approximately 0.4% of men will be diagnosed with testicular cancer during their lifetime; the median age at diagnosis is 33 years.

The incidence (new cases) and prevalence (all cases) of testicular cancer, based on the 2011–2015 SEER data, is 5.7 per 100 000 men per year, and 257 823 men, respectively. In 2018, the estimated total number of new cases and mortality (cancer deaths) of testicular cancer is 9310 and 400 respectively.

The overall five-year survival rate of all testicular cancer cases combined is 95.3%. The distribution of cases over the disease stages and corresponding relative survival is shown in Table 18. Over the past 10 years, the incidence testicular cancer has been increasing, on average 0.8% per year. However, in the same timeframe, the mortality of testicular cancer has been stable (Figure 19) [49].

T A B L E   1 8 .   **Testicular cancer stages:** distribution and five-year relative survival.

Stage of cancer	Patients diagnosed in stage (%)	5-year survival (%)
Local	68	99.2
Regional	19	96.0
Distant	12	73.7
Un-staged	1	79.7

Local – cancer is confined to the testicle.
Regional – cancer is spread outside of but remains close to the testicle.
Distant – cancer is outside of and distant from the testicle.
Un-staged – cancer stage is unknown.

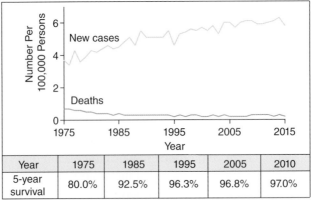

New Cases, Deaths, and 5-year Relative Survival

Year	1975	1985	1995	2005	2010
5-year survival	80.0%	92.5%	96.3%	96.8%	97.0%

SEER 9 Incidence & U.S. Mortality 1975–2015, All Races, Males. Rates are Age-Adjusted

Figure 19. Graph shows 40-year trends of new cases (light gray) and deaths (dark gray) of testicular cancer. Table shows five-year survival by year. Data from *SEER 9 Incidence & 5-year Relative Survival Percent from 1975–2010, and U.S. Mortality 1975–2015*. All Races, males. Rates are age-adjusted. *Source*: https://www.cancer.gov

Gynecologic Cancers

Endometrial Uterine Cancer

Endometrial cancer originates in the lining of the uterus. Approximately 75% of primary endometrial cancers are endometrioid cancers, which are adenocarcinomas. The remaining, less common, endometrial cancers consist of uterine papillary serous (<10%), mucinous (1%), clear cell (4%), squamous cell (<1%), and mixed (10%) [50].

Endometrial cancer is the ninth most commonly diagnosed cancer, which makes it the most common gynecologic cancer. Risk factors for endometrial cancer include obesity, hypertension, and diabetes mellitus. Approximately 2.9% of women will be diagnosed with endometrial cancer during their lifetime; the median age at diagnosis is 62 years.

The incidence (new cases) and prevalence (all cases) of endometrial cancer, based on the 2011–2015 SEER data, is 26.0 per 100000 women per year, and 727200 women, respectively. In 2018, the estimated total number of new cases and mortality (cancer deaths) of endometrial cancer is 63230 and 11350 respectively.

The overall five-year survival rate of all endometrial cancer cases combined is 81.1%. The distribution of cases over the disease stages and corresponding relative survival is shown in Table 19. Over the past 10 years, the incidence and mortality of endometrial cancer have been increasing, on average 1.3 and 1.6% per year, respectively (Figure 20) [51].

Ovarian Cancer

The most common ovarian cancers are epithelial tumors that originate in the tissue covering the ovary. Germ cell tumors, which originate in the germ (egg) cells, are less common. Ovarian low malignant potential tumors have abnormal cells that may become a precursor to ovarian cancer [52, 53].

Ovarian cancer is relatively rare, representing about 1.3% of new cancers. The main risk factor for ovarian cancer is a positive family history. Approximately

TABLE 19. **Endometrial cancer stages:** distribution and five-year relative survival.

Stage of cancer	Patients diagnosed in stage (%)	5-year survival (%)
Local	67	94.9
Regional	21	68.6
Distant	9	16.3
Un-staged	3	52.0

Local – cancer is confined to the uterus.
Regional – cancer is spread outside of but remains close to the uterus.
Distant – cancer is outside of and distant from the uterus.
Un-staged – cancer stage is unknown.

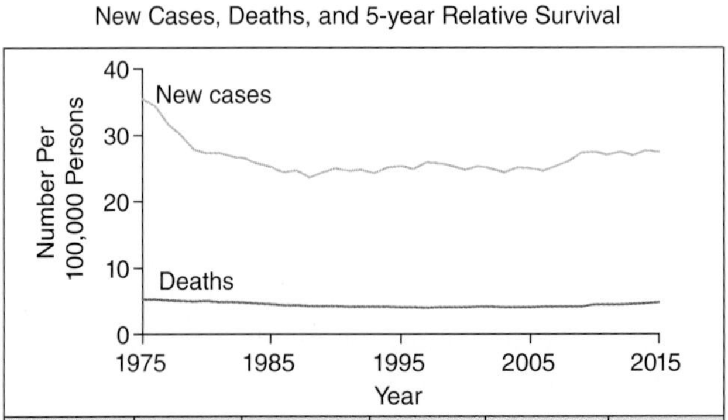

New Cases, Deaths, and 5-year Relative Survival

Year	1975	1985	1995	2005	2010
5-year survival	87.8%	82.8%	84.1%	82.9%	83.4%

SEER 9 Incidence & U.S. Mortality 1975–2015, All Races, Females. Rates are Age-Adjusted

Figure 20. Graph shows 40-year trends of new cases (light gray) and deaths (dark gray) of uterine cancer. Table shows five-year survival by year. Data from *SEER 9 Incidence & 5-year Relative Survival Percent from 1975–2010, and U.S. Mortality 1975–2015*. All Races, females. Rates are age-adjusted. *Source*: https://www.cancer.gov

1.3% of women will be diagnosed with ovarian cancer during their lifetime; the median age at diagnosis is 63 years.

The incidence (new cases) and prevalence (all cases) of ovarian cancer, based on the 2011–2015 SEER data, is 11.6 per 100 000 women per year, and 224 940 women, respectively. In 2018, the estimated total number of new cases and mortality (cancer deaths) of ovarian cancer is 22 240 and 14 070 respectively.

The overall five-year survival rate of all ovarian cancer cases combined is 47.4%. The distribution of cases over the disease stages and corresponding relative survival is shown in Table 20. Over the past 10 years, the incidence and mortality of ovarian cancer have been declining, on average 1.5 and 2.3% per year, respectively (Figure 21) [54].

TABLE 20. **Ovarian cancer stages:** distribution and five-year relative survival.

Stage of cancer	Patients diagnosed in stage (%)	5-year survival (%)
Local	15	92.3
Regional	20	74.5
Distant	59	29.2
Un-staged	6	24.8

Local – cancer is confined to the ovary.
Regional – cancer is spread outside of but remains close to the ovary.
Distant – cancer is outside of and distant from the ovary.
Un-staged – cancer stage is unknown.

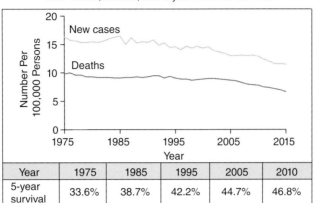

New Cases, Deaths, and 5-year Relative Survival

Year	1975	1985	1995	2005	2010
5-year survival	33.6%	38.7%	42.2%	44.7%	46.8%

SEER 9 Incidence & U.S. Mortality 1975–2015, All Races, Females. Rates are Age-Adjusted

Figure 21. Graph shows 40-year trends of new cases (light gray) and deaths (dark gray) of ovarian cancer. Table shows five-year survival by year. Data from *SEER 9 Incidence & 5-year Relative Survival Percent from 1975–2010, and U.S. Mortality 1975–2015*. All Races, females. Rates are age-adjusted. *Source*: https://www.cancer.gov

Cervical Uterus Cancer

Most cervical cancers originate at the squamous-columnar junction, the transition zone between squamous cells on the outer cervix (about 90%) and columnar, glandular cells on the inner cervix (10%). Therefore, cervical cancer can involve squamous cells, glandular cells, or both. Cervical intraepithelial neoplasia (CIN), (irregular cells in the tissue lining the cervix) or adenocarcinoma *in situ*, may be a precursor of the cancer [55].

Compared to other cancers, cervical cancer is relatively rare. Risk factors for cervical cancer include HPV infection, high parity, smoking, and long-term oral contraceptives. Approximately 0.6% of women will be diagnosed with cervical cancer during their lifetime; the median age at diagnosis is 50 years.

The incidence (new cases) and prevalence (all cases) of cervical cancer, based on the 2011–2015 SEER data, is 7.4 per 100 000 women per year, and 257 524 women, respectively. In 2018, the estimated total number of new cases and mortality (cancer deaths) of cervical cancer is 13 240 and 4170 respectively.

The overall five-year survival rate of all cervical cancer cases combined is 66.2%. The distribution of cases over the disease stages and corresponding relative survival is shown in Table 21. Over the past 10 years, the incidence of cervical cancer has been stable. However, in the same time frame, the mortality of cervical cancer has been declining, on average 0.7% per year (Figure 22) [56].

TABLE 21. **Cervical cancer stages:** distribution and five-year relative survival.

Stage of cancer	Patients diagnosed in stage (%)	5-year survival (%)
Local	45	91.7
Regional	36	56.0
Distant	15	17.2
Un-staged	4	50.0

Local – cancer is confined within the cervix.
Regional – cancer is spread outside of but remains close to the cervix.
Distant – cancer is outside of and distant from the cervix.
Un-staged – cancer stage is unknown.

New Cases, Deaths, and 5-year Relative Survival

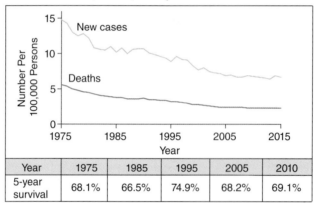

Year	1975	1985	1995	2005	2010
5-year survival	68.1%	66.5%	74.9%	68.2%	69.1%

SEER 9 Incidence & U.S. Mortality 1975–2015, All Races, Females. Rates are Age-Adjusted

Figure 22. Graph shows 40-year trends of new cases (light gray) and deaths (dark gray) of cervical cancer. Table shows five-year survival by year. Data from *SEER 9 Incidence & 5-year Relative Survival Percent from 1975–2010, and U.S. Mortality 1975–2015*. All Races, females. Rates are age-adjusted. *Source:* https://www.cancer.gov

Vulvar Cancer

About 90% of vulvar cancers are squamous cell carcinomas, meaning they origi-nate in the epithelial lining of the vagina. Vulvar cancer most often affects the outer vaginal lips. Less often, cancer affects the inner vaginal lips, clitoris, or vaginal glands. Vulvar cancer usually forms slowly over a number of years. Abnormal cells can grow on the surface of the vulvar skin for a long time. Vulvar intraepithelial neoplasia (VIN), irregular cells in the tissue lining the cervix, may be a precursor of the cancer [57].

Compared to other cancers, vulvar cancer is rare. Risk factors for vulvar cancer include VIN, HPV infection, and genital warts. Approximately 0.3% of women will be diagnosed with cervical cancer during their lifetime; the median age at diagnosis is 68 years.

The incidence (new cases) of vulvar cancer, based on the 2011–2015 SEER data, is 2.5 per 100000 women per year. In 2018, the estimated total number of new cases and mortality (cancer deaths) of cervical cancer is 6190 and 1200 respectively.

The overall five-year survival rate of all vulvar cancer cases combined is 71%. The distribution of cases over the disease stages and corresponding relative survival is shown in Table 22. Over the past 10 years, the incidence and mortality of vulvar cancer have been increasing, on average 0.6 and 1.2% per year, respectively (Figure 23) [58].

TABLE 22. **Vulvar cancer stages:** distribution and five-year relative survival.

Stage of cancer	Patients diagnosed in stage (%)	5-year survival (%)
Local	59	86.3
Regional	30	53.3
Distant	6	18.6
Un-staged	5	58.7

Local – cancer is confined within the vulva.
Regional – cancer is spread outside of but remains close to the vulva.
Distant – cancer is outside of and distant from the vulva.
Un-staged – cancer stage is unknown.

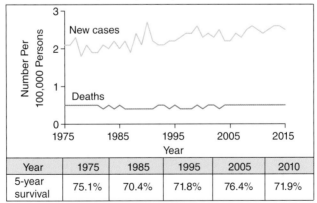

Year	1975	1985	1995	2005	2010
5-year survival	75.1%	70.4%	71.8%	76.4%	71.9%

SEER 9 Incidence & U.S. Mortality 1975–2015, All Races, Females. Rates are Age-Adjusted

Figure 23. Graph shows 40-year trends of new cases (light gray) and deaths (dark gray) of vulvar cancer. Table shows five-year survival by year. Data from *SEER 9 Incidence & 5-year Relative Survival Percent from 1975–2010, and U.S. Mortality 1975–2015*. All Races, females. Rates are age-adjusted. *Source:* https://www.cancer.gov

Hematologic Cancers

Leukemia

Leukemia begins in the blood-forming tissue, such as the bone marrow, and causes large numbers of abnormal blood cells to be produced. The abnormal blood cells have an extended lifespan, and may overwhelm the normal blood cells and prevent them from fulfilling their functions, such as transporting oxygen to tissues, controlling bleeding and fighting infections. The four main types of leukemia are acute lymphoblastic leukemia (ALL), acute myeloid leukemia (AML), chronic lymphocytic leukemia (CLL), and chronic myeloid leukemia (CML) [59].

Leukemia is the 10th most commonly diagnosed cancer. It is a common childhood cancer, but it most often occurs in older adults and is slightly more common in men. Risk factors for leukemia include previous cancer treatment, certain genetic disorders (such as Down syndrome), some blood disorders, and exposure to chemicals such as benzene. Approximately 1.5% of men and women will be diagnosed with leukemia during their lifetime; the median age at diagnosis is 66 years.

The incidence (new cases) and prevalence (all cases) of leukemia, is 13.8 per 100 000 men and women per year, and 405 815 people, respectively. In 2018, the estimated total number of new cases and mortality (cancer deaths) of leukemia is 60 300 and 24 370, respectively.

The five-year overall survival of leukemia is 61.4%. There is no standard staging system for leukemia. Leukemia is described as untreated, in remission, or recurrent. Over the past 10 years, the incidence of leukemia has been increasing, on average 0.3% per year. However, in the same interval, the mortality of leukemia has been declining, on average 1.5% per year (Figure 24) [60].

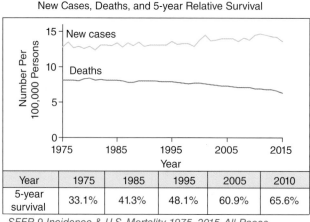

New Cases, Deaths, and 5-year Relative Survival

Year	1975	1985	1995	2005	2010
5-year survival	33.1%	41.3%	48.1%	60.9%	65.6%

SEER 9 Incidence & U.S. Mortality 1975–2015, All Races, Both sexes. Rates are Age-Adjusted

Figure 24. Graph shows 40-year trends of new cases (light gray) and deaths (dark gray) of leukemia. Table shows five-year survival by year. Data from *SEER 9 Incidence & 5-year Relative Survival Percent from 1975–2010, and U.S. Mortality 1975–2015*. All races, both sexes. Rates are age-adjusted. *Source:* https://www.cancer.gov

Non-Hodgkin Lymphoma

Non-Hodgkin lymphoma is classified into two prognostic groups (groups with distinct, likely outcomes): the indolent (slowly progressing) lymphomas and the aggressive lymphomas. Indolent lymphomas include follicular lymphoma (the most common), lymphoplasmacytic lymphoma, marginal zone lymphoma, splenic marginal zone lymphoma, and primary cutaneous anaplastic large cell lymphoma. Aggressive lymphomas include diffuse large B-cell lymphoma (the most common), mediastinal large B-cell lymphoma, follicular large cell lymphoma, and anaplastic large cell lymphoma, among many others [61].

Non-Hodgkin lymphoma is the seventh most commonly diagnosed cancer, and is the eighth leading cause of cancer death. It is more common in men, advanced age, and among the Caucasian race. Risk factors for non-Hodgkin lymphoma include inherited immune disorders, autoimmune disease, Epstein–Barr virus, and taking immunosuppressant drugs after an organ transplant. Approximately 2.1% of men and women will be diagnosed with non-Hodgkin lymphoma during their lifetime; the median age at diagnosis is 67 years.

The incidence (new cases) and prevalence (all cases) of non-Hodgkin lymphoma, is 19.4 per 100 000 men and women per year, and 686 042 people, respectively. In 2018, the estimated total number of new cases and mortality (cancer deaths) of non-Hodgkin lymphoma is 74 680 and 19 910, respectively.

The overall five-year survival rate of all non-Hodgkin lymphoma cases combined is 71.4%. The distribution of cases over the disease stages and corresponding relative survival is shown in Table 23. Over the past 10 years, the incidence and mortality of non-Hodgkin lymphoma have been declining, on average 0.7 and 2.2% per year, respectively (Figure 25) [62].

TABLE 23. **Non-Hodgkin lymphoma stages:** distribution and five-year relative survival.

Stage of cancer	Patients diagnosed in stage (%)	5-year survival (%)
Stage I	25	81.8
Stage II	14	75.3
Stage III	16	69.1
Stage IV	34	61.7
Un-staged	11	76.4

Stage I – cancer only in originating layer of cells.
Stage II – cancer is confined to the primary site.
Stage III – cancer is spread outside of but remains close to primary site / regional lymph nodes.
Stage IV – cancer is outside of and distant from the primary site.
Un-staged – cancer stage is unknown.

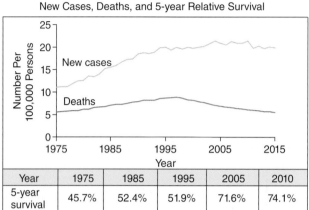

New Cases, Deaths, and 5-year Relative Survival

Year	1975	1985	1995	2005	2010
5-year survival	45.7%	52.4%	51.9%	71.6%	74.1%

SEER 9 Incidence & U.S. Mortality 1975–2015, All Races, Both sexes. Rates are Age-Adjusted

Figure 25. Graph shows 40-year trends of new cases (light gray) and deaths (dark gray) of non-Hodgkin lymphoma. Table shows five-year survival by year. Data from *SEER 9 Incidence & 5-year Relative Survival Percent from 1975–2010, and U.S. Mortality 1975–2015*. All races, both sexes. Rates are age-adjusted. *Source*: https://www.cancer.gov

Myeloma

Myeloma, also known as multiple myeloma, begins in plasma cells (white blood cells that produce antibodies); when a plasma cell becomes abnormal, it is a myeloma cell. Eventually myeloma cells collect in the bone marrow of several bones, which is called multiple myeloma [63].

Myeloma is relatively rare, although it is more common in men and in the African American race. Significant risk factors for myeloma include prior exposure to radiation or certain chemicals, and a history of monoclonal gammopathy of undetermined significance (MGUS). Approximately 0.8% of men and women will be diagnosed with myeloma during their lifetime; the median age at diagnosis is 69 years.

The incidence (new cases) and prevalence (all cases) of myeloma, is 6.7 per 100 000 men and women per year, and 124 733 people, respectively. In 2018, the estimated total number of new cases and mortality (cancer deaths) of myeloma is 30 770 and 12 770, respectively.

The overall five-year survival rate of all myeloma cases combined is 50.7%. The distribution of cases over the disease stages and corresponding relative survival is shown in Table 24. Over the past 10 years, the incidence of myeloma has been rising, on average 0.9% per year. However, in the same interval, the mortality of myeloma has been declining, on average, 0.5% per year (Figure 26) [64].

Hodgkin Lymphoma

Hodgkin lymphoma is a cancer of the immune system that begins in the lymph system. Hodgkin lymphomas (HL) have two categories: the most common is

TABLE 24. **Myeloma stages:** distribution and five-year relative survival.

Stage of cancer	Patients diagnosed in stage (%)	5-year survival (%)
Local	5	72.0
Distant	95	49.6

Local – cancer is confined to the primary site.
Distant – cancer is outside of and distant from the primary site.

New Cases, Deaths, and 5-year Relative Survival

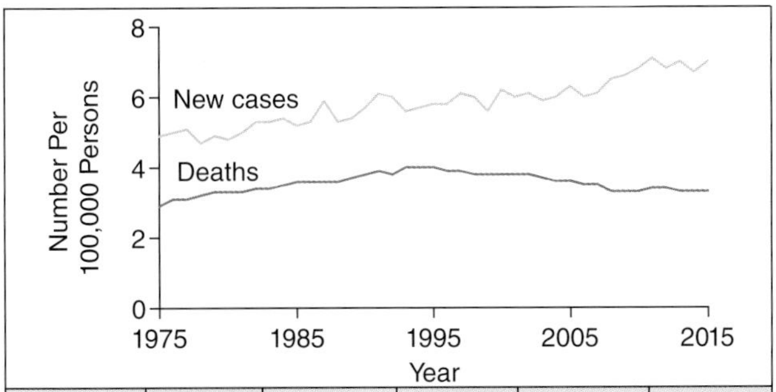

Year	1975	1985	1995	2005	2010
5-year survival	26.3%	27.1%	33.3%	46.6%	53.0%

SEER 9 Incidence & U.S. Mortality 1975–2015, All Races, Both sexes. Rates are Age-Adjusted

Figure 26. Graph shows 40-year trends of new cases (light gray) and deaths (dark gray) of myeloma. Table shows five-year survival by year. Data from *SEER 9 Incidence & 5-year Relative Survival Percent from 1975–2010, and U.S. Mortality 1975–2015*. All races, both sexes. Rates are age-adjusted. *Source*: https://www.cancer.gov

classical Hodgkin lymphoma, which is subdivided based on the structure of the characteristic Reed-Sternberg cells (giant cells found in HL), into four subtypes: nodular sclerosis HL, mixed-cellularity HL, lymphocyte depletion HL, and lymphocyte-rich classical HL. The second category is the nodular lymphocyte-predominate HL [65].

Hodgkin lymphoma is rare. It is more common in men and in young adults. Risk factors for HL include a positive family history with a first-degree relative and a history of Epstein–Barr virus infection. Approximately 0.2% of men and women will be diagnosed with Hodgkin lymphoma during their lifetime; the median age at diagnosis is 39 years.

The incidence (new cases) and prevalence (all cases) of Hodgkin lymphoma, is 2.5 per 100000 men and women per year, and 208805 people, respectively. In 2018, the estimated total number of new cases and mortality (cancer deaths) of Hodgkin lymphoma is 8500 and 1050, respectively.

The overall five-year survival rate of all Hodgkin lymphoma cases combined is 86.6%. The distribution of cases over the disease stages and corresponding relative survival is shown in Table 25. Over the past 10 years, the incidence and mortality of Hodgkin lymphoma have been declining, on average 1.8 and 2.8% per year, respectively (Figure 27) [66].

TABLE 25. **Hodgkin lymphoma stages:** distribution and five-year relative survival.

Stage of cancer	Patients diagnosed in stage (%)	5-year survival (%)
Stage I	15	92.3
Stage II	40	93.4
Stage III	21	83.0
Stage IV	20	72.9
Un-staged	4	82.7

Stage I – cancer only in originating layer of cells.
Stage II – cancer is confined to the primary site.
Stage III – cancer is spread outside of but remains close to primary site / regional lymph nodes.
Stage IV – cancer is outside of and distant from the primary site.
Un-staged – cancer stage is unknown.

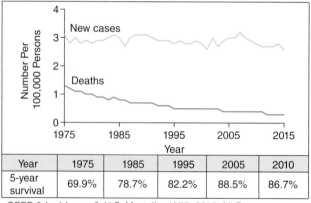

New Cases, Deaths, and 5-year Relative Survival

Year	1975	1985	1995	2005	2010
5-year survival	69.9%	78.7%	82.2%	88.5%	86.7%

SEER 9 Incidence & U.S. Mortality 1975–2015, All Races, Both sexes. Rates are Age-Adjusted

Figure 27. Graph shows 40-year trends of new cases (light gray) and deaths (dark gray) of Hodgkin lymphoma cancer. Table shows five-year survival by year. Data from *SEER 9 Incidence & 5-year Relative Survival Percent from 1975–2010, and U.S. Mortality 1975–2015*. All races, both sexes. Rates are age-adjusted. *Source:* https://www.cancer.gov

Pediatric Oncology

The most common cancers in children (0–14 years) include leukemia (31%), brain and other CNS cancer (26%), neuroblastoma (6%), and non-Hodgkin lymphoma (5%). In adolescents (15–19 years), the most common cancers include brain and other CNS cancer (22%), leukemia (13%), thyroid cancer (11%), Hodgkin lymphoma (11%), and gonadal germ cell tumors (10%) [67].

Cancer is the second leading cause of death in children aged 5–14 years, following accidents. White children have the highest incidence (new cases) of childhood cancers; African American children have the lowest cancer incidence. American Indian/Alaska Native children have the lowest cancer mortality (deaths) of all racial/ethnic groups. Unlike many adult cancers, childhood and adolescent cancer incidence is not consistently higher among populations with lower socioeconomic status. Risk factors for childhood cancer include ionizing radiation, both high and low birth weights, and parental smoking [68].

The incidence (new cases) of childhood cancer, based on the 1975–2015 SEER data, is 177.2 per 1 000 000 per year. The prevalence of childhood cancer in children and adolescents (0–20 years), in 2015, was 118 794. The total number of people (of all ages) alive in 2015 that have been diagnosed with childhood cancer is 359 890. The top three cancers among childhood cancer survivors are ALL, brain and other CNS cancers, and lymphoma [67].

Childhood cancers are most often treated with chemotherapy and sometimes radiation, and both therapies can have long-term effects. Children should have follow-up care to assess for short- and long-term complications, late effects of therapy, and to detect recurrent and secondary cancers. For example, children who received chest irradiation have an increased risk for lung cancer, which will increase 20-fold if there is also a history of smoking [69].

REFERENCES

1. SEER Training Modules: Cancer as a Disease. National Cancer Institute, Bethesda, MD (2018). Available from: http://training.seer.cancer.gov/disease/categories (accessed 05/28/2018).
2. Cancer Types: Metastatic Cancer. National Cancer Institute, Bethesda, MD (2018). Available from: https://www.cancer.gov/types/metastatic-cancer (accessed 05/29/2018).
3. Cancer Types: Breast Cancer Treatment (PDQ®)–Health Professional Version. National Cancer Institute, Bethesda, MD (2018). Available from: https://www.cancer.gov/types/breast/hp/breast-treatment-pdq (accessed 05/28/2016).
4. Cancer Types: Male Breast Cancer Treatment (PDQ®)–Health Professional Version. National Cancer Institute, Bethesda, MD (2018). Available from: https://www.cancer.gov/types/breast/hp/male-breast-treatment-pdq (accessed 05/29/2018).
5. SEER Stat Fact Sheets: Breast Cancer. National Cancer Institute, Bethesda, MD (2018). Available from: http://seer.cancer.gov/statfacts/html/breast.html (accessed 05/28/2018).
6. Cancer Types: Skin Cancer Treatment (PDQ®)–Health Professional Version. National Cancer Institute, Bethesda, MD (2018). Available from: https://www.cancer.gov/types/skin/hp/skin-treatment-pdq (accessed 05/29/2018).

7. Cancer Types: Melanoma Treatment (PDQ®)–Health Professional Version. National Cancer Institute, Bethesda, MD (2018). Available from: https://www.cancer.gov/types/skin/hp/melanoma-treatment-pdq (accessed 05/29/2018).

8. SEER Stat Fact Sheets: Melanoma of the Skin. National Cancer Institute, Bethesda, MD (2018). Available from: http://seer.cancer.gov/statfacts/html/melan.html (accessed 11/26/2018)

9. Cancer Types: Adult Central Nervous System Tumors Treatment (PDQ®)–Health Professional Version. National Cancer Institute, Bethesda, MD (2018). Available from: https://www.cancer.gov/types/brain/hp/adult-brain-treatment-pdq (accessed 05/29/2018).

10. Cancer Types: Childhood Brain and Spinal Cord Tumors Treatment Overview (PDQ®)–Health Professional Version. National Cancer Institute, Bethesda, MD (2018). Available from: https://www.cancer.gov/types/brain/hp/child-brain-treatment-pdq (accessed 05/29/2018).

11. SEER Stat Fact Sheets: Brain and Other Nervous System Cancer. National Cancer Institute, Bethesda, MD (2018). Available from: http://seer.cancer.gov/statfacts/html/brain.html (accessed 05/29/2018).

12. Cancer Types: Thyroid Cancer Treatment (Adult) (PDQ®)–Health Professional Version. National Cancer Institute, Bethesda, MD (2018). Available from: https://www.cancer.gov/types/thyroid/hp/thyroid-treatment-pdq (accessed 05/29/2018).

13. SEER Stat Fact Sheets: Thyroid Cancer. National Cancer Institute, Bethesda, MD (2018). Available from: http://seer.cancer.gov/statfacts/html/thyro.html (accessed 05/29/2018).

14. Cancer Types: Osteosarcoma and Malignant Fibrous Histiocytoma of Bone Treatment (PDQ®)–Health Professional Version. National Cancer Institute, Bethesda, MD (2018). Available from: https://www.cancer.gov/types/bone/hp/osteosarcoma-treatment-pdq (accessed 05/30/2018).

15. Cancer Types: Ewing Sarcoma Treatment (PDQ®)–Health Professional Version. National Cancer Institute, Bethesda, MD (2018). Available from: https://www.cancer.gov/types/bone/hp/ewing-treatment-pdq (accessed 05/30/2018).

16. SEER Stat Fact Sheets: Bone and Joint Cancer. National Cancer Institute, Bethesda, MD (2018). Available from: http://seer.cancer.gov/statfacts/html/bones.html (accessed 05/30/2018).

17. Cancer Types: Non-Small Cell Lung Cancer Treatment (PDQ®)–Health Professional Version. National Cancer Institute, Bethesda, MD (2018). Available from: https://www.cancer.gov/types/lung/hp/non-small-cell-lung-treatment-pdq (accessed 05/29/2018).

18. Cancer Types: Small Cell Lung Cancer Treatment (PDQ®)–Health Professional Version. National Cancer Institute, Bethesda, MD (2018). Available from: https://www.cancer.gov/types/lung/hp/small-cell-lung-treatment-pdq (accessed 05/29/2018).

19. SEER Stat Fact Sheets: Lung and Bronchus Cancer. National Cancer Institute, (2018). Available from: http://seer.cancer.gov/statfacts/html/lungb.html (accessed 05/29/2018).

20. Cancer Types: Lip and Oral Cavity Cancer Treatment (Adult) (PDQ®)–Health Professional Version. National Cancer Institute, Bethesda, MD (2018). Available from: https://www.cancer.gov/types/head-and-neck/hp/adult/lip-mouth-treatment-pdq (accessed 05/30/2018).

21. Cancer Types: Oropharyngeal Cancer Treatment (Adult) (PDQ®)–Health Professional Version. National Cancer Institute, Bethesda, MD (2018). Available from: https://www.cancer.gov/types/head-and-neck/hp/adult/oropharyngeal-treatment-pdq (accessed 05/30/2018).

22. SEER Stat Fact Sheets: Oral Cavity and Pharynx Cancer. National Cancer Institute, Bethesda, MD (2018). Available from: http://seer.cancer.gov/statfacts/html/oralcav.html (accessed 05/30/2018).

23. Cancer Types: Laryngeal Cancer Treatment (Adult) (PDQ®)–Health Professional Version. National Cancer Institute, Bethesda, MD (2018). Available from: https://www.cancer.gov/types/head-and-neck/hp/adult/laryngeal-treatment-pdq (accessed 05/30/2018).

24. SEER Stat Fact Sheets: Laryngeal Cancer. National Cancer Institute, Bethesda, MD (2018). Available from: http://seer.cancer.gov/statfacts/html/laryn.html (accessed 05/30/2018).

25. Cancer Types: Colon Cancer Treatment (PDQ®)–Health Professional Version. National Cancer Institute, Bethesda, MD (2018). Available from: https://www.cancer.gov/types/colorectal/hp/colon-treatment-pdq (accessed 05/31/2018).

26. Cancer Types: Rectal Cancer Treatment (PDQ®)–Health Professional Version. National Cancer Institute, Bethesda, MD (2018). Available from: https://www.cancer.gov/types/colorectal/hp/rectal-treatment-pdq (accessed 05/31/2018).

27. SEER Stat Fact Sheets: Colon and Rectum Cancer. National Cancer Institute, Bethesda, MD (2018). Available from: http://seer.cancer.gov/statfacts/html/stomach.html (accessed 05/31/2018).

28. Cancer Types: Pancreatic Cancer Treatment (PDQ®)–Health Professional Version. National Cancer Institute, Bethesda, MD (2018). Available from: https://www.cancer.gov/types/pancreatic/hp/pancreatic-treatment-pdq (accessed 05/31/2018).

29. SEER Stat Fact Sheets: Pancreatic Cancer. National Cancer Institute, Bethesda, MD (2018). Available from: https://seer.cancer.gov/statfacts/html/pancreas.html (accessed 05/31/2015).

30. Cancer Types: Adult Primary Liver Cancer Treatment (PDQ®)–Health Professional Version. National Cancer Institute, Bethesda, MD (2018). Available from: https://www.cancer.gov/types/liver/hp/adult-liver-treatment-pdq (accessed 05/31/2018).

31. Cancer Types: Bile Duct Cancer (Cholangiocarcinoma) Treatment (PDQ®)–Health Professional Version. National Cancer Institute, Bethesda, MD (2018). Available from: https://www.cancer.gov/types/liver/hp/bile-duct-treatment-pdq (accessed 05/31/2018).

32. SEER Stat Fact Sheets: Liver and Intrahepatic Bile Duct Cancer. National Cancer Institute, Bethesda, MD (2018). Available from: https://seer.cancer.gov/statfacts/html/livibd.html (accessed 05/31/2018).

33. Cancer Types: Gastric Cancer Treatment (PDQ®)–Health Professional Version. National Cancer Institute, Bethesda, MD (2018). Available from: https://www.cancer.gov/types/stomach/hp/stomach-treatment-pdq (accessed 05/31/2018).

34. SEER Stat Fact Sheets: Stomach Cancer. National Cancer Institute, Bethesda, MD (2018). Available from: https://seer.cancer.gov/statfacts/html/stomach.html (accessed 05/31/2018).

35. Cancer Types: Esophageal Cancer Treatment (PDQ®)–Health Professional Version. National Cancer Institute, Bethesda, MD (2018). Available from: https://www.cancer.gov/types/esophageal/hp/esophageal-treatment-pdq (accessed 05/31/2018).

36. SEER Stat Fact Sheets: Esophagus Cancer. National Cancer Institute, Bethesda, MD (2018). Available from: https://seer.cancer.gov/statfacts/html/esoph.html (accessed 05/31/2018).

37. Cancer Types: Small Intestine Cancer Treatment (PDQ®)–Health Professional Version. National Cancer Institute, Bethesda, MD (2018). Available from: https://www.cancer.gov/types/small-intestine/hp/small-intestine-treatment-pdq (accessed 05/31/2018).

38. SEER Stat Fact Sheets: Small Intestine Cancer. National Cancer Institute, Bethesda, MD (2018). Available from: https://seer.cancer.gov/statfacts/html/smint.html (accessed 06/01/2018).

39. Cancer Types: Anal Cancer Treatment (PDQ®)–Health Professional Version. National Cancer Institute, Bethesda, MD (2018). Available from: https://www.cancer.gov/types/anal/hp/anal-treatment-pdq (accessed 06/01/2018).

40. SEER Stat fact Sheets: Anal Cancer. National Cancer Institute, Bethesda, MD (2018). Available from: https://seer.cancer.gov/statfacts/html/anus.html (accessed 05/31/2018).

41. Cancer Types: Prostate Cancer Treatment (PDQ®)–Health Professional Version. National Cancer Institute, Bethesda, MD (2018). Available from: https://www.cancer.gov/types/prostate/hp/prostate-treatment-pdq (accessed 06/01/2018).

42. SEER Stat Fact Sheets: Prostate Cancer. National Cancer Institute, Bethesda, MD (2018). Available from: https://seer.cancer.gov/statfacts/html/prost.html (accessed 06/01/2018).

43. Cancer Types: Bladder Cancer Treatment (PDQ®)–Health Professional Version. National Cancer Institute, Bethesda, MD (2018). Available from: https://www.cancer.gov/types/bladder/hp/bladder-treatment-pdq (accessed 06/01/2018).

44. Cancer Types: Bladder Cancer. National Cancer Institute, Bethesda, MD (2018). Available from: https://seer.cancer.gov/statfacts/html/urinb.html (accessed 06/01/2018).

45. Cancer Types: Renal Cell Cancer Treatment (PDQ®)–Health Professional Version. National Cancer Institute, Bethesda, MD (2018). Available from: https://www.cancer.gov/types/kidney/hp/kidney-treatment-pdq (accessed 06/04/2018).

46. Cancer Types: Transitional Cell Cancer of the Renal Pelvis and Ureter Treatment (PDQ®)–Health Professional Version. National Cancer Institute, Bethesda, MD (2018). Available from: https://www.cancer.gov/types/kidney/hp/transitional-cell-treatment-pdq (accessed 06/04/2018).

47. SEER Stat Fact Sheets: Kidney and Renal Pelvis Cancer. National Cancer Institute, Bethesda, MD (2018). Available from: https://seer.cancer.gov/statfacts/html/kidrp.html (accessed 06/04/2018).

48. Cancer Types: Testicular Cancer Treatment (PDQ®)–Health Professional Version. National Cancer Institute, Bethesda, MD (2018). Available from: https://www.cancer.gov/types/testicular/hp/testicular-treatment-pdq (accessed 06/04/2018).

49. SEER Stat Fact Sheets: Testicular Cancer. National Cancer Institute, Bethesda, MD (2018). Available from: https://seer.cancer.gov/statfacts/html/testis.html (accessed 06/04/2018).

50. Cancer Types: Endometrial Cancer Treatment (PDQ®)–Health Professional Version. National Cancer Institute, Bethesda, MD (2018). Available from: https://www.cancer.gov/types/uterine/hp/endometrial-treatment-pdq (accessed 06/04/2018).

51. SEER Stat Fact Sheets: Uterine Cancer. National Cancer Institute, Bethesda, MD (2018). Available from: https://seer.cancer.gov/statfacts/html/corp.html (accessed 06/04/2018).

52. Cancer Types: Ovarian Epithelial, Fallopian Tube, and Primary Peritoneal Cancer Treatment (PDQ®)–Health Professional Version. National Cancer Institute, Bethesda, MD (2018). Available from: https://www.cancer.gov/types/ovarian/hp/ovarian-epithelial-treatment-pdq (accessed 06/05/2018).

53. Cancer Types: Ovarian Low Malignant Potential Tumors Treatment (PDQ®)–Health Professional Version. National Cancer Institute, Bethesda, MD (2018). Available from: https://www.cancer.gov/types/ovarian/hp/ovarian-low-malignant-treatment-pdq (accessed 06/05/2018).

54. SEER Stat Fact Sheets: Ovarian Cancer. National Cancer Institute, Bethesda, MD (2018). Available from: https://seer.cancer.gov/statfacts/html/ovary.html (accessed 06/05/2018).

55. Cancer Types: Cervical Cancer Treatment (PDQ®)–Health Professional Version. National Cancer Institute, Bethesda, MD (2018). Available from: https://www.cancer.gov/types/cervical/hp/cervical-treatment-pdq (accessed 06/05/2018).

56. SEER Stat Fact Sheets: Cervical Cancer. National Cancer Institute, Bethesda, MD (2018). Available from: https://seer.cancer.gov/statfacts/html/cervix.html (accessed 06/05/2018).

57. Cancer Types: Vulvar Cancer Treatment (PDQ®)–Health Professional Version. National Cancer Institute, Bethesda, MD (2018). Available from: https://www.cancer.gov/types/vulvar/hp/vulvar-treatment-pdq (accessed 06/05/2018).

58. SEER Stat Fact Sheets: Vulvar Cancer. National Cancer Institute, Bethesda, MD (2018). Available from: https://seer.cancer.gov/statfacts/html/vulva.html (accessed 06/05/2018).

59. Cancer Types: Leukemia Treatment (PDQ®)–Health Professional Version. National Cancer Institute, Bethesda, MD (2018). Available from: https://www.cancer.gov/types/leukemia/hp (accessed 06/05/2018).

60. SEER Stat Fact Sheets: Leukemia. National Cancer Institute, Bethesda, MD (2018). Available from: https://seer.cancer.gov/statfacts/html/leuks.html (accessed 06/05/2018).

61. Cancer Types: Adult Non-Hodgkin Lymphoma Treatment (PDQ®)–Health Professional Version. National Cancer Institute, Bethesda, MD (2018). Available from: https://www.cancer.gov/types/lymphoma/hp/adult-nhl-treatment-pdq (accessed 06/05/2018).

62. SEER Stat Fact Sheets: Non-Hodgkin Lymphoma. National Cancer Institute, Bethesda, MD (2018). Available from: https://seer.cancer.gov/statfacts/html/nhl.html (accessed 06/05/2018).

63. Cancer Types: Plasma Cell Neoplasms (Including Multiple Myeloma) Treatment (PDQ®)–Health Professional Version. National Cancer Institute, Bethesda, MD (2018). Available from: https://www.cancer.gov/types/myeloma/hp/myeloma-treatment-pdq (accessed 06/05/2018).

64. SEER Stat Fact Sheets: Myeloma. National Cancer Institute, Bethesda, MD (2018). Available from: https://seer.cancer.gov/statfacts/html/mulmy.html (accessed 06/05/2018).

65. Cancer Types: Adult Hodgkin Lymphoma Treatment (PDQ®)–Health Professional Version. National Cancer Institute, Bethesda, MD (2018). Available from: https://www.cancer.gov/types/lymphoma/hp/adult-hodgkin-treatment-pdq (accessed 06/05/2018).

66. SEER Stat Fact Sheets: Hodgkin Lymphoma. National Cancer Institute, Bethesda, MD (2018). Available from: https://seer.cancer.gov/statfacts/html/hodg.html (accessed 06/05/2018).

67. A.M. Noone, N. Howlader, M. Krapcho (eds), et al. SEER Cancer Statistics Review, 1975–2015: Section 29: Childhood Cancer by the ICCC. National Cancer Institute, Bethesda, MD (based on November 2017 SEER data submission, posted to the SEER web site, April 2018). Available from: https://seer.cancer.gov/csr/1975_2015/results_merged/sect_29_childhood_cancer_iccc.pdf (accessed 06/13/2018).

68. Cancer Types: Cancer in Children and Adolescents. National Cancer Institute, Bethesda, MD (2017). Available from: https://www.cancer.gov/types/childhood-cancers/child-adolescent-cancers-fact-sheet (accessed 06/13/2018).

69. Cancer Types: Childhood Cancers. National Cancer Institute, Bethesda, MD (2017). Available from: https://www.cancer.gov/types/childhood-cancers (accessed 06/13/2018).

10

CANCER EPIDEMIOLOGY: IDENTIFYING CANCER RISK

Bernard F. Cole[1], Brian L. Sprague[2], Thomas P. Ahern[2],
and Judy R. Rees[3,4]

[1] *Department of Mathematics and Statistics, College of Engineering and Mathematical Sciences, University of Vermont, Burlington, VT, USA*
[2] *Department of Surgery, Larner College of Medicine, University of Vermont, Burlington, VT, USA*
[3] *Department of Epidemiology, Geisel School of Medicine, Dartmouth College, Hanover, NH, USA*
[4] *Norris Cotton Cancer Center, Lebanon, NH, USA*

INTRODUCTION

This chapter introduces basic cancer epidemiology and related areas of clinical research. Our goal is to provide the lay reader with a foundational understanding of research concepts that will help make the cancer-research literature, as well as other chapters in this book, more accessible.

The field of cancer epidemiology encompasses many facets of scientific exploration, but the most important one is the ascertainment of factors that are associated with cancer risk. There are many examples. Smoking and asbestos exposure have been linked to lung cancer. Certain diets have been associated with colorectal cancer. Family history has been linked to colorectal cancer and breast cancer. Specific genes have been identified as predictors of breast cancer risk.

Cancer: Prevention, Early Detection, Treatment and Recovery, Second Edition. Edited by Gary S. Stein and Kimberly P. Luebbers.
© 2019 John Wiley & Sons, Inc. Published 2019 by John Wiley & Sons, Inc.

By ascertaining risk factors for disease, medical practitioners and public health officials are able to better control the disease by promoting lifestyles that avoid known risk factors, or by identifying people who have an elevated risk of developing cancer, making it more likely that the disease can be detected earlier and thus more-successfully treated.

Notice right away that we use the words "are associated with" rather than the stronger term "cause." The reason for this is that proving a causal relationship is often very difficult or impossible in an observational study (as opposed to an interventional study). Smoking provides an excellent example. We may say that smoking causes lung cancer because we have decades of data from a variety of sources that all point to a causal relationship. However, this was not always so clear, and for many years smoking was considered safe. Even after seeing increases in cancers among smokers, many still doubted that smoking was the cause. In epidemiology, we are often confronted with this problem. In particular, we might see an association but not necessarily a cause for any increased risk. Nevertheless, understanding risk factors that are associated with cancer is an important step in developing strategies for prevention. In this chapter, we will introduce basic epidemiological methods that cancer researchers utilize to identify risk factors and promote prevention and control of the disease.

A related area, also covered in this chapter, is clinical trials research, where investigators evaluate new treatments for cancer in an interventional study. Specifically, we discuss the randomized clinical trial, which is widely considered to be the best tool available for determining the effectiveness of experimental treatments.

PRIMARY AND SECONDARY PREVENTION

Prevention of cancer is a national priority in the United States. There are two overarching strategies that are employed. These are primary and secondary prevention. Primary prevention is focused on minimizing cancer risk among people who have never had the disease. Examples of interventions along these lines include educational initiatives to reduce cancer risk (e.g., weight reduction, smoking cessation, exercise), and decreasing or eliminating exposure to hazardous materials that have been linked to cancer. The goal of secondary prevention is to lessen the risk of cancer recurrence or progression in people with cancer or a history of cancer. Cancer screening is a key strategy for secondary prevention. The goal of screening is to detect cancers earlier in their development, when they can be more effectively treated. Mammography and colonoscopy are two major examples of screening procedures. Another type of secondary prevention involves using drugs that have been shown to reduce risk. For example, low-dose aspirin has been shown to lower colorectal cancer risk in people with a history of colon polyps.

CONFOUNDING AND BIAS

Confounding and bias are the two most important sources of error in epidemiologic studies. The presence of either one can render a study uninterpretable. Cancer epidemiologists must go to great lengths to avoid these errors in order to produce meaningful data. Bias is a systematic error that can occur in many ways. The two most major, potential sources of bias are in the selection of study participants ("selection bias") and in the way that data are collected from those participants ("measurement bias"). For example, suppose we are interested in studying "vaping" (the use of electronic cigarettes), specifically what proportion of the population are regular users and whether usage differs between men and women. To do this study, we visit some college campuses and randomly ask people to answer a few questions about how often they vape. This study will be biased because the population being sampled will be younger, better educated, and have a higher standard of living on average than the general population. As a result, we would not be able to use the data from this experiment to make any statements about the vaping habits of the general population, including whether vaping rates differ for males as compared to females. However, if we had intended to study the same question in college students, the study described above would not have suffered from selection bias.

The way in which data are collected can also result in bias. A simple example of measurement bias is when a piece of equipment is not properly calibrated, such as when a scale consistently under-reports a person's weight. In this example, the rate of obesity in the study population will be underestimated, as will the average body mass index. Measurement bias is also common when the data collection relies on human judgment or memory, such as when a patient is asked about prior risky behaviors, or when a nurse assesses a patient's mobility, or when a pathologist looks at a tissue sample and renders an opinion. In each of these examples, there may be external factors that influence the responses. A patient might underreport drug and alcohol use to avoid the doctor's disapproval. Mobility might be over- or underestimated depending on how healthy the patient otherwise looks. Any knowledge the pathologist might have about the patient's history, including any medications or prior illnesses, might affect her opinion about the tissue sample.

It is difficult to completely eliminate bias; however, with careful scientific technique it can be minimized to an acceptable level. Selection bias is avoided by carefully defining the study population of interest and seeking participants in such a way that the characteristics of the participants mimic those of the population as closely as possible. Selection bias can be completely avoided by randomly selecting participants from the general population of interest; however, such an approach is usually not practical and may even be impossible. For one thing, we cannot force people to participate! The next best thing is to recruit participants from sources that are considered to be generally representative of the population. Selection bias can also be minimized by recruiting participants from multiple and diverse settings.

Measurement bias can also be avoided with careful scientific technique that is applied uniformly. All equipment used to make measurements must be regularly calibrated. Protocols for collecting opinion-based data ("rating") need to be developed and tested for accuracy, including comparisons across different raters. Where possible, "blinding" is a highly effective tool for minimizing bias. In a blinded study, those investigators who are providing opinion-based assessments are blinded to some or all of the participant's characteristics. For example, the pathologist can be asked to render an opinion without knowing the identity of the patient from whom the tissue sample was obtained.

Confounding is another source of error in epidemiologic studies and occurs when an apparent association between a risk factor and disease status is actually due to some other risk factor. Confounding can also cause the masking of a risk factor, making it harder to detect. Age is a common confounder. Older age is probably the number one risk factor for most diseases, and therefore, before we can say that some other risk factor is associated with disease, we first need to be sure that age is not a confounder. Let us consider a simple hypothetical example. Suppose we are interested in the association between climate and melanoma risk. We randomly sample people from eastern Massachusetts and southern Florida and compare the prevalence of (i.e., percent of people with) a history of melanoma. We might see that the prevalence in southern Florida is much higher than that in eastern Massachusetts. Is this due to climate? Perhaps some of the effect is due to climate; however, we also note that the population in eastern Massachusetts is probably somewhat younger on average when one considers that the region has a large number of college campuses, and southern Florida is a popular retirement destination. In this hypothetical example, the true effect of climate is confounded by age. The actual difference in risk between the two climates is probably less than this study would suggest. Fortunately, it is relatively straightforward to control confounding due to age. The investigators could match participants according to age, or if the extent of the confounding is not too severe, could use statistical approaches to estimate the climate effect after adjusting for any age differences.

While age is almost always an important potential confounder in epidemiologic studies, there are many other ways that confounding can result in error. Sophisticated statistical approaches are often employed to correct for confounding; however, these only work for confounders that have been measured. There will always be the potential that some unknown confounder might explain an apparent association.

STATISTICAL ANALYSIS

Statistics plays a critically important role in epidemiology, and it is important to understand a few of the most important statistical concepts in order to interpret the findings of any study. We will concern ourselves in this chapter with the general concepts rather than the details of any specific analytical method. In a nutshell, the statistical method is concerned with differentiating a true effect or association from random variability. Consider a simple example, where we flip a coin 10 times. If the coin were fair, we would expect to see five heads and five tails. Of course, by random chance, we could easily see some other outcome. Suppose we see eight heads and two tails. Is this evidence that the coin is more likely to come up heads rather than tails, or was the result merely random variability? Similar questions arise in biomedical research, where we might see apparently higher risk of disease in a subgroup of people, or we might observe that a particular medicine appears to be more effective than others at treating a disease. Statistical analysis provides a structured way to address these questions. The material below summarizes the key aspects of statistical analysis, which we illustrate later in the chapter in a number of case studies.

There are two basic modes of statistical analysis: estimation and testing. Estimation consists of using data to best approximate an unknown characteristic of a population. A simple example is the result of a political poll. The characteristic of interest is the percentage of voters who plan to support a particular candidate. In cancer research, we are interested in characteristics such as the percent of women in the United States living with breast cancer (or any other oncologic disease), or how much a new treatment prolongs life among patients diagnosed with lung cancer.

All estimates are subject to error due to random variability, much the way that a totally fair coin can come up heads 8 times in 10 tosses. Random variability is often addressed by presenting the estimate along with a range of plausible values reflecting the potential random error. In the biomedical literature, the range is called a "confidence interval." For example, a 95% confidence interval is expected to capture the true population characteristic 95% of the time. In the biomedical literature, statistics are often presented with 95% confidence intervals. The accuracy of an estimate is determined by several factors. The two most important of these are the amount of variation in the data (i.e., how spread out the data are) and the sample size, with larger sample sizes resulting in narrower confidence intervals and thus more overall accuracy.

There are a great many different kinds of statistical estimates related to risk of disease – far too many to be covered in this chapter; however, we will mention some of the more commonly seen ones. A very common summary measure is the relative risk. This is a ratio describing the risk in an exposed group relative to the risk in an unexposed group. If the ratio is greater than one, it indicates that risk is higher in the exposed group. If the ratio is less than one, it indicates that risk is lower in the exposed group (i.e., the exposure is "protective"). There are several forms of relative risk, referred to by terms such as risk ratio, odds ratio, hazard ratio, and rate ratio, but the idea is the same. Relative risks are often shown with 95% confidence intervals to provide some assessment of the accuracy of the estimate. Another common summary in cancer research is actually shown in a graph called a "survival curve," which illustrates the percent of study participants free of an event (e.g., disease relapse/progression or death) over the time since study enrollment. The curves all start at 100% at time zero and gradually decrease over time as events are observed. By comparing the survival curves of two groups, an investigator can evaluate differences in risk.

In statistical testing, we are concerned with using available data to answer a specific yes-or-no question about the general population. For example, we might be interested to know whether risk of cancer differs for two groups of individuals. Investigators will first use statistical analysis to estimate the risk in each group. In a testing scenario, they will then use these estimates to form a conclusion, either that the risk differs between the two groups or that there is no evidence of an association. There are two possible errors that can be made. The investigators might conclude that there is a difference in risk when, in fact, there is no association at all. This is called a false-positive result or a type-I error. Alternatively, the investigators might fail to detect a difference in risk when one actually exists. This is called a false-negative result or a type-II error.

While neither of these errors can be eliminated completely, they can be controlled. By convention, the type-I error rate is fixed to occur with some low probability, usually 5 or 1%. This can be done by specifying an appropriate criterion for concluding that an association exists. For example, in the coin-flipping example, we might conclude that the coin is unfair only if we see less than 2, or more than 8 heads from 10 tosses. If the coin were, in fact, fair, the probability of seeing less than two or more than eight heads, and thus incorrectly concluding that it is unfair, would be about 2% – a very unlikely outcome. A similar approach is used in epidemiology and clinical research to control the rate at which false-positive conclusions are made. The type-II error rate is typically controlled by the sample size. A larger sample size will decrease the false-negative error rate, all else being equal. Thus, in many investigations, the sample size is selected to be large enough that the type-II error rate is reduced to some acceptable level, usually below 20%.

Often, the result of a statistical test is summarized with something called a p-value, which might be displayed in tables, text, or figures. The p-value, always a number between 0 and 1, can be thought of as a measure of how consistent the sample data are with the assumption that no association exists in the general population. If the p-value is small, say less than 0.05 or 0.01, the investigators will conclude with "statistical significance" that an association exists (in the general population). Formally, the p-value is defined as the probability of obtaining the observed data, or more extreme data, under the assumption that there is no association. This formal definition can be difficult to wrap one's head around; however, for our purposes, it is fine to think of a p-value simply as a measure of consistency between the data and the assumption of no association.

At this point, it is helpful to make an analogy between statistical testing and a jury trial. The jury needs to decide on a verdict, and there are two kinds of errors that can be made. The jury can find the defendant guilty when the defendant is actually innocent, or the jury can find the defendant not guilty when the defendant is, in fact, guilty. These errors are respectively analogous to type-I and type-II errors in statistical testing. In the US judicial system, the criterion for finding a defendant guilty is to find "guilt beyond a reasonable doubt," which is analogous to setting the type-I error rate at a specific level (e.g., 5%). The second error in a jury trial is controlled by putting resources into criminal investigations, enabling the police and prosecutors to develop the evidence needed for a conviction. These resources are analogous to the sample size in statistical testing. If we strengthen the criterion for conviction, say to "guilt beyond a shadow of a doubt," there would be more cases where a guilty defendant goes free. On the other hand, if the criterion were weakened, say to "more likely guilty than not," there would be more cases where an innocent defendant is convicted. This tradeoff in errors is similar to that in the statistical testing paradigm.

COHORT AND CASE-CONTROL STUDIES

There are two primary types of studies that epidemiologists use to evaluate potential risk factors for cancer, the cohort study and the case-control study. A cohort study is one in which the investigator gathers a sample of individuals from a particular population. If the study is for primary prevention, the individuals will have no prior history of the disease (e.g., cancer). Otherwise, if the study is for secondary prevention, all of the participants will have had cancer at some prior point in their lives. The investigator will then follow the study participants, often for many years or decades, and watch for the occurrence of new cancers. The main advantage of a cohort study is that the investigator can look at how baseline factors, as well as changes in these factors over time, are associated with subsequent cancer risk. For example, in a study of colorectal cancer risk, one can measure factors such as age, body mass index, diet, smoking behavior, and alcohol consumption at baseline as well as over the course of follow-up. Then, when cancers occur, the investigator can analyze whether any of these risk factors, or changes in them, is associated with the occurrence of cancer. Such data are very useful for identifying people with an elevated cancer risk as well as developing preventative strategies or interventions. Because the risk factors are measured at baseline, a cohort study can inform us about how altering these factors might lessen risk. As a result, the cohort study is something of a gold standard in epidemiology. The primary disadvantage of cohort studies is that they usually require many participants (sometimes in the tens of thousands) and many years to complete. Moreover, it is nearly impossible to conduct a cohort study for a rare disease because, even with thousands of participants, there will be too few cancers observed to provide meaningful data on risk association.

A case-control study attempts to ascertain risk factors "after the fact" so to speak. The investigator will assemble two groups of study participants. The first group will consist of people who have been diagnosed with the disease being studied. These are the "cases." The second group – called the "controls" – will consist of people who have no history of the disease. Often, the control group is obtained by randomly sampling people from the general population in such a way that the controls are matched to the cases on certain factors, usually age and gender. Once the investigator has assembled the two groups, she will measure certain exposures and characteristics (e.g., history of smoking, alcohol consumption, hair color, chemical exposure, genetic traits) in the two groups and look for any differences. If a certain exposure causes the disease (or conversely is preventative) the investigator should see a difference in the exposure levels between the cases and controls in the study.

There are two key advantages of case-control studies. The first is that they can be completed far more quickly than a cohort study because participants do not need to be followed over time. The second is that rare diseases can be efficiently studied in the case-control setting. However, these studies have several important limitations as well. Measurement of exposures can be difficult and subject to "recall bias." This happens when people are asked about their prior behaviors, such as alcohol consumption and diet. They might not accurately recall their behaviors, especially if they are being asked about their exposures over a long period of time. Even very risky behaviors such as cigarette smoking typically result in cancer only after lengthy exposure. Thus, inaccuracy in the recall of prior exposures and behaviors can make it difficult for a case-control study to correctly identify risk factors. If not done carefully, the selection of the control group can also cause bias in case-control studies. For example, seeking participants through motor vehicle registrations, or randomly dialing telephone numbers, or through the internet will likely fail to obtain a representative control group because not all cases might drive cars, have a dedicated telephone, or use computers. Bias in the control group selection will likely result in false risk factors being detected.

CLINICAL TRIALS

A clinical trial is a study involving human participants who are given an experimental treatment or a control treatment. We can think of clinical trials as a very special kind of cohort study – one where the investigator actually determines the exposure, in this case administration of a drug. Participants, often patients with a specific disease, are treated and followed over time to assess clinical effectiveness and safety of the new therapy.

Clinical trials are conducted under very rigorous and highly regulated conditions. Studies of new drugs cannot begin until several conditions are met. There must be sufficient preclinical (i.e., laboratory and animal) evidence demonstrating potential for use in humans. In the United States, the investigator must apply for and be granted approval for an investigational new drug by the US Food and Drug Administration (FDA). All studies involving humans are regularly monitored for safety by an institutional review board (IRB), which consists of independent experts who have no stake in the outcome of the trial. These boards are called "institutional" because they typically consist of experts employed by the investigator's home institution. All serious and unexpected adverse events must be reported to the IRB and FDA in real time. Larger clinical trials often involve an external review board that provides a further level of protection. These boards consist of experts who are external to the investigator's home institution. Most studies undergo annual or biannual review by these boards as well as provide real-time reporting of serious, unexpected adverse events. Finally, all participants in any clinical trial must provide written informed consent to be included in the study. The consent document must clearly outline, in easy-to-understand language, what is involved in trial participation and what the possible risks and benefits of participation are. Only after all of these protections are approved and in place may a clinical trial commence.

A clinical trial might be stopped early for several different reasons related to safety, or effectiveness, or concerns regarding study progress. For example, if too many adverse events are observed, the study will be stopped or suspended until additional protections are in put place. These might involve more-frequent monitoring of patients and dose reductions. If early indicators of effectiveness suggest that the drug is not as effective as thought, the trial might be stopped early. On the other hand, if the early indications are clear that the drug is highly effective, the trial might be stopped early so that other patients can benefit from the treatment. Finally, studies are sometimes stopped early for other reasons such as poor accrual of participants or lack of compliance with treatment protocols.

There are many different kinds of clinical trials. In this chapter, we introduce the most common types called phase 1, phase 2 and phase 3, which are typically used to test new drug therapies.

The goal of a phase 1 study is to determine an appropriate dosage and to assess side effects. These studies are usually small in size, often involving fewer than 18 participants. In a typical phase I study, the investigator will gradually increase the dosage, in groups of three participants, until side effects become too severe. The maximum tolerated dose is then taken to be the largest dose where the side effects are tolerable. Once an appropriate dosage is established, a phase 2 study can be conducted. This is the first primary evaluation of the drug's effectiveness in patients with a specific disease. Phase 2 trials are designed to be completed in a relatively short period of time. They often involve moderate numbers of patients, usually at least 24 and less than 50, but sometimes more. The goal of a phase 2 trial is to ascertain the drug's activity in terms of reducing disease burden. In cancer studies involving solid tumors, investigators will evaluate the drug's ability to shrink the tumor by at least 50% (called a "partial response") or cause it to disappear entirely (called a "complete response"). A successful phase 2 trial will demonstrate a response rate sufficiently high to justify further testing in patients, and might lead to a phase 3 trial.

Phase 3 clinical trials are the most important tool available for determining the safety and effectiveness of a new treatment. A phase 3 trial is comparative in the sense that it compares a new, experimental treatment with an existing approach, called the control (which might be a currently available drug, or a placebo, or supportive care). Patients are randomly assigned to receive either the new treatment or the control. Randomization of participants is very important because it is the only way to ensure that the only difference between the two groups is the treatment received. In other words, randomization ensures that no other patient factors likely to influence outcome (e.g., age or disease burden) are associated with treatment assignment. These trials typically involve many more patients than phase 2 studies, often hundreds, and in some cases, thousands of patients. The primary goal of a phase 3 trial is to evaluate the effectiveness of the new treatment in terms of an important clinical outcome. In cancer clinical trials, phase 3 trials are often designed to compare treatments in terms of survival time or the time until disease relapse or progression. Of course, adverse events are also rigorously monitored in phase 3 trials.

Because the number of patients needed to run a phase 3 trial is so large, investigators often need to recruit study participants from several different medical institutions. The key advantage of this approach is that it improves the ability of the study to recruit a diverse sample of patients, one which better reflects the characteristics of the general patient population.

A phase 3 trial with a successful outcome, demonstrating improvement in clinical outcome with a new treatment, has the potential to change clinical practice at the national, even international, level. In the United States, the FDA requires that such a study be conducted before it will grant approval for the drug to be used to treat a specific condition and for the drug's sponsor (e.g., a pharmaceutical company) to market the treatment. The FDA's review process is also very rigorous and requires that the sponsor provide all data regarding the new drug to the agency. FDA scientists will reevaluate these data and form their own conclusions regarding the safety and effectiveness of the new drug. In cases where concerns remain about granting approval, the FDA will sometimes employ an external advisory committee to review the information and advise the agency on approval. For new cancer treatments, the Oncologic Drugs Advisory Committee (ODAC) fills this role. All committee meetings are open to the public, and all committee members undergo a comprehensive review of potential conflicts of interest. Information about ODAC is freely available from the FDA's website.

A randomized phase 3 trial is generally considered the "gold standard" for evaluating new treatments. A rigorously designed and executed phase 3 trial will have minimal bias when estimating the treatment effect. If the sample size is large (as is typical for such studies) the chance of any confounding factors is negligible owing to the randomization of participants. Nevertheless, clinical trials do have certain limitations. From a practical perspective, clinical trials can take a very long time to complete, often several years, and require significant resources, sometimes in the tens of millions of dollars. The most important scientific limitation is the ability of the trial results to be generalized to the overall population of patients with the disease being studied. Selection bias can still be a problem in a randomized clinical trial. Patients who enroll in clinical trials might have quite different characteristics from the general population. A well-run study will mitigate this limitation by recruiting patients from a diversity of settings so that the study sample will be as representative of the general population as possible.

A second important scientific limitation relates to the ascertainment of patient outcomes, especially those that rely on clinical opinion, such as when diagnosing a disease progression. Some clinical trials attempt to minimize bias along these lines by using placebo controls and blinding participants, and providers to the treatment assignment. In other words, both the study participants and their health-care providers do not know whether the patient is taking the active drug or an identical-looking substance that lacks the active ingredient (such a substance is called a "placebo"). Such studies are called "double-blind, placebo-controlled." The main advantage of these studies is that any assessment of how well a patient is doing will not be affected by knowledge of what treatment the patient is getting. However, it is often not possible to conduct a blinded study, especially if the treatment is known to have specific side effects, or administration of a placebo would be too invasive to be ethical (e.g., intravenous administration). In trials where blinding and placebo controls are not possible, and where ascertainment of the primary clinical endpoint involves investigator opinion, there is always significant concern that investigator bias (however unintentional) might affect study results. In cancer clinical trials, this concern can be addressed by several means. A common approach is to employ an independent expert panel to review clinical documentation for each participant and provide a second opinion. Despite limitations, the randomized controlled clinical trial is still considered by most experts to be the best way to evaluate a new therapy.

CASE STUDY: CLINICAL TRIAL OF ASPIRIN TO PREVENT COLORECTAL ADENOMAS

Colorectal cancer is a significant public-health issue in developed nations. In the United States, the number new diagnoses of colorectal cancer in 2017 is expected to be over 135 000, while the number of people who die from the disease is expected to be over 50 000. Most colorectal cancers develop slowly over time and begin as a polyp on the inner lining of the colon or rectum. The most common type of polyp leading to cancer is the adenomatous polyp. As a result, there has been interest in finding ways to prevent the occurrence of these lesions in the hopes of reducing the incidence of colorectal cancer. Because these lesions can usually be visualized and removed during a colonoscopy, regular colonoscopic screening is one approach. There has also been interest in whether they can be prevented with drugs or dietary supplements. The Aspirin/Folate Polyp Prevention Study was a double-blind, placebo-controlled, randomized clinical trial of aspirin, folic acid, or both to prevent colorectal adenomas in people having a history of adenomas [1]. In this chapter, we will confine ourselves to the aspirin component of the study. Aspirin was studied in two doses, 81 and 325 mg/day. Study participants were randomly assigned to receive aspirin in dosages of 81, 325 mg/day or identical-looking placebo tablets for up to three years, at which time participants underwent colonoscopic follow-up. The study randomized 1121 patients with a recent history adenomas to receive placebo (372 patients), 81 mg of aspirin (377 patients), or 325 mg of aspirin (372 patients) daily. The study showed that low-dose aspirin is effective for preventing colorectal adenomas. The incidence of one or more adenomas (after three years of follow-up) was 47% in the placebo group, 38% in the group given 81 mg of aspirin per day, and 45% in the group given 325 mg of aspirin per day. The null hypothesis for the study was that the risk of adenoma incidence after three years of treatment was the same for all three treatment groups. This null hypothesis was rejected with a p-value of 0.04. Confidence intervals were then used to compare each aspirin group versus the placebo group. The relative risks of any adenoma occurrence (as compared with the placebo group) were 0.81 in the 81-mg group (95% confidence interval, 0.69–0.96) and 0.96 in the 325-mg group (95% confidence interval, 0.81–1.13). Based on these results, the investigators concluded that low-dose aspirin had a moderate beneficial effect for preventing adenomas in the large bowel.

CASE STUDY: CANCER SURVIVORSHIP AND RACIAL DISPARITIES

An important area of cancer research is survivorship, which looks at how patients do over time following their diagnosis. Survivorship research has many facets and addresses a wide array of health-related dimensions. These include physical and psychological symptoms, quality of life, economic impact of cancer, health behaviors, and identification of special populations. For example, patients who were treated with chemotherapy sometimes complain of "chemo brain," which is a decrease in cognitive function. Patients usually describe it as being "in a fog" or not having as good a memory as before treatment. Researchers began studying cognitive ability in cancer survivors in comparison to similarly aged control subjects. A number of studies concluded that chemotherapy was associated with measurable cognitive deficits, leading to the development of supportive care options designed to help patients who suffer from chemo brain.

Survivorship research includes the identification of special populations of survivors who are at higher risk of poor clinical outcomes (e.g., disease progression or early death). In particular, researchers have been evaluating potential racial disparities among cancer survivors in the United States, usually comparing outcomes for African American patients to white patients. Hypothesized causes of racial disparities include social and cultural factors as well as economic factors, accessibility of high-quality healthcare and biological factors. A major data source for such studies is the Surveillance, Epidemiology, and End Results (SEER) program of the US National Cancer Institute. The SEER data set is a comprehensive collection of information obtained from all cancer patients in 18 representative regions of the United States. The program was begun in the early 1970s, and the latest SEER report contains critical information about cancer trends between 1975 and 2014. Patients in the SEER database were followed for survival, and researchers have utilized this information to ascertain special populations of cancer survivors who are at higher risk of early death. For example, one study found that African American survivors of ovarian cancer are at elevated risk of early death as compared to white survivors [2]. In this study, researchers utilized data from ovarian cancer patients diagnosed between 2001 and 2009 in 37 states across the United States. They obtained these data from SEER and other state cancer registries. A total of 172 848 ovarian cancers were diagnosed in this time period, with about 8% of them were diagnosed in African American women. The investigators found that African American women had consistently worse survival compared to white women despite having similar disease severity. Among the African American women, the estimated five-year survival rate was about 31%, whereas for white women it was 42%. These estimates were adjusted for age and differing background mortality risks; as a result, the disparity is not explained by differing patient age or higher background mortality rate. Studies such as this are used to inform public health policy, often leading to new initiatives to ensure that all patients receive appropriate, evidence-based treatment. They might also lead to new research into better methods for detecting cancer in its early stages, when it can be more effectively treated.

CASE STUDY: DIESEL MOTOR EXHAUST EXPOSURE AND LUNG CANCER RISK

Diesel motor exhaust is a common air pollutant, especially in urban areas and in European countries where a higher percentage of vehicles are diesel powered. Concerns about exposure to diesel exhaust have grown significantly in both Europe and the United States following the Volkswagen emissions scandal, where certain so-called "clean diesel" automobiles were equipped with a defeat device that enabled the vehicles to pass emission standards during regulatory testing but to emit up to 40 times the allowable amount of nitrogen oxides during real-world driving (nitrogen oxides are major contributors to smog and acid rain) [3]. Diesel exhaust has been classified as a carcinogen by the International Agency for Research on Cancer (IARC) based on studies that showed an association between exposure and increased risk of lung cancer; however, results were not consistent between these studies. To evaluate this issue further, researchers conducted a case-control study in Swedish men [4]. They enrolled 993 lung cancer cases and 2359 controls who were matched to the lung-cancer cases by age and year of enrollment in the study. Ascertaining exposure to diesel exhaust is not directly possible in a retrospective study such as this. Instead, the investigators used a surrogate marker for exposure based on occupation. Based on information from other studies, they were able to estimate each man's exposure intensity based on his occupation, even accounting for job changes and whether the man worked part time or full time. Since lung cancer is known to be associated with smoking history, asbestos exposure, radon exposure, and exposure to air pollution (other than diesel exhaust), the investigators also collected data on these exposures. They then utilized a mathematical model to assess the association between occupational exposure to diesel exhaust and lung cancer risk after adjusting for the other exposures known to be linked to lung cancer. Overall, the investigators found that men who were occupationally exposed to any diesel exhaust had 15% higher odds of having lung cancer as indicated by an odds ratio of 1.15. However, this result was not statistically significant. The 95% confidence interval for the odds ratio was from 0.94–1.41, which includes "no association" (i.e., an odds ratio of 1.0). The investigators did find that men in the highest quartile of exposure duration (≥ 34 years) were at significantly higher risk as compared to men who were never occupationally exposed to diesel exhaust. The odds ratio for this comparison was 1.66 with a 95% confidence interval from 1.08 to 2.56 (p = 0.03), which excludes "no association" (i.e., an odds ratio of 1). They concluded that they "found no convincing association between exposure intensity and lung cancer risk" but that long-term exposure to diesel exhaust was associated with an increased risk of lung cancer.

CONCLUSION

In this chapter, we provided a brief summary of the major types of epidemiologic studies used to evaluate cancer risk factors, and we summarized the advantages and limitations of each study design. A basic understanding of statistical methods is necessary for interpreting the results of any study, so we included an introductory discussion of the statistical-testing paradigm that is common in biomedical research. Finally, we illustrated the concepts with three case studies that illustrate the variety of study designs covered. While this chapter will not turn the reader into a practicing epidemiologist or statistician, we hope that this brief introduction is helpful for broadly understanding and interpreting the scientific literature related to cancer epidemiology.

REFERENCES

1. Baron, J.A., Cole, B.F., Sandler, R.S. et al. (2003). A randomized trial of aspirin to prevent colorectal adenomas. *N. Engl. J. Med.* 348 (10): 891–899.
2. Stewart, S.L., Harewood, R., Matz, M. et al. (2017). Disparities in ovarian cancer survival in the United States (2001–2009): findings from the CONCORD-2 study. *Cancer* 123 (Suppl 24): 5138–5159.
3. Barrett, S.R.H., Speth, R.L., Eastham, S.D. et al. (2015). Impact of the Volkswagen emissions control defeat device on US public health. *Environ. Res. Lett.* 10 (11): 114005/1–114005/10.
4. Ilar, A., Plato, N., Lewne, M. et al. (2017). Occupational exposure to diesel motor exhaust and risk of lung cancer by histological subtype: a population-based case-control study in Swedish men. *Eur. J. Epidemiol.* 32 (8): 711–719.

VI

TUMOR TYPES

11

BREAST CANCER: UNIQUE CHARACTERISTICS AS TARGETS FOR THERAPY

Ogheneruona Apoe, Ted James, Gabriella Szalayova, Amrita Pandit, and Marie E. Wood

University of Vermont Medical Center, Burlington, VT, USA

OVERVIEW

Breast cancer is the most common type of cancer found in women, and the second leading cause of cancer deaths. One in eight women will develop breast cancer in her lifetime, and it is estimated that there will be 266 120 new cases of breast cancer in 2018, with 40 920 deaths [1]. While breast cancer is more prevalent in women, it also occurs in men, although much less commonly. Male breast cancer accounts for less than 1% of all cases of breast cancer, and it is estimated that there will be 2550 new cases in 2018 [1].

The incidence of breast cancer is higher in Caucasian women than in African American women, however, the mortality rate is higher in African Americans. Some of this is attributed to the higher incidence of triple negative breast cancer cases in African American women, although socioeconomic factors also play a role. Breast cancer incidence is significantly lower in Native American women compared to other ethnicities. The highest rate of breast cancer in the world is noted in women living in North America.

Cancer: Prevention, Early Detection, Treatment and Recovery, Second Edition. Edited by Gary S. Stein and Kimberly P. Luebbers.
© 2019 John Wiley & Sons, Inc. Published 2019 by John Wiley & Sons, Inc.

RISK FACTORS

There are several well-known risk factors for the development of breast cancer including, age, menstrual and pregnancy history, genetic mutations, abnormalities in cells, breast density, obesity, and radiation therapy to the chest area. However, up to 75% of women have no identifiable risk factors [2].

Age: Breast cancer risk increases with age, and although it can occur in women of a younger age, it is relatively uncommon under the age of 40. Most breast cancers occur in women who are aged 50 years or older.

Menstrual and Pregnancy History: Early age at menarche (<12 years old), late age at first pregnancy (>30 years old), never having been pregnant, and late age at menopause (>55 years old) are all associated with an increased risk of breast cancer [3]. Having a period before the age of 12 confers a 20% higher risk of developing breast cancer compared to those who have their first period after age 14.

Genetic Mutations: The most well-known genes associated with risk for breast cancer are BRCA1 and BRCA2. These genes account for about 10% of all breast cancers and 20–25% of hereditary breast cancers [4] and are associated with a high risk of cancer development. The likelihood of carrying a BRCA mutation is increased by the presence of the following features: early onset breast cancer, male breast cancer, bilateral breast cancer, breast and ovarian cancer in the same woman or family, and Ashkenazi Jewish heritage [2, 5].

The BRCA1 gene mutation is associated with a lifetime breast cancer risk of 55–65%, and also carries an increased lifetime risk of ovarian cancer (39%) [4]. The BRCA2 gene mutation is associated with a lifetime breast cancer risk of 45%, and an increased risk of ovarian cancer (11–17%) [4]. Other genes have been identified and are associated with a moderate or high risk of breast cancer [6]. Additional highly penetrant genes associated with high breast cancer risk include: TP53 (associated with Li-Fraumeni syndrome), PTEN (associated with Cowden syndrome), STK-11 (associated with Peutz-Jeghers syndrome), CDH1 (associated with hereditary diffuse gastric cancer), and PALB2. Moderately penetrant genes (associated with lower but still elevated risk) include ATM, CHEK2, and others.

Cellular Abnormalities: The presence of atypical or abnormal cells on breast biopsy is associated with an increased risk of invasive of breast cancer. These conditions include ductal hyperplasia (slightly increased risk, 1.5–2 times higher), atypical ductal/ lobular hyperplasia (moderately increased, 4–5 times higher risk), and lobular carcinoma *in situ* (higher risk, 8–11 times increased).

Breast Density: There is an association between breast density and breast cancer, with the risk of breast cancer increasing with the amount of breast density. Women who have a breast density of 75% or greater are four to six times more likely to develop breast cancer when compared to women with a breast density of 10% or less [7]. Recent legislation has been enacted by several states around mandatory reporting of breast density. This has caused much confusion and anxiety for both women and their providers. Further information regarding breast density and the notification laws can be found at the following website (http://www.breastdensity.info).

Obesity: Obesity appears to be a risk factor for the development of several cancers including breast cancer. There is a modestly increased risk for breast cancer associated with obesity especially for postmenopausal breast cancer. Fat tissue produces an excess amount of estrogen, and this may influence the growth of breast cancer cells.

Radiation Therapy: Mantle field radiation (extended field radiation therapy to the neck, chest, and underarm areas used to treat Hodgkin lymphoma) before the age of 30, but especially before the age of 20, is associated with an increased risk of breast cancer. This risk is about 20–30 times that of the general population. The younger a woman was when she received radiation to the chest area, the higher her risk of developing breast cancer throughout her lifetime. Smaller doses of radiation in the form of chest x-rays and mammograms do not carry this extent of risk.

Hormone Replacement Therapy: Hormone replacement therapy (HRT) has been associated with an increased risk for developing breast cancer. This is true for the combination of estrogen and progesterone therapy but not for estrogen alone (which is used in women who have had a hysterectomy) [8, 9]. The risk begins to increase after four years of use and decreases after the therapy is stopped.

RISK ASSESSMENT

There are a few tools that can be used to help clinicians understand an individual's risk for breast cancer. These tools use a variety of factors to calculate risk. The most widely used tool is the Gail Model (https://www.cancer.gov/bcrisktool) [10]; however, this model does not use extensive family history or breast density to calculate risk. The Claus model uses family history (up to two first and/or second degree relatives) [11] to calculate short and longer term risk, but does not include other important risk factors such as atypia seen on breast biopsy in the calculation of risk. The Tyrer-Cuzick model uses a variety of risk factors to calculate risk [12]. This model has recently been updated to include mammographic density [13]. Each of these models has strengths and limitations. It is important that the provider using them understand the limitations to ensure that the most accurate risk assessment is provided.

Genetic testing is another way to asses risk for those with a strong family history or known mutation in the family. Genetic testing should always be performed with counseling to ensure that the individual being testing understands the test being ordered, the limitations of testing, and the implications of both a positive and negative test result.

SCREENING

The purpose of screening is to detect breast cancer at an early stage, which is generally associated with a better prognosis. Breast imaging in the form of a mammogram is the usual modality for breast cancer screening. Early detection of breast cancer through screening mammography in women age 40–74 has been shown to decrease mortality by 31%, in addition to decreasing the incidence of stage II and higher breast cancers by 25% [14].

Screening mammogram is usually performed yearly beginning between the ages of 40 and 50 in women who are of average risk. In women who are high risk (due to strong family history, genetic mutations, history of chest radiation, or abnormal biopsies), screening mammograms should be started by age 30 [15]. In addition, screening with breast MRI and ultrasound have been shown to improve outcomes in women who are at high risk.

There is no upper age limit at which breast cancer screening should be stopped, but it is recommended that annual screening should be stopped when life expectancy is less than five to seven years, or the results obtained from the mammogram would not be acted on based on age or underlying medical conditions [15].

One drawback to mammograms is the potential of identifying a breast cancer at such an early stage that if it was otherwise undetected and left alone it may not even have had the opportunity to cause problems during the expected natural lifetime of the patient. This phenomenon has been termed overdiagnosis. The extent to which this type of "indolent" cancer is actually detected by mammogram varies based on different interpretations of the data and differing opinion over our still somewhat limited understanding of the biological complexities of breast cancer. Women ultimately should make an informed decision about screening practices based on personal risk factors and discussions with their care provider.

PREVENTION

Breast cancer prevention efforts can be divided into several groups: lifestyle modification, chemoprevention, and surgical prevention.

Lifestyle Modification: This refers to any set of behavioral changes that are geared toward mitigating the effects of known breast cancer risk factors. This pertains mainly to obesity and includes physical activity/exercise, and dietary changes to aid in weight loss. Smoking cessation and alcohol reduction have also been suggested as forms of lifestyle modification for breast cancer risk reduction.

Chemoprevention: This refers to any pharmacological substance that is utilized to decrease the incidence of breast cancer.

In clinical trials, two classes of drugs have been shown to decrease the incidence of breast cancer. These are selective estrogen receptor modulators

(SERMs) (e.g., Tamoxifen and Raloxifene), and aromatase inhibitors (AI) (e.g., Arimidex and Exemestane).

SERMs: Studies utilizing SERMs in both pre- and postmenopausal women have shown benefit [16–18]. Tamoxifen was shown to decrease the incidence of breast cancer by 49%, and Raloxifene was shown to be just as effective in postmenopausal women. Significant adverse effects, such as increased risk of endometrial cancer, strokes, and blood clot formation, have limited the widespread use of SERMs.

AI: This class of drug is only clinically beneficial in postmenopausal women, but has shown more significant decrease in the incidence of breast cancer than in studies using SERMs [19, 20]. The side effects of AIs such as muscle/joint pain and osteoporosis are less severe than SERMs.

Surgical Prevention: This pertains to the surgical removal of nonaffected organs in order to prevent breast cancer, and can be considered in women who are known to have a very high risk of developing breast cancer (e.g., BRCA mutation carrier)

Prophylactic/Risk-Reducing Mastectomy: This is the surgical removal of all breast tissue in women who do not have breast cancer, and can reduce the risk of cancer development by 95%.

Prophylactic/Risk-Reducing Salpingo-oophorectomy: This is the surgical removal of normal ovaries and fallopian tubes in premenopausal women. For women who carry the BRCA gene mutation, this procedure results in a significant reduction in the risk of breast cancer development by about 50%.

DETECTION

There are multiple ways that breast cancer may present. A woman can sometimes detect breast cancer simply by observing that something is not right with her breast. Some common changes include the presence of unusual lumps or bumps that can be felt, breast swelling, irritation of the nipple or skin of the breast, skin dimpling, redness of the breast, fluid discharge from the nipple (clear or bloody discharge), turning inward of the nipple, swollen lymph nodes under the arm or above the collar bone, unusual breast pain, change in the shape of the breast, or a rash/crusting involving the nipple. Detection of any of the above-mentioned breast changes should be reported as soon as possible to a care provider. Standard image techniques for detecting breast cancer include mammogram, ultrasound, or magnetic resonance image (MRI). If findings on imaging or physical exam are suspicious for breast cancer, the next usual step will be a biopsy.

Core needle biopsy has become the biopsy choice of preference as it typically provides adequate tissue for a comprehensive diagnosis and may allow patients to avoid having to undergo a full surgical excision of the area in question if the biopsy is conclusively negative [21]. Core-needle biopsies are typically performed in the office or clinic, using local anesthesia (i.e., numbing medication injected into the breast). Core needle biopsies are often performed using ultrasound or other breast imaging techniques (e.g., MRI or mammography/x-rays) to visualize and guide the needle into the correct area of concern in the breast. However, if the area of concern is easily felt, the core-needle biopsy can be done by hand while feeling the lump (i.e., palpation-guided core-needle biopsy).

The removed breast tissue is then evaluated by a pathologist. If breast cancer is present the next step is to determine how advanced the cancer is. There are various imaging and lab tests, which are sometimes needed to evaluate for the presence of cancer spreading outside the breast area (also known as metastasis). Furthermore, the cancer cells will be analyzed for the presence or absence of certain markers (estrogen and progesterone receptors and HER2 gene amplification) that are important in the treatment of breast cancer and prevention of recurrence after the cancer is removed.

Multiple research studies have demonstrated that performing a core-needle biopsy for the diagnosis of breast cancer before surgery decreases the chances of having to have multiple operations to appropriately treat the cancer. Otherwise women end up undergoing one surgery for the biopsy, waiting a few days to get the results, and then may require another surgery to specifically address the diagnosed cancer (i.e., getting enough of a margin around the cancer and assessing lymph nodes for spread of cancer). Knowing the cancer diagnosis in advance of surgery also allows women to make choices between lumpectomy and mastectomy, with or without reconstruction. There are rare circumstances where a needle biopsy is not feasible or the results are inconclusive and a surgical biopsy is necessary; however, these are the exception to the rule. Only under very rare circumstances should an open surgical biopsy be performed instead of a core-needle biopsy.

Following the biopsy, it typically takes a few days to get the results while the pathologist analyzes the tissue specimen. Even if the biopsy does not demonstrate cancer, certain benign breast conditions such as atypical ductal hyperplasia can still increase the risk of developing breast cancer and may need to be addressed by a specialist.

TYPES OF BREAST CANCER

Breast cancer can be divided into different types as outlined below.

Non-invasive Breast Cancer: This occurs when the cancerous cells remain confined to the specific area from which they originated, e.g., the cancerous cells remain within the ducts or lobules and do not extend beyond these structures.

Invasive Breast Cancer: This occurs when the cancerous cells extend beyond the lining of the specific area from which they originate and spread into the surrounding tissues. Most breast cancers are invasive.

Invasive Ductal Carcinoma (IDC), also known as infiltrating ductal carcinoma, is the most common type of breast cancer, and accounts for up to

70–80% of invasive breast cancers. The cancer starts in the breast ducts (the tube that connects the milk-producing gland to the nipple), and spreads to the surrounding tissues.

Invasive Lobular Carcinoma (ILC), also known as infiltrating lobular carcinoma, accounts for about 10–15% of invasive breast cancers. It starts in the breast lobules (the milk-producing glands) and spreads to surrounding tissues. Treatment for lobular cancer is similar to ductal cancer.

Other histologic types of breast cancer include metaplastic, tubular, mucinous, and medullary. These types are rare and the treatment of them is not particularly different to that for invasive ductal cancer. However the prognosis maybe different, e.g., tubular carcinoma is associated with a very low risk of recurrence and medullary cancer is more commonly associated with BRCA1 mutations.

Inflammatory Breast Cancer (IBC): This is a relatively uncommon but aggressive form of breast cancer, and accounts for 1–3% of all breast cancers. IBC usually affects the skin overlying the breast and can cause it to appear red, warm, thickened, tender or itchy, which can be confused with an infection. IBC can also cause the skin to have a pitted appearance like an orange peel. Despite the name, the appearance of this type of cancer is not caused by inflammation or infection, but rather by cancer cells blocking the lymph vessels in the breast. Treatment of IBC starts with chemotherapy.

Paget's disease of the Breast/ Nipple: This is a fairly uncommon type of breast cancer (1–4% of breast cancer cases) that typically involves the skin of the nipple and/or areola. The symptoms can often be mistaken for eczema, dermatitis, or infection, and as such, patients may often have the disease for a few months before being diagnosed. Skin changes include scaling/crusting of the nipple, redness, rash, itching, tingling/ burning sensation, or inversion of the nipple. There can also be a bloody or straw colored nipple discharge. While the disease may appear to be confined to the skin surface, almost all cases are associated with an underlying cancer in the breast tissue.

CLINICAL CHARACTERISTICS, HISTOLOGY AND PROGNOSTIC MARKERS

Histology refers to the cellular makeup of tissues, while prognostic markers are properties of the cancer that determine the overall prognosis and are used to guide treatment, e.g., hormone receptor and HER2 status.

Grade refers to how the abnormal cells appear under the microscope, and how much they differ from the normal appearance of that particular type of cell. The more abnormal the cells appear, the higher the grade of the tumor. The higher the grade, the more likely the cells are to grow rapidly or spread. Low-grade tumors typically grow more slowly and have a less likelihood of spreading, though it is still possible.

STAGING

The stage gives information on the extent of the disease. The purpose of staging is to determine how far the cancer has spread, which in turn helps to guide treatment and determine prognosis.

The most commonly used method of staging for breast cancer is the TNM system, which looks at three different features of the primary tumor. T (tumor) takes into account the size of the tumor, N (node) looks at whether there is any spread of cancer to the nearby lymph nodes, and M (metastasis) looks at spread of cancer to distant sites beyond the primary tumor. With this information the cancer can be assigned a stage ranging from 0 to IV.

Stage 0 is the earliest stage of cancer and is indicative of a tumor that is confined to the cells lining a specific area, i.e., they have not invaded surrounding tissues. Ductal carcinoma *in situ* (DCIS) is stage 0 breast cancer (the earliest form of breast cancer) and occurs when the cancerous cells are confined to the lining of the ducts. This typically has a good prognosis.

Stage IV is the most advanced stage and is indicative of an invasive cancer that has not only gone beyond the tissues of the breast, but has also spread to other organs or sites distant from the original tumor, such as the brain, lungs, liver, or bones. This is typically associated with a poorer prognosis.

Different tests are used to help in staging breast cancer, some of which include: lymph node evaluation, chest X-ray, CT scan, PET scan, or bone scan.

TREATMENT

Treatment is often directed by the stage of the disease as well as the histological makeup and prognostic markers of the cancer itself.

- Surgery
- Radiation
- Systemic treatments including hormonal therapy, chemotherapy, and biologic agents

Surgery

Surgery has progressed from radical resection to minimally invasive procedures for breast cancer, and it is understood that "less is more" when it comes to breast surgery [22, 23]. The lumpectomy (also known as partial mastectomy or breast conserving surgery), a procedure where the breast tumor is removed, and the remaining nipple, skin, and breast tissue is preserved, has become the dominant treatment for early stage breast cancer. New techniques have been pioneered that sample a few lymph nodes in the axilla (sentinel node biopsy) rather than removing them all (axillary node dissection). These breast conserving surgeries and minimally invasive procedures have been established as the new standard of care [24–27].

The surgical management of breast cancer often includes various options that the surgeon and patient must navigate including decisions regarding the specific type of surgery to remove the breast cancer, use of radiation, and reconstructive considerations. The ultimate treatment decision for surgery is often influenced as much by patient's personal preference as any other medical factor. The decision for the type of breast cancer surgery should ideally involve a process of shared decision-making where the surgeon and patient partner together to explore options, weigh pros and cons, and ultimately reach a mutual decision for the best surgical option for the individual patient.

The type of breast surgery that a patient undergoes will depend on the size of the breast cancer relative to the breast, the extent of tumor presence, features of the particular cancer, characteristics of the patient, and the patient's treatment priorities. For early-stage breast cancer, the most common procedure selected is a partial mastectomy. This is an ideal procedure for removing a relatively small, isolated breast cancer; especially in a patient who can receive radiation following surgery. The goal of the partial mastectomy (also known as lumpectomy) is to remove the entire tumor with a surrounding rim (i.e., margin) of normal breast tissue. Ensuring a margin of normal breast tissue provides a safeguard that all of the known cancer was indeed removed.

In many cases the tumor is discovered by mammogram before it can be felt. In these cases of non-palpable breast cancer, the surgeon must use another technique to localize the tumor at the time of surgery. This can be done using mammographic or ultrasound guidance. Ultrasound can also be used in the operating room to direct the surgeon to the location to perform the lumpectomy. One potential drawback to the lumpectomy procedure is the possibility of tumor ultimately being found at the edge of the breast specimen removed. This is called a positive margin and if one or more margins are found to be positive, then a decision must be made about returning for further surgery, which may simply consist of a wider lumpectomy or reexcision of the concerning breast margin.

Finally, the lumpectomy procedure may or may not be performed along with a sentinel lymph node biopsy depending on the type and stage of breast cancer. Extensive research concludes that there is no difference in survival between lumpectomy and mastectomy for breast cancer. Similarly, recurrence rates for early-stage breast cancer (stage I–II) and for even some stage III breast cancers are essentially equivalent between mastectomy and lumpectomy.

A mastectomy implies removal of all of the breast tissue. This can be performed as a total or simple mastectomy (which means removal of all of the breast tissue, overlying breast skin, nipple and areola, without the removal of lymph nodes) or a skin sparing mastectomy or a modified radical mastectomy (MRM). A skin sparing mastectomy preserves as much of the breast skin as possible and involves the removal of nipple and areola. This procedure facilitates the ability to perform an immediate breast reconstruction with an implant or using natural tissue from another part of the body (e.g., abdominal tissue) with a more favorable cosmetic outcome. A MRM involves removal of all breast tissue, overlying skin, nipple, and areola as well as most of the lymph nodes in the axilla (i.e., under the armpit). The major difference between MRM and radical mastectomy, which is an obsolete procedure today, is that MRM spares the pectoralis muscle of the chest. This enables the women to have a better functional outcome. It also provides a better cosmetic outcome; especially if one wishes to undergo reconstruction.

Any invasive breast cancer has a risk of involving the lymph nodes draining the area. The sentinel lymph node procedure is a minimally invasive alternative to an axillary lymph node dissection and involves identification and removal of the first few lymph nodes (i.e., sentinel nodes) to which a breast cancer may spread. In order to do the procedure, a radioactive solution or blue dye is injected into the breast and is followed as it drains though lymphatic vessels into the sentinel nodes. The sentinel nodes can then be identified by detecting radioactivity or the appearance of blue coloration. These nodes (typically two to three) are removed and sent to a pathologist for further examination. The expertise of pathologist helps to determine if the sentinel lymph nodes are positive (i.e., contain spread of breast cancer cells) or are negative (i.e., do not contain spread of breast cancer cells). If no cancer is detected in the sentinel nodes, then the risk of the remaining nodes being involved with cancer is so low that they can usually be left alone. This procedure is not performed if cancer is already known to be in the nodes. This procedure has replaced axillary lymph node dissection (i.e., complete removal of most or all of the lymph nodes in the axilla) as the procedure of choice for assessing lymph nodes in people with early-stage invasive breast cancer. The risk of chronic swelling of the arm (lymphedema) and nerve injury is significantly reduced with sentinel node biopsy compared to a complete axillary lymph node dissection. Although the sentinel node biopsy has become a common procedure, it requires a great deal of skill. It should be done only by a surgeon who has experience with this technique.

Axillary Lymph Node Dissection involves removal of most of the lymph nodes in the axilla (i.e., under the armpit). This procedure may be performed if cancer has been demonstrated or is highly suspected to be already present in the lymph nodes. The risk of chronic arm swelling and nerve injury are much higher in axillary dissections compared to sentinel node biopsies.

Radiation

Radiation plays an essential role in the management of breast cancer. Radiation therapy is administered by a radiation oncologist; a physician specialized in the use of radiation to treat disease. Most patients undergoing breast conserving surgery will receive breast radiation. Studies have demonstrated that omitting radiation after breast conserving surgery leads to higher risk of the cancer returning in the breast. Radiation is usually not required if a mastectomy is performed, unless the tumor is more advanced (e.g., larger in size, involving lymph nodes, inflammatory breast cancer). Radiation is delivered directly to the breast and not to the entire body. The radiation is usually emitted from an external source and directed as a beam to the breast. Care is taken not to have the radiation delivered to the lungs or heart. Major side effects from radiation include fatigue and skin irritation. It is very rare for therapeutic breast radiation to cause cancer; however, there is a small risk of this occurring. The duration of radiation therapy for breast cancer varies and is traditionally between four to six weeks. This traditional approach delivers radiation to the entire breast. There are newer techniques of "partial" breast radiation where the radiation is delivered just to the area of the breast where the lumpectomy was performed. The rationale is based on data indicating that most breast cancer recurrences occur in the breast near the site where the tumor was removed. Partial breast radiation can be delivered in a few as one to three weeks. Whether whole breast or partial breast radiation is used, the delivery is typically 5-days per week (Monday–Friday). Fortunately, each session only requires approximately 10–15 minutes.

Systemic Treatment

Systemic treatment for breast cancer involves treatment with pills or injections and can be done to either prevent cancer from developing (as discussed above) or prevent cancer from reoccurring. This treatment can be given either neoadjuvant therapy (before surgery), or adjuvant therapy (after surgery), or to treat metastatic cancer. A discussion of how, when, and which agent would be used is beyond the scope of this chapter. We have attempted to give the reader a brief view into how these agents work and how they might be used in the treatment of breast cancer.

Hormonal Therapy: Breast cancer found to have receptors for estrogen or progesterone on the surface of the cell are labeled as estrogen receptor positive (ER+) or progesterone receptor positive (PR+). Tumors that do not have either ER or PR receptors are called either estrogen receptor negative (ER-) or progesterone negative (PR-). Women with either ER or PR positive tumors can be treated with hormonal therapies (such as tamoxifen or AI). These therapies can be given after surgery, radiation, and/or chemotherapy is complete with the goal of preventing recurrence. For women with receptor positive metastatic breast cancer, hormonal therapies are the mainstay of treatment and are utilized until the cancer becomes resistant to this treatment, often switching from one agent to another. Sometimes chemotherapy may be used first in women with metastatic receptor positive breast cancer, if the disease appears to be progressing rapidly.

Chemotherapy: This involves the use of pharmacological agents to destroy cancer cells. Because these drugs typically target cells that grow rapidly, other normal cells of the body that normally replicate quickly may also be affected (such as hair, intestines, and bone marrow). Chemotherapy can be used to prevent a recurrence or treat metastatic breast cancer. Chemotherapy is the mainstay of treatment for ER and PR- breast cancer. Chemotherapy can be added to hormonal therapy to further reduce the risk of recurrence or when receptor positive breast cancer becomes resistant to hormonal therapy. There are many chemotherapy agents with known activity in breast cancer. They can be used alone or in combination. It is important that patients receive teaching regarding the side effects of chemotherapy and know when it is important to call (for example, a fever when bone marrow function is at its lowest).

Biological Therapy: This therapy involves the use of pharmacological agents to destroy cancer cells based on specific biological markers of the cells themselves. The most common biologic target in breast cancer involves the Her2 gene, which is amplified (overexpressed) in about 20% of breast cancer. Trastuzumab (or Herceptin) targets this Her2 amplification and in doing so can cause cell death. There are other agents that target Her2 amplification and other biologic agents and many agents being studied in clinical trials.

Clinical Trials

Clinical trials are an important way of developing new advances and finding new, best therapies. Participation by women in clinical trials is how we have come to achieve many of the advances outlined above. Participation in clinical trials may offer individuals newer therapies not conventionally available. Individuals are encouraged to ask if they may be eligible for clinical trials at any stage of disease. Information regarding all clinical trials can be found at https://clinicaltrials.gov. Individuals can search their disease and location for trials close to them.

SURVIVORSHIP

The advances discussed above fortunately have led to more patients surviving breast cancer and thus the need to address issues related to survivorship. These issues include, but are not limited to, functional issues related to surgery, long-term side effects from chemotherapy (e.g., cardiac problems and nerve injury/peripheral neuropathy), issues with sexual function, body image concerns, economic issues related to the cost of treatment, and emotional/psychosocial considerations.

Staying healthy after breast cancer treatment is an essential factor for increasing quality of life, and also helps to improve cancer survival rates. For example, maintaining a healthy body weight following breast cancer treatment not only reduces the risk of cardiac disease, but also lowers the risk of breast cancer recurrence. Eating the right types of food, maintaining physical activity, and keeping emotionally/spiritually healthy are all important recommendations for breast cancer survivors.

Adjusting to life after breast cancer treatment may be difficult for some patients. Fear of recurrence and the possibility of death can overwhelm people. It is important to realize the severity of worry, and to acknowledge that worry can cause harm. Excessive worry and stress can rob patients of quality of life and even create health issues through depression and negative effects of stress on the immune system. This can be debilitating, and in some cases present a worse problem than the cancer itself. Programs and specialists are available to help patients who are having trouble living in the aftermath of cancer and managing the risk of cancer recurrence.

It is important to recognize that despite optimal treatment; breast cancer can return. Regular follow-up with medical management and continued screening are essential after treatment ends. Recurrence of breast cancer can be in the breast where it was initially treated (local recurrence), within the adjacent lymph nodes (regional recurrence), or elsewhere in the body (distal recurrence). The most common site of breast cancer recurrence is the bone, but breast cancer can also recur in the nodes, skin, lungs, liver, and rarely the brain. Patients treated for breast cancer that experience symptoms consistent with a recurrence (e.g., new breast mass, swelling of adjacent nodes, unusual headaches, respiratory problems, or bone pain) need to be appropriately evaluated for breast cancer recurrence. This may involve body scans and/or blood work. Obtaining these tests in patients who do not have symptoms is controversial, as studies have not conclusively demonstrated any survival benefit for routine testing in previously treated breast cancer patients without complaints. Blood work is not always sensitive or specific for breast cancer recurrence (i.e., some patients with recurrence will have normal blood work, and other patients may have abnormal blood work for reasons unrelated to cancer). Disadvantages of routine scans include the risk of false positives (i.e., identifying something that is not real – leading to more scans, tests, and invasive procedures). Other disadvantages include increased cost, inconvenience to the patient, and possible anxiety. Whether or not to undergo routine testing will be based on risk factors and should be a mutual decision between the patient and oncologist.

Ideally, all breast cancer patients should receive an individualized survivorship care plan after treatment that includes guidelines for monitoring and maintaining their health. Many groups have now developed various types of care plans to help improve the quality of care of survivors as they move from treatment and into long-term recovery.

Cancer organizations such as the American Cancer Society and Komen Foundation provide tips on staying active and healthy after cancer treatment, information about living with the possibility of cancer recurrence, and approaches survivors can take to increase their chances for survival and quality of life after breast cancer therapy. Inspiration and hope from stories about other people whose lives have been touched by breast cancer are also available.

Resources for breast cancer survivors:

- http://www.cancer.org/treatment/survivorshipduringandaftertreatment
- http://ww5.komen.org/BreastCancer/aftertreatment.html

REFERENCES

1. Siegel, R.L., Miller, K.D., and Jemal, A. (2018). Cancer statistics, 2018. *CA Cancer J. Clin.* 68 (1): 7–30. https://doi.org/10.3322/caac.21442.
2. Pazdur, R., Lawrence, R.C., Hoskins, W.J. et al. (2007-2008). *Cancer Management: A Multidisciplinary Approach*, 10e. CMP Medica.
3. National Cancer Institute. (2012, 24 September 2012). Breast Cancer Risk in American Women. Retrieved from http://www.cancer.gov/cancertopics/types/breast/risk-fact-sheet
4. National Cancer Institute. (2015, 1 April 2015). BRCA1 and BRCA2: Cancer Risk and Genetic Testing. Retrieved from http://www.cancer.gov/cancertopics/causes-prevention/genetics/brca-fact-sheet
5. Harris, J., Lippman, M., Morrow, M. et al. (2010). *Diseases of the Breast*, 4e. Philadelphia, PA: Lippincott, Williams & Wilkins.
6. Economopoulou, P., Dimitriadis, G., and Psyrri, A. (2015). Beyond BRCA: new hereditary breast cancer susceptibility genes. *Cancer Treat. Rev.* 41 (1): 1–8. https://doi.org/10.1016/j.ctrv.2014.10.008.
7. Boyd, N.F. (2013). Mammographic density and risk of breast cancer. *Am. Soc. Clin. Oncol. Educ. Book* https://doi.org/10.1200/EdBook_AM.2013.33.e57.
8. Anderson, G.L., Limacher, M., Assaf, A.R. et al. (2004). Effects of conjugated equine estrogen in postmenopausal women with hysterectomy: the Women's Health Initiative randomized controlled trial. *JAMA* 291 (14): 1701–1712. https://doi.org/10.1001/jama.291.14.1701.
9. Rossouw, J.E., Anderson, G.L., Prentice, R.L. et al. (2002). Risks and benefits of estrogen plus progestin in healthy postmenopausal women: principal results from the women's health initiative randomized controlled trial. *JAMA* 288 (3): 321–333.
10. Parmigiani, G., Berry, D., and Aguilar, O. (1998). Determining carrier probabilities for breast cancer-susceptibility genes BRCA1 and BRCA2. *Am. J. Hum. Genet.* 62 (1): 145–158.
11. Claus, E.B., Risch, N., and Thompson, W.D. (1994). Autosomal dominant inheritance of early-onset breast cancer. Implications for risk prediction. *Cancer* 73 (3): 643–651.
12. Amir, E., Evans, D.G., Shenton, A. et al. (2003). Evaluation of breast cancer risk assessment packages in the family history evaluation and screening programme. *J. Med. Genet.* 40 (11): 807–814.
13. Brentnall, A.R., Harkness, E.F., Astley, S.M. et al. (2015). Mammographic density adds accuracy to both the Tyrer-Cuzick and Gail breast cancer risk models in a prospective UK screening cohort. *Breast Cancer Res.* 17 (1): 147. https://doi.org/10.1186/s13058-015-0653-5.
14. Tabar, L., Fagerberg, C.J., Gad, A. et al. (1985). Reduction in mortality from breast cancer after mass screening with mammography. Randomised trial from the breast cancer screening working group of the Swedish national board of health and welfare. *Lancet* 1 (8433): 829–832.
15. Lee, C.H., Dershaw, D.D., Kopans, D. et al. (2010). Breast cancer screening with imaging: recommendations from the society of breast imaging and the ACR on the use of mammography, breast MRI, breast ultrasound, and other technologies for the detection of clinically occult breast cancer. *J. Am. Coll. Radiol.* 7 (1): 18–27. https://doi.org/10.1016/j.jacr.2009.09.022.
16. Cuzick, J., Powles, T., Veronesi, U. et al. (2003). Overview of the main outcomes in breast-cancer prevention trials. *Lancet* 361 (9354): 296–300.
17. Fisher, B., Costantino, J.P., Wickerham, D.L. et al. (1998). Tamoxifen for prevention of breast cancer: report of the national surgical adjuvant breast and bowel project P-1 study. *J. Natl. Cancer Inst.* 90 (18): 1371–1388.
18. Vogel, V.G., Costantino, J.P., Wickerham, D.L. et al. (2006). Effects of tamoxifen vs raloxifene on the risk of developing invasive breast cancer and other disease outcomes: the NSABP study of tamoxifen and raloxifene (STAR) P-2 trial. *JAMA* 295 (23): 2727–2741. https://doi.org/10.1001/jama.295.23.joc60074.

19. Cuzick, J., Sestak, I., Forbes, J.F. et al. (2014). Anastrozole for prevention of breast cancer in high-risk postmenopausal women (IBIS-II): an international, double-blind, randomised placebo-controlled trial. *Lancet* 383 (9922): 1041–1048. https://doi.org/10.1016/S0140-6736(13)62292-8.

20. Goss, P.E., Ingle, J.N., Alés-Martínez, J.E. et al. (2011). Exemestane for breast-cancer prevention in postmenopausal women. *N. Engl. J. Med.* 364 (25): 2381–2391. https://doi.org/10.1056/NEJMoa1103507.

21. James, T.A., Mace, J.L., Virnig, B.A. et al. (2012). Preoperative needle biopsy improves the quality of breast cancer surgery. *J. Am. Coll. Surg.* 215 (4): 562–568. https://doi.org/10.1016/j.jamcollsurg.2012.05.022.

22. Sakorafas, G.H. and Safioleas, M. (2009). Breast cancer surgery: an historical narrative. Part I. From prehistoric times to Renaissance. *Eur. J. Cancer Care (Engl)* 18 (6): 530–544. https://doi.org/10.1111/j.1365-2354.2008.01059.x.

23. Winchester, D.P., Trabanino, L., and Lopez, M.J. (2005). The evolution of surgery for breast cancer. *Surg. Oncol. Clin. N. Am.* 14 (3): 479–498. vi. doi:https://doi.org/10.1016/j.soc.2005.04.006.

24. Cotlar, A.M., Dubose, J.J., and Rose, D.M. (2003). History of surgery for breast cancer: radical to the sublime. *Curr. Surg.* 60 (3): 329–337. https://doi.org/10.1016/S0149-7944(02)00777-8.

25. Fisher, B., Anderson, S., Bryant, J. et al. (2002). Twenty-year follow-up of a randomized trial comparing total mastectomy, lumpectomy, and lumpectomy plus irradiation for the treatment of invasive breast cancer. *N. Engl. J. Med.* 347 (16): 1233–1241. https://doi.org/10.1056/NEJMoa022152.

26. Krag, D.N., Anderson, S.J., Julian, T.B. et al. (2010). Sentinel-lymph-node resection compared with conventional axillary-lymph-node dissection in clinically node-negative patients with breast cancer: overall survival findings from the NSABP B-32 randomised phase 3 trial. *Lancet Oncol.* 11 (10): 927–933. https://doi.org/10.1016/S1470-2045(10)70207-2.

27. Rostas, J.W. and Dyess, D.L. (2012). Current operative management of breast cancer: an age of smaller resections and bigger cures. *Int. J. Breast Cancer* 2012: 516417. https://doi.org/10.1155/2012/516417.

12

PROSTATE CANCER

David W. Sobel and Mark K. Plante

Division of Urology, Department of Surgery, University of Vermont Medical Center, Burlington, VT, USA

WHAT IS A PROSTATE?

The prostate is a male sex gland that assists reproductive function by contributing nutrients to the ejaculate. It is composed of glands that empty into ducts supported by fat, muscle, and fibrous cells. When ejaculation occurs the prostate contracts providing fluid that combines with fluid from the seminal vesicles and sperm from the testes. The combination is then expelled into the urethra. The urethra runs from the bladder, through the prostate and to the outside through the penis.

As men age, the prostate grows and can restrict the flow of urine from the bladder, making it harder to urinate. This process of slow growth with aging has been termed "benign prostate hyperplasia" and is accompanied by symptoms such as increased frequency of urination during the day or night, urgency, hesitancy, intermittency of urination, and decreased stream. Slow, orderly growth of the prostate forms what is called an adenoma, a benign tumor. When the growth becomes disordered, due to many factors including DNA damage from the environment and genetic predispositions, prostate cancer arises.

Cancer: Prevention, Early Detection, Treatment and Recovery, Second Edition. Edited by Gary S. Stein and Kimberly P. Luebbers.
© 2019 John Wiley & Sons, Inc. Published 2019 by John Wiley & Sons, Inc.

HOW COMMON IS PROSTATE CANCER AND ARE THERE ANY RISK FACTORS?

After skin cancer, prostate cancer is the most commonly diagnosed cancer in men. Approximately 240 000 men are diagnosed annually. Men have a 1 in 7 lifetime risk of developing prostate cancer, and a 1 in 38 risk of dying from prostate cancer. As such, most men with prostate cancer will not die from it; nonetheless, it remains the second most common cause of cancer death in men, behind only lung cancer [1]. If Prostate cancer is so prevalent, how do we go about preventing it? What are the risk factors for prostate cancer? All in all, we have learned over recent years that one's risk of prostate cancer is a complex interplay of age, race, family history, and other factors such as smoking and diet.

While the genetic basis of prostate cancer continues to be studied, it has been shown that a family history positive of prostate cancer results in an increased risk of developing prostate cancer. In fact, the risk nearly doubles if you have any affected family member and nearly triples if your brother has prostate cancer [2]. Age is also a risk factor for prostate cancer with the average age at diagnosis being 66 years [1]. Race has also been established as a risk factor. African Americans have the highest incidence of prostate cancer worldwide with higher death rates as well [1, 3]. Links between smoking and prostate cancer are not as clear as those with lung cancer. That said, there does appear to be a correlation between smoking at the time of diagnosis and worse prostate cancer outcomes, including prostate specific mortality. Further, some association with smoking has been seen with more aggressive cancer and an increased risk of recurrence after treatment [4].

CAN IT BE PREVENTED?

Are there any drugs that have been shown to prevent prostate cancer? In 1993, the Prostate Cancer Prevention trial (PCPT) was started to see whether a drug called finasteride could prevent prostate cancer from developing. Finasteride is in a class of drugs called "5-alpha reductase" inhibitors that stops the enzyme that converts testosterone into a more potent form called dihydrotestosterone. Finasteride had been used for many years as a treatment for benign prostatic hyperplasia as it causes the prostate to "shrink." It was hypothesized that since prostate cancer is a testosterone influenced disease, perhaps finasteride could prevent it from developing. The PCPT was completed in 2003 and showed that in "low risk disease" (more about that later) finasteride reduced the overall relative risk of prostate cancer by 25% [5]. Great news, however, it also showed that there might be more "high risk" prostate cancer in the participants who received the drug. As a result of this, the FDA refused to approve the drug for prostate cancer prevention, and in fact, finasteride now comes with a warning depicting this. All in all, as is true in many areas of medicine, more than one trial would be necessary to prove with certainty these findings and as such, finasteride remains a drug only useful for conditions related to benign prostatic growth. Another trial looked at the other available "5-alpha reductase" inhibitor, dutasteride, and, unfortunately, again, the results did not support its use as a preventative agent for prostate cancer.

Other supplements have been identified as potential cancer risk reducers, but unfortunately none have borne out. The Selenium and Vitamin E Cancer prevention Trial (SELECT) randomized participants to placebo, vitamin E, selenium, or both vitamin E and selenium. After eight years the study showed no protective value for prostate cancer development. Even more concerning, vitamin E was correlated with an increased risk of prostate cancer [6]. Vitamin C has similarly not shown any reduction in prostate cancer risk.

WHAT TESTS EXIST TO TRY TO SCREEN FOR IT?

The problem with prostate cancer is that in the vast majority of men diagnosed with it, it is asymptomatic. In order to diagnose it in the early stages, physicians and patients need rely on a controversial blood test, prostate specific antigen (PSA) and digital rectal examination (DRE). PSA is an enzyme (specifically, a serine protease) that liquefies the seminal fluid upon ejaculation in order to help sperm motility. It is a secreted by the prostate and found in both blood and semen. PSA began as a subject of study in the 1960s, but it wasn't until the late 1980s that it was measured quantitatively in the blood, becoming a tumor marker for prostate cancer [7]. In the ensuing years, its use as a tumor marker was popularized and in conjunction with a digital rectal exam, it became the screening test of choice for prostate cancer [8].

Much controversy exists, however, regarding the use of PSA as screening test for prostate cancer. In 2012, the US Preventative Services Task Force (USPSTF) recommended against PSA-based screening for prostate cancer. This statement was first published in the Annals of Internal Medicine in 2012 and was met with considerable skepticism by urologists nationwide. The argument was that many men may have asymptomatic prostate cancer diagnosed as a result of PSA screening that would never have become symptomatic during their lifetime. Thus, these men may, in turn, be subjected to unnecessary further testing and treatment. In fact, the USPSTF gave PSA as a screening test a "D" (failing) rating stating, "There is moderate or high certainty that this service has no net benefit or that the harms outweigh the benefits" [9]. Much controversy [10] exists about this statement and how it was arrived at, ranging from reports of flawed calculations regarding the number of patients that need to have a prostate biopsy in order to find one man with prostate cancer, to overestimation of the risk and complications of treatment. Most urologists agree that without a better screening tool, PSA has some utility in detecting prostate cancer in its earliest stage and that since its appearance in the clinical arena, the number of deaths resulting from prostate cancer have steadily declined in the absence of significant treatment advances.

Over the years, numerous different PSA modalities have been developed and then evaluated to better determine who might have clinically important prostate cancer. For instance, PSA velocity, the change in PSA over time, can be more indicative of potential aggressive cancer if it is sharply rising. PSA density, which is a measure of PSA relative to the volume of the prostate that is measured at the time of biopsy, is another measure of potential risk of cancer but like PSA velocity it is imperfect in its ability to discriminate in all cases. Age adjusted PSA is another concept whereby it is well known that PSA rises in men as they age given that the prostate also grows with age.

Digital rectal examination remains a crucial tool in screening for prostate cancer as most prostate cancers develop in what is termed the "peripheral zone" or outside edge of the prostate. This peripheral zone can be felt during an exam where a physician inserts a gloved, lubricated finger into the rectum to palpate the prostate to reveal nodular cancer. Also, it needs be remembered that not all prostate cancers will have an elevated PSA.

Many new and novel ancillary tests have entered the clinical landscape, including the Prostate Health Index (a blood test), PCA3 (a urine test), and the 4Kscore (a blood test) aiming to answer such questions as "What should be done in the case of an elevated PSA and a negative biopsy?" and "What is the likelihood of finding an aggressive cancer on biopsy?"

The American Urologic Association (AUA) has issued the following recommendations for screening [11]:

---

1. PSA screening in men under age 40 years is not recommended.
2. Routine screening in men between ages 40–54 years at average risk is not recommended.
3. For men ages 55–69 years, the decision to undergo PSA screening involves weighing the benefits of preventing prostate cancer mortality in 1 man for every 1000 men screened over a decade against the known potential harms associated with screening and treatment. For this reason, shared decision-making is recommended for men age 55–69 years that are considering PSA screening, and proceeding based on patients values and preferences.
4. To reduce the harms of screening, a routine screening interval of two years or more may be preferred over annual screening in those men who have participated in shared decision-making and decided on screening. As compared to annual screening, it is expected that screening intervals of two years preserve the majority of the benefits and reduce over diagnosis and false positives.
5. Routine PSA screening is not recommended in men over age 70 or any man with less than a 10–15 year life expectancy

---

HOW IS IT DIAGNOSED?

Based on a patient's risk factors, PSA, and digital rectal exam, a urologist may recommend a prostate biopsy. A prostate biopsy is a procedure done in the office under ultrasound guidance and often with local anesthesia and antibiotics taken before and after to prevent infection. Patients will likely need to stop any anticoagulation such as Coumadin® (Warfarin), Lovenox® (enoxaparin), or Xarelto® (rivaroxaban) prior to the biopsy. During the procedure, the patient lies on their side and the urologist introduces the ultrasound probe into the rectum and injects lidocaine into the nerves surrounding the prostate. With the block completed, a hollow core needle is then sequentially directed into the prostate to acquire the biopsy cores, typically 12 – six from each side of the prostate – in the area most common to harbor cancer, the peripheral zone. These cores are then sent to pathology for analysis. The process itself takes approximately 10 minutes and patients can drive themselves to and from the procedure.

Pain during the procedure is usually minimal thanks to lidocaine anesthetic injected prior to biopsy. Minor bleeding from the rectum or blood in the urine is quite common but self-limited. Rare but important is if patients develop a fever and infection after the biopsy, it is best treated with antibiotics after reevaluation with their doctor. Again, uncommon, systemic infections occur in less than 1 in 100 patients [12].

A pathologist then examines the collected biopsy tissue under a microscope in order to interpret the architecture of cells in the prostate tissue. If cancer is seen, it is graded using a system first described in 1974 by the pathologist Donald Gleason. The Gleason score is a number that reflects the aggressiveness of the prostate cancer based on the architectural pattern created by the cancer cells in the tissue. Previously, this was based on the most common and second most common pattern in each biopsy specimen being given a primary and secondary grade from 1 to 5. The two grades are then added together for a total score out of 10, depicted as such: "3 + 4 = 7" or "4 + 4 = 8." Gleason grade 6 through 10 is represents malignant patterns, with grades 1 to 5 rarely reported. Due to the confusion of Gleason grade 6 actually being low risk prostate cancer, pathologists have begun using a new 5-grade classification system that incorporates the Gleason patterns representing malignancy but assigns them a grade 1 through 5, 5 being the highest risk. As this is still being implemented in most pathology labs, typically reports will include both the classic Gleason score ("3 + 4 = 7") and the new group grade ("Grade 2").

According to the AUA, higher Gleason grade, higher PSA levels, and more advanced tumor stage have all been associated with increased prostate cancer mortality. Based on these criteria, the AUA has risk-stratified patients into the following categories [13]:

Low risk: PSA<10 ng/ml and a Gleason score of 6 or less and clinical stage
   T1c or T2a
Intermediate risk: PSA >10 to 20 ng/ml or a Gleason score of 7 or clinical
   stage T2b but not qualifying for high risk
High risk: PSA>20 ng/ml or a Gleason score of 8–10 or clinical stage T2c

It should also be noted that the usual 12-core biopsy might miss clinically significant cancer up to 20–30% of the time. In certain instances, a urologist may suggest a patient undergo a re-biopsy with possible sampling of other locations in the prostate. The use of Magnetic Resonance Imaging (MRI), has evolved significantly in recent years and may be used to try to help identify suspicious areas in the prostate to target biopsies.

Once the diagnosis of cancer is made, the urologist may order (as clinically indicated) a number of other imaging tests in order to rule out prostate cancer spread beyond the gland itself. A bone scan may be indicated when the PSA is >20, the clinical stage is T3 or T4, the Gleason score is ≥8, the serum alkaline phosphatase is elevated, or if symptoms of new bone pain are present suggestive of metastatic disease. A pelvic CT or MRI may also be performed to look for cancer extension to outside the prostate in the form of enlarged lymph nodes or other organ spread, less commonly seen than cancer spread to bone. MRI uses a magnetic field to create images that reflect tissue composition and anatomic detail not seen with other imaging modalities. For the prostate, MRI is becoming increasingly useful for both diagnosis, as already discussed, and for more accurate tumor staging, including for those patients following an active surveillance protocol.

Prostate cancer is staged according to the American Joint Committee on Cancer using the "TNM" system: [14]

T stands for tumor, which describes the size and extent of the primary
   tumor.
   T1a: Cancer is found incidentally during a transurethral resection of the
      prostate (TURP). Cancer is in no more than 5% of the tissue removed.
   T1b: Cancer is found during a TURP but is in more than 5% of the tissue
      removed
   T1c: Cancer is found by needle biopsy that was done because of an
      increased PSA.
   T2a: The cancer is palpable on digital rectal exam (DRE) in one half or less
      of only one side (left or right) of your prostate.
   T2b: The cancer on DRE is in more than half of only one side (left or right)
      of your prostate.
   T2c: The cancer on DRE is in both sides of your prostate.
   T3a: The cancer extends outside the prostate but not to the seminal
      vesicles.
   T3b: The cancer has spread to the seminal vesicles.

T4: The cancer has grown into tissues next to your prostate (other than the seminal vesicles), such as the urethral sphincter (muscle that helps control urination), the rectum, the bladder, and/or the wall of the pelvis.

N stands for nodes.
   N0: no positive regional lymph nodes.
   N1: metastasis in regional nodes.

M stands for metastases.
   M0: no distant metastasis.
   M1a: non-regional lymph nodes.
   M1b: Bony metastasis.
   M1c: Other sites with or without bone disease

WHAT TREATMENTS ARE AVAILABLE FOR CLINICALLY LOCALIZED DISEASE?

Clinically localized prostate cancer is cancer that is contained to the prostate within its capsule. Several different management options exist for patients with this including to do no more than monitor the cancer.

How do prostate cancer patients choose between monitoring, termed active surveillance, and treatment for their cancer and if the choice is for treatment, among the various options for treatment? Many facets need be considered including risk stratification, life expectancy, quality of life concerns, and also patient preference. Certain treatments may not be preferable for high risk prostate cancer patients. A patient might decide that the risk and type of certain side effects unique to a specific modality would be untenable. Further, for those patients with advanced age or other significant health issues, they may elect to abstain from treatment until and if it becomes symptomatic. Active surveillance, surgery performed to remove the entire prostate and its capsule, radiation delivered using different modalities, cryotherapy using the ability of freezing to kill tissue, and hormonal therapy are the presently accepted standard of care treatment options.

It is well established that not all men with the diagnosis of prostate cancer will die from it. As such, active surveillance has been increasingly adopted as a means to monitor patients with asymptomatic low risk prostate cancer, allowing for "timely intervention for cases demonstrating tumor progression" [15]. Active surveillance allows for minimizing the morbidity of treatment for patients with lower risk disease until and if cancer progression occurs, suggesting intervention may be necessary. The eligibility for active surveillance has been changed in recent years, however, most urologists agree that it should be considered for patients with low risk disease by defined by the AUA and National Comprehensive Cancer Network (NCCN) guidelines (PSA $\leq 10\,$ng/ml, Gleason score ≤ 6, Clinical stage $\leq$T2a, ≤ 3 positive biopsy cores, $\leq 50\%$ cancer in each core) [16]. The Prostate Cancer Intervention vs Observation Trial (PIVOT) [17] showed that in men with low to intermediate risk disease, there was no difference in death from prostate cancer or death from any cause at 12 years. However, 90% of the men studied were > 60 years old and, in this age group, no difference in survival was shown by being on an active surveillance protocol. It should be noted, however, that many advances for the treatment of other diseases and conditions have been made in recent years. Therefore, with extending survival for patients with them, care needs be exercised before applying dated trial results to decisions regarding treatment for present day populations of prostate cancer patients.

Active surveillance entails close monitoring of patients with serial digital rectal examination (DRE), serum PSA testing, as well as consideration of repeat prostate biopsies every one to three years. As discussed, MRI is increasingly being considered to either supplant or help guide repeat biopsies. Patients may elect to discontinue active surveillance and seek treatment for progressive and accelerated increases of PSA, repeat biopsies that demonstrate pathologic progression to intermediate or high-grade cancer, or, in some cases, anxiety associated with nonaction.

Surgery for prostate cancer is termed "radical prostatectomy" and entails removal of the entire prostate, seminal vesicles, and, often times, local pelvic lymph nodes. The traditional open retropubic radical prostatectomy, pioneered by Walsh at Johns Hopkins in 1982 [18], involves making an incision below the belly button, removing the prostate located between the bladder and the urethra, and reconnecting the bladder neck to the urethra. Hospital stays are usually two to four days following surgery with a Foley catheter left draining the bladder for one to two weeks in order to allow the newly created connection between the bladder and the urethra to heal.

Today in the United States, the vast majority (>80%) of radical prostatectomies are performed laparoscopically with the help of a robotic surgical platform [19]. The abdomen is filled with CO_2 gas and robot arms/instruments and a camera are passed through five small incisions that are used to complete the removal of the prostate. Robot assisted surgery for prostate cancer has resulted in shorter hospital stays compared to the traditional approach, usually between one and two days following surgery, as well as reduced overall blood loss during surgery. However, patients are still requiring for a catheter to be left for one to two weeks and to date, no long-term improvements in cancer outcomes have been shown including cancer recurrence rates and mortality [20, 21].

Complication rates after surgery remain largely similar with both techniques and include blood loss requiring blood transfusion, blood clot formation in the legs, pain, and infection. These complications are not unique to prostate surgery and are typical of many different surgical procedures. Specific to radical prostatectomy, potential complications that relate to quality of life include urinary incontinence and erectile dysfunction. The urinary incontinence seen after this surgery is typically stress-induced, meaning that leakage occurs with coughing, sneezing, or during physical exertion. Typically it is worst immediately after surgery once the catheter is removed and usually improves over 6 to 12 months.

For those with severe and continuing incontinence following radical prostatectomy, both nonsurgical and surgical options exist to help ameliorate it including placement of an artificial urinary sphincter where necessary. Erectile dysfunction incidence after prostatectomy depends on several factors including the patient's age and erection function before surgery, and whether the nerves adjacent to the prostate were preserved during the procedure [22]. Numerous effective treatments exist to help treat this complication.

Radiation energy has been used to treat prostate cancer patients for decades with many advancements in both the method as well as the precision of delivery. Administered by radiation oncologists, external Beam Radiation Therapy (EBRT) is where the energy is delivered to the prostate and pelvis from outside the body. It can be used alone for patients with low risk prostate cancer, or in conjunction with either short or long courses of hormone therapy androgen deprivation therapy (ADT) for patients with intermediate or high risk disease. Prior to the 1970s, radiation treatments consisted of delivering the radiation with a very limited ability to focus it at the prostate and pelvis. With the advent of enhanced abilities to image the inside of the body, notably using computed tomography (CT), radiation can now be directed much more precisely toward the prostate. Conformal Radiation Therapy (CRT) where the radiation beam "conforms" to the shape of the prostate is now the standard of care. This advancement has reduced the amount of radiation that is unnecessarily delivered to other organs close to the prostate such as the bladder and rectum and has reduced the complications seen as a result. Intensity Modulated Radiation Therapy (IMRT) is another advancement whereby the radiation dose delivered to the prostate and pelvis is optimized. Treatment plans usually consist of up to 50 separate daily doses of radiation delivered to the prostate, five days a week. Proton or neutron beam radiation uses a different radioactive source to the standard gamma radiation source used in most centers. It is much more costly overall and, to date, no research has been completed supporting its superiority for both treatment outcomes and complication rates reported by many of the centers that offer it [23]. Despite the continuing technological advances, radiation therapy continues to have associated side effects and complications including erection difficulties and lower urinary tract symptoms, such as frequency, urgency, and burning with urination. For patients with higher risk disease, improved outcomes have been seen with the addition of ADT from 6 to 24 months depending on the disease risk category.

Brachytherapy is another form of radiation energy delivery to the prostate whereby small radioactive "seeds" are inserted directly into the prostate using ultrasound guidance. It can be used alone as monotherapy for low or intermediate risk prostate cancer patients or in combination with external beam radiation for patients with higher risk disease. The procedure is typically performed on an outpatient basis. Patients who have previously undergone a TURP for benign enlargement, in most cases, will not be able to undergo brachytherapy given the higher rates of urinary complications [24]. After brachytherapy, irritative lower urinary tract symptoms such as frequency and burning with urination are not uncommon but thankfully resolve over time. There is a risk of patients being unable to urinate after the procedure requiring for a catheter to be left short term. Erectile dysfunction, while possible, is typically seen less frequently than with other treatment modalities [25].

For those patients with low or intermediate risk prostate cancer deemed poor candidates for either surgery for any form of radiation therapy, cryotherapy may be an option [26]. Cryotherapy involves placement of probes in the prostate using ultrasound guidance similar to that used for brachytherapy. Argon gas is then pumped into the probes causing the adjacent tissue temperature to lower to approximately −40 °C. The resultant freezing creates an "ice ball" with both immediate and delayed cell death then taking place [27]. A catheter is placed in the urethra to the bladder during the procedure allowing for warming and thereby preventing damage to the urethra that previously resulted in severe complications. The freezing procedure is completed and then repeated once more to maximize the cellular destruction for cancer eradication. Urethral damage can nonetheless occur despite the warming catheter, resulting in urinary retention, incontinence, pain, and fistula formation. The fistulas seen are abnormal communications between the urethra and the rectum and, again, are much less commonly seen, occurring in <1% of cases. Unfortunately, most men will report erectile dysfunction following cryotherapy. Patients with large prostates may not be good candidates as it is more difficult to reliably achieve uniform freezing of all the tissue in larger glands. The procedure is typically performed on an outpatient basis, again, similar to brachytherapy and unlike brachytherapy, most patients are left with a Foley catheter for one week.

Finally, for those patients with a limited life expectancy, typically considered to be less than 10 years, due to age or other health issues, the question often and appropriately becomes "Why treat something that is asymptomatic and likely will not result in death?" Watchful waiting is thus an option whereby intervention is only considered for only symptomatic disease progression and in the form of palliative intervention as opposed to curative intent.

WHAT IF THE CANCER RECURS AFTER TREATMENT?

Despite the "curative" intent options discussed, prostate cancer can nonetheless recur. Specifically, high risk disease is most likely to recur, including those patients with a high pretreatment PSA, high Gleason grade, and/or higher stage tumors. Positive margins and positive lymph nodes after surgery, suggesting not all the cancer was removed, as well as smoking and obesity have also been associated with higher rates of recurrence.

The definition of recurrence is different based on which treatment was performed. With radical prostatectomy, the goal is for all prostate cells, including prostate cancer cells, to be removed. Thus, no detectable PSA should be seen after surgery. For those patients with a tumor that extends beyond the surgical margins or outside the prostate, the PSA may either never become undetectable or be found to rise after surgery. These patients may be considered for subsequent external beam radiation treatment in order to try again to eradicate the cancer. If the PSA was never undetectable, it is called adjuvant radiation. If it became undetectable but subsequently rose, it is called salvage radiation.

For those patients who underwent radiation therapy, biochemical recurrence is defined differently. As the prostate was never removed, the PSA may never become undetectable. Different thresholds for how much the PSA must rise from the lowest level after radiation exist as suggested by different professional organizations. However, at present, the most widely accepted threshold is a rise of 2 ng/ml or more above the lowest recorded PSA. If prostate cancer recurs after radiation, a patient may be eligible for either surgery in the form of a salvage radical prostatectomy or salvage cryotherapy. Both treatments, given that they are salvage procedures, are technically more difficult and are associated with higher complication rates than if they are performed in the primary treatment setting.

WHAT ARE THE TREATMENTS FOR LOCALLY ADVANCED AND METASTATIC DISEASE?

Locally advanced disease refers to prostate cancer that extends outside the prostate capsule to the surrounding tissues whereas metastatic disease refers to cancers that have spread beyond prostate and its surrounding tissues.

If a cancer is locally advanced with extension outside the prostate capsule, it is likely not amenable to radical prostatectomy given the likelihood all the cancer may not be able to be removed by surgical means. Not infrequently, radiation therapy is felt to be the more appropriate choice given its ability to treat tissues outside the prostate and, in combination with ADT, effect good long-term cancer control even if not curative.

Metastatic prostate cancer, having by definition spread to distant locations, most commonly the lymph nodes and/or bone, is treated primarily with ADT. This allows for the destruction of cancer deposits everywhere in the body by removing the influence of testosterone which stimulates their growth. Often times it is asymptomatic but at times may cause bone pain, kidney obstruction, anemia, fatigue, and other generalized physical manifestations.

ADT, which is primarily the cessation of testosterone production by the testes, can be achieved either by surgical means, their removal, or medically using either of a variety of injected medications. Drugs such as Lupron® (leuprolide), Trelstar® (triptorellin), or Zoladex® (goserelin), to name a few, belong to a class of drugs that stop production of testosterone by the testicles by way of the action in the brain. With the first injection, the body's response is to make more androgens, created a so-called "testosterone flair." That is why, for the first injection, they are typically paired with another oral medication that blocks the testosterone receptors in order to block the influence of the flair on the cancer. The most commonly used medication is Casodex® (bicalutamide). Within a month after the first injection, the serum testosterone level is usually lowered to castration level (<50 ng/dl). If the cancer cells are responsive to the ADT, the tumor burden will shrink and symptoms improve with lowered PSA as well. To summarize, ADT is seldom if ever used as a monotherapy for clinically localized disease, but can be used in conjunction with radiotherapy and less commonly surgery for more advanced local disease and is the first line treatment for metastatic disease.

It must be noted that there are significant side effects associated with ADT and they compound with the duration of treatment. Physical changes can include gynecomastia, the benign growth of breast tissue, as well as atrophy of the testis, loss of skeletal muscle mass and overall weight gain. In addition, most men will experience hot flashes, fatigue, loss of libido, and erectile dysfunction. Longer-term risks include osteoporosis, weakening of the bones as well as the development of cardiac issues [28].

Many advanced and metastatic prostate cancer patients will have their disease controlled with ADT, however, prostate cancer cells are always at risk of developing the ability to evade the effects of androgen deprivation. These cells will then regain the ability to grow and divide, leading to disease progression termed "castration resistant" prostate cancer. This nonresponse of the tumor to ADT is when other treatment options need be considered, many of which have been developed within the last decade. Typically urologists will turn to medical oncologists for assistance in managing these patients with respect to which agents will be considered in addition to continuing ADT.

Within the newly available treatment options that now exist for castration resistant prostate cancer are oral medications that further reduce both the production and action of androgens, Zytiga® (abiraterone) and Xtandi® (enzalutamide). Numerous chemotherapy drugs have been researched over the last two decades. Finally, some specific medications have shown success in treating this very difficult stage of the disease and have even been shown to potentially be of use earlier in the treatment of patients before their cancers become castrate resistant. Therapies using the immune system as a way to eradicate tumor cells have also been developed with one, Provenge® (Sipuleucel-T), using a patient's own immune cells, obtained from a simple blood draw, to then be reinjected in combination with a medication to stimulate the immune action. For patients with metastatic prostate cancer to bone who have associated bone pain, external radiation to those specific lesions can palliate some of the symptoms. Injectable radioactive materials have also been developed that target the prostate cancer deposits in the bone, the most recently available, Xofigo® (Radium 223).

The common saying, "Today is the best time in history to have prostate cancer" is certainly true in many respects given the recent advances for treatment of this common male cancer. With better tools for screening and detection being developed and better treatments for advanced disease available, curing and controlling prostate cancer should be more successful than ever before. Research also continues in finding ways to reduce prostate cancer risk as well as working to improve presently available surgical and nonsurgical therapy treatment options.

REFERENCES

1. American Cancer Society. "What are the key statistics about prostate cancer?" *Prostate Cancer.* 12 March 2015. Web. 22 July 2015.

2. Bruner, D.W., Moore, D., Parlanti, A. et al. (2003). Relative risk of prostate cancer for men with affected relatives: systematic review and meta-analysis. *Int. J. Cancer* 107 (5): 797–803.

3. Powell, I.J., Bock, C.H., Ruterbusch, J.J. et al. (2010). Evidence supports a faster growth rate and/or earlier transformation to clinically significant prostate cancer in black than in white American men, and influences racial progression and mortality disparity. *J. Urol.* 183: 1792–1796.

4. Kenfield, S.A., Stampfer, M.J., Chan, J.M., and Giovannucci, E. (2011). Smoking and prostate cancer survival and recurrence. *JAMA* 305 (24): 2548–2555.

5. Thompson, I.M., Goodman, P.J., Tangen, C.M. et al. (2003). The influence of finasteride on the development of prostate cancer. *N. Engl. J. Med.* 349: 215–224.

6. Lippman, S.M., Klein, E.A., Goodman, P.J. et al. (2009). Effect of selenium and vitamin E on risk of prostate cancer and other cancers: the selenium and vitamin E cancer prevention trial (SELECT). *JAMA* 301 (1): 39–51.

7. Rao, A.R., Motiwala, H.G., Karim, O.M. et al. (2008). The discovery of prostate-specific antigen. *BJU Int.* 101: 5–10.

8. Catalona, W.J., Smith, D.S., Ratliff, T.L. et al. (1991). Measurement of prostate-specific antigen in serum as a screening test for prostate cancer. *N. Engl. J. Med.* 324: 1156–1161.

9. Moyer, V.A. and Force USPST (2012). Screening for prostate cancer: U.S. preventive services task force recommendation statement. *Ann. Intern. Med.* 157 (2): 120–134.

10. Carlsson, S., Vickers, A.J., Roobol, M. et al. (2012). Prostate cancer screening: facts, statistics, and interpretation in response to the US preventive services task force review. *J. Clin. Oncol.* 30 (21): 2581–2584.

11. Carter, H.B., Albertsen, P.C., Barry, M.J. et al. (2013). *Early Detection of Prostate Cancer: AUA Guideline*. AUA Education and Research, Inc.

12. Rodríguez, L.V. and Terris, M.K. (1998). Risks and complications of transrectal ultrasound guided prostate needle biopsy: a prospective study and review of the literature. *J. Urol.* 160 (6 Pt 1): 2115–2120.

13. Thompson, I., Thrasher, B.J., Aus, G. et al. (2007). Guideline for the management of clinically localized prostate cancer: 2007 update. *J. Urol.* 177: 2106–2131.

14. American Cancer Society. "The AJCC TNM staging system." *Prostate Cancer.* 12 March 2015. Web. 31 July 2015.

15. Glass, A., Punnen, S., and Carroll, P.R. (2013). Active surveillance for prostate cancer. *AUA Updat. Ser.* 32: 130–135.

16. Reese, A.C., Landis, P., Han, M. et al. (2013). Expanded criteria to identify men eligible for active surveillance of low risk prostate cancer at Johns Hopkins: a preliminary analysis. *J. Urol.* 190: 2033–2038.

17. Wilt, T.J., Brawer, M.K., Jones, K.M. et al. (2012). Radical prostatectomy versus observation for localized prostate cancer. *N. Engl. J. Med.* 367: 203–213.

18. Walsh, P.C. (1998). Anatomic radical prostatectomy: evolution of the surgical technique. *J. Urol.* 160: 2418–2424.

19. Lowrance, W.T., Eastham, J.A., Savage, C. et al. (2012). Contemporary open and robotic radical prostatectomy practice patterns among urologists in the United States. *J. Urol.* 187: 2087–2092.

20. Barocas, D.A., Salem, S., Kordan, Y. et al. (2010). Robotic assisted laparoscopic prostatectomy versus radical retropubic prostatectomy for clinically localized prostate cancer: comparison of short-term biochemical recurrence-free survival. *J. Urol.* 183: 990–996.

21. Krambeck, A.E., DiMarco, D.S., Rangel, L.J. et al. (2009). Radical prostatectomy for prostatic adenocarcinoma: a matched comparison of open retropubic and robot-assisted techniques. *BJU Int.* 103: 448–453.

22. Alemozaffar, M., Sanda, M.G., Kaplan, I.D. et al. (2011). Prediction of erectile function following treatment for prostate cancer. *JAMA* 306: 1205.

23. Russell, K., Caplan, R., Laramore, G. et al. (1994). Photon versus fast neutron external beam radiotherapy in the treatment of locally advanced prostate cancer: results of a randomized prospective trial. *Int. J. Radiat. Oncol. Biol. Phys.* 28: 47–54.

24. Blasko, J.C., Grimm, P.D., and Raghe, H. (1993). Brachytherapy and organ preservation in the management of carcinoma of the prostate. *Semin. Radiat. Oncol.* 3: 240–249.

25. Sanda, M.G., Dunn, R.L., Michalski, J. et al. (2008). Quality of life and satisfaction with outcome among prostate-cancer survivors. *N. Engl. J. Med.* 358: 1250.

26. Babaian, R.J., Donnelly, B., Bahn, D. et al. (2008). Best practice statement on cryosurgery for the treatment of localized prostate cancer. *J. Urol.* 180 (5): 1993–2004.

27. Baust, J.G. and Gage, A.A. (2005). The molecular basis of cryosurgery. *BJU Int.* 95: 1187.

28. Daniell, H.W. (2001). Osteoporosis due to androgen deprivation therapy in men with prostate cancer. *J. Urol.* 58 (suppl 2A): 101.

13

GASTROINTESTINAL CANCER

Kristina Guyton, Baddr Shakhsheer, and Neil Hyman

Department of Surgery, University of Chicago Medicine, Chicago, IL, USA

INTRODUCTION

Gastrointestinal cancer ranks among the commonest malignancies worldwide. These cancers can present in a variety of ways within a population; presentation depends on whether population-based screening modalities are available and utilized or whether an individual diagnosis is predicated upon an evaluation of symptoms. While early colon cancer in the United States and early gastric cancer in Japan may be detected in the asymptomatic phase with screening endoscopy, most cases of gastrointestinal malignancies are diagnosed based on suggestive symptomatology. These signs most commonly include overt or occult gastrointestinal blood loss or consequences of the associated mass effect.

The location of the cancer within the gastrointestinal tract dictates general oncologic behavior and therefore the determined treatment and prognosis. As such this chapter is divided into three sections: esophageal cancer, gastric cancer, and colorectal cancer. Modes of presentation, detection, and treatment are discussed for each type.

Cancer: Prevention, Early Detection, Treatment and Recovery, Second Edition. Edited by Gary S. Stein and Kimberly P. Luebbers.
© 2019 John Wiley & Sons, Inc. Published 2019 by John Wiley & Sons, Inc.

ESOPHAGEAL CANCER

Incidence, Risk Factors, Prevention

Diagnoses of esophageal cancer are notoriously feared by both patients and practitioners because of the complex management and poor prognosis with advanced disease. Approximately 17 000 new cases of esophageal cancer are diagnosed annually in the United States; worldwide there are approximately 500 000 new cases of esophageal cancer detected annually. Most esophageal cancers are of two pathological subtypes: squamous cell cancer (SCC) and adenocarcinoma. Anatomically, SCC predominately presents in the middle and upper third of the esophagus. Adenocarcinoma predominates in the lower third of the esophagus and is thought to arise from Barrett's esophagus (intestinal metaplasia), a consequence of gastroesophageal reflux disease (GERD).

The incidence of esophageal cancer varies globally, with the highest incidence of disease found in East Asia (China, Singapore), Africa, and Iran. These locations suffer almost exclusively from SCC. Risk factors in this part of the world include malnutrition and poor fruit and vegetable intake. In the United States, however, the major risk factors for SCC are tobacco and alcohol use. A history of specific "benign" esophageal disorders may predispose to SCC as well, including achalasia or strictures, especially those caused by caustic ingestions in childhood. In contrast, adenocarcinoma most frequently arises from areas of Barrett's esophagus: intestinal metaplasia of the lower esophagus that arises from chronic GERD. Barrett's esophagus increases the risk of esophageal cancer 30-fold.

Prevention of esophageal cancer is best achieved by avoidance of the risk factors, namely alcohol and tobacco usage. The recent decrease in incidence of SCC in the United States correlates with a decline in national smoking rates. Patients with a diagnosis of Barrett's esophagus should be screened endoscopically at regular, prescribed intervals in order to identify any progression toward adenocarcinoma.

Detection

Esophageal cancer often presents at an advanced stage, with just over half of diagnoses presenting as incurable locally advanced or metastatic disease. Presenting symptoms of disease commonly include dysphagia to solid foods and weight loss. The dysphagia is progressive and an indicator of long-standing disease as the lumen must be significantly narrowed by the tumor in order to result in symptoms. Patients may also experience retrosternal pain and possibly vocal changes if the recurrent laryngeal nerve is involved. Friable tumors that bleed can cause iron-deficiency anemia.

Early cancers in the United States are typically detected by endoscopy, either incidentally or as a result of a surveillance program in those diagnosed with

Barrett's esophagus. Barium studies may also suggest the presence of an esophageal cancer, demonstrating a tapering or irregularity in the esophageal wall; however, endoscopy with tissue biopsy remains the standard method of diagnosis.

Computed tomography (CT) scan of the chest, abdomen, and pelvis is used for staging. However, this modality is poor at evaluating depth of invasion and involvement of local lymph node basins; these characteristics are best evaluated by endoscopic ultrasound (EUS). Positron emission tomography (PET) scans may be used to evaluate for smaller metastatic lesions that a CT scan alone cannot visualize. The most metastatic sites include the liver, lungs, bone, and adrenal glands. Metastatic patterns differ with histology as SCC metastases tend to occur locally in the thorax while adenocarcinoma metastases tend to be found intraabdominally.

Treatment

Treatment of esophageal cancer differs widely based on tumor characteristics. If EUS shows that the tumor is confined to the esophageal mucosa (T1a tumors), endoscopic mucosal resection can be employed to remove the tumor and confirm the precise depth of invasion. Other local therapies for T1a tumors include photo-dynamic therapy, radiofrequency ablation, laser therapy, cryotherapy, and argon plasma coagulation. These therapies avoid the morbidity associated with esophagec-tomy, which is the standard of care for more advanced tumors or in centers where advanced endoscopic techniques are not available. There are early indications that some of these therapies may also be applicable in submucosal tumors (T1b) in lieu of esophagectomy; however, definitive data is still unavailable.

The majority of esophageal tumors present at later stages with tumors that invade through the esophageal wall (T3 or greater) and include nodal involvement. In the United States, most of these patients receive neoadjuvant chemotherapy and radiation and then are restaged to determine the appropriateness of proceeding with esophagectomy.

Esophagectomy remains a mainstay of treatment in the curative treatment of esophageal cancer. However, complications of this surgery are significant and can occur in up to 30% of patients, these include conduit necrosis (death of the portion of the stomach brought into the chest to replace the esophagus), anastomotic leak, and aspiration. Many studies have demonstrated that this operation is best per-formed in high volume centers in order to minimize perioperative morbidity and mortality. Despite advances in care, the perioperative death rate remains approxi-mately 2% even in the best of centers. Following surgery, long-term side effects of esophagectomy include dysphagia, chronic cough, and reflux.

Indications for chemotherapy for esophageal cancer consist of locally advanced disease: full-thickness esophageal wall involvement (T3 or greater), nodal involve-ment, and/or metastatic disease. Standard regimens include ECF (epirubicin, cisplatin, 5-fluorouracil), DCF (docetaxel, cisplatin, 5-fluorouracil), and more recently EOX (epirubicin, oxaliplatin, and capecitabine). However, there is no consensus as to the best regimen; regardless of choice, most contemporary regimens incorporate combinations of agents. Most recently, esophageal tumors have been tested for HER2 overexpression in order to determine candidacy for trastuzumab.

Endoscopic stenting may be used as a palliative treatment for dysphagia in patients who are not candidates for surgery.

Recovery

Recovery after treatment for esophageal cancer focuses on detection of recurrence and management of treatment morbidities. Most recurrences are detected within one year of treatment and may present near the primary tumor site or as distant metastases. There is no standard posttherapy surveillance strategy and the curability of recurrent cancer is quite low. In patients with recurrent dysphagia endoscopy can be used to assess the esophagus, employing dilation for stenosis or stenting for mass effect. Nutritional counseling is often a key component of care for patients as they commonly require active monitoring of their nutritional status.

GASTRIC CANCER

Incidence, Risk Factors, Prevention

Gastric cancer remains one of the most prevalent cancers in the world. It was the leading cause of cancer death worldwide until the 1980s when it was overtaken by lung cancer. This change in incidence was partially due to risk factor mitigation and may also be attributed to treatment of Helicobacter pylori (H. pylori) on a global scale; however the initial decline in incidence preceded the discovery of H. pylori. Domestically, there are approximately 25 000 cases of gastric cancer diagnosed annually. Gastric cancer is more common in men than women. Rates of gastric cancer incidence vary geographically, with the highest rates in East Asia, Eastern Europe, and South America; by comparison, it is far less common in North America. More than 70% of gastric cancers occur in developing nations.

There are two common histological variants of gastric cancer that have distinct biological behaviors and pathologic outcomes: intestinal type and diffuse (or infiltrative) type. Intestinal gastric cancer is more commonly found in the elderly and in males while diffuse type cancer, with its more dire prognosis, is most prevalent in younger age groups and has equal incidence among genders.

Anatomic location is an important factor both for surgical planning as well as prognosis. Proximal gastric tumors share many clinical and histopathologic features with distal esophageal cancers (mostly adenocarcinoma) and may indeed share a common etiology. Distal tumors, which have decreased in incidence over time, are more likely to be associated with chronic gastritis.

Certain risk factors for the intestinal type cancer provide clues as to its pathogenesis. Atrophic gastritis – an autoimmune disease that leads to gastric epithelial atrophy as well as loss of chief and parietal cells – is associated with an increased risk of gastric cancer. In populations where atrophic gastritis is prevalent, gastric cancer rates are higher. Intestinal metaplasia, which can be caused by *H. pylori* infection and bile reflux, is also a risk factor for intestinal type gastric cancer. Dietary risk factors include salt intake and foods high in nitrites such as processed meats; places were these foods are common, such as East Asia, have substantially higher rates of gastric cancer. Group A blood type is associated with a higher likelihood of gastric cancer for unknown reasons. Other frequently reported risk factors include obesity, smoking, and *H. pylori* infection.

Progress toward preventing gastric cancer has been made through risk factor modification. Efforts at primary prevention through eradicating *H. pylori* have resulted in some suggested benefit. Promoting diets with a high intake of fruits and vegetables is known to be beneficial in protecting populations against gastric cancer.

Detection

Screening in high risk populations remains controversial. In countries with high incidences of gastric cancer such as Japan, routine screening in patients older than 40 years of age facilitates early diagnoses, either via barium studies or endoscopy. The cost-benefit analysis and precise effectiveness of such programs have yet to be definitively determined.

In the United States most patients present with advanced disease. The most common early symptoms are weight loss and persistent abdominal pain. Proximal tumors can also present with dysphagia. Diffuse cancers can cause nausea and early satiety due to an inability of the stomach to distend. Bleeding tumors can potentially be detected by positive fecal occult tests, anemia, or more overt manifestations of gastrointestinal blood loss.

The first step in the diagnostic workup for suspected gastric cancer is endoscopy and biopsy. Any suspect ulcer should be biopsied several times at the ulcer margin and base. For diffuse type gastric cancer, endoscopic biopsies may generate false-negative results: these tumors infiltrate the submucosa or deeper levels of the stomach wall and spread along those planes.

Staging of gastric cancer is performed via CT scan. Because the peri-gastric lymph node basins are numerous, local lymph node metastases that are geographically distant from the tumor are not necessarily indicative of unresectability. EUS can determine the depth of invasion and can be used to biopsy lymph nodes. A staging laparoscopy is sometimes performed in tumors more advanced than T1, as approximately one-quarter of patients with negative CT scans will have otherwise occult metastases evident on laparoscopy with peritoneal washings.

Treatment

The treatment of gastric cancer depends on stage. The earliest gastric cancers, those confined to the mucosa (T1), can be managed endoscopically by mucosal resection if the patient lacks evidence of lymph node involvement. In the absence of metastases or in centers that do not have access to advanced endoscopic techniques, gastric cancer is managed by surgical gastrectomy. Surgical planning depends upon the anatomic location of the lesion; most physicians now treat tumors close to the esophagogastric junction as esophageal cancers.

Surgical resectability is the strongest predictor of survival. Chemotherapy and radiotherapy may be administered for patients with nodal disease or tumors that are T3 or greater; chemotherapy for a T2N0 cancer remains controversial. The two predominant regimens are perioperative chemotherapy (as a result of the MAGIC trial) or adjuvant chemoradiotherapy (as a result of the American Intergroup trial, also known as the Macdonald protocol). First line therapy in the MAGIC protocol includes epirubicin, cisplatin, and 5-fluorouracil (ECF). The Macdonald protocol utilizes leucovorin-modulated 5-fluorouracil with radiotherapy. As with esophageal tumors, recent attention has been given to the HER2 status of gastric tumors for possible trastuzumab therapy.

Recovery

Nutritional issues plague this patient population, especially those who undergo total gastrectomy. Patients can suffer from caloric malnutrition as a result of early satiety and gastroparesis; additionally other complications of surgery and surgical reconstruction can result in dumping syndromes and reflux gastritis. The dietary intake and weight status of these patients are concerning and require frequent monitoring and assessment.

There is insufficient evidence to recommend a specific posttreatment surveillance regimen for patients treated for gastric cancer. Pragmatically, most clinicians utilize a combination of regular physical exam, endoscopy, and imaging.

COLORECTAL CANCER

Incidence, Risk Factors, Prevention

Colorectal cancer affects over 136000 Americans per year, making it the third most commonly diagnosed cancer in the United States. One in every 20 Americans (5%) will develop colon or rectal cancer over the course of their lifetime. In incidence it follows lung cancer and prostate cancer or breast cancer, in men and women, respectively. Annually 50000 Americans will die of their disease, making colorectal cancer the third most lethal cancer in either sex. Within the United

States, colorectal cancer incidence is about 25% higher in men and 20% higher in African American populations. Approximately 90% of colorectal cancers are diagnosed in patients over the age of 50.

Although the United States has one of the highest incidences in the world, it also has one of the highest five-year survival rates: 64% of those treated for colorectal cancer are long-term survivors. Fortunately, both the incidence and overall mortality associated with colon and rectal cancer have begun to decline, by 2–3% per year over the last 15 years. Although the reasons for the declines are unclear, it is hypothesized to be due to increased screening, removal of at risk lesions (precursor lesions that may become malignant in the future), and detection of cancers at earlier stages.

The incidence of colorectal cancer varies over 10-fold around the world, with the highest incidence in developed regions such as Australia, New Zealand, North America, and Europe and the lowest in Africa and South-Central Asia. The geographic variation is thought to be primarily a result of environmental and dietary factors in addition to background genetic vulnerability. This assertion is supported by the fact that individuals who migrate from a low risk to a higher risk region are observed to take on the higher risk rate that is reflective of their new environment.

Other factors increasing the risk of colorectal cancer development include low socioeconomic status, inactivity, smoking, alcohol intake, and diets high in animal fats and low in fiber, fruits, and vegetables. However, there really is no dominant risk factor or causative agent as seen in other cancers (e.g., cigarette smoking and lung cancer). Supplements that have shown promise in their ability to decrease the risk of colorectal cancer development include calcium, aspirin, and nonsteroidal anti-inflammatories (NSAIDS). However, again, in the absence of a primary or small group of established predisposing factors, the impact of the interventions is relatively modest at best.

Around 5% of colorectal cancers are associated with inherited colorectal cancer syndromes. The two most prominent include FAP (familial adenomatous polyposis) and HNPCC (hereditary nonpolyposis colorectal cancer). FAP is characterized by the development of hundreds to thousands of adenomatous polyps carpeting the colonic and rectal mucosa beginning in the teenage years with nearly inevitable progression to cancer. In comparison, those affected by HNPCC develop relatively few polyps, but they tend to be located in the right colon and progress rapidly into cancers.

Even outside defined genetic syndromes, family history of colon or rectal cancer is an important risk factor. The presence of colorectal cancer or adenomatous polyps in a first-degree relative confers twice the risk of developing colorectal cancer as compared to the general population. The risk is further elevated if there are multiple affected family members or there is evidence of early onset of disease. Inflammatory bowel disease, particularly ulcerative colitis, is associated with an increased risk of colorectal cancer that progressively increases further over time; the cancer risk increases with the extent, duration, and activity of ulcerative colitis.

More recently, it has become apparent that Crohn's colitis also increases the risk of malignancy to a similar extent as ulcerative colitis if the distribution of disease and longevity are similar in magnitude.

The classic paradigm for the development of colorectal cancer is the adenoma to carcinoma progression. With years of research and the advent of advanced molecular biology techniques, we have refined the understanding of this complex multistep process. We know that through genetic losses and mutations, normal mucosa becomes hyperproliferative, develops into an adenoma, progresses to carcinoma, and ultimately acquires the ability to metastasize. Should a cell progress along this pathway it will gradually acquire genetic changes, gaining oncogenes and losing tumor-suppressor gene activity.

Detection

Early detection of colorectal cancer results in a much higher likelihood of cure. Research suggests that colonic or rectal adenomas are present for 8–10 years prior to their progression to cancer, affording ample opportunity for early intervention. This makes colorectal cancer an attractive target for secondary prevention: removal of the precursor lesion (the adenomatous polyp) may prevent the later occurrence of an overt carcinoma.

Most patients with premalignant lesions or early-stage malignancy have no discernable symptoms. However, proactive screening provides the opportunity to diagnose colorectal cancer early when the prognosis is excellent, or even to prevent its occurrence altogether. Consequently, population-based screening of patients prior to symptom development has achieved considerable traction over recent decades. The National Polyp Study (NPS) found that screening colonoscopy with removal of encountered polyps reduced colorectal cancer risk by 76–90% over the subsequent 10 years. Earlier and more frequent screening should be considered in "high-risk" individuals with the following: a personal history of colorectal cancer or adenomatous polyp, a genetic syndrome predisposing to colorectal cancer (HNPCC, FAP), one or more first-degree relatives with colorectal cancer, two or more second-degree relatives with colorectal cancer, or inflammatory bowel disease (IBD) causing pancolitis or longstanding (>8–10 years) active disease. In individuals with "average risk," screening should begin at age 50 and an emerging consensus suggests that screening should end at age 75 to avoid overuse.

A surveillance plan should be developed for each patient based on effectiveness, safety, cost, and availability of the screening tests. Options for screening strategies in average-risk patients include a complete colonoscopy at 10-year intervals; a CT colonography or flexible sigmoidoscopy every five years; or fecal occult blood testing on three samples annually. There has been a gradual shift over many decades in colon cancer location from the distal large intestine, with incidences now found more proximally in the right colon. For this reason, screening colonoscopy, which affords visualization of the entire colon, has been encouraged. Another advantage of colonoscopy includes the ability to biopsy and tattoo unresectable lesions, allowing for a definitive diagnosis and easier surgical localization of nonpalpable lesions.

There are distinct presenting symptoms characteristic of the right versus left colon cancers. Tumors of the left colon or rectum are more likely to present with a change in bowel habits or blood per rectum. Altered bowel habits can result from narrowing of the bowel lumen by the tumor, thereby resulting in a narrowing of the stool caliber or obstructive symptoms such as cramping, abdominal distension, nausea, and vomiting. A bleeding left-sided tumor will usually manifest as blood on the surface of the stool and is more likely to be noticeable to the patient. Conversely, the right colon has a wide diameter and contains liquid stool. Obstructive symptoms are therefore less common and any blood lost into the lumen is mixed into the stool, and less likely to be noticed by the patient. Right colon cancers typically present with fatigue, iron deficiency anemia, and/or a positive fecal occult blood test. Digital rectal exam can detect distal rectal tumors and abdominal exam may reveal a palpable mass for large tumors.

In patients diagnosed with colon and rectal cancer a thorough history and physical examination is performed to identify the direct consequences of the tumor (e.g., obstructive symptoms), search for evidence of metastatic disease, and determine the patient's suitability for surgical resection. Evaluation of the entire colon, either with colonoscopy or less commonly with barium enema or CT colonography, is performed; approximately 5% of patients will have more than one cancer in the colon and synchronous polyps may be removed endoscopically or addressed at subsequent resection. CT scan of the abdomen and pelvis can be used to evaluate the local extent of the colon cancer and identify metastatic deposits, occurring most often in the liver. If CT is not performed preoperatively, intraoperative ultrasound may be used to evaluate the liver parenchyma.

Patients with rectal cancer usually require a more thorough imaging evaluation owing to the multidisciplinary nature of its disease management. Since precise localization is key to treatment planning, flexible, or rigid sigmoidoscopy is usually performed to identify the location in reference to the pelvic floor and determine the suitability for sphincter salvage. Similarly, transrectal ultrasound or, increasingly, pelvic magnetic resonance imaging (MRI) is of great value in developing an individualized, patient-specific treatment recommendation. A CT scan of the chest, abdomen, and pelvis becomes an important tool in rectal cancer treatment planning to evaluate for distant metastases. A useful tool for evaluating therapeutic effect and tumor recurrence, carcinoembryonic antigen (CEA) is a serum tumor-marker elevated in two-thirds of all colorectal cancers. If elevated preoperatively, CEA should return to normal levels when all disease is resected and subsequently can function as a tool to monitor for recurrence.

As emphasized earlier, stage at diagnosis is the most important predictor of survival. The current TNM (Tumors/Nodes/Metastasis) system describes the extent of the disease and can be further categorized into stages. A stage I lesion involves only partial thickness of the bowel wall. Stage II denotes a transmural lesion without lymph node or distant metastases. Stage III disease signifies lymph node metastases, without distant disease. Stage IV indicates metastatic disease. Complete cure is possible in 90% or more of patients with local disease. The likelihood of five year survival is 70% for cases of regional disease spread and 12% for distant disease spread.

Treatment

Surgery is the mainstay of colorectal cancer treatment. If there is no evidence of metastatic disease, as an isolated modality surgery has the capacity to completely cure an individual. However, approximately 20% of patients present with distant metastatic disease at the time of diagnosis. With the development of increasingly aggressive surgical procedures, adjunctive ablative modalities, and improvements in chemotherapy, some of these patients may also be treated for cure. Even when cure is not possible, surgery remains useful as a way to palliate advanced local disease since many patients will suffer tumor-specific complications long before they succumb to systemic disease.

Reflective of the understanding of colorectal cancer spread, surgical resection is based on two principles: removal of local disease and removal of the lymph nodes draining that portion of the intestine. In general, lymphatic drainage follows the venous flow toward large mesenteric veins; patterns of vascular and lymphatic drainage will determine the area of resection. For colon cancer, proximal, and distal negative margins are easily obtainable as the lymphatic drainage area is wider than the area needed to control local disease. A large area of resection in the colon usually has minimal effect on overall bowel function.

Complete resection of rectal cancer, due to the confined location within the bony pelvis, can be difficult to obtain; nevertheless, the critical impact of surgical technique on long-term outcomes, especially local control, has been clearly demonstrated. The traditional approach to resection of rectal cancer is known as an abdominoperineal resection (APR) where the rectum and anus are completely removed and a permanent colostomy is created. This remains the treatment for most distal rectal cancers. However, tumors of the upper and middle portions of the rectum are managed with a low anterior resection (LAR). This procedure involves removal of the upper rectum, the sigmoid colon, and the lymph nodes associated with the inferior mesenteric vessels, with subsequent connection of the descending colon to the rectum. The frontiers of sphincter salvage continue to be pushed forward with new reconstructive surgical techniques and the potential contributions of laparoscopic and robotic access. Superficial low rectal cancers may be appropriate for local, transrectal excision of only the tumor, leaving the rectum otherwise intact.

In colorectal cancer, chemotherapy is an important addition to surgery for many patients and remains the primary treatment for most patients with metastatic disease. The goal of adjuvant (after surgical resection) chemotherapy is to eradicate occult micrometastases that were not detectable at the time of resection and is commonly used to treat intermediate stage colon cancer. Neoadjuvant (before surgical resection) chemotherapy, typically along with pelvic radiation, is frequently used as part of the multidisciplinary management of rectal cancer to downstage the tumor, thereby facilitating margin-negative resection. With colon cancer, recurrence after resection is more likely at sites distant to the primary site (e.g., liver and lung) while in rectal cancer, locoregional recurrence is nearly as likely as distant metastasis. Thus, adjuvant chemotherapy is an important component of colon cancer treatment, while neoadjuvant chemotherapy and radiation therapy for local tumor control are more commonly used with rectal cancer.

Regimens with fluorouracil (5-FU) and leucovorin have been shown to reduce disease recurrence and improve survival in patients with locally advanced colon cancer, both by 30%. Since 2000, the therapeutic repertoire has expanded to include irinotecan, oxaliplatin, and humanized monoclonal antibodies that inhibit blood vessel growth and cancer cell growth by targeting vascular endothelial growth factor (VEGF) and epidermal growth factor receptor (EGFR) respectively. The best way to combine and sequence these agents is still under investigation. Radiation therapy plays a pivotal role in rectal cancer and may be valuable in colon cancer when it is locally advanced. In an effort to identify those patients who will most benefit from the addition of chemotherapy and radiation therapy recent years have brought an intense focus on personalized or customized individual treatment plans. New paradigms of care are goals of active investigation and, simultaneously, the subject of heated debate among practitioners.

Recovery

Following surgery and adjuvant therapy, the focus of care turns to monitoring for disease recurrence or progression. Nearly 40% of patients who initially present with stage II or stage III disease will experience a recurrence following primary therapy. As with initial detection, early recognition of recurrence or metastasis may improve five-year survival. Although follow-up regimens clearly identify a small subset of patients with recurrent disease that may be retreated with curative intent, the optimal modalities and follow-up intervals are uncertain. As discussed earlier, the CEA level, if elevated with the initial tumor, can be a useful marker of recurrence or progression. Screening colonoscopy is recommended within one year of resection and at three-to-five-year intervals in this population as additional cancers in the colon are seen in up to 3% of patients in this group. For rectal cancer, proctoscopic monitoring of the anastomotic site for local recurrence is commonly performed. CT scans of the chest, abdomen, and pelvis assist in monitoring for local recurrence or systemic disease, most often found in the liver and/or lung.

CONCLUSION

Gastrointestinal tract cancers continue to be a major source of cancer related morbidity and mortality worldwide. Recent advances in screening, surgical technique, and adjuvant therapy have improved outcomes and disease-free survival. Continued research toward prevention and treatment is active and ongoing.

14

LUNG CANCER

Farrah B. Khan[1], C. Matthew Kinsey[2], and Garth Garrison[2]

[1]Division of Hematology and Oncology, University of Vermont Medical Center, Burlington, VT, USA
[2]Division of Pulmonary and Critical Care Medicine, University of Vermont Medical Center, Burlington, VT, USA

LUNG CANCER OVERVIEW

Lung cancer is one of the most commonly diagnosed cancers and continues to be the leading cause of cancer-related death for both women and men in the United States. Overall, close to 6% of men and women will be diagnosed with lung cancer during their lives. Lung cancer is associated with a poor prognosis as only approximately 18% of those diagnosed with lung cancer survive five years although there is a wide variety in prognosis based on stage at diagnosis [1]. Most, but not all, lung cancers can be attributed to cigarette smoking.

EPIDEMIOLOGY AND RISK FACTORS

Epidemiology

Lung cancer is now the second most frequently diagnosed malignancy following breast cancer. In 2017, it is estimated that there will be 222 500 new cases identified

Cancer: Prevention, Early Detection, Treatment and Recovery, Second Edition. Edited by Gary S. Stein and Kimberly P. Luebbers.
© 2019 John Wiley & Sons, Inc. Published 2019 by John Wiley & Sons, Inc.

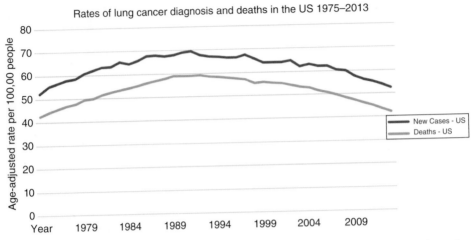

Figure 1. Rates of lung cancer diagnosis and attributable deaths in SEER database. (See insert for color representation.)

and 155870 deaths attributed to the disease [2]. Lung cancer is the leading cause of cancer-related deaths in people of all ethnicities in the United States. The rate of lung cancer is higher in men than women and his particularly high in African American men. Close to 80% of new lung cancer cases are diagnosed in people who are between the ages of 55 and 80. Overall the rate of new diagnoses has declined since the 1990s, mostly driven by a decline of the rate in men (Figure 1). The rate in women increased until the early 2000s and has since remained relatively stable.

Risk Factors

Exposures to carcinogens are the largest risk factor for development of lung cancer. The most common carcinogen causing lung cancer is cigarette smoking. Radon and asbestos exposures are other well-known causative agents. Diesel exhaust, arsenic, silica, and chromium have also been associated with increased lung cancer risk. As with most cancers, the risk of developing lung cancer increases with age, particularly following the sixth decade. People with a prior history of lung cancer or a history of lung cancer in a first degree relative have a higher risk of lung cancer.

Cigarettes

Cigarette smoking is the leading risk factor for lung cancer in the United States. Cigarette smoke contains thousands of identified substances including over 20 of which are known to cause lung cancer. These toxic substances include polycyclic aromatic hydrocarbons (PAH), nicotine-derived nitrosamine ketone (NNK), ethylene oxide, cadmium, and a radioactive isotope of polonium (polonium-210) [3]. These compounds cause damage to DNA which over time can result in mutations to tumor suppressor genes like *TP53* and genes important to cell signaling like *KRAS*. These mutations can then lead to the unregulated growth of cells characteristic of cancer.

Radon

Radon is thought to be the second leading cause of lung cancer in the United States. It is a naturally occurring odorless radioactive gas found in soil. Over 20 000 cases of lung cancer per year can be linked to radon exposure [4]. Coexposure with cigarette smoking can lead to a nearly 10-fold increase in risk for lung cancer. The average concentration of in the United States is around 0.4 picocuries per liter (pCi/L) and can be much higher in some regions. The concentration within homes and buildings can increase significantly in areas with poor ventilation such as basements. As a radioactive substance, radon can cause DNA damage to cells in the lung ultimately leading to cancer-causing mutations.

Asbestos

Asbestos is another significant environmental exposure that can increase risk for lung cancer. It has also been associated with, among other things, chronic lung disease, and malignant mesothelioma in addition to lung cancer. Asbestos is a general term for fiber shaped silicates that had been used extensively in the past as a building material and as electrical and thermal insulation. While use of asbestos has been banned in many countries, it is still used in some regions and older buildings containing asbestos can be found throughout the world. Of the numerous types of asbestos, crocidolite is considered the most carcinogenic (though Libby vermiculite is the most carcinogenic). Chrysotile has been the most commonly used type of asbestos and is likely carcinogenic but appears less potent than others. The mechanism of how asbestos causes cancer is not fully understood but may involve the chronic activation of the immune system with the formation and release of reactive oxygen species. There is a significant increase in lung cancer risk when asbestos exposure is combined with cigarette smoking.

PREVENTION AND SCREENING

Prevention

Preventing and/or reducing exposures to carcinogens, most importantly cigarette smoke, is the most effective method of reducing the risk of developing lung cancer. In 1965, 41% of US adults smoked, which declined to below 16% in 2015. This has been associated with a decline in lung cancer rate from a peak of

69.2 per 100000 people in 1992 to 52.4 per 100000 people in 2014. While risk may always remain elevated above never-smokers, the risk of lung cancer declines over time following smoking cessation and patients of any age are encouraged to quit smoking. Radon mitigation and restrictions on the use and handling of asbestos may provide additional benefits in reducing lung cancer rates. Diet and exercise have been suggested as additional strategies to reduce risk but there is limited data on their efficacy.

Screening

Screening is an important strategy to reduce lung cancer deaths since a risk for lung cancer, although much smaller, exists even after smoking cessation. It is a strategy not to reduce lung cancer rates but rather to reduce lung cancer deaths. Early-stage lung cancer is typically asymptomatic making it challenging to diagnose. However, given the well-established risk factors, it is possible to identify a population at high risk for developing the disease. Efforts to develop an effective screening test have been ongoing since the 1960s.

Screening with Low Dose CT

Screening at-risk populations of cigarette smokers with periodic chest X-rays with or without sputum analysis has not been shown to be beneficial. With improvement in computed tomography (CT) technology including the development of rapid multislice systems, obtaining high resolution images of the lungs and detection of small nodules has become more feasible. Using an annual low-dose CT scan in high-risk smokers, the International Early Lung Cancer Action Program (I-ELCAP) study showed an increase in the detection of early stage cancers. The detected early stage cancers were associated with a 10-year survival of 88% [5].

In 2011, the National Lung Screening Trial (NLST) was published and demonstrated a significant reduction in lung cancer related deaths when a strategy of annual low-dose chest CT scans was performed. In this study, people 55–74 years old with a 30 pack-year smoking history and who were actively smoking or quit within the last 15 years were randomly assigned to a strategy of annual chest X-ray versus annual low-dose chest CT scan. In this study, there was a 20% reduction in lung cancer related death when the strategy of annual CT was used [6]. Largely based on this study, lung cancer screening has been recommended by numerous professional organizations and government organizations including the United States Preventive Services Task Force. Screening of populations with additional risk factors including other exposures and/or family history has been suggested but does not have the same level of evidence to support.

Screening Controversies

Several ongoing controversies exist regarding lung cancer screening including high aggregate cost for annual CT scans in a large population, high false positive rate, and risk of overdiagnosis. In the NLST, 39% of those screened had a positive result of which only 4% were shown to be malignant. False positive results may cause unnecessary anxiety which could impair quality of life and impact participation in screening. False positive results may also lead to unnecessary biopsy procedures. Overdiagnosis refers to diagnosing cancers that would not have developed in to clinically significant problems. The rate of overdiagnosed lung cancers in the NLST may be as high as 18% [7].

SYMPTOMS

In many cases lung cancer can grow undetected in the chest when it is in its early stages. Most individuals have extra lung capacity (e.g., the output of the heart determines maximum exercise) so the disease may not be detected until more prominent symptoms due to involvement of the main airways or chest wall occur hence the importance of CT screening for lung cancer in at risk individuals (e.g., those without symptoms). The most common symptoms of lung cancer are a cough that does not go away or gets worse with time. Coughing up blood occurs in between 5 and 20% of cases of lung cancer and is a symptom that always warrants immediate evaluation by a medical professional [8, 9]. In some cases, direct involvement of the chest wall or a significant tumor blocking the airway can result in chest pain. This pain is often worse with deep breathing, coughing, or laughing. Shortness of breath is a common symptom and can be associated with wheezing. Some individuals also develop infections such as bronchitis or pneumonia.

It is well recognized that cancer can result in weight loss. Sometimes this is due to a loss of appetite. Lung cancer may also be associated with symptoms of fatigue, tiredness, or weakness. If the cancer spreads to distant organs it can result in bone pain or nervous system changes including headache, weakness, dizziness, difficulty walking, and seizures. Involvement of the lymph nodes can result in swollen glands or lumps that you can feel under the skin.

DIAGNOSIS

The approach to the diagnosis of lung cancer is highly dependent on where the suspected cancer is located. Regardless, it generally requires obtaining cancer cells. The goal is to obtain enough cells to perform all the needed pathology tests to determine what type of cancer it is and what potential therapies a patient may be a candidate for. An additional requirement prior to therapy is the determination of the stage of the cancer, or how extensive it is. Ideally, meeting all of these requirements would only require a single procedure.

If there is evidence of spread outside of the chest to other areas of the body, the recommended approach would be to perform a biopsy of that area. This is generally a percutaneous (through the skin) needle biopsy that is performed under ultrasound or CT guidance. The patient is given pain medications, and medications to make them feel sleepy during the procedure. Sometimes ultrasound or CT scan is not needed. If there is a suspicious lesion under the skin for instance, a needle biopsy can sometimes be performed with just local numbing medication.

If fluid has built up between the lung and chest wall (e.g., a pleural effusion) due to lung cancer, cancer cells may potentially be obtained from the fluid by a simple needle procedure (thoracentesis). This is an ideal location to attempt to obtain cancer cells since it also provides information about where the cancer may have spread. However, if the cancer has indeed involved the inside of the chest wall and is causing fluid build-up, cancer cells are detectable in the fluid only about 60% of the time [10]. If a thoracentesis has been performed and there are no cancer cells obtained, biopsies of the inside of the chest wall (e.g., the pleura) may be recommended. This procedure is performed by an interventional pulmonologist or surgeon. During this procedure, referred to as pleural biopsy, the patient is sedated and a small incision (<1 cm) is made in the chest to allow entry of a camera. The cancer on the inside of the chest wall may then be visualized using the camera and directly biopsied. This is considered the "gold standard" procedure and has a very high diagnostic yield [11]. Therapies to treat the fluid buildup, such as pleurodesis or tunneled pleural catheter placement, may be performed at the same time.

If there is a nodule or mass close to the middle of the chest or that has blocked off a breathing tube, a bronchoscopy may be recommended. The patient is given pain medications and medications to make them feel sleepy during the procedure and a camera (bronchoscope) is placed into the airway. The suspected cancer can then be identified and biopsies performed. If the suspected cancer is not inside the airway but next to a large airway a bronchoscope with a small ultrasound on it may be used to guide needle biopsies (endobronchial ultrasound, EBUS). This approach has the advantage that "invasive" staging of the lymph nodes lying along the airway may be performed at the same time as biopsies of the cancer by taking small needle biopsies of these lymph nodes. Imaging modalities such as positron emission tomography PET scanning (a way to label sugar and determine what structures actively take up the sugar) only detect cancer in these lymph nodes when it is actually present approximately 7 times out of 10. Thus, invasive staging is recommended when the cancer is large or closer to the middle of the chest [12].

If there is a nodule or mass in the outer part of the lung, a percutaneous (through the skin) needle biopsy may be recommended. This procedure is generally performed under CT scan guidance. Similar to percutaneous biopsies of other parts of the body, the patient is given pain medications and medications to make them feel sleepy during the procedure.

STAGING

Staging is the process of determining how far a cancer might have spread. There are very detailed guidelines, based on data from hundreds of thousands of patients, which tell us how to determine the final stage [13]. Once the lung cancer type and stage are determined, then a treatment plan can be implemented. Staging of the lung cancer is based on characteristics of the tumor itself (T stage), extent lymph node involvement (N stage), and presence of cancer outside the lung (metastasis, M stage). The T stage is predominantly determined by the size and location of the primary tumor (e.g., where it started) on a CT scan or potentially after it has been surgically removed (if a treatment option). Lymph node staging in lung cancer is based on the location (not number) of lymph nodes involved with cancer. It is determined after careful review of the CT scan, PET scan, and the results of lymph node biopsies. The M stage is also determined through review of imaging and biopsy. A final overall stage (I, II, III, or IV) can then be determined and appropriate treatment options discussed (detailed later).

TYPES OF LUNG CANCER

Determining the specific type of lung cancer is extremely important, as it greatly informs treatment decision-making. There are two main categories of lung cancer: non-small cell and small cell. Of lung cancers 85–90% are within the non-small cell group.

Non-small cell lung cancers are further divided into: adenocarcinoma (50% of non-small cell lung cancers); squamous cell (30%), with the remaining approximately 20% comprising of a mix of different cell types, such as large cell, adenosquamous, poorly differentiated.

Adenocarcinomas: The most common type of lung cancer in both smoker and nonsmokers. These may be slow-growing cancers that arise in the outer portions of the lung.

Squamous cell: Strongly associated with smoking and often arises from the large airways.

Small cell lung cancers are rare in nonsmokers. These tend to be faster growing tumors than non-small cell lung cancers and are often diagnosed with metastatic disease.

TREATMENT

Non-small Cell Lung Cancer

The stage of non-small cell lung cancer often determines the nature of treatment. The stage is calculated by: the size of the tumor, the location of any involved lymph nodes, and if there are metastases. Also, the cell type of non-small cell lung cancer (adenocarcinoma, squamous cell, etc.) can play a large part in determining the best treatment.

Treatment by stage:
Stage I: surgery or radiation (if not a surgical candidate).

Stage II: surgery or radiation (if not a surgical candidate). If there are aggressive features to the tumor, namely lymph nodes involved or large tumor, chemotherapy is given after surgery. Sometimes radiation may be needed to the surgical site.

Stage III: Based on the size and extent of the cancer spread, surgery plus postoperative chemotherapy, or treatment with chemotherapy and radiation (e.g., chemoradiation) are considered. In some instances, giving chemoradiation therapy upfront can shrink a tumor enough to have it surgically removed.

Stage IV: Systemic treatment. Radiation can be used in short courses to treat symptomatic metastases (e.g., lung tumor causing difficulties with breathing; pain from a bone metastasis). It can also be given as whole brain irradiation or focused treatment to an area for brain metastases. In the event of a solitary brain metastases, surgical resection, followed by surgical cavity irradiation may be possible. Generally, stage IV lung cancer cannot be cured.

In earlier stages of cancer, when chemoradiation is used, a fairly standard combination is radiation for about six to seven weeks, five days a week, plus chemotherapy given weekly for eight weeks. The chemotherapy duration may be longer, if given before surgery. One of the most common regimens is with carboplatin and paclitaxel. In certain Stage III cancers treated with chemoradiotherapy, there is role to give immunotherapy for a year after to decrease the risk of the cancer coming back [14].

Treatment according to cell type:
With stage IV cancer, treatment can be affected by the cell type, as follows:
Adenocarcinoma
When it comes to Stage IV cancer, more information is beneficial to identify the optimal treatment. Thus, when a diagnosis of adenocarcinoma is made, the sample is sent off for molecular marker testing, also known as non-genomic sequencing (NGS). The frequency of these markers are low, approximately 10–15% percent of

tumors. For example, a study of targeted agent erlotinib versus chemotherapy in first treatment for Stage IV cancer showed significantly improved response rates with erlotinib in individuals whose cancer expressed the EGFR mutation [15]. Comparable findings have been shown for other targetable mutations.

Some examples of targets and their medication:

> EGFR (15% of adenocarcinomas) – erlotinib, afatinib
>
> ALK (4%) – crizotinib, alectinib, ceritinib
>
> ROS1, RET (<4%) – cabozantinib

Another test that is performed on tumor tissue is PD-L1 testing. Positivity for this indicates a good response to immunotherapy with pembrolizumab. In individuals whose tumor was highly positive for PD-L1, pembrolizumab in first-line treatment improved survival and produced long-lasting responses in comparison to first-line chemotherapy [16].

There is still a role for chemotherapy as first-line treatment for metastatic cancer. The molecular marker testing can take approximately three weeks to result, and if an individual has a large tumor burden, or is very symptomatic, treatment needs to start sooner with chemotherapy. Chemotherapy also has the advantage of more quickly killing cancers cells than targeted therapies/immunotherapies, which often take one to two months to start improving symptoms. There are a number of chemotherapy options and one that is common for adenocarcinoma is carboplatin or cisplatin plus pemetrexed, a chemotherapy medicine that is especially active against adenocarcinoma. This is typically given once every three weeks for four to six cycles. There is also a role to continue every three week pemetrexed after finishing the combination treatment as "maintenance therapy," i.e., to maintain the reduction in tumor after the initial therapy. Maintenance with pemetrexed has been shown to be well tolerated and yielded an improvement in progression free survival of approximately two months [17].

Squamous cell

This cell type is not associated with expressing targets for specific agents and thus, the first line remains chemotherapy, typically carboplatin or cisplatin plus one of many other chemotherapeutic agents, for example, paclitaxel. This regimen is given once every three weeks for four to six cycles. Maintenance chemotherapy with medicines from the paclitaxel family have been studied, but without the benefit seen with pemetrexed in adenocarcinoma, and much more treatment limiting toxicity. There is no meaningful role for pemetrexed in squamous cell carcinoma.

Recurrent Non-Small Cell Lung Cancer

In both non-squamous and squamous type non-small cell lung cancers, there is a role for immunotherapy with nivolumab if cancer recurs after first-line chemotherapy. In a head-to-head comparison of nivolumab with what had previously been the "gold standard" second-line treatment, taxotere, response rates and progression-free survival rates almost doubled with nivolumab. There was also an improvement in overall survival with nivolumab. Based on these findings, nivolumab is now the considered second line agent [18, 19].

Large cell and other variants of non-small lung cancer

These are usually treated akin to squamous cell cancers.

Small Cell Lung Cancer

With small cell lung cancers, the key to deciding about treatment rests on whether the volume of the lung tumor can be treated with radiation, if so, treatment comprises a combination of chemotherapy and radiation. A typical treatment regimen would be radiation five days a week for seven weeks along with a combination of two chemotherapy medicines given for three days in a row every three weeks. The three week period is considered a "cycle" of treatment. The chemotherapy continues after the radiation is done. Typically, four to six cycles of chemotherapy are administered. The typical regimen consists of either carboplatin or cisplatin plus etoposide. Atezolizumab has been approved for first line treatment in combination with carboplatin and etoposide after the IMpower133 Phase 3 trial of triplet therapy versus chemotherapy plus placebo showed a statistically significant improvement in median overall survival (12.3 months versus 10.3 months) and progression free survival (5.3 months versus 4.2 months) in the atezolizumab containing arm [20].

If the volume of the lung cancer is too large to be treated with radiation, or if there are other sites of cancer within the body, treatment is with chemotherapy alone. In this setting it is still given three days in a row every three weeks, with four to six cycles administered.

The risk of small cell recurring in the brain is estimated to be 50% at two years from diagnosis. The risk increases with increased length of time out from diagnosis [21]. As local and systemic therapies continue to improve survival rates, the importance of trying to reduce the risk of brain metastases becomes more and more important. Radiation to the brain may be offered after completing radiation and/or chemotherapy to reduce this risk. This is called prophylactic cranial irradiation (PCI), and a review of a number of studies with strict entry criteria demonstrated a significant reduction in development of brain metastases and an improvement in survival [22].

The risk of short-term memory loss and other difficulties with cognition is present with PCI, but hard to quantify, with increased risk associated with increased age at the time of PCI. The dose of radiation or the length can also contribute. Additionally, chemotherapy may contribute to neurocognitive difficulties (also known as "chemobrain") that make it hard to tell what is purely from PCI. The risk can be reduced by sparing the hippocampus of the brain.

Of individuals diagnosed 10 to 15% with small lung cancer have brain metastases at the time of diagnosis. Depending on the number of locations of the metastases in the brain, these can be treated by surgery, whole brain irradiation, or stereotactic radiosurgery. When surgically removed, the cavity that remains is treated with radiation to prevent the risk of local recurrence. Whole brain irradiation is typically a higher dose than PCI, as it is being used to treat existing metastases, rather than prevent them. A drug called memantine was studied in patients receiving whole brain irradiation, resulting in significantly improved mental processing speed, long-term memory, and efficiency at performing tasks [23]. So, this is often offered during whole brain radiation for a duration of 24 weeks.

Although small cell lung cancer is very responsive to upfront treatment with radiation and/or chemotherapy, unfortunately the rate of recurrence is very high, and difficult to treat in this context. To date, the standard second treatment option remains a chemotherapy called topotecan, with response rates ranging about 15–25%. There are also other chemotherapies that can be used at the time of recurrence. Immunotherapy with a medicine called nivolumab, either alone or in combination with another immunotherapy called Ipilimumab has been studied for recurrent small cell lung cancer, with improved response rates that are lasting anywhere from four to greater than six months, with the average not having been reached yet in the study population [24]. There are studies that are ongoing that compare immunotherapy to chemotherapy. Currently, the decision to proceed with chemotherapy versus immunotherapy after recurrence is based on prior treatments, medical comorbidities, and amount of tumor.

IMPORTANT ADDITIONAL TREATMENTS IN NON-SMALL AND SMALL CELL LUNG CANCERS

Zoledronic Acid and Denosumab

These are medications used to help heal damage from metastases in the bone, and prevent fractures, and associated morbidity and mortality related to such fractures. Significant improvements in overall survival have been seen in individuals with non-small cell lung cancer on these agents [25].

Palliative Care

Palliative Care is a valuable resource, regardless of cancer stage in lung cancer, for assistance in coming to terms with the diagnosis, symptom management, and support in end-of-life discussions and care. Early institution of palliative care in addition to standard of care treatment in individuals with newly diagnosed metastatic cancer resulted in significantly lower rates of depression, a reduction of overly aggressive end-of-life treatment, and improved median survival versus those who did not receive Palliative Care [26].

Exercise During Cancer Treatment

Regular exercise during treatment helps with treatment related fatigue; alleviates stress, and supports muscle strength and joint flexibility to maintain stamina. Ideally, it is a mix of aerobic exercise to raise the heart rate and strength training.

The NCCN Clinical Practice Guidelines in Oncology recommendation is for 30 minutes of exercise a day, five days a week [27]. This can be broken up into three 10-minute exercise sets if there is insufficient strength to do it all in one block. Many cancer centers and physical therapy offices offer supervised cancer-directed programs. The University of Vermont Medical Center has the Steps to Wellness Oncology program. Always check with your Oncologist or Primary Care Provider before starting a new exercise regimen when on treatment to ensure it is safe for you.

SURVIVORSHIP AND SURVEILLANCE

Surveillance

The general guidelines for surveillance of non-small and small cell lung cancers is established by the NCCN Guidelines, and modified according to an individual's scan results and clinical symptoms (if any). The current recommendations are for five years of active surveillance with an Oncologist, after which time, surveillance can be transitioned back to the individual's Primary Care Provider and incorporated into an annual physical exam.

Non-small Cell

A common surveillance regimen after treatment would be a physical examination, symptom review, and CT scan every four to six months, with transition to annual visit and CT as appropriate to the stage and the individual.

Small Cell

A common surveillance regimen after treatment would be a physical examination, symptom review, and CT scan every three months, with subsequent transition to every six months. Typically small cell is followed at closer intervals for longer, given its strong propensity to recur or metastasize.

Survivorship

Some of the more common potential long-lasting sequelae of treatment for lung cancer may include:

  Numbness or tingling in the hands and feet called neuropathy or parethesias (chemotherapy);
  Ringing in the ears called tinnitus, or decreased hearing (chemotherapy or brain irradiation);

Shortness of breath (chest radiation);

Difficulty swallowing called dysphagia (chest radiation);

Permanent hair loss (whole brain radiation).

Survivorship care plans are now being prepared after receiving treatment for a number of Hematologic and Oncologic malignancies. These are personalized documents that give the names and doses of all chemotherapies, radiation treatments, and surgeries an individual might have received as part of their treatment. It highlights potential associated longer term side effects for which an individual and their Oncologist and Primary Care Provider should look. Examples of survivorship documents can be found on the American Cancer Society [28] and American Society of Clinical Oncology [29] websites.

CONFLICTS OF INTEREST

The authors of this manuscript report no financial or other conflicts of interest related to the work.

REFERENCES

1. Siegel, R.L., Miller, K.D., and Jemal, A. (2015). Cancer statistics, 2015. *CA: A Cancer Journal for Clinicians* 65 (1): 5–29. https://doi.org/10.3322/caac.21254.

2. https://seer.cancer.gov/csr/1975_2014, accessed 17 January 2018

3. Hecht, S.S. (2012). Lung carcinogenesis by tobacco smoke. *International Journal of Cancer* 131 (12): 2724–2732. https://doi.org/10.1002/ijc.27816.

4. Lantz, P.M., Mendez, D., and Philbert, M.A. (2013). Radon, smoking, and lung cancer: the need to refocus radon control policy. *American Journal of Public Health* 103 (3): 443–447. https://doi.org/10.2105/AJPH.2012.300926.

5. International Early Lung Cancer Action Program Investigators, Henschke, C.I., Yankelevitz, D.F. et al. (2006). Survival of patients with stage I lung cancer detected on CT screening. *The New England Journal of Medicine* 355 (17): 1763–1771. https://doi.org/10.1056/NEJMoa060476.

6. National Lung Screening Trial Research Team, Aberle, D.R., Adams, A.M. et al. (2011). Reduced lung-cancer mortality with low-dose computed tomographic screening. *The New England Journal of Medicine* 365 (5): 395–409. https://doi.org/10.1056/NEJMoa1102873.

7. Patz, E.F., Pinsky, P., Gatsonis, C. et al. (2014). Overdiagnosis in low-dose computed tomography screening for lung cancer. *JAMA Internal Medicine* 174 (2): 269–274. https://doi.org/10.1001/jamainternmed.2013.12738.

8. Colice, G.L. (1997). Detecting lung cancer as a cause of hemoptysis in patients with a normal chest radiograph: bronchoscopy vs CT. *Chest* 111: 877–884. https://doi.org/10.1378/chest.111.4.877.

9. Corner, J. (2005). Is late diagnosis of lung cancer inevitable? Interview study of patients' recollections of symptoms before diagnosis. *Thorax* 60 (4): 314–319. https://doi.org/10.1136/thx.2004.029264.

10. Hooper, C., Lee, Y.C.G., Maskell, N. et al. (2010, August). Investigation of a unilateral pleural effusion in adults: British thoracic society pleural disease guideline 2010. *Thorax* 65 (2): 4–17. https://doi.org/10.1136/thx.2010.136978.

11. Noppen, M. (2010). The utility of thoracoscopy in the diagnosis and management of pleural disease. *Seminars in Respiratory and Critical Care Medicine* 31 (6): 751–759. https://doi.org/10.1055/s-0030-1269835.

12. Silvestri, G.A., Gonzalez, A.V., Jantz, M.A. et al. (2013). Methods for staging non-small cell lung cancer: diagnosis and management of lung cancer, 3rd ed: American college of chest physicians evidence-based clinical practice guidelines. *Chest* 143 (5 Suppl): e211S–e250S. https://doi.org/10.1378/chest.12-2355.

13. https://www.iaslc.org/research-education/staging

14. Antonia, S.J., Villegas, A., Daniel, D. et al. (2017). Durvalumab after chemoradiotherapy in stage III non-small-cell lung cancer. *The New England Journal of Medicine* 377 (20): 1919–1929. https://doi.org/10.1056/NEJMoa1709937.

15. Kobayashi, S., Ji, H., Yuza, Y. et al. (2005). An alternative inhibitor overcomes resistance caused by a mutation of the epidermal growth factor receptor. *Cancer Research* 65 (16): 7096–7101. https://doi.org/10.1158/0008-5472.CAN-05-1346.

16. Reck, M., Rodríguez-Abreu, D., Robinson, A.G. et al. (2016). Pembrolizumab versus chemotherapy for PD-L1-positive non-small-cell lung cancer. *The New England Journal of Medicine* 375 (19): 1823–1833. https://doi.org/10.1056/NEJMoa1606774.

17. Paz-Ares, L., De Marinis, F., Dediu, M. et al. (2012). Maintenance therapy with pemetrexed plus best supportive care versus placebo plus best supportive care after induction therapy with pemetrexed plus cisplatin for advanced non-squamous non-small-cell lung cancer (PARAMOUNT): a double-blind, phase 3, randomised controlled trial. *The Lancet Oncology* 13 (3): 247–255. https://doi.org/10.1016/S1470-2045(12)70063-3.

18. Borghaei, H., Paz-Ares, L., Horn, L. et al. (2015). Nivolumab versus Docetaxel in advanced nonsquamous non-small-cell lung cancer. *The New England Journal of Medicine* 373 (17): 1627–1639. https://doi.org/10.1056/NEJMoa1507643.

19. Brahmer, J., Reckamp, K.L., Baas, P. et al. (2015). Nivolumab versus Docetaxel in advanced squamous-cell non-small-cell lung cancer. *The New England Journal of Medicine* 373 (2): 123–135.

20. Horn, L., Mansfield, A.S., Szczęsna, A. et al. (2018). First-line atezolizumab plust chemotherapy in extensive stage small cell lung cancer. *NEJM* 379: 2220–2229.

21. Yang, G.Y. and Matthews, R.H. (2000). Prophylactic cranial irradiation in small-cell lung cancer. *The Oncologist* 5 (4): 293–298.

22. Aupérin, A., Arriagada, R., Pignon, J.P. et al. (1999). Prophylactic cranial irradiation for patients with small-cell lung cancer in complete remission. Prophylactic Cranial Irradiation Overview Collaborative Group. *The New England Journal of Medicine* 341 (7): 476–484. https://doi.org/10.1056/NEJM199908123410703.

23. Brown, P.D., Pugh, S., Laack, N.N. et al. (2013). Memantine for the prevention of cognitive dysfunction in patients receiving whole-brain radiotherapy: a randomized, double-blind, placebo-controlled trial. *Neuro-Oncology* 15 (10): 1429–1437. https://doi.org/10.1093/neuonc/not114.

24. Antonia, S.J., López-Martin, J.A., Bendell, J. et al. (2016). Nivolumab alone and nivolumab plus ipilimumab in recurrent small-cell lung cancer (CheckMate 032): a multicentre, open-label, phase 1/2 trial. *The Lancet Oncology* 17 (7): 883–895. https://doi.org/10.1016/S1470-2045(16)30098-5.

25. Hirsh, V., Major, P.P., Lipton, A. et al. (2008). Zoledronic acid and survival in patients with metastatic bone disease from lung cancer and elevated markers of osteoclast activity. *Journal of Thoracic Oncology: Official Publication of the International Association for the Study of Lung Cancer* 3 (3): 228–236. https://doi.org/10.1097/JTO.0b013e3181651c0e.

26. Temel, J.S., Greer, J.A., Muzikansky, A. et al. (2010). Early palliative care for patients with metastatic non-small-cell lung cancer. *The New England Journal of Medicine* 363 (8): 733–742. https://doi.org/10.1056/NEJMoa1000678.

27. https://www.nccn.org/patients/resources/life_with_cancer/exercise.aspx

28. https://www.cancer.net/survivorship/follow-care-after-cancer-treatment/asco-cancer-treatment-and-survivorship-care-plans

29. https://www.cancer.org/treatment/survivorship-during-and-after-treatment/survivorship-care-plans.html

15

PEDIATRIC ONCOLOGY

Jessica L. Heath

Pediatric Hematology-Oncology, University of Vermont, Burlington, VT, USA

Childhood cancer is in many ways quite different from cancer in the adult population. The types of cancers that are commonly seen, the treatments that are used, and the outcomes all differ significantly. In addition, there are unique challenges to overcome for children and families facing a pediatric cancer diagnosis. This chapter will cover cancer prevention, detection, treatment, and recovery in the context of treating the child with cancer as part of a whole family. It will additionally provide some exploration of the issues and challenges that are specific to children and families.

All forms of childhood cancers are rare diseases. The incidence of childhood cancer in the United States is 0.17%, meaning that about 15 000 children are diagnosed with cancer each year. The rarity of pediatric cancer has implications for how it is studied and treated, as will be discussed in detail below. Childhood cancer is a significant problem in the United States and globally. It is the leading disease related cause of death of children under the age of 14 in the United States. Overall, success rates for curing many forms of childhood cancer have increased dramatically over the past 50 to 60 years, largely as a result of cooperative group efforts and multicenter clinical trials. Other pediatric cancers remain harder to treat, and ongoing research into better treatment options for these aggressive tumor types is happening around the world.

Cancer: Prevention, Early Detection, Treatment and Recovery, Second Edition. Edited by Gary S. Stein and Kimberly P. Luebbers.
© 2019 John Wiley & Sons, Inc. Published 2019 by John Wiley & Sons, Inc.

WHY DO KIDS GET CANCER?

In the vast majority of cases, we don't know why some children get cancer and others don't. The development of cancer in a child may be the result of a complex interplay of genetic risk and environmental exposure. Childhood cancer may affect anyone, regardless of race, gender, socioeconomic status, or zip code. Unfortunately, it appears as though the incidence of childhood cancer is on the rise, with a 1–3% annual increase in incidence. This increase is largely reflective of an increasing incidence of cancer in children less than one year of age. The force behind this increase is likely multifactorial, and is poorly understood.

There is some thought that toxic environmental exposures may increase the risk of childhood cancer. There have been few comprehensive studies done in this area, but there is some emerging data on some early childhood exposures. For example, exposure to tobacco products prenatally or during infancy and early childhood has been associated with an increased risk of developing leukemia. There have also been some studies done looking at a possible relationship between prenatal or early childhood insecticide exposure and the development of leukemia. At this point, studies regarding the risk of childhood cancer from environmental exposures are preliminary in nature, and do not provide hard and fast evidence that these exposures cause cancer in children, but do merit further study.

There is a special environmental concern regarding the possible contribution of radiation exposure to the risk of developing pediatric cancer. While it is well established that high dose radiation therapy given for the treatment of cancer does put one at risk for the development of a subsequent cancer, it is harder to predict the risk of developing cancer due to infrequent exposure by the occasional X-ray or CT scan, and even harder to predict risk due to ambient or background radiation. That being said, most hospitals with a pediatric specialty do have specific protocols in place to minimize radiation exposure from routine radiologic procedures like CT scans.

There are some preexisting medical conditions that put children at higher risk for certain cancers. Some children are born with genetic changes that put them in jeopardy for developing cancer at some point later in life, either in childhood or as an adult. These cancer predisposition syndromes are sometimes inherited from a parent, and other times are found "de novo," or new, in that child. If a child is found to have a cancer predisposition syndrome, their doctor may recommend additional specialized care. This can include routine testing to detect cancer early, such as periodic bone marrow biopsy or whole body MRI. Children with some cancer predisposition syndromes, such as those that predispose to leukemias, will undergo bone marrow transplantation in an attempt to abort the potential development of leukemia. Children with Down Syndrome are at increased risk for the development of leukemia compared with other children; however, leukemia is still rare in this population, and routine screening is not recommended.

Outside these rare cases, including cancer predisposition syndromes and certain exposures such as chemotherapy or radiation therapy for a prior cancer, it is impossible to predict which children will suffer from these very rare diseases. This makes efforts at early detection in the general population challenging, as early detection efforts often rely on identifying a specific "at risk" population, or having an easy, reliable, and inexpensive test for a common disease. In the absence of these conditions, there is a high risk of false positive or false negative screening results. In the case of false positives, a child would theoretically undergo additional, possibly invasive, testing in order to confirm/refute a cancer diagnosis. The financial burden on the healthcare system, the emotional burden on the family, and the medical burden of unnecessary testing in the case of false positive results makes any type of cancer screening program in childhood not feasible. There are therefore no screening or early detection programs in place. Cancer detection relies instead on the clinical acumen of physicians on the frontlines of medicine – in the primary care offices, urgent care centers, and emergency rooms – where pediatric cancer patients are most likely to present.

These same issues make efforts at prevention of pediatric cancer especially difficult. In nearly all cases of childhood cancer, there is nothing that a parent or child has done to result in that cancer, and there is nothing that could have been done to prevent it from happening. However, efforts are in place that are aimed to encourage healthy lifestyle adoption and at the prevention of adult cancers. These include providing guidance to patients and caregivers at routine well-child visits regarding the dangers of tobacco products, alcohol, and other drugs, giving advice regarding nutrition and obesity avoidance, and offering the human papillomavirus (HPV) vaccine.

WHAT KINDS OF CANCER IMPACT CHILDREN?

Children are impacted by a myriad of solid tumors and leukemias. The types of cancer for which children are at highest risk vary by age (Figure 1). Leukemias are the most common cancer in children under the age of 15. There are two principle types of leukemia, acute lymphoblastic leukemia (ALL), and acute myeloid leukemia (AML). Together, ALL and AML comprise 30% of all childhood cancers, with ALL making up 23% of diagnoses. Brain tumors as a group comprise the next most common category of cancers afflicting children. Infants and very young children may experience cancers that are unique to that age group and not typically found in older children or adults. These include retinoblastoma, hepatoblastoma, neuroblastoma, and nephroblastoma (or Wilms' tumor).

There is a change in cancer epidemiology as children age. In adolescence and young adulthood (ages 15–35), there is a virtual absence of the more developmentally immature cancers (retinoblastoma, hepatoblastoma, nephroblastoma). Lymphomas and germ cell tumors become the most dominant cancers in this age group. We also begin to see the emergence of more "adult-type" tumors such as melanoma, breast cancer, and cervical cancer as children age into their early 20s.

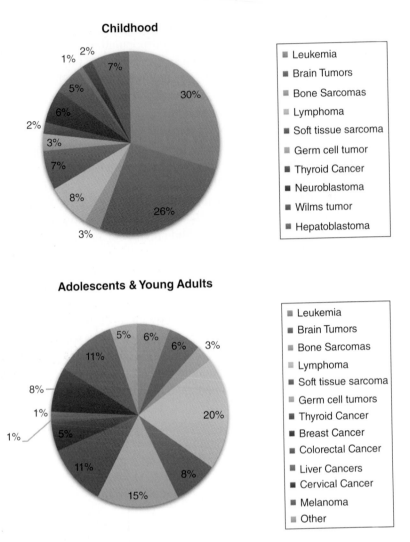

Figure 1. Most Common Cancers in Children and Adolescents. The cancers seen most commonly in children under the age of 15 years are shown in the top panel. Leukemia and brain tumors are by far the most common cancers seen in this age group. The cancers most commonly seen in adolescents and young adults (AYAs) (ages 15–29 years) are shown in the lower panel. The most frequently seen cancers in this age group differ from what is seen in younger children, and include lymphoma and germ cell tumors (including testicular cancer). (See insert for color representation.)

There are some tumor types that are seen across childhood, adolescence, and into young adulthood. These include leukemia, lymphoma, and brain tumors as mentioned above, as well as bone and soft tissue sarcomas. Interestingly, even a cancer type that may be seen both in children and adults is often markedly different in childhood compared with what is seen in adulthood. This is true in terms of the genetics and biology of the tumor, as well as in the treatment response. There is ongoing scientific inquiry as to what may underlie these differences; currently it is not well understood.

TREATMENTS FOR CHILDHOOD CANCER

Childhood cancers are treated with varying combinations of surgery, chemotherapy, radiation therapy, and/or hematopoietic stem cell transplantation (bone marrow transplant), depending on the specific cancer type. In contrast to adult cancer patients, with a minority treated on clinical trials, the majority of children with cancer are enrolled on a clinical trial. Because childhood cancer is so rare, there are very few institutions that are large enough to conduct their own independent clinical trials. Thus, many clinical trials in pediatric oncology are conducted through multicenter cooperative groups. The Children's Oncology Group (COG) is one such group and is composed of over 9000 experts in pediatric cancer at over 200 institutions throughout North America, Europe, Australia, and New Zealand. This group is responsible for orchestrating and carrying out many of the clinical trials on which children with cancer are treated. Approximately 60% of children with cancer in the United States are treated on one of these trials, and many more are enrolled on patient registries and tumor banking studies.

Cooperative group clinical trials are an important driving force behind the improvements in cure rates that have been seen in many childhood cancers over the past 50 to 60 years. The most dramatic example is found in the most common type of childhood cancer, ALL. In the 1950s, cure rates for childhood ALL were close to 0%. Today, children with some types of ALL have cure rates of 97%. Many other pediatric cancers have also seen substantial improvements in outcome as a result of these clinical trials.

Unfortunately, the successes that we have seen in the treatment of childhood cancer have come at a high price. Most childhood cancers are still treated with combinations of traditional chemotherapy, often administered at high doses, and radiation therapy. These treatments typically work by killing rapidly dividing cells (they are "cytotoxic"), rather than specifically targeting the cancer cells themselves. This leads to a high burden of toxicity, and side effects that may be seen in the short or long term. Long-term side effects of cancer therapy are termed "late effects," and are particularly prominent in children, as survivors of childhood cancer are expected to live many decades beyond the end of their cancer-directed therapy. In fact, approximately 60% of children treated for cancer will experience at least one late effect as a result of their treatment; many survivors experience multiple toxicities. These late effects can range from organ damage such as to the kidneys, liver, or heart, to neurocognitive effects, to infertility, to long-term psychological effects resulting from the trauma of the experience.

A preferable alternative to using broadly cytotoxic medicines would be to use therapies that are targeted toward each individual tumor. Targeted therapies may allow for increased cure rates, but also decreased short- and long-term toxicities, including late effects as described above. Targeted therapies are being developed to a great degree in a variety of cancers that are prominent in adults, but less so in pediatric oncology. There are multiple factors contributing to this discrepancy. First, many targeted therapies are directed toward specific gene mutations found in the cancer cells. Pediatric cancers, in general, carry a much lower mutational burden than do adult cancers, resulting in fewer "targetable" lesions. Second, because of the rarity of pediatric cancers, it is difficult to conduct early phase clinical trials with enough patients to obtain meaningful and informative results. For example, renal cell carcinoma is a very difficult tumor to treat. It is typically treated with surgery, and is not responsive to radiation therapy, or to any traditional chemotherapy agent. Clinical trials to test novel targeted therapies would be incredibly helpful and informative in this cancer; however, only seven to eight children per year in the United States are diagnosed with this rare tumor, making it extremely difficult to accumulate enough data to make strong determinations of the efficacy of a new therapy.

Despite these difficulties, some targeted therapies have been used with tremendous success in pediatric cancers. One example is the case of a particularly aggressive form of ALL, called Ph+ ALL. This leukemia is characterized by the Philadelphia chromosome (hence the Ph+), which is an abnormal rearrangement of genetic material between two chromosomes in the leukemia cell. This rearrangement results in an overactive protein (an enzyme called a kinase). Tyrosine kinase inhibitors have been developed to block this overactive protein, and thereby kill the leukemia cells. Inclusion of tyrosine kinase inhibitors into chemotherapy regimens has increased the survival for this type of leukemia by 50%! This is an example of the outstanding potential of targeted therapies.

Some targeted therapies seek to harness the power of the immune system to kill off tumor cells. This strategy has also been used effectively in pediatric cancer; a great example is seen in a subset of high risk neuroblastoma. Neuroblastoma is typically seen in young children, under the age of five, and is an abnormal proliferation of the very early nerve cells. These neuroblastoma cells express a protein on their surface that is unique to the tumor cells, called disialoganglioside, or GD2. GD2 is only expressed on a few types of normal cells, such as some cells of the nervous system. Antibodies have been designed to recognize GD2 on the cell surface, and trigger an attack by the immune system. The inclusion of anti-GD2 directed immunotherapy to an intensive regimen of chemotherapy, radiation therapy, surgery, and stem cell transplantation has more than doubled survival rates for children suffering from this aggressive cancer. There are several similar therapies currently in clinical trials for other pediatric cancers, attempting to harness the power of the immune system to directly target tumor cells, and leave noncancerous cells intact.

The healthcare of survivors of childhood cancer is therefore best undertaken by physicians who are knowledgeable in the field of late effects and experienced in the care of survivors. As the treatments for childhood cancers have improved, the number of adult survivors of childhood cancer in the United States is steadily growing, and is expected to be over 500 000 by the year 2020. Currently, there are very few programs that are well versed in the care of adult survivors of childhood cancer, and the transition of adult long-term survivors from their pediatric oncology clinic to a new medical home is a great challenge. With the increasing numbers of survivors due to improved cancer care, this is an area that will need to be addressed in coming years.

WHAT IS THE IMPACT OF A CANCER DIAGNOSIS ON A CHILD?

A cancer diagnosis in childhood carries with it a myriad of effects, not only on the physical health of the child, but also on the psychological health, social well-being, and educational outcomes for that child. Like physical late effects of cancer treatment described above, children can also sustain psychological late effects as a result of having gone through the traumatic experience of cancer. Children and adolescents with cancer are at increased risk for depression, anxiety, chronic worrying, and post-traumatic stress disorder during or after treatment. The adverse physical effects of cancer therapy certainly contribute to this, and interact with social determinants of a child's psychological health. Cancer treatment in general carries obvious short-term side effects, including severe fatigue, nausea, vomiting, weight loss, and loss of hair. Some children have surgeries, and may endure scarring, loss of limb, or other physical sequelae. Children treated with chemotherapy have compromised immune systems, and are at a great increased risk for infection. These overt physical impacts of treatment alter the ways in which children are able to interact with their peers. For children who are visibly different from their peers as a result of their treatment, social interactions can be quite challenging. Many children with cancer report feeling self-conscious and isolated from their peer groups due to changes in their physical appearance. This sense of isolation is magnified by actual absences from school and extracurricular activities. Most children with cancer miss some part of school, due to the aforementioned side effects, as well as frequent visits to the clinic or prolonged hospitalizations. For many children, the school environment is where friendships are developed, and the bulk of their social interactions had previously taken place.

Missing school also impacts the child's ability to achieve educational and vocational success. Even though individualized educational plans through school, or home tutoring is made available, children with cancer are at increased risk for needing to repeat a grade, or failing individual subjects. This is most notable in math and science classes, where ordered, sequential thinking and significant active working memory are necessary for learning. Challenges to success are seen in all pediatric cancer patients for the above reasons. Children with brain tumors and leukemia, i.e., those children receiving chemotherapy or radiation therapy that is specifically directed at their brain and central nervous system are at particularly high risk for having difficulty in school, both in the short and long term. Neurocognitive effects of cancer therapy are well described, and our understanding of how chemotherapy and radiation therapy affect the developing brain and nervous system is continuing to evolve. Negative effects on memory, higher order executive function, attention, and processing speed are well documented. These effects may be temporary, and may improve with time; however a substantial percentage of children will be left with lasting negative effects in these domains, hindering academic success.

While this paints a relatively toxic picture of the psychological health of survivors of childhood cancer, the reality is that many survivors are thriving. In general, survivors of childhood cancer self-report health-related quality of life and life satisfaction scores that are comparable to their siblings, and to the general population. We often think of the negative psychological impact of a cancer diagnosis on a child, and it is counterintuitive to consider that there may also be a positive impact. However, there is some evidence emerging that survivors may develop positive traits as a result of their experience, including increased resilience.

The myriad of challenges encountered by the child with cancer does necessitate a multidisciplinary team in order to address the physical health issues and also the psychosocial needs of the child. A comprehensive care team for the pediatric cancer patient therefore includes physicians, nurses, advanced practice providers, licensed clinical social workers, child life specialists, neurocognitive and learning evaluations, physical and occupational therapists, nutritionists, and chaplains or spiritual advisors.

HOW DOES A CHILDHOOD CANCER DIAGNOSIS AFFECT THE FAMILY?

A diagnosis of cancer in a child has far reaching consequences, not only for that child, but also for the family and social system in which that child exists (Figure 2). The parents and siblings of children with cancer have been interviewed and studied the most within this context; however, it is not unreasonable to anticipate that the extended family and social circles around that child are also impacted. Parents of

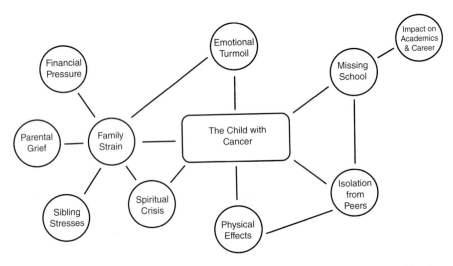

Figure 2. The Impact of Childhood Cancer. A cancer diagnosis in a child has far reaching impact, modulated by the family and social system in which that child exists.

children with cancer experience a number of significant emotional, psychological, financial, and social stresses throughout the cancer diagnosis, treatment, posttreatment surveillance, and beyond. Parents report a variety of very strong emotions contributing to psychological strain, including overwhelming sadness, fear, guilt, helplessness, and loss of control. Grief is a nearly universal experience, whether it is grief over loss of life; grief regarding pain, sadness or other distress of the child; or grief for the loss of their child's health, or a life not shadowed by cancer. Parental strain can be exacerbated by feelings of isolation. Because childhood cancer is so rare, there are very few other parents who can relate to what these families are going through. However, there has been a growth of online support systems for parents, some providing information in "nonmedical speak," others in the form of forums where parents can gain emotional support from others who have traveled this road.

A cancer diagnosis proves a significant financial strain for families. Although children can be covered by their parents' health insurance or Medicaid, there are significant costs not covered by insurance. The cost of prescriptions, clinic visits, emergency room (ER) visits, and hospitalizations adds up quickly. These costs vary depending on the treatment regimen, some of which require daily clinic visits for a week, others mandate a hospitalization every two weeks, or carry significant side effects resulting in frequent trips to the ER. In addition, the cost of traveling to receive care can be significant. As previously mentioned, almost all children with cancer are treated at centers that specialize in the care of pediatric oncology patients. This means that some families will travel three to four hours to reach that clinic and hospital. Gas, tolls, parking, and overnight stays in hotels for these trips also add up quickly. Finally, many children with cancer are unable to attend daycare or school for at least some period of time as a result of the side effects of cancer and its treatment. This impacts the ability of one or both parents to work, which imposes further financial strain on the family.

A cancer diagnosis in a child also has a significant impact on that child's siblings. Siblings of all ages may be scared for their affected brother/sister. They may not have a full understanding of the implications of a cancer diagnosis, and may be worried about the health and mortality of their sibling. Siblings may also feel isolated by the experience, as parental and familial attention is focused on the ill child. Siblings are sometimes shuttled between the households of family and friends during times when their parent(s) are staying in the hospital or in clinic with the ill child. Many pediatric oncology centers have programs focused on inclusion and support of siblings. These programs are often coordinated by program social workers, child psychologists, and child life specialists.

ADOLESCENTS AND YOUNG ADULTS: A SPECIAL POPULATION

Adolescents and young adults (AYAs) with cancer are increasingly recognized as a special population, with unique biology, needs, and challenges compared with younger children and older adults. The AYA age group comprises individuals between the approximate ages of 15 and 39, although definitions vary. One of the alarming aspects of AYA cancer care is that improvements in cure rates that have been seen in younger children and older adults with cancer, have not been mirrored in the AYA population. A number of reasons are thought to contribute to this problem, and include biological, financial, societal, and personal issues. In 2006 the National Cancer Institute and Live Strong Young Adult Alliance (a program of the Lance Armstrong Foundation) cosponsored a peer review group to focus on this specific issue. That group of experts identified 11 key areas on which to focus clinical, research, and advocacy efforts in order to improve outcomes in this population and issued five priority recommendations as listed in Figure 3.

The most common cancers seen in adolescence (aged 15–24) include lymphomas, leukemias, testicular cancer, and thyroid cancer. Increasingly seen in this age group is melanoma, likely due to increased environmental exposure to harmful ultraviolet light sources. Young adults (ages 25–39) are most likely to be affected by lymphomas, bone and soft tissue sarcoma, melanoma, and (in 35–39 year olds) breast cancer. This differs significantly from the epidemiology of cancer in older adults, where breast, colorectal, prostate, and lung cancer predominate. For unclear reasons, even in cancer types that are shared between AYAs and young children, AYAs are more likely to have unfavorable biological factors associated with their tumor, such as high risk genetic mutations. The biological underpinnings of the differences inherent in AYA cancers are an area of active research in the scientific community and will hopefully shed light on differences in cure rates.

Recommendation	Example(s)
Identify the characteristics that distinguish the unique cancer burden in the AYA oncology patient.	• Define biological differences in cancers that affect AYAs • Explore health care disparities in this population
Provide education, training, and communication to improve awareness, prevention, access, and quality cancer care to AYAs.	• Raise awareness of the issues surrounding AYA oncology • Provide education targeting patients, families and healthcare providers
Create the tools to study the AYA cancer problem.	• Develop national registries/databases • Improve availability of clinical trials
Ensure excellence in service delivery across the cancer control continuum (i.e., prevention, screening, diagnosis, treatment, survivorship, and end of life).	• Develop and distribute standards of care for this patient population
Strengthen and promote advocacy and support of the AYA cancer patient.	• Develop resources to specifically address the psychosocial needs of AYAs with cancer

Figure 3. Priority Recommendations of the National Cancer Institute's adolescents and young adult (AYA) Progress Review Group. In 2006, the National Cancer Institute convened a working group jointly with the LiveStrong Young Adult Alliance (part of the Lance Armstrong Foundation) to examine the unique issues within the field of AYA oncology, and provide recommendations around how to improve care for this population. These are the priority recommendations that emerged from that summit. Source: Adapted from "Closing the Gap; Research and Care Imperatives for Adolescents and Young Adults with Cancer: Report of the Adolescent and Young Adult Oncology Progress Review Group."

Another factor impacting outcomes in this group is the lack of awareness among patients and healthcare providers about the warning signs and symptoms of cancer in this age group. Most young adults have not experienced a cancer diagnosis in their peer group, and are unaware of the risk. In addition, these individuals are less aware of "red flag" symptoms of various cancers, are more likely to assume that symptoms are likely due to a sports injury or less serious illness, and are less likely to seek medical attention independently. Educational and awareness programs directed at both patients/parents and primary care physicians are necessary to improve timely referral to a specialist and diagnosis.

Access to healthcare services is an additional challenge – one that is constantly in flux as the political climate and laws around healthcare and health insurance change. Many AYAs benefit from continuing on their parents' insurance and in the absence of being able to do so, are more likely to become uninsured in their late teenage years and early 20s. These uninsured young adults are less likely to seek medical care or advice outside of an emergency. This not only contributes to delays in diagnosis, but can present additional challenges navigating the tremendous expense of cancer therapy.

An additional interesting point regarding access to healthcare lies in where AYAs receive their cancer care. Nearly all children with cancer receive their cancer-related care at an academic medical center, where their treating oncologist is a member of a national cooperative group. This leads to improved access to experts in the field, clinical trials, and the newest therapies. However, only 20% of 15–19 year olds, and less than 10% of 20–30 year olds with cancer are treated at these cooperative group institutions, and even fewer are enrolled on a clinical trial. Increased efforts to provide AYAs with optimal cancer care must include conversations about the most appropriate location and care team, specifically including availability of, and access to, clinical trials.

Finally, any discussion of AYA oncology care must include the specific psychosocial needs of this age group. AYAs are at unique developmental stages, which helps to understand the origin of some of the specific difficulties of a cancer diagnosis at this age. Erik Erikson was a renowned developmental psychologist who in the 1950s published a theory of human psychosocial development composed of eight stages (Figure 4). Essentially, during adolescence (age 12–18 years), an individual is in the "identity versus role confusion" stage. The major developmental goals of an individual at this stage are centered on defining who you are as a person. This necessitates the graduated development of independence from parents, identifying with peer groups, and making decisions about who you are and who you want to be. All of these tasks are hampered by a cancer diagnosis, where, as an adolescent they are increasingly dependent on the care of a guardian. Adolescents with cancer rely on their parents/guardians for financial support and health insurance, accompaniment to clinic appointments, surgeries, and during hospitalizations, medication management, and navigation of an incredibly complex medical system with intricate and high-stakes medical decision-making. A cancer diagnosis also makes it far more difficult for many adolescents to form strong peer relationships, as they become so different from their peers. This is true not only in

Age Range	Developmental Stage	Developmental Task
Birth–18 months	Trust vs Mistrust	Develop a trust of caregivers
18 months–3 years	Autonomy vs Shame/Self-doubt	Gain a sense of control over oneself and independence
3–5 years	Initiative vs Guilt	Gain a sense of control over one's environment
6–11 years	Industry vs Inferiority	Cope with social and academic demands, gain a sense of competence
12–18 years	Identity vs Role Confusion	Establish one's identity and sense of self
19–40 years	Intimacy vs Isolation	Form intimate, reciprocal relationships with peers
40–65 years	Generativity vs Stagnation	Create something that will have a lasting and positive impact, possibly through having children
65 years–death	Ego Integrity vs Despair	Find fulfillment in examining one's life

Figure 4. Erikson's Stages of Development. Erik Erikson's stages of development are shown here. It can be helpful to consider the developmental stage of a patient as they navigate a chronic illness like cancer.

terms of physical appearance (loss of hair, surgical scarring, loss of limb), but also in terms of their life experiences and perspectives.

Erikson considered the development of young adults (age 19–30 years) to be centered on the primary tasks around "intimacy versus isolation," with the goals of this stage of development focused on forming intimate, reciprocal romantic, and peer relationships. Many of the same issues around forming peer groups in adolescence apply here to the development of deep and intimate relationships. Sexual drive and sexual function may also be impacted by the illness and its treatments. At the same time, individuals at this stage of life are also often developing their careers, and considering having children. A cancer diagnosis at this stage of life is uniquely disruptive. It can be hard to start or maintain a career in the face of intensive chemotherapy, not feeling well, and many days of missed work. This can have an impact not only on current and future career prospects, but also on financial security and health insurance. Finally, the impact of potential infertility as a result of treatment is felt particularly acutely at this stage of life.

While there are many unique challenges surrounding a cancer diagnosis in the AYA population, there is hope that care for these patients will improve over the coming decades. Recognition of these particular challenges has sparked an advocacy movement, including patients, healthcare providers, and scientists dedicated to increasing survival and quality of life in this special patient population.

WHERE DO WE GO FROM HERE?

The past several decades have seen enormous growth in the field of pediatric oncology. We have a greater understanding of the biological basis of various cancers that affect children, and have made tremendous strides in the treatment of many childhood cancers.

The continued improvement in the care of children, adolescents, and young adults with cancer relies on a multifaceted approach to the problem of pediatric cancers. Better therapies require a better understanding of each individual disease. Only by attaining a deep grasp of the biological basis for the development of an individual type of cancer, can we develop and test new targeted therapies that may be more effective and carry a lower risk of short- and long-term sequelae.

Understanding that a cancer diagnosis in a child impacts the entire family is crucial to providing complete and compassionate care to that child and their family both during and after treatment. Multidisciplinary care teams who are expertly trained in the field of pediatric oncology are required to provide optimal and comprehensive care to these children and their families.

16

LEUKEMIA, LYMPHOMA, AND MYELOMA

Alan Rosmarin

Wolters Kluwer Health, Waltham, MA, USA

INTRODUCTION

Leukemia, lymphoma, and myeloma are cancers of blood forming cells. These blood cancers, or hematologic malignancies, occur when a single cell from bone marrow, lymph nodes, or spleen undergoes malignant transformation, forms a clone of cancerous cells, and grows out of control. Cancers can form in almost any type of cell. Some solid tumors, such as lung or breast cancer, can spread to bone or other blood-bearing organs, but only cancers that originate in blood-forming cells are referred to as hematologic malignancies.

Mature blood cells have different specialized functions, such as red blood cells (RBCs), which carry oxygen to the body's cells; platelets, which initiate blood clot formation; and various white blood cells (WBCs) that provide immune and inflammatory responses to protect against infections. Blood cells develop in the bone marrow – a spongy tissue that lies within the pelvis, vertebrae, ribs, and other bones. All blood cell types ultimately arise from hematopoietic stem cells (HSC), a rare cell type in the bone marrow that generates rapidly dividing progenitor cells to meet the enormous, ongoing daily needs for blood cell formation.

Cancer: Prevention, Early Detection, Treatment and Recovery, Second Edition. Edited by Gary S. Stein and Kimberly P. Luebbers.
© 2019 John Wiley & Sons, Inc. Published 2019 by John Wiley & Sons, Inc.

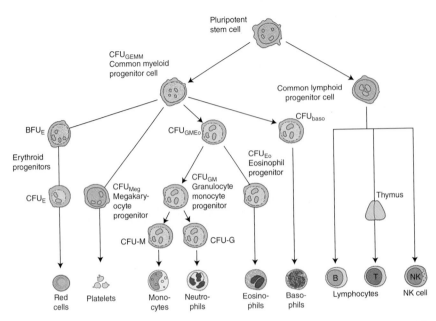

Figure 1. Diagrammatic representation of the bone marrow pluripotent stem cell and the cell lines that arise from it. Note: Baso, basophil; BFU, burst-forming unit; CFU, colony-forming unit; E, erythroid; Eo, eosinophil; GEMM, granulocyte, erythroid, monocyte, and megakaryocyte; GM, granulocyte, monocyte; Meg, megakaryocyte; NK, natural killer. *Source*: Hoffbrand and Moss, 2011. Reproduced with permission of John Wiley and Sons.

There are two major categories of progenitor cells: the myeloid progenitor cell and the lymphoid progenitor cell (Figure 1). Myeloid progenitor cells gives rise to RBCs, platelets, and various WBCs that include neutrophils, monocytes, eosinophils, and basophils, which are responsible for innate immunity – the initial defense against microbes. Acute myelogenous (myeloid) leukemia (AML) and chronic myelogenous (myeloid) leukemia (CML) arise from malignant transformation of a myeloid progenitor cell or a HSC, and can spread in the bone marrow, the peripheral blood, and other organs.

Lymphoid progenitor cells give rise to lymphocytes, a group of WBCs that are responsible for adaptive or specific immunity. These WBCs include T lymphocytes, B lymphocytes, and plasma cells. These lymphoid cells produce receptors that recognize different microbial antigens and generate a specific immune response. The primary lymphoid organs where lymphocytes develop and mature are the bone marrow for B cells and plasma cells, and the thymus for T cells (Figure 2). Malignant transformation of a lymphoid progenitor cell or more mature B or T lymphocytes gives rise to acute lymphoblastic (lymphoid) leukemia (ALL) or chronic lymphocytic (lymphoid) leukemia (CLL).

B lymphocytes mature in the bone marrow and circulate in the bloodstream until they recognize an antigen, which activates it to develop into memory B cells or plasma cells. Plasma cells are mature B cells that secrete antibodies (immunoglobulins), which are highly specific proteins that bind antigens and lead to the destruction

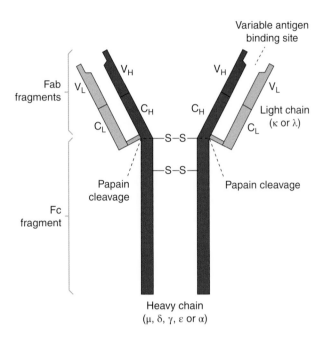

Variable antigen
binding site

Figure 2. Basic structure of an immunoglobulin molecule. Each molecule is made up of two light (κ or λ) (blue areas) and two heavy (purple) chains, and each chain is made up of variable (V) and constant (C) portions, the V portions including the antigen-binding site. The heavy chain (μ, δ, γ, ε or α) varies according to the immunoglobulin class. IgA molecules form dimers, while IgM forms a ring of five molecules. Papain cleaves the molecules into an Fc fragment and two Fab fragments. *Source*: Hoffbrand and Moss, 2011. Reproduced with permission of John Wiley and Sons. (See insert for color representation.)

of invading microbes (Figure 3). Each plasma cell makes a unique antibody that provides a "fingerprint," which identifies that cell and all of its progeny.

Specific immune responses are generated in secondary lymphoid organs, including lymph nodes, the spleen, and lymphoid tissues of the digestive and respiratory tracts (Figure 2). Lymph nodes are small, bean-shaped organs that are clustered in the neck, armpits, chest, abdomen, pelvis, and groin. Lymph fluid contains B and T lymphocytes and it drains from tissues throughout the body into lymph nodes through a spidery network of lymph vessels. Antigens from viruses and bacteria are concentrated in lymph nodes, where they encounter lymphocytes. When B lymphocytes recognize specific antigens, they become activated and mature within lymph nodes to become memory B cells or antibody-secreting plasma cells. When T lymphocytes recognize specific antigens in the lymph nodes, they become activated and mature in the lymph node, but then enter the bloodstream to populate sites of inflammation or infection in the body.

Hodgkin lymphoma (HL) and most non-Hodgkin lymphoma (NHL) are cancers of the B lymphocytes; less commonly, NHL corresponds to T lymphocytes. Multiple myeloma is a cancer of the plasma cells, in which one clone of malignant plasma cells grows out of control in bone marrow and other organs.

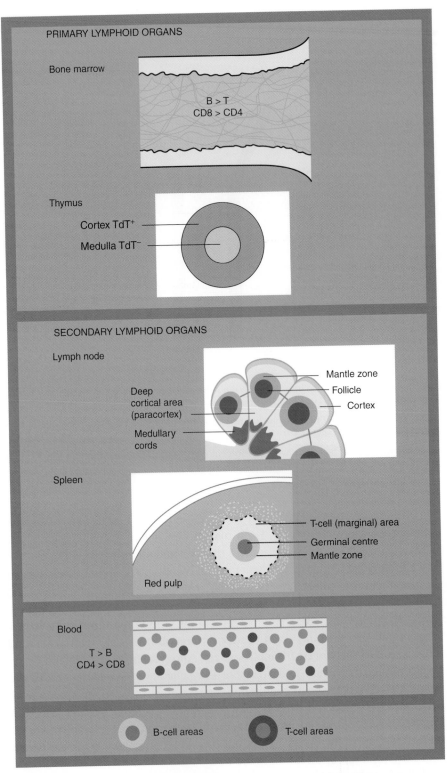

Figure 3. Primary and secondary lymphoid organs and blood. TdT, terminal deoxynucleotidyl.
Source: Hoffbrand and Moss, 2011. Reproduced with permission of John Wiley and Sons.

In lymphomas, cancerous lymphocytes are mostly found in lymph nodes, but can spread to the spleen and other secondary lymphoid tissues, and to other organs. In leukemias, malignant cells are predominantly found in the bone marrow and circulating blood. However, the distinction between leukemias and lymphomas can sometimes be blurred. For example, in some cases of lymphoma, cells can be found in the blood and bone marrow and, conversely, sometimes leukemic cells can be found in the lymph nodes.

Leukemias, lymphomas, and myelomas make up about 10% of all new cancer cases in the United States. Some kinds of solid tumors, e.g., breast cancer, can have a strong pattern of inheritance, and some lung cancers are known to be caused by smoking or other noxious agents. In contrast, hematologic malignancies are only rarely considered familial, and no specific cause of malignant transformation is identified for most patients affected by hematologic malignancies. However, there is an increased association of certain types of leukemia with extensive exposure to benzene or other organic solvents, and some lymphomas are thought to be caused by herbicides, such as Agent Orange.

Some hematologic malignancies are associated with certain abnormalities of chromosomes, such as loss or duplication of parts of chromosomes or entire chromosomes. In other cases, two distinct chromosomes may join to form an abnormal "fusion" chromosome. Such abnormalities can help to confirm the identity of a particular hematologic malignancy, such as the characteristic fusion of chromosomes 9 and 22 in CML. Identification of the genes associated with such abnormal chromosomes has provided important clues about the nature of these cancers and, in many cases, has offered new and more effective ways to treat these disorders. It is important to understand that such chromosome abnormalities arise during the lifetime of the affected individual, and they are not generally inherited by their children.

Because hematologic malignancies arise in cells that can travel in the bloodstream, bone marrow, and various lymphoid organs, the malignant cells may spread – even before the cancer is discovered. Thus, in contrast to certain solid tumors in which surgery can be curative, removal of the tumor mass is only rarely able to cure a lymphoma, leukemia, or myeloma. Rather, hematologic malignancies are generally treated with systemic therapies, such as chemotherapy or immunotherapy, or with regional treatments, such as radiation therapy. Many hematologic malignancies are very responsive to systemic treatments, and several types have high rates of cure. For some hematologic malignancies, stem cell transplantation may offer an effective treatment or even cure.

There are two general categories of stem cell transplants: autologous and allogeneic. In an autologous stem cell transplant (auto transplant), the transplanted HSCs come from the patient. This approach allows the patient to receive higher-dose chemotherapy than would be otherwise feasible, because higher doses of chemotherapy may kill cancer cells more effectively, but would also severely damage the bone marrow. An auto transplant repopulates the HSCs in the bone marrow after high-dose chemotherapy, allowing normal, healthy blood cells to regenerate. In an allogeneic stem cell transplant (allo transplant), the donated bone marrow comes from another person who is genetically similar to the patient. Importantly, in an allo transplant, the donor's transplanted stem cells may recognize remaining cancer cells as foreign and destroy them – an effect known as graft-versus-disease effect. However, this comes with the potential side effect of graft-versus-host disease, in which the transplanted stem cells may also recognize normal cells in the recipient as foreign and attack them, causing effects on the skin, GI tract, and other organs.

LEUKEMIAS

Leukemias account for more than 50 000 cases per year in the United States, or approximately 3% of all cancers. There are four major categories of leukemia, and they differ significantly in regard to who is likely to be affected, the expected outcome (prognosis), types of treatment, and possible complications. The first factor in classifying leukemia refers to the microscopic appearance of the leukemic cells, i.e., do they resemble normal, mature blood cells, or do they appear more like immature stem cells? When the cells are immature in appearance, they are described as acute leukemia, because they typically are fast growing and will progress quickly without treatment. In contrast, the more mature appearing cells of chronic leukemias tend to accumulate slowly and the diseases typically progress over months or years. The other factor used in classification is whether the cells arose in the lymphoid lineage or in the myeloid lineage. Thus, there are four main categories of leukemia: AML, CML, ALL, and CLL. In some cases, leukemia can be further classified based on characteristic chromosome or genetic changes, some of which are described below.

The clinical presentation of a hematologic malignancy may be subtle, or may even be asymptomatic – presenting simply as a laboratory abnormality. For others, symptoms may include fatigue and shortness of breath from anemia, bleeding and bruising from low platelet counts, or recurrent infections. Some may note swollen lymph nodes, abdominal fullness, bone pain, unexplained fevers, drenching sweats, or weight loss. The diagnosis of leukemia is typically made when a bone marrow aspirate and biopsy is performed to investigate such symptoms, or to evaluate abnormal cells in the peripheral blood. Lymphomas or myeloma may be diagnosed with a biopsy of an involved lymph node, bone marrow, or other organ.

ACUTE MYELOGENOUS LEUKEMIA (AML)

AML is the most common leukemia in adults, with more than 20 000 new cases annually in the United States and an average age at diagnosis of 65. As leukemic cells replace normal blood cells in the bone marrow, patients typically present with fatigue, weakness, pallor, recurrent infections, bruising, or abnormal bleeding. The WBC count can be abnormally high or low, and leukemic blast cells may be found in the peripheral blood. The diagnosis may also be suspected when there is unexplained

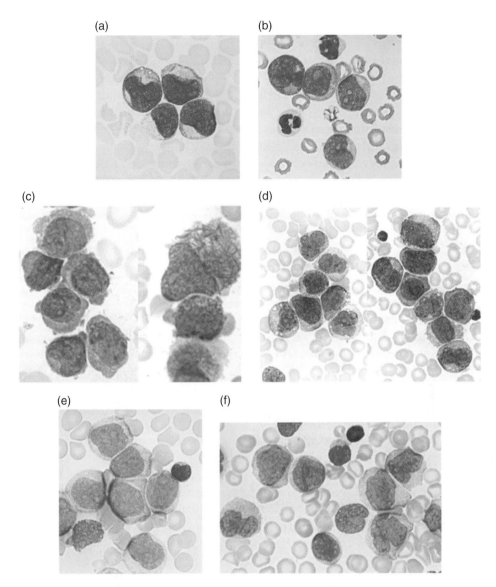

Figure 4. Morphological examples of acute myeloid leukemia. (a) Blast cells without differentia-tion show few granules but may show Auer rods, as in this case; (b) cells in differentiation show multiple cytoplasmic granules; or (c) M_3 blast cells contain prominent granules or multiple Auer rods; (d) myelomonocytic blasts have some monocytoid differentiation; (e) monoblastic leukemia in which >80% of blasts are monoblasts; (f) monocytic with <80% of blasts monoblasts. Source: Hoffbrand and Moss, 2011. Reproduced with permission of John Wiley and Sons.

anemia or thrombocytopenia (low platelet count). Bone marrow aspiration and biopsy will show that more than 20% of the cells in the bone marrow are myeloid blasts, often with cellular inclusions, called Auer rods (Figure 4). The cells are evaluated by flow cytometry to characterize specific protein markers, by cytogenetics to identify abnormal chromosomes, and with molecular analyses. Increasingly, the specific genetic signature allows for separation of AML into favorable, intermediate, and unfavorable risk categories, with 25%, 65%, and 10% of patients falling into each category respectively.

Treatment of AML involves chemotherapy, with the goal of inducing a complete remission (less than 5% blasts in the bone marrow), and hopefully long-term cure. This intensive induction chemotherapy, typically involving an anthracycline (such as daunorubicin or idarubicin) and cytarabine, may not be tolerable by the frail or elderly, and alternative treatments may be offered that instead focus on relief of symptoms. Approximately two-thirds of adults will attain complete remission after intensive induction therapy. Although this may reduce the leukemic cells 1000-fold or more, most will suffer a relapse of AML within months if no further treatment is administered. For this reason, postremission consolidation chemotherapy to further eradicate remaining leukemia cells may be offered, which can include high dose cytarabine, auto transplant, or allo transplant. Because of the severity of the underlying acute leukemia and the intensity of therapy, these disorders should be treated at a center with extensive experience.

A specific subtype of AML, acute promyelocytic leukemia (APL) deserves particular mention. Prior to the development of more specific and effective therapies, APL was the most lethal form of AML. However, with contemporary approaches to treatment, APL now has the highest cure rate among subtypes of AML. This type of acute leukemia usually harbors the t(15;17) translocation, in which the *PML* gene on chromosome 15 is fused to the retinoic acid receptor α gene, *RARA*, on chromosome 17. The resultant fusion protein – PML-RARA – causes arrested differentiation of the malignant promyelocyte cells. In contrast to other forms of AML for which intensive induction chemotherapy is the standard of care, the mainstay of treatment for APL is all-*trans* retinoic acid (ATRA), which binds to the abnormal PML-RARA protein and permits the arrested malignant cells to differentiate into normal, mature neutrophils. There is hope that an increased understanding of the underlying molecular defects in other forms of AML will lead to effective novel targeted therapies, as has occurred with APL.

ACUTE LYMPHOBLASTIC LEUKEMIA (ALL)

ALL is the most common childhood malignancy in the United States, accounting for nearly one-third of all childhood cancers. The peak incidence is age three to seven, but there is also a secondary peak after age 40. There are over 6000 new cases of ALL annually, of which around 3000 will be diagnosed in children. Accumulation of lymphoblasts in the bone marrow may cause failure of normal blood cell development and lead to fatigue, pallor, fever, and bleeding. ALL may also cause swelling of lymph nodes and the spleen, and infiltration of the central nervous system (CNS) by leukemic blasts. Eighty-five percent of ALL is of B-cell lineage (B-ALL), and the remainder is of T-cell origin (T-ALL). The most common chromosomal abnormality in childhood B-ALL, t(12;21), which forms the TEL-AML1 fusion protein, is associated with a good prognosis in ALL. Much like AML, the specific genetic signature allows for stratification of childhood ALL into low, average, high, and very high risk categories, with 15%, 36%, 25%, and 24% of patients falling into each category respectively.

Treatment of ALL consists of induction chemotherapy with vincristine, asparaginase, and dexamethasone, with or without an anthracycline (such as daunorubicin or doxorubicin). After achieving remission, consolidation chemotherapy or allo transplant may be offered, depending on the child's prognosis. Most childhood ALL patients are treated in standardized clinical trial protocols, which include preventive chemotherapy for the CNS. Maintenance therapy with lower intensity treatment that further enhances the cure rate may extend for two or more years. Remarkably, these approaches have converted this formerly lethal disease to one with a cure rate of 85–90%. Although contemporary treatment cures most children with ALL, therapy can cause long-term effects on growth, maturation, and cognition. Thus, there is interest in identifying treatment that is less toxic, but still curative, for children with good prognosis disease. Those with poor prognostic features should receive more intensive therapy, with the goal of increasing the cure rate.

In contrast to the successes of childhood ALL, fewer than 40% of adults with ALL are free of leukemia after five years, and less than 5% of adults over the age of 70 survive for five years. Therefore, new treatments for adult ALL are in development to prevent relapse after an initial response to induction therapy. Blinatumomab is a type of immunotherapy that uses the patient's own immune system to kill ALL cells by inducing the patient's own normal T lymphocytes to attack residual leukemia. Other approaches include investigational use of CAR-T cells (chimeric antigen receptor T-cells) that can destroy residual leukemic cells. These and other approaches that utilize the immune system to destroy leukemic cells offer new treatment strategies in situations where conventional chemotherapy has not yet achieved the successes seen in childhood ALL.

CHRONIC MYELOGENOUS LEUKEMIA (CML)

Chronic leukemias, which are classified as either myeloid or lymphoid, are distinguished by the more mature appearance of the leukemic cells, and their slower rate of progression. CML arises from an abnormal hematopoietic pluripotent stem cell that undergoes rearrangement of chromosomes 9 and 22 (the so-called Philadelphia chromosome; Ph') to create a fusion of two genes, known as BCR-ABL (Figure 5). The Ph' chromosome and BCR-ABL were the first specific chromosomal and genetic abnormalities to be associated with a particular human cancer. BCR-ABL is an uncontrolled, overactive tyrosine kinase protein that drives CML by adding phosphate groups to other cellular proteins. Development of imatinib (Gleevec®)

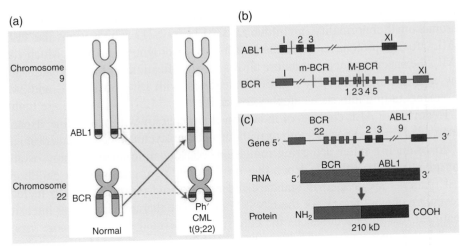

Figure 5. The Philadelphia chromosome (Ph'). (a) There is translocation of part of the long arm of chromosome 22 to the long arm of chromosome 9 and reciprocal translocation of part of the long arm of chromosome 9 to chromosome 22 (the Philadelphia chromosome). This reciprocal translocation brings most of the *ABL* gene into the *BCR* region on chromosome 22 (and part of the *BCR* gene into juxtaposition with the remaining portion of *ABL* on chromosome 9). (b) The breakpoint in *ABL* is between exons 1 and 2. The breakpoint in *BCR* is at one of the two points in the major breakpoint cluster region (M-BCR) in chronic myeloid leukemia (CML) or in some cases of Ph+acute lymphoblastic leukemia (ALL). (c) This results in a 210k-Da fusion protein product derived from the *BCR-ABL* fusion gene. In other cases of Ph+ALL, the breakpoint in *BCR* is at a minor breakpoint cluster region (m-BCR) resulting in a smaller *BCR-ABL* fusion gene and a 190-kDa protein. *Source*: Hoffbrand and Moss, 2011. Reproduced with permission of John Wiley and Sons. (See insert for color representation.)

and other targeted agents against BCR-ABL represented a paradigm shift in the treatment of cancer, and proved that attacking specific molecular abnormalities could achieve effective and durable remissions. Numerous targeted agents have since been developed that attack the underlying molecular defects in other types of cancer.

There are over 6600 cases of CML diagnosed annually, accounting for about 15% of all leukemias. Most patients are diagnosed between the ages of 40 and 60, but it can also occur in children. The only known risk factor is ionizing radiation, such as occurred in atomic bomb survivors in Japan. Patients may present with fatigue, weight loss, night sweats, or an enlarged spleen, but many are diagnosed incidentally when a routine blood count shows an abnormally high WBC. Typically, the peripheral blood shows an increased number of both immature and mature myeloid cells, including neutrophils, basophils, eosinophils, and monocytes. Chromosomal testing of peripheral blood and bone marrow will detect the Ph' chromosome, and molecular studies confirm the presence of the BCR-ABL fusion gene. Most patients (85–90%) are identified in chronic phase of CML. Before the development of highly effective tyrosine kinase inhibitors (TKIs), the chronic phase CML would typically progress within four to five years to the more aggressive accelerated phase or blast phase, which resembles an aggressive acute leukemia. These latter stages of CML are characterized by an increased percentage of myeloblasts in the peripheral blood or bone marrow, and worsening symptoms of fever, fatigue, and weight loss.

Initial treatment of chronic phase CML consists of daily treatment with oral TKIs that specifically block BCR-ABL activity. Although these drugs are highly effective, some patients develop resistance or do not tolerate the TKI. However, newer TKIs provide a good response in many of these patients with chronic phase CML, including some who have developed mutations of BCR-ABL that render it resistant to imatinib. In patients with a suboptimal response to multiple TKIs, allo transplant is the only proven curative treatment option.

CHRONIC LYMPHOCYTIC LEUKEMIA (CLL)

CLL is the most common leukemia of the lymphoid lineage, with over 14 000 cases diagnosed annually in the United States. It has a peak incidence between 60 and 80 years of age. CLL arises from a clone of malignant B cells that resemble mature lymphocytes, but they also aberrantly express T-cell proteins (e.g., CD5), in addition to the usual complement of B cell proteins. These abnormal lymphocytes accumulate in the blood, bone marrow, spleen, and lymph nodes. Most patients are diagnosed when a routine complete blood count shows an elevated WBC and lymphocyte count. A peripheral blood smear shows a large number of lymphocytes, including so-called "smudge cells" (Figure 6). The most frequent clinical sign is enlargement of lymph nodes in the neck, armpits, groin, or other

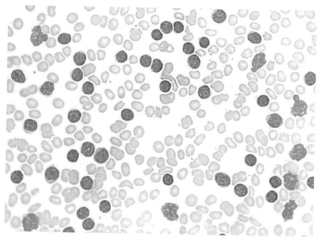

Figure 6. Chronic lymphocytic leukemia: peripheral blood smear showing lymphocytes with thin rims of cytoplasm, coarse condensed nuclear chromatin and rare nucleoli. Typical smudge cells are present. *Source*: Hoffbrand and Moss, 2011. Reproduced with permission of John Wiley and Sons.

sites, or abdominal fullness from an enlarged spleen. The diagnosis of CLL is confirmed when flow cytometry studies show that the abnormal cells express both B cell antigens and the CD5 antigen. Various classification systems are used to stage CLL and establish the prognosis, based on whether the patient has lymphadenopathy, an enlarged spleen, anemia, or thrombocytopenia.

Because most cases of CLL progress slowly, patients may be observed for months or years ("watchful waiting,") before treatment is needed. In other words, the mere presence of CLL is not reason enough to initiate treatment, because early treatment will not affect the long-term outcome. Generally, treatment is begun when the patient has, or is likely to soon have, significant symptoms, such as an enlarging spleen, or troublesome or rapidly growing lymph nodes that interfere with the activities of daily living. Other indications for treatment include development of constitutional, or "B symptoms," i.e., fatigue, fevers, drenching night sweats, or weight loss; a rapid increase in the peripheral blood lymphocyte count; or progressive anemia or thrombocytopenia. Some may develop abnormal antibodies that attack their own red blood cells or platelets, causing anemia or thrombocytopenia, and these conditions may be indications for treatment. Because CLL may affect development of normal lymphocytes, some patients may suffer recurrent bacterial infections due to inadequate levels of normal immunoglobulins (antibodies); such patients may benefit from periodic infusions of immunoglobulins.

As with other forms of leukemia, chromosome and molecular testing can stratify patients into high-risk and standard-risk disease and offer important prognostic information. The choice of therapy is influenced by the patient's age and medical condition, as well as various prognostic factors. Because the malignant lymphocytes express the B lymphocyte antigen, CD20, they are vulnerable to immunotherapy that utilizes rituximab and related monoclonal antibodies. Treatment may also include steroids, with or without oral chemotherapy, combination chemotherapy, or various forms of intensive chemoimmunotherapy. In recent years, new targeted agents have been developed for CLL, and there are ongoing clinical trials to identify effective, less toxic therapies for initial treatment, and for relapsed disease. High-risk patients are encouraged to enroll in clinical trials, and to have evaluation for possible stem cell transplantation. Nevertheless, insights into the molecular mechanism that cause CLL will continue to provide new and more effective avenues of attack against this form of leukemia.

LYMPHOMAS

Lymphoma is the most common category of hematologic malignancy, with more than 80 000 new cases diagnosed each year in the United States. There are two main categories of lymphoma: Hodgkin Lymphoma (HL, also known as Hodgkin Disease), and NHL. HL and NHL differ in regard to treatment and prognosis. Lymphomas are primarily cancers of B lymphocytes, but a small percentage is of T lymphocyte origin. These disorders can range from slowly growing indolent disorders that require no therapy, to rapidly progressive lethal disorders. The malignant lymphocytes are mostly found in lymph nodes, but they can spread via the blood or lymphatic vessels to the spleen, bone marrow, peripheral blood, and other organs. Patients may feel swelling of lymph nodes, or a vague fullness in the abdomen due to enlargement of the spleen. Lymphomas can also cause anemia, thrombocytopenia, immune deficiencies, and constitutional symptoms of fever, drenching sweats, and weight loss.

Diagnosis of both HL and NHL is usually made by biopsy of a suspicious lymph node. Although this may be performed as a fine needle aspirate (in which a small number of cells are aspirated from a lymph node), Hodgkin lymphoma usually requires a core needle biopsy or excision of an involved lymph node to detect the characteristic appearance. Both HL and NHL are more common in patients with autoimmune disorders, such as rheumatoid arthritis, lupus, or Sjögren's syndrome, and in immunosuppressed patients, such as those with HIV or following a solid organ transplant.

HODGKIN LYMPHOMA

Thomas Hodgkin was a nineteenth-century British anatomist and pathologist who first described Hodgkin lymphoma in 1832. Seventy-five years later, an Austrian pathologist named Carl Sternberg, and an American pediatrician and pathologist named Dorothy Reed, independently described the characteristic "owl-eye" lymphocyte that is diagnostic of HL and now bears their names: the Reed-Sternberg (RS) cell (Figure 7).

HL is less common than NHL, with only about 9000 new cases seen annually in the United States. It has a bimodal age distribution with a peak in young adults between ages 15 and 35 and a second peak in adults over the age of 50. HL has a 2:1 male predominance, and patients present typically with painless, firm lymph node swelling in the neck, armpits, or groin. Others are found to have a mass in the chest on routine chest X-ray. Some patients may also have systemic "B" symptoms, including unexplained fever, drenching night sweats, unexplained weight loss (>10% body weight over six months), severe itching, or pain in lymph nodes when drinking alcohol.

HL is divided into two main categories. Classical Hodgkin lymphoma (CHL) includes 95% of patients with HL and is defined by diagnostic RS cells. Only 5% of HL patients have nodular lymphocyte-predominant Hodgkin lymphoma (NLPHL). This uncommon variant does not have typical RS cells but instead contains large, abnormal B cells, called lymphohistiocytic or "popcorn" cells, based on their microscopic appearance.

Prognosis in HL is influenced by the histology (microscopic appearance) and by the extent of disease in the body, which is defined by staging (Figure 8). A whole body computed tomography (CT) scan or (positron emission tomography)

(a)

(b)

(c)

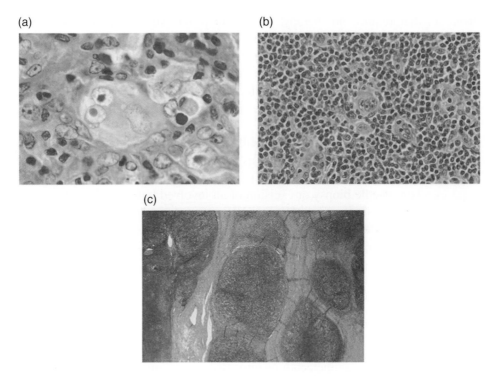

Figure 7. Hodgkin lymphoma: (a) high-power view of a lymph node biopsy showing two typical multinucleate Reed-Sternberg cells, one with a characteristic owl eye appearance, surrounded by lymphocytes, histiocytes, and an eosinophil; (b) mixed cellularity; and (c) nodular sclerosing Hodgkin lymphoma. *Source*: Hoffbrand and Moss, 2011. Reproduced with permission of John Wiley and Sons. (See insert for color representation.)

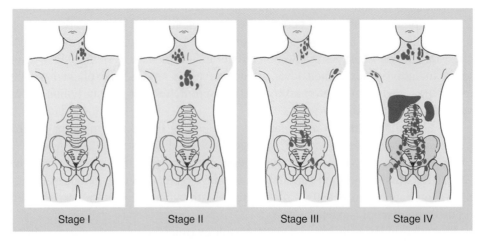

Stage I   Stage II   Stage III   Stage IV

Figure 8. Staging of Hodgkin and non-Hodgkin lymphoma. *Source*: Hoffbrand and Moss, 2011. Reproduced with permission of John Wiley and Sons. (See insert for color representation.)

PET/CT is used to determine the extent of lymphoma involvement throughout the body. Stage I indicates lymphoma involvement in only one lymph node region, and stage II indicates lymphoma involving two or more lymph node areas confined to one side of the diaphragm. Patients are also classified as A or B according to whether or not B symptoms (fever, night sweats, weight loss) are present. Stage I and stage II disease without B symptoms or bulky disease (a lymph node area > 10 cm in diameter), is considered early stage HL. Stage III has involvement in lymph node areas above and below the diaphragm, and stage IV indicates widespread disease outside the lymph node areas including the liver, lung, or bone marrow. Advanced stage HL includes stage III and IV disease, as well as patients with stage I and II disease with bulky disease and/or B symptoms.

Over 80% of patients with HL are cured with standard treatment and, unlike advanced stage disease in many other forms of cancer, even stage IV lymphomas can be cured. Early stage HL is typically treated with combination chemotherapy (usually adriamycin, bleomycin, vinblastine, and dacarbazine; ABVD), sometimes followed by radiation therapy to involved lymph nodes. Advanced stage HL is treated with more cycles of ABVD chemotherapy, with or without radiation therapy. Newer agents that directly target the RS cells are being tested in resistant disease, although high dose chemotherapy followed by auto transplant is also effective.

NON-HODGKIN LYMPHOMA

NHL is the most common hematologic malignancy, with over 71 000 new cases diagnosed each year, making up about 5% of all cancers. It is the sixth most common cause of cancer in both men and women in the United States. The incidence of NHL increases with age, and median age at diagnosis is 64. NHL is not a single disease; rather, the World Health Organization classifies more than 60 types of NHL! They are broadly divided into four main categories: indolent (slow-growing) or aggressive lymphomas of either B-cell or T-cell origin. B-cell lymphomas make up 85% of all NHL, while the remainder of NHL are T-cell lymphomas. Aggressive lymphomas commonly present with a rapidly growing mass and B symptoms, and they can be rapidly fatal within weeks, if untreated. Examples of aggressive lymphomas include diffuse large B-cell lymphoma (DLBCL), Burkitt lymphoma, and adult T-cell lymphoma. Indolent lymphomas typically present with slow growing lymphadenopathy that may wax and wane for years, even without treatment. Examples of indolent lymphomas include follicular lymphoma and marginal zone lymphoma. Patients with B-cell lymphomas benefit from rituximab (Rituxan®), which is directed against the B-cell protein, CD20; rituximab was the first therapeutic monoclonal antibody approved for cancer treatment. Nearly all B-cell lymphomas express CD20, therefore rituximab and related monoclonal antibodies have become a mainstay of treatment in B-cell lymphomas.

DLBCL is the most common type of NHL, accounting for 30–40% of all NHL. It is an aggressive lymphoma with 60% of patients presenting with advanced stage disease. DLBCL is typically treated with six or eight cycles of rituximab together with cyclophosphamide, vincristine, doxorubicin, and prednisone (known as R-CHOP) chemotherapy. Despite its aggressive nature, most cases of DLBCL can be cured. For those patients with refractory or relapsed DLBCL, high-dose chemotherapy with auto l transplant is the preferred treatment.

Follicular lymphoma is the most common indolent NHL and the second most common NHL, representing around 25% of all NHL cases. Most people with limited stage follicular lymphoma (stage I and stage II with non-bulky disease) can be effectively treated with radiation therapy, and a portion of these patients will be cured. However, many patients are managed with a "watchful waiting" approach, and are treated only when the disease is progressing and symptoms develop. Most people present with advanced stage follicular lymphoma (stage II bulky disease, stage III, and stage IV disease). Because this is not curable with conventional chemotherapy and radiation, treatment is usually initiated after development of symptoms, steady progression of disease, or when extensive bone marrow involvement causes anemia or thrombocytopenia. Treatment may involve a variety of chemotherapy agents, typically combined with rituximab. Although most patients will have either a partial or complete response, the majority are expected to ultimately develop recurrent disease. Thus, patients may be treated with continued rituximab therapy to maintain the remission. Once patients develop recurrent disease that requires treatment, therapy typically includes rituximab with one of a variety of chemotherapy regimens. High-dose chemotherapy followed by auto transplant is an option in patients who relapse after a short initial response.

Peripheral T-cell lymphomas (PTCL) comprise around 15% of all NHL, and are generally more aggressive lymphomas with a worse prognosis than B-cell NHL. Because T-cell lymphomas do not express the B-cell protein, CD20, rituximab is not used for treatment. The most common T-cell lymphomas include PTCL, anaplastic large cell lymphoma, and angioimmunoblastic T-cell lymphoma. Initial treatment for most PTCL includes more aggressive chemotherapy, but many patients will have relapsed or refractory disease. High dose chemotherapy followed by auto or allo transplant has been the standard treatment, but newer targeted agents are currently being evaluated for these disorders. Other T-cell lymphomas, such as Mycosis Fungoides that initially involve the skin, grow slowly and may be managed with watchful waiting or localized treatment to troublesome sites of disease.

MYELOMA

Multiple myeloma is a cancer of plasma cells. Plasma cells are mature B cells that secrete immunoglobulin proteins (antibodies), which bind to antigens and activate the humoral and cellular immune response. There are five subclasses of antibodies:

immunoglobulin G (IgG), IgA, IgM, IgD, and IgE. Because the multiple myeloma plasma cells grow out of control, they secrete one specific immunoglobulin – known as a monoclonal (M) protein – with IgG being the most common subtype. The myeloma cells divide and multiply in bones and in the bone marrow, crowding out and overtaking normal plasma cells and other blood cells. Secretion of large amounts of M protein can lead to kidney failure, anemia, bone fractures, recurrent infections, and high levels of calcium in the blood. Approximately 60% of patients present with bone pain, particularly in the back or chest from vertebral fractures. Anemia is present in over 70% of myeloma patients at diagnosis that can result in generalized fatigue and weakness.

There are an estimated 26 000 new cases of multiple myeloma annually in the United States, making up 1.6% of all new cancer cases. It is generally a disease of older adults, with a median age at diagnosis of 69 years. Myeloma is two to three times more common in African Americans and blacks in Africa, compared to whites, and has a 1.5 to 1 predominance in men compared to women. Five-year median overall survival in all patients with myeloma approaches 50%.

Diagnosis of myeloma is made by biopsy of bone marrow, bone, or soft tissue that shows ≥10% clonal plasma cells. However, there must also be evidence of end-organ damage with anemia, high calcium, renal failure, or lytic bone lesions to confirm the diagnosis of multiple myeloma. Treatment of myeloma is tailored depending upon patient factors such as age and overall health, and upon the unique biology or genetic composition of each patient's multiple myeloma. In general, in younger patients in reasonably good health, the standard of care is intensive combination chemotherapy followed by auto transplant. For older patients or for those with multiple medical comorbidities who are unlikely to tolerate an auto transplant, combination chemotherapy is continued until there is evidence of stable disease or until treatment is no longer tolerated.

Specific genetic abnormalities place each patient into high risk, intermediate risk, or standard risk myeloma. This risk stratification helps guide the selection of initial chemotherapy, and predicts outcome after treatment. In intermediate and high risk patients, maintenance chemotherapy is often used after transplant, with the goal of preventing disease relapse. Within the last 15 years, there have been seven new drugs approved by the FDA for treatment of myeloma that has significantly changed the treatment paradigm. Immunotherapy has been the most significant advancement in myeloma with immunomodulatory agents such as lenalidomide, and proteosome inhibitors such as bortezomib now used as the backbone of initial combination chemotherapy regimens. However, despite these advances, multiple myeloma remains an incurable disease and additional treatment options are needed. Promising treatments include monoclonal antibodies that target the immune system to attack myeloma cells. Elotuzumab is a promising drug that activates natural killer (NK) cells to specifically kill myeloma cells. Daratumumab is another monoclonal antibody that binds to the CD38 protein on myeloma cells and activates the immune system to destroy the myeloma cells.

CONCLUSION

Continued advances in gene sequencing and other molecular testing have led to a better understanding of the pathogenesis of hematologic malignancies. This knowledge drives development of new treatments – sometimes referred to as "targeted therapy," "personalized medicine," or "precision medicine" – that specifically targets a patient's unique cancer biology. The era of targeted therapy in oncology began nearly two decades ago with imatinib in CML and rituximab in B-cell lymphoma changing the treatment paradigm of those diseases. Today, new targeted therapies for leukemias, lymphomas, and myelomas continue to set the pace for the field of oncology by improving outcome and minimizing toxicity to the patient.

17

GU ONCOLOGY: KIDNEY, UROTHELIAL, AND PENILE CANCER

Scott Perrapato

University of Vermont Larner College of Medicine, Burlington, VT, USA

INTRODUCTION

This section will cover the epidemiology, prevention strategies, early detection, treatment summaries, and recovery (survivorship) programs for kidney, urinary collecting systems (to include renal pelvis, ureter, bladder, and the urethra), and male penile cancers.

Kidney, renal pelvic, and ureteral cancers are especially interesting because they have well defined hereditary and environmental causes. Knowing these, an individual can work toward prevention and early detection to identify these cancers in their earliest and most curable forms. Some of these cancers have well-understood biochemical abnormalities that have led to specific and focused treatments.

Kidney cancer, or cancer of the kidney tissue, has many associated risk factors that can be controlled to reduce an individual's risk of developing cancer (Table 1).

Urothelial cancer, which originates in the renal pelvis, ureter, bladder, or urethra, affects an increasing number of men and women each year [2]. Many of these individuals develop cancer from preventable environmental exposures such as cigarette smoke, chronic infections of the urinary tract, and chemical exposures.

Cancer: Prevention, Early Detection, Treatment and Recovery, Second Edition. Edited by Gary S. Stein and Kimberly P. Luebbers.

TABLE 1. Epidemiology of kidney cancer.

Environmental	Hereditary/Individual
Cigarette Smoking	Von Hippel Lindau Disease
Cadmium	Hereditary Papillary Renal Carcinoma
Phenacetin analgesics	Hypertension
	Obesity
	Long-term dialysis

Source: Adapted from Chow (2010) [1].

Penile cancer is commonly related to chronic infection and inflammatory disorders, most specifically to infection with Human Papilloma Virus (HPV). Preteen boys are now able to get the HPV vaccination, which should significantly decrease the incidence of penile cancer [3].

KIDNEY CANCER

Epidemiology

The kidneys are a paired set of solid organs that filter impurities from the body while regulating the body's electrolyte and water balance. They are located in the retroperitoneum. One in 750 individuals is born with only one kidney. Tumors of the kidney most commonly arise from the proximal renal collecting tubules and are called adenocarcinomas but there are other types and other origins of kidney tumors (Table 2).

Kidney cancer can be sporadic, meaning that there is not always a known genetic abnormality. However there are some epidemiological and genetic risk factors associated with it. Cigarette smoking has been implicated in a higher incidence of kidney cancer. Research is also identifying more and more genetic changes associate with kidney cancer.

Prevention

Prevention of kidney cancer involves both identifying and minimizing an individual's risk factors as well as early detection. Von Hippel Lindau (VHL) Disease and hereditary papillary renal carcinoma result in genetic changes that lead to an increased chance of developing kidney cancer. VHL Disease is a result of mutation or change in chromosome 3p and not only leads to an increased chance of developing kidney cancer, but also the development of tumors in other locations such as the retina and cerebellum. Hereditary papillary renal carcinoma results in changes in the kidney that make an individual more likely to develop kidney cancer.

TABLE 2. Renal cancer cell types.

Benign		
	Oncocytoma	
	Angiomyolipoma (hamartoma)	
	Juxtaglomerula	
	Leiomyoma	
	Hemangioma	
Malignant		
	Adenocarcinoma	
		Conventional Clear Cell
		Papillary
		Chromophobe
		Collecting Duct
		Neuroendocrine
		Sarcomatoid
	Metastatic tumors (tumors that originate in other organs and spread to the kidney)	
Pediatric		
	Nephroblastoma (Wilms' Tumor)	

Long-term hemodialysis has been shown to be associated with the development of abnormal kidney cysts with cells that are more likely to develop into tumors over time. Individuals with renal problems and renal failure can lessen their risk of kidney cancer by maintaining renal health by controlling diabetes and hypertension.

Early Detection

Kidney tumors are often initially painless and because of the location of the kidney within the retroperitoneum, the affected individual might not notice the early development of a kidney cancer. Therefore early detection involves a person knowing if they are at risk. Those at risk will have periodic imagining of the kidney (such as renal ultrasound [US] or magnetic resonance imaging [MRI]) to identify kidney tumors at their smallest size so they can be effectively treated. When diagnosed early kidney cancer is a very curable disease. However once kidney cancer has spread, it can only be temporarily controlled and only rarely cured.

Currently kidney tumors are analyzed (with biopsy or surgical excision) and classified by morphology (appearance) and cytogenetics (genetic analysis) to differentiate benign tumors from potentially malignant and metastatic tumors to the kidney (Table 2). This information will help identify the individual's treatment options as well as determine if their family members have an increased risk of developing kidney cancer so they can enter early detection programs.

The standard tumor, node, metastasis (TNM) grading and staging system is utilized to characterize the tumor and determine what treatment guidelines/options are available for a particular patient [4].

Treatment

The treatment of kidney cancer involves primary surgical excision or ablative therapies. If a kidney tumor is found early before it has grown to a large size or spread beyond the kidney, local excision (i.e., partial nephrectomy, cryosurgery [freezing the tumor]) or ablative therapies (i.e., radiofrequency ablation [destroying the tumor with high frequency energy]) can be used and offer a chance at maintaining renal function while effectively curing the cancer [5, 6]. Use of this targeted treatment has a greater chance of saving renal function and lessening the chance of renal failure requiring dialysis in later life. Traditionally small renal tumors were treated with radical nephrectomy, during which the entire kidney and associated adrenal gland are removed. This treatment is now reserved for large tumors or tumors that are positioned close to the renal blood supplies and would make local excision dangerous.

Kidney cancer that spreads beyond the kidney, either by the lymphatic system of the bloodstream, often manifests as tumors within the lymph nodes, liver, lungs, or bone. Biologic research of the cellular pathways of kidney cancer is rapidly developing and leading to many innovative and effective treatments for the individual with metastatic kidney cancer. Metastatic kidney cancer can be treated with biologic response modifiers (i.e., interferon and interleukin) and targeted therapies (i.e., inhibitors of vascular endothelial growth factors [VEGF]).

There is a subset of patients with kidney cancer who also develop a limited number (i.e., one or two) of tumors outside the kidney. These tumors can often be treated with pinpoint radiotherapy, surgical excision, or the ablative treatments mentioned above. Therapy for these tumors yields very positive long-term results.

Molecular targeted therapy utilizing tyrosine kinase inhibitors (i.e., Sunitinib) is a revolutionary "bench to bedside" therapeutic success story for kidney cancer patients. Tyrosine kinase inhibitors target and disrupt tumor cell angiogenesis (VEGF receptors) causing both primary and metastatic kidney tumors to regress with minimal side effects. "[A]ntiangiogenic tyrosine kinase inhibitors dominate the landscape of treatment, representing four of the seven Food and Drug Administration approved therapies for [renal cell carcinoma]" [7].

Recovery

Recovery for individuals with kidney cancer is dependent on the primary therapy for localized kidney cancer and the possibility of systemic treatment if metastatic disease is present. Recovery following kidney cancer treatment of the primary tumor involves the body healing after the surgical or ablative procedure as well as the psychological readjustment to the cancer diagnosis.

There is considerable research showing that insuring the best uncomplicated recovery, including psychological readjustment, may lead to longer disease-free outcomes [8].

Developing an understanding of a tumor's specific epigenetic and biochemical abnormalities will lead to targeted systemic therapy including immunotherapy and targeted biologic therapy and may lead to longer periods of remission. A new molecular level genetic and functional classification for kidney cancer mutations is being developed for the most common type of kidney cancer: clear cell carcinoma (ccRCC). "These discoveries provide insight into ccRCC development and set the foundation for the first molecular genetic classification of the disease, paving the way for subtype-specific therapies" [9].

UROTHELIAL CANCER

Epidemiology

The bladder is the most common site for the development of urothelial cancer accounting for approximately 95% of all urothelial cancers. However urothelial cancer can arise anywhere in the urinary tract that is lined with transitional cells or urothelium (renal pelvis, ureters, bladder, and parts of the urethra). The majority of bladder cancers are transitional cell cancers that develop from the natural lining within the bladder. These transitional cells can differentiate into squamous cell carcinomas (1–2%) and adenocarcinomas (<1%) [10]. There are multiple known causes of bladder cancer, the most common association being exposure to cigarette smoke (smokers are four times more likely of developing bladder cancer than nonsmokers). Because of the easy migration of people and transportation of merchandize and foods between countries we are seeing more unique causes of urothelial cancer such as aristolochia use or schistosomiasis infection. See Figure 1 for more causes of urothelial cancer.

Cigarette smoke
Male gender
Aniline dyes
Benzene derivatives
Phenacetin analgesic toxicity
Parasitic infections (i.e. schistosomiasis)
Aristolochia (herbal supplement)
Hereditary factors (i.e. Lynch Syndrome)

Figure 1. Causes of urothelial cancer.

Prevention

Limiting or eliminating any and all cancer-inducing exposures are excellent preventive strategies for all urothelial cancers. However because of the long latency from risk factor exposure to cancer development, urothelial cancer risk may remain for life even for individuals who stop smoking and remove themselves from chemical exposures and other risk factors.

Early Detection

Early detection of urothelial cancer involves knowing a person's particular risk factors and a heightened level or surveillance or check-ups. Typically the first indicator that a person has developed urothelial cancer is blood in the urine (either seen under a microscope during routine urine analysis or gross hematuria seen by the individual). Usually there are no other signs or symptoms.

Individuals who experience hematuria require a standard evaluation comprised of radiologic evaluation, direct visualization, and cellular analysis of the urinary tract.

Radiologic evaluation of the urinary tract, usually in the form of a CT scan of the abdomen and pelvis with intravenous (IV) contrast examines the kidneys, renal pelvis, ureters, bladder, and urethra for small masses. During an in-office cystoscopy, the urologist will examine the lining of the bladder with a thin, flexible camera inserted through the urethra. And finally urine cytology will provide information about the presence of blood, atypical cells, and special urinary tumor markers.

Treatment

Treatment of urothelial cancers is dependent on the location and stage of the cancer at the time it is diagnosed (Figure 2). Superficial urothelial cancer involves the lining of the renal pelvis, ureter, bladder, or urethra and may extend into the tissue immediately below the lining. It will not invade in the muscle tissue around the bladder. In most people, superficial urothelial cancer can be treated with local ablation or removal techniques or minimally invasive surgery (Table 2). Intravesical treatments with immunotherapeutic or chemotherapeutic agents may be used to convert abnormal bladder lining to normal in a large percentage of patients (Table 2).

Once urothelial cancer invades into or through the deep muscle layer covering the bladder, it is considered invasive cancer and needs to be treated with a systemic/whole-body technique. Invasive urothelial cancer will metastasize in 50% of patients. Invasive cancer is best treated with a combination of systemic chemotherapy and radiation or systemic chemotherapy and surgical excision. Urothelial cancer that has metastasized elsewhere in the body (i.e., lungs, liver, bones) may be controlled by systemic chemotherapy and immunotherapy but it is only cured in a small percentage of patients.

Superficial:
  Ablation:
    Electrodessication, Laser therapy ("light amplification by stimulated emission of radiation" i.e. Neodymium YAG)

  Topical therapies:
    Immunotherapy (Bacillus Calmette-Guerin (BCG), Interferon, Interleukin)
    Chemotherapy (valrubicin, mitomycin, gemcitabine)

  Systemic immunotherapy: (pembrolizumab/PD-1 blockade)

  Invasive:
    Neoadjuvant systemic chemotherapy with partial/total surgical excision or external radiotherapy

Figure 2. Treatment of urothelial cancer (Bladder, Ureter, Renal Pelvis).

Because of high rates of recurrence and spread within the urinary tract, physicians may recommend removal of the kidney, ureter, or bladder where the site of cancer is located. This discussion is multifaceted and is not an option for every patient.

Recovery

Patients treated for urothelial cancer will need to undergo surveillance cystoscopy and urinalysis every 3 to 6 months for several years (interval frequency and duration are physician and patient dependent). Patients who undergo removal of their bladders to control the spread of urothelial cancer in the bladder may have surgery to redirect urine outside the body through a conduit made of small bowel. Another option following removal of the bladder is bladder reconstruction with a pouch of small bowel.

PENILE CANCER

Epidemiology

Penile cancer is exceedingly rare in the United States. It is associated with infection with STDs like HPV as well as poor hygiene. It accounts for less than 1% of cancers in men in the United States, but is more common in parts of Asia, Africa, and South America. The most common type of penile cancer is squamous cell carcinoma with about 95% of cases. Other types of penile cancer are basal cell carcinoma and melanoma, sarcoma of the smooth muscle and connective tissue within the penis, and adenocarcinoma of the sweat glands in the skin of the penis.

Prevention

Penile cancer is rare but offers one of the best opportunities for cancer prevention. There are several etiologies of penile cancer. HPV is one of the most common causes of squamous cell carcinoma of the penis. The HPV vaccination is now available for young boys and has the potential to dramatically decrease the incidence of penile cancer [3, 11, 12]. Additionally chronic inflammation due to poor hygiene and environmental causes (tropical and desert climates) has also been implicated in causing penile cancer. Rates of penile cancer are significantly lower in men who are circumcised than men who are uncircumcised [12]. Circumcision allows for greater ease of maintaining genital hygiene.

Early Detection

Regular self-examinations are recommended for the assessment of any skin cancer, including penile cancer.

Treatment

When penile cancer is detected early, it can be treated by local surgical excision, radiation, or laser/cryosurgery. These treatments can all be curative.

Penile cancer that is discovered after it is locally invasive has a significant chance of spreading to local lymph nodes and then metastasizing to the rest of the body. Inguinal sentinel node sampling techniques help determine the extent of metastasis through the lymphatic system and helps guide therapy.

Penile cancer that has metastasized or spread throughout the body may not be curable but can be placed into remission and controlled utilizing chemotherapy.

Recovery

Localized penile cancer is usually managed with "penile preserving" therapy that decreases the physical and psychological morbidity and long-term problems associated with penile cancer treatment. However locally advanced, regionally spread, and widely spread/metastasized penile cancers all require lifelong therapy. Patients undergoing surgical removal of penile cancerous tissue may experience decreased sexual potency and rarely may experience decreased quality of urinary stream with difficulty fully emptying the bladder [10].

REFERENCES

1. Chow, W.-H., Dong, L.M., and Devesa, S.S. (May 2010). Epidemiology and risk factors for kidney cancer. *Nature Reviews Urology* 7 (5): 245–257.
2. Munoz, J.J. and Ellison, L.M. (November 2000). Upper tract urothelial neoplasms: incidence and survival during the last 2 decades. *The Journal of Urology* 164 (5): 1523–1525.
3. "HPV Vaccine Is Recommended for Boys." Centers for Disease Control and Prevention. August 4, 2015. Accessed August 4, 2015.
4. "Cancer Staging." National Cancer Institute. January 6, 2015. Accessed August 1, 2015.
5. Blitstein, J. and Ghavamian, R. (June 2008). Laparoscopic partial nephrectomy in the treatment of renal cell carcinoma: a minimally invasive means to nephron preservation. *Expert Review of Anticancer Therapy* 8 (6): 921–927.
6. Weizer, A.Z., Palella, G.V., Montgomery, J.S. et al. (April 2011). Robot-assisted retroperitoneal partial nephrectomy: technique and perioperative results. *Journal of Endourology* 25 (4): 553–557.
7. Lee, C.-H. and Motzer, R.J. (August 2014). Sunitinib as a paradigm for tyrosine kinase inhibitor development for renal cell carcinoma. *Urologic Oncology* 33: 275–279.
8. Straatman, J., Cuesta, M., de-Lange-deKlerk, E., and van Der Pert, D. (2016). Long-Term Survival After Complications Following Major Abdominal Surgery. *Journal of Gastrointestinal Surgery.* 20: 1036–1041.
9. Brugarolas, J. (May 2014). Molecular genetics of clear-cell renal cell carcinoma. *Journal of Clinical Oncology* 32 (18): 1968–1976.
10. "What is bladder cancer" American Cancer Society. February 25, 2015. Accessed August 5, 2015.
11. "HPV and Men Fact Sheet." Centers for Disease Control and Prevention." January 28, 2015. Accessed August 4, 2015.
12. "Can penile cancer be prevented?" American Cancer Society. April 20, 2015. Accessed August 13, 2015.

18

LIVER CANCER: HEPATOCELLULAR CARCINOMA

Jennifer LaFemina[1], Bradley Switzer[2], and Giles F. Whalen[1]

[1]*Division of Surgical, UMass Memorial Health Center, University of Massachusetts Medical School, Worcester, MA, USA*
[2]*Division of Medical Oncology, UMass Memorial Health Center, University of Massachusetts Medical School, Worcester, MA, USA*

GENERAL CONSIDERATIONS

The colloquial term "Liver Cancer" encompasses many different types of malignant tumors in the liver. In the United States and Western Europe the most common type of malignant tumor in the liver is a metastasis from another area. The most common sources of these metastatic tumors in the liver are primary lung cancers, breast cancers, and colorectal and other gastrointestinal cancers such as pancreas or stomach. However, it is not unusual to see metastases to the liver from less common cancers like melanomas, kidney cancers, or some sarcomas and endocrine cancers. All of these "liver cancers" are typically managed by the systemic treatment algorithms for advanced metastatic disease of whatever the primary site is. Occasionally, this includes liver directed local-regional therapy such as partial liver resections, thermal ablations, embolization of a liver tumor's blood supply or chemotherapy administered selectively into the hepatic artery.

Additionally, there are a number of different kinds of malignant tumors that start in the liver, which are termed "liver cancer." These include malignancies that arise from the epithelium lining the bile ducts anywhere from main ducts to small biliary ductules (Cholangiocarcinomas – also colloquially called "Klatskin tumors" when they develop at the bifurcation of the main right and left bile ducts) or from the epithelium lining biliary mucinous cysts ("cystadenocarcinomas"). Malignancies that arise from the supporting vascular stroma such as "epithelioid hemangioenothelioma" and its clinically more aggressive cousin "hemangiosarcoma" (which has a known association with exposure to an older radiologic contrast agent called "thorotrast" and environmental exposure to polyvinyl chloride manufacture) are even more unusual but also occur.

However, all of these primary malignant tumors of the liver are rare in comparison to the primary cancers that arise from hepatocytes themselves. These primary cancers have some variability in epidemiology, degree of differentiation, clinical patterns, and behavior, which is denoted by their pathologic names. Hepatoblastoma occurs in pediatric patients. Fibrolamellar Hepatocellular Cancer (HCC) tends to develop in younger patients in the absence of antecedent liver inflammation, and typically behaves somewhat less aggressively than the common form of HCC. The common form of HCC is often simply called "Hepatoma" and may occasionally have a mixed histologic pattern ("cholangiohepatocellular cancer"). It is usually associated with underlying chronic hepatitis and cirrhosis, and owing to the prevalence of viral hepatitis in the developing world, is one of the most common malignant solid tumors afflicting human beings on earth. It presents a number of treatment challenges in both developed and underdeveloped countries and remained a quite lethal problem in 2015. For all of these reasons, HCC is the focus of this chapter on "liver cancer."

DEMOGRAPHICS

Hepatocellular carcinoma (HCC or hepatoma) is the most common primary liver malignancy. Worldwide, HCC is the fifth most common cancer and the third most common cause of cancer death, with approximately 748 300 new liver cancer cases and 695 900 cancer related deaths diagnosed in 2008 [1, 2]. While it is a less common entity in the United States, recent reports suggest that the incidence and mortality rates associated with this cancer continue to increase in the United States due to the increase in Hepatitis B (HBV) and C (HCV) virus infection [3–6]. Globally, men are diagnosed more commonly than women, and more than half of these cases and deaths were believed to occur in China [2]. Furthermore, HCC is the most common cause of death in cirrhotics. Metastatic disease to the liver, rather than HCC, is the most common type of cancer of the liver in the United States.

RISK FACTORS

Chronic hepatic inflammation is the major driving factor in the development of HCC. For this reason up to 90% of patients diagnosed with HCC do so within a background of chronic liver disease [7, 8]. Risk factors vary by regional exposures. Chronic HBV is the primary risk factor in developing countries in eastern Asia and sub-Saharan Africa and is identified in more than 50% of cases. Additional risk factors, particularly in developing countries, include alfatoxin B1 exposure and parasitic infections (e.g., schistosomiasis, liver flukes) [7]. Hepatitis C virus (HCV) and heavy alcohol use are the most common risk factors in patients in industrialized countries such as the United States. However, other processes that are associated with chronic hepatocellular injury and HCC development include: iron storage diseases (e.g., hemochromatosis, Wilson's disease), α1-antitrypsin, hereditary tyrosinemia, type I glycogen storage disease, familial polyposis coli, Budd-Chiari syndrome, biliary cirrhosis, diabetes, and obesity [7, 9, 10]. Alcohol works synergistically with HBV and HCV to increase the risk of carcinogenesis. For example, HCV and alcohol usage in the setting of cirrhosis can increase the risk of developing HCC by nearly 17-fold.

PREVENTION

As chronic HBV infection is the most common risk factor leading to HCC development, HBV vaccination is one option to reduce the risk of HCC, particularly in high-risk areas. Population-based analyses have demonstrated that a universal HBV vaccination program in Taiwan reduced the incidence of HCC in children [11]. As there is no vaccine currently available for HCV, prevention relies on avoidance of transmission of contaminated blood. Antiviral medications can interrupt the transformation of acute to chronic infection, which ultimately is the primary way to reduce the incidence of HCC.

SCREENING

In industrialized nations in which intervention is feasibility, early disease standard screening protocols have been designed to diagnosis early stage HCC and reduce death due to disease. The American Association of Liver Disease (AASLD) recommends a screening liver ultrasound every six months for high-risk individuals [12]. Ultrasound is the preferred means of screening as it is safe, widely available, and has acceptable sensitivity (up to 80%) and specificity (greater than 90%) [13]. High risk individuals are defined as patients with HCV or cirrhosis from any cause. Routine use of alpha-fetoprotein (AFP) has not been recommended as its sensitivity is limited (approximately 60%), in part because nearly 20% of patients with HCC will not produce this tumor marker.

CLINICAL PRESENTATION

For patients in whom screening has been initiated, HCC may be diagnosed in its earliest stages, when symptoms are absent or minimal. However, for patients in whom the tumor is diagnosed at later stages, signs and symptoms may include abdominal pain, unintentional weight loss, ascites, or a palpable right upper quadrant mass. Portal hypertension or jaundice may be present, but the latter is generally a late sign of the disease. Rarely, tumor rupture may lead to a diagnosis.

DIAGNOSTIC WORKUP

Radiology

If a mass is detected on surveillance ultrasound, further evaluation is based on the size of the lesion. For lesions <1 cm, diagnosis can be unreliable, and HCC will be identified only about 50% of cases [14]. For this reason, close surveillance with repeat ultrasound scans every three months is recommended. If the lesion remains stable, the patient can return to standard surveillance. However, if the lesion is enlarging, then the patient can transition to the diagnostic algorithm for lesions >1 cm.

If the mass detected on ultrasound screening is >1 cm, a four-phase liver CT or dynamic contrast enhanced MRI should be performed. The classic features of HCC include venous or delayed phase "washout" and arterial hypervascularization (Figures 1–6). If these are demonstrated, then no further diagnostic workup is

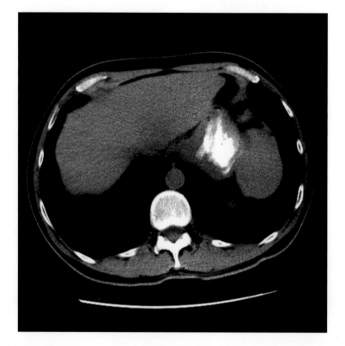

Figure 1. CT scan of the liver in a patient without IV contrast. Note no clear defined tumor.

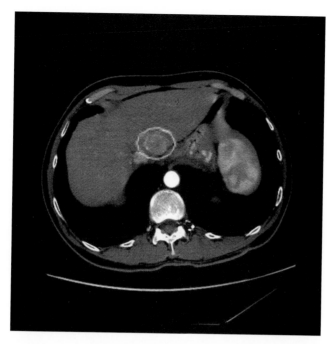

Figure 2. The same CT as in Image 1 but with the Arterial phase of contrast administration, shown inside the red circle. Note clear definition of the tumor with the administration of contrast. (See insert for color representation.)

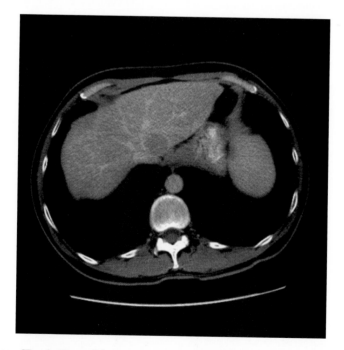

Figure 3. Same CT as in Figure 2 but representative image demonstrates the venous phase. The CT is repeated one minute after the IV bolus of contrast is given, this allows for the contrast to be "washed out" of the arteries (and tumor) and catches it when it is concentrated in the veins of the liver. This is an example of the classic "wash out" seen in HCC and is not seen in other tumors found in the liver helping to distinguish this from other growths.

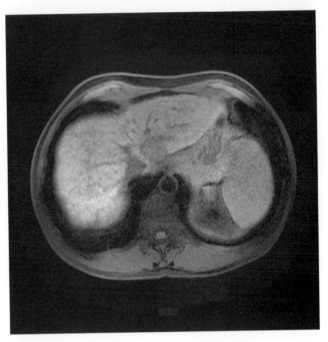

Figure 4. T1 weighted MRI of the Liver in the same patient without contrast. The tumor is diffi-cult to visualize.

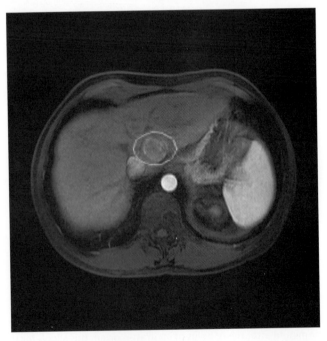

Figure 5. T1 MRI of the liver with IV contrast in the arterial phase. You can see both the hepatic artery and tumor enhance with contrast. The tumor is circled in red for clarification. (See insert for color representation.)

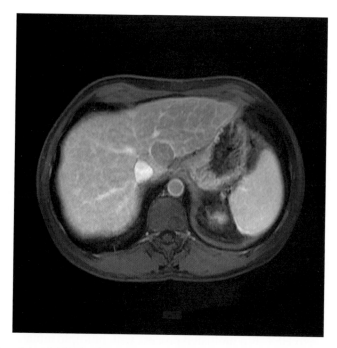

Figure 6. T1 weighted MRI in the same patient with HCC in the venous phase. The MRI is repeated three minutes after contrast administration to time the images with contrast concentrating in the venous system. Notice that the tumor has "washed out" all the contrast and the liver venous system is enhancing around it.

needed to confirm the diagnosis. However, if imaging fails to demonstrate these classic HCC features, then proceeding with the alternate imaging modality or a biopsy is required.

While AFP is not employed for screening purposes up to 90% of patients with HCC will have an elevated AFP reading. However, it is important to note that AFP is neither particularly sensitive nor specific for the diagnosis of HCC, and it can be falsely elevated in other conditions, such as cirrhosis. However, the combination of AFP >200 ng/ml and classic imaging findings yields a sensitivity of approximately 100% for the diagnosis of HCC.

Pathology

Microscopic Evaluation

Since the tumor is commonly diagnosed radiographically, many of the pathologic specimens are procured during resection or explant. Only when imaging is not "classic" is a biopsy obtained to confirm diagnosis. Tumors in livers without cirrhosis typically present as a solitary mass and may occasionally have satellite nodules surrounding the primary tumor. Tumors in livers with cirrhosis are often multifocal with several discrete synchronous lesions. The tumors tend to have scant stroma and central necrosis because of poor internal tumor vascularization.

The neoplastic cells and microarchitecture of HCC resemble normal liver. They can form trabeculae, cords, nests, and may contain bile pigment in cytoplasm similar to normal liver tissue. It can be difficult to differentiate normal versus abnormal and there is a spectrum ranging from well-differentiated (normal looking) to poorly-differentiated (loss of nearly all normal cellular architecture and function). As the tumor becomes more abnormal, it shows signs of cellular atypia, thickened cords, abnormal trabeculi, and ultimately mitotic complexes.

Multiple subtypes of HCC exist, and multiple subtypes may be found within the same tumor or patient. However, the variations have little impact on outcome and treatment. The exception to this rule is fibrolamellar hepatocellular carcinoma. Fibrolamellar HCC is a distinct variant of HCC recognized by characteristic histologic changes, prolonged survival relative to conventional HCC, and an increased incidence in adolescents or young adults.

Molecular Pathogenesis

The progression to HCC is a multistep process with multiple mutations implicated in pathogenesis. *TP53* and β-catenin have most commonly been implicated in tumor progression, cited in 25–40% and 25% of HCC cases, respectively [7]. Chromosomal amplifications and deletions, epigenetic alterations, and miRNA alterations have all been implicated and ultimately these changes induce changes in cell survival and proliferation pathways. Notable cascade pathways that have been implicated include the EGFR and Ras (more than 50% of cases), mTOR (up to 50% of cases), IGF1R (20%), HGF and cMET pathways, and the Wnt pathway (up to one-third of cases) [7, 15–18]. It is believed that pathways related to angiogenesis are also instrumental in tumor development.

STAGING AND PROGNOSIS

In decades past, survival was universally dismal. With the advent of screening protocols, up to 40% of patients will be diagnosed at an early stage, allowing for optimal treatment with potentially curative options [14]. Staging systems have been developed to reflect treatment options and prognosis related to HCC. The single greatest predictor of long-term survival for patients with HCC is the ability to perform surgical resection of their cancer. Thus, patients are often staged and subsequently referenced as operable or inoperable.

One of the most difficult things about measuring prognosis and establishing set treatment options for HCC is that these patients often have two separate and distinct life-threatening diseases. In the United States, most patients with HCC have advanced cirrhosis making the traditional staging models insufficient to predict survival. To date, there are more than six staging systems for HCC (including the

TABLE 1. American Joint Commission on Cancer staging for liver tumors (excluding intrahepatic cholangiocarcinoma).

Primary Tumor (T)	
T0	No primary tumor
T1	Solitary tumor, no vascular involvement
T2	Solitary tumor, vascular involvement *or* multiple tumors <5 cm
T3a	Multiple tumor, >5 cm
T3b	Single/multiple tumor(s), any size, and involvement of major portal/hepatic vein branch
T4	Tumor(s), direct invasion into adjacent organs other than gallbladder *or* perforation of visceral peritoneum
Nodes (N)	
N0	No node metastases
N1	Positive node metastases
Distant Metastasis (M)	
M0	No distant metastasis
M1	Distant metastasis present
Stages	
I	T1N0M0
II	T2N0M0
IIIA	T3aN0M0
IIIB	T3bN0M0
IIIC	T4N0M0
IVA	Any TN1M0
IVB	AnyTAnyNM1

Okuda, CLIP, and BCLC systems), but not one has been adopted universally. The American Joint Committee on Cancer (AJCC) is an organization that publishes a unified classification system with the goal of selecting the most effective treatment, determining prognosis, and continuing evaluation of cancer control measures. The current, seventh edition AJCC staging, as shown in Table 1, has been independently validated by multiple institutions in multiple countries.

TREATMENT

Overview

Worldwide, HCC is traditionally found in younger patients with chronic HBV, but in the United States, HCC often arises in the setting of cirrhosis making the potential treatment algorithm significantly different and dependent upon the patient's medical comorbidities. If HCC is a result of cirrhosis, then special consideration is needed in treatment of both diseases since both are potentially life-ending. For this reason, the management of patients with HCC often involves multiple teams including hepatology, radiology (diagnostic and interventional), surgery, medical oncology, nutrition, social work, and drug/alcohol counseling.

As with many cancers, surgery is a mainstay of treatment and offers a chance for long-term survival benefit. Aside from thermal or chemical ablation, which may be as effective as partial liver resection for small (<2 cm) HCC, other available treatments are primarily offered to assist with symptoms or disease stabilization rather than a cure.

Surgical Resection

Surgical removal of the tumor, either by partial hepatectomy or by total hepatectomy with orthotopic liver transplant (OLT) is the most effective treatment for most patients with localized HCC. However, the presence of underlying cirrhosis and the limited availability of donor livers for transplantation significantly limit the applicability of this treatment for many patients with HCC.

Cirrhotic livers are fibrotic and poorly functioning, sometimes lacking the ability to compensate or regenerate, even for small surgeries. To determine if a patient is a candidate for resection, liver function must be assessed. The Child-Turcotte-Pugh score is the principal means by which to assess underlying liver function [19]. Included parameters include serum bilirubin, serum albumin, INR, presence of ascites, and degree of encephalopathy. While this system has been widely validated, it does lack HCC-specific parameters, which some of the aforementioned staging systems have incorporated. In general, Child-Pugh Class A patients have satisfactory liver function and may tolerate hepatic resection; those with class C disease are at a significant risk of complications and death following even minor hepatic resections.

When resection is considered, it is no longer believed that a 1 cm margin is necessary. However, the entire tumor must be completely removed. A microscopically positive margin or satellites nodules increase the risk of recurrence [20]. The pattern of spread for HCC includes local extension (diaphragm, adjacent organs), regional (peritoneal surfaces and lymph nodes in the hepatic hilum), and vascular dissemination, to distant sites (via the portal and hepatic veins).

Extrahepatic metastatic disease is a contraindication to surgical resection because it will not change the oncologic outcome. Additional contraindications for surgical resection without transplantation include severe liver dysfunction and inadequate future liver remnant (FLR) because the patient cannot recover from the operation. Portal venous embolization (PVE) of the part of the liver to be resected may be employed preoperatively to grow the volume of the postoperative liver remnant to an adequate size prior to the stress of operative hepatectomy. This strategy may convert an unresectable circumstance into one that can be undertaken with a greater margin of safety. However, a cirrhotic liver may not retain the capacity to respond to this stimulus.

With margin-negative resection, median survival is approximately 30–40 months, with a five-year survival of 30–40%. Recurrence in the remaining liver develops in up to 70% of cases, and it is believed to at least be in part related to a tendency of the remaining liver to generate new HCCs in addition to liver metastases from the originally resected primary or regrowth at the margins. Vascular invasion and severe hepatic fibrosis are the strongest predictors of survival and recurrence, though size >5 cm, multiple tumors, high mitotic rate, lack of a tumor capsule, and an unfavorable differentiation status are also associated with worse prognosis [21].

Orthotopic Liver Transplantation

After a partial hepatic resection, the most common pattern of recurrence is intrahepatic. Some if not most of these recurrences are new primary HCCs, which develop in the chronically damaged and inflamed remnant liver. Total removal of the patient's diseased liver with OLT not only addresses that problem, but also cures the patient's underlying cirrhosis and liver failure. Consequently, many patients who could not have a resectable localized HCC removed because their underlying liver dysfunction made the operative risk prohibitive would now become eligible for a curative treatment of their cancer. The problem with using OLT for the treatment of HCC is the relative scarcity of donor livers compared to the number of patients who have end-stage liver disease, and the longer survival of patients who undergo OLT for reasons other than HCC (since those patients do not typically have an aggressive cancer prone to recurrence and metastasis in an immunosuppressed host). In order for the scarce resource of donor livers to be ethically used for the treatment of patients with HCC it is important to select HCC patients who are less likely to develop metastatic HCC in the near term – so that the donor liver is not "wasted." Various prognostic clinical-pathologic criteria have been employed to accomplish this selection.

The Milan Criteria were ultimately chosen to identify those patients who were likely to have the greatest benefit, and lowest risk of recurrence of HCC, following OLT. The Milan criteria include: one tumor <5 cm, or up to three tumors <3 cm, with no associated vascular invasion or metastatic spread [22]. With total hepatectomy and OLT in these highly selected patients outcomes are excellent and better than partial hepatectomy: five-year survival can exceed 60%. There are several other large centers, such as the University of California at San Francisco, with their own transplantation requirements, but to date the United Network for Organ Sharing (UNOS) has adopted the Milan criteria as the standard for distributing donor livers for transplantation in the treatment of HCC.

When a patient is found to be a candidate for liver transplantation, their name is placed on the UNOS transplant list, and they await organ availability. The liver transplant waiting list contains cirrhotic patients both with and without cancer, and the wait time on the list can be several years. In order to prioritize the allocation of donor livers, the Model for End Stage Liver Disease (MELD) score has been widely accepted. This score takes into account a patient's INR, bilirubin, and creatinine. To reduce the risk of cancer progression while on the waiting list (especially if the patient's liver failure is not quite end stage), extra points are awarded to the MELD score to improve allocation of donor livers to the HCC patient whose tumor or tumors fall within the Milan criteria.

Systemic Therapies

Cytotoxic Chemotherapy

Traditional cytotoxic chemotherapy has only a limited role in HCC. Since the worldwide prevalence of HCC is exceedingly high, much research has been conducted in an attempt to find the ideal chemotherapy. Unfortunately, this has been associated with little success. Many agents, even in combination, have been tried and are not particularly effective or have significant toxicities. Further studies have been performed employing immunomodulating therapies such as interferon, which in combination with cytotoxic chemotherapy has demonstrated some promising results.

There are several proposed mechanisms for the relative resistance of HCC to chemotherapy. First, it is believed that the tumors themselves are chemo-refractory. There tends to be a high rate of expression for several multidrug resistance (MDR) genes in HCC. Examples are p-glycoprotein, glutathione-S-transferase, heat shock proteins, and mutations in p53. All of these are processes that cells have developed to resist the effects of chemotherapy either by pumping chemotherapy out of the cell or by inactivating it. Second, concomitant liver failure affects the ability of patients to tolerate chemotherapy, resulting in a greater number of toxicities and dose reductions. This greatly affects patient outcomes when patients are treated with traditional chemotherapy. Finally, there is a significant heterogeneity of patients and of underlying causes of HCC. The exact reason for the relative resistance to chemotherapy is not known but is likely a combination of the above. Interestingly, studies done in Asia have demonstrated greater benefits associated with cytotoxic chemotherapy than those studies in the United States, which is possibly related to a more favorable patient profile in Asian countries (younger, fewer patients with cirrhosis).

Targeted Therapies

Sorafenib is a small molecule that inhibits tumor cell proliferation and tumor angiogenesis, and increases the rate of apoptosis in a wide range of models. It has been termed a multi-targeted tyrosine kinase inhibitor (TKI) and affects several more intracellular proteins than the tyrosine kinases. It acts by inhibiting the serine–threonine kinases, Raf-1 and B-Raf, the receptor tyrosine kinase activity of vascular endothelial growth factor receptors (VEGFRs) one to three and platelet-derived growth factor receptor β (PDGFR-β) [23]. By affecting the above proteins and signaling pathways, it blocks blood vessel growth affectively starving the tumor of energy and initiates pathways that lead to apoptosis or programmed cell death. Targeted chemotherapy is typically better tolerated with less side effects because it "targets" the damaged or cancerous cells leaving most normal cells minimally affected.

A randomized trial demonstrated that well-selected patients with preserved liver function and advanced HCC who were treated with sorafenib had a significant (three months) improvement in survival [23]. There is ongoing research to determine if this agent is useful in the adjuvant setting and if it improves outcomes of surgery.

Locoregional Therapies

Many HCC tumors are limited to the liver at the time of diagnosis. If the person is not a surgical candidate, if the tumor is locally unresectable, or if the patient is awaiting OLT, locoregional treatments can be adopted in an attempt to keep the tumor from progressing, and occasionally cure it. These treatments are not only reasonable options in selected patients with well-compensated liver disease, they are the most commonly used treatments for HCC worldwide.

CHEMICAL ABLATION
Tumors are injected directly with cytotoxic chemicals, such as ethanol (the most common) or acetic acid, via image guidance with ultrasound or CT scan. The agents are directly toxic to both the normal and cancerous tissues. This technique can have a 90–100% necrosis rate for tumors <2 cm, but its benefit decreases significantly with larger tumors. Drawbacks include multiple injections over several days and difficulties of imaging small tumors that are deep within the liver.

THERMAL ABLATION
Not unlike chemical ablation, specialized needle probes can deliver thermal energies to kill tumors. Tumors are localized and targeted with a needle using image guidance. This can be performed percutaneously or surgically via laparoscopy or laparotomy. These probes can be super cooled with liquid nitrogen (cryoablation) so that the tumors are frozen *in situ*, or more commonly, heat energy is delivered to boil the tumor tissue *in situ* (Microwave ablation and Radiofrequency ablation). Radiofrequency ablation (RFA) is the best studied and most commonly used now. An RFA probe is threaded into the tumor, and tissue is ablated using the heat generated from the high frequency alternating current. This technique usually requires only one setting and also has a 90–100% necrosis rate that decreases significantly with increasing size of tumor >2 cm [24, 25]. Microwave ablation provides a somewhat wider zone of necrosis and may be better than RFA for slightly larger lesions [26–28]. While recurrence is higher with larger tumors, RFA in well-selected patients with small tumors yields an overall survival rate similar to resection [29–32]. However, for a medically-suitable patient with a larger HCC that meets Milan Criteria, surgical resection may be preferable to RFA due to a lower recurrence and improved survival [33].

Transarterial Chemoembolization (TACE)

TACE involves both intraarterially infused chemotherapy as well as hepatic artery occlusion. The catheter is inserted into the femoral artery and ultimately directed into the hepatic artery via angiography. Chemotherapy can be infused either prior to embolization or on gelatin sponge scaffold. Lipiodol, an iodized ethyl ester of fatty acids of poppy seed oil, has been effective because its half-life is extended in hepatic tumors, increasing its duration of treatment. In well-selected patients with preserved liver function (Child-Pugh Class A and some Bs) and unresectable tumors, TACE has been shown to have significant benefit in overall survival compared to best supportive care [34, 35]. With modern vascular imaging and intervention, this technique has evolved to embolize only the vessels feeding the HCC tumor ever more selectively and accurately, and with use of different sizes of drug eluting beads to diminish toxicity.

Radiation Therapy

Conventional external beam radiotherapy plays a limited role in patients with HCC. However, the use to of yttrium-90 to perform radioembolization may have utility. It appears to be similar to TACE in terms of efficacy for local control of unresectable HCC and for downstaging and bridging to OLT, but also appears to have a lower toxicity profile [36]. Stereotactic Body Radiation Treatment (SBRT), which is a form of gated CT-guided radiation surgery, delivers focused lethal doses of radiation to a particular target volume in the liver in just a few fractions and appears to been effective in controlling disease in unresectable candidates [37, 38]. Since multiple parts of the liver receive some of the beams that are focused on the target volume from multiple angles, toxicity depends upon the size of the tumor being treated and its location within the liver. SBRT combined with TACE may be more effective controlling suitable HCC tumors in 3–5 cm range than either one alone.

Palliative Care

Since many of the patients with HCC are not surgically resectable and ultimately will succumb to their diseases, effective palliation of symptoms is an important component of their care. Pain is always an important symptom to control and often can be difficult in patients with cirrhosis since their liver does not clear opiates as well as in healthy individuals. Cirrhotic patients often struggle with hepatic encephalopathy and control of their circulating ammonia levels is crucial for mental clarity. Patients with a dysfunctional liver have poor production of albumin leading to a low intravascular oncotic pressure. This can result in a struggle with peripheral edema and ascites. The aggressive use of diuretics, and when necessary therapeutic abdominal paracentesis, can improve a patient's quality of life in the settings of advanced liver disease and HCC. Medical management of a patient's cirrhosis can also improve survival from their HCC.

Immunotherapy

Emerging data suggests that immunomodulation might reduce the risk of HCC recurrence after resection. A Japanese trial examined autologous lymphocytes activated *in vitro* with recombinant IL2 and antibody to CD3 and found an 18% reduction in recurrence, but not overall survival, following hepatectomy [39]. Another trial from Japan has shown that postoperative interferon appears to reduce the risk of late recurrence in HCV patients who adhere to treatment [40].

Emerging Therapies

As previously discussed, multiple signaling cascades have been implicated in the pathogenesis of HCC. After success with sorafenib, there is ongoing work evaluating the utility of targeted agents in advanced HCC. For instance, everolimus (inhibitor of the mTOR pathway), tivantinib (inhibitor of MET kinase), and brivanib, sunitinib, and linifanib (inhibitors of antiangiogenic multikinase) are being evaluated both independently and in combination with cytotoxic chemotherapy, TACE, and inhibitors of other signaling pathways. Furthermore, as discussed, HBV vaccination strategies have reduced the risk of HCC in children in Taiwan. Ultimately, a vaccine against HCV might serve as a viable means by which to reduce the risk of HCC in the United States.

REFERENCES

1. Ferlay, J., Shin, H.R., Bray, F. et al. (Dec 15 2010). Estimates of worldwide burden of cancer in 2008: GLOBOCAN 2008. *International journal of cancer/Journal International du Cancer* 127 (12): 2893–2917.
2. Jemal, A., Bray, F., Center, M.M. et al. (Mar-Apr 2011). Global cancer statistics. *CA: A Cancer Journal for Clinicians* 61 (2): 69–90.
3. Altekruse, S.F., McGlynn, K.A., and Reichman, M.E. (Mar 20 2009). Hepatocellular carcinoma incidence, mortality, and survival trends in the United States from 1975 to 2005. *Journal of Clinical Oncology: Official Journal of the American Society of Clinical Oncology* 27 (9): 1485–1491.
4. Altekruse, S.F., Henley, S.J., Cucinelli, J.E. et al. (Apr 2014). Changing hepatocellular carcinoma incidence and liver cancer mortality rates in the United States. *The American Journal of Gastroenterology* 109 (4): 542–553.
5. Davis, G.L., Alter, M.J., El-Serag, H. et al. (Feb 2010). Aging of hepatitis C virus (HCV)-infected persons in the United States: a multiple cohort model of HCV prevalence and disease progression. *Gastroenterology* 138 (2): 513–521. 521 e511-516.
6. Tanaka, Y., Kurbanov, F., Mano, S. et al. (Mar 2006). Molecular tracing of the global hepatitis C virus epidemic predicts regional patterns of hepatocellular carcinoma mortality. *Gastroenterology* 130 (3): 703–714.

7. Forner, A., Llovet, J.M., and Bruix, J. (Mar 31 2012). Hepatocellular carcinoma. *Lancet* 379 (9822): 1245–1255.
8. Sherman, M. (Feb 2010). Hepatocellular carcinoma: epidemiology, surveillance, and diagnosis. *Seminars in Liver Disease* 30 (1): 3–16.
9. El-Serag, H.B. (Sep 22 2011). Hepatocellular carcinoma. *The New England Journal of Medicine* 365 (12): 1118–1127.
10. El-Serag, H.B., Tran, T., and Everhart, J.E. (Feb 2004). Diabetes increases the risk of chronic liver disease and hepatocellular carcinoma. *Gastroenterology* 126 (2): 460–468.
11. Chang, M.H., Chen, C.J., Lai, M.S. et al. (Jun 26 1997). Universal hepatitis B vaccination in Taiwan and the incidence of hepatocellular carcinoma in children. Taiwan Childhood Hepatoma Study Group. *The New England Journal of Medicine* 336 (26): 1855–1859.
12. Bruix, J. and Sherman, M. (Nov 2005). Practice guidelines committee AAftSoLD. Management of hepatocellular carcinoma. *Hepatology* 42 (5): 1208–1236.
13. Singal, A., Volk, M.L., Waljee, A. et al. (Jul 2009). Meta-analysis: surveillance with ultrasound for early-stage hepatocellular carcinoma in patients with cirrhosis. *Alimentary Pharmacology & Therapeutics* 30 (1): 37–47.
14. Llovet, J.M., Burroughs, A., and Bruix, J. (Dec 6 2003). Hepatocellular carcinoma. *Lancet* 362 (9399): 1907–1917.
15. Villanueva, A., Chiang, D.Y., Newell, P. et al. (Dec 2008). Pivotal role of mTOR signaling in hepatocellular carcinoma. *Gastroenterology* 135 (6): 1972–1983. 1983 e1971-1911.
16. Villanueva, A., Newell, P., Chiang, D.Y. et al. (Feb 2007). Genomics and signaling pathways in hepatocellular carcinoma. *Seminars in Liver Disease* 27 (1): 55–76.
17. Sahin, F., Kannangai, R., Adegbola, O. et al. (Dec 15 2004). mTOR and P70 S6 kinase expression in primary liver neoplasms. *Clinical Cancer Research: An Official Journal of the American Association for Cancer Research* 10 (24): 8421–8425.
18. Farazi, P.A. and DePinho, R.A. (Sep 2006). Hepatocellular carcinoma pathogenesis: from genes to environment. *Nature Reviews. Cancer* 6 (9): 674–687.
19. Child, C.G. and Turcotte, J.G. (1964). Surgery and portal hypertension. *Major Problems in Clinical Surgery* 1: 1–85.
20. Poon, R.T., Fan, S.T., Ng, I.O. et al. (Apr 2000). Significance of resection margin in hepatectomy for hepatocellular carcinoma: a critical reappraisal. *Annals of Surgery* 231 (4): 544–551.
21. Contrereas, C.M., Choi, E.A., and Abdalla, E.K. (2012). Chapter 12: hepatobiliary cancers. In: *The MD Anderson Surgical Oncology Handbook*, 5e (ed. B.W. Feig and C.D. Ching), 429–430. Philadelphia, PA: Lippincott, Williams & Wilkins.
22. Mazzaferro, V., Regalia, E., Doci, R. et al. (Mar 14 1996). Liver transplantation for the treatment of small hepatocellular carcinomas in patients with cirrhosis. *The New England Journal of Medicine* 334 (11): 693–699.
23. Llovet, J.M., Ricci, S., Mazzaferro, V. et al. (Jul 24 2008). Sorafenib in advanced hepatocellular carcinoma. *The New England Journal of Medicine* 359 (4): 378–390.
24. Livraghi, T., Meloni, F., Di Stasi, M. et al. (Jan 2008). Sustained complete response and complications rates after radiofrequency ablation of very early hepatocellular carcinoma in cirrhosis: is resection still the treatment of choice? *Hepatology* 47 (1): 82–89.
25. Poulou, L.S., Botsa, E., Thanou, I. et al. (May 18 2015). Percutaneous microwave ablation vs radiofrequency ablation in the treatment of hepatocellular carcinoma. *World Journal of Hepatology* 7 (8): 1054–1063.
26. Liang, P., Yu, J., Lu, M.D. et al. (Sep 7 2013). Practice guidelines for ultrasound-guided percutaneous microwave ablation for hepatic malignancy. *World Journal of Gastroenterology: WJG* 19 (33): 5430–5438.
27. Lloyd, D.M., Lau, K.N., Welsh, F. et al. (Aug 2011). International multicentre prospective study on microwave ablation of liver tumours: preliminary results. *HPB: The Official Journal of the International Hepato Pancreato Biliary Association* 13 (8): 579–585.

28. Simon, C.J., Dupuy, D.E., and Mayo-Smith, W.W. (Oct 2005). Microwave ablation: principles and applications. *Radiographics: A Review Publication of the Radiological Society of North America, Inc.* 25 (Suppl 1): S69–S83.

29. Wang, Y., Luo, Q., Li, Y. et al. (2014). Radiofrequency ablation versus hepatic resection for small hepatocellular carcinomas: a meta-analysis of randomized and nonrandomized controlled trials. *PLoS One* 9 (1): e84484.

30. Chen, X., Chen, Y., Li, Q. et al. (May 1 2015). Radiofrequency ablation versus surgical resection for intrahepatic hepatocellular carcinoma recurrence: a meta-analysis. *The Journal of Surgical Research* 195 (1): 166–174.

31. Chen, M.S., Li, J.Q., Zheng, Y. et al. (Mar 2006). A prospective randomized trial comparing percutaneous local ablative therapy and partial hepatectomy for small hepatocellular carcinoma. *Annals of Surgery* 243 (3): 321–328.

32. Hong, S.N., Lee, S.Y., Choi, M.S. et al. (Mar 2005). Comparing the outcomes of radiofrequency ablation and surgery in patients with a single small hepatocellular carcinoma and well-preserved hepatic function. *Journal of Clinical Gastroenterology* 39 (3): 247–252.

33. Huang, J., Yan, L., Cheng, Z. et al. (Dec 2010). A randomized trial comparing radiofrequency ablation and surgical resection for HCC conforming to the Milan criteria. *Annals of Surgery* 252 (6): 903–912.

34. Llovet, J.M., Real, M.I., Montana, X. et al. (May 18 2002). Arterial embolisation or chemoembolisation versus symptomatic treatment in patients with unresectable hepatocellular carcinoma: a randomised controlled trial. *Lancet* 359 (9319): 1734–1739.

35. Lo, C.M., Ngan, H., Tso, W.K. et al. (May 2002). Randomized controlled trial of transarterial lipiodol chemoembolization for unresectable hepatocellular carcinoma. *Hepatology* 35 (5): 1164–1171.

36. Abdelfattah, M.R., Al-Sebayel, M., Broering, D. et al. (Mar 2015). Radioembolization using yttrium-90 microspheres as bridging and downstaging treatment for unresectable hepatocellular carcinoma before liver transplantation: initial single-center experience. *Transplantation Proceedings* 47 (2): 408–411.

37. Price, T.R., Perkins, S.M., Sandrasegaran, K. et al. (Jun 15 2012). Evaluation of response after stereotactic body radiotherapy for hepatocellular carcinoma. *Cancer* 118 (12): 3191–3198.

38. Yoon, S.M., Lim, Y.S., Park, M.J. et al. (2013). Stereotactic body radiation therapy as an alternative treatment for small hepatocellular carcinoma. *PLoS One* 8 (11): e79854.

39. Takayama, T., Sekine, T., Makuuchi, M. et al. (Sep 2 2000). Adoptive immunotherapy to lower postsurgical recurrence rates of hepatocellular carcinoma: a randomised trial. *Lancet* 356 (9232): 802–807.

40. Mazzaferro, V., Romito, R., Schiavo, M. et al. (Dec 2006). Prevention of hepatocellular carcinoma recurrence with alpha-interferon after liver resection in HCV cirrhosis. *Hepatology* 44 (6): 1543–1554.

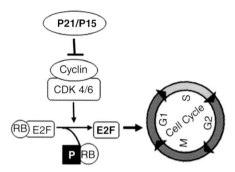

Figure 2.7 Role of RB in the regulation of E2F transcription factor activity during cell cycle. RB sequesters E2F and blocks E2F-mediated expression of genes necessary for the progression of cells from G_1 to S phase. Whereas cyclin-CDKs phosphorylate RB and restore E2F activity to promote cell proliferation, p21 and p15 inhibit cyclin-CDKs and block E2F activation. The arrow indicates activation and "T" indicates inhibition.

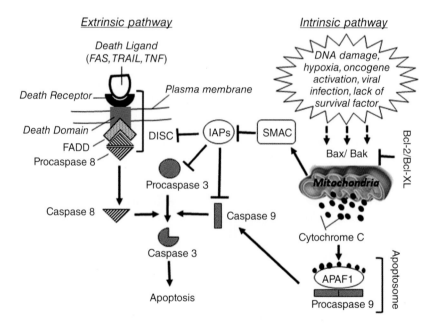

Figure 2.9 Apoptotic pathways showing proteolytic cascade of caspases. Initiator caspases 8 and 9 activated by extrinsic and intrinsic signals, respectively, converge on caspase 3 to execute apoptosis. The arrow indicates activation and "T" indicates inhibition.

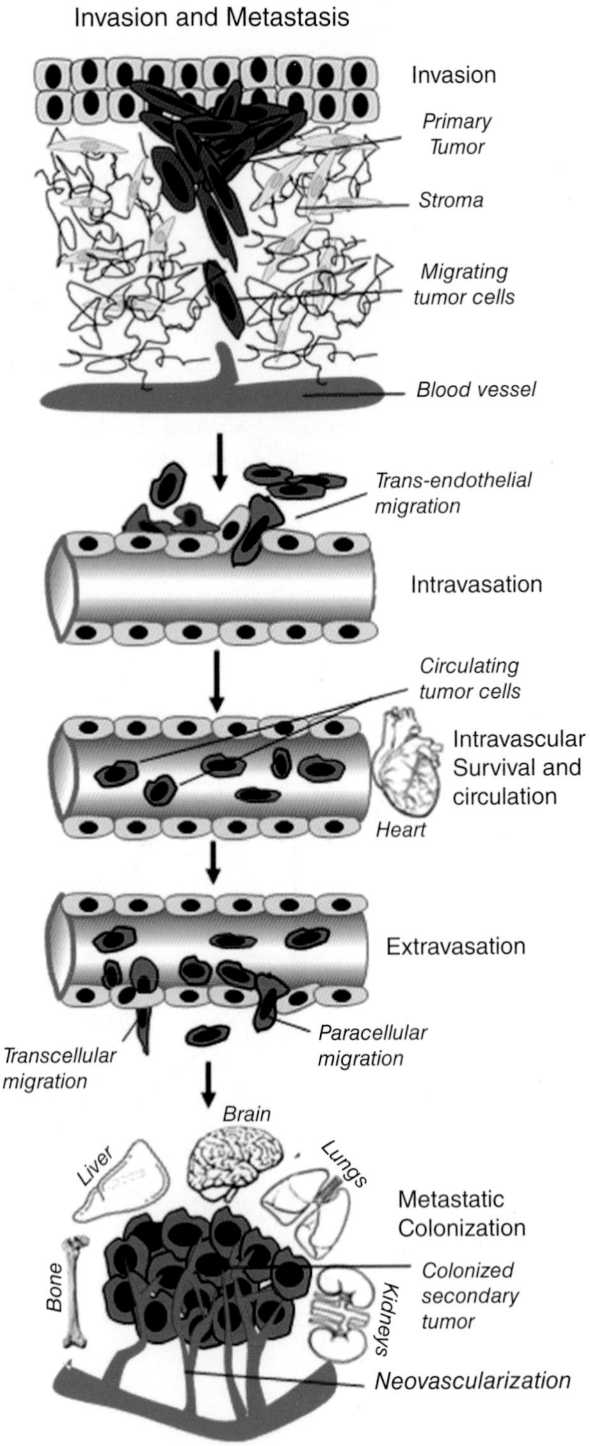

Figure 2.10 Biological processes associated with the spreading of cancer from its primary site to other organs in body. *Source: From invasion to metastasis.*

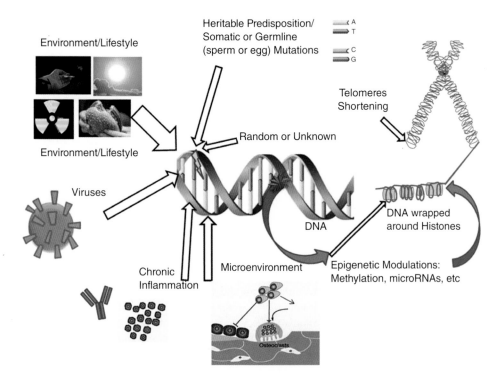

Figure 3.1 The causes of cancer. Cancerous mutations can be caused by numerous factors. Mutations of the DNA can result from the environment (pollutants, carcinogens, UV exposure, radiation), choices we make and our lifestyles, random or unknown mutations that are not corrected properly, epigenetic alterations, viruses, or even from changes in the local, noncancerous microenvironment of the cell. Mutations can also be inherited from our parents and can be enhanced by inflammatory signals. Shortening of telomeres during aging can also contribute to cancer.

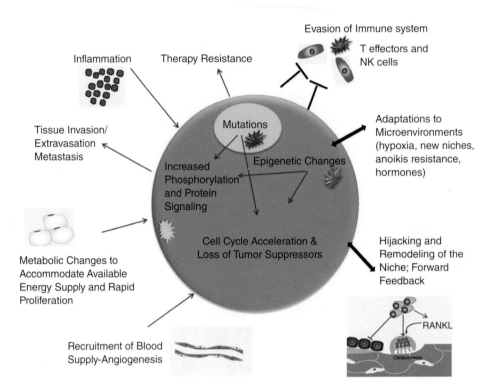

Figure 3.2 Drivers of cancer. Genetic and epigenetic mutations or alterations typically lead to altered signaling within a cell that alter cell proliferation, cell cycle progression, apoptosis, or cell death, such that the cell becomes immortal. Inflammation, metabolic changes, immune evasion, and signals from the local microenvironment can also contribute to tumorigenesis and disease progression. These mutations then lead to therapy resistance, increased survival ability, blood vessel recruitment (angiogenesis), invasion and metastasis, and modulation of the local noncancerous environment to further support the tumor through positive feedback loops.

Figure 7.1 Breast cancer case study molecular diagnostics: immunohistochemistry (IHC) and chromogenic in situ hybridization (CISH). All assays were applied to the same tumor. (a) Hematoxylin and eosin (H&E) stained section of the breast tumor. (b) IHC for breast cancer standard of care marker estrogen receptor (ER): >90% of tumor cell nuclei show are stained; strong intensity on average. (c) IHC for standard of care marker progesterone receptor (PgR): 31–40% of tumor cell nuclei are stained; moderate intensity on average. (d) IHC for HER2 using an FDA approved IHC assay (antibody clone 4B5 [Table 1]) ~90% tumors cells show uniform intense complete membrane staining demonstrating HER2 overexpression. (e) ISH for *HER2* DNA (INFORM dual ISH probe [Table 2]). Red CISH signals indicate signals from the centromere of chromosome 17; the black signals represent the *HER2* DNA locus and are generated by the enzymatic deposition of silver atoms (silver-enhanced ISH [SISH]). The excess of black signals relative to red indicates *HER2* DNA amplification. (f) An experimental ISH assay for *HER2* RNA (RNAscope®, Advanced Cell Diagnostics, Newark, CA) showing abundant *HER2* RNA expression (dot signals). (g) IHC for Ki67 (mouse monoclonal antibody MIB-1). Ki67 is a cell proliferation marker often used in diagnostic assessments; however, because of a lack of consensus about scoring criteria Ki67 is not recommended for routine breast IHC. (h) RNAscope ISH for *MKI67* for comparison with the protein stain. Scale bar: 20 μm.

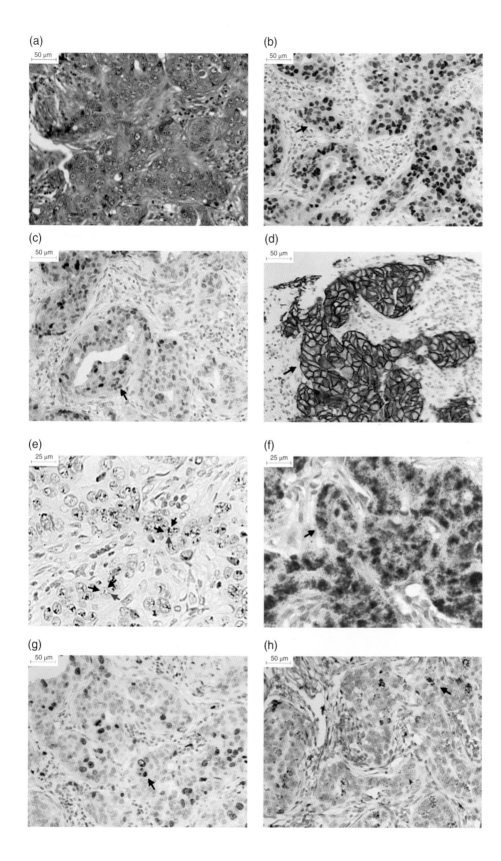

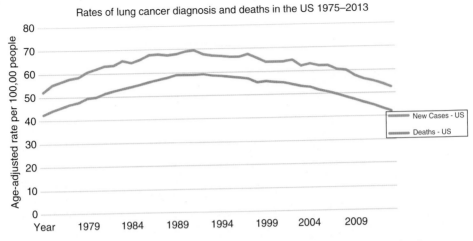

Figure 14.1 Rates of lung cancer diagnosis and attributable deaths in SEER database.

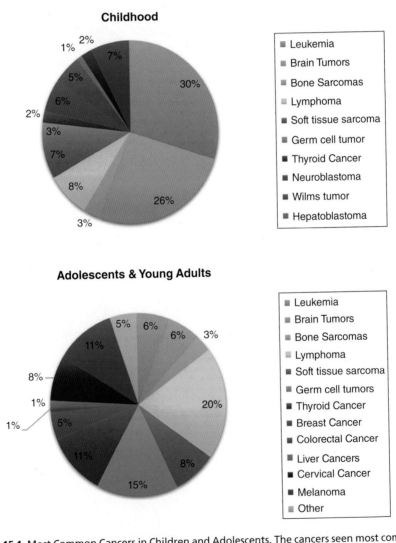

Figure 15.1 Most Common Cancers in Children and Adolescents. The cancers seen most commonly in children under the age of 15 years are shown in the top panel. Leukemia and brain tumors are by far the most common cancers seen in this age group. The cancers most commonly seen in adolescents and young adults (AYAs) (ages 15–29 years) are shown in the lower panel. The most frequently seen cancers in this age group differ from what is seen in younger children, and include lymphoma and germ cell tumors (including testicular cancer).

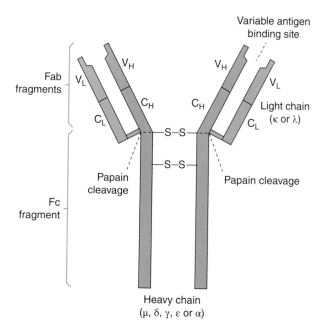

Figure 16.2 Basic structure of an immunoglobulin molecule. Each molecule is made up of two light (κ or λ) (blue areas) and two heavy (purple) chains, and each chain is made up of variable (V) and constant (C) portions, the V portions including the antigen-binding site. The heavy chain (μ, δ, γ, ε or α) varies according to the immunoglobulin class. IgA molecules form dimers, while IgM forms a ring of five molecules. Papain cleaves the molecules into an Fc fragment and two Fab fragments. Source: Hoffbrand and Moss, 2011. Reproduced with permission of John Wiley and Sons.

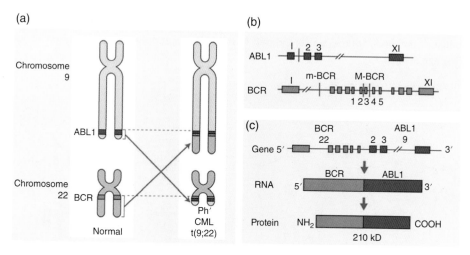

Figure 16.5 The Philadelphia chromosome (Ph'). (a) There is translocation of part of the long arm of chromosome 22 to the long arm of chromosome 9 and reciprocal translocation of part of the long arm of chromosome 9 to chromosome 22 (the Philadelphia chromosome). This reciprocal translocation brings most of the *ABL* gene into the *BCR* region on chromosome 22 (and part of the *BCR* gene into juxtaposition with the remaining portion of *ABL* on chromosome 9). (b) The breakpoint in *ABL* is between exons 1 and 2. The breakpoint in *BCR* is at one of the two points in the major breakpoint cluster region (M-BCR) in chronic myeloid leukemia (CML) or in some cases of Ph+acute lymphoblastic leukemia (ALL). (c) This results in a 210k-Da fusion protein product derived from the *BCR-ABL* fusion gene. In other cases of Ph+ALL, the breakpoint in *BCR* is at a minor breakpoint cluster region (m-BCR) resulting in a smaller *BCR-ABL* fusion gene and a 190-kDa protein. *Source*: Hoffbrand and Moss, 2011. Reproduced with permission of John Wiley and Sons.

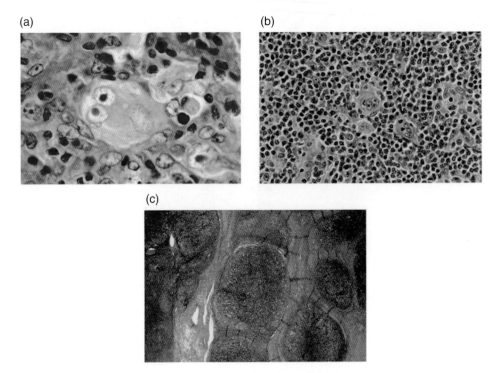

Figure 16.7 Hodgkin lymphoma: (a) high-power view of a lymph node biopsy showing two typical multinucleate Reed-Sternberg cells, one with a characteristic owl eye appearance, surrounded by lymphocytes, histiocytes, and an eosinophil; (b) mixed cellularity; and (c) nodular sclerosing Hodgkin lymphoma. Source: Hoffbrand and Moss, 2011. Reproduced with permission of John Wiley and Sons.

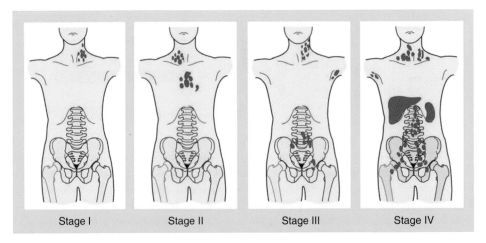

Stage I   Stage II   Stage III   Stage IV

Figure 16.8 Staging of Hodgkin and non-Hodgkin lymphoma. Source: Hoffbrand and Moss, 2011. Reproduced with permission of John Wiley and Sons.

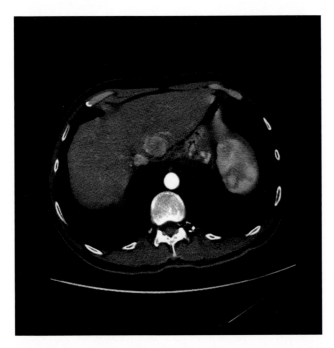

Figure 18.2 The same CT as in Image 1 but with the Arterial phase of contrast administration, shown inside the red circle. Note clear definition of the tumor with the administration of contrast.

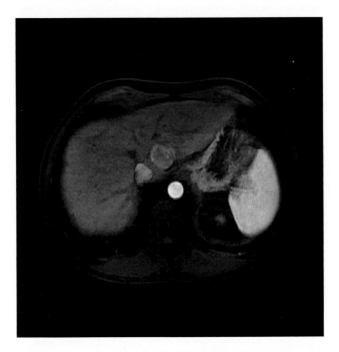

Figure 18.5 T1 MRI of the liver with IV contrast in the arterial phase. You can see both the hepatic artery and tumor enhance with contrast. The tumor is circled in red for clarification.

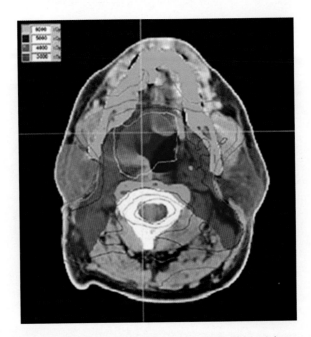

Figures 21.6 Intensity Modulated Radiation Therapy (IMRT). This axial computed tomograph shows the planned dose administration for this patient's IMRT. High dose areas are in red and purple, while low dose zones are designated in green.

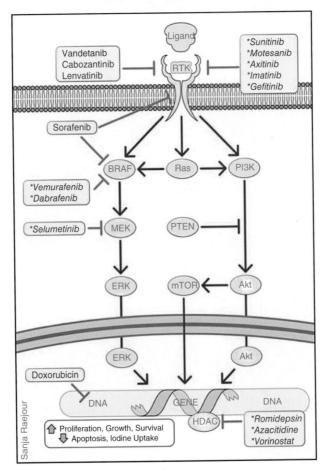

Figure 23.1 Intracellular signaling pathways in thyroid cancer. Therapies currently in use and key molecular targets that are inhibited. * indicates drugs approved for use in other cancers. Source: Illustration by Sanja Raejour.

(a)

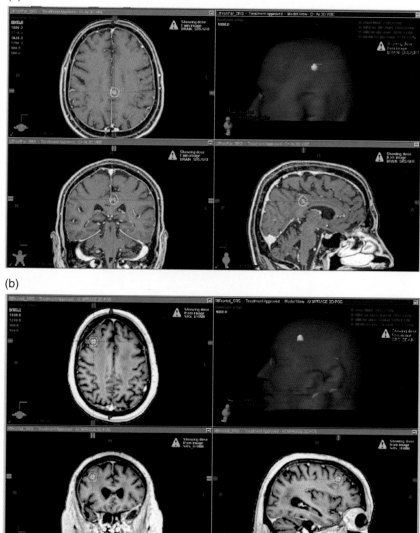

(b)

Figure 24.1 (a) and (b). Radiosurgery treatment of a central nervous system metastasis. Source: Courtesy of the University of Massachusetts Medical School, Department of Radiation Oncology.

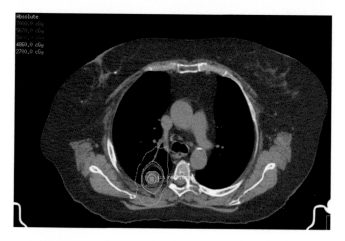

Figure 25.3 Isodose plan for treatment of lung tumor.

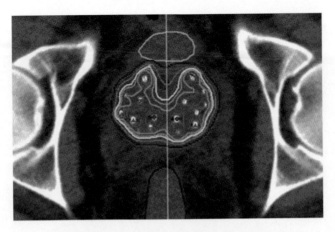

Figure 25.11 Prostate seed brachytherapy plan.

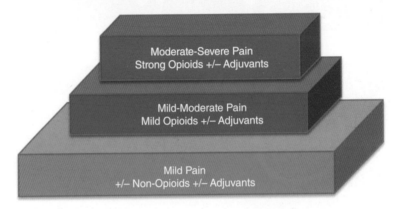

Figure 26.1 World Health Organization (WHO) analgesic ladder.

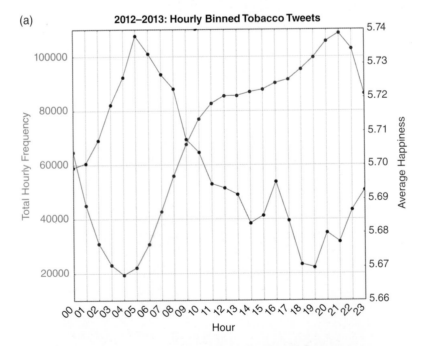

Figure 31.3 (a) (Left) Average number of tobacco related tweets captured per local hour of occurrence spanning 2012–2013. (b) (right) Wordshift graph illustrating the words influencing a shift in positivity from 10:00 in comparison to 3:00.

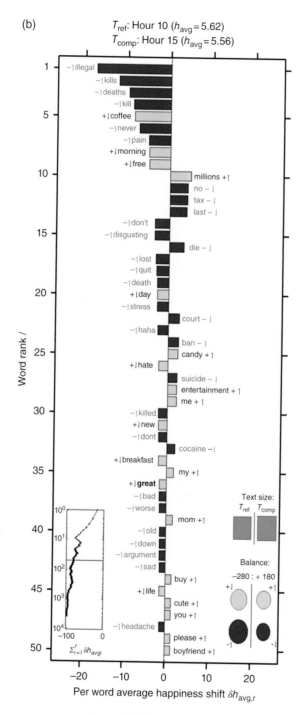

Figure 31.3 (*Continued*)

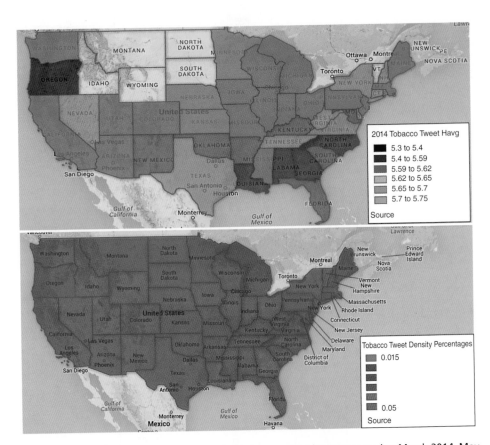

Figure 31.5 Approximately 40 000 geo-tagged tobacco related tweets spanning March 2014–May 2014. (Top) The density of tobacco tweets is computed by finding the ratio of tobacco related tweets to the total number of geo-located tweets per state appearing in the geo-stamped subset of twitter (~1% of all tweets). (Bottom) The states are recolored based on their average happiness scores of their tobacco-related tweets, uncolored states have less than 500 tweets.

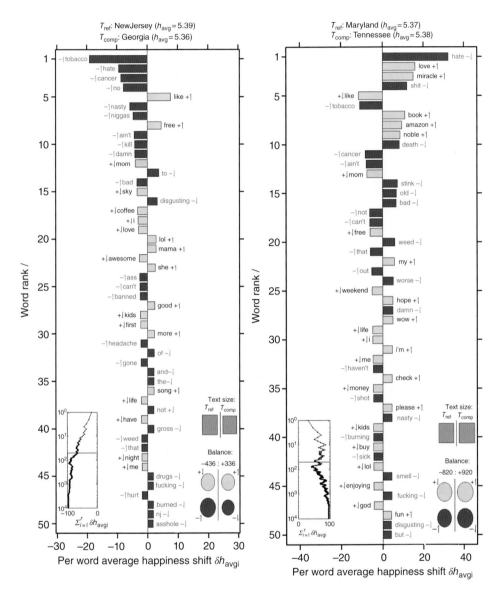

Figure 31.6 (Left) The positivity of geo-tagged tweets from New Jersey are compared to Georgia. A slightly negative shift in positivity is due to an increase in words like "smoke," "headache," and "cancer," as indicated by the up arrow (more frequent) and minus sign (negative sentiment). (Right) Maryland Tobacco Tweets are compared to Tennessee. A slightly positive shift occurs due to an increase of "love" and decrease of "hate."

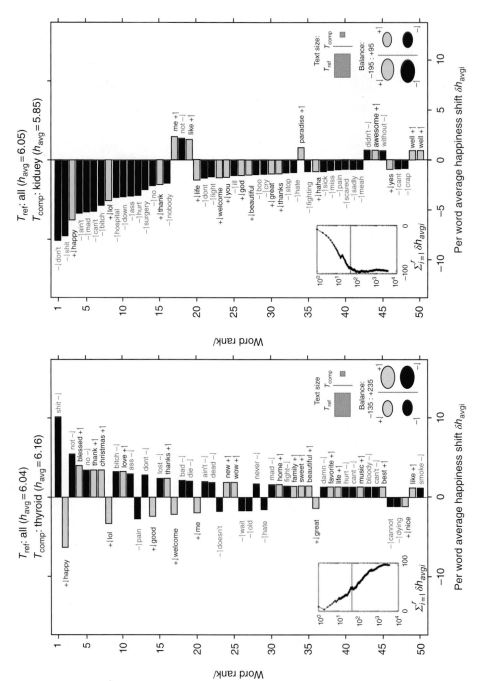

Figure 31.7 Wordshift graphs for the happiest cancer patient tweet-set and the saddest cancer patient tweet-set.

19

GYN ONCOLOGY: UNIQUE CHARACTERISTIC TARGETS FOR THERAPY IN GYNECOLOGIC CANCERS

Celeste Straight[1], Leslie Bradford[2], and Susan Zweizig[3]

[1]UMass Memorial Health Care, Central Mass Ob/Gyn,
Shrewsbury, MA, USA
[2]Maine Medical Center, Scarborough, ME, USA
[3]University of Massachusetts, UMass Memorial Health Care,
Worcester, MA, USA

Gynecologic cancers are a diverse group of tumors arising from several different tissue types and are controlled by a variety of signaling pathways. In this chapter we will discuss mechanisms by which therapy can be targeted in cancers of gynecologic origin, including ovarian, primary peritoneal, fallopian tube, uterine, cervical, vaginal, and vulvar cancers. Specifically, we will define the cellular characteristics of each tumor type, the biologic pathways in tumorigenesis (tumor development and growth), and molecular targets that have helped us to better understand disease pathogenesis, prognosis, and treatment.

CERVIX CANCER

Cervix cancer is the most common gynecologic cancer worldwide, and is both preventable and curable. Yet it remains one of the leading causes of death among women, especially in underdeveloped nations. Cervical cancer provides an excellent model for understanding the progression of dysplasia to cancer, strategies for prevention and screening, and novel approaches to therapy.

The most common type of cervical cancer is squamous cell carcinoma. Squamous cancers of the cervix also share similarities with squamous cell cancers of the head and neck and those of the vagina and vulva. This is because these cancers share a common origin. Of cervical carcinomas 98% are a result of a persistent infection with one of nine oncogenic, or "high-risk," viral types of human papillomavirus (HPV) [1, 2].

HPV is a group of more than 150 related viruses. It is named for the papillomas, or warts, that some HPV types can cause. Each HPV virus is given a number, referred to as its HPV type. There are over 40 HPV types that can infect the genital areas of males and females.

SQUAMOUS CARCINOMAS AND HUMAN PAPILLOMA VIRUS INFECTION

HPVs establish infection only in the keratinocytes of the skin or mucus membranes. The virus infects, or binds to the basement membrane, when it is exposed through microabrasions or epithelial trauma. HPV lesions then arise through proliferation of these basal keratinocytes. The viral particles are subsequently released through desquamation of cells.

Thus, HPV DNA is frequently present in anal, vulvar, vaginal, and cervical carcinomas as well as their precursors. As such, it is a natural target for therapy against these cancers, but also plays a key role in prevention.

Most squamous cell carcinomas of the cervix are caused by a persistent infection with "high-risk" subtypes of HPV. HPV is also associated with cancer of the vulva, vagina, anus, and the oropharynx. Infection of tissue with high-risk HPV is usually spontaneously cleared by the host. However, cases of persistent viral infection will cause production of proteins (E6 and E7) that work together to cause unregulated growth of cells resulting in invasive cancer. Some of this unregulated growth can be held in check by protective proteins that exist within normal cells. These proteins include P53 and the retinoblastoma, or RB, protein. If these proteins are defective, inactive, or absent then the cells will be more likely to develop cancer. Underlying immune disorders can also facilitate the development of invasive cancer in HPV infected individuals [2]. Persistent HPV infection then can lead to elaboration of the E6 and E7 protein products of the human papilloma virus that work to inhibit tumor suppressor gene activity and thereby immortalize the cells that are infected with oncogenic human papilloma virus. The HPV E7 protein prevents the activity of the retinoblastoma or Rb tumor suppressor protein, while HPV E6 is a different protein that inactivates P53, an important tumor suppressor gene. When it is functional, P53 expression allows cells to repair DNA damage and prevent carcinogenesis. By affecting P53 expression, as well as through direct effects, the E6 and E7 proteins lead to unregulated cell growth and progression first to cancer precursors and then finally invasive carcinoma.

IMMUNOSUPPRESSION

The risk of persistent infection with a high-risk HPV is increased in men and women that have immunosuppression. Decreased immunity can occur as a result of medications (such as chronic steroid use), primary immune deficiency diseases, or acquired immune deficiency illnesses such as HIV infection. The severity and duration of infection can be decreased by normalization of the immune function. In general, if women are less likely to clear their HPV infection spontaneously because their immune system is compromised, or weakened, they are at an increased risk for developing a cervical or vaginal cancer, or in some cases, vulvar cancer [3].

PREVENTION, SCREENING, AND TREATMENT

Because of its role in tumorigenesis, HPV and its proteins are targets for prevention, screening, and therapy. Our understanding of the role of HPV in the development of gynecologic cancers has resulted in the development of a vaccine against these high-risk oncogenic human papilloma viruses. Specifically, these vaccines are formulated against the viral capsid protein, which is the envelope that surrounds the HPV virus.

There are several vaccines in use that have been found to be effective in preventing infection and reducing the incidence of cervical cancer precursors. These include a bivalent vaccine that targets HPV subtypes 16 and 18, a quadrivalent vaccine that protects previously unexposed women against the four major high risk HPV types (6, 11, 16, 18), and a newly FDA-approved nanovalent vaccine designed to prevent infection by HPV subtypes (6, 11, 16, 18, 31, 33, 45, 52, and 58) [4]. This is found to be at least as effective as the quadrivalent vaccine. These vaccines are formulated against the viral capsid protein which is the envelope that surrounds the HPV virus [5].

Multiple prospective randomized controlled trials have studied the efficacy of a quadrivalent vaccine that protects patients against HPV 6, 11, 16, and 18 and the efficacy with prevention of external lesions as well as precancerous changes and cancer of the cervix [6]. In these trials, the quadrivalent vaccine has been found to be 99% effective in preventing precancerous or cancerous lesions in women who had not been exposed to HPV at their initial evaluation on enrollment to the study. The efficacy of the vaccine is found to change somewhat with age and for this reason, at the current time, administration of this vaccine is recommended in women under the age of 26 who have not been exposed to high-risk HPV subtypes [7]. Although it is too early to see a great effect in the numbers of invasive cancer, there appears to be a substantial decrease in the number of precancers of the cervix. The vaccines have also been studied in the treatment of patients who have already been infected with HPV. Although these vaccines have been shown to be most effective in women and men who have not yet been exposed to one of the onco-genic viral subtypes of HPV, they have also been shown to be effective in women who have undergone treatment for precancerous HPV related conditions of the cervix [6, 8, 9]. Further studies with the use of these vaccines in already infected individuals are ongoing.

The burden of HPV-related disease is a significant problem worldwide and current research is focusing on strategies for large-scale administration of this vaccine to attempt to decrease morbidity and mortality from these diseases.

In addition to vaccine-related immunity, the cervix itself is anatomically accessible for screening and for treatment. Treatment can consist of the application of topical medications or excisional techniques for treatment of precancerous changes of the cervix. Similarly, the cervix and uterus are sensitive targets for radiation therapy. Radiation therapy, for instance, is the primary treatment modality for locally advanced cervical carcinoma. Extrapolating from the head and neck data, the use of concurrent chemotherapy using cisplatin has improved the efficacy of radiation treatment. Radiation can be given externally in a focused beam through the abdominal wall and also can be delivered directly to the cervical tissues internally in very high doses using a form that is inserted into the cervix and vagina (brachytherapy).

An exciting molecular target in HPV-induced cancers is vascular endothelial growth factor (VEGF). VEGF promotes the growth of small blood vessels that supply cancer cells. This phenomenon has been found to be important in the progression of cervical carcinoma. When cancer is established or recurrent, it may be treated with chemotherapy, a biologic agent known as bevacizumab. This is a monoclonal antibody against VEGF that has been shown to improve disease-free survival in cervical carcinoma that is recurrent, persistent, or metastatic when it is added to combination therapy [10].

CANCER OF THE VULVA

Cancers of the vulva share similarities with cancers of the cervix in that over 90% of vulvar cancers are of squamous histology. Similarly, HPV infection is a major risk factor, as is chronic immunosuppression. These cancers arise from one of two pathogenic mechanisms that correlate with age. Younger women tend to have vulvar cancers that are associated with HPV infection, whereas older women can have keratinizing squamous carcinomas associated with inflammatory vulvar skin diseases. Therefore many squamous carcinomas in older women are not associated with HPV infection [11].

The vulva itself is an external site that is readily available to the use of topical medications, excision of affected skin, and irradiation. Similar to that risk found in women with cervix cancer, women who have immunosuppression related to human immunodeficiency virus infection, transplant recipients, or women on chronic immunosuppression therapy have a higher risk of persistent HPV infection, cervical carcinoma, and vulvar cancers [3]. HPV infection has been shown to be associated with over 80% of precancerous lesions of the vulva [11].

In general, cancers of the vulva are treated by excision of the primary vulvar tumor with evaluation for lymph node metastasis and surgery can be combined with radiotherapy. In addition to surgical procedures, Imiquimod is a topical medication commonly used for venereal warts. It can also be used to treat squamous and melanocytic precancerous conditions of the vulva. Imiquimod is applied to the vulva and acts as an immune cell modulator that stimulates the T cell immune response of vulvar skin to malignant or premalignant cells [12].

Treatment of vaginal carcinomas is very similar to that of treatment of the cervix and includes excision, irradiation, and prevention with HPV vaccination and treatment of premalignant lesions.

ENDOMETRIAL CANCER

The uterus is lined by the endometrium and generally is a readily accessible organ whose biological characteristics provide unique targets for therapy of this gynecologic cancer.

American Cancer Society statistics predict that in 2015, there will be an estimated 54 870 new diagnoses of uterine cancer and 10 170 deaths from this disease [13].

Uterine cancer is generally thought to have an excellent prognosis with 70% of cases being Stage I at diagnosis and having a five-year survival of 90% [14]. However the incidence and mortality from this disease continue to rise. This increase in mortality is likely because of the aging population and an accompanying increase in the number of women with poor prognosis histologic subtypes.

Uterine cancer is divided into two main types. Type I is mostly mediated by estrogen and progesterone stimulation. These tumors typically present at early stage and are associated with generally good overall survival. Common genetic changes seen in Type I cancers include microsatellite instability (MSI) and inactivation of PTEN (phosphatase with tensin homology), which is present in 83% of endometrial carcinomas [15]. Type II endometrial cancers are more common in

TABLE 1. Type II versus Type II endometrial cancer.

	Type I	Type II
Clinical Features		
Risk factors	Unopposed estrogen	Age
Race	White > black	No difference
Cellular differentiation	Well differentiated	Poorly differentiated
Histology	Endometrioid	Non-endometrioid
Stag of diagnosis	Early (I/II)	
Prognosis	Favorable	
Molecular features		
K-ras overexpression	Yes	Yes
Her2/neu overexpression	No	Yes
P53 overexpression	No	Yes
PTEN mutations	Yes	No
MSI (microsatellite instability)	Yes	No

older individuals and include the papillary serous carcinomas, clear cell carcinomas, and uterine carcinosarcomas. These tumors are seen in a background of atrophic endometrium and are estrogen independent. Serous carcinomas are generally modulated by lack of P53 expression (Table 1).

Type I endometrial adenocarcinomas EAC are estrogen dependent and can arise when there is excessive estrogen stimulation of the lining of the uterus without adequate modulation of growth by progesterone. For this reason, endometrial cancer is more common in obese women where adrenal hormones are converted to estrogen in peripheral adipose tissue. Endometrial cancer is also more common in women who are prone to anovulation because of this lack of progesterone steroid production. Excessive estrogen stimulation leads to thickening and hyperplastic changes of the endometrium or endometrial hyperplasia. Although most endometrial hyperplasias are benign, some are found to have atypia and these are the precursor lesions to endometrial cancer. The effect of continuous estrogen stimulation or unopposed estrogen replacement therapy given without adequate progesterone supplementation has also been correlated with the development of uterine cancer (Figure 1). Administration of progesterone has been shown to directly counteract the neoplastic effects of estrogen on endometrial cells and also to down regulate expression of estrogen receptors. Progesterone administration leads to cell death in endometrial glands and stroma through the transforming growth factor beta (TGF-beta). For this reason, significant attention has been given to prevention of uterine cancer in susceptible individuals and to treatment of uterine cancer with administration of progesterone. Traditionally, this hormone was administered orally; however progesterone containing intrauterine devices have been shown to be effective in direct delivery of drug to the affected tissue [16, 17].

Type II endometrial cancers are estrogen independent tumors that are associated with an advanced stage at the time of diagnosis and poor overall survival. Although they account for only 15% of uterine cancers they are responsible for significant

(a)

	Type I	Type II
CLINICAL FEATURES		
Risk factors	Unopposed estrogen	Age
Race	White > black	No difference
Cellular Differentiation	Well differentiated	Poorly differentiated
Histology	Endometrioid	Non-endometrioid
Stage at Diagnosis	Early (I/II)	Late (III/IV)
Prognosis	Favorable	Poor
MOLECULAR FEATURES		
K-ras overexpression	Yes	Yes
Her2/neu overexpression	No	Yes
P53 overexpression	No	Yes
PTEN mutations	Yes	No
MSI (microsatellite instability)	Yes	No

(b)

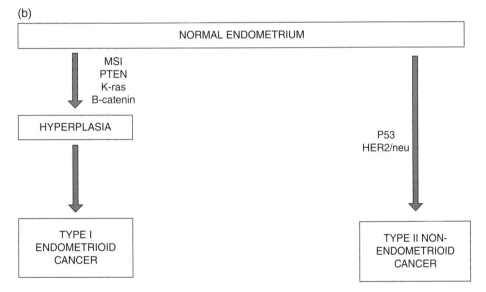

Figure 1. Molecular pathogenesis of Type I and Type II endometrial cancers.

mortality from this disease [18]. Histologies include papillary serous, clear cell, and poorly differentiated endometrial cancer. Because they cannot be prevented or controlled by hormonal mediation, significant attention has been directed toward molecular targets for therapy in this group of tumors (Figure 1).

Papillary serous carcinoma, for instance, is a Type II endometrial cancer that accounts for only 5–10% of uterine cancers and is an aggressive subtype of uterine cancer with 40–70% of cases having disease outside the uterus at the time of presentation. The five-year overall survival of 27–55% compared to 80% for Type I cancers. Papillary serous carcinoma is more commonly seen in older women and those with a lower body mass index (BMI). This aggressive cancer accounts for 39% of deaths from uterine cancers. Its molecular profile is well characterized with alterations in P53, PIK3/AKT/mTOR, and Her2neu pathways and therefore is an excellent candidate for targeted therapies [18].

PIK3CA/AKT/mTOR PATHWAY

The PIK3CA/AKT/mTOR signaling pathway is critical to endometrial cell proliferation, tumor invasion and metastasis, and cell survival [19].

This pathway can be activated by gene amplification of HER2 (Human epidermal growth factor 2) or by oncogenic mutations of the PIK3CA/AKT genes. PIK3CA gene mutations are found in 40–50% of all types of endometrial cancer [20]. Phosphorylation of HER2 upregulates the PIK3CA/AKT/mTOR pathway. In endometrial cancers, overexpression of HER2 has been seen in 4–69% of cases and is associated with poor prognosis [21].

The c-erbB2 gene, a member of the tyrosine kinase receptor family, encodes HER2, which then affects PIK3CA downstream. Both HER2 gene amplification and PIK3CA gene mutations are found in biologically aggressive endometrial cancers with 29% of serous cancers demonstrating HER2 amplification and as many as 22% carrying PIK3CA mutations with or without HER2 amplification [20].

Another mode of activation in non-HER2-overexpressing tumors is the loss of PTEN. P53 regulates transcription of PTEN, which regulates cell growth (in opposition to PIK3CA). When p53 and PTEN are mutated or inactivated subsequent activation of the PIK3 pathway leads to increased phosphorylation of intracellular tyrosine kinase residues and resultant tumorigenesis [22]. mTor inhibitors are a class of medications currently most frequently used in the treatment of kidney disease that have shown some effect in the treatment of recurrent endometrial cancer. Further investigation into these treatments is planned [23].

METFORMIN IN EAC

Metformin is best known as a first line medication used to treat insulin resistant diabetes; it decreases circulating insulin and glucose levels and has been found to have antitumor properties in patients with gynecologic malignancies including ovarian and endometrial adenocarcinomas (EAC). Evidence suggests that Metformin lowers all cancer risk, and reduces cancer incidence and deaths among diabetic patients [24]. Metformin has been shown to improve outcomes in diabetic patients with endometrial cancer with improved recurrence free and overall survival demonstrated. Metformin is thought to inhibit tumor growth and proliferation through its effect on metabolism in neoplastic cells. It has both indirect effects through improvement in insulin resistance and decreasing insulin and glucose levels and direct effects by activating AMPK and inhibition of the mTOR pathway [25].

As previously described, in EAC, genetic mutations often include loss of the PTEN tumor suppressor gene, and dysregulation of the PI3K/mTOR pathway. Activation of mutations in PI3K leads to loss of inhibition of mTOR, which promotes unregulated protein synthesis and cell division. Metformin acts as a secondary mTOR inhibitor via activating an upstream regulator called AMPK, a protein kinase that regulates cell metabolism and energy use. When activated, AMPK signals cells to conserve energy and decrease production and thus inhibits growth and proliferation signaling cascades including mTOR. This pathway affects tumor survival in several ways including inhibiting tumor growth, decreasing overall energy sources, inhibiting invasion of tumor cells and increasing tumor cell sensitivity to chemotherapeutic drugs such as cisplatin and paclitaxel [26, 27]. Metformin has been found in retrospective studies to be associated with improved overall survival and recurrence-free survival in groups of women taking Metformin as treatment for diabetes over cohorts that are not. In other models, Metformin appears to work synergistically when in combination with hormonal treatment for EAC with progesterone analogs. It is unclear whether Metformin will be universally effective in cancer treatment and prevention or whether this benefit will be limited to the insulin resistant population. The relative importance of Metformin's various effects alone or in combination including direct effect on tumor cells versus its indirect effect of improvement of the metabolic milieu on antitumor activity is unclear, but is an active area of current clinical trial investigation [28].

OVARIAN CANCER

Ovarian cancer is well known as the most lethal gynecologic cancer and is the fifth leading cause of cancer related deaths in women in the United States. The American Cancer Society national cancer statistics predict that in 2015, approximately 21 290 women will be diagnosed with ovarian cancer and 14 800 will die of this disease [13, 29]. Three-quarters of these diagnoses will be made at advanced stages that carry a five-year survival of 30% compared to the 90% five-year survival that is possible when the cancer is confined to the ovary at time of diagnosis [30]. The survival of patients with ovarian cancer has not changed much in the last 30 years, prompting interest in tumorigenesis, molecular signatures, and exploration of possible targets for therapy of this disease.

Histology and Genesis of Ovary Cancer

Ovarian cancer is divided by its histologic (cellular) characteristics into epithelial and non-epithelial carcinomas. Epithelial ovarian cancer (EOC) is the most common type of ovary cancer and is most commonly found in women in the fifth decade of life. The non-epithelial tumors include germ call and sex cord stromal tumors that will be discussed separately. Epithelial ovarian, fallopian tube, and peritoneal cancers are similar in their histologic appearance, patterns of spread, and treatment. Evidence from molecular studies supports a common origin of these tumors from the distal portion of the fallopian tube or endometriosis implants [31, 32]. Therefore it is important to note that not all ovarian cancers originate in the ovary itself. Within this group of epithelial cancers there are two separate groups of tumors that have different clinical behaviors, different histologies, and different molecular signatures that we will describe as low grade and high grade tumors.

LOW GRADE AND BORDERLINE EOC

Low grade EOCs include tumors with cellular origin of low grade serous, endometrioid, and mucinous background. These tumor types make up only approximately 25% of the malignant epithelial ovarian tumors and are typically found to be slow growing, less aggressive, less invasive, and more likely to be diagnosed at an earlier stage with a better prognosis [33]. They arise in mullerian epithelium such as the ovary, or in endometriosis, and exhibit overall genetic stability and progression to malignancy by mutations in multiple genes including *KRAS, BRAF, PTEN*, ERBB2, and *beta-catenin* [34]. Although these low grade tumors are much more likely to be diagnosed at earlier stages, and thus have higher overall survival rates, they are less responsive to chemotherapeutic treatment because of their genetic makeup and slower rate of cellular division [35, 36].

Borderline Ovarian Tumors

Borderline ovarian tumors are defined by having malignant potential via abnormal cellular growth without evidence of invasion. Borderline tumors are most often confined to the ovary at time of diagnosis, indolent in nature, and have a good prognosis even in cases where disease has spread outside of the ovary. They are treated by surgery alone. The most commonly found tumors are serous, making up approximately 65% of all borderline tumors. Evidence of serous borderline tumors link to low grade serous Type I cancer includes similar KRAS and BRAF mutations with otherwise genetically stable characteristics [37]. Serous borderline tumors are bilateral in approximately half of cases and have potential for micropapillary invasion, which can lead to invasive disease and peritoneal implants [38].

Mucinous borderline tumors however, are unilateral in over 90% of cases. They are not found to have invasive peritoneal implants and are almost always diagnosed at stage I. They exhibit the same KRAS mutation seen in both benign mucinous cystadenoma and low grade mucinous ovarian carcinoma [39].

Endometrioid Ovarian Cancer

Endometrioid ovarian cancer is thought to arise from the uterine endometrium and endometriosis transported to the ovaries by retrograde flow through the fallopian tubes. Endometrioid ovarian cancer arises from endometriosis after loss of tumor suppressor genes such as PTEN in endometrial type cells [40, 41].

Mucinous Ovarian Cancer

Mucinous ovarian cancers are typically slow growing, unilateral, and large at the time of diagnosis. The histology of mucinous tumors includes very well-differentiated cells with low cellular division activity and normal appearing nuclei [36]. They are only known to have one oncogenic genetic mutation of KRAS. This mutation turns on growth factors and increases cellular division and thereby tumor growth [42].

Low Grade Serous Ovarian Cancer

Low grade serous ovarian cancers typically behave in a benign manner. Their cell growth is slow and they have low grade cellular and nuclear changes, with uniformity to the cells [36, 43]. Low grade serous tumors (LGSC) express KRAS, BRAF, and ERBB2 oncogene mutations as seen in their borderline counterparts.

In summary, low grade EOCs are all characterized by relative genetic stability and are associated with mutations of tumor suppressor PTEN genes and oncogenes KRAS, BRAF, ERBB2, which may provide targets for detection and treatment in future practices. Although their lower rates of cell division and growth makes them less susceptible to treatment with chemotherapy these tumors have a generally better prognosis and overall survival as they are typically diagnosed at an earlier stage than high grade EOC.

High Grade Ovarian Cancers

This group of EOC includes high grade serous ovarian, fallopian, peritoneal cancer, carcinosarcoma, or malignant mixed mesodermal tumors and undifferentiated carcinomas. These tumors are universally known to be very aggressive, fast growing, genetically unstable, and poorly differentiated. They are associated with poor prognosis and a lower overall survival than the low grade tumors and are typically diagnosed at later stages with significant metastasis at the time of diagnosis. Precursor lesions of these tumors are frequently found in the fallopian tube. Because they are rapidly dividing cancer cells, these cancers are more susceptible to chemotherapeutic agents.

All of the above mentioned high grade histologies are known to be highly genetically unstable and have TP53 genetic mutation or loss. As described earlier in this chapter, P53 controls cell death and division and therefore when this gene is lost or mutated cells can divide in an unregulated fashion and do not undergo apoptosis (cell death) as they should when their genes become damaged [36, 44].

Molecular evidence in the genesis of ovarian cancer has changed the understanding of the origin and progression of high grade ovarian, fallopian tube, and peritoneal cancers. Evidence suggests that ovarian cancers may not only develop in or on the ovarian epithelium but also from histologically similar fallopian and peritoneal tissues [13]. Lesions at the distal fallopian tube that are related to, if not a precursor of, high grade serous ovarian cancer are called serous tubal intraepithelial neoplasias (STIN) and serous tubal intraepithelial carcinoma (STIC) and carry identical P53 mutations to high grade Type II EOCs [45, 46]. Additionally, STIC is incidentally found in surgical specimens of women with high grade serous carcinoma of the ovary without known BRCA mutation, pointing toward a likely correlation between STIC and advanced disease [47, 48].

Clear Cell Carcinoma of the Ovary

Clear Cell ovarian cancer is also thought to originate from endometriosis but unlike endometrioid ovarian cancers clear cell tumors are likely to be high grade at the time of diagnosis and have a significantly worse prognosis when diagnosed at an advanced stage. Clear cell and endometrioid tumors both exhibit MSI while serous and mucinous types do not [13, 36, 40].

Non-epithelial

There are a broad variety of non-epithelial ovarian cancer types including germ cell tumors, such as dysgerminoma, sex cord stromal tumors such as granulosa or Sertoli-Leydig cell tumors, sarcomas, and rare lipoid tumors that make up approximately 10% of all ovarian cancers. Most germ cell and sex cord stromal tumors are benign with only a small percentage having malignant transformation. When these cancers are diagnosed in young women it is important to keep in mind that fertility sparing surgical treatment can be curative. Patients diagnosed at advanced stages or with recurrence also have good prognosis as these tumors are highly chemosensitive and respond well to therapy. Targeted molecular therapy with antiangiogenic and tyrosine kinase inhibitors is currently under study in these tumors [49].

Genetics

The lifetime risk of diagnosis of ovary cancer in American women is 1 in 72 and is significantly elevated in women with a family history of ovarian cancer. Investigation into inheritance patterns and incidence of cancer diagnosis has led to the discovery of inherited genetic mutations that play a role in Hereditary Breast and Ovarian Cancer syndrome as discussed earlier in this book. Specific genes involved in the biologic steps of tumor formation in ovarian cancer include the breast cancer susceptibility (BRCA I and II) genes and mismatch repair genes MLH1, MSH2, MSH6 found in patients affected by Lynch Syndrome. These are inherited in an autosomal dominant manner and therefore can be inherited from a male or female parent.

BRCA is a tumor suppressor gene that plays a protective role in cell regulation by monitoring and repairing double stranded DNA breaks. Mutation of BRCA I and II significantly increases a woman's overall lifetime risk of breast and ovarian cancer specifically ER/PR and HER2 (triple) negative breast cancer and EOC. In general women with EOC associated with these mutations have a better prognosis than women who are not mutation carriers. The importance of taking a family history in women who have cancer cannot be underestimated since the identification of BRCA mutation in affected relatives also allows for early genetic testing and the possibility for early preventive intervention. Currently, the study of BRCA function and molecular pathways is an active area of investigation with development of screening protocols, risk reducing surgery and successful new models for therapy targeted toward steps in DNA repair [38, 50].

Lynch Syndrome or hereditary non-polyposis colorectal cancer syndrome (HNPCC) involves another group of genetic mutations in several possible DNA mismatch repair genes including, but not limited to MLH1, MSH2, and MSH6, which significantly increase the risk of endometrial (40–60%), ovarian (12%), and gastrointestinal cancer (50%) in carriers [30, 37, 51]. MSH6 is associated with the highest risk of endometrial cancer, whereas MLH1 and MSH2 are more commonly mutated in women with Lynch Syndrome who are at risk for ovarian cancer. Women with Lynch syndrome are most likely to have concurrent endometrial carcinoma at time of diagnosis of ovarian cancer, and again – a family history of any of the above cancers and screening for uterine cancer at time of evaluation are imperative [30, 37].

CURRENT THERAPY

Surgical Cytoreduction

The mainstay of treatment in all ovarian cancers is surgical resection and minimization of tumor burden. In EOC the goal of surgery is complete cytoreduction with no residual tumor. A multitude of studies have proven that this results in the best survival and progression free disease intervals. The completeness of surgical resection is further classified by descriptive definitions including complete cytoreduction with no gross residual at the end of the surgical procedure, optimal cytoreduction where the maximum diameter of gross residual disease measures less than or equal to 1 cm, and suboptimal where tumor implants greater than 1 cm in diameter remain at the end of surgical resection.

Complete cytoreduction is associated with the best overall survival and patients who have disease that is cytoreduced with surgery to less than 1 cm residual have improved survival compared to those patients where this degree of disease removal cannot be accomplished (suboptimal cytoreductive surgery) [52].

Patients that present with a disease burden that is not likely to be completely removed surgically at the time of initial diagnosis are candidates for neoadjuvant chemotherapy (chemotherapy given before surgery) in an attempt to shrink the tumor burden prior to surgery.

In borderline tumors and malignant germ cell tumors without evidence of metastatic disease where the prognosis is excellent, fertility sparing surgery with close follow-up is acceptable [49].

Chemotherapy

High grade ovarian cancers are frequently diagnosed at advanced stage with significant dissemination of intraabdominal disease and require a systemic approach to therapy with chemotherapy. This can be given in adjuvant form (after surgery) or as a neoadjuvant treatment (before surgery). First line chemotherapy for high grade epithelial, fallopian, and peritoneal carcinoma is a combination of platinum-based treatment (typically carboplatin, cisplatin) with a taxane (paclitaxel). Treatment can be administered intravenously or by direct instillation into the peritoneal cavity (intraperitoneally). High grade EOC is typically responsive to the drugs because of its rapid rate of cellular division and genetic instability. Platinum drugs are alkylating agents that cross link DNA strands interfering with DNA replication and causing cell death; taxanes interfere with microtubule function, preventing cell division.

Chemotherapeutic regimens for non-epithelial ovarian carcinomas can vary and may include bleomycin, etoposide, and cisplatin [49]. Chemotherapy is not used in borderline tumors [53].

Targeted Therapy for Ovarian Cancer

PARP Inhibition

Targeted therapy for high grade EOC has focused on the frequent P53 mutations as well as BRCA mutations. BRCA mutations can be inherited but a significant number of patients who do not harbor a BRCA mutation in their genome will have a mutation in the tumor itself. In BRCA mutated tumors double stranded repair is impaired and therefore these tumors rely on single strand repair to survive. PARP inhibitors block this single strand DNA repair. Exposure of a BRCA mutated cell to such an inhibitor leads to lack of DNA repaired and consequent cell death, i.e., synthetic lethality [54]. Therefore PARP inhibitors are an important class of agents that are effective toward tumor cells with absent BRCA capability, whether this is inherited or not. PARP inhibitors are an FDA approved treatment for recurrent ovarian cancer in BRCA mutation carriers. This class of agents is currently under investigation in the treatment of newly diagnosed and recurrent ovarian cancer in women with and without a BRCA mutation [55].

Angiogenesis

Angiogenesis is the formation of new blood vessels mediated by several molecular pathways and is instrumental in tumor formation, maintenance, and growth as blood vessels are aberrantly recruited to support abnormal cell division. VEGF inhibitors have been well studied in targeted therapy for cancer treatment. Specifically the monoclonal antibody Bevacizumab blocks VEGF receptors and has been shown to slow tumor progression, increasing the progression free interval, but not overall survival in patients with suboptimally cytoreduced ovarian cancer when used in combination with a platinum/taxane chemotherapy regimen [56, 57].

This combination of Bevacizumab added to standard chemotherapy may have an overall survival benefit in poor prognosis ovarian cancer patients including stage IV, inoperable stage III, and suboptimally cytoreduced patients [58].

Very recently research of targeted molecular therapy and antiangiogenic medications has resulted in a breakthrough of not only progression free survival but also overall survival for the first time in women treated with VEGFR Tyrosine Kinase inhibitor cediranib [55]. Several other pathways in angiogenesis targeting endothelial growth factor and tyrosine kinase receptors are currently being studied but without sufficient evidence to guide treatment recommendations [55, 59].

MEK Inhibitors

Low grade serous carcinomas (LGSC) and mucinous tumors with their poor responses to standard chemotherapies are an attractive group for development of targeted therapies. These tumors have multiple known genetic mutations specific to their histologic group, specifically BRAF and KRAS mutations that affect the mitogen activated protein kinase (MAPK) pathway. This pathway is usually activated by growth factors and promotes cell growth and survival. When there is a mutation in the KRAS or BRAF genes this pathway is activated without any inhibition by usual mediators. MEK1/2 is an enzyme product of this activation and drives cell proliferation and inhibition of cell death. MEK inhibitors are a class of drugs currently under study for treatment of LGSC. The MEK inhibitor selumetinib has shown some effect and continues to be investigated [55, 60].

PI3K/mTOR, PTEN

In clear cell and endometrioid type EOCs PTEN mutations have been found and the previously discussed PIK3/mTOR pathway is being studied as a potential therapeutic target. mTOR inhibitors are currently being combined with taxanes in some clinical trials and have shown effect in patients with ovarian cancer [61].

Metabolism

Metformin has been shown to inhibit oncogenic pathways and decrease tumor growth by several metabolic pathways including inhibiting ATP production through a regulatory protein AMPK activation, inhibited protein synthesis by blocking of the mTOR pathway, and inhibition of adipocyte induced proliferation and migration. Metformin has been correlated with decreased cancer risk and improved survival and response to chemotherapy for multiple solid tumors in retrospective studies. Further research and application to human ovarian models is needed [62, 63].

IMMUNE THERAPY

In efforts to find novel treatments for EOC, focus has shifted to the immune system and its potential role as cancer therapy. Specifically, in ovarian cancer, the level of activity of particular immune cells have been correlated with both increased progression free survival and can be triggered by chemotherapy to increase sensitivity, recognition of, and attack on tumor cells, thus rendering chemotherapeutic agents more effective. Increasing data are showing that the immune system plays an important role in EOC response to surgery and chemotherapy, showing that when cytotoxic T cells infiltrate ovarian tumors there is a delay in disease progression and improved overall survival [64].

CONCLUSION

Cancers of gynecologic origin make up a substantial part of the total incidence of cancers diagnosed yearly, and as highlighted in this chapter, are diverse in not only origin, but in tumor behavior and biologic characteristics. Effective therapy relies on a broad and thorough understanding of the differences and similarities of each tumor type. Although the current mainstay of treatment for most gynecologic cancers includes surgical debulking, chemotherapy, and radiation, ongoing studies focused on targeted therapy for respective tumor types will continue to guide our management.

REFERENCES

1. Ndiaye, C., Mena, M., Alemany, L. et al. (2014). HPV DNA, E6/E7 mRNA, and p16INK4a detection in head and neck cancers: a systematic review and meta-analysis. *The Lancet Oncology* 15: 1319–1331.

2. Schiffman, M., Castle, P.E., Jeronimo, J. et al. (2007). Human papillomavirus and cervical cancer. *Lancet* 370: 890–907.

3. Engels, E.A., Pfeiffer, R.M., Fraumeni, J.F. et al. (2011). Spectrum of cancer risk among U.S. solid organ transplant recipients. *JAMA* 306 (17): 1891–1901.

4. Ault, K.A. (2007). Effect of prophylactic human papillomavirus L1 virus-like-particle vaccine on risk of cervical intraepithelial neoplasia grade 2, grade 3, and adenocarcinoma in situ: a combined analysis of four randomised clinical trials. *Lancet* 369 (9576): 1861–1868.

5. Petrosky, E., Bocchini, J.A., Hariri, S. et al. (2015). Use of 9-valent human papillomavirus (HPV) vaccine: updated HPV vaccination recommendations of the advisory committee on immunization practices. *MMWR. Morbidity and Mortality Weekly Report* 64 (11): 300–304.

6. FUTURE II Study Group (2007). Quadrivalent vaccine against human papillomavirus to prevent high-grade cervical lesions. *The New England Journal of Medicine* 356: 1915–1927.

7. ACOG Committee Opinion Number 641 September 2015 Timothy J Perren

8. Joura, A.E., Guiliano, A.R., Iversen, O.E. et al. (2015). A 9-valent HPV vaccine against infection and intraepithelial neoplasia in women. *The New England Journal of Medicine* 372: 711–723.

9. Joura, E.A., Leodolter, S., Hernandez-Avila, M. et al. (2007). Efficacy of a quadrivalent prophylactic human papillomavirus (types 6, 11, 16, and 18) L1 virus-like-particle vaccine against high-grade vulval and vaginal lesions: a combined analysis of three randomised clinical trials. *Lancet* 369: 1693–1702.

10. Tewari, K.S., Sill, M.W., Long, H.J. 3rd et al. (2014). Improved survival with Bevacizumab in advanced cervical cancer. *The New England Journal of Medicine* 370 (8): 734–743. https://doi.org/10.1056/NEJMoa1309748.

11. Zweizig, S., Korets, S., and Cain, J. (October, 2014). Key concepts in the management of vulvar cancer. *Clinical Obstetrics and Gynaecology* 28 (7): 959–966.

12. van Seters, M., van Beurden, M., ten Kate, F.J. et al. (2008). Treatment of vulvar intraepithelial neoplasia with topical imiquimod. *The New England Journal of Medicine* 358 (14): 1465–1473.

13. Siegel, R.L., Miller, K.D., and Jemal, A. (2015). Cancer statistics, 2015. *Cancer Journal for Clinicians* 65 (1): 5–29. https://doi.org/10.3322/caac.21254.

14. Creutzberg, C.L., Van Putten, W.L., Koper, P.C. et al. (2000). Surgery and postoperative radiotherapy versus surgery alone for patients with stage-1 endometrial carcinoma: multicentre randomized trial. PORTEC study group. *Lancet* 355: 1404–1411.

15. Hecht, J.L. and Mutter, G.L. (2006). Molecular and pathologic aspects of endometrial carcinogenesis. *Journal of Clinical Oncology* 24: 4783–4791.

16. Gallos, I.D., Shehmar, M., Thangaratinam, S. et al. (2010). Oral progestogens vs levonorgestrel-releasing intrauterine system for endometrial hyperplasia: a systematic review and metaanalysis. *American Journal of Obstetrics and Gynecology* 203 (6): 547.e1.

17. Kalogera, E., Dowdy, S.C., and Bakkum-Gamez, J.N. (2014). Preserving fertility in young patients with endometrial cancer: current perspectives. *International Journal of Women's Health* 6: 691–701.

18. Del Carmen, M., Birrer, M., and Schorge, J.O. (2012). Uterine papillary serous cancer: a review of the literature. *Gynecologic Oncology* 127: 651–661.

19. Engelman, J.A. (2009). Targeting PI3K signalling in cancer: opportunities, challenges and limitations. *Nature Reviews Cancer* 9: 550–562. https://doi.org/10.1038/nrc2664.

20. Kandoth, C., McLellan, M.D., Vandin, F. et al. (2013). Mutational landscape and significance across 12 major cancer types. *Nature* 502: 333–339. https://doi.org/10.1038/nature12634.

21. Slomovitz, B.M., Broaddus, R.R., Burke, T.W. et al. (2004). Her-2/neu overexpression and amplification in uterine papillary serous carcinoma. *Journal of Clinical Oncology* 22 (15): 3126–3132.

22. Oda, K., Stokoe, D., Takentani, Y. et al. (2005). High frequency of coexistent mutations of PIK3CA and PTEN genes in endometrial carcinoma. *Cancer Research* 65: 10669–10673.

23. Husseinzadeh, N. and Husseinzadeh, H.D. (2014). mTOR inhibitors and their clinical application in cervical, endometrial and ovarian cancers: a critical review. *Gynecologic Oncology* 133 (2): 375–381.

24. Evans, J.M., Donnelly, L.A., Emslie-Smith, A.M. et al. (2005). Metformin and reduced risk of cancer in diabetic patients. *BMJ* 330 (7503): 1304–1305.

25. Ko, E., Walter, P., Jackson, A. et al. (2014). Metformin is associated with improved survival in endometrial cancer. *Gynecologic Oncology* 132 (2014): 438–442.

26. Hadad, S., Iwamoto, T., Jordan, L. et al. (2011). Evidence for biological effects of metformin in operable breast cancer: a pre-operative, window-of-opportunity, randomized trial. *Breast Cancer Research and Treatment* 128 (3): 783–794.

27. Hanna, R.K., Zhou, C., Malloy, K.M. et al. (2012). Metformin potentiates the effects of paclitaxel in endometrial cancer cells through inhibition of cell proliferation and modulation of the mTOR pathway. *Gynecologic Oncology* 125 (2): 458–469. https://doi.org/10.1016/j.ygyno.2012.01.009.

28. Ko, E., Franasiak, J., Sink, K. et al. (2011). Obesity, diabetes, and race in Type 1 and Type 2 endometrial cancers. *Journal of Clinical Oncology* 29 (15-suppl): 5111.

29. Chi, D., Eisenhauer, E., Lang, J. et al. (2006). What is the optimal goal of primary cytoreductive surgery for bulky stage IIIC epithelial ovarian carcinoma (EOC)? *Gynecologic Oncology* 103 (2): 559–564.

30. Jones, M.B. (2006). Borderline ovarian tumors: current concepts for prognostic factors and clinical management. *Clinical Obstetrics and Gynecology* 49 (3): 517–525.

31. Kurman, R.J. and Shih, I. (2011). Molecular pathogenesis and extraovarian origin of epithelial ovarian cancer—shifting the paradigm. *Human Pathology* 42 (7): 918–931.

32. Lim, R., Lappas, M., Ahmed, N. et al. (2011). 2D-PAGE of ovarian cancer: analysis of soluble and insoluble fractions using medium-range immobilized pH gradients. *Biochemical and Biophysical Research Communications* 406 (3): 408–413.

33. Gershenson, D.M., Sun, C.C., Lu, K.H. et al. (2006). Clinical behavior of stage II-IV low-grade serous carcinoma of the ovary. *Obstetrics and Gynecology* 108 (2): 361–368.

34. Mayr, D., Hirschmann, A., Löhrs, U. et al. (2006). KRAS and BRAF mutations in ovarian tumors: a comprehensive study of invasive carcinomas, borderline tumors and extraovarian implants. *Gynecologic Oncology* 103 (3): 883–887.

35. Cass, I., Baldwin, R.L., Varkey, T. et al. (2003). Improved survival in women with BRCA-associated ovarian carcinoma. *Cancer* 97 (9): 2187–2195.

36. Kurman, R.J. and Shih, I. (2010). The origin and pathogenesis of epithelial ovarian cancer- a proposed unifying theory. *The American Journal of Surgical Pathology* 34 (3): 433–443.

37. Watson, P., Bützow, R., Lynch, H.T. et al. (2001). The clinical features of ovarian cancer in hereditary nonpolyposis colorectal cancer. *Gynecologic Oncology* 82 (2): 223–228.

38. Vang, R., Shih, I., and Kurman, R.J. (2009). Ovarian low-grade and high-grade serous carcinoma: pathogenesis, clinicopathologic and molecular biologic features, and diagnostic problems. *Advances in Anatomic Pathology* 16 (5): 267–282.

39. Watson, P., Vasen, H.F., Mecklin, J. et al. (2008). The risk of extra-colonic, extra-endometrial cancer in the lynch syndrome. *International Journal of Cancer* 123 (2): 444–449.

40. Colombo, N., Peiretti, M., Garbi, A. et al. (2012). Non-epithelial ovarian cancer: ESMO clinical practice guidelines for diagnosis, treatment and follow-up. *Annals of Oncology: Official Journal of the European Society for Medical Oncology* 23 (Suppl 7): vii20–vii26.

41. Prowse, A.H., Manek, S., Varma, R. et al. (2006). Molecular genetic evidence that endometriosis is a precursor of ovarian cancer. *International Journal of Cancer* 119 (3): 556–562.

42. Rossner, P. Jr., Gammon, M.D., Zhang, Y. et al. (2009). Mutations in p53, p53 protein overexpression and breast cancer survival. *Journal of Cellular and Molecular Medicine* 13 (9b): 3847–3857.

43. Shih, I. and Kurman, R.J. (2004). Ovarian tumorigenesis: a proposed model based on morphological and molecular genetic analysis. *The American Journal of Pathology* 164 (5): 1511–1518.

44. Wethington, S.L., Park, K.J., Soslow, R.A. et al. (2013). Clinical outcome of isolated serous tubal intraepithelial carcinomas (STIC). *International Journal of Gynecological Cancer* 23 (9): 1603–1611.

45. Gayther, S.A., Warren, W., Mazoyer, S. et al. (1995). Germline mutations of the BRCA1 gene in breast and ovarian cancer families provide evidence for a genotype-phenotype correlation. *Nature Genetics* 11: 428–433.

46. Kaye, S.B., Lubinski, J., Matulonis, U. et al. (2012). Phase II, open-label, randomized, multicenter study comparing the efficacy and safety of olaparib, a poly (ADP-ribose) polymerase inhibitor, and pegylated liposomal doxorubicin in patients with BRCA1 or BRCA2 mutations and recurrent ovarian cancer. *Journal of Clinical Oncology: Official Journal of the American Society of Clinical Oncology* 30 (4): 372–379.

47. Li, H., Lu, Z., Shen, K. et al. (2014). Advances in serous tubal intraepithelial carcinoma: correlation with high grade serous carcinoma and ovarian carcinogenesis. *International Journal of Clinical and Experimental Pathology* 7 (3): 848.

48. Zeppernick, F., Meinhold Heerlein, I., and Shih, I. (2015). Precursors of ovarian cancer in the fallopian tube: serous tubal intraepithelial carcinoma–an update Ovarian cancer and the fallopian tube. *Journal of Obstetrics and Gynaecology Research* 41 (1): 6–11.

49. Gilks, C.B., Irving, J., Kobel, M. et al. (2015). Incidental nonuterine high-grade serous carcinomas arise in the fallopian tube in most cases: further evidence for the tubal origin of high-grade serous carcinomas. *The American Journal of Surgical Pathology* 39 (3): 357–364.

50. Martin, L.P., Hamilton, T.C., and Schilder, R.J. (2008). Platinum resistance: the role of DNA repair pathways. *Clinical Cancer Research* 14 (5): 1291–1295.

51. Kurman, R.J., Visvanathan, K., Roden, R. et al. (2008). Early detection and treatment of ovarian cancer: shifting from early stage to minimal volume of disease based on a new model of carcinogenesis. *American Journal of Obstetrics and Gynecology* 198 (4): 351–356.

52. Elattar, A., Bryant, A., Winter-Roach, B.A. et al. (2011). Optimal primary surgical treatment for advanced epithelial ovarian cancer. *Cochrane Database of Systematic Reviews* 10 (8): CD007565.

53. Shih, K.K., Zhou, Q.C., Aghajanian, C. et al. (2010). Patterns of recurrence and role of adjuvant chemotherapy in stage II–IV serous ovarian borderline tumors. *Gynecologic Oncology* 119 (2): 270–273.

54. Dunlop, M.G., Farrington, S.M. et al. (1997). Cancer risk associated with germline DNA mismatch repair gene mutations. *Human Molecular Genetics* 6 (1): 105–110.

55. Liu, J. and Matulonis, U.A. (2014). New strategies in ovarian cancer: translating the molecular complexity of ovarian cancer into treatment advances. *Clinical Cancer Research* 20 (20): 5150–5156.

56. Kurman, R.J. and Shih, I. (2008). Pathogenesis of ovarian cancer: Lessons from morphology and molecular biology and their clinical implications. *International Journal of Gynecological Pathology* 27 (2): 151–160.

57. Liu, J.F., Konstantinopoulos, P.A., and Matulonis, U.A. (2014). PARP inhibitors in ovarian cancer: current status and future promise. *Gynecologic Oncology* 133 (2): 362–369.

58. Oza, A.M., Cook, A.D., Pfisterer, J. et al. (2015). Standard chemotherapy with or without bevacizumab for women with newly diagnosed ovarian cancer (ICON7): overall survival results of a phase 3 randomised trial. *The Lancet Oncology* 16 (8): 928–936.

59. Black, D. and Barakat, R.R. (2006). Fallopian tube cancer. In: *Textbook of Uncommon Cancer*, 477–484. Wiley.

60. Farley, J., Brady, W.E., Vathipadiekal, V. et al. (2013). Selumetinib in women with recurrent low-grade serous carcinoma of the ovary or peritoneum: an open-label, single-arm, phase 2 study. *The Lancet Oncology* 14 (2): 134–140.

61. Mabuchi, S., Kuroda, H., and Takahashi, R. (2015). The PI3K/AKT/mTOR pathway as a therapeutic target in ovarian cancer. *Gynecologic Oncology* 137 (1): 173–179.
62. Lengyel, E., Litchfield, L.M., Mitra, A.K. et al. (2014). Metformin inhibits ovarian cancer growth and increases sensitivity to paclitaxel in mouse models. *American Journal of Obstetrics and Gynecology* 212 (4): 479.e1–479.e10.
63. Nevadunsky, N.S., Van Arsdale, A., Strickler, H.D. et al. (2014). Metformin use and endometrial cancer survival. *Gynecologic Oncology* 132 (1): 236–240.
64. Wefers, C., Lambert, L.J., Torensma, R. et al. (2015). Cellular immunotherapy in ovarian cancer: targeting the stem of recurrence. *Gynecologic Oncology* 137 (2): 335–342.

20

TESTICULAR CANCER

Christopher B. Allard and Michael L. Blute

Massachusetts General Hospital, Boston, MA, USA

INTRODUCTION

Testicular cancer is rare, with an annual incidence of 5 to 10 cases per 100 000 men in the United States. However, due to the low incidence of other cancers in young men, testicular cancer is the most common solid cancer in men under age 35. Improved treatments over the past few decades have dramatically improved prognosis for patients with testicular cancer. Today, the vast majority of patients with testicular cancer can expect a high likelihood of cure [1].

A wide range of treatments may be used for testicular cancer, from a simple outpatient surgery with subsequent surveillance to a combination of major surgery or radiation therapy and chemotherapy. Treatment strategies depend primarily on the subtype of cancer and on the extent – or stage – of the disease.

Unlike many solid cancers, biopsies are seldom performed for testicular cancer and the primary tumor is removed even if metastases are present. The extent of disease is determined by a combination of imaging and measurement of serum tumor markers, which provide important diagnostic and prognostic information.

Cancer: Prevention, Early Detection, Treatment and Recovery, Second Edition. Edited by Gary S. Stein and Kimberly P. Luebbers.
© 2019 John Wiley & Sons, Inc. Published 2019 by John Wiley & Sons, Inc.

HISTOLOGIC SUBTYPES

The histologic subtype of testicular cancer is determined microscopically after surgical removal of the testicle (radical orchiectomy). Germ cell tumors (GCT) comprise 90–95% of testicular tumors and will be the focus of this chapter. Determination of the particular GCT subtype(s) is necessary to help decide which additional treatments, if any, will be recommended after radical orchiectomy.

GCTs arise from sperm cell precursors and may differentiate into one or several subtypes. These subtypes are classified broadly as seminoma or non-seminomatous germ cell tumors (NSGCT). NSGCTs include the malignant subtypes embryonal, yolk-sac, and choriocarcinoma, and the benign subtype teratoma. Most often, NSGCTs contain mixed elements with multiple subtypes. Mixed tumors containing elements of both seminoma and NSGCT are classified as NSGCT for the purposes of treatment.

Seminomas are the most common GCT in men aged 40 to 50 while NSGCT occur more often in men aged 20 to 40. All GCTs are extremely rare after age 50; in fact, after age 50, lymphoma is the most common cause of testicular masses. The broad classification of GCTs into seminoma or NSGCT forms the basis of treatment considerations for many GCTs owing to differences in prognosis and response to treatments.

Testicular tumors not arising from germ cells are rare and include the sex cord stromal tumors (including Leydig cell tumors, Sertoli cell tumors, granulosa cell tumors, and gonadoblastomas). Rarely, malignancies originating in other organs may metastasize to the testicles. The classification of testicular tumor subtypes according to the World Health Organization (WHO) is shown in Table 1.

RISK FACTORS

Several conditions confer an increased risk of testicular cancer. The most common is the presence of a congenital undescended testicle (cryptorchidism). During normal embryologic development, the testicles descend into the scrotum from the back of the abdomen (the retroperitoneum). Cryptorchidism, which occurs in 2% of newborns, results from incomplete descent of one or both testicles.

Cryptorchid testes may descend after birth, but seldom descend after one year. When they fail to spontaneously descend, cryptorchid testes require surgical correction in order to facilitate testicular self-examination and possibly decrease the risk of cancer. The risk of testicular cancer among people with cryptorchidism is increased two to four times. Importantly, the contralateral normal testicle is also thought to be at increased risk, although a lower risk than the cryptorchid testicle. In spite of the increased risk, it is important to remember that the vast majority of patients with a history of cryptorchidism will not develop testicular cancer [3, 4].

TABLE 1. Histologic classification of testicular tumors based on the World Health Organization (simplified) [2].

Germ cell tumors

    A. Precursor lesions

        1. Intratubular germ cell neoplasia

    B. Tumors of one histologic type

        1. Seminoma

        2. Embryonal carcinoma

        3. Teratoma

        4. Choriocarcinoma

        5. Yolk sac tumor

    C. Malignant mixed germ cell tumor

Sex cord/stromal tumors

    D. Leydig cell tumor

    E. Sertoli cell tumor

    F. Granulosa cell tumor

Mixed germ cell and sex cord/stromal tumors

    G. Gonadoblastoma

    H. Unclassified tumors

Other risk factors include a family history of testicular cancer in a close relative. A personal history of testicular cancer in one testicle is associated with an increased risk of developing cancer in the other. Less common risks include some disorders of sexual differentiation in which the gonads are abnormal. Infertility has also been associated with testicular tumors [5, 6].

SCREENING

Cancer screening involves testing individuals without signs or symptoms of a disease in order to identify cancers before they become clinically apparent. The main objective of screening programs is to diagnose cancers early in order to optimize cure rates. There is presently no widely accepted means of screening asymptomatic individuals for testicular cancer. While some testicular cancers produce elevated levels of certain tumor markers in the blood, many do not, and some people may have elevated markers due to other causes, making serum blood or urine tests unreliable and expensive.

Experts debate the possible benefits of routine testicular self-examination as a means of screening. The United States Preventive Services Task Force (USPSTF) presently recommends against screening for testicular cancer (by self-examination or physician examination), citing the low incidence of testicular cancer, a lack of evidence for a survival benefit from screening, as well as the high cure rate associated with even advanced forms of testicular cancer. While there is little doubt that self-examination will detect some testicular cancers, critics of screening argue that these tumors would have been detected eventually, even without self-examination, and that overall survival would be no worse in late-detected compared to early-detected cancers (due to the high survival rates for testicular cancer in general).

However, many physicians still recommend self-examination on the basis that it is noninvasive, cost-free, and has the potential to identify cancers at earlier stages when treatments can be less aggressive, thereby minimizing immediate and long-term side effects. While a "false-positive" result may cause a patient anxiety, a specialist can usually easily determine, using physical examination, and ultrasound, whether the lesion in question is a tumor or a benign entity. Invasive investigations such as biopsy are not typically performed, and there is little risk of unnecessary treatment of benign conditions.

If one chooses to perform testicular self-examination, there should be gentle palpation of both testicles on a monthly basis. A testicular tumor is typically much harder than the normal testicle and is often painless and non-tender to touch. Other anatomic structures may be confused for tumors, including cysts (spermatoceles), fluid (hydroceles), and enlarged veins (varicoceles). If a mass is felt, this should be evaluated by a physician. Some diagnoses can be easily made on the basis of physical examination alone, while others will require an ultrasound.

SIGNS AND SYMPTOMS

The most common presenting complaint of patients with testicular tumors is a painless scrotal mass. Testicular tumors are slightly more common in the right testicle as a result of the increased rate of cryptorchidism on this side. Occasionally the mass may be painful or tender. Less commonly, patients may develop breast growth (gynecomastia) related to hormones produced from the tumor.

Men with metastatic testicular cancer may experience a variety of symptoms depending on the extent and location(s) of the metastases. Testicular cancer usually spreads in a predictable pattern through lymphatics to the retroperitoneum before spreading to organs. Left-sided tumors typically spread to the left para-aortic and preaortic lymph nodes (the lymph nodes in proximity with the aorta below the level of the kidneys) and seldom cross over to the right side. Right-sided tumors may spread to the interaortocaval and precaval lymph nodes (the lymph nodes in proximity with the inferior vena cava and to the right of the aorta) and can cross over to the left (pre- and para-aortic) nodes.

When testicular cancer metastasizes beyond the retroperitoneal lymph nodes, the commonly affected sites include the lungs, liver, brain, and bones. Symptoms may be nonspecific, including back pain, shortness of breath, weight loss, nausea/vomiting, and neurologic symptoms.

DIAGNOSTIC WORKUP AND STAGING

When testicular cancer is suspected based on a patient's history and physical examination, a scrotal ultrasound is performed to visualize the primary tumor and to ensure that the contralateral testicle appears normal. Removal of the affected testicle (radical orchiectomy) is then performed, facilitating pathologic confirmation of cancer and determination of the histologic subtype(s).

Abdominal imaging (usually by computed tomography [CT] scan) will be performed to assess for the presence of lymphatic or organ metastases and chest CT or X-ray will assess for lung metastases. Other investigations, such as a brain MRI or bone scan, are performed when signs or symptoms suggest metastases in these regions. Blood will be drawn to assess for elevated levels of the tumor markers: alpha-fetoprotein (AFP), Beta human chorionic gonadotropin (B-hCG), and lactate dehydrogenase (LDH). These tests will be performed before orchiectomy to establish baseline levels and several weeks after orchiectomy. Persistently elevated tumor markers after orchiectomy is usually a sign of metastases, which may be microscopic if imaging findings are normal.

While all testicular tumors are initially managed by radical orchiectomy, subsequent treatment options depend on the histologic subtype and clinical stage of the disease as revealed by the diagnostic workup. Testicular tumor staging is defined by the TNM (tumor, node, metastasis) staging system as shown in Table 2. The testicular cancer TNM staging system is unique for its inclusion of tumor markers as an integral component of the classification system.

The TNM stage is assigned based on each of the tumor, node, metastases, and serum tumor marker components. For example, a patient with a tumor limited to the testicle without lymphovascular invasion that has negative serum tumor markers and no metastases seen on imaging would have clinical stage T1N0M0S0. Alternatively, a patient with a primary tumor invading the spermatic cord who has multiple lymph nodes of size 2–5 cm, lung metastases, normal LDH, AFP = 900, and B-hCG = 30000 would have stage T3N2M1aS2.

After defining the TNM stage, testicular cancer is next defined according to its stage grouping. The two examples above would be classified as stage groups 1a and 3b, respectively. The stage grouping system for testicular cancer is shown in Table 3.

MANAGEMENT OF TESTICULAR TUMORS

The management of testicular tumors begins with radical orchiectomy. Subsequent options depend largely on the histologic subtype and the tumor stage group (Table 3). In many cases, multiple options may be offered, and management decisions will be based on individual prognosis and patient preferences based on an understanding of the benefits and risks of each approach.

TABLE 2. The testicular cancer TNM staging system of the American Joint Committee of Cancer and the International Union Against Cancer (simplified) [7].

Primary Tumor (T)	Definition of stage
T0	No evidence of primary tumor
Tis	Intratubular germ cell neoplasia
T1	Limited to the testicle/epididymis; no lymphovascular invasion
T2	Limited to the testicle/epididymis with lymphovascular invasion or tumor extending through into tunica vaginalis
T3	Invades spermatic cord
T4	Invades the scrotum
Regional Lymph Nodes (N)	
N0	No regional lymph node metastasis
N1	Metastasis with a lymph node mass <2 cm or multiple lymph nodes, none more than 2 cm
N2	Metastasis with a lymph node mass 2–5 cm or multiple lymph nodes, any mass 2–5 cm
N3	Metastasis with a lymph node mass >5 cm
Distant Metastases (M)	
M0	No evidence of distant metastases
M1a	Non-regional lymph node metastases or lung metastasis
M1b	Distant metastasis other than M1a
Serum Tumor Markers (S)	
S0	Markers within normal limits
S1	LDH <1.5 × normal
	hCG <5000 mlu/ml
	AFP <1000 ng/ml
S2	LDH 1.5–10 × normal
	hCG 5000–50 000 mlu/ml
	AFP 1000–10 000 ng/ml
S3	LDH >10 × normal
	hCG >50 000 mlu/ml
	AFP >10 000 ng/ml

TABLE 3. AJCC stage grouping of testicular cancer [7].

Stage 0	Tis	N0	M0	S0
Stage 1a	T1	N0	M0	S0
Stage 1b	T2-T4	N0	M0	S0
Stage 1s	Any T	N0	M0	S1–3
Stage 2a	Any T	N1	M0	S0–1
Stage 2b	Any T	N2	M0	S0–1
Stage 2c	Any T	N3	M0	S0–1
Stage 3a	Any T	Any N	M1a	S0–1
Stage 3b	Any T	Any N	M0	S2
	Any T	Any N	M1a	S2
Stage 3c	Any T	Any N	M1b	Any S
	Any T	Any N	Any M	S3

Radical Orchiectomy

Radical orchiectomy is the surgical removal of the testicle and spermatic cord through an inguinal incision. This procedure is performed soon after confirmation of a testicular mass and is curative in many patients with localized (N0M0S0) disease.

This procedure is typically performed in an outpatient setting. The testicle is removed through an inguinal (groin) incision, rather than a scrotal incision, because a scrotal incision may alter the otherwise predictable lymphatic spread of testicular tumors and because this incision facilitates removal of the entire spermatic cord along with the testicle. An example of a radical orchiectomy specimen is shown in Figure 1. Complications may include local bleeding (hematoma) and nerve injury.

Sperm banking should be considered since treatments may affect fertility potential. However, in the usual setting of a normal contralateral testicle, testosterone levels will remain normal after treatment, and there will be no need for testosterone replacement.

Important prognostic and staging information obtained by pathologic assessment of the radical orchiectomy specimen include the extent of local tumor invasion, histology, and the presence or absence of lymphovascular invasion.

Active Surveillance

Patients with low-risk, clinically localized (e.g., stage 1A) testicular tumors (seminoma or NSGCT) are increasingly being managed with active surveillance after orchiectomy [8–11]. This strategy consists of close observation and no further

Figure 1. Radical orchiectomy specimen showing testicle (bivalved) with mixed non-seminomatous germ cell tumor replacing most of the testicular parenchyma.

treatments unless relapses are detected. Patients who develop relapse are treated with cisplatin-based chemotherapy. This strategy is only appropriate for patients without evidence (imaging or tumor markers) of lymphatic or metastatic disease.

The rationale for this approach is twofold: 1) Survival rates are nearly 100% for any management approach, including active surveillance, since chemotherapy is highly effective at treating relapse, and 2) Active surveillance prevents overtreatment in the majority of patients who are cured by radical orchiectomy alone.

Approximately 75% of patients with NSGCT and 80–85% with seminoma managed by active surveillance will not relapse, and will therefore not experience side effects from unnecessary treatments [12]. Among patients who do relapse, the majority will occur in the first year, although later relapses can occur [13, 14].

Among all patients managed by active surveillance, 25–30% with NSGCT and 15–20% with seminoma develop relapses. Patients at increased risk for relapse include those with lymphovascular invasion and/or a high proportion of embryonal histology in the orchiectomy specimen for NSGCT [15, 16], and tumors larger than 4 cm or with rete testis invasion for seminoma [17]. Even with risk factors present, the majority of patients with stage 1A NSGCT or seminoma will be cured by radical orchiectomy.

Surveillance protocols vary by institution and consist of periodic history, physical examination, serum tumor markers, chest X-ray or CT, and CT of the abdomen and pelvis [18]. Follow-up is intensive initially (every few months) and gradually decreases in frequency. Adherence to an intensive follow-up protocol is of paramount importance in detecting relapses early; active surveillance is not appropriate for patients who cannot commit to this schedule or who will experience excessive anxiety as a consequence.

The primary disadvantages of active surveillance are the need for frequent follow-up with exposure to significant radiation from multiple CT scans, which may increase the risk of developing future malignancies [19], and an increased risk of relapse [20]. A further consideration is that relapses will be treated with chemotherapy, while primary retroperitoneal lymph node dissection (RPLND) for stage 1 NSGCT or radiotherapy for stage 1 seminoma decrease the likelihood of requiring chemotherapy compared to surveillance.

Retroperitoneal Lymph Node Dissection

RPLND may be performed in the setting of negative tumor markers to decrease the risk of relapse in stage 1 NSGCT, or to remove enlarged lymph nodes in stage 2A/B NSGCT [21, 22]. RPLND is not routinely used for the initial management of seminoma, but is occasionally necessary if patients have residual tumors after chemotherapy.

RPLND generally involves a vertical abdominal incision from the pubic bone to the xiphoid process; bowel is then mobilized to reveal the retroperitoneum, and all lymphatic tissue below the kidneys is then carefully excised and sent for pathologic evaluation [23]. While some surgeons are performing this procedure through minimally invasive means (laparoscopic or robotic-assisted) [24, 25], an open approach involving a large incision is still considered the standard of care.

When performed for stage 1 NSGCT (i.e., with no evidence of enlarged lymph nodes or metastases), this procedure provides superior staging information over CT scans in order to determine if any cancer resides in the retroperitoneum. The primary advantage of RPLND over active surveillance for stage 1 disease is a decreased risk of relapse; RPLND decreases relapse rates from 20–30% (for active surveillance) to 5%, thereby reducing the frequency of abdominal imaging necessary during follow-up. When relapses do occur, they tend to be outside the retroperitoneum.

Compared to chemotherapy for stage 1 or 2 GCT, RPLND is associated with lower risks of infertility and other long-term side effects. Furthermore, surgical excision is the only method that successfully treats teratoma. While teratoma is a benign histology, it can enlarge and compress adjacent structures or occasionally transform into a malignant subtype. Teratoma may be present in the retroperitoneum even when it is absent in an orchiectomy specimen.

Disadvantages of RPLND compared to chemotherapy include the failure to treat any microscopic disease outside the retroperitoneum, which may result in relapse requiring chemotherapy [26]. Patients with extensive lymph node involvement in RPLND specimens may also require chemotherapy in addition to RPLND. Finally, RPLND is a major operation with risks of significant morbidity. Previously, failure of semen emission (but not erectile dysfunction) was a common side effect. This has become rare with improved "nerve-sparing" techniques. Other complications, such as injury to organs or blood vessels, may occur. Bowels may be slow to recover function (ileus) resulting in a prolonged hospital stay, and rarely patients may develop a leak of lymphatic fluid (chylous ascites), which may take weeks or months to resolve.

Radiotherapy

Unlike NSGCT, Seminomas are highly sensitive to radiation. Hence, the role for external beam radiotherapy (EBRT) to the retroperitoneum is comparable to that of RPLND in NSGCT; it is an appropriate strategy for stage 1–2B testicular tumors to minimize the risk of relapse (when there is no evidence of disease) or to treat low-volume lymph node-positive disease [27]. Unlike RPLND, EBRT can be utilized when serum markers are slightly elevated (TNM stage 1S).

Compared to active surveillance, primary EBRT reduces the risk of relapse from 15–20% to 5%, with most relapses occurring outside the retroperitoneum. Side effects of radiotherapy include fatigue, nausea, and bone marrow suppression. Long-term toxicities may include bowel inflammation, infertility, and secondary malignancies. In young patients with a long life expectancy, EBRT significantly increases the risk of developing other cancers later in life as well as heart disease [28, 29].

Chemotherapy

The introduction of cisplatin-based chemotherapy regimens has resulted in dramatic improvements in prognosis for patients with testicular cancer. The availability of highly effective chemotherapy has been the driving force behind increased utilization of active surveillance, since relapses can so often be effectively cured [30].

Chemotherapy is an option for clinical stage 1 tumors, reducing relapse rates to under 5% [31, 32]. In stage 2 disease, chemotherapy can effectively treat retroperitoneal lymph nodes [22]. In patients with extensive lymph node involvement or stage 3 (metastatic) disease, as well as any patient with elevated post-orchiectomy tumor markers (except stage 1S seminoma, as mentioned above), chemotherapy is standard.

Chemotherapy agents used for NSGCT consist of a combination of bleomycin, etoposide, and cisplatin (BEP). Generally, two cycles of BEP are given for stage 1 and 3 cycles for stage 2. For patients with stage 3 disease, the number of cycles depends on the International Germ Cell Cancer Collaborative Group (IGCCCG) risk classification, as shown in Table 4. Patients with "good risk" metastatic disease are generally treated with three cycles of BEP and those with "intermediate" or "poor risk" receive four cycles.

Stage 1 seminoma may be treated with single-agent carboplatin with one or two cycles. Higher stages are managed with BEP with three or four cycles depending on the IGCCCG risk classification as shown in Table 4. For seminoma, no patients are classified as "poor risk."

In spite of the clear benefits of chemotherapy in terms of decreasing relapse and treating known disease, as well as treating microscopic disease residing outside

TABLE 4. International Germ Cell Cancer Collaborative Group (IGCCCG) risk classification of advanced germ cell tumors [33].

	NSGCT	Seminoma
Good risk	Testicular or retroperitoneal primary *and*	Any primary site *and*
	No non-pulmonary visceral metastases *and*	No non-pulmonary visceral metastases *and*
	Post-orchiectomy markers ≤ S1	Normal AFP, any hCG, any LDH
Intermediate risk	Testicular or retroperitoneal primary *and*	Any primary site *and*
	Non non-pulmonary visceral metastases *and*	Non-pulmonary visceral metastases *and*
	Post-orchiectomy markers = S2	Normal AFP, any hCG, any LDH
Poor risk	Mediastinal primary tumor *or*	No seminoma patients classified as poor prognosis
	Non-pulmonary visceral metastases *or*	
	Post-orchiectomy markers = S3	

the retroperitoneum, long-term side effects may occur. Potential side effects include lung inflammation/fibrosis (interstitial pneumonitis and pulmonary fibrosis), atherosclerosis, bone marrow suppression, infertility, secondary cancers, and heart disease [34, 35].

SUMMARY

Testicular cancer is the most common malignancy in young men. It is usually detected when a patient feels a painless hard lump on their testicle. The usual management involves removal of the testicle and staging investigations, including imaging and measurement of serum tumor markers, aimed at identifying the subtype and extent of disease. Further treatments will depend on the type (histology) of tumor as well as the presence and location(s) of any metastases. In patients without evidence of metastatic disease, active surveillance is an appropriate option in most motivated patients, since it is associated with excellent long-term outcomes and reduced need for aggressive treatments. Fortunately, the vast majority of patients with testicular cancer are curative with modern treatment strategies.

REFERENCES

1. Carver, B.S., Serio, A.M., Bajorin, D. et al. (2007). Improved clinical outcome in recent years for men with metastatic nonseminomatous germ cell tumors. *J. Clin. Oncol.* 25 (35): 5603–5608. https://doi.org/10.1200/JCO.2007.13.6283.
2. Eble, J.N., Sauter, G., Epstein, J.L. et al. (2004). *World Health Organization Classification of Tumours. Pathology and Genetics of Tumours of the Urinary System and Male Genital Organs.* Lyon: IARC Press.
3. Wood, H.M. and Elder, J.S. (2009). Cryptorchidism and testicular cancer: separating fact from fiction. *J. Urol.* 181 (2): 452–461. https://doi.org/10.1016/j.juro.2008.10.074.
4. Akre, O., Pettersson, A., and Richiardi, L. (2009). Risk of contralateral testicular cancer among men with unilaterally undescended testis: a meta analysis. *Int. J. Cancer* 124 (August 2008): 687–689. https://doi.org/10.1002/ijc.23936.
5. Lambert, S.M. and Fisch, H. (2007). Infertility and testis cancer. *Urol. Clin. N. Am.* 34: 269–277. https://doi.org/10.1016/j.ucl.2007.02.002.
6. Hemminki, K., Chen, B., Schettler, T. et al. (2006). Familial risks in testicular cancer as aetiological clues. *Int. J. Androl.* 29: 205–210. https://doi.org/10.1111/j.1365-2605.2005.00599.x.
7. American Joint Committee on Cancer AJCC (2009). *American Joint Committee on Cancer (AJCC) Staging Manual*, 7e. New York, NY: Springer-Verlag.
8. Krege, S., Beyer, J., Souchon, R. et al. (2008). European consensus conference on diagnosis and treatment of germ cell cancer: a report of the second meeting of the European Germ Cell Cancer Consensus group (EGCCCG): part I. *Eur. Urol.* 53: 478–496. https://doi.org/10.1016/j.eururo.2007.12.024.
9. Wood, L., Kollmannsberger, C., Jewett, M. et al. (2010). Canadian consensus guidelines for the management of testicular germ cell cancer. *Can. Urol. Assoc. J.* 4 (2): e19–e38.
10. Nguyen, C.T., Fu, A.Z., Gilligan, T.D. et al. (2010). Defining the optimal treatment for clinical stage I nonseminomatous germ cell testicular cancer using decision analysis. *J. Clin. Oncol.* 28 (1): 119–125. https://doi.org/10.1200/JCO.2009.22.0400.

11. Nichols, C.R., Roth, B., Albers, P. et al. (2013). Active surveillance is the preferred approach to clinical stage I testicular cancer. *J. Clin. Oncol.* 31 (28): 3490–3493. https://doi.org/10.1200/JCO.2012.47.6010.

12. Mortensen, M.S., Lauritsen, J., Gundgaard, M.G. et al. (2014). A nationwide cohort study of stage I seminoma patients followed on a surveillance program. *Eur. Urol.* 66 (6): 1172–1178. https://doi.org/10.1016/j.eururo.2014.07.001.

13. Carver, B.S., Motzer, R.J., Kondagunta, G.V. et al. (2005). Late relapse of testicular germ cell tumors. *Urol. Oncol.* 23: 441–445. https://doi.org/10.1016/j.urolonc.2005.06.003.

14. Baniel, J., Foster, R.S., Einhorn, L.H. et al. (1995). Late relapse of clinical stage I testicular cancer. *J. Urol.* 154: 1370–1372. https://doi.org/10.1097/00005392-199510000-00031.

15. Albers, P., Siener, R., Kliesch, S. et al. (2003). Risk factors for relapse in clinical stage I non-seminomatous testicular germ cell tumors: results of the German testicular cancer study group trial. *J. Clin. Oncol.* 21 (8): 1505–1512. https://doi.org/10.1200/JCO.2003.07.169.

16. Vergouwe, Y., Steyerberg, E.W., Eijkemans, M.J.C. et al. (2003). Predictors of occult metastasis in clinical stage I nonseminoma: a systematic review. *J. Clin. Oncol.* 21: 4092–4099. https://doi.org/10.1200/JCO.2003.01.094.

17. Warde, P., Specht, L., Horwich, A. et al. (2002). Prognostic factors for relapse in stage I seminoma managed by surveillance: a pooled analysis. *J. Clin. Oncol.* 20 (22): 4448–4452. https://doi.org/10.1200/JCO.2002.01.038.

18. Rustin, G.J., Mead, G.M., Stenning, S.P. et al. (2007). Randomized trial of two or five computed tomography scans in the surveillance of patients with stage I nonseminomatous germ cell tumors of the testis: Medical Research Council Trial TE08, ISRCTN56475197 – the National Cancer Research Institute Testis Cancer Clinical Study Groups. *J. Clin. Oncol.* 25 (11): https://doi.org/10.1200/JCO.2006.08.4889.

19. Brenner, D.J. and Hall, E.J. (2007). Computed tomography – an increasing source of radiation exposure. *N. Engl. J. Med.* 357: 2277–2284. https://doi.org/10.1056/NEJMra072149.

20. Kollmannsberger, C., Tandstad, T., Bedard, P.L. et al. Patterns of relapse in patients with clinical stage I testicular cancer managed with active surveillance. *J. Clin. Oncol.* 2014: 1–7. https://doi.org/10.1200/JCO.2014.56.2116.

21. Hotte, S.J., Mayhew, L.A., Jewett, M. et al. (2010). Management of stage I non-seminomatous testicular cancer: a systematic review and meta-analysis. *Clin. Oncol.* 22 (1): 17–26. https://doi.org/10.1016/j.clon.2009.09.005.

22. Stephenson, A.J., Bosl, G.J., Motzer, R.J. et al. (2007). Nonrandomized comparison of primary chemotherapy and retroperitoneal lymph node dissection for clinical stage IIA and IIB nonsemi-nomatous germ cell testicular cancer. *J. Clin. Oncol.* 25 (35): 5597–5602. https://doi.org/10.1200/JCO.2007.12.0808.

23. Jewett, M. and Groll, R.J. (2007). Nerve-sparing retroperitoneal lymphadenectomy. *Urol. Clin. N. Am.* 34: 149–158. https://doi.org/10.1016/j.ucl.2007.02.014.

24. Janetschek, G., Hobisch, A., Peschel, R. et al. (2000). Laparoscopic retroperitoneal lymph node dissection for clinical stage I nonseminomatous testicular carcinoma: long-term outcome. *J. Urol.* 163 (June): 1793–1796.

25. Nelson, J.B., Chen, R.N., Bishoff, J.T. et al. (1999). Laparoscopic retroperitoneal lymph node dissection for clinical stage I nonseminomatous. *Urology* 54 (6): 1064–1067.

26. Albers, P., Siener, R., Krege, S. et al. (2008). Randomized phase III trial comparing retroperito-neal lymph node dissection with one course of bleomycin and etoposide plus cisplatin chemo-therapy in the adjuvant treatment of clinical stage I nonseminomatous testicular germ cell tumors: AUO Trial AH 01/94. *J. Clin. Oncol.* 26 (18): 2966–2972. https://doi.org/10.1200/JCO.2007.12.0899.

27. Chung, P., Mayhew, L.A., Warde, P. et al. (2010). Management of stage I seminomatous testicu-lar cancer: a systematic review. *Clin. Oncol. (R. Coll. Radiol.)* 22: 6–16. https://doi.org/10.1016/j.clon.2009.08.006.

28. Travis, L.B., Curtis, R.E., Storm, H. et al. (1997). Risk of second malignant neoplasms among long-term survivors of testicular cancer. *J. Natl. Cancer Inst.* 89 (19): 1429–1439.

29. Huddart, R.A., Norman, A., Shahidi, M. et al. (2003). Cardiovascular disease as a long-term complication of treatment for testicular cancer. *J. Clin. Oncol.* 21 (8): 1513–1523. https://doi. org/10.1200/JCO.2003.04.173.

30. Feldman, D.R., Bosl, G.J., Sheinfeld, J. et al. (2008). Medical treatment of advanced testicular cancer. *JAMA* 299 (6): 672–684.

31. Tandstad, T., Dahl, O., Cohn-Cedermark, G. et al. (2009). Risk-adapted treatment in clinical stage I nonseminomatous germ cell testicular cancer: the SWENOTECA management program. *J. Clin. Oncol.* 27 (13): 2122–2128. https://doi.org/10.1200/JCO.2008.18.8953.

32. Studer, U.E., Burkhard, F.C., and Sonntag, R.W. (2000). Risk adapted management with adjuvant chemotherapy in patients with high risk clinical stage i nonseminomatous germ cell tumor. *J. Urol.* 163 (June 1987): 1785–1787.

33. Group IGCCC (1997). International germ cell consensus classification: a prognostic factor-based staging system for metastatic germ cell cancers. *J. Clin. Oncol.* 15 (2): 594–603.

34. Meinardi, B.M.T., Gietema, J.A., van der Graaf, W.T. et al. (2014). Cardiovascular morbidity in long-term survivors of metastatic testicular cancer. *J. Clin. Oncol.* 18: 1725–1732.

35. Kaufman, M.R. and Chang, S.S. (2007). Short- and long-term complications of therapy for testicular cancer. *Urol. Clin. N. Am.* 34: 259–268. https://doi.org/10.1016/j.ucl.2007.02.011.

21

HEAD AND NECK CANCER

Kumar G. Prasad[1,2], Marion Couch[1], and Michael Moore[3]

[1]*Department of Otolaryngology–Head and Neck Surgery, Indiana University School of Medicine, Indianapolis, IN, USA*
[2]*Department of Otolaryngology–Head and Neck Surgery, Meritas Health, North Kansas City, MO, USA*
[3]*Arilla Spence DeVault Professor of Otolaryngology–Head and Neck Surgery, Chief, Division of Head and Neck Surgery, Indiana University School of Medicine, Indianapolis, IN, USA*

INTRODUCTION

Head and neck cancers are a relatively rare group of diseases that affect approximately 59 340 Americans every year, resulting in an estimated 12 290 deaths [1]. For the purposes of this chapter, head and neck cancer refers to malignancies arising from the upper aerodigestive tract and its supporting structures as well as salivary gland cancer. Thyroid and skin malignancies will be covered elsewhere.

Over 90% of cancers of the upper aerodigestive tract are squamous cell carcinomas, arising from squamous epithelium that makes up its lining. Other tumors include lymphomas from lymphoid tissue, salivary gland malignancies from the major or minor salivary glands, and cancers of the underlying stromal components such as the muscle, bone, and nerves. Malignancies developing from these mesenchymal structures are termed sarcomas.

Cancer: Prevention, Early Detection, Treatment and Recovery, Second Edition. Edited by Gary S. Stein and Kimberly P. Luebbers.
© 2019 John Wiley & Sons, Inc. Published 2019 by John Wiley & Sons, Inc.

ANATOMY

The upper aerodigestive tract begins at the junction of the skin and mucosa of the lips and nose and continues down to the level of the cricoid cartilage, which separates the larynx from the cervical trachea and the hypopharynx from the cervical esophagus. Anatomic subsites include the nasal cavity, paranasal sinuses, nasopharynx, oral cavity, oropharynx, larynx, and hypopharynx.

A physiologically intact upper aerodigestive tract is critical for vital functions such as olfaction (smell), mastication (chewing), articulation, phonation, deglutition (swallowing), and maintenance of a patient airway. Additionally, the human craniofacial anatomy provides the facial structure that is central to one's identity. Consequently, when malignancies affect this region, care must be made to not only eradicate disease but also to preserve function and cosmesis.

The cervical lymphatic system is also particularly relevant when managing head and neck malignancies. The lymph nodes of the neck are distributed diffusely across the anterior and lateral neck, with a few nodes also being present

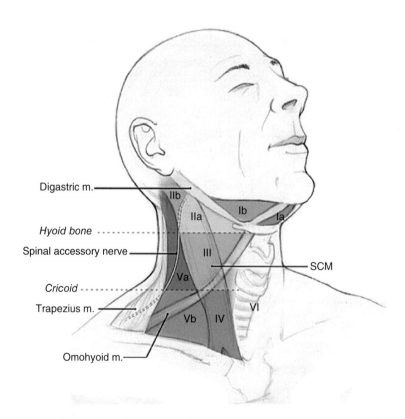

Figure 1. The cervical lymph node basins. This diagram demonstrates the number of cervical lymphatic basins of the neck. *Source:* Reprinted by permission from Springer Nature, *Normal Cervical Lymph Node Appearance and Anatomic Landmarks in Neck Ultrasound*, Lisa A Orloff and Masie L. Shindo, 2017 [2].

T A B L E 1. Lymph node drainage patterns of the head and neck.

Primary Site	Lymph Node Drainage
Oral Cavity	I (floor of mouth, often bilateral), II–III, with IV at risk for lateral tongue cancers
Oropharynx	II–IV (often bilateral for tongue base, soft palate and posterior pharyngeal wall)
Nasopharynx	Retropharyngeal LNs, II, V
Larynx/Hypopharynx	II–IV (rare for early glottis cancers, high rate for supraglottic cancer)

posteriorly. These lymphatics are subdivided into six nodal basins according to their anatomic location (see Figure 1).

Depending on the location of the primary tumor, certain lymph nodes will be at higher risk for metastatic involvement. Table 1 provides a brief summary of the lymphatic basins most associated with each primary site [1].

EPIDEMIOLOGY AND RISK FACTORS

While head and neck malignancies can occur spontaneously, the vast majority are due to exposure to environmental carcinogens. Mucosal cancers are 2.7 times more common in men, while females are more likely to develop thyroid cancer. In the United States, the most common risk factor for mucosal cancer is exposure to tobacco, with approximately 85% of patients with mouth and throat cancers having a history of regular use. The mechanism for carcinogenesis was first described by Slaughter et al. as "field cancerization" in 1953 [3] which results from genetic insults that are acquired through regular exposure to known carcinogens. The host cell's DNA is affected and then passed on to their progeny. Once enough insults have been sustained (the so-called "Multi-hit hypothesis") an inability to regulate cell growth and death may predispose cells to malignant transformation [4]. This concept has been further expanded to demonstrate that all mucosal surfaces exposed to a known carcinogen are at risk for malignant growth, and has been used to explain the presence of multiple synchronous cancers throughout the head, neck, lung, and esophagus. The oral cavity is the first point of contact for carcinogens such as tobacco and alcohol and therefore has been shown to have the highest rate of second primary tumors [5]. Heavy alcohol use has also been demonstrated to have a direct correlation with numerous upper aerodigestive tract cancers, but when used along with cigarette smoking, the risks are synergistic [6]. Additional risk factors include exposure to UV rays in the form of sunlight (cutaneous cancers), prior head and neck radiation, heavy metals and sawdust (sinonasal malignancies), and prior infection with the Ebstein-Barr virus (nasopharyngeal cancer).

Over the last three decades, there has been a significant decline in the rate of diagnosis of tobacco-related cancers secondary to public awareness initiatives, health policy, and societal pressures. Over this same period, however, there has

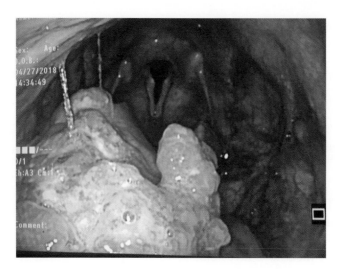

Figure 2. Transnasal fiberoptic laryngoscopy. This is a photograph taken from a flexible fiberoptic pharyngolaryngoscopy demonstrating a large exophytic tumor of the right base of tongue.
Beyond the tumor, the normal appearing vocal folds and hypopharynx can be seen.

been a significant increase in the incidence of oropharyngeal cancer [7]. Developing
research suggests that this new spike in oropharyngeal cancer is caused by prior
infection with high-risk subtypes of the human papillomavirus (HPV 16 and 18),
which has been shown to be an independent risk factor for cancers of the tonsils
and tongue base [8]. The significance of these findings is that HPV+ cancers tend
to have a better prognosis and survival rate than patients with similar non-HPV-
related cancers [9–12]. Additionally, with the development of approved HPV vaccines,
infections that predispose to these cancers may be preventable, thus decreasing the
disease burden in our population (see Figure 2).

PREVENTION

The primary approach to prevention of head and neck cancer is avoidance of
high-risk exposures such as tobacco and alcohol. Diets rich in fruits and vegetables, especially those with large amounts of vitamin A and beta carotenes, have
shown to reduce the risk of developing mucosal head and neck cancers while consumption of large amounts of meats and animal products increases risk. Unfortunately,
there are no proven options for chemoprevention of head and neck cancer.

   The development of HPV-related oropharyngeal cancer has been linked to increased
numbers of sexual partners, early age of onset of sexual activity, and participation in oral
sex [13, 14]. As a result, safe sexual practices may reduce the risk of contracting the
disease. In addition, the recent institution of HPV vaccines approved by the US Food
and Drug Administration is an exciting opportunity to provide primary prevention to the
at risk population. Currently there are three commercially available vaccines, Gardisil-4

and -9 (Merck & Co) and Cervarix (GlaxoSmithKlein). Gardisil-4 is a quadrivalent vaccine active against HPV subtypes 6, 11, 16, and 18 and is approved for use in girls and boys aged 9 to 45. Cervarix is a bivalent vaccine active against HPV subtypes 16 and 18 and is approved in women ages 9 to 26 [15]. Gardisil-9 has the same coverage as Gardisil-4 and provides additional protection against subtypes 31, 33, 45, 52, and 58. Each is given as a three IM shot series, with only two shots being needed if it is started before the fifteenth birthday. Preliminary data suggests that the use of these vaccines is safe and effective.

CLINICAL PRESENTATION

Initial evaluation of head and neck cancer centers around doing a focused, yet thorough history, including the chief complaint, history of present illness, past medical and surgical history, medications, allergies, social history (including tobacco and alcohol use as well as sexual history), and review of systems. The most common signs and symptoms depend on the size and location of the disease process, as well as the behavior of the tumor itself. Cancers of the oral cavity, due to their accessible location, are often recognized by the patient as a tongue or mouth sore that worsens instead of improving over time. In more advanced disease, difficulty with speech and swallowing may develop. When the adjacent bone of the mandible or maxilla is involved worsening pain typically occurs and the patient may develop numbness on the involved lower lip (due to extension into the inferior alveolar nerve) or cheek (due to involvement of the infraorbital nerve).

Pharyngeal malignancies may present with sore throat on the side of the tumor made worse with swallowing. Additional symptoms include ear pain (referred through contributions of cranial nerves IX and X), difficulty swallowing, muffled speech, hoarseness, and even difficulty breathing with more advanced tumors. Due to their inconspicuous location and abundant lymphatic drainage of the pharynx, tumors in these locations often may initially present with a painless neck mass. This represents a regional metastasis from an underlying primary tumor that has yet to be identified (a so called unknown primary). In fact, this is the most common presentation of patients with HPV-related cancer of the oropharynx [16]. Nasopharyngeal cancers may present with a neck mass and/or a unilateral middle-ear effusion as a result of obstruction of the ipsilateral Eustachian tube orifice.

EXAMINATION

After obtaining a thorough history, a detailed physical examination of the head and neck should be performed to assess the breathing status and voice, ears, nose, oral cavity, oropharynx, and neck (including palpation of the parotid and thyroid glands). An assessment of cranial nerves 2–12 should also be done to look for any deficits.

At the completion of the exam, a transnasal fiberoptic laryngoscopy provides additional evaluation of the mucosal surfaces of the nasal cavity, nasopharynx, oropharynx, hypopharynx, and larynx (see Figure 2). In addition to looking for mucosal lesions/masses, the movement of the vocal folds with phonation and deep inspiration should be assessed. Finally, a determination should be made based on this exam as to the best way to manage the airway in the event general anesthesia is required in the future.

ADDITIONAL WORKUP

Following the history and physical examination, it is often necessary to obtain additional imaging in order to evaluate both the primary tumor and cervical lymph node basins. Computed tomography scan (CT) is most commonly used due to the fact that it is quick and relatively inexpensive. However, there is an associated exposure to ionizing radiation and the view of some areas (i.e., oral cavity) may be obscured due to dental artifacts. Despite this, CT with intravenous contrast is the imaging of choice for an initial workup of head and neck cancer, especially when there is question of involvement of adjacent bony structures such as the mandible.

Magnetic resonance imaging (MRI) is another modality that can be considered. MRI is superior to CT in the assessment of tongue base involvement, as well as to determine invasion of the mandibular bone marrow space or perineural spread of tumor [17] but does take longer to complete. Thus, these images can be more subject to motion artifact, especially when evaluating the larynx.

In addition to assessing the head and neck, it is important to rule out distant metastases in patients with advanced disease. A formal metastatic workup includes imaging of the chest, which can be accomplished via chest X-ray or non-contrasted CT. A positron emission tomography with CT (PET/CT) may also utilized to rule out metastatic disease but is recommended only for those patients with clinical stage III and IV disease [18].

TISSUE BIOPSY/EXAMINATION UNDER ANESTHESIA

Probably the most critical aspect of the initial workup of a suspected head and neck cancer is a tissue biopsy. The vast majority of oral cavity lesions are accessible for biopsy in the clinic setting using local anesthesia. For most tumors of the pharynx and larynx, a specimen must be obtained in the operating room under general anesthesia for assessment of the entirety of the upper aerodigestive tract, otherwise known as a direct laryngoscopy. The primary lesion is evaluated for its size and location as well as its proximity to the midline and other critical structures such as the mandible, maxilla, and larynx. A sample is typically sent from the junction between the normal and abnormal tissue in an effort to both maintain orientation and confirm invasion of the basement membrane. While the patient is under anesthesia, an esophagoscopy and/or bronchoscopy can be performed to screen for additional primary lesions.

STAGING

Staging criteria for head and neck cancer is based on subsite of origin of the primary tumor, as well as a number of other factors. Each tumor is given a TNM stage, representing the size of the primary tumor ("T"), the nature of cervical nodal metastasis ("N"), and any evidence of distant metastatic disease ("M"). The staging combines all information gained from the clinician's history, physical exam, imaging, and biopsy results. Some regions (i.e., oral cavity and oropharynx) are staged based on the size of the primary tumor, while others depend on the subsites involved and the function of the affected organ (i.e., larynx). Still others combine the two approaches, with the true stage being the highest stage reached by either method (i.e., hypopharynx).

The recently published AJCC Eighth Edition staging system also incorporates certain disease aspects to better provide prognostic differentiation due to staging [19]. Examples of changes include adding depth of invasion in addition to size in the staging of oral cavity cancer, with deeper tumors receiving a higher "T" classification. Moreover, a separate staging system has also been developed for HPV-related oropharyngeal cancers, reflecting the different biologic behavior and responsiveness to treatment of these tumors. Finally, modifications have been made to the nodal staging system, creating both a clinical and pathologic staging system and also incorporating extracapsular spread in HPV-negative cancer, recognizing the survival detriment that comes with this pathologic finding.

The TNM stage is then combined to form an overall comprehensive stage designated as I–IVc. Cancers that are stage I or II are classified as early stage disease and thus are typically managed with single modality therapy (either surgery or radiation therapy [RT]). Stage III and IVa tumors are considered advanced stage and are treated with multimodality therapy (surgery with postoperative adjuvant radiation ± chemotherapy, or upfront chemoradiation therapy with surgery for treatment failure). Stage IVb tumors are considered surgically unresectable and are therefore usually managed with chemoradiation therapy. Those that are stage IVc have demonstrated distant metastatic disease and are typically treated with palliative chemotherapy, adding radiation only in an effort to improve locoregional control.

While an exhaustive review of all head and neck cancer staging is beyond the scope of this chapter, a brief outline is provided. As with cancer of other regions of the body, head and neck malignancies are staged for three main reasons:

1. To allow clinicians to accurately and succinctly put the patient's disease into clinical context when communicating to other practitioners.
2. To predict the patient's prognosis based on their disease and to guide their treatment.
3. To assist in stratification of patients and their disease in clinical trials and other areas of scientific investigation.

MULTIDISCIPLINARY APPROACH TO HEAD AND NECK CANCER CARE

Successful management of head and neck cancer involves contributions from many members of a multidisciplinary team. The primary disciplines include head and neck oncologic and reconstructive surgery, medical oncology, radiation oncology, pathology, diagnostic radiology, dentistry, oromaxillofacial prosthodontics, and speech and swallowing therapy. Additional teams that may be involved are neuro-surgery (for patients with skull base tumors), oculoplastic surgery (for tumors around the eye), and vascular surgery (for tumors encasing the carotid artery). These disciplines meet on a regular basis and present patients prospectively in the form of a tumor board. At this conference, input is obtained from all members of the team in order to find the optimal therapy and to discuss options for enrollment in clinical trials.

MANAGEMENT

The primary modalities used to treat head and neck cancer are surgery and RT, either used individually or in combination with each other. Chemotherapy, while not an effective treatment in and of itself for head and neck cancer, can enhance the effects of RT and can also provide some disease control and palliation for patients with distant metastases or inoperable disease.

A number of factors must be taken into consideration when determining the most appropriate strategy for management. The typical approach is to choose the treatment that will provide the best chance of cancer control, while still attempt-ing to minimize the morbidity of the therapy. Certain tumor locations are better served by one treatment approach over another. An example would be oral cavity cancer, where upfront surgery is the gold standard and the addition of postoperative (adjuvant) RT or chemoradiation therapy may be warranted based on surgical and histopathological findings. Conversely, the majority of cancers of the pharynx and larynx are often treated with upfront "organ preserving" nonsurgical therapy, reserving surgery only for tumors unresponsive to radiation ± chemotherapy ("salvage surgery").

SURGICAL MANAGEMENT OF THE PRIMARY TUMOR

Surgical extirpation of cancer involves removing the primary lesion with a cuff of normal surrounding tissue, typically in an *en bloc* fashion. The approach is dictated by the location and size of the tumor as well as its involvement of, or proximity to, adjacent structures. For tumors of the oral cavity, early stage lesions can usually be removed transorally. A similar approach can be utilized for select cancers of the oropharynx, hypopharynx, and larynx, often using rigid endoscopes to assist in exposing the tumor. For these lesions, cancer removal is typically accomplished through the use of a laser and an operative microscope (i.e., transoral laser micro-surgery, TLM) or by implementing a surgeon-controlled robotic operating system (i.e., transoral robotic surgery, TORS, see Figure 3). In each instance, the involved or at risk cervical nodal basins must also be addressed through the use of a con-comitant or staged neck dissection.

For more advanced tumors, it is often necessary to employ an open surgical approach. In these instances, a neck incision is typically combined with a lip inci-sion and division of the mandible (mandibulotomy) in order to achieve adequate exposure to the primary lesion and surrounding structures (see Figure 4a and b).

This is of particular importance to large tumors of the pharynx, as the lateral extent of these advanced lesions is often in close proximity to the carotid arterial

(a)   (b)

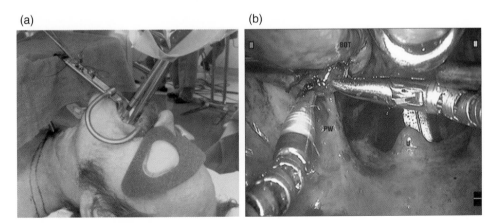

Figure 3. Transoral robotic surgery for cancer of the pharynx: (a) shows the robot with instru-ment ports placed transorally. A bedside assistant helps with counter tension and suctioning and also monitors any pressure on the lips and teeth; (b) shows a transoral view for robotic tumor resection. Here, a grasping instrument is used along with a cautery to remove this left oropharyn-geal tumor with a margin of normal surrounding tissue.

(a)   (b)

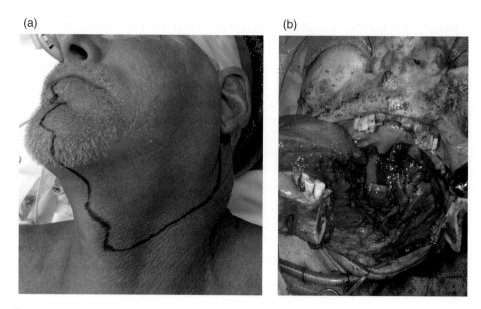

Figure 4. Lip split with mandibulotomy to approach base of tongue and pharynx: (a) shows the view following the initial soft tissue exposure through a lip-split incision. The mandible is pre-plated and subsequently divided with a saw; (b) shows the wide exposure to the oropharynx provided by this approach. Notice the controlled view of the neck vessels in relation to the resection bed of the tumor.

(a)   (b)

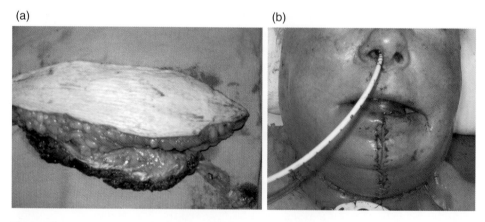

Figure 5. Reconstruction following a transmandibular approach to the pharynx: (a) shows a musculocutaneous anterolateral thigh free flap used to repair the oral and pharyngeal defect. The mandible is then rotated back to its native position, plated, and the soft tissue incisions are closed to yield the final result seen in (b).

system. For such approaches, it then becomes necessary to perform reconstructive surgery in the same setting in order separate the contamination of the aerodigestive tract from the sterile environment of the neck. Figure 5 shows how an anterolateral thigh muscle and skin free flap can be used to repair such a complex defect of the oral cavity and pharynx.

Advances in reconstructive surgery over the past four decades have dramatically increased the ability of surgeons to functionally rehabilitate patients following major tumor resections. Options range from simple to complex and include: healing by secondary intention, primary closure, skin grafting, allografting, local rotational flaps, regional pedicled flaps, and free tissue transfer. Free tissue transfer involves the harvesting of a composite tissue graft from a distant part of the body, designed to be perfused by a single arterial and venous system via microsurgical anastomosis to native blood vessels in the neck. There are a large variety of donor sites available to choose from, allowing the surgeon to select an option that best replaces the tissue and function lost as part of the ablation.

SURGICAL MANAGEMENT OF CERVICAL LYMPHATICS

Resection of the primary tumor is often accompanied by a so-called "neck dissection." This component of the procedure involves removal of the clinically involved and/or at risk cervical nodal basins at the time of the primary surgery (selective neck dissection). The original description of the radical neck dissection by Crile in 1906 involved the removal of all fat and lymph nodes in levels I through V of the neck, along with the ipsilateral sternocleidomastoid muscle, internal jugular vein, and spinal accessory nerve. While oncologically effective, this procedure is associated with significant morbidity to the patient including shoulder dysfunction and pain, swelling of the face and neck, and a cosmetic deformity due to the removed soft tissue bulk. These techniques have been largely abandoned in modern day head and neck surgery, with its use only when these structures are clinically involved with cancer.

NONSURGICAL THERAPY

The primary nonsurgical option for treating head and neck cancer employs external beam RT alone or in conjunction with chemotherapy. This technique has a well-established role in the treatment of head and neck cancer as definitive therapy or as adjuvant to primary surgical treatment. Since this technique, by definition, uses an energy source that intentionally causes tissue damage in an effort to ablate the tumor, adjacent tissue can also be affected, producing both short- and long-term side effects. Over the past 30 years there have been many technological advances in radiation delivery methods, allowing for better targeting and delivery of RT in an effort to minimize toxic effects to surrounding normal tissue.

Probably the most notable advance, to date, has been intensity-modulated radiation therapy (IMRT). This technique allows for better sparing of normal surrounding tissues by modulating the radiation beam in multiple small beamlets, while at the same time adequately overlapping in the area of the tumor volume to optimize the tumoricidal effect. Using this technology, clinicians can determine the treatment

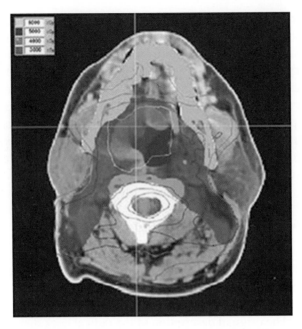

Figure 6. Intensity Modulated Radiation Therapy (IMRT). This axial computed tomograph shows the planned dose administration for this patient's IMRT. High dose areas are in red and purple, while low dose zones are designated in green. (See insert for color representation.)

field three dimensionally by using CT, MRI, and/or PET/CT images obtained just prior to treatment planning (see Figure 6).

As mentioned above, chemotherapy has not been demonstrated to be an effective single modality therapy for head and neck cancer, but has been shown to play an important role in many treatment regimens by acting as a radiosensitizing agent. Concomitant chemotherapy administered at the same time as RT, is the most commonly utilized approach. In this instance, a drug(s) is(are) administered to potentiate the effects of RT in order to achieve improved locoregional control and organ preservation. This strategy has been found to have particular application in treating moderately advanced cancers of the oropharynx, as well as the larynx and hypopharynx [20]. In these instances, concomitant chemoradiation has been found to provide improved locoregional control and laryngectomy-free survival [20, 21]. Platinum-based agents, such as cisplatin and carboplatin, are the most commonly used drugs, typically administered on days 1, 22, and 43 of radiotherapy.

ADJUVANT THERAPY

In some patients with head and neck cancer undergoing up-front surgery, it is necessary to employ radiation or chemoradiation in the postoperative setting. This therapy is referred to as adjuvant therapy and can play a critical role in disease control of patients with advanced disease. Such treatment is recommended to start within six weeks of the date of surgery, unless significant medical or wound related complications occur, in order to minimize the risk of tumor recurrence. Important considerations in planning include the location and extent of the primary tumor as well as associated pathologic lymph nodes. Additional factors include the status of the resection margins, extension to soft tissue/bone, perineural or lympho-vascular invasion at the primary site, size/location/number of involved lymph nodes, and the presence or absence of extra-capsular spread (ECS). Patients with positive surgical margins and/or extracapusular spread seen in their cervical lymph nodes are recommended to have adjuvant chemotherapy added to RT [22, 23].

IMMUNOTHERAPY

Over the past decade, there have been exciting advances in treatment approaches through the implementation of a drug class referred to as immunotherapies. The mechanism of these medications is complex and is based on the fact that many cancers in part affect the host by turning off the local immune response, thus allowing for more unchecked tumor growth and progression. These new medications work to block this inhibition, thus allowing for a more robust immune response against the cancer, at times with dramatic results. Currently, there are two approved immunotherapies for use in head and neck cancer, Pembrolizumab and Nivolumab. Such agents are approved for the treatment of recurrent/metastatic cancers as well as those that have been refractory to platinum-based chemotherapies. Because of their encouraging results, these and other agents are currently the subject of many active clinical trials to better define their role in treatment of head and neck cancers.

SURVIVORSHIP AND SURVEILLANCE

For all head and neck cancer patients, long-term clinical follow-up is necessary for disease surveillance. Patients remain at risk not only for disease recurrence after treatment but also for second primary tumors in the head and neck, esophagus, and lungs. There is no surveillance method that has been proven to be superior. However, a typical approach includes an office visit with history and head and neck physical examination, including fiberoptic laryngoscopy every two to three months during the first year, every three to six months during the second year, and every four to eight months thereafter until the fifth year. Beyond that an annual surveillance is typically adequate. Development of concerning symptoms such as progressive pain, dysphagia, voice change, ear pain, weight loss, or the development of any head and neck masses should prompt a visit prior to the scheduled appointment. Baseline imaging (PET/CT, CT, or MRI) is obtained around three months after the completion of treatment. For patients with up-front nonsurgical therapy, it has been shown that a posttreatment PET/CT scan is an effective way of looking for residual disease in cervical lymph nodes [24–26]. For those who continue to smoke, chest imaging should be obtained on an annual basis.

In addition to the head and neck surgeon, visits with other disciplines such as radiation oncology, medical oncology, primary care, dental, and speech/swallow pathology may be indicated. Common long-term side effects of RT to the head and neck include dry mouth, altered taste, neck stiffness due to fibrosis, and swelling/lymphedema. Swallow function should be assessed prior to, during, and after completion of treatment and may require long-term therapy due to the associated fibrosis and dysphagia. Thyroid function should be assessed at least annually by checking a blood thyroid stimulating hormone level. Dry mouth resulting from alteration of mucosal minor salivary gland function is managed with salivary replacements and frequent hydration. Smoking and alcohol cessation counseling and interventions should be continued if not previously successful. Meticulous dental care and surveillance is also necessary due to the increased risk of dental caries resulting from radiation xerostomia. Moreover, dental extractions should be avoided following radiation, when possible, in an effort to minimize the risk of osteoradionecrosis.

SUMMARY

While head and neck cancer overall represents roughly only 6% of all malignancies, the diagnosis and treatment of afflicted patients remains a challenge for healthcare providers. The vast majority of these tumors are related to exposure to carcinogens such as tobacco and alcohol. However, the HPV has emerged as an independent risk factor in certain subsites such as the base of tongue and tonsil. Disease prevention centers around the avoidance of carcinogens, safe sex practices, and, when appropriate, HPV vaccination.

Diagnostic challenges result from the inherent difficulty of examining the upper aerodigestive tract in a primary care setting. Delays in diagnosis of head and neck cancer contribute to the advanced tumor staging upon presentation, which in turn creates difficult decisions to be made regarding treatment strategies. The generally accepted principle is to utilize single-modality therapy in the form of surgery or radiation for early stage cancers, while reserving multimodality therapy for advanced stage disease.

While great debate has existed over the years regarding the appropriate therapeutic modality in individual cases, the overarching theme has been to maximize both treatment response and disease-free survival while preserving a functional upper aerodigestive tract with respect to respiration, phonation, and deglutition. This principle has been the impetus for advances in the development of effective chemotherapeutic drugs and radiation delivery strategies for possible organ preservation. Likewise, the desire to improve the functional status of patients undergoing surgery has resulted in an increased interest in advanced microsurgical reconstructive options for head and neck cancer surgeons.

Successfully caring for the head and neck cancer patient is truly a unique challenge and requires the coordinated efforts of primary care providers, head and neck oncologic and reconstructive surgeons, medical oncologists, radiation oncologists, pathologists, radiologists, dentistry, and speech pathologists. The importance of the role of each member of the multidisciplinary team is not to be understated in our goal to provide these patients with the best possible care.

REFERENCES

1. Shah, J.P. (1990). Patterns of cervical lymph node metastasis from squamous carcinomas of the upper aerodigestive tract. *Am. J. Surg.* 160: 405–409.
2. Lisa A. Orloff, Maisie L. Shindo. Normal Cervical Lymph Node Appearance and Anatomic Landmarks in Neck Ultrasound. (2017)
3. Slaughter, D.P., Southwick, H.W., and Smejkal, W. (1953). Field cancerization in oral stratified squamous epithelium; clinical implications of multicentric origin. *Cancer* 6: 963–968.
4. Califano, J. et al. (1996). Genetic progression model for head and neck cancer: implications for field cancerization. *Cancer Res.* 56: 2488–2492.
5. Haughey, B.H., Gates, G.A., Arfken, C.L., and Harvey, J. (1992). Meta-analysis of second malignant tumors in head and neck cancer: the case for an endoscopic screening protocol. *Ann. Otol. Rhinol. Laryngol.* 101: 105–112.
6. Mehanna, H., Paleri, V., West, C.M., and Nutting, C. (2010). Head and neck cancer–part 1: epidemiology, presentation, and prevention. *BMJ* 341: c4684.
7. Sturgis, E.M. and Cinciripini, P.M. (2007). Trends in head and neck cancer incidence in relation to smoking prevalence: an emerging epidemic of human papillomavirus-associated cancers? *Cancer* 110: 1429–1435.
8. Zandberg, D.P., Bhargava, R., Badin, S., and Cullen, K.J. (2013). The role of human papillomavirus in nongenital cancers. *CA Cancer J. Clin.* 63: 57–81.
9. Chaturvedi, A.K., Engels, E.A., Anderson, W.F., and Gillison, M.L. (2008). Incidence trends for human papillomavirus-related and -unrelated oral squamous cell carcinomas in the United States. *J. Clin. Oncol.* 26: 612–619.
10. Ang, K.K. et al. (2010). Human papillomavirus and survival of patients with oropharyngeal cancer. *N. Engl. J. Med.* 363: 24–35.
11. Marur, S., D, Souza, G., Westra, W.H., and Forastiere, A.A. (2010). HPV-associated head and neck cancer: a virus-related cancer epidemic. *Lancet Oncol.* 11: 781–789.
12. O, Rorke, M.A. et al. (2012). Human papillomavirus related head and neck cancer survival: a systematic review and meta-analysis. *Oral Oncol.* 48: 1191–1201.
13. D, Souza, G., Agrawal, Y., Halpern, J. et al. (2009). Oral sexual behaviors associated with prevalent oral human papillomavirus infection. *J. Infect. Dis.* 199: 1263–1269.
14. Gillison, M.L. (2004). Human papillomavirus-associated head and neck cancer is a distinct epidemiologic, clinical, and molecular entity. *Semin. Oncol.* 31: 744–754.
15. Markowitz, L.E. et al. (2014). Human papillomavirus vaccination: recommendations of the advisory committee on immunization practices (ACIP). *MMWR Recomm. Rep.* 63: 1–30.
16. McIlwain, W.R., Sood, A.J., Nguyen, S.A., and Day, T.A. (2014). Initial symptoms in patients with HPV-positive and HPV-negative oropharyngeal cancer. *JAMA Otolaryngol. Head Neck Surg.* 140: 441–447.

17. Caldemeyer, K.S., Mathews, V.P., Righi, P.D., and Smith, R.R. (1998). Imaging features and clinical significance of perineural spread or extension of head and neck tumors. *Radiographics* 18: 97–110; quiz 147.

18. Fogh, S.E. et al. (2012). Value of fluoro-2-deoxy-D-glucose-positron emission tomography for detecting metastatic lesions in head and neck cancer. *Am. J. Clin. Oncol.* 35: 311–315.

19. Amin, M., Edge, S., Greene, F.L. et al. (eds.) (2017). *AJCC Cancer Staging Manual*, 8e. New York: Springer Nature.

20. Forastiere, A.A. et al. (2003). Concurrent chemotherapy and radiotherapy for organ preservation in advanced laryngeal cancer. *N. Engl. J. Med.* 349: 2091–2098.

21. Forastiere, A.A. et al. (2013). Long-term results of RTOG 91-11: a comparison of three nonsurgical treatment strategies to preserve the larynx in patients with locally advanced larynx cancer. *J. Clin. Oncol.* 31: 845–852.

22. Bernier, J. et al. (2004). Postoperative irradiation with or without concomitant chemotherapy for locally advanced head and neck cancer. *N. Engl. J. Med.* 350: 1945–1952.

23. Cooper, J.S. et al. (2004). Postoperative concurrent radiotherapy and chemotherapy for high-risk squamous-cell carcinoma of the head and neck. *N. Engl. J. Med.* 350: 1937–1944.

24. Nayak, J.V. et al. (2007). Deferring planned neck dissection following chemoradiation for stage IV head and neck cancer: the utility of PET-CT. *Laryngoscope* 117: 2129–2134.

25. Koshkareva, Y., Branstetter, B.F., Gaughan, J.P., and Ferris, R.L. (2014). Predictive accuracy of first post-treatment PET/CT in HPV-related oropharyngeal squamous cell carcinoma. *Laryngoscope* 124: 1843–1847.

26. Passero, V.A. et al. (2010). Response assessment by combined PET-CT scan versus CT scan alone using RECIST in patients with locally advanced head and neck cancer treated with chemoradiotherapy. *Ann. Oncol.* 21: 2278–2283.

22

BRAIN AND SPINAL CORD TUMORS

Lawrence Recht

Department of Neurology & Clinical Neurosciences, Stanford University School of Medicine, Stanford, CA, USA

INTRODUCTION

Cancers of the central nervous system(CNS; involving either brain or spinal cord) are among the most challenging neoplasms in terms of clinical decision-making. First, they arise within a critical, highly complex environment wherein small perturbations can result in a major functional impact. Second, their location within a closed structure (i.e., the skull and spinal canal) results in early dysfunction due to mass effect. Finally, the existence of both a specialized cerebrospinal fluid (CSF) circulation as well as a physiological barrier between the systemic circulation and neural tissue (i.e., the blood brain barrier, BBB) complicates treatment decisions.

Because few absolutes exist concerning treatment for this diverse set of cancers, patients and their families often are quickly overwhelmed and frequently voice frustration about not having either the time or the resources to make critical decisions. This chapter will provide the necessary foundation from which to make such decisions. I will do this in three parts. First, I will offer a brief primer on the CNS, especially how it pertains to brain and spinal cord tumors. Second, I will describe the diagnostic and therapeutic toolkit, since in effect, no matter how complex the nosology and molecular biologic subtyping is becoming, there exist only a few

Cancer: Prevention, Early Detection, Treatment and Recovery, Second Edition. Edited by Gary S. Stein and Kimberly P. Luebbers.
© 2019 John Wiley & Sons, Inc. Published 2019 by John Wiley & Sons, Inc.

reliable treatments. Third, with this background, I will discuss the more important categories of these cancers, both those that arise primarily within the brain and spinal cord as well as those that spread (metastasize) to the CNS.

AN ONCOLOGIC PRIMER ON THE CNS

While space does not permit an exhaustive or even a general outline of CNS functions, there are certain general properties that need to be addressed to better understand the unique issues that arise when tumors develop within this crucial organ.

The CNS is a highly complex structure that is composed of two major neural subtypes: the *neurons*, which generally transmit chemical information at synapses to other neurons after electrical activation and the *glia*, which are highly specialized cells that maintain the highly controlled milieu that enables neuronal function. Glial cells can be subclassified into *astrocytes*, which stabilize the environment and help nourish neurons and *oligodendrocytes*, which synthesize the myelin that enables quicker transmission impulses between neurons. Smaller more specialized classes of glia also exist, the most important one from a neuro-oncologic standpoint being the *ependymal cells* that insulate brain tissue from the CSF.

Primary brain tumors often resemble one or more of these cell types and the pathologic nomenclature (which is still used) starts with their cellular appearance, i.e., astrocytoma, oligodendroglioma, ependymoma, etc. This convention persists even though it is known that neurons and glia arise from common progenitor cells and that intermediate forms of these cell types have been identified.

At the organ level, the brain and spinal cord are organized into regions that are differentially vulnerable to injury and manifest very different levels of dysfunction when injured. This results in the location of a tumor being a major determinant of morbidity and treatment options. For example, tumors that arise in the frontal cortex are amenable to large resections and often grow to large sizes before producing symptoms while tumors located on the motor strip more posteriorly are extremely difficult to remove without causing permanent paralysis.

The brain also possesses a unique circulatory system that is designed to ensure a stable environment so as to maintain cellular function. For instance, a barrier exists between the arterial circulation and the brain environment that restricts movement of molecules based on their polarity and size (i.e., the BBB); this barrier contains specialized receptors (multidrug resistance receptors or MDR) that actively extrude many chemotherapeutic agents. Furthermore, to provide a cushion for the brain as well to provide access to a number of important molecules, another circulatory system exists within the CNS in which CSF, produced within the lateral ventricles via an active filtration of blood within in the *choroid plexus*, bathes the CNS via the ventricular system. Tumors frequently are located in areas that can block these pathways early; such interruption of flow results in *hydrocephalus*.

Finally, the last important principle with specialized oncologic ramifications concerns mass effect. The CNS, being enclosed within a rigid skull and spinal canal, is very vulnerable to even slight changes in mass. Under normal circumstances, brain volume represents the sum of the brain cells and extracellular space in addition to what is present in the vasculature and CSF. This creates a baseline pressure that represents an optimum between arterial and venous pressure (measured as the CSF pressure) and ensures optimal tissue function. There are compensatory mechanisms wherein increasing mass can be compensated by decreasing one or more of these normal components so that the volume remains stable and the CSF pressure does not change; this property is called *compliance*. As volume increases, however, the compensatory mechanisms become overwhelmed with a resultant rise in CSF pressure that at least temporarily prevents disastrous movement of tissue components. This pressure increase generally results in clinical manifestations of headache and cognitive impairment and results in swelling of the optic nerve head, which can be observed on examination (i.e., *papilledema*). If left untreated, compensatory mechanisms all eventually fail resulting in movement of tissue from a high to lower pressure compartment (i.e., herniation), which can have lethal consequences and constitutes a medical emergency. Mass effect and herniation was once a common early cause of death in brain tumor patients; current treatments (discussed later) have for the most part prevented this cause of early death.

THE NEURO-ONCOLOGIC DIAGNOSTIC AND TREATMENT TOOLKIT

The clinical approach to a new CNS lesion consists of stabilization of symptoms, establishment of a diagnosis, and then a specific therapeutic plan directed toward the tumor. In real time, these approaches often overlap and a specific approach may have value for multiple steps (such as bevacizumab which can function as both an important symptom stabilizer in addition to a specific antitumor agent). In this section, therefore, I will introduce the reader to some of the more important diagnostic and therapeutic modalities before proceeding to a discussion of specific tumor types.

Diagnostic Modalities

Imaging

Magnetic resonance (MR) imaging is the single best diagnostic test available for detecting and following CNS tumors. Considering its general overall safety, including lack of radiation exposure, it is the preferred imaging modality in all but those patients who have in place therapeutic metal hardware (i.e., pacemakers, defibrillators) that can be magnetized. It functions by rapid, repeated magnetization and relaxation of molecules with odd numbers of protons, the most common of which by far is hydrogen. Thus, for most imaging sequences, the high-quality resolution is based on differences primarily in water content. For the purposes of detecting and following brain and spinal cord tumors, contrast agents containing paramagnetic materials such as gadolinium are used to detect breaches in the BBB (through which they ordinarily cannot pass due to their polarity).

MR imaging enables the clinician to detect mass lesions with very high sensitivity, although specificity is rather low. Thus, although it can detect essentially every tumor, it is often difficult to distinguish the type of tumor or even if the lesion is neoplastic. Furthermore, it is often hard to be certain that changes occurring during the clinical course after a tumor are diagnosed are due to tumor growth or treatment effect. Specialists have proposed criteria with which these distinctions can be reliably made [1], but they still rely heavily on clinical judgment.

While MR imaging remains the most useful diagnostic tool in neuro-oncology, other ancillary tests can provide ancillary information and are frequently utilized. These include both positron emission tomography (PET) using flurodeoxyglucose (FDG) and magnetic resonance spectroscopy (MRS), which have both been advanced as being useful in distinguishing whether an MR lesion is neoplastic in addition to its aggressiveness.

Pathologic Analysis

Examination of tissue obtained at surgery is the gold standard in terms of establishing the diagnosis, assessing prognosis, and guiding specific treatment decisions.

Although most CNS cancers can be detected with standard histologic methods, difficulties can arise in certain situations, especially when there are only small samples for analysis (i.e., when resections are not possible). For example, it is sometimes difficult for the pathologist to distinguish slow growing tumors from reactive gliosis (which results from non-neoplastic processes). In such cases, pathologists can assess tissue for the presence of mutant p53 or isocitrate dehydrogenase (IDH1 and 2). Alternatively, it sometimes can be challenging to distinguish lymphoma from an inflammatory process. In this instance, the key to establishing a cancer diagnosis is establishing clonality (i.e., inflammatory lesions consist of mixtures of T and B lymphocytes whereas tumors are composed predominantly or exclusively of only one type); this distinction can be made using either histochemical stains or flow cytometry.

Once neoplasm is confirmed, the next step is to characterize tumor type as either primary (arising within the brain or its adjacent structures) or metastatic (arising due to spread from another type of cancer). This is generally possible using standard histochemical techniques. However, while distinguishing primary from metastatic tumor is generally straightforward, it can often be very challenging pinpointing the site of origin for metastatic tumors when a prior history of cancer is not present. In these cases, the pathologist is helped by the availability of numerous markers that are highly correlated with the organ of origin.

Primary CNS tumors are characterized by the predominant cell type present. This convention dates back to the classic studies of Bailey and Cushing, in the 1920s [2], who viewed brain tumors as aberrations of development and developed a nomenclature that persists for the most part to the present, forming the foundation of the World Health Organization (WHO) classification system. Once the cell type is determined, primary tumors are then "graded" according to their aggressiveness. This grading depends on both the cell's morphology as well as the tissue architecture. Regarding the latter, the two most important features in terms of determining prognosis are *necrosis* and *vascular proliferation*; when both are present tumors are considered of the highest grade.

Once a tumor type and grade is determined, there are frequently a number of other tests that can be performed to help the clinician gauge prognosis and more importantly, treatment options. At the current time, however, only a few are particularly helpful and only three will be discussed here: 1p/19q deletion analysis, methyl guanine methyltransferase MGMT promoter methylation, and IDH mutation.

The presence of codeletions (or more accurately a translocation) on 1p and 19q characterizes the oligodendroglioma [3]. Their presence is important because it conveys in general a better prognosis compared to patients with other astrocytic tumors as well as being better treatment responsiveness [4, 5]. At the current time, the prevailing opinion is that unless a tumor contains these deletions, they should not be considered oligodendrogliomas [6, 7].

Methylation of the MGMT gene promoter results in the cessation of transcription of this important enzyme that repairs breaks caused by alkylating agents such as temozolomide and CCNU, the two most commonly used chemotherapeutic agents in primary tumors [8–10]. The presence of methylation of this gene can be detected using PCR and when methylation is noted correlates with better responsiveness to alkylating agents and better prognoses for patients with malignant gliomas [5, 11, 12].

The isocitrate dehydrogenase enzyme (IDH) mutation was identified as being associated with secondary glioblastomas in the Cancer Genome Project [13]. Subsequent studies have demonstrated that the presence of a mutation is a powerful predictor of both better prognosis and is correlated with MGMT promoter methylation and 1p/19q codeletion status [14, 15]. Since 90% of mutations can be detected using standard immunohistochemical techniques, assessment of this marker is being incorporated in the standard pathological assessment of glioma patients in many centers throughout the world.

While other cancer markers such as EGFR amplification are often assessed in selected pathologic laboratories, they add little to diagnostic and therapeutic decision-making. In addition, similar to other cancer types, there is currently much interest in performing comprehensive molecular analyses so as to develop personalized targeted therapies. At the current time, however, there is no evidence that basing therapy on the results of these analyses improves outcome.

Therapeutic Modalities

Surgery

Surgery plays a crucial role both in diagnosis and treatment of brain tumors. While biopsy establishes the diagnosis and can be performed with little risk, it is clear that outcome improves with greater surgical extent, even with infiltrative tumors such as gliomas [16, 17]. Furthermore, larger resections provide immediate relief of mass effect, often enabling patients to be weaned from steroids quickly. Therefore, the prevailing opinion is that an attempt at removing as much tumor as safely possible should be the first step in tumor treatment.

Several technological advances, including awake mapping, MR guided surgical platforms, and functional MR imaging have enabled surgeons to remove greater amounts of tumor tissue without increasing morbidity, even in brain areas associated with the eloquent cortex. While complete removal is ideal, it is generally agreed that this should not be done at the cost of functional impairment.

Radiation Therapy

Many cancers that involve brain and spinal cord are treated with radiation therapy, which is probably the single most effective agent for most tumors. As is described elsewhere in this book, radiation involves treating tumor fields with ionizing particles that produce tissue damage, including breakage of DNA. Such particles are usually administered via external sources although there have been many studies over the years assessing local implantation of heavier particles (i.e., brachytherapy).

Because tumor cells have less capacity to repair DNA breaks, treatment is usually divided into fractions (i.e., *fractionated*) so that normal cells would have a chance to repair while dose levels build up in tumors. This generally results in the maximally tolerated radiation dose being built up over several weeks of treatment. Recently however, the development of stereotactic radiosurgery has improved targeting so that higher doses can be administered with less normal tissue exposure. While such radiation dosing schemes have not yet replaced standard fractionation, these techniques are being used more frequently in brain and spinal cord tumors in order to minimize treatment time and improve dosing to tumor relative to normal tissues.

Based on the premise that tumor hypoxia within tumors impairs radiation effectiveness (which depends on creating oxygen radicals), several radiosensitizers such as misonidazole have been evaluated over past decades. To date, however, none of them have proved efficacious.

Chemotherapy

As mentioned earlier, delivering drugs to CNS tumors is more difficult compared to other organs because of the presence of the BBB that excludes many agents based on polarity and size, usually via MDRs and increased oncostatic pressure [18].

This restricts the number of agents that can be used in CNS tumors compared to what is generally available in systemic cancers.

Two of the most frequently used chemotherapeutic agents currently used by neuro-oncologists are temozolomide and nitrosoureas such as lomustine (CCNU). Since they are both small, lipophilic molecules, they readily penetrate the brain and brain tumors and have the advantage of being administered orally. Temozolomide is preferred because of its more favorable toxicity profile, especially in terms of marrow sparing, and forms a major component of the standard framework for glioblastoma treatment.

Both temozolomide and CCNU work more effectively when the MGMT promoter methylation is present. However, even when the promoter methylation is not present, the absence of other available effective agents results in their being used universally in these patients.

Although combinations of chemotherapy regimens are often utilized in other cancers, it is interesting that there has never been a clinical study supporting the use of more than one chemotherapeutic agent at a time in glioma, the most common primary CNS tumor. Nevertheless, the combination of CCNU, procarbazine, and vincristine (PCV) is often used by neuro-oncologists; its use being based on studies dating back to the 1980s, studies that empirically assessed its efficacy versus anaplastic oligodendrogliomas [19, 20].

While chemotherapy is not particularly effective versus the most common primary brain tumors, it is efficacious and potentially curative in CNS lymphoma (see below), where administration of high doses of methotrexate, with or without other agents, can result in cure without using irradiation. Unlike the situation in glioma, however, it is fairly well accepted that similar to the situation in systemic lymphoma, combination therapy with more than one agent results in better efficacy [21].

Targeted Agents and Immunotherapies

Considering the mediocre results achieved with conventional chemotherapeutic approaches in most brain tumors, much research has been directed toward designing agents that target specific pathways that either promote tumor growth or enhance immune responsiveness. With the discovery that there are multiple potential pathways that can be differentially activated, this concept closely aligns with a parallel interest in elucidating vulnerable pathways in a particular patient, i.e., *personalized* cancer treatment.

The most effective targeted agent in neuro-oncology is arguably bevacizumab, a monoclonal antibody directed against vascular derived growth factor (VEGF), a very important growth factor involved in the stimulation of tumor blood vessel development. Although very specific in the sense that it only binds one molecule (i.e., VEGF), this agent has two important applications in brain tumor treatment. The first concerns its "antiangiogenic" effect and closely parallels a general oncologic treatment goal, i.e., the inhibition of vessel formation. Its consistent ability to produce a significant albeit transient response in recurrent malignant gliomas led to its provisional approval by the FDA in 2009 [22, 23] for this application. In addition, however, linked with the fact that new tumor vessels are "leaky," bevacizumab has a profound effect on tumor swelling (edema), rendering it the first true "steroid sparing" agent for chronic peritumoral edema.

Another important target is the epidermal growth factor receptor (EGFR) for which there exist a number of targeted agents, the most commonly utilized in brain tumor being erlotinib. While there was much enthusiasm for this agent as recently as a decade ago, the clinical results in glioblastoma have been disappointing in contrast to the experience in lung cancer [24]. This probably relates to the fact that the specific mutation that conveys sensitivity in lung cancers is not commonly encountered in brain.

Other targeted therapies are limited to specific situations, outside trials. Everolimus for example, which exerts its actions via the mTOR axis, is very effective at shrinking the subependymal giant cell astrocytoma (SEGA) that arises in the setting of the familial disease, tuberous sclerosis [25], although its impact on other brain tumors is much less dramatic. Rituxumab, a monoclonal antibody targeting B-cells, improves treatment of CNS lymphomas when added to cytoreductive treatments. Recently, reports have started appearing indicating that patients with unusual variants of gliomas such as pleomorphic xanthoastrocytoma and epithelioid glioblastoma containing specific a specific mutation in the Braf inhibitor may respond to targeted agents such as vemurafenib [26].

It is also important to note that recent experience indicates that systemic cancers that are responsive to targeted agents (such as EGFR mutant lung cancers and Braf mutant melanomas) often remain sensitive to these agents when they metastasize to the brain. This often results in brisk responses of CNS metastases when used as single agents, sometimes even obviating the need to radiation.

Using drugs to alter the immune system so as to create an unfavorable environment for tumor growth is another way to specifically target tumors. Building on decades of work, there has been recently much interest in this approach, which can be subdivided into two general approaches. The first approach involves the induction of antitumor immune responses by promoting the development of cells specifically targeting novel antigens on tumor cells. This can be accomplished by vaccine approaches that either induce immunity to a specific epitope, such as the truncated EGFrviii that is expressed by approximately 30% of GBMs (glioblastoma multiforme) but not on normal cells or by introducing a patient's immune cells to an extract of tumor antigens isolated from the resected tumor. Studies are currently underway examining both of these approaches [27, 28].

An alternative approach is to administer therapies that block tumor-induced immunotolerance (checkpoint inhibitors), a complex process in which tumors evade immune surveillance. This approach has been very effective in melanoma wherein the administration of the anti-CTLA4 agent, ipilumamab, can have profound antitumor effects. Such approaches are currently being assessed in brain tumors in numerous clinical trials. The fact that such agents are often associated with brisk, inflammatory responses will make it difficult to use in brain tumors, however, where, as mentioned above, small increases in mass effect can result in profound neurological dysfunction.

Supportive Therapies

Although not in the strict sense anticancer agents, the optimal utilization of treatments for both mass effect and seizures are important for maintaining quality and even quantity of life for the neuro-oncologic patient.

Steroids have been until recently the only effective method for treating peritumoral edema, a major component of mass effect. The most commonly used steroid, dexamethasone, is a fluorinated corticosteroid that has been available since the late 1950s and is generally preferred by most clinicians treating brain tumors because of the impression that it causes less severe side effects. Although its mechanism of action in reducing edema remains unknown, it is especially effective in decreasing swelling that results from extracellular water that accumulates because of leaky vessels (*vasogenic edema*). This effect requires doses far beyond that which saturates steroid receptors, however, which requires patients to be treated long term with supraphysiological doses. Although they are well tolerated for short periods of time, multiple side effects, including immunosuppression, skin friability, myopathy, and diabetes, arise after as little as two weeks of use and can be very disabling, resulting in the very frustrating scenario of patients being more disabled from the ravages of steroids than their brain tumors.

Despite an obvious need and decades of research, an effective steroid substitute has remained elusive. Recently, however, and quite serendipitously, it has been noted that bevacizumab is just as, if not more, effective at treating tumor edema [29].

Brain tumors also are frequently accompanied by seizures, which can be the most significant symptom impacting quality of life. Although there is no one superior anti-seizure agent, certain important guidelines should be adhered to in patients with brain tumors: (i) there is no need to administer them prophylactically (i.e., before seizures are noted); and (ii) although effective anti-epileptics, agents such as phenytoin, carbamazepine, and phenobarbital should not be prescribed as first line agents because of their tendency to induce P450 liver enzymes that interfere with chemotherapy and steroid metabolism; generally, levetiracetam is used because of its excellent side effect profile and no enzyme inducing capacity.

Seizures may sometimes not be well controlled with one or even more antiepileptics. In fact, there is evidence that if one agent does not work, adding more is unlikely to be very effective (while producing added toxicity). In selected cases, therefore, such as the low grade gliomas (see below), refractory seizures are often better treated with specific anti-tumor therapies such as surgery, radiation, or chemotherapy.

A SURVEY OF PRIMARY AND METASTATIC CNS TUMORS

While a complete pathologic description of brain and spinal cord tumors is far beyond the scope of this chapter, there are certain general organizing principles that can help orient the reader. In this section, we will address several categories of tumors according to whether the tumor arises from CNS tissue itself (i.e., *primary* tumors) or has spread from elsewhere (*metastatic* tumors).

Primary Tumors

Etiology/Demographics

Unlike many other cancers, there are few clear risk factors for the development of CNS tumors. The only one that has enough evidence to support a definite linkage is therapeutic radiation [30, 31]. Many other environmental toxins, including pesticides and chemical exposures (and cell phone usage, which I believe is extremely unlikely) have been suggested as causative agents, but the linkage remains tenuous.

CNS tumors are highly prevalent in the phakomatoses, a group of uncommon familial genetic syndromes, including neurofibromatosis (gliomas, acoustic neuromas, meningiomas, schwannomas), von Hippel Lindau disease (hemangioblastoma), and tuberous sclerosis (SEGA). Primary brain tumors, especially gliomas, are also commonly encountered in Li-Fraumeni (p53 mutant) syndrome [32].

Because of their relative low incidence in the general population and the cost of diagnostic testing (i.e., MR), there does not exist a rationale for screening the general population. In patients with familial syndromes and who have been exposed to therapeutic radiation, periodic MRs on a yearly or bi-annual basis can be justified.

Glioma

Although often used in conversations and papers, there is no pathological definition of glioma. Nevertheless, practitioners often use this term to denote some or all of the group of tumors that pathologically contain at least one cellular component of the class of glial cells (see above for which cell types this constitutes).

While there are many reasons for carefully classifying tumors, the ultimate goal lies in informing practice decisions. In this regard, since the brain is sensitive to the long-term effects of treatment, the clinician has to carefully consider how aggressive to be up-front, i.e., if survival is expected to be very long without therapy, it is preferable to either reduce treatment intensity or defer treatment entirely. Unfortunately, as the WHO classification system has evolved, there has been a tendency to subclassify tumors according to subtle differences in cellular composition despite the fact that these finer distinctions complicate clinical decision-making. Furthermore, as biological advances have forced pathologists to add molecular alterations into the classification, awkward situations develop where certain Grade III tumors have much better prognoses than those graded II. Therefore, because even a concise review of the many tumor types is beyond the scope of this chapter (readers who are interested are referred to the last WHO classification monograph [33]), I will present what I consider the four meaningful subdivisions of this class of brain tumors.

Glioblastoma. Glioblastomas (GB), the most common glioma, is the most malignant type. They are defined by the presence of microvascular proliferation and pseudopalisading necrosis (termed secondary structures) rather than a particular cellular morphology or proliferation rate and are highly correlated with the presence of enhancement on MR scanning. They comprise 50% of gliomas and have an annual incidence of approximately 4–5 per 100 000 people in the United States.

Over the years, there has been a trend among pathologists to label a tumor as GB even in the absence of frank necrosis if vascular proliferation is present; this grade inflation creates a problem when trying to compare studies performed in different eras. Thus, patients who were considered anaplastic astrocytomas in the first randomized study proving the value of RT in malignant brain tumor [34] would almost certainly be classified as GB today.

The significance of a GB diagnosis is its aggressiveness; although survivals are increasing, the two-year survival is still significantly less than 50%. Therefore, an up-front aggressive therapeutic approach is justified from the perspective of controlling tumor growth.

The first step in treatment is to remove as much tumor as safely possible after which adjuvant treatment in the form of local maximally tolerated radiation therapy (up to 60 Gy) is administered over six weeks along with daily temozolomide, followed by at least six cycles of temozolomide administered on a 5-days-on–23-days-off schedule [35]. Even with maximal therapy, recurrence rates of 75% per year are noted during the first two years.

Many attempts have been made to subclassify GB into prognostic groups using molecular and biochemical approaches, but they have added little to treatment decision-making. At present, the most useful marker in this regard is the presence of IDH mutation; even when present, however, prognoses are not better enough to change treatment decisions. On the other hand, the most valuable adjunct prognostic biomarker has been the extent of methylation of the promoter element of the DNA repair gene MGMT. As mentioned above, increased methylation effectively inhibits repair of DNA breaks that occur with administration of alkylating agents such as temozolomide. Promoter methylation correlates with improved outcomes when patients are treated with the standard framework of radiation and temozolomide and is highly correlated with the phenomenon of pseudoprogression, which is also correlated with good outcome [11]. Even if the tumor is not promoter methylated, however, patients are still treated within the same framework outside clinical studies. The only general exception to this rule is in the older patient. Although there is no specific cutoff where one becomes old, the short-term risks of irradiation can outweigh the minimal gains in survival after 70 years of age, and there are numerous studies demonstrating that these patients do as well when RT is omitted [36].

Although its precise role remains unclear, the one other agent that has impacted outcome for the GB patient is bevacizumab. Provisionally approved in 2009 in the United States for use in recurrent disease, this well-tolerated monoclonal agent was widely adopted by practitioners because of its dramatic initial effects on clinical status and imaging with minimal side effects. When assessed for its ability to improve survival when incorporated into the standard framework, however, no benefit was noted [37, 38], which has certainly diminished enthusiasm for this agent for academic neuro-oncologists (although it has not diminished its use by clinicians who continue to find it very useful for patients with recurrent disease).

Pilocytic astrocytoma/gangliocytoma/Dysembryoplastic neuroectodermal tumor (DNET). On the other end of the glioma spectrum, pilocytic astrocytomas and DNETs are infrequently occurring tumors with very low growth potentials that present mostly in childhood.

The pilocytic astrocytoma most frequently occur in children, and is the most frequent primary brain tumor in patients under 19 years. It is characterized pathologically by microcystic changes; these cysts can become confluent, producing larger ones that can become quite symptomatic and require urgent surgery to relieve mass effect. They are also distinguished pathologically by the presence of Rosenthal fibers, which is a pathological finding associated in general with slowly evolving processes. They may arise anywhere within the CNS, but most frequently occur in the cerebellum and optic pathway (the latter frequently seen in association with neurofibromatosis).

DNETs and gangliocytomas are tumors characterized by the presence of large, plump neuronal appearing cells (ganglion cells). They arise most frequently in the temporal lobes during childhood and often are associated with seizures.

Because of their slow growth potential and tendency to arise in young patients, surgery is the preferred sole therapeutic modality. Adjuvant therapy in the form of radiation or chemotherapy is generally not indicated, even when removals are incomplete, since growth rates are very slow and 10-year survivals exceed 90% [39].

1p/19q deleted oligodendroglioma. Studies dating back to the mid-1980s indicated that oligodendrogliomas, gliomas that contain cells with a characteristic appearance of immature oligodendrocytes, were more responsive to chemotherapy than other gliomas [4, 5]. Subsequent work over the next several years linked this sensitivity with the presence of an unbalanced translocation of t(1;19)(q10;p10) with loss of the derivative chromosome (termed codeletion of 1p/19q) [3]. While it remains unclear both why this deletion is so strongly correlated with the oligodendroglial phenotype and how it confers treatment sensitivity and better survival, the presence of this biomarker is such a powerful predictor that in my opinion, it justifies being treated as a separate subgroup of gliomas.

Because of the strong correlation with oligodendroglioma, a case can be made for testing only for this deletion (which requires an *in situ* hybridization analysis) when at least some component of oligodendroglioma is present in the pathologic

TABLE 1. List of "The Rest" of gliomas ranked in approximate order of time to transform into glioblastoma multiforme (GBM).

Glioma name
• Anaplastic astrocytoma
• Anaplastic oligoastrocytoma
• Anaplastic ependymoma
• Diffuse astrocytoma
• Oligoastrocytoma
• Gliomatosis cerebri
• Ependymoma
• Glioneuronal tumors
• Astroblastoma
• Angiocentric glioma
• Pleomorphic xanthoastrocytoma
• Subependymal giant cell astrocytoma
• Ganglioglioma
• Neurocytoma
• Subependymoma

specimen. This applies even to tumors containing the characteristic features of the GB, since general experience has indicated that those tumors will behave much more indolently and are best not considered GB.

"The Rest." The remaining group which comprises 25–30% of gliomas, encompasses a wide range of histologies (Table 1) that share two important clinical characteristics: (i) the evolution over time to GB; and (ii) the relative treatment resistance to both radiation and chemotherapy.

Although grouping these various histologies together is far from customary, it clarifies why treatment decisions are difficult for the clinician (and patients). First, although evolution toward GB is common, it can vary widely (from months to years) and there is no evidence that it can be altered with treatment. Considering that these tumors tend to occur earlier in life than the GB (with a median age around 40 compared to 60 for GB) and are often associated with lengthy periods of stability, controversy still exists as to when to administer treatment (i.e., at diagnosis or at a later time when tumor growth becomes apparent). Furthermore, it remains uncertain in these patients what constitutes adequate treatment. Up until recently, it has been customary to first administer either radiation or chemotherapy alone. The recent emerging data from pivotal unpublished RTOG (Radio Therapy Oncology Group) studies, however suggest that combination therapy with both radiation and chemotherapy is more effective than RT alone. Thus, the timing of therapy is crucial to balance maximal tumor control with minimal long-term toxicity.

These tumors occur throughout life and can be located anywhere within the CNS. When they arise in the cerebral cortex, they usually present with seizures, the intractability of which can be the most pressing issue. They generally do not enhance after contrast administration. At the time of imaging diagnosis, surgical decisions are based on accessibility. When located in polar, accessible regions, complete resections can often result in long-term stability. If total or near total resection is not possible, it is recommended that at least a biopsy be performed, if only to confirm the diagnosis of glioma.

Once diagnosis is established, the decision must be made whether to or when to administer further treatment. Criteria have been proposed to help clinicians decide whether to treat immediately based on a number of features including extent of resection, age of the patient, symptoms, degree of anaplasia in tissue sample [40], and although it still remains to be confirmed, emerging biomarkers such as the IDH1 mutant may also prove valuable in decision making [41, 42].

In the near future, it is hoped that treatment issues will be clarified, especially as it relates to timing and intensity. In this group of gliomas, since there is an evolution toward GB that can be very variable, a real key will be in distinguishing those tumors destined to do this quickly. One can make a cogent argument that these patients would do best if treated similarly to those with GB.

CNS Lymphoma

Although relatively uncommon (with an incidence only 10% that of GB), primary CNS lymphomas (PCNSL) are an important subset of primary malignant brain tumors that can arise in both immunocompetent patients and in the setting of chronic immunosuppression.

PCNSLs are intraparenchymal tumors that generally present with signs of generalized CNS dysfunction and one or more enhancing masses on MR. They are distinguished from lymphomas presenting elsewhere in that there is no (or at most minimal) systemic evidence of lymphoma at time of presentation; in lymphoma parlance, this represents stage IE disease.

Diagnosis is generally made at the time of biopsy. Although there is a general impression that steroids should be held prior to tissue diagnosis because of their oncolytic effects on lymphoma cells, this is probably overstated and these agents should not be withheld if necessary to control mass effect [43]. Furthermore, although it has been advocated that no more than a biopsy be performed on these patients; recent findings have suggested that survivals are better in those who undergo resections [44].

The vast majority (95%) is diffuse large B cell lymphomas. At the time of diagnosis, workup including CSF examination, ophthalmological examination, and a search for concomitant systemic disease should be performed.

Unlike the GB, where cures are extremely rare, long-term survivals in these patients are possible for approximately 25% of these patients with intensive treatment with high doses of methotrexate (at least 3 g/M^2) and another chemotherapeutic (several agents have been used, most frequently cytarabine and ifosfamide), often followed by autologous stem cell transplantation [45]. Whole brain radiation therapy (WBRT) was once an established part of the treatment regimen, but recent data supports withholding this in the upfront treatment [21].

Although most CNS lymphomas occur in immunocompetent patients at the present time, they also arise in the immunosuppressed patient. This was a much more pressing problem in the late 1980s in the AIDS population, but has since virtually disappeared in the United States since the development of HAART (highly active antiretroviral therapy) regimens. Currently, it occurs mainly in transplant patients as part of the post-transplant lymphoproliferative disorder spectrum (PTLD). Treatment in these patients is particularly complicated since there is often a need to maintain immunosuppression for transplant viability.

Meningioma

Meningiomas are probably the most common primary brain tumor, although many are diagnosed serendipitously and require no treatment. Cytologically, they resemble arachnoid cells are almost always attached to a dural (i.e., the covering surface of the brain) element.

Most meningiomas grow slowly and do not invade into brain tissue or metastasize (Grade I). However, they can demonstrate more rapid growth rates, which can be visualized pathologically based on higher mitotic indices (Grade II) or sarcomatous elements (Grade III). Irrespective of grade, they can also produce major morbidity by dint of location.

Treatment is primarily surgical; complete removals are generally associated with long progression free survivals and cure, even in Grade II meningiomas. Radiation therapy is often used to control growth when resection is not complete or when growth recurs. Chemotherapy is only occasionally effective and is not recommended as frontline treatment. Because of their robust expression of various growth factor receptors, including EGFR, estrogen, and progesterone receptors and somatostatin receptors, agents targeting these sites have been explored but have not demonstrated efficacy.

Medulloblastoma

Another group of primary brain tumors assume the appearance of very primitive neuronal precursors and are classified as primitive neuroectodermal tumors (PNET). They are uncommon tumors with an annual incidence in the range of one to two per million that occur primarily in younger age groups. Although they can arise anywhere in the CNS, they most commonly arise in the cerebellum, where they are called *medulloblastomas*.

Medulloblastomas are very aggressive neoplasms that often present with signs of increased intracranial pressure (which usually produces a scenario of a young child who is frequently waking up with headaches or nausea and vomiting). Approximately 10% of patients present with disseminated disease (most frequently in the CSF pathways but systemic metastases to bone and other sites can occur). Careful staging is therefore necessary after diagnosis and tumors are classified into standard and high risk depending on both extent of residual disease and dissemination. While medulloblastomas are classified into five categories according to histologic and molecular characteristics, this has little impact on clinical decision-making.

The first step in treatment is radical excision, since the best survivals are associated with the least residual disease. Although very malignant neoplasms, five-year survival rates exceed 75% for average risk patients with craniospinal irradiation. Radiation dose is very important in producing cures [46]. Because of the attendant morbidity with high dosages, especially in children, lower doses that are combined with chemotherapy generally provide equivalent results [47]. High risk patients are generally treated with a combination of intensive chemotherapy and full dose radiation; even with this intensive treatment, however, five-year survival rates are only in the range of 20 to 40%.

CNS Involvement by Systemic Cancer

Numerically, the number of patients who develop CNS tumors due to spread from another site (i.e., metastasize) is at least an order of magnitude greater than those who develop primary ones. Furthermore, although these tumors frequently occur late in the disease course, they are also a common early or even presenting manifestation.

Brain Metastases

Metastases to the brain represent a common manifestation of cancer that can occur in 5–10% of patients during the course of their cancer. It occurs most often in the setting of lung and breast cancer because these are the most common cancers. However, any malignancy can be a source of brain metastases and it is a very frequent occurrence in some of the less common cancers, such as melanoma and choriocarcinoma. Brain MR is therefore a required screening test in the workups of several cancers including certain forms of lung cancer, melanoma, and hematopoietic cancers.

Most brain metastasis result from hematogenous spread of the cancer in which emboli lodge in end vessels; thus, they are most frequently located at the junction of gray and white matter. Although they characteristically produce much more swelling relative to primary brain tumors, it is often hard to distinguish them from primary malignant brain tumors on MR. Since the lung is generally the most likely primary source of these metastases, either as a primary or a site of additional metastases, chest CTs are the most important screening test [48] in these situations. Nevertheless, even when comprehensive workups are launched, the primary site often cannot be detected. At this point, a brain biopsy or resection becomes necessary.

Even though eradication of brain metastases does not guarantee improved survival, optimal treatment remains an important therapeutic issue in oncology because of the significant quality of life issues caused by neurological morbidity.

Controversy exists as to the best management, especially how to balance surgery and radiation therapy. Early work, dating back to the 1970s, established whole brain irradiation as an effective treatment [49], albeit with cognitive consequences [50]. With the establishment in the 1990s of stereotactic radiosurgery as an effective way to deliver higher dosages to tumor with relative sparing of normal tissue, there has therefore been a trend toward sparing normal brain and irradiating only tumor sites in order to decrease neurological toxicities, primarily in the cognitive spheres [51]. Recent studies suggest that hippocampal sparing whole brain irradiation may also result in less cognitive impairment and a trial to assess this modality is currently underway.

Brain metastases may respond to chemotherapy, especially in cases where the tumor is known to be sensitive (such as lymphomas). Nevertheless, this modality is only occasionally used for brain metastases, most often when irradiation fails. On the other hand, targeted therapies that work on systemic tumors also work on metastases. Thus, numerous reports exist of lung cancers with mutated EGFR and ALK responding to erlotinib and crizotinib, respectively, and similar results have been found in melanoma as well.

Leptomeningeal Cancer

Leptomeningeal cancer (LMD) describes a condition where cancers metastasize primarily to the CSF pathways. Tumor cells are generally embedded just beneath the ependymal cell layer; symptoms are produced by altering CSF flow dynamics or impinging nerves as they exit the brain.

While it can be associated with brain metastases, there are several characteristics of LMD that distinguish it from ICMs. First, the route into the CNS is probably different. Thus, unlike ICMs, which invade via the blood supply, LM results from either local spread from brain, bone, bone marrow, or via nerve fascicles. In leukemias, spread into the CNS occurs early via overflow from the bone marrow. In contrast, LMD is generally a late manifestation of solid cancers, which tend to involve bone and bone marrow late in their course.

Although LMD is generally associated with very poor outcomes in solid tumors, it has different implications in lymphocytic leukemias and lymphomas, where seeding frequently occurs early. Depending on the type of leukemia or aggressive lymphoma, early spread into the CNS can be present at time of diagnosis and failure of initial treatment can occur if CNS disease is not detected and treated. Therefore, spinal fluid examination is part of staging and prophylactic therapy, generally in the form of intracompartmental chemotherapy, is often used.

Diagnosis is generally made on CSF examination, although if suspicion is high, a diagnosis of LMD can be made on imaging and clinical grounds. Treatment consists of either craniospinal irradiation or chemotherapy injected directly in the CSF, either via lumbar puncture or an inserted Ommaya reservoir. Leukemias and lymphomas are much more responsive to conventional intrathecal agents, i.e., methotrexate or cytosine arabinoside, that can be placed in this space and is preferred to irradiation due to the lower long-term morbidity.

Solid cancers, where LMD occurs late in the disease course, are generally insensitive to methotrexate and cytosine arabinoside (breast cancer being the one exception). Treatment options are therefore limited and survivals generally short once diagnosis is made. I generally recommend focal irradiation be administered for symptomatic control with systemic treatment to try to control the remainder of the disease.

REFERENCES

1. Wen, P.Y., Macdonald, D., Reardon, D.A. et al. (2010). Updated response assessment criteria for high-grade gliomas: response assessment in neuro-oncology working group. *Journal of Clinical Oncology* 28: 1963–1972.
2. Bailey, P. and Cushing, H. (1926). *A Classification of Tumors of the Glioma Group*. Philadelphia, PA: Lippincott.
3. Jenkins, R.B., Blair, H., Ballman, K.V. et al. (2006). A t(1;19)(q10;p10) mediates the combined deletions of 1p and 19q and predicts a better prognosis of patients with oligodendroglioma. *Cancer Research* 66 (20): 9852–9861.
4. Kaloshi, G., Benouaich-Amiel, A., Diakite, F. et al. (2007). Temozolomide for low-grade gliomas: predictive impact of 1p/19q loss on response and outcome. *Neurology* 68 (21): 1831–1836.
5. Levin, N., Lavon, I., Zelikovitsh, B. et al. (2006). Progressive low-grade oligodendrogliomas: response to temozolomide and correlation between genetic profile and O6-methylguanine DNA methyltransferase protein expression. *Cancer* 106 (8): 1759–1765.
6. Laigle-Donadey, F., Martin-Duverneuil, N., Lejeune, J. et al. (2004). Correlations between molecular profile and radiologic pattern in oligodendroglial tumors. *Neurology* 63 (12): 2360–2362.
7. Van Den Bent, M.J., Looijenga, L.H., Langenberg, K. et al. (2003). Chromosomal anomalies in oligodendroglial tumors are correlated with clinical features. *Cancer* 97 (5): 1276–1284.
8. Esteller, M., Garcia-Foncillas, J., Andion, E. et al. (2000). Inactivation of the DNA-repair gene *MGMT* and the clinical response of gliomas to alkylating agents. *New England Journal of Medicine* 343 (19): 1350–1354.
9. Jaeckle, K.A., Eyre, H.J., Townsend, J.J. et al. (1998). Correlation of tumor O^6 methylguanine-DNA methyltransferase levels with survival of malignant astrocytoma patients treated with bis-chloroethylnitrosourea: a southwest oncology group study. *Journal of Clinical Oncology* 16 (10): 3310–3315.
10. Hegi, M.E., Diserens, A.-C., Gorlia, T. et al. (2005). *MGMT* gene silencing and benefit from temozolomide in glioblastoma. *The New England Journal of Medicine* 352: 997–1003.
11. Brandes, A.A., Franceschi, E., Tosoni, A. et al. (2008). MGMT promoter methylation status can predict the incidence and outcome of pseudoprogression after concomitant radiochemotherapy in newly diagnosed glioblastoma patients. *Journal of Clinical Oncology* 26: 2192–2197.
12. Gerstner, E.R., Yip, S., Wang, D.L. et al. (2009). MGMT methylation is a prognostic biomarker in elderly patients with newly diagnosed glioblastoma. *Neurology* 73: 1509–1510.
13. Yan, H., Parsons, D.W., Jin, G. et al. (2009). IDH1 and IDH2 mutations in gliomas. *The New England Journal of Medicine* 360 (8): 765–773.
14. Yang, H., Ye, D., Guan, K. et al. (2012). IDH1 and IDH2 mutations in tumorigenesis: mechanistic insights and clinical perspectives. *Clinical Cancer Research* 18: 5562–5571.
15. Sanson, M., Marie, Y., Paris, S. et al. (2009). Isocitrate dehydrogenase I codon 132 is an important prognostic biomarker in gliomas. *Journal of Clinical Oncology* 27: 4150–4154.
16. Keles, G.E., Lamborn, K.R., and Berger, M.S. (2001). Low-grade hemispheric gliomas in adults: a critical review of extent of resection as a factor influencing outcome. *Journal of Neurosurgery* 95: 735–745.

17. Bloch, O., Han, S.J., Cha, S. et al. (2012). Impact of extent of resection for recurrent glioblastoma on overall survival: clinical article. *Journal of Neurosurgery* 117 (6): 1032–1038.
18. Muldoon, L.L., Soussain, C., Jahnke, K. et al. (2007). Chemotherapy delivery issues in central nervous system malignancy: a reality check. *Journal of Clinical Oncology* 25: 2295–2305.
19. Cairncross, J.G., Macdonald, D.R., and Ramsay, D.A. (1992). Aggressive oligodendroglioma: a chemosensitive tumor. *Neurosurgery* 31 (1): 78–82.
20. Cairncross, G., Macdonald, D., Ludwin, S. et al. (1994). Chemotherapy for anaplastic oligoden-droglioma. *Journal of Clinical Oncology* 12: 2013–2021.
21. Thiel, E., Karfel, A., Martus, P. et al. (2010). High-dose methotrexate with or without whole brain radiotherapy for primary CNS lymphoma (G-PCNSL-SG-1): a phase 3, randomised, non-inferiority trial. *Lancet Oncology* 11: 1036–1047.
22. Vredenburgh, J.J., Desjardins, A., Herndon, J.E. et al. (2007). Phase II trial of bevacizumab and irinotecan in recurrent malignant glioma. *Clinical Cancer Research* 13: 1253–1259.
23. Friedman, H.S., Prados, M.D., Wen, P.Y. et al. (2009). Bevacizumab alone and in combination with irinotecan in recurrent glioblastoma. *Journal of Clinical Oncology* 28: 4733–4740.
24. Van Den Bent, M.J., Brandes, A.A., Rampling, R. et al. (2009). Randomized phase II trial of erlotinib versus temozolomide or carmustine in recurrent glioblastoma: EORTC brain tumor group study 26034. *Journal of Clinical Oncology* 8: 1268–1274.
25. Franz, D.N., Belousova, E., Sparagana, S. et al. (2014). Everolimus for subependymal giant cell astrocytoma in patients with tuberous sclerosis complex: 2-year open-label extension of the ran-domised EXIST-1 study. *The Lancet. Oncology* 15 (13): 1513–1520.
26. Robinson, G.W., Orr, B.A., and Gajjar, A. (2014). Complete clinical regression of a BRAF V600E-mutant pediatric glioblastoma multiforme after BRAF inhibitor therapy. *BMC Cancer* 14: 258.
27. Swartz, A.M., Li, Q.J., and Sampson, J.H. (2014). Rindopepimut: a promising immunotherapeu-tic for the treatment of glioblastoma multiforme. *Immunotherapy* 6 (6): 679–690.
28. Hdeib, A. and Sloan, A.E. (2015). Dendritic cell immunotherapy for solid tumors: evaluation of the DCVax((R)) platform in the treatment of glioblastoma multiforme. *CNS Oncology* 4 (2): 63–69.
29. Levin, V.A., Bidaut, L., Hou, P. et al. (2010). Randomized double-blind placebo-controlled trial of bevacizumab therapy for radiation necrosis of the central nervous system. *International Journal of Radiation Oncology, Biology, Physics* 79 (5): 1487–1495.
30. Neglia, J.P., Meadows, A.T., Robison, L.L. et al. (1991). Second neoplasms after acute lympho-blastic leukemia in childhood. *The New England Journal of Medicine* 325: 1330–1336.
31. Reiling, M.V., Rubnitz, J.E., Rivera, G.K. et al. (1999). High incidence of secondary brain tumours after radiotherapy and antimetabolites. *Lancet* 354: 34–39.
32. Li, F.P., Fraumeni, J.F. Jr., Mulvihill, J.J. et al. (1988). A cancer family syndrome in twenty-four kindreds. *Cancer Research* 48 (18): 5358–5362.
33. Louis DN. *WHO Classification of Tumours of the Central Nervous System.* World Health Organization (WHO), (2007).
34. Walker, M.D., Green, S.B., Byar, D.P. et al. (1980). Randomized comparisons of radiotherapy and nitrosoureas for the treatment of malignant glioma after surgery. *The New England Journal of Medicine* 303 (23): 1323–1329.
35. Stupp, R., Mason, W.P., Van Den Bent, M. et al. (2005). Radiotherapy plus concomitant and adju-vant temozolomide for glioblastoma. *The New England Journal of Medicine* 352: 987–996.
36. Wick, W., Platten, M., Meisner, C. et al. (2012). Temozolomide chemotherapy alone versus radi-otherapy alone for malignant astrocytoma in the elderly: the NOA-08 randomised, phase 3 trial. *The Lancet. Oncology* 13 (7): 707–715.
37. Gilbert, M.R., Dignam, J.J., Armstrong, T.S. et al. (2014). A randomized trial of bevacizumab for newly diagnosed glioblastoma. *The New England Journal of Medicine* 370 (8): 699–708.
38. Chinot, O.L., Wick, W., Mason, W. et al. (2014). Bevacizumab plus radiotherapy-temozolomide for newly diagnosed glioblastoma. *The New England Journal of Medicine* 370 (8): 709–722.

39. Burkhard, C., Di Patre, P.L., Schuler, D. et al. (2003). A population-based study of the incidence and survival rates in patients with pilocytic astrocytoma. *Journal of Neurosurgery* 98 (6): 1170–1174.

40. Gorlia, T., Wu, W., Wang, M. et al. (2013). New validated prognostic models and prognostic calculators in patients with low-grade gliomas diagnosed by central pathology review: a pooled analysis of EORTC/RTOG/NCCTG phase III clinical trials. *Neuro-Oncology* 15 (11): 1568–1579.

41. Metellus, P., Coulibaly, B., Colin, C. et al. (2010). Absence of IDH mutation identifies a novel radiologic and molecular subtype of WHO grade II gliomas with dismal prognosis. *Acta Neuropathologica* 120 (6): 719–729.

42. Dubbink, H.J., Taal, W., Van Marion, R. et al. (2009). IDH1 mutations in low-grade astrocytomas predict survival but not response to temozolomide. *Neurology* 73 (21): 1792–1795.

43. Porter, A.B., Giannini, C., Kaufmann, T. et al. (2008). Primary central nervous system lymphoma can be histologically diagnosed after previous corticosteroid use: a pilot study to determine whether corticosteroids prevent the diagnosis of primary central nervous system lymphoma. *Annals of Neurology* 63: 662–667.

44. Weller, M., Martus, P., Roth, P. et al. (2012). Surgery for primary CNS lymphoma? Challenging a paradigm. *Neuro-Oncology* 14 (12): 1481–1484.

45. Rubenstein, J.L., Hsi, E.D., Johnson, J.L. et al. (2013). Intensive chemotherapy and immunotherapy in patients with newly diagnosed primary CNS lymphoma: CALGB 50202 (Alliance 50202). *Journal of Clinical Oncology: Official Journal of the American Society of Clinical Oncology* 31 (25): 3061–3068.

46. Berry, M.P., Jenkin, R.D., Keen, C.W. et al. (1981). Radiation treatment for medulloblastoma. A 21-year review. *Journal of Neurosurgery* 55 (1): 43–51.

47. Packer, R.J., Gajjar, A., Vezina, G. et al. (2006). Phase III study of craniospinal radiation therapy followed by adjuvant chemotherapy for newly diagnosed average-risk medulloblastoma. *Journal of Clinical Oncology: Official Journal of the American Society of Clinical Oncology* 24 (25): 4202–4208.

48. Mavrakis, A.N., Halpern, E.F., Barker, F.G. 2nd et al. (2005). Diagnostic evaluation of patients with a brain mass as the presenting manifestation of cancer. *Neurology* 65 (6): 908–911.

49. Borgelt, B., Gelber, R., Kramer, S. et al. (1980). The palliation of brain metastases: final results of the first two studies by the Radiation Therapy Oncology Group. *International Journal of Radiation Oncology, Biology, Physics* 6 (1): 1–9.

50. Sun, A., Bae, K., Gore, E.M. et al. (2011). Phase III trial of prophylactic cranial irradiation compared with observation in patients with locally advanced non-small-cell lung cancer: neurocognitive and quality-of-life analysis. *Journal of Clinical Oncology: Official Journal of the American Society of Clinical Oncology* 29 (3): 279–286.

51. Kocher, M., Soffietti, R., Abacioglu, U. et al. (2011). Adjuvant whole-brain radiotherapy versus observation after radiosurgery or surgical resection of one to three cerebral metastases: results of the EORTC 22952-26001 study. *Journal of Clinical Oncology: Official Journal of the American Society of Clinical Oncology* 29 (2): 134–141.

23

THYROID CANCER

Frances E. Carr

*University of Vermont Cancer Center, Department of Pharmacology,
University of Vermont Larner College of Medicine, Burlington, VT, USA*

OVERVIEW

In this chapter, we will discuss thyroid cancer, the types and classifications, incidence and frequency of occurrence, known causes, risk factors, diagnostics, classification and staging, therapies, and ongoing innovational studies.

INTRODUCTION

Thyroid cancer is the most common malignancy of the endocrine system and the incidence has been increasing faster globally than that reported for other solid tumors over the past few decades [1–4]. An estimated 56 870 new cases were predicted for 2017 in the United States with an expected overall 96.6% survival rate after five years [5]. The prognosis for patients with early detection of thyroid cancer and receiving standard therapies is generally good; however, the outcome for patients with resistant or recurrent disease is poor. Due to the lack of effective therapies, patients with advanced or metastatic thyroid cancer, have a risk of higher

Cancer: Prevention, Early Detection, Treatment and Recovery, Second Edition. Edited by Gary S. Stein and Kimberly P. Luebbers.
© 2019 John Wiley & Sons, Inc. Published 2019 by John Wiley & Sons, Inc.

mortality than all other endocrine cancers combined. Since the 1980s, the rate has tripled with an increase in tumors of all sizes, particularly for women. The increase in incidence has been attributed to advances in diagnostic technologies as well as access to healthcare, but environmental and lifestyle factors and prior cancers are increasingly recognized as contributing causes [6–11]. Studies have illustrated an increase in large and small tumors in both rural and urban environments, further implicating that factors beyond increased surveillance contribute to the rise in thyroid cancer incidence [9, 12, 13]. This chapter will review the thyroid gland, types of thyroid cancer, risk factors, diagnostics, therapies, and advances in genomic medicine for detection and treatment.

GENERAL INFORMATION

The thyroid gland is a butterfly-shaped tissue, with two lobes, located at the front of the neck at the trachea. Thyroid hormones and calcitonin are synthesized in two distinct thyroid cell types; epithelial (or follicular cells) and parafollicular (or "C" cells), respectively. The epithelial cells form follicular structures around a lumen; these functional units are responsible for the synthesis, storage, and secretion of thyroid hormones (triiodothyronine, T_3 and thyroxine, T_4). The thyroid gland actively takes up iodine, a mineral found in foods and iodized salt, which is necessary to make T_3 and T_4. The iodinated protein thyroglobulin serves as the backbone matrix for the synthesis of T_3 and T_4, and it also serves as a marker for thyroid function. A network of transcription factors, including Titf1, Hhex, Pax8, and Foxe1, are critical for thyroid cell development, proliferation, differentiation, and survival [14]. In addition, thyroid hormones act primarily in the nucleus regulating gene expression by interacting with a thyroid hormone receptor (TR), which is a DNA binding protein. Thyroid hormones can also bind to plasma membrane receptors and affect changes in signaling pathways in the cytoplasm and cause rapid cell changes. Disruption of these developmental regulators can alter cell function and facilitate tumor growth, as seen in thyroid cancer.

TYPES OF THYROID CANCER

The development of nodules in the thyroid is common in adults, with only 8–15% of the nodules malignant [15]. Thyroid cancers are heterogeneous and range in virulence from the most common well-differentiated papillary thyroid cancer (PTC) and follicular thyroid cancer (FTC) to the less common poorly differentiated (PDTC) and undifferentiated aggressive anaplastic thyroid cancer (ATC). These tumors are distinguished by molecular profile, pathophysiology, biological behavior, morphology, and clinical presentation. Histological and cytological features are used to classify follicular cell derived tumors; PTC and FTC are generally well differentiated and retain characteristics of normal thyroid cells, whereas poorly differentiated and ATC bear little resemblance to thyroid cells. PTC and FTC may progress to poorly differentiated tumors in advanced disease. ATC is a dedifferentiated thyroid cancer. Aggressiveness and lethality of these tumors are correlated with loss of differentiation and responsiveness to treatment. Approximately 20–30% of differentiated tumors can develop diffuse aggressive metastases to bone and lung [16, 17].

Papillary Thyroid Carcinoma (PTC)

PTC is the most common type, accounting for 75–85% of thyroid tumor cases. It is considered to be primarily a slow-growing well-differentiated tumor that is usually localized to the thyroid gland and proximal lymph nodes [18]. A subset of these tumors may be more aggressive presenting with local tissue and vascular invasion as well as metastasis. Aggressive PTC have higher recurrence rates and decreased 10-year survival rates with or without dedifferentiation [19]. PTC is commonly diagnosed between the ages of 30 and 60 although it can occur at any age. Females are affected three to four times more often than males. Variants of PTC include tumors composed of neoplastic follicles with nuclear features of PTC often termed PTCFV. The main subtypes include infiltrative (nonencapsulated) and encapsulated. Overall, the prognosis for PTC is very good [20].

Papillary Microcarcinoma (PMC)

PMC of the thyroid is a lesion 1.0 cm or less in size most often found incidentally. It is the fastest growing subtype of thyroid cancer in patients of all ages [21]. PMC is a tumor that is classically considered relatively innocuous without significant clinical relevance. However, PMC has a 20–50% incidence of metastases to cervical lymph nodes at the time of diagnosis and the probability may reach 90% if micrometastases are present [22, 23]. Although the mortality rate in patients harboring this tumor is less than 1%, a subset of these PMC will exhibit aggressive behavior that negatively affects prognosis and survival. This observation suggests two subcategories; incidental and non-incidental with different clinical outcomes [23–26]. Currently, there are no concrete predictive or prognostic parameters to assist in clinical and surgical management of patients with this tumor although studies have stratified PMC to inform disease management [25, 27–29].

Follicular Thyroid Carcinoma (FTC)

FTC is the next most common type of differentiated thyroid cancer. It accounts for 10–15% of differentiated thyroid cancers in areas where dietary iodine is readily available and 25–40% of FTC in areas of iodine deficiency [30, 31]. The overall incidence of FTC has declined while that of PTC has increased [31]. The decrease in the incidence of FTC may be correlated with an increase in dietary iodine supplementation [32] although there is not yet a consensus on this observation. FTC presents as a larger tumor generally than PTC, with distinct cellular characteristics. FTC does not have the distinct nuclear features of PTC, nor does it contain amyloid and calcitonin, as does PTC. Most classifications of this type of thyroid cancer are based upon the degree of invasiveness and the relationship of the tumor or lesion to the surrounding tissues including vascular invasion. FTC generally occurs later in life and is considered to be a more aggressive tumor type. Depending upon the capsular and vascular invasion, follicular thyroid tumors may metastasize to lung and bone. A subtype of FTC, Hürthle or oxyphilic, represents less than 3% of thyroid cancers. It is also less differentiated and often difficult to detect [33, 34].

Follicular Adenoma (FA)

FA has cellular characteristics that are similar to FTC but is a common benign neoplasm. A FA cannot be distinguished from an FTC based upon cytologic features alone. The determination of FTC as distinct from FA is primarily based upon the tumor characteristics notably morphologic evidence of vascular invasion [35]. The development of molecular biomarkers is now significant in classifying a FA or FTC as discussed later.

Poorly Differentiated Thyroid Cancer (PDTC)

PDTC is a rare malignancy most often considered to be the bridge between well-differentiated cancers (PTC and FTC) and dedifferentiated thyroid cancer with distinct morphology and aggressive cell behavior [36, 37]. PDTCs are usually large tumors with extrathyroidal extensions and identified histologically. While this tumor type is not yet a distinct category for classification, it is critical for distinguishing the tumor from the most aggressive undifferentiated thyroid cancer with more treatment options. The development of novel biomarkers can be critical for diagnosis of PDTC to determine appropriate management.

Anaplastic Thyroid Carcinoma (ATC)

ATC is one of the most aggressive solid tumors in humans. It is a rare type of undifferentiated thyroid cancer accounting for 2–5% of thyroid lesions with a quick onset and rapid metastases throughout the body. It is widely invasive often replacing most of the thyroid gland and surrounding soft tissue. In contrast to differentiated thyroid cancer, ATC cells do not display any of the cellular features or biological functions of normal follicular thyroid cells such as uptake of iodine and thyroid hormone synthesis [37]. The spectrum of ATC reflects major histological categories defining the main differential diagnoses. It is critical to distinguish ATC from PDTC as there are more treatment options for the responsive PDTC. The prognosis is poor for ATC with a mortality rate over 90% and a mean survival of six months after diagnosis.

Medullary Thyroid Carcinoma (MTC)

MTC is derived from parafollicular C cells, which produce the hormone calcitonin. MTC accounts for only 1–2% of thyroid tumors and may be sporadic (80%) or familial (20%). The inherited forms of MTC can be associated with other endocrine tumors in Multiple Endocrine Neoplasia (MEN) 2A and (MEN) 2B syndromes. The familial form is characterized by a point mutation in the germline DNA encoding a RET (rearranged during transfection translocation) oncogene [38]. As a less differentiated cancer that metastasizes to lymph nodes and other organs, it is more aggressive than PTC or FTC. It is most often associated with high levels of circulating calcitonin and carcinoembryonic antigen (CEA), which serve as biomarkers for the disease and disease progression [37, 39].

RISK FACTORS

The primary risk factors for thyroid cancer include gender, age, and exposure to radiation including radiation from medical diagnostics. There is emerging recognition of other influences such as lifestyle and environmental impacts [40, 41]. The tumor may not emerge until 5 to 20 or more years after exposure complicating identification of tumor initiating events. The American Thyroid Association (ATA), a professional international organization of healthcare providers and scientists with expertise in thyroid function and disease, issues specific guidelines on risk factors, classification of thyroid tumors, and recommendations for clinical management of benign and malignant thyroid tumors. Additional medical and scientific groups that develop clinical practice guidelines include the American Association of Clinical Endocrinologists, National Comprehensive Cancer Network, European Thyroid Association, and Society for Nuclear Medicine and Molecular Imaging among others. For the purposes here, the ATA guidelines will be the primary reference.

Radiation

Exposure to radiation is well documented as the primary risk factor for the development of thyroid cancer. Comprehensive studies of atomic bomb survivors in Japan; populations exposed to radiation from Chernobyl as well as other epidemiological analyses have clearly demonstrated the risk of development of thyroid cancer after exposure to external radiation [42–45]. Critically, the thyroids of children are particularly sensitive to external radiation. Head or neck radiation treatments during childhood in particular are risk factors for later development of thyroid cancer; the risk decreases with increased age at initial exposure. In the decades before the 1960s, children were treated with low doses of radiation for a wide range of disorders including tonsillitis and acne. However, even low dose exposure increases the risk of thyroid cancer development, which continues for 30 to 50 years after exposure [46, 47]. The increased use of imaging tests such as X-rays and computerized tomography (CT) scans in medical diagnostics represents the largest source of man-made radiation exposure. The rising use of CT for diagnostics, particularly in children, has raised concerns about the risk of thyroid cancer with this exposure [48]. A correlation with exposure to dental X-rays as well as other diagnostic X-rays is associated with an increased thyroid cancer risk [49, 50]. The most vulnerable population is children who should have minimal exposure to diagnostic radiation sources.

Gender

Thyroid cancer occurs in women approximately three to four times more often than in men. Although the reasons for this are not entirely clear, thyroid cancer incidence increases at puberty and decreases after menopause [12]. Estrogen is a potent growth promoter of thyroid cells and tissues [51–54]. In contrast to other cancers, how this critical hormone acts to initiate or advance thyroid tumors is still not understood. Another risk factor may actually be a prior breast cancer. In the United States one in eight women are expected to develop breast cancer during their lifetimes. With improved survival, the rates of secondary cancers have revealed an unexpected co-occurrence of breast and thyroid cancer. Recent studies suggest that women with thyroid cancer have a greater risk for development of breast cancer and women with breast cancer are at increased risk for development of thyroid cancer; these findings suggest a common etiology [53, 55, 56, 57]. The co-occurrence of these cancers is most pronounced when either tumor type is diagnosed at younger ages implicating contributing hormonal factors in promoting or inhibiting tumor growth [56]. Identifying the factors similarly contributing to these cancers is important for early detection as well as development of effective treatments.

Age

The age at which a thyroid cancer occurs has been a factor considered in classification and risk as reflected in ATA guidelines [20, 58]. Thyroid tumors are most common between the ages of 25 and 65 with a higher incidence for women at a younger age and for men at an older age. A recent analysis investigating the influence of age on thyroid cancer risk determined that while the prevalence of nodules in the thyroid increases with age, the risk of a malignancy decreases. However if cancer is present, a more aggressive disease is more likely. Early identification of thyroid cancers in older patients is important to achieve the best outcome [59]. Thyroid cancer in children is relatively rare accounting for 0.5–5% of total thyroid cancer incidence. The pediatric cancers are usually differentiated and may present at a more advanced stage but carry an excellent prognosis with long-term survival greater than 95%. However, therapy recommended for an adult may not be appropriate for a child who is at higher risk from overaggressive treatment [60].

Familial Factors

Although thyroid cancer is primarily a sporadic disease, a smaller percentage (20–25% MTC; 5% PTC) may arise due to familial syndromes [61, 62]. Familial non-medullary thyroid cancer (FNMTC) may be associated with other syndromes including Cowden's syndrome, Werner's syndrome, familial adenomatous polyposis, Carney complex, Pendred syndrome, and familial multinodular goiter. Of note, FNMTC appears to be more common in patients with other cancers including breast cancer in women and prostate cancer in men, but the reasons for the association are not understood. Although somatic mutations (DNA changes in cells not associated with inheritance) have been identified for sporadic thyroid tumors, the same mutations do not seem to be associated with familial thyroid cancer [63]. While FNMTC has been reported as a more-aggressive disease with lymph metastases than sporadic cancers, there is not agreement on these observations [64]. In contrast, the genetic events associated with familial medullary thyroid cancer (FMTC) and the relationship to cellular characteristics and histopathology are well-established. FMTC, a dominant inheritance pattern disease, is most often associated with multiple endocrine neoplasia, MEN2A or MEN2B syndromes. Most patients with FMTC have specific mutations in the RET proto-oncogene allowing for directed therapeutic interventions [65–69].

Thyroid Disease

Thyroid disease, most notably goiter, is both a familial and sporadic thyroid cancer risk factor. A goiter is a term that simply refers to an abnormal enlargement of the thyroid gland. This growth may arise with an overproduction (hyperthyroidism), underproduction (hypothyroidism), or normal production (euthyroidism) of TH. A primary cause of hypothyroid goiter is global iodine deficiency. The incidence in the United States has dramatically declined with the advent of iodine supplementation such as iodized salt. Iodine is a critical component of TH and a deficiency results in an underproduction of TH. This causes a compensatory increase in thyroid-stimulating hormone (TSH) from the pituitary gland in the brain. TSH stimulates thyroid gland growth as well as TH production. Excess TSH causes an abnormal increase in thyroid tissue during the process of stimulating thyroid cells to produce TH. A goiter may also be caused by antibodies that attack the thyroid gland destroying the tissue and reducing TH production. This autoimmune disease, Hashimoto's thyroiditis, is the most common cause of hypothyroidism in the United States. It impacts women at least 10 times more often than men and occurs primarily between 45 and 65 years of age [70–72]. Associated with the damaged tissue and decreased TH production is pituitary secretion of TSH to stimulate TH production and growth of the thyroid gland. Conversely, antibodies may attack the thyroid and increase tissue growth and hormone production resulting in an enlarged gland and hyperthyroidism. This autoimmune disease, Grave's Disease, is the most common cause of hyperthyroidism in the United States and has a genetic predisposition although no specific gene has been identified to date [73–75].

The direct relationship between thyroid diseases and subsequent development of thyroid cancer has been controversial. When patients undergo treatment for thyroid cancer, there are often cellular changes associated with Hashimoto's thyroiditis. However, it is unclear whether these changes occur before or after thyroid cancer [71]. More recent comprehensive epidemiological studies have found a higher risk of thyroid cancer in patients with toxic multinodular goiter [76] as well as subsequent to either Hashimoto's or Grave's autoimmune diseases [70, 71, 77–81].

Iodine and Micronutrients

Iodine deficiency has long been recognized as a risk factor for development of thyroid disease but whether this dietary deficiency is linked to development of thyroid cancer was unclear. A recent comprehensive review of available evidence suggests iodine deficiency is a clear risk factor for thyroid cancer particularly FTC and possibly ATC [82]. Conversely, excess dietary iodine has been implicated as a risk factor for PTC [83]. Further studies are needed to clarify the overall relationship. Beyond iodine, whether dietary factors are also associated with thyroid cancer risk is still unknown. Foods containing substances that disrupt the production of thyroid hormones by interfering with iodine uptake in the thyroid gland are referred to as goitrogens. The over-consumption of uncooked cruciferous vegetables shows a correlation with thyroid cancer risk. Other dietary factors including nitrate and nitrite show a positive association with thyroid cancer as well [84]. Other studies suggest a protective effect of vitamins and dietary factors notably by calcium intake [85]. However, reported results have been inconsistent and further investigations are clearly warranted [86].

Obesity and Environmental Factors

Obesity has been indicated as a risk factor for a number of cancer types [87–89] including thyroid cancer [90–92]. The increasing prevalence of obesity not only affects the incidence of cancer but also the prognosis for cancer survivors. More refined measures, that is, considerations beyond body mass index, and more comprehensive studies are needed to determine the specific relationships between adiposity and cancer initiation [93]. Nevertheless, the finding that reduction in body mass correlates with better health outcomes and improved response to cancer therapies emphasizes the overall relationship between obesity and cancer risk. The thyroid gland and TH are integral to metabolic processes in the body and may be associated with Metabolic syndrome. In a large cohort study of Metabolic syndrome, there was a correlation with thyroid cancer incidence notably for women [94].

In the past decades, human exposure to environmental toxins such as asbestos, pesticides, plastic derivatives, phthalates, and polychlorinated biphenyls (PCBs) among others has increased [95]. Endocrine disrupting chemicals (EDCs), substances in the environment that mimic or disrupt hormonal systems, are now widely recognized as public health concerns [96, 97]. Numerous studies have linked exposure to derivatives of plastics such as phthalates, bisphenol A, pesticides and other chemicals to a variety of cancers notably breast and prostate [98, 99]. The increase in thyroid cancer incidence has been attributed, in part, to a wide variety of factors including exposure to EDCs. Although studies are currently sparse, they reveal a potential link of EDCs with thyroid cancer [100]. Interestingly, the rise in obesity has also been linked to exposure to EDCs suggesting that the risk for thyroid cancer may be through multiple routes [101, 102]. Human studies are made complicated by the length of time needed to gather cancer incidence data and establish linkages to EDC exposure, which may have occurred decades prior to tumor development. Comprehensive animal and human cell culture studies have provided insights into possible biomarkers, which can be used to link exposure and cancer incidence [103, 104]. The development of biomarker panels and other factors that can be monitored in the blood or biopsy tissues are now emerging. These advances will provide new tools to link exposure to environmental toxins and thyroid cancer [105].

Prevention

Most people with thyroid cancer have no known risk factors. Imaging tests such as X-rays and CT scans particularly for children are limited and use the lowest dose available. For FMTC, testing for gene mutations (e.g., RET proto-oncogene) can facilitate early identification of risk and subsequent intervention such as removal of the thyroid gland. Maintaining a healthy lifestyle and minimizing toxin exposure continue to be the best preventive measures.

DIAGNOSIS AND CLASSIFICATION OF THYROID CANCER

Thyroid cancer may be detected through routine physical examinations. Most early cancers are found when patients have a lump in the neck, trouble breathing, swallowing, or hoarseness. Ultrasound tests can detect changes in the thyroid gland including the development of nodules, which are abnormal overgrowths of tissue that are most often benign. These nodules may range in size from less than 1 cm, a microcarcinoma, to greater than 2 cm or larger with or without multiple nodules. Nodules are often benign (noncancerous) and may be assessed by imaging tests including radioiodine scans, CT, magnetic resonance imaging (MRI), and positron emission tomography (PET). CT, MRI, and PET are used particularly for determining potential metastases. Radioiodine scans utilize a small amount of radioactive iodine (I^{131}), which is taken up preferentially by thyroid cells. If there is low detection of the radioactivity, there is a higher likelihood of a malignancy associated with the nodule. Symptoms are not usually caused by the nodules but by changes to the function of the thyroid gland resulting in overproduction or underproduction of TH. Changes in thyroid gland function can be monitored through TH, TSH, and calcitonin and CEA (markers for MTC) in the blood.

Assessment of a nodule may be conducted using a sampling of the nodule cells by a biopsy procedure, such as fine-needle aspiration (FNA), or surgical biopsy. Differential diagnoses of thyroid cancers have largely been through histological examination of cells. Thyroid cancers reflect a spectrum of cellular changes, altered nuclear architecture, distinct molecular profiles that correlate with tumor aggression and metastases. As studies have elucidated more of the pathogenesis and histology of thyroid cancer, guidelines have changed, as have the prognostic implications. Guidelines for diagnosis, classification, and recommended treatments for management of thyroid cancer in the United States are summarized by the ATA [20, 39, 58, 106] with specific guidelines for children who have thyroid nodules and differentiated cancers [60].

Molecular Diagnostics

Thyroid cancers reflect a spectrum of cellular changes with distinct molecular profiles that correlate with tumor aggression and metastases. With a greater understanding of the biology underlying the development and progression of thyroid cancers, novel biomarkers including genetic mutations, epigenomic profiling (changes in genes and function that do not involve DNA sequence), as well as expression of tumor suppressor or tumor promoters (oncogenes) are now used to guide diagnosis and treatment plans. Advances in the diagnosis and treatment of differentiated and undifferentiated thyroid cancers have led to clinical management tailored to an individual's risk factors for response to treatment, disease recurrence, and mortality. The evolution of this comprehensive approach to personalized medicine has arisen with the advent of newer molecular diagnostic and targeted therapies in addition to advances in standardized interpretation of cytopathology [29]. It is important to note that most thyroid nodules are benign, thus the challenge lies in identifying the malignant and most particularly the aggressive tumors that warrant invasive treatments and subsequent care. Without precise identification of the malignancy, the risk of overtreatment, and subsequent morbidity is high.

DNA Mutations

The most common mutations identified in thyroid cancers include BRAF, RAS, RET/PTC, and NTRK1, which alter specific cellular signaling pathways. Activation of mitogen-activated protein kinase (MAPK) due to mutations in one or more of these enzymes is common in PTC [107]. FTC is characterized by either RAS mutations or PAX8/PPARγ rearrangements reflecting changes in the phosphatidylinositol-3-kinases-AKT serine/threonine kinase 1 (PI3K-Akt) pathway [108, 109]. Additional mutations are characteristic of PDTC and ATC [109–111]. The development of a gene expression classifier based upon these changes advanced the accuracy and sensitivity of cancer detection [112–115]. MTC, both familial and sporadic, frequently carry point mutations in one or both of the RET and RAS genes [116, 117]. However, using the power of next generation sequencing (high throughput technology that enables rapid DNA or RNA sequence determination), confirmation of the known gene mutations combined with additional DNA sequence changes provides a more sensitive assay for not only identifying thyroid cancer but also the type and subtypes [118, 119]. Recently, additional somatic (acquired, non-inherited) DNA mutations have been identified that provide information beyond the type of thyroid cancer. BRAF is a gene that encodes the oncoprotein B-Raf, a serine/threonine protein kinase, which helps to transmit signals from outside the cell (via a receptor tyrosine kinase (RTK) to the nucleus through the RAS/MAPK signaling pathway (Figure 1). When this protein is overactive, it contributes to cancer growth by allowing cell growth and division (proliferation), cell movement (migration), inhibition of programmed cell death (apoptosis), and cell-specific development (differentiation). The most common BRAF gene mutation, V600E, is present in a number of cancers including thyroid. Another somatic mutation in the enzyme Telomerase Reverse Transcriptase (TERT) allows cells to avoid apoptosis and to become immortal as cancer cells. TERT is responsible for adding nucleotides to the ends of telomeres, ends of chromosomes, which protect against deterioration. Telomere length is important for cell health. Mutations in TERT, BRAF, or both are now recognized as critical predictors of disease aggression, response to therapies, and thyroid cancer outcome [120–127]. By integrating and validating these additional DNA mutations, a six-genotype genetic prognostic model for poor outcomes of PTC has been established [128].

Epigenetic Alterations

Beyond somatic DNA mutations, additional cellular changes occur to advance thyroid tumor growth. Changes in the levels of key proteins in the cell including molecules that act as tumor suppressors or tumor promoters may occur through epigenetic changes. Epigenetics refers to changes in gene expression by mechanisms other than changes in the DNA sequence itself. Given the extensive changes in the nuclear architecture noted particularly in PDTC and ATC, it is not surprising

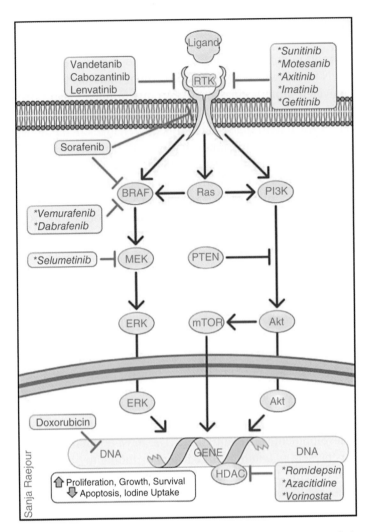

Figure 1. Intracellular signaling pathways in thyroid cancer. Therapies currently in use and key molecular targets that are inhibited. * indicates drugs approved for use in other cancers. *Source*: Illustration by Sanja Raejour. (See insert for color representation.)

to note that global modifications of DNA, such as changes in methylation and acetylation, are extensive and linked to altered expression of tumor-associated genes. For example, hypomethylation leads to instability in the genome, reduction in DNA repair, and activation of oncogenes. Hypermethylation is usually associated with gene silencing; particularly those of tumor suppressors, but the mechanisms by which specific regions of the genome or specific gene target are silenced are not known.

Some of the most important epigenetic regulators of gene expression are small RNA molecules transcribed from DNA that are not translated into proteins. These RNA molecules are collectively referred to as noncoding RNAs (ncRNAs) and are divided into two main groups; short ncRNAs (<30 nucleotides) and long ncRNAs (>200 nucleotides). These groups include microRNAs (miRNAs), small interfering RNAs (siRNAs), enhancer RNAs (eRNAs), and piwi associated RNAs (piRNAs) now recognized as central to many biological processes including cell proliferation, differentiation as well as cell death in part through the regulation of gene expression networks [129]. MiRNAs typically down-regulate gene expression by binding to specific RNA sequences and blocking translation or causing mRNA cleavage and degradation, which ultimately decreases protein levels. Conversely, miRNAs may activate translation through interaction at different regions of the mRNA targets to increase protein levels [130]. The results of these regulatory changes may decrease tumor suppressors resulting in a direct or indirect increase in oncogene expression levels and/or specific proteins in signaling pathways that control cell proliferation, migration, and cell death.

Not surprisingly, recent studies have shown that miRNAs are altered in various cancers implicating these ncRNAs in initiation and progression of tumors [131, 132]. Overexpression of miRNAs could lead to repression of tumor suppressor genes while the decrease in miRNA expression could enable an increase in tumor promoter and oncogene expression. During thyroid tumor progression, miRNA expression is anomalously regulated resulting in altered expression of tumor regulating genes and loss of regulation of intracellular signaling pathways. Importantly, changes in miRNA levels can be detected in serum samples ("liquid biopsies"), which bring the possibility of expanding these potential diagnostic biomarkers to support clinical management [133–135]. MiRNAs may eventually be novel therapeutic target molecules. In thyroid cancer, differential miRNA expression has been the focus of recent studies. Specific miRNA profile changes have been correlated with PTC [136, 137], FTC [138], PDTC [139, 140], ATC [140] and MTC [141, 142] with emerging evidence for diagnostic utility when aligned with clinicopathological features [143–145]. In particular, distinguishing PDTC and ATC directly impacts clinical management and overall prognosis [140]. Thus, differential expression of select miRNAs may be critical in identifying aggressive metastatic disease and distinguishing benign and malignant lesions [138, 143, 146–150]. The emergence of miRNA profiles for diagnostics is particularly important for distinguishing indolent versus aggressive PMC. Cytopathology characteristics are not different for these PMCs. Patients are particularly at risk for aggressive overtreatment as well as undertreatment without molecular tools to support the diagnostics. MiRNA profiles characteristic of PMCs are just emerging [151]. Of note, a recent report identified a miRNA signature distinguishing indolent from aggressive PMCs further advancing this potential diagnostic tool [152]. Serum miRNA profiles are additional tools to monitor response to therapy in thyroid cancer follow-up as well as indicators of recurrent disease [153–155]. Finally, miRNA expression profiles may eventually also reveal linkages between risk factors and later development of thyroid cancers as recently suggested for Graves' disease and PTC [156].

Molecular Classification

The ATA has recently issued specific guidelines on the application of molecular profiling for thyroid nodules and the impact on therapeutic monitoring and perioperative care [119, 157, 158]. Next-generation sequencing is a method that can be used to detect multiple genetic alterations in large regions of the genome [159]. Understanding the genomic variants, DNA mutations, and epigenomic modifications now provides a framework for classification of thyroid cancers into molecular subtypes to reflect the underlying signaling and properties of the tumors. This integrated data should better inform the management of the disease [160]. The comprehensive shared data atlas for PTC should also be extended to other types of thyroid cancer. Molecular profiling panels have been established for thyroid cancer and they strengthen risk prediction by the ATA or other systems [118, 161, 162].

STAGING OF THYROID CANCER

The type and stage of thyroid cancer is diagnosed histologically. Generally, cells are obtained by fine-needle aspiration biopsy (FNAB) or from surgical tissue specimens. The classifications for FNA biopsy results are as follows: nondiagnostic, suspicious for malignancy, suspicious for neoplasm, lesion of undetermined significance, malignant, or benign [20, 58, 115, 163]. The thyroid cells are assessed by cytopathology and classified as differentiated (PTC, PMC, FTC), poorly differentiated, or undifferentiated (MTC and ATC). After classification of thyroid cancer type, the tumor is staged. The American Joint Committee on Cancer (AJCC) has designated thyroid cancer staging by the tumor, node, metastasis (TNM) classification system [20, 58, 164] and additional risk stratification that encompass other variables thought to be predictors of prognosis guide treatment options and potential for recurrent disease [165]. Thyroid cancer stages are summarized in Table 1.

Recurrent Thyroid Cancer

Patients treated for differentiated thyroid cancer should be followed carefully based on individual risk for recurrent disease [166]. Approximately 10–30% of patients thought to be disease free after initial treatment will develop recurrence to residual and/or metastases. Of these patients, approximately 80% develop recurrence with disease in the neck alone, and 20% develop recurrence with distant metastases. The most common site of distant metastasis is the lung. The advent of serum biomarkers should facilitate this active monitoring.

T A B L E 1. Thyroid cancer stages; National Cancer Institute and American Thyroid Association.

Type of Thyroid Cancer	Stage (TNM)	Description
Differentiated Thyroid Cancer (DTC) • Papillary (PTC) • Follicular (FTC)	Stage I Stage II	Aged <55 • DTC is localized to the thyroid gland • DTC has local invasion
Differentiated Thyroid Cancer (DTC) • Papillary (PTC) • Follicular (FTC)	Stage I Stage II Stage IIIa Stage IIIb	Aged>55 • DTC is localized to the thyroid gland • DTC>2 cm and <4 cm; limited to the thyroid gland • DTC >4 cm but limited to the thyroid • Tumor any size with gross extrathyroidal invasion or positive • lymph nodes
Differentiated Thyroid Cancer (DTC) • Papillary (PTC) • Follicular (FTC)	Stage IV	• Tumor extends beyond the thyroid capsule with local lymph node or distant metastases • Lung and bone metastases
Medullary (MTC)	Stage 0 Stage II Stage III Stage IV-A Stage IV-B Stage IV-C	• Biochemical/genetic assessment • Tumor <2 cm • Tumor >2 cm but <4 cm with no metastases • Advanced without distant metastases • Very advanced with or without lymph metastases, without distant metastases • Distant metastases
	Advanced and metastatic	Targeted therapy • TKIs • DNA modifiers Palliative chemotherapy
Dedifferentiated Thyroid Cancer Anaplastic (ATC)	Stage IV-A Stage IV-B Stage IV-C	• Tumor within the thyroid • Extrathyroidal extension with or without lymph node metastases • Distant metastases
Papillary Microcarcinoma (PMC)	Indolent Aggressive	Surveillance (surgery) Surgery

Source: Adapted from by Mary Ann Liebert, Inc. (2016), (2012).

TREATMENT OPTIONS

Standard Therapeutic Strategies

Based upon the stage and type of thyroid cancer, overall treatment options are established as summarized in Table 2 [20, 58, 106, 167, 168]. Clinical management for thyroid cancer includes observation, surgery, radioiodine therapy, external

T A B L E 2. Current treatment recommendations for thyroid cancers; National Cancer Institute and American Thyroid Association.

Type of Thyroid Cancer	Stage (TNM)	Treatment Options
Papillary (PTC) and Follicular (FTC)	Stages I and II	Surgery with or without I^{131} treatment • Lobectomy (partial) (tumor<1 cm) • Total thyroidectomy (tumor>1 cm) • I^{131} • Thyroid suppression 　Treatment with TH (thyroxine) to decrease 　　TSH (thyroid stimulating hormone) Active surveillance (watchful waiting)
Papillary (PTC) and Follicular (FTC)	Stage III	Surgery with or without I^{131} treatment EBRT
Papillary (PTC) and Follicular (FTC)	Stage IV	Iodine sensitive: • Radiation therapy (I^{131} ablation) 　Primary and metastatic thyroid cells that uptake 　　iodine may be ablated by therapeutic I^{131} Iodine resistant: • Targeted therapy 　Tyrosine kinase inhibitors (TKIs) 　DNA modifiers • Surgery • External-beam radiation therapy (EBRT) 　Localized tumors that are unresponsive to I^{131} • Thyroid suppression
Recurrent Papillary (PTC) and Follicular (FTC)		Targeted therapy External-beam radiation therapy (EBRT) Chemotherapy Additional surgery
Medullary (MTC)	Localized	Total thyroidectomy External-beam radiation therapy (EBRT)
	Advanced and metastatic	Targeted therapy • TKIs • DNA modifiers Palliative chemotherapy
Anaplastic (ATC)		Surgery EBRT Chemotherapy (clinical trials)
Papillary Microcarcinoma (PMC)	Indolent Aggressive	Surveillance (surgery) Surgery

Source: Adapted from by Mary Ann Liebert, Inc. (2016), (2012). Notes: T = primary tumor;
　M = distant metastases; N = regional lymph nodes.

beam radiation therapy, and chemotherapy. Generally, the origin and characteristics of the cancer influence both disease progression and response to therapies. Most thyroid cancers are very treatable with an excellent prognosis and overall low mortality. Treatment options include surgery, radioactive iodine (I^{131}) ablation, and targeted therapies. Surgical options for primary tumors include partial thyroid removal (hemithyroidectomy), near total thyroidectomy, and total thyroidectomy depending upon the stage and risk stratification. I^{131} has been used in combination with thyroidectomy to completely ablate the thyroid gland and eradicate possible residual cancer. I^{131} is taken up by the thyroid cells via sodium iodide transporters, emits short-wavelength beta rays, and causes acute cell death. The disease site, tumor characteristics, and metastases are taken into consideration in order to develop a treatment plan for cancer patients [20, 39, 58, 60, 106].

External Beam Radiation Therapy

External beam radiation therapy (EBRT) is generally used for the palliative treatment of patients with inoperable or advanced disease. It may also be used for patients with a high likelihood of residual disease after surgery or for tumors that are unresponsive to I^{131} treatment [163].

Hormone Therapy

The pituitary hormone TSH increases thyroid gland growth and production of TH. After treatment for thyroid cancer that involves complete removal of the thyroid gland, patients take replacement TH (levothyroxine-T4) to avoid hypothyroidism and suppress TSH. TH supplements are critical for normal metabolism but also for suppression of TSH. The lower TSH level reduces the possibility of continued growth of any remaining malignant thyroid cells and recurrence or progression of thyroid cancer.

Targeted Therapies

Targeted cancer therapies are substances that interfere with molecules involved specifically in cancer cell growth and survival. By contrast, traditional or broad spectrum chemotherapy drugs act against all actively dividing cells. To be effective then, targeted therapies require the knowledge of factors that play a key role in cancer cell growth and survival. Thus, options for the treatment of thyroid cancers are based upon the understanding of the underlying cellular disruptions caused by gene mutations, gene modifications (epigenetics), and identification of key tumor suppressors and oncogenes. Targeted approved cancer therapies include agents that prevent cell growth signaling, interfere with tumor blood vessel development, promote cancer cell death, and stimulate the immune system to destroy cancer cells.

The major pathways and current approved pharmacological therapies for thyroid cancers are summarized in Figure 1. Additional signaling pathways and potential targets are involved in the progression of thyroid tumorigenesis [169]. Therapies by type of thyroid cancer and stage are listed in Table 2. Key cellular signaling pathways disrupted in thyroid cancers include the tyrosine kinase signaling pathways MAPK and PI3K. Activation of these pathways may be linked directly to DNA mutations and/or to epigenetic modifications of key regulators. For example, the BRAF V600E mutation, common in PTC, leads to the constitutive activation of a protein kinase, and stimulates MAPK signaling, and subsequent disruption of cell regulation. RAS family mutations are primarily associated with FTC and they lead to a constitutively activated state of this protein and stimulation of the PI3K-Akt pathway. Mutations in the RET oncogene, common in FMTC, activates both MAPK and PI3K-Akt. Drug targets inhibit activity of proteins in these pathways that regulate cell growth, movement, and death (apoptosis).

Tyrosine kinase inhibitors (TKIs): Agents that inhibit key enzymes in cellular pathways include TKIs. Growth factors bind to receptor molecules at the cell membrane and transmit signals through activation of enzymes, receptor-tyrosine kinases (RTK). The TKIs may act specifically on one enzyme or on multiple targets in a pathway. These agents are primarily used for I^{131}-refractory disease, locally recurrent disease, or metastatic and aggressive disease.

DNA modulators and chemotherapeutic agents: Therapeutic options for PDTC, ATC, and advanced MTC are limited. Pharmaceutical molecules that target disruption of DNA replication and subsequent protein synthesis are of considerable interest for these thyroid cancers. Advances in the understanding of the disrupted cellular signaling reveal novel molecular targets [170]. Progressive dedifferentiation is relatively common in ATC. Molecular diagnosis of specific mutations, such as of BRAF and RAS, may allow targeted therapies such as BRAF inhibitors. Cytotoxic chemotherapy such as doxorubicin and cisplatin target DNA replication to inhibit rapidly dividing and growing cells [106, 171, 172].

Survival Rates

The survival rates of patients with thyroid cancer depend upon a number of variables including age and other risk factors. Specific histotype (e.g., PTC, FTC, MTC) and the degree of tumor cell differentiation of the tumor cells are critical factors. Overall, the prognosis is excellent for PTC and FTC with 95 and 80% survival rates, respectively, after 35–40 years. After 10 years, the survival rate for MTC is lower at 65%. With increased dedifferentiation the prognosis is less favorable with a survival rate lower than 20% for PDTC after five years and less than 10% for ATC after six months subsequent to diagnosis [173].

Management of Low-Risk Differentiated and Microcarcinomas of the Thyroid

Effective management of low-risk thyroid cancers is controversial. There has been a recognition that there are risks of overtreatment as well as undertreatment. Treatment guidelines and diagnostics are evolving with advances in understanding of underlying causes for thyroid cancers and responses to evolving therapies

[20, 39, 58, 106, 174–176]. Active surveillance (watchful waiting) as well as minimal surgery (lobectomy) are important therapeutic options [177]. There is clearly a need for systematic studies to address these issues [178].

INNOVATIONS

Ongoing studies are revealing novel molecular events underlying the initiating events of thyroid cancers, new serum diagnostic tools, and promising new therapeutics. Highlights of the many discoveries include inducing aggressive cells to return to a state with more normal thyroid cell characteristics (redifferentiation), identification of tumor initiating cells (TICs)(stem cells), novel tumor suppressors, and miRNAs in diagnostics and monitoring response to treatment. Ultimately, understanding the disrupted biology in thyroid cancers should reveal factors that contribute to the higher incidence in women than men.

Redifferentiation

There has been exciting progress in the understanding of the cellular changes associated with thyroid cancers. The promise of advances in identifying molecular alterations reveals potential prognostic markers, diagnostic markers, and therapeutic targets. These molecular discoveries now provide opportunities for the development of novel molecular-based treatment strategies for patients with all stages of thyroid cancer. For example, differentiated thyroid cancers are generally more responsive to treatments than undifferentiated tumors. One strategy to enhance treatments for advanced tumors, particularly ATC, is redifferentiation; stimulation of the cells to again acquire characteristics of normal thyroid cells monitored by molecular markers [179, 180]. By reestablishing expression of tumor suppressors, aggressive tumorigenesis is inhibited.

Stem Cells

Tremendous progress has been made in understanding the initiating events of thyroid cancer. Putative cancer-forming stem cells appear to drive tumorigenesis through the ability to undergo self-renewal, develop into cancer cell types, and mediate metastases. This exciting discovery could potentially lead to the development of new strategies particularly for advanced stage disease, PDTC, and ATC [181]. Identifying gene mutations, epigenetic changes, and selective cell pathway activation should reveal therapeutic targets in this subpopulation of cells in thyroid tumors. Importantly, these stem cells or TICs may be a clue to resistance to treatment and/or recurrent disease [182].

Novel Tumor Suppressors

Unraveling the cellular signaling pathways inevitably reveals novel tumor suppressors which may be targets for reexpression or activation. These factors are potential key regulators in cell pathways that control cell growth, proliferation, differentiation, and programmed death, all of which are important for normal cell function. The exponential increase in the findings of novel tumor regulators will advance the development of methods to reactivate or reintroduce these critical factors. A detailed review of these molecules and pathways is beyond the scope of this summary. However, disruption of thyroid hormone signaling in the thyroid gland is of particular importance. Silencing or mutations of TRs is common in a number of endocrine-related cancers including thyroid cancers [183, 184]. Reintroduction of these factors reduces tumor growth in animal models [185, 186]. Unraveling the mechanisms of this tumor suppression can reveal interventions that may restore cell differentiation and physiological regulation [187–190].

Advanced Molecular Diagnostics and Therapies

An emerging field of study is the effect of cannabinoids on endocrine-related cancers. Cannabinoids, the chemical compounds in cannabis plants, are a large class of more than 70 compounds that have been used to provide relief to an array of symptoms (including nausea, pain, and inflammation) related to cancer therapy and other disorders. In humans, activation of the endogenous endocannabinoid system has been linked to decreased tumor growth [191–194]. Identifying novel therapies with low toxicity particularly for aggressive thyroid cancers including ATC is of paramount importance. Further studies are needed to establish the mechanisms by which cannabinoids may inhibit tumor growth to realize novel treatments.

As important as the discovery of new molecules is the review and repositioning of pharmacological agents, currently in use for treatment of noncancer diseases, for the development of anticancer treatments. This emerging area of study includes drugs used for metabolic disorders. The signaling pathways disrupted in a variety of diseases may be similar to those in certain cancer types. Thus, the application of these pharmaceutical agents as antitumor treatments holds much promise.

Advances in molecular diagnostics particularly based upon epigenetic changes, are revealing important innovations for therapies as well. Noncoding RNA, including both miRNA and lncRNA, are now recognized as a new level of gene regulation with potential use as diagnostic, prognostic, and therapy response markers in FNAB and serum [195–198]. MiRNA profiles may distinguish aggressive from nonaggressive thyroid tumors as well as thyroid cancer type [144, 152, 196, 197, 199–203]. Specific miRNAs likely participate in thyroid cancer development and progression by directly affecting the expression of genes responsible for differentiation, tumor invasion, and metastases formation [143, 145, 147, 150]. Modifying individual miRNA expression levels may reveal specific therapeutic molecules. While genomic medicine is now commonly used for diagnostic or therapeutic decision-making, epigenetic changes including miRNA expression profiles or selected miRNAs are only recently incorporated for thyroid cancer [204, 205]. Nevertheless, miRNA-based interventions are on the horizon. Taking advantage of the comprehensive thyroid cancer data including genomic and epigenomic profiles will revolutionize the diagnosis and treatment of thyroid cancers across the spectrum of types and stages of disease.

REFERENCES

1. Jemal, A., Bray, F., Center, M.M. et al. (2011). Global cancer statistics. *CA: A Cancer Journal for Clinicians* 61: 69–90.
2. Pellegriti, G., Frasca, F., Regalbuto, C. et al. (2013). Worldwide increasing incidence of thyroid cancer: update on epidemiology and risk factors. *Journal of Cancer Epidemiology* 2013: 1–10.
3. Zhu, C., Zheng, T., Kilfoy, B.A. et al. (2009). A birth cohort analysis of the incidence of papillary thyroid cancer in the United States, 1973–2004. *Thyroid* 19: 1061–1066.
4. Lim, H., Devesa, S.S., Sosa, J.A. et al. (2017). Trends in thyroid cancer incidence and mortality in the United States, 1974–2013. *JAMA* 317: 1338–1348.
5. Thyroid cancer facts and figures. 2017 National Cancer Institute Surveillance, Epidemiology, and End Results program web site. http://seer.cancer.gov/statfacts/html/thyro.html.
6. Davies, L. and Welch, H.G. (2006). Increasing incidence of thyroid cancer in the United States, 1973–2002. *JAMA* 295: 2164–2167.
7. Morris, L.G., Sikora, A.G., Tosteson, T.D. et al. (2013). The increasing incidence of thyroid cancer: the influence of access to care. *Thyroid* 23: 885–891.
8. Sprague, B.L., Warren Andersen, S., and Trentham-Dietz, A. (2008). Thyroid cancer incidence and socioeconomic indicators of health care access. *Cancer Causes & Control* 19: 585–593.
9. Hanley, J.P., Jackson, E., Morrissey, L.A. et al. (2015). Geospatial and temporal analysis of thyroid cancer incidence in a rural population. *Thyroid* 25: 812–822.
10. Enewold, L., Zhu, K., Ron, E. et al. (2009). Rising thyroid cancer incidence in the United States by demographic and tumor characteristics, 1980–2005. *Cancer Epidemiology, Biomarkers & Prevention* 18: 784–791.
11. Pazaitou-Panayiotou, K., Iliadou, P.K., Chrisoulidou, A. et al. (2013). The increase in thyroid cancer incidence is not only due to papillary microcarcinomas: a 40-year study in 1,778 patients. *Experimental and Clinical Endocrinology & Diabetes* 121: 397–401.
12. Li, N., Du, X.L., Reitzel, L.R. et al. (2013). Impact of enhanced detection on the increase in thyroid cancer incidence in the United States: review of incidence trends by socioeconomic status within the surveillance, epidemiology, and end results registry, 1980–2008. *Thyroid* 23: 103–110.
13. Zahnd, W.E., James, A.S., Jenkins, W.D. et al. (2017). Rural-urban differences in cancer incidence and trends in the United States. *Cancer Epidemiology, Biomarkers & Prevention* 27 (11): 1265–1274.
14. De Felice, M. and Di Lauro, R. (2011). Minireview: intrinsic and extrinsic factors in thyroid gland development: an update. *Endocrinology* 152: 2948–2956.
15. Kwong, N., Medici, M., Angell, T.E. et al. (2015). The influence of patient age on thyroid nodule formation, multinodularity, and thyroid cancer risk. *The Journal of Clinical Endocrinology and Metabolism* 100: 4434–4440.

16. Durante, C., Haddy, N., Baudin, E. et al. (2006). Long-term outcome of 444 patients with distant metastases from papillary and follicular thyroid carcinoma: benefits and limits of radioiodine therapy. *The Journal of Clinical Endocrinology and Metabolism* 91: 2892–2899.

17. Muresan, M.M.O.P., Leclere, J., Sirveaux, F. et al. (2008). Bone metastases from differentiated thyroid carcinoma. *Endocrine-Related Cancer* 15: 37–49.

18. Albores-Saavedra, J., Henson, D.E., Glazer, E. et al. (2007). Changing patterns in the incidence and survival of thyroid cancer with follicular phenotype--papillary, follicular, and anaplastic: a morphological and epidemiological study. *Endocrine Pathology* 18: 1–7.

19. Mete, O. and Asa, S.L. (2011). Pathological definition and clinical significance of vascular invasion in thyroid carcinomas of follicular epithelial derivation. *Modern Pathology* 24: 1545–1552.

20. Haugen, B.R., Alexander, E.K., Bible, K.C. et al. (2016). 2015 American thyroid association management guidelines for adult patients with thyroid nodules and differentiated thyroid cancer. *Thyroid* 26: 1–133.

21. Hughes, D.T., Haymart, M.R., Miller, B.S. et al. (2011). The most commonly occurring papillary thyroid cancer in the United States is now a microcarcinoma in a patient older than 45 years. *Thyroid* 21: 231–236.

22. Sakorafas, G.H., Sampanis, D., and Safioleas, M. (2010). Cervical lymph node dissection in papillary thyroid cancer: current trends, persisting controversies, and unclarified uncertainties. *Surgical Oncology* 19: e57–e70.

23. Noguchi, S., Yamashita, H., Uchino, S. et al. (2008). Papillary microcarcinoma. *World Journal of Surgery* 32: 747–753.

24. Mehanna, H., Al-Maqbili, T., Carter, B. et al. (2014). Differences in the recurrence and mortality outcomes rates of incidental and nonincidental papillary thyroid microcarcinoma: a systematic review and meta-analysis of 21,329 person-years of follow-up. *The Journal of Clinical Endocrinology and Metabolism* 99: 2834–2843.

25. Sugitani, I., Toda, K., Yamada, K. et al. (2010). Three distinctly different kinds of papillary thyroid microcarcinoma should be recognized: our treatment strategies and outcomes. *World Journal of Surgery* 34: 1222–1231.

26. Ito, Y., Miyauchi, A., Inoue, H. et al. (2010). An observational trial for papillary thyroid microcarcinoma in Japanese patients. *World Journal of Surgery* 34: 28–35.

27. Pisanu, A., Reccia, I., Nardello, O. et al. (2009). Risk factors for nodal metastasis and recurrence among patients with papillary thyroid microcarcinoma: differences in clinical relevance between nonincidental and incidental tumors. *World Journal of Surgery* 33: 460–468.

28. Ito, Y., Higashiyama, T., Takamura, Y. et al. (2007). Prognosis of patients with benign thyroid diseases accompanied by incidental papillary carcinoma undetectable on preoperative imaging tests. *World Journal of Surgery* 31: 1672–1676.

29. Niemeier, L.A., Kuffner Akatsu, H., Song, C. et al. (2012). A combined molecular-pathologic score improves risk stratification of thyroid papillary microcarcinoma. *Cancer* 118: 2069–2077.

30. De Crea, C., Raffaelli, M., Sessa, L. et al. (2014). Actual incidence and clinical behaviour of follicular thyroid carcinoma: an institutional experience. *Scientific World Journal* 2014: 7.

31. La Vecchia, C., Malvezzi, M., Bosetti, C. et al. (2015). Thyroid cancer mortality and incidence: a global overview. *International Journal of Cancer* 136: 2187–2195.

32. Burgess, J.R., Dwyer, T., McArdle, K. et al. (2000). The changing incidence and spectrum of thyroid carcinoma in Tasmania (1978–1998) during a transition from iodine sufficiency to iodine deficiency. *The Journal of Clinical Endocrinology and Metabolism* 85: 1513–1517.

33. Kushchayeva, Y., Duh, Q.Y., Kebebew, E. et al. (2004). Prognostic indications for hurthle cell cancer. *World Journal of Surgery* 28: 1266–1270.

34. Ahmadi, S., Stang, M., Jiang, X.S. et al. (2016). Hürthle cell carcinoma: current perspectives. *OncoTargets and Therapy* 9: 6873–6884.

35. Yoon, J.H., Kim, E.-K., Youk, J.H. et al. (2014). Better understanding in the differentiation of thyroid follicular adenoma, follicular carcinoma, and follicular variant of papillary carcinoma: a retrospective study. *International Journal of Endocrinology* 2014: 9.

36. Bongiovanni, M., Fadda, G., and Faquin, W.C. (2018). Poorly differentiated thyroid carcinoma. In: *The Bethesda System for Reporting Thyroid Cytopathology: Definitions, Criteria, and Explanatory Notes* (ed. S.Z. Ali and E.S. Cibas), 177–188. Cham: Springer International Publishing.

37. De Lellis, R.A., Lloyd, R.V., Heitz, P.U. et al. *Pathology and Genetics of Tumors of Endocrine Organs*. Lyons, France: IARC Press.

38. Eng, C., Mulligan, L.M., Smith, D.P. et al. (1995). Mutation of the RET protooncogene in sporadic medullary thyroid carcinoma. *Genes, Chromosomes and Cancer* 12: 209–212.

39. Wells, S.A. Jr., Asa, S.L., Dralle, H. et al. (2015). Revised American Thyroid Association guidelines for the management of medullary thyroid carcinoma. *Thyroid* 25: 567–610.

40. Marcello, M.A., Cunha, L.L., Batista, F.A. et al. (2014). Obesity and thyroid cancer. *Endocrine-Related Cancer* 21: T255–T271.

41. Pazaitou-Panayiotou, K., Polyzos, S.A., and Mantzoros, C.S. (2013). Obesity and thyroid cancer: epidemiologic associations and underlying mechanisms. *Obesity Reviews* 14: 1006–1022.

42. Ron, E., Lubin, J.H., Shore, R.E. et al. (1995). Thyroid cancer after exposure to external radiation: a pooled analysis of seven studies. *Radiation Research* 141: 259–277.

43. Shore, R.E. (1992). Issues and epidemiological evidence regarding radiation-induced thyroid cancer. *Radiation Research* 131: 98–111.

44. Moysich, K.B., Menezes, R.J., and Michalek, A.M. (2002). Chernobyl-related ionising radiation exposure and cancer risk: an epidemiological review. *The Lancet Oncology* 3: 269–279.

45. Sinnott, B., Ron, E., and Schneider, A.B. (2010). Exposing the thyroid to radiation: a review of its current extent, risks, and implications. *Endocrine Reviews* 31: 756–773.

46. Veiga, L.H., Holmberg, E., Anderson, H. et al. (2016). Thyroid cancer after childhood exposure to external radiation: an updated pooled analysis of 12 studies. *Radiation Research* 185: 473–484.

47. Cardis, E., Kesminiene, A., Ivanov, V. et al. (2005). Risk of thyroid cancer after exposure to ^{131}I in childhood. *Journal of the National Cancer Institute* 97: 724–732.

48. Brenner, D.J. and Hall, E.J. (2007). Computed tomography — an increasing source of radiation exposure. *The New England Journal of Medicine* 357: 2277–2284.

49. Neta, G., Rajaraman, P., Berrington de Gonzalez, A. et al. (2013). A prospective study of medical diagnostic radiography and risk of thyroid cancer. *American Journal of Epidemiology* 177: 800–809.

50. Han, M.A. and Kim, J.H. (2017). Diagnostic x-ray exposure and thyroid cancer risk: systematic review and meta-analysis. *Thyroid*.

51. Rajoria, S., Suriano, R., Shanmugam, A. et al. (2010). Metastatic phenotype is regulated by estrogen in thyroid cells. *Thyroid* 20: 33–41.

52. Santin, A.P. and Furlanetto, T.W. (2011). Role of estrogen in thyroid function and growth regulation. *Journal of Thyroid Research* 2011: 875125.

53. Derwahl, M. and Nicula, D. (2014). Estrogen and its role in thyroid cancer. *Endocrine-Related Cancer* 21: T273–T283.

54. Dong, W., Zhang, H., Li, J. et al. (2013). Estrogen induces metastatic potential of papillary thyroid cancer cells through estrogen receptor alpha and beta. *International Journal of Endocrinology* 2013: 941568.

55. Kuo, J.H., Chabot, J.A., and Lee, J.A. (2016). Breast cancer in thyroid cancer survivors: an analysis of the surveillance, epidemiology, and end results-9 database. *Surgery* 159: 23–29.

56. Bolf, E.L., Sprague, B.L., and Carr, F.E. (2018). A linkage between thyroid and breast cancer: A common etiology? *Cancer Epidemiology, Biomarkers and Prevention*. DOI: 10.1158/1055-9965. EPI-18-0877 [Epub ahead of print].

57. Nielsen, S.M., White, M.G., Hong, S. et al. (2016). The breast-thyroid cancer link: a systematic review and meta-analysis. *Cancer Epidemiology, Biomarkers & Prevention* 25: 231–238.

58. Haugen, B.R. (2017). 2015 American Thyroid Association management guidelines for adult patients with thyroid nodules and differentiated thyroid cancer: what is new and what has changed? *Cancer* 123: 372–381.

59. Kwong, N., Medici, M., Angell, T.E. et al. (2015). The influence of patient age on thyroid nodule formation, multinodularity, and thyroid cancer risk. *The Journal of Clinical Endocrinology and Metabolism* 100: 4434–4440.

60. Francis, G.L., Waguespack, S.G., Bauer, A.J. et al. (2015). American Thyroid Association: management guidelines for children with thyroid nodules and differentiated thyroid cancer. *Thyroid* 25: 716–759.

61. Kebebew, E., Greenspan, F.S., Clark, O.H. et al. (2005). Extent of disease and practice patterns for medullary thyroid cancer. *Journal of the American College of Surgeons* 200: 890–896.

62. Sturgeon, C. and Clark, O.H. (2005). Familial nonmedullary thyroid cancer. *Thyroid* 15: 588–593.

63. Griffith, C.C. and Seethala, R.R. Familial non-medullary thyroid cancer: an update on the genetic and pathologic features. *Diagnostic Histopathology* 22: 101–107.

64. Navas-Carrillo, D., Rios, A., Rodriguez, J.M. et al. (2014). Familial nonmedullary thyroid cancer: screening, clinical, molecular and genetic findings. *Biochimica et Biophysica Acta* 1846: 468–476.

65. Nose, V. (2008). Familial non-medullary thyroid carcinoma: an update. *Endocrine Pathology* 19: 226–240.

66. Dotto, J. and Nose, V. (2008). Familial thyroid carcinoma: a diagnostic algorithm. *Advances in Anatomic Pathology* 15: 332–349.

67. Nose, V. (2011). Familial thyroid cancer: a review. *Modern Pathology* 24 (Suppl 2): S19–S33.

68. LiVolsi, V.A., Baraban, E., and Baloch, Z.W. (2017). Familial thyroid carcinoma: the road less traveled in thyroid pathology-an update. *Diagnostic Histopathology* 23: 366–377.

69. Richards, M.L. (2010). Familial syndromes associated with thyroid cancer in the era of personalized medicine. *Thyroid* 20: 707–713.

70. Lai, X., Xia, Y., Zhang, B. et al. (2017). A meta-analysis of Hashimoto's thyroiditis and papillary thyroid carcinoma risk. *Oncotarget* 8: 62414–62424.

71. Lee, J.-H., Kim, Y., Choi, J.-W. et al. (2013). The association between papillary thyroid carcinoma and histologically proven Hashimoto's thyroiditis: a meta-analysis. *European Journal of Endocrinology* 168: 343–349.

72. Caturegli, P., De Remigis, A., and Rose, N.R. (2014). Hashimoto thyroiditis: clinical and diagnostic criteria. *Autoimmunity Reviews* 13: 391–397.

73. Smith, T.J. and Hegedus, L. (2016). Graves' disease. *The New England Journal of Medicine* 375: 1552–1565.

74. Ross, D.S., Burch, H.B., Cooper, D.S. et al. (2016). 2016 American thyroid association guidelines for diagnosis and management of hyperthyroidism and other causes of thyrotoxicosis. *Thyroid* 26: 1343–1421.

75. Burch, H.B. and Cooper, D.S. (2015). Management of Graves' disease: a review. *JAMA* 314: 2544–2554.

76. Smith, J.J., Chen, X., Schneider, D.F. et al. (2013). Cancer after thyroidectomy: a multi-institutional experience with 1,523 patients. *Journal of the American College of Surgeons* 216: 571–577.

77. Resende de Paiva, C., Grønhøj, C., Feldt-Rasmussen, U. et al. (2017). Association between Hashimoto's thyroiditis and thyroid cancer in 64,628 patients. *Frontiers in Oncology* 7: 53.

78. Chen, Y.-K., Lin, C.-L., Chang, Y.-J. et al. (2013). Cancer risk in patients with Graves' disease: a nationwide cohort study. *Thyroid* 23: 879–884.

79. Belfiore, A., Russo, D., Vigneri, R. et al. (2001). Graves' disease, thyroid nodules and thyroid cancer. *Clinical Endocrinology* 55: 711–718.

80. Ren, M., Wu, M.C., Shang, C.Z. et al. (2014). Predictive factors of thyroid cancer in patients with Graves' disease. *World Journal of Surgery* 38: 80–87.

81. Boutzios, G., Vasileiadis, I., Zapanti, E. et al. (2014). Higher incidence of tall cell variant of papillary thyroid carcinoma in Graves' disease. *Thyroid* 24: 347–354.

82. Zimmermann, M.B. and Galetti, V. (2015). Iodine intake as a risk factor for thyroid cancer: a comprehensive review of animal and human studies. *Thyroid Research* 8: 8.

83. Leung, A.M. and Braverman, L.E. (2014). Consequences of excess iodine. *Nature Reviews. Endocrinology* 10: 136–142.

84. Choi, W.J. and Kim, J. (2014). Dietary factors and the risk of thyroid cancer: a review. *Clinical nutrition research* 3: 75–88.

85. Cho, Y.A., Lee, J., and Kim, J. (2016). Association between nutrient intake and thyroid cancer risk in korean women. *Nutrition Research and Practice* 10: 336–341.

86. O'Grady, T.J., Kitahara, C.M., DiRienzo, A.G. et al. (2014). The association between selenium and other micronutrients and thyroid cancer incidence in the NIH-AARP diet and health study. *PLoS One* 9: e110886.

87. Renehan, A.G., Tyson, M., Egger, M. et al. (2008). Body-mass index and incidence of cancer: a systematic review and meta-analysis of prospective observational studies. *Lancet* 371: 569–578.

88. Arnold, M., Leitzmann, M., Freisling, H. et al. (2016). Obesity and cancer: an update of the global impact. *Cancer Epidemiology* 41: 8–15.

89. Arnold, M., Pandeya, N., Byrnes, G. et al. (2015). Global burden of cancer attributable to high body-mass index in 2012: a population-based study. *The Lancet Oncology* 16: 36–46.

90. Rinaldi, S., Lise, M., Clavel-Chapelon, F. et al. (2012). Body size and risk of differentiated thyroid carcinomas: findings from the epic study. *International Journal of Cancer* 131: E1004–E1014.

91. Xu, L., Port, M., Landi, S. et al. (2014). Obesity and the risk of papillary thyroid cancer: a pooled analysis of three case-control studies. *Thyroid* 24: 966–974.

92. Kitahara, C.M., Platz, E.A., Freeman, L.E.B. et al. (2011). Obesity and thyroid cancer risk among U.S. Men and women: a pooled analysis of five prospective studies. *Cancer Epidemiol Biomarkers Prevention: A Publication of the American Association for Cancer Research, Cosponsored by the American Society of Preventive Oncology* 20: 464–472.

93. Schmid, D., Ricci, C., Behrens, G. et al. (2015). Adiposity and risk of thyroid cancer: a systematic review and meta-analysis. *Obesity Reviews* 16: 1042–1054.

94. Almquist, M., Johansen, D., Björge, T. et al. (2011). Metabolic factors and risk of thyroid cancer in the metabolic syndrome and cancer project (Me-Can). *Cancer Causes & Control* 22: 743–751.

95. Zhang, Y., Guo, G.L., Han, X. et al. (2008). Do polybrominated diphenyl ethers (PBDEs) increase the risk of thyroid cancer? *Bioscience Hypotheses* 1: 195–199.

96. Gore, A.C., Chappell, V.A., Fenton, S.E. et al. (2015). Edc-2: the Endocrine Society's second scientific statement on endocrine-disrupting chemicals. *Endocrine Reviews* 36: E1–E150.

97. Diamanti-Kandarakis, E., Bourguignon, J.-P., Giudice, L.C. et al. (2009). Endocrine-disrupting chemicals: an Endocrine Society scientific statement. *Endocrine Reviews* 30: 293–342.

98. Soto, A.M. and Sonnenschein, C. (2010). Environmental causes of cancer: endocrine disruptors as carcinogens. *Nature Reviews. Endocrinology* 6: 363–370.

99. Di Donato, M., Cernera, G., Giovannelli, P. et al. (2017). Recent advances on Bisphenol-A and endocrine disruptor effects on human prostate cancer. *Molecular and Cellular Endocrinology* 457: 35–42.

100. Zhou, Z., Zhang, J., Jiang, F. et al. (2017). Higher urinary Bisphenol A concentration and excessive iodine intake are associated with nodular goiter and papillary thyroid carcinoma. *Bioscience Reports* 37: https://doi.org/10.1042/BSR20170678.

101. Tang-Peronard, J.L., Andersen, H.R., Jensen, T.K. et al. (2011). Endocrine-disrupting chemicals and obesity development in humans: a review. *Obesity Reviews* 12: 622–636.

102. Braun, J.M. (2016). Early-life exposure to edcs: role in childhood obesity and neurodevelopment. *Nature Reviews. Endocrinology* 13: 161.

103. Mughal, B.B. and Demeneix, B.A. (2017). Flame retardants and increased risk of thyroid cancer. *Nature Reviews. Endocrinology* 13: 627.

104. Hoffman, K., Lorenzo, A., Butt, C.M. et al. (2017). Exposure to flame retardant chemicals and occurrence and severity of papillary thyroid cancer: a case-control study. *Environment International* 107: 235–242.

105. Lee, H.-S., Kang, Y., Tae, K. et al. (2018). Proteomic biomarkers for Bisphenol A – early exposure and women's thyroid cancer. *Cancer Research and Treatment* 50: 111–117.

106. Smallridge, R.C., Ain, K.B., Asa, S.L. et al. (2012). American Thyroid Association guidelines for management of patients with anaplastic thyroid cancer. *Thyroid* 22: 1104–1139.

107. Adeniran, A.J., Zhu, Z., Gandhi, M. et al. (2006). Correlation between genetic alterations and microscopic features, clinical manifestations, and prognostic characteristics of thyroid papillary carcinomas. *The American Journal of Surgical Pathology* 30: 216–222.

108. Nikiforova, M.N., Lynch, R.A., Biddinger, P.W. et al. (2003). RAS point mutations and PAX8-PPARγ rearrangement in thyroid tumors: evidence for distinct molecular pathways in thyroid follicular carcinoma. *The Journal of Clinical Endocrinology and Metabolism* 88: 2318–2326.

109. Hou, P., Liu, D., Shan, Y. et al. (2007). Genetic alterations and their relationship in the phosphatidylinositol 3-Kinase/AKT pathway in thyroid cancer. *Clinical Cancer Research* 13: 1161–1170.

110. Kondo, T., Ezzat, S., and Asa, S.L. (2006). Pathogenetic mechanisms in thyroid follicular-cell neoplasia. *Nature Reviews. Cancer* 6: 292.

111. Landa, I., Ibrahimpasic, T., Boucai, L. et al. (2016). Genomic and transcriptomic hallmarks of poorly differentiated and anaplastic thyroid cancers. *The Journal of Clinical Investigation* 126: 1052–1066.

112. Ohori, N.P., Nikiforova, M.N., Schoedel, K.E. et al. (2010). Contribution of molecular testing to thyroid fine-needle aspiration cytology of "follicular lesion of undetermined significance/atypia of undetermined significance". *Cancer Cytopathology* 118: 17–23.

113. Moses, W., Weng, J., Sansano, I. et al. (2010). Molecular testing for somatic mutations improves the accuracy of thyroid fine-needle aspiration biopsy. *World Journal of Surgery* 34: 2589–2594.

114. Chudova, D., Wilde, J.I., Wang, E.T. et al. (2010). Molecular classification of thyroid nodules using high-dimensionality genomic data. *The Journal of Clinical Endocrinology and Metabolism* 95: 5296–5304.

115. Alexander, E.K., Kennedy, G.C., Baloch, Z.W. et al. (2012). Preoperative diagnosis of benign thyroid nodules with indeterminate cytology. *The New England Journal of Medicine* 367: 705–715.

116. de Groot, J.W.B., Links, T.P., Plukker, J.T.M. et al. (2006). RET as a diagnostic and therapeutic target in sporadic and hereditary endocrine tumors. *Endocrine Reviews* 27: 535–560.

117. Moura, M.M., Cavaco, B.M., Pinto, A.E. et al. (2011). High prevalence of RAS mutations in RET-negative sporadic medullary thyroid carcinomas. *The Journal of Clinical Endocrinology and Metabolism* 96: E863–E868.

118. Nikiforova, M.N., Wald, A.I., Roy, S. et al. (2013). Targeted next-generation sequencing panel (ThyroSeq) for detection of mutations in thyroid cancer. *The Journal of Clinical Endocrinology and Metabolism* 98: E1852–E1860.

119. Nikiforov, Y.E., Carty, S.E., Chiosea, S.I. et al. (2015). Impact of the multi-gene ThyroSeq next-generation sequencing assay on cancer diagnosis in thyroid nodules with atypia of undetermined significance/follicular lesion of undetermined significance cytology. *Thyroid* 25: 1217–1223.

120. Tavares, C., Melo, M., Cameselle-Teijeiro, J.M. et al. (2016). Endocrine tumours: genetic predictors of thyroid cancer outcome. *European Journal of Endocrinology* 174: R117–R126.

121. Vuong, H.G., Duong, U.N., Altibi, A.M. et al. (2017). A meta-analysis of prognostic roles of molecular markers in papillary thyroid carcinoma. *Endocrine connections* 6: R8–R17.

122. Xing, M., Alzahrani, A.S., Carson, K.A. et al. (2015). Association between BRAF V600E mutation and recurrence of papillary thyroid cancer. *Journal of Clinical Oncology* 33: 42–50.

123. Xing, M., Liu, R., Liu, X. et al. (2014). BRAF V600E and TERT promoter mutations cooperatively identify the most aggressive papillary thyroid cancer with highest recurrence. *Journal of Clinical Oncology* 32: 2718–2726.

124. Melo, M., da Rocha, A.G., Vinagre, J. et al. (2014). TERT promoter mutations are a major indicator of poor outcome in differentiated thyroid carcinomas. *The Journal of Clinical Endocrinology and Metabolism* 99: E754–E765.

125. Russo, M., Malandrino, P., Nicolosi, M.L. et al. (2014). The BRAF(V600E) mutation influences the short- and medium-term outcomes of classic papillary thyroid cancer, but is not an independent predictor of unfavorable outcome. *Thyroid* 24: 1267–1274.

126. Gandolfi, G., Ragazzi, M., Frasoldati, A. et al. (2015). TERT promoter mutations are associated with distant metastases in papillary thyroid carcinoma. *European Journal of Endocrinology* 172: 403–413.

127. Xing, M., Alzahrani, A.S., Carson, K.A. et al. (2013). Association between BRAF V600E mutation and mortality in patients with papillary thyroid cancer. *JAMA* 309: 1493–1501.

128. Shen, X., Liu, R., and Xing, M. (2017). A six-genotype genetic prognostic model for papillary thyroid cancer. *Endocrine-Related Cancer* 24: 41–52.

129. Gurtan, A.M. and Sharp, P.A. (2013). The role of miRNAs in regulating gene expression networks. *Journal of Molecular Biology* 425: 3582–3600.

130. Stack, B.C. Jr., Ferris, R.L., Goldenberg, D. et al. (2012). American Thyroid Association consensus review and statement regarding the anatomy, terminology, and rationale for lateral neck dissection in differentiated thyroid cancer. *Thyroid* 22: 501–508.

131. Calin, G.A. and Croce, C.M. (2006). MicroRNA-cancer connection: the beginning of a new tale. *Cancer Research* 66: 7390–7394.

132. Iorio, M.V. and Croce, C.M. (2012). MicroRNA dysregulation in cancer: diagnostics, monitoring and therapeutics. A comprehensive review. *EMBO Molecular Medicine* 4: 143–159.

133. Zen, K. and Zhang, C.Y. (2012). Circulating microRNAs: a novel class of biomarkers to diagnose and monitor human cancers. *Medicinal Research Reviews* 32: 326–348.

134. Krützfeldt, J. (2016). Strategies to use microRNAs as therapeutic targets. *Best Practice & Research. Clinical Endocrinology & Metabolism* 30: 551–561.

135. Celano, M., Rosignolo, F., Maggisano, V. et al. (2017). MicroRNAs as biomarkers in thyroid carcinoma. *International Journal of Genomics* 2017: 11.

136. Aragon Han, P., Weng, C.H., Khawaja, H.T. et al. (2015). MicroRNA expression and association with clinicopathologic features in papillary thyroid cancer: a systematic review. *Thyroid* 25: 1322–1329.

137. Chruscik, A. and Lam, A.K. (2015). Clinical pathological impacts of microRNAs in papillary thyroid carcinoma: a crucial review. *Experimental and Molecular Pathology* 99: 393–398.

138. Wojtas, B., Ferraz, C., Stokowy, T. et al. (2014). Differential miRNA expression defines migration and reduced apoptosis in follicular thyroid carcinomas. *Molecular and Cellular Endocrinology* 388: 1–9.

139. Dettmer, M.S., Perren, A., Moch, H. et al. (2014). MicroRNA profile of poorly differentiated thyroid carcinomas: new diagnostic and prognostic insights. *Journal of Molecular Endocrinology* 52: 181–189.

140. Landa, I., Ibrahimpasic, T., Boucai, L. et al. (2016). Genomic and transcriptomic hallmarks of poorly differentiated and anaplastic thyroid cancers. *The Journal of Clinical Investigation* 126: 1052–1066.

141. Abraham, D., Jackson, N., Gundara, J.S. et al. (2011). MicroRNA profiling of sporadic and hereditary medullary thyroid cancer identifies predictors of nodal metastasis, prognosis, and potential therapeutic targets. *Clinical Cancer Research* 17: 4772–4781.

142. Mian, C., Pennelli, G., Fassan, M. et al. (2012). MicroRNA profiles in familial and sporadic medullary thyroid carcinoma: preliminary relationships with RET status and outcome. *Thyroid* 22: 890–896.

143. Nikiforova, M.N., Tseng, G.C., Steward, D. et al. (2008). Microrna expression profiling of thyroid tumors: biological significance and diagnostic utility. *The Journal of Clinical Endocrinology and Metabolism* 93: 1600–1608.

144. Nikiforova, M.N., Chiosea, S.I., and Nikiforov, Y.E. (2009). MicroRNA expression profiles in thyroid tumors. *Endocrine Pathology* 20: 85–91.

145. Wojcicka, A., Kolanowska, M., and Jazdzewski, K. (2016). Mechanisms in endocrinology: microRNA in diagnostics and therapy of thyroid cancer. *European Journal of Endocrinology* 174: R89–R98.

146. Stokowy, T., Wojtas, B., Krajewska, J. et al. (2015). A two miRNA classifier differentiates follicular thyroid carcinomas from follicular thyroid adenomas. *Molecular and Cellular Endocrinology* 399: 43–49.

147. Boufraqech, M., Klubo-Gwiezdzinska, J., and Kebebew, E. (2016). MicroRNAs in the thyroid. *Best Practice & Research. Clinical Endocrinology & Metabolism* 30: 603–619.

148. Graham, M.E.R., Hart, R.D., Douglas, S. et al. (2015). Serum microRNA profiling to distinguish papillary thyroid cancer from benign thyroid masses. *Journal of Otolaryngology - Head & Neck Surgery* 44: 33.

149. Marini, F., Luzi, E., and Brandi, M.L. (2011). MicroRNA role in thyroid cancer development. *Journal of Thyroid Research* 2011: 407123.

150. Faam, B., Ghaffari, M.A., Ghadiri, A. et al. (2015). Epigenetic modifications in human thyroid cancer. *Biomedical Reports* 3: 3–8.

151. Shen, R., Liyanarachchi, S., Li, W. et al. (2012). MicroRNA signature in thyroid fine needle aspiration cytology applied to "atypia of undetermined significance" cases. *Thyroid* 22: 9–16.

152. Sidiropoulos, N., Crothers, J., Balla, A. et al. (2015). Molecular pathogenesis of papillary thyroid microcarcinoma. *Thyroid* 25 (Suppl 1): A356–A383.

153. Rosignolo, F., Sponziello, M., Giacomelli, L. et al. (2017). Identification of thyroid-associated serum microRNA profiles and their potential use in thyroid cancer follow-up. *Journal of the Endocrine Society* 1: 3–13.

154. Zhang, Y., Xu, D., Pan, J. et al. (2017). Dynamic monitoring of circulating microRNAs as a predictive biomarker for the diagnosis and recurrence of papillary thyroid carcinoma. *Oncology Letters* 13: 4252–4266.

155. Lee, J.C., Zhao, J.T., Clifton-Bligh, R.J. et al. (2013). MicroRNA-222 and microRNA-146b are tissue and circulating biomarkers of recurrent papillary thyroid cancer. *Cancer* 119: 4358–4365.

156. Pohl, M., Grabellus, F., Worm, K. et al. (2017). Intermediate microRNA expression profile in Graves' disease falls between that of normal thyroid tissue and papillary thyroid carcinoma. *Journal of Clinical Pathology* 70: 33–39.

157. Yip, L., Wharry, L.I., Armstrong, M.J. et al. (2014). A clinical algorithm for fine-needle aspiration molecular testing effectively guides the appropriate extent of initial thyroidectomy. *Annals of Surgery* 260: 163–168.

158. Ferris, R.L., Baloch, Z., Bernet, V. et al. (2015). American Thyroid Association statement on surgical application of molecular profiling for thyroid nodules: current impact on perioperative decision making. *Thyroid* 25: 760–768.

159. Metzker, M.L. (2010). Sequencing technologies - the next generation. *Nature Reviews. Genetics* 11: 31–46.

160. The Cancer Genome Atlas (2014). Integrated genomic characterization of papillary thyroid carcinoma. *Cell* 159: 676–690.

161. Alzahrani, A.S., Alsaadi, R., Murugan, A.K. et al. (2016). TERT promoter mutations in thyroid cancer. *Hormones and Cancer* 7: 165–177.

162. Song, Y.S., Lim, J.A., Choi, H. et al. (2016). Prognostic effects of TERT promoter mutations are enhanced by coexistence with BRAF or RAS mutations and strengthen the risk prediction by the ATA or TNM staging system in differentiated thyroid cancer patients. *Cancer* 122: 1370–1379.

163. Cooper, D.S., Doherty, G.M., Haugen, B.R. et al. (2009). Revised American Thyroid Association management guidelines for patients with thyroid nodules and differentiated thyroid cancer. *Thyroid* 19: 1167–1214.

164. Edge, S.B. and Compton, C.C. (2010). The American Joint Committee on cancer: the 7th edition of the AJCC cancer staging manual and the future of TNM. *Annals of Surgical Oncology* 17: 1471–1474.

165. Omry-Orbach, G. (2016). Risk stratification in differentiated thyroid cancer: an ongoing process. *Rambam Maimonides Medical Journal* 7: e0003.

166. Tufano, R.P., Clayman, G., Heller, K.S. et al. (2015). American Thyroid Association: management of recurrent/persistent nodal disease in patients with differentiated thyroid cancer: a critical review of the risks and benefits of surgical intervention versus active surveillance. *Thyroid* 25: 15–27.

167. National Cancer Institute. Thyroid cancer treatment (pdq). 2014; http://cancer.gov/cancertopics/pdq/treatment/thyroid/HealthProfessional. Accessed January 15, 2018.

168. Wells, S.A. Jr., Asa, S.L., Dralle, H. et al. (2015). American Thyroid Association Guidelines Task Force on Medullary Thyroid C. Revised American Thyroid Association guidelines for the management of medullary thyroid carcinoma. *Thyroid* 25: 567–610.

169. Xing, M. (2013). Molecular pathogenesis and mechanisms of thyroid cancer. *Nature Reviews Cancer* 13: 184.

170. Cabanillas, M.E., Zafereo, M., Gunn, G.B. et al. (2016). Anaplastic thyroid carcinoma: treatment in the age of molecular targeted therapy. *Journal of Oncology Practice/American Society of Clinical Oncology* 12: 511–518.

171. Smallridge, R.C. and Copland, J.A. (2010). Anaplastic thyroid carcinoma: pathogenesis and emerging therapies. *Clinical Oncology (Royal College of Radiologists)* 22: 486–497.

172. Viola, D., Valerio, L., Molinaro, E. et al. (2016). Treatment of advanced thyroid cancer with targeted therapies: ten years of experience. *Endocrine-Related Cancer* 23: R185–R205.

173. Elisei, R. and Pinchera, A. (2012). Advances in the follow-up of differentiated or medullary thyroid cancer. *Nature Reviews. Endocrinology* 8: 466–475.

174. Tuttle, R.M., Haugen, B., and Perrier, N.D. (2017). Updated American Joint Committee on cancer/tumor-node-metastasis staging system for differentiated and anaplastic thyroid cancer (eighth edition): what changed and why? *Thyroid* 27: 751–756.

175. Oda, H., Miyauchi, A., Ito, Y. et al. (2017). Comparison of the costs of active surveillance and immediate surgery in the management of low-risk papillary microcarcinoma of the thyroid. *Endocrine Journal* 64: 59–64.

176. Ghaznavi, S.A. and Tuttle, R.M. (2017). Disease-related and treatment-related unfavorable events in the management of low-risk papillary microcarcinoma of the thyroid by active surveillance versus immediate surgery. *Annals of Thyroid* 2: 16.

177. Tarasova, V.D. and Tuttle, R.M. (2017). Current management of low risk differentiated thyroid cancer and papillary microcarcinoma. *Clinical Oncology* 29: 290–297.

178. Haymart, M.R., Esfandiari, N.H., Stang, M.T. et al. (2017). Controversies in the management of low-risk differentiated thyroid cancer. *Endocrine Reviews* 38: 351–378.

179. Reddi, H.V., Driscoll, C.B., Madde, P. et al. (2013). Redifferentiation and induction of tumor suppressors mir-122 and mir-375 by the PAX8/PPARγ fusion protein inhibits anaplastic thyroid cancer: a novel therapeutic strategy. *Cancer Gene Therapy* 20: 267.

180. Vivaldi, A., Miasaki, F.Y., Ciampi, R. et al. (2009). Re-differentiation of thyroid carcinoma cell lines treated with 5-aza-2′-deoxycytidine and retinoic acid. *Molecular and Cellular Endocrinology* 307: 142–148.

181. Lin, R.Y. (2011). Thyroid cancer stem cells. *Nature Reviews. Endocrinology* 7: 609–616.

182. Gao, Y.J., Li, B., Wu, X.Y. et al. (2014). Thyroid tumor-initiating cells: increasing evidence and opportunities for anticancer therapy (review). *Oncology Reports* 31: 1035–1042.

183. Kim, W.G. and SY, C. (2013). Thyroid hormone receptors and cancer. *Biochimica et Biophysica Acta (BBA) – General Subjects* 1830: 3928–3936.

184. Aranda, A., Martinez-Iglesias, O., Ruiz-Llorente, L. et al. (2009). Thyroid receptor: roles in cancer. *Trends in Endocrinology and Metabolism* 20: 318–324.

185. Kim, W.G., Zhao, L., Kim, D.W. et al. (2014). Inhibition of tumorigenesis by the thyroid hormone receptor beta in xenograft models. *Thyroid* 24: 260–269.

186. Martinez-Iglesias, O., Garcia-Silva, S., Tenbaum, S. et al. (2009). Thyroid hormone receptor β1 acts as a potent suppressor of tumor invasiveness and metastasis. *Cancer Research* 69: 501–509.

187. Carr, F.E., Tai, P.W., Barnum, M.S. et al. (2016). Thyroid hormone receptor-β (TRβ) mediates runt-related transcription factor 2 (RUNX2) expression in thyroid cancer cells: a novel signaling pathway in thyroid cancer. *Endocrinology* 157: 3278–3292.

188. Guigon, C.J., Kim, D.W., Zhu, X. et al. (2010). Tumor suppressor action of liganded thyroid hormone receptor beta by direct repression of beta-catenin gene expression. *Endocrinology* 151: 5528–5536.

189. Park, J.W., Zhao, L., Willingham, M.C. et al. (2017). Loss of tyrosine phosphorylation at y406 abrogates the tumor suppressor functions of the thyroid hormone receptor β. *Molecular Carcinogenesis* 56: 489–498.

190. Zambrano, A., García-Carpizo, V., Gallardo, M.E. et al. (2014). The thyroid hormone receptor β induces DNA damage and premature senescence. *The Journal of Cell Biology* 204: 129–146.

191. Chakravarti, B., Ravi, J., and Ganju, R.K. (2014). Cannabinoids as therapeutic agents in cancer: current status and future implications. *Oncotarget* 5: 5852–5872.

192. Hermanson, D.J. and Marnett, L.J. (2011). Cannabinoids, endocannabinoids and cancer. *Cancer Metastasis Reviews* 30: 599–612.

193. Bifulco, M., Malfitano, A.M., Pisanti, S. et al. (2008). Endocannabinoids in endocrine and related tumours. *Endocrine-Related Cancer* 15: 391–408.

194. Kushchayeva, Y., Jensen, K., Burman, K.D. et al. (2014). Repositioning therapy for thyroid cancer: new insights on established medications. *Endocrine-Related Cancer* 21: R183–R194.

195. Di Leva, G. and Croce, C.M. (2010). Roles of small RNAs in tumor formation. *Trends in Molecular Medicine* 16: 257–267.

196. Yip, L., Kelly, L., Shuai, Y. et al. (2011). MicroRNA signature distinguishes the degree of aggressiveness of papillary thyroid carcinoma. *Annals of Surgical Oncology* 18: 2035–2041.

197. Yoruker, E.E., Terzioglu, D., Teksoz, S. et al. (2016). MicroRNA expression profiles in papillary thyroid carcinoma, benign thyroid nodules and healthy controls. *Journal of Cancer* 7: 803–809.

198. Yu, S., Liu, Y., Wang, J. et al. (2012). Circulating microRNA profiles as potential biomarkers for diagnosis of papillary thyroid carcinoma. *The Journal of Clinical Endocrinology and Metabolism* 97: 2084–2092.

199. Li, X., Abdel-Mageed, A.B., Mondal, D. et al. (2013). MicroRNA expression profiles in differentiated thyroid cancer, a review. *International Journal of Clinical and Experimental Medicine* 6: 74–80.

200. Dettmer, M., Perren, A., Moch, H. et al. (2013). Comprehensive microRNA expression profiling identifies novel markers in follicular variant of papillary thyroid carcinoma. *Thyroid* 23: 1383–1389.

201. Celano, M., Rosignolo, F., Maggisano, V. et al. (2017). MicroRNAs as biomarkers in thyroid carcinoma. *International Journal of Genomics* 2017: 6496570.

202. Mancikova, V., Montero-Conde, C., Perales-Paton, J. et al. (2017). Multilayer OMIC data in medullary thyroid carcinoma identifies the STAT3 pathway as a potential therapeutic target in RETM918T tumors. *Clinical Cancer Research* 23: 1334–1345.

203. Sasanakietkul, T., Murtha, T.D., Javid, M. et al. (2017). Epigenetic modifications in poorly differentiated and anaplastic thyroid cancer. *Molecular and Cellular Endocrinology*.

204. Hsiao, S.J. and Nikiforov, Y.E. (2018). Utilization of molecular markers in the diagnosis and management of thyroid nodules. In: *Thyroid and Parathyroid Ultrasound and Ultrasound-Guided FNA* (ed. D.S. Duick, R.A. Levine and M.A. Lupo), 465–487. Cham: Springer International Publishing.

205. Nikiforov, Y.E. (2017). Role of molecular markers in thyroid nodule management: then and now. *Endocrine Practice* 23: 979–988.

VII

TREATMENT OPTIONS

24

RADIATION THERAPY

Thomas J. FitzGerald[1,2], Maryann Bishop-Jodoin[2,3], Fran Laurie[2,3], Christopher Riberdy[1], Jesse N. Aronowitz[1], Elizabeth Bannon[1], Bruce A. Bornstein[1], Carla D. Bradford[1], Harry Bushe[1], Maria G. Cicchetti[1,2], Linda Ding[1], Jonathan M. Glanzman[1], David J. Goff[1], Beth B. Herrick[1], Jessica R. Hiatt[1], I-Lin Kuo[1], Yuan-Chyuan Lo[1], Janaki Moni[1,2], Richard S. Pieters[1], Paul S. Rava[1], Allison Sacher[1], Jonathan Saleeby[1], Shirin Sioshansi[1], Kenneth Ulin[1], John M. Varlotto[1], and Tao Wang[3]

[1]*Department of Radiation Oncology, University of Massachusetts Medical School; UMass Memorial Healthcare, Worcester, MA, USA*
[2]*IROC Rhode Island, Lincoln, RI, USA*
[3]*Department of Radiation Oncology, University of Massachusetts Medical School, Worcester, MA, USA*

Radiation therapy is an important component to the care of the patient with cancer. Nearly 60% of all patients afflicted with cancer will require radiation therapy at some time point during the course of their disease with a large majority of patients on treatment now treated with curative intent. This chapter will detail the history of radiation therapy, identify essential components to a modern department of radiation therapy, and focus on salient areas of improvement in patient outcome for the next generation of cancer patients.

Cancer: Prevention, Early Detection, Treatment and Recovery, Second Edition. Edited by Gary S. Stein and Kimberly P. Luebbers.
© 2019 John Wiley & Sons, Inc. Published 2019 by John Wiley & Sons, Inc.

HISTORY

Soon after the discovery of X-rays by Roentgen in 1895, it was identified that long-term exposure to X-rays during one or multiple sessions would result in visible injury to the skin including erythema and ulceration [1, 2]. Because many dermal lesions were then being treated by caustic agents including chemicals and electricity, X-rays were quickly assimilated into the care of individuals with dermal lesions and other abnormalities now made more visible through the discovery of imaging [3]. X-rays identified other sites of disease (chest), and, because X-rays were thought to be bactericidal, within a few years there were many published reports concerning the use of X-rays in treating rodent ulcer, epithelioma, tuberculosis, and other maladies [4]. The enthusiasm of practitioners quickly diminished when injuries from long protracted exposure to low energy X-rays became more visible and the successes described in scientific papers were thought to be exaggerated. By 1920, many deleterious public health effects of protracted X-ray exposure were being published including the deaths of factory workers at a radioluminsecent watch factory and the death of industrialist Eben Byers due to radium poisoning from Radithor, radium water used for medicinal purpose [5].

A second wave of enthusiasm paralleled the discovery of X-rays with the discovery of radium by Madame Curie. Radium emanation was passed into both glass- and platinum-based metal tubes. The tubes could be applied in a number of strategic locations including the oral cavity, esophagus, rectum, vagina/cervix, and other body areas through surgical incisions. Investigators began to use radium for well-defined local problems including skin lesions and keloid formation with X-rays used for areas requiring larger volumes for treatment [6].

A significant step forward for modern radiotherapy occurred in 1922 when Henri Coutard, working at the Curie Institute, presented evidence at the International Congress of Oncology that laryngeal cancer could be treated without disastrous effects and by 1934 presented a protracted, fractionated strategy for treatment, which has served as the foundation of fractionated radiation treatment of today. His radiation treatment plan created a significant but recoverable acute mucosal reaction within head and neck tissue. His new assumption was that cancer cells had similar regenerative capacity to normal cells and recovery for normal cells would be improved with less radiation daily dose delivered over a more extended period of time. He was inspired by the work of Claudius Regaud who observed that single large doses of X-ray to rabbits would create significant dermal injury but the same dose applied as a series of smaller doses would not injure tissue to the same degree. By 1935, his strategy for radiation care was adopted by many hospitals and the era of modern radiotherapy moved to an enterprise level [7].

Although equipment for the application of radiation therapy improved with optimum X-ray filtration and protection for the investigator, treatments continued to be delivered with low energy units measured in thousands of electron volts, essentially delivered at X-ray energies consistent with diagnostic radiology. These low energies would deliver higher radiation doses at the skin/dermal surface in order to achieve full dose at the depth of the tumor, hence leading to sequela to tissues of limited self-renewal including bone and cartilage with nonhealing ulcers the by-product of injury to underlying support tissues and stroma. During this era, radiologists largely performed both roles of diagnostic imaging and therapeutic treatment. The American Board of Radiology (ABR) was incorporated in January 1934 and began to develop guidelines for training in both diagnostic radiology and therapeutic radiology/radiation oncology. The ABR certifies all radiologists and radiation oncologists practicing in today in the United States.

As is often the case in the development of technology, improvements in radar and microwave cyclotron function during World War II became the infrastructure for the development of the modern clinical linear accelerator. The advent of the microwave power tube became the source to accelerate atomic particles to the speed of light, thus capable of generating X-ray energies measured in millions of electron volts. This created a new opportunity for X-ray biology and patient care as high energy X-rays were absorbed in tissue uniform to the atomic number of the material treated, thus creating far more consistent and predictable treatment effects in tissue. The first microwave-based linear accelerator for modern radiation treatment was at Hammersmith hospital in London, England with the first patient treated in 1953. In 1957, Henry Kaplin utilized the linear accelerator at Stanford Hospital, then in San Francisco CA, to treat the left eye of a two-year-old child with retinoblastoma whose right eye was surgically removed because of disease. The patient survived with vision intact for the rest of his life. The cyclotron was moved to Palo Alto, CA when Stanford Hospital relocated to its current location. The first accelerator could not rotate around the patient therefore the patient had to be positioned for the beam. The first 360° isocentric linear accelerator was built in 1960. Adjustments in standing wave accelerator technology in 1968 significantly reduced the cost of the accelerator and at this time patient care could move forward with advanced X-ray treatment at an enterprise function now at a more affordable cost to healthcare institutions. The ability to integrate multiple higher energy X-ray technology into a single accelerator was developed in 1972 by the Varian Corporation. Today, full volumetric real time imaging is completely integrated with computer driven four-dimensional treatment planning for execution of daily patient treatment with tens of thousands of medical accelerators distributed worldwide for patient care. Treatments now take minutes with full-image validation of the target. Particle therapy with protons has become more practical with smaller and more cost-effective cyclotrons available for patient care. With less normal tissue exposed to treatment, more compressed treatment schedules including single fraction treatments with accelerated fractionation patterns are now being revisited for patient care. Training programs in radiation oncology are attracting very talented and outstanding student investigators. Coupled with strong integration of basic and clinical translational science into our training programs, the future of radiation oncology remains exceptionally strong from an academic and patient-care perspective.

RADIATION BIOLOGY AND SENSITIZATION

Interactions in both tumor tissue and normal cells serve as the basis for modern radiation biology. The absorption of X-rays in tissue creates a variety of changes in tumor and normal tissue cell biology. Many of these changes can be repaired however exposure to X-ray leaves an invisible footprint on the status of cells exposed to radiation. In this section the focus will be on salient features associated with the tissue biology of exposure to X-rays.

Irreversible and irreparable X-ray damage to tissue, both in tumor and normal cells, is lethal if exposure leads to cell death. A human being exposed to total body radiation has a 50% likelihood of death if the total body exposure is greater than 2.0 Gy assuming no peripheral support. If supportive care such as fluids and blood products are available, human beings may be able to survive exposures of 6.0 Gy. There are rare survival reports of nuclear workers exposed during a radiation accident of slightly greater exposure. Sublethal damage is often repaired in hours of exposure however sublethal events can become lethal damage if more exposure is delivered in a protracted fractionated manner as during traditional radiation treatment fractionation strategies. Damage is considered potentially lethal if conditions after exposure can be modified to facilitate cell death [6].

Repair of damage caused by exposure to radiation is initiated immediately after exposure and is generally thought to be complete within hours of treatment. There is direct damage from radiation at the DNA level. Exposure to radiation also generates highly reactive oxygen species from water molecules, which also create intracellular damage. The molecules that cause intracellular damage are short lived and rapidly interact with various biomolecules housed within a cell including interactions in the nucleus. Much of the science directed at enhancing sensitivity to radiation by chemotherapy and other targeted agents is directed at inhibition of intracellular tumor repair post radiation treatment. There are also alternative strategies for synergistic damage to tumor tissue in combination with radiation therapy making the cell kill additive in nature. Investigators have shown in prostate cancer that cell adhesion molecules called integrins, are selectively expressed in cancerous tissue and not expressed in normal prostate epithelium [8, 9]. It has been demonstrated that hormone therapy directly affects tumor adhesion molecules and radiation therapy inhibits tumor cell integrin expression at high doses, therefore adhesion cell modifiers may play an important role in synergistically increasing cell kill with radiation therapy [8, 9]. Strategies such as this will become important adjuncts to treatments of the future as targeted therapies become more commonplace in patient care.

Both tumor cells and normal tissue exhibit varied differential sensitivity during the cell cycle [6, 10]. Cells are more sensitive to radiation therapy during the G2-M phase of the cell cycle and are more resistant to therapy in the DNA synthesis phase of the cell cycle. This provides an important therapeutic window of opportunity for protracted treatments as cells will redistribute into other phases of the cell cycle during a fractionated course of therapy. Modern targeted therapies are now being directed cell cycle check point inhibition and these are now being tested as vehicles to synchronize cell cycle sensitivity with radiation treatment.

Both tumor cells and normal tissue repopulate during a protracted phase of radiation management. Injury and cell death create an accelerated response by tissues for re-growth and cell proliferation. This is visible in acute responding normal tissue such as skin. It is postulated that extending a treatment course due to inter-current issues beyond an established time expectation for treatment may lead to tumor regrowth. This is thought to be especially important in cancers of the head and neck and cervix as treatment delays in these disease sites have been associated with worse patient outcome [6, 11–13]. It is also thought that clonogenic tumor cell populations may be less sensitive to radiation therapy and may populate tumor more quickly than anticipated under traditional circumstances and accelerate tumor cell repopulation [6]. Altered radiation treatment fractionation has been evaluated in several tumor systems to determine if different treatment strategies with multiple daily treatments and selective increased dose to targets improves patient outcome [14].

As tumor dies and debris is reabsorbed, active tumor will be in closer approximation to blood vessels and as such become better oxygenated. Tumor is more sensitive to radiation in an oxygenated environment; therefore a larger oxygen component to the tumor microenvironment should have a direct positive effect on tumor cell kill. The converse is true. Tumor is less sensitive to radiation therapy in a hypoxic environment. There have been many attempts with modest success, to increase oxygen levels in tumor as well as develop chemical modifiers of radiation treatment to improve cell kill in regions of hypoxia [6].

As early as 1906, Bergonie and Tribondeau recognized that cells were especially sensitive to radiation treatment when they were rapidly dividing and undifferentiated. We have come to recognize that various tumor cell systems have differential sensitivity to radiation treatment ranging from very sensitive (leukemia, lymphoma) to more resistant (melanoma, glioblastoma). The more resistant tumors appear to have a higher capacity for repairing sublethal events and become good candidates to evaluate biological strategies for radiation sensitization with inhibitors of repair [15]. Likewise the converse is true. Investigators have demonstrated that most leukemia cell lines and human bone marrow progenitor cells have little to no capacity to repair sublethal events, thus explaining dose rate independent survival to X-ray exploiting a biologic advantage to bone marrow transplantation and total body radiation therapy [16–22].

Various compounds have been used to enhance tumor cell kill and promote protection of normal tissue from radiation therapy. The areas of emphasis have included cell signaling pathways, cell cycle specific pathways, angiogenesis, and cell adhesion molecules. Many agents are used as part of the standard of care for patients treated with radiation therapy. These include alkylating agents such as Temodar for treatment of patients with glioblastoma and cis-platinum for patients with cancers of the head and neck and cervix. Although it is not perfectly clear whether these agents synergize cell kill with radiation therapy or whether their effect is simply additive in nature, it is clear from many publications that integrated treatments in many disease sites improve patient outcome and a greater emphasis should be placed on drug/X-ray interaction as we move into the new era of targeted therapy for cancer care [23].

Radiation protectors are compounds that are designed to reduce damage in normal tissues by X-ray treatment and need to be *in situ* during or shortly after exposure to treatment. These compounds are generally antioxidants and sulfhydryl groups that work to scavenge free radicals promoted after exposure to X-ray. These agents need to be selective in protecting normal tissue and not tumor and also must be delivered with minimal toxicity to normal tissue [23]. Amifostine may be the best example of a compound used clinically to achieve this objective as it serves to protect mucosa during exposure to X-ray treatment. Mitigators are compounds that may be able to modify potential damage from radiation treatment and are delivered after exposure. Examples of this include the use of varied oxygen species including oxygen saturated scaffolding applied to tissue in order to promote healing of tissues exposed to radiation therapy [24]. Investigators have demonstrated the improvement in the dermal healing of mice exposed to 50 Gy radiation therapy with the use of oxygen saturated gauze applied to the skin surface in a strategic manner. The same investigators have demonstrated that although heavily irradiated tissues remain oxygenated, absolute blood vessel counts are decreased as part of a late effect of management, therefore more susceptible to long-term damage if there is a secondary injury or infection [24, 25]. Various small molecule inhibitors of transforming growth factor beta are under evaluation to determine if they can ameliorate damage done by radiation therapy thought to be aggravated through this mechanism [26].

The science of radiation biology continues to expand as radiation therapy continues to increase in demand as a patient care option. As we extend treatment strategies including altered fractionation, biology will become an increasingly important vehicle for patient care. We are at the forefront of our understanding of molecular fingerprints to individual patient treatment and we will adjust treatment strategies based on tumor specific and patient specific molecular biology moving forward.

MODERN RADIATION THERAPY

Modern radiation therapy patients treated with curative intent are uniformly evaluated as part of a multidisciplinary format where colleagues from all oncology-based disciplines including diagnostic radiology and pathology are available for patient care discussion. In this venue, patient care decisions can be made in a team format with primary discussion carried out with the patient and the family by selected team members. The format achieves many objectives. The group can review all materials at the same time and discussions can be managed with each member reviewing the identical objects. The format also serves the purpose of harmonizing the multidisciplinary team and building mutual trust between the individual disciplines. A good working relationship becomes excellent when individual disciplines understand problems and limitations faced by other disciplines. The relationship becomes outstanding when the recognition of the limitations becomes the foundation of process improvements for the team. This includes opportunities for research in quality assurance and clinical-translational science. The interactions between the disciplines minimize the risk of communication gaps among healthcare disciplines as well as limit miscommunication to the patient and family.

Once a strategy for potential radiation treatment is established between involved providers, the patient is seen in consultation by the radiation oncologist to discuss and develop a radiation-therapy care plan. A comprehensive review of patient oncology history and medical comorbidities is conducted in order to provide a balanced review of the risks and benefits of radiation treatment, obtain consent for treatment, and move forward with a treatment planning strategy and planning arrangements.

The simulation process has changed significantly over the past decade. Fluoroscopic-based simulators have been largely replaced with volumetric imaging based simulators mostly utilizing computer tomography. There are a few institutions that have magnetic resonance (MR) simulators. MR-based units have importance in translational science as metabolic events can be monitored during treatment as part of a study. An immobilization device is constructed for the patient that will permit daily reproducibility of the treatment plan. Devices are now uniformly lightweight, facile to construct, and permit the patient to reproduce the same position on daily basis. Commonly used devices include custom head/shoulder molds, vacuum locked torso casts, and individualized custom devices used for extremity treatment.

Treatment planning systems and clinical processes have radically changed over the past decade. Many senior radiation oncologists were exclusively trained using fluoroscopy and two-dimensional planning techniques. Today, all targets are drawn using computer tomography tools with drawn targets superimposed on skin rendering objects for placement of tattoos and digital radiographs for target validation. Four-dimensional planning is conducted for target areas such as lung and upper abdomen where motion management influences target reproducibility. Computer planning now depicts targets with volume definition and can provide dose volume histogram analysis indicating the specific dose to a defined volume of normal tissue. This permits planning strategies to be compared and the radiation oncologist may choose a plan that provides less radiation dose to a specific target organ at risk for injury. This is important in all body regions for therapy. A lung cancer patient may have an underlying cardiomyopathy, therefore the radiation oncologist may intentionally choose a plan that places more radiation dose in subcutaneous tissue in order to avoid dose to the cardiac structures. The planning process provides opportunities to fuse many diagnostic images into the planning images to facilitate accurate drawing of targets and develop strategic process for the addition of expanded target to accommodate for subclinical disease, motion, and limitations of establishing accurate daily positioning. Images commonly used for fusion in order to enhance target definition include positron emission tomography and MR imaging. Most fusion software recognizes bony anatomy as points of fusion. This means that fusion of imaging from the central nervous system is generally highly accurate due to the multiple irregular features of skull anatomy. Registration of images from other sites (chest/abdomen) can be less accurate and subject to variation due to motion and challenges of registration due to more limited bony anatomy in these regions. The radiation oncologist needs to be aware of these problems and understand that limitations of registration can influence the accuracy of target location. The radiation oncologist must provide the treatment planning team with normal tissue radiation therapy dose constraints. The constraints set limits on the volume of normal tissue that can be exposed to treatment. The constraints are based on established standards as well as individual strategic points driven by specific patient needs. This sets a process in motion for the planning team to follow with the radiation oncologist acknowledging incidences where requested constraints cannot be met. The radiation oncologist also must define what is optimal for daily image guidance for each specific patient. Modern radiation therapy equipment has incorporated outstanding diagnostic imaging capacity into the treatment units including both kilovoltage (KV) units and cone beam computer tomography. The KV units are helpful for identifying and tracking fiducials (metal markers), which are placed in and about targets. The cone beam units define both the volume treated and the potential motion of normal tissue into the target region. The imaging information can be acquired in rapid sequence and validate daily position more accurately than dermal landmarks.

Nearly all radiation therapy treatment plans are now built using three-dimensional computer models juxtaposing tumor with normal tissue. Once a plan is finalized and validated by the physics team, most radiation treatments today are executed through intensity modulation (IMRT). The most common form of IMRT treatment delivery is dynamic motion of multileaf collimators housed within the gantry of the treatment unit. Thousands of leaf motions can move in synchrony with gantry motion to optimize radiation therapy treatment planning to offer full dose to tumor targets and minimize, as best as possible, radiation dose to normal tissue. Multileaf motion coupled with treatment table rotation and gantry motion can provide exquisite coverage of the treatment target excluding as much normal tissue as possible with near uniform X-ray doses through the target. Both image guidance and motion management, however, become extremely important in the use of intensity modulation. Tools for intensity modulation predate the development of tools for image guidance. As such, it was likely that ultra-conformal radiation therapy may have not provided a full dose to targets in the preimage guided era of radiation oncology as targets may have been in motion, which was not detected by MV imaging used on traditional accelerators. Therefore, as treatment has become more precise, the need for image guidance becomes very important. IMRT treatment plans improve dose distribution in all body regions compared to more traditional treatment methods and can be validated through dose volume histogram analysis in all areas of interest.

BRACHYTHERAPY

As early as 1901, investigators including Pierre Curie suggested that radioactive isotopes could be used to treat malignancies. Multiple publications of single cases were described, however it was generally recognized that outcomes were influenced by the technical inability to accurately place intracavitary and interstitial radioisotopes within a tumor and define the target of interest that was intended to receive a full dose. Interest in brachytherapy became reinvigorated when remote after-loading systems were developed that permitted radioactive materials to be stored safely before use, which significantly decreased X-ray exposure to health-care workers. Today, most brachytherapy applications are performed within the department using high dose rate after-loading devices for nearly all intracavitary and interstitial applications. The devices and catheters that support the use of the radioactive element are generally placed in the operating room. The patient is brought to the radiation oncology department post recovery and undergoes volumetric imaging to evaluate placement of the catheters and the potential relationship of normal tissue to the radioactive element. The modern high dose rate devices permit alteration of dwell times of specific regions within the catheters in order to adjust radiation dose appropriately to tumor targets and avoid normal tissue tissues as much as possible. In the care of the gynecology patient, there are opportunities in patients treated for definitive cervix cancer to strategically place packing or similar device into the vagina to move both the bladder and rectum away from the applicators and decrease radiation dose to these structures. Studies have and continue to demonstrate the importance of brachytherapy in gynecologic malignancies. The procedure permits amplification of radiation dose at a dose rate that is forgiving to normal tissues yet increases radiation dose to curative levels in patients with gross tumor. Although there are published reports suggesting that teletherapy and external therapy can be used as an alternative to brachytherapy, implant therapy continues to be associated with our best clinical outcomes. Brachytherapy is used to supplement care in our surgically resected high-risk endometrial patients largely at the vaginal apex where tumor can flow in a retrograde manner to the upper vagina through cervical and paracervical tissue. Brachytherapy can also be used as the sole point of care in patients with medically inoperable endometrial cancer. The essential aspect of using brachytherapy for gynecologic cancers is the unique ability to augment dose to tumor and place sharp radiation dose gradients to normal tissues that otherwise would receive higher dose with standard techniques of external therapy. Strategic use of radiation dose rate also permits protection of normal tissue from treatment while augmenting radiation dose to tumor.

Brachytherapy can be used to supplement radiation dose to close surgical resection margins or residual gross tumor as part of an operative procedure. This has been done in numerous disease sites including tumors of the central nervous system, head and neck, thorax, abdomen, pelvis, and extremity therapy. The approach can be with either permanent seeds placed independently or with a mesh or with the placement of secure remote after-loading catheters to be used for postoperative radiation treatment. Both approaches permit amplifying radiation therapy to targets with less exposure to normal tissue.

One of the most common forms of brachytherapy today is for both definitive and integrated care for the patient with prostate carcinoma. Brachytherapy can be used as a sole modality of care for patients with low risk factors for recurrence and as part of an integrated care plan combined with external radiation therapy for patients with intermediate and high risk features for tumor recurrence. Patients with prostate glands of average size with minimum or modest symptom scores are very good candidates for this form of treatment [26]. The introduction of real time image guidance with intraoperative ultrasound significantly moved this field forward. Pre-implant planning with radioactive seeds placed in a durable position has significantly reduced sequela from brachytherapy and improved implant geometry [27].

RADIOSURGERY AND STEREOTACTIC RADIATION THERAPY

With the advent of advanced radiation therapy technologies, treatment of small lesions initially in the central nervous system with compressed treatment schedules became possible via radiosurgery (Figure 1a and b). Today, an extraordinary portfolio of technologies is capable of providing treatment to exceptionally small targets with full image validation of the target coupled with real time evaluation of target motion management. Radiosurgery with doses of 2000 cGy to lesions in the central nervous system has been very successful and has now prompted clinical trials evaluating deferred radiation therapy to the whole brain in lieu of radiosurgery as a sole modality of initial therapy for metastatic disease and selected primary diseases of the central nervous system, both benign and malignant in origin. The success of radiosurgery to the central nervous system has permitted the development of similar techniques with compressed fractionation schemes to multiple body regions including lung, liver, and selected other regions in which high dose to normal tissue can be avoided. Modern equipment has cone beam computer tomography integrated with the ability to perform modulated arc therapy for treatment execution. This is an important feature as it permits dose to be adjusted to critical structures including motion management and limit dose to structures such as the rib in patients with disease is close approximation to the chest wall and other structures. Comparison of selected cases demonstrates a sig-

(a)

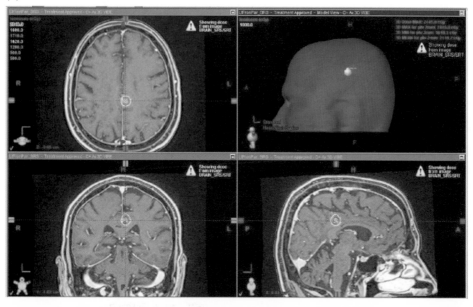

(b)

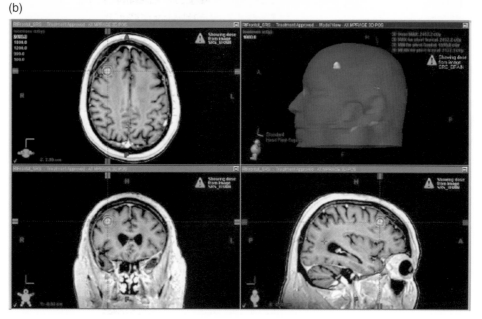

Figure 1. (a) and (b). Radiosurgery treatment of a central nervous system metastasis. *Source:* Courtesy of the University of Massachusetts Medical School, Department of Radiation Oncology. (See insert for color representation.)

nificant decrease in rib/chest wall dose in patients with disease within 2 cm of the chest wall [14]. Traditional static field radiosurgery generates significant more dose to rib and chest wall relative to modulated arc therapy whereas multileaf collimators can move in a synchronous manner relative with gantry motion to contour radiation dose in a more conformal manner to the rib, chest wall, and other critical structures. With compressed fractionation schedules, we are at risk of injury to tissue of limited self-renewal potential. This has been seen in multiple publications as rib fractures, chest wall pain, and injury to the diaphragm are now being seen with stereotactic body radiosurgery [14, 28, 29]. Accelerated fractionation with part of the initial management of patients with radiation therapy dating to 1960–1970 and reports of injury to tissues of limited self-renewal were identified. These injuries did not easily heal and left patients with chronic pain to bone and soft tissue. We are beginning to see these injuries again with the use of compressed fractionation; therefore we need to remain disciplined in our readoption of these techniques and make every effort to limited exaggerated areas of radiation dose inhomogeneity and the volume of normal tissue receiving prescription dose radiation therapy. Radiosurgery and stereotactic radiation therapy are important areas of research as cell kill from high dose therapy may be exceptionally proficient. Our ability to exclude normal tissue from therapy has significantly improved and high radiation dose to small targets appears to offer durable tumor control. These techniques will offer a significant advantage to patients moving forward.

FUTURE DIRECTIONS

Our technology has advanced to an extraordinary level and our discipline can now routinely perform tasks we could only imagine in the past. We are capable of both documenting and minimizing normal tissue in the radiation therapy treatment field. We can manage both patient and target motion and validate that the target is correctly treated on a daily basis. We can treat small targets to exceptionally high dose using compressed treatment schedules in many disease sites. From a clinical perspective, we are functioning at a level well beyond the expectations of our leaders from the past.

We need to focus on areas of weakness to further improve our clinical care processes. Although we have exceptional technology, the technology is not uniformly applied through our discipline and this creates problems in quality assurance. Fairchild and colleagues reviewed data from multiple important clinical trials and found that the quality of care on clinical trials directly influenced patient outcome [30]. In the HeadSTART clinical trial evaluating the role of Tirapazamine in locally advanced squamous cell carcinoma of the head and neck, patient survival was directly associated with compliance to study guidelines. Although our computational methods have become more standardized, the application of imaging to the target by radiation physicians remains nonuniform. If targets are excluded from the radiation therapy treatment field or additional normal tissue is unnecessarily included in the field, outcomes are influenced [31].

A statistically significant decrease in patient failure was demonstrated when patients were not treated to protocol guidelines in an intermediate and advanced risk Hodgkin lymphoma clinical trial [32]. This issue was identified as part of a quality assurance review of the imaging and radiation therapy objects after publication of the primary study endpoint. This raises several points of concern as the original message in the primary paper was that radiation therapy did not add to patient survival. In actuality, there was a significant improvement in survival if the patient was treated with radiation therapy but only if they were treated to study recommendations. Issues such as this will need to be addressed moving forward otherwise a significant quantity of information will be unintentionally misinterpreted by well-meaning study coordinators and patient care specialists.

Particle therapy with protons will become more available in the next five years. Cyclotrons are becoming more compact and can be fitted into therapy vaults once only reserved for photon accelerators. Protons have an advantage of decreasing radiation exit dose and decreasing integral dose to patients. Protons are especially important to the pediatric patient as more normal tissue can be spared for growth and development in selected patients. Tumors of the central nervous system comprise 25% of the pediatric oncology population and selected patients from this group may benefit the most from proton therapy as a smaller dose reaches normal tissue.

This is an exceptionally exciting time for the science of radiobiology. Targeted therapies are progressing and defining improvements in patient care and coupled with there are improvements in science in nearly all oncology related disease sites. For example, cell adhesion molecules called integrins have been shown to be expressed in prostate cancer and not in normal prostate epithelium. As such, they have become a selective target to integrate anti-integrin therapy. Investigators have shown that only high dose radiation therapy can eliminate integrin expression and hormone therapy may play a strong role in limiting the efficacy of cell adhesion molecules, therefore making tumors more vulnerable to hormone therapy in this disease. Prostate cancer cells resistant to the effects of radiation therapy demonstrate epithelial mesenchymal transformation (EMT). Sensitivity to radiation therapy and reversal of EMT transition can be reversed with ERK 1 and 2 knockdown as well as selective inhibition of BCL-2. Neuroendocrine differentiation is a defined feature in GU disease especially resistant to traditional therapies. Treatment with anti-neuroendocrine receptor therapy in experimental settings has sensitized cell lines with these features to radiation therapy. It is clear that soon this knowledge will enter clinical care [9].

Radiation therapy has a bright future and will remain an integral part of the care of the patient with cancer.

REFERENCES

1. Gould, G. and Lloyd, J. (1898). X-ray burns. Amputation of the thigh, x-ray burns, their nature and treatment. *The Phila. Med. J.* 1: 354.
2. Metzenbaum, M. (1905). Radium: its value in the treatment of lupus, rodent ulcer, and epithelioma with reports of cases. *Int. Clin.* 4: 21–31.

3. Aronowitz, J.N. (2007). Ethereal fire: antecedents of radiology and radiotherapy. *AJR Am. J. Roentgenol.* 188: 904–912.

4. Singer, H. (1914). Radiation emanation. *Md. Med. J.* 57: 7 24–7 25.

5. Macklis, R.M. (1993). The great radium scandal. *Sci. Am.* 269: 94–99.

6. Hall, E.J. and Giaccia, A.J. (2011). *Radiobiology for the Radiologist*, 7e. Philadelphia: Lippincott Williams & Wilkins.

7. Coutard, H. (1937). The results and methods of treatment of cancer by radiation. *Ann. Surg.* 106: 584–598. https://doi.org/10.1097/00000658-193710000-00010.

8. Simon, E.L., Goel, H.L., Teider, N. et al. (2005). High dose fractionated ionizing radiation inhibits prostate cancer cell adhesion and beta(1) integrin expression. *Prostate* 64: 83–91.

9. Wang, T., Languino, L.R., Lian, J. et al. (2011). Molecular targets for radiation oncology in prostate cancer. *Front. Oncol.* 1: 17. https://doi.org/10.3389/fonc2011.00017.

10. Hendry, J.H. (1985). The cellular basis of long term marrow injury after irradiation. *Radiother. Oncol.* 3: 331–338.

11. D'Ambosio, D.J., Li, T., Horwitz, E.M. et al. (2008). Does treatment duration affect outcome after radiotherapy for prostate cancer? *Int. J. Radiat. Oncol. Biol. Phys.* 72: 1402–1407.

12. Suwinski, R., Sowa, A., Rutkowski, T. et al. (2003). Time factor in postoperative radiotherapy: a multivariate locoregional control analysis in 868 patients. *Int. J. Radiat. Oncol. Biol. Phys.* 56: 399–412.

13. Deacon, J., Peckham, M.J., and Steel, G.G. (1984). The radioresponsiveness of human tumours and the initial slope of the cell survival curve. *Radiother. Oncol.* 2: 317–323.

14. Ding, L., Lo, Y.C., Kadish, S. et al. (2013). Volume modulated arc therapy (VMAT) for pulmonary stereotactic body radiotherapy (SBRT) in patients with lesions in close approximation to the chest wall. *Front. Oncol.* 3: 12. https://doi.org/10.3389/fonc.2013.00012.

15. Hall, E.J., Marchese, M.J., Astor, M.B. et al. (1986). Response of cells of human origin, normal and malignant, to acute and low dose rate irradiation. *Int. J. Radiat. Oncol. Biol. Phys.* 12: 655–659.

16. FitzGerald, T.J., McKenna, M., Rothstein, L. et al. (1986). Radiosensitivity of human bone marrow granulocyte-macrophage progenitor cells and stromal colony forming cells: effect of dose rate. *Radiat. Res.* 107: 205–215.

17. FitzGerald, T.J., McKenna, M., Kase, K. et al. (1986). Effect of X-irradiation dose rate on the clonagenic survival of human and experimental animal hematopoietic tumor cell lines: evidence of heterogeneity. *Int. J. Radiat. Oncol. Biol. Phys.* 12: 69–73.

18. Greenberger, J.S., FitzGerald, T.J., Klassen, V. et al. (1988). Alteration in hematopoietic stem cell seeding and proliferation by both high and low dose rate irradiation of bone marrow stromal cells in vitro. *Int. J. Radiat. Oncol. Biol. Phys.* 14: 85–94.

19. Greenberger, J.S., FitzGerald, T.J., Rothstein, L. et al. (1986). Long term culture of human granulocytes and granulocyte progenitor cells. *Prog. Clin. Biol. Res.* 211: 159–185.

20. Greenberger, J.S., Klassen, V., Kase, K. et al. (1984). Effects of low dose rate irradiation on plateau phase bone marrow stromal calls in vitro: demonstration of a new form of non-lethal, physiologic damage to support of hematopoetic stem cells. *Int. J. Radiat. Oncol. Biol. Phys.* 10: 1027–1037.

21. Naparstek, E., FitzGerald, T.J., Sakakeeny, M.A. et al. (1986). Induction of malignant transformation of cocultivated hematopoietic stem cells by X-irradiation of murine bone marrow stromal cells in vitro. *Cancer Res.* 46: 4677–4684.

22. FitzGerald, T.J., Santucci, M.A., Haryigaya, K. et al. (1988). Radiosensitivity of human bone marrow stromal cell lines: effect of dose rate. *Int. J. Radiat. Oncol. Biol. Phys.* 15: 1153–1159.

23. Citron, D., Cotrim, A.P., Hyodo, F. et al. (2010). Radioprotectors and mitigators of radiation-induced normal tissue injury. *Oncologist* 15: 360–371.

24. Chin, M.S., Freniere, B.B., Lo, Y.C. et al. (2012). Hyperspectral imaging for early detection of oxygenation and perfusion changes in irradiated skin. *J. Biomed. Opt.* 17: 026010. https://doi.org/10.1117/1.JBO.17.2.026010.

25. Brizell, D.M., Murphy, B.A., Rosenthal, D.I. et al. (2008). Phase II study of palifermin and concurrent chemoradiation in head and neck squamous cell carcinoma. *J. Clin. Oncol.* 26: 2489–2496.

26. Anscher, M.S., Thrasher, B., Zgonjanin, L. et al. (2008). Small molecule inhibitor of transforming growth factor-beta protects against development of radiation-induced lung injury. *Int. J. Radiat. Oncol. Biol. Phys.* 71: 829–837.

27. Lafata, K.J., Bushe, H., and Aronowitz, J. (2014). A simple technique for the generation of institution-specific nomograms for permanent prostate cancer brachytherapy. *J. Contemp. Brachytherapy* 6: 293–296.

28. Bongers, E.M., Haasbeek, C.J., Lagerwaard, F.J. et al. (2011). Incidence and risk factors for chest wall toxicity after risk-adapted stereotactic radiotherapy for early-stage lung cancer. *J. Thorac. Oncol.* 6: 2052–2057.

29. Sioshansi, S., Rava, P.S., Karam, A.R. et al. (2014). Diaphragm injury after liver stereotactic body radiation therapy. *Pract Radiat Oncol.* 4: e227–e230.

30. Fairchild, A., Straube, W., Laurie, F. et al. (2013). Does quality of radiation therapy predict outcomes of multicenter cooperative group trials? A literature review. *Int J Radiat Oncol Biol Phys.* 87 (2): 246–260.

31. Peters, L.J., O'Sullivan, B., Giralt, J. et al. (2010). Critical impact of radiotherapy protocol compliance and quality in the treatment of advanced head and neck cancer: results from TROG 02.02. *J. Clin. Oncol.* 28: 2996–3001.

32. FitzGerald, T.J., Urie, M., Ulin, K. et al. (2008). Processes for quality improvements in radiation oncology clinical trials. *Int. J. Radiat. Oncol. Biol. Phys.* 71: S76–S79.

25

RADIATION THERAPY – THE PATIENT EXPERIENCE

Harold James Wallace, III

Division of Radiation Oncology, Larner College of Medicine, University of Vermont, Burlington, VT, USA

There have been dramatic changes in radiation treatment since its first use over a century ago. A patient being treated in the modern era would not recognize the procedures initially utilized shortly after the discovery of X-rays by Roentgen in 1895. Low-energy treatment devices, a limited understanding of the mechanism of radiation injury, and a lack of formalized training for individuals interested in radiation treatment limited its systematic use for many years. Modern radiation treatment is carried out by a highly trained team of professionals; radiation doses are calculated with sophisticated computerized treatment planning systems and delivered using equipment that can pinpoint highly conformal dose clouds of radiation to any target within the body. While radiation therapy has its risks and associated side effects, the understanding of the biology of cancer, the physics of radiation delivery, and the expertise of the cancer care team in symptom management assures patients in the twenty-first century that the ratio of potential for benefit to the risk of toxicity has never been better.

Cancer: Prevention, Early Detection, Treatment and Recovery, Second Edition. Edited by Gary S. Stein and Kimberly P. Luebbers.
© 2019 John Wiley & Sons, Inc. Published 2019 by John Wiley & Sons, Inc.

THE RADIATION TEAM

Radiation oncologists are physicians who have completed specialty training in the use of radiation therapy in the care of individuals with cancer. Radiation oncologists complete five years of training after having graduated from medical school, four of those years exclusively dedicated to the study of radiation oncology. Radiobiology, radiation physics, as well as clinical radiation oncology training, are included in the education process. Board certification requires the successful completion of written and oral exams. While some radiation oncologists complete additional training in a subspecialty fellowship, this is not required. Many radiation oncologists in large practices focus their interest and care for individuals with specific types of cancers; however those in smaller or solo practices maintain a broad knowledge regarding the use of radiation in all disease sites and can expertly care for all patients who require radiation treatment.

Radiation therapists are the individuals who deliver radiation treatment. Their training can be within two- or four-year programs and includes instruction on the biology of cancer, anatomy, medical physics, and technical instruction on equipment operation. They spend much of their last year in radiation oncology clinics gradually taking more responsibility for patient's care. Radiation therapists become registered upon completion of their education and successfully completing a national registry exam and are licensed by individual states. They have the opportunity to interact with patients daily, frequently for many weeks and develop close and supportive relationships with those under treatment.

Medical physicists are responsible for every aspect of radiation treatment delivery [1]. Medical physicists complete a minimum of two years of additional education after college and have either a master's degree or a PhD like radiation oncologists, board certification is obtained after completing postgraduate education, doing a hospital-based residency program, and successfully completing written and oral exams. While the multiple computer and radiation producing systems available for treatment planning and delivery enhance the likelihood that a patient can get appropriate and potentially curable treatment with fewer side effects than ever before, these multiple systems must be tested, calibrated, and checked on a continual basis to ensure the prescribed radiation dose is delivered to the correct location. Prior to a patient's first treatment a medical physicist checks and confirms all aspects of the planned treatment. All patients' treatment charts are reviewed by a medical physicist on an ongoing basis during treatment. All new software, equipment, and techniques involved in treatment planning or delivery must be carefully reviewed by a medical physicist before it can be used clinically. Medical physicists are responsible for the safe use and storage of any radioactive isotopes used in a radiation oncology department and are subject to routine review by the Nuclear Regulatory Agency or other state agencies.

Medical dosimetrists in collaboration with the radiation oncologist and medical physicist design a treatment plan that is specific to each patient. The radiation oncologist reviews with the dosimetrist the goal of treatment and, in collaboration with the dosimetrist, develops a treatment plan that is designed to deliver the most appropriate dose of radiation to the intended target (usually the tumor or site where there is suspicion that tumor may exist) while minimizing radiation dose, and risk of radiation injury, to normal structures or organs that are in proximity to the target. Medical dosimetrists may have formalized training within a dedicated university-based training program or as part of an apprentice-like program at a radiation oncology center. Dosimetrists become certified after successful completion of a national certification exam.

Nurses in radiation oncology departments have specific expertise in educating patients about the practical nature of the prescribed treatment and in managing the side effects that can be associated with cancer therapies. These nurses help coordinate the multifaceted care a given patient may need and serve as common points of contact for patients, physicians, radiation therapists, social workers, and others outside the department of radiation oncology.

Social workers are integral to the care of many patients with cancer and frequently work within radiation oncology departments. The challenges facing patients who need radiation treatment are not usually limited to the symptoms of their disease or the side effects of their treatment. Social workers are uniquely trained to help with many of the other issues one may face during the course of treatment for cancer. Their understanding of specific patients and their family concerns, patient assistance programs, transportation options, insurance issues, and many others can relieve the patient from some of the stressors associated with cancer and its treatment.

The entire team within a radiation oncology department is dedicated to providing a professional and comforting experience for the individual who requires therapy. In addition to this highly functioning and integrated team, many patients who require radiation therapy are cared for by others with different areas of expertise in the treatment of those with cancer.

Multidisciplinary Teams

Many cancers are treated with more than one of the classic modalities of cancer care (surgery, chemotherapy, and radiation therapy), and the importance of additional complimentary therapies is becoming better understood by patients and the medical community. In order to facilitate the most seamless and expeditious approach to the care of a patient with cancer, many healthcare facilities have developed multidisciplinary teams of physicians and other healthcare providers. These teams are usually composed of medical, surgical, and radiation oncologists, and frequently include nurse navigators, research nurses, dietitians, and social workers. Diagnostic radiologists, surgical pathologists, medical geneticists/counselors, and others may also play important roles in these teams. The prevalence of these dedicated teams varies widely but they are present at most larger institutions. These teams are usually disease-site specific and often dedicated to the care of patients with breast, lung, gastrointestinal, genitourinary, and gynecologic tumors. These teams meet to discuss specific patients within the context of a tumor board and review the clinical circumstance as well as the related radiologic and pathologic findings. Many institutions have dedicated clinics staffed by all of the appropriate cancer specialists so a patient can see all of them in a single setting. This is not only efficient for the patient, but engenders direct communication with all involved in a patient's care. Nutritional counseling, rehabilitation services, as well as complimentary therapies may be available through these multidisciplinary programs. These groups frequently serve as the foundation for disease-specific research programs and encourage enrollment in local, regional, and national clinical trials.

THE EQUIPMENT

Radiation treatment is dependent on accurate delivery of high doses of high energy radiation to a specific target. The equipment in a radiation oncology facility is specifically designed for this purpose.

Most radiation treatment is delivered using high energy X-rays. These X-rays have a predictable behavior but that behavior depends on the nature of the tissues within the treated area. Denser tissues such as bone absorb more radiation, while less dense tissue including air in the lungs or sinuses absorb very little. Radiation therapy computer planning systems can account for these variations in X-ray absorption however, an accurate representation of each patient's tumor and surrounding normal structures is required to allow for these complicated calculations. Computerized tomography (CT) scans provide an ideal way for the radiation oncologist to visualize these structures and for the dosimetrist and physicist using these planning systems to plan and calculate the radiation dose. The process of obtaining these images is called a simulation, and the machine used is called a simulator. CT scanners are most commonly used as simulators (Figure 1) however, magnetic resonance imaging (MRI) scanners and CT/Positron Emission Tomography (PET) scanners are used at some institutions. Unlike imaging equipment used for diagnostic purposes, the precise location of the patient and the target within needs to be determined. Laser lights are used to project lines on the patient during the time of simulation. These lines are oriented in a specific fashion and cross on specific locations on the patient's skin. These crossing points are usually permanently marked on the patient using tattoos and are used in the radiation treatment rooms where an identical set of lasers is utilized (Figure 2).

Radiation treatment plans are designed on specialized computer systems. These systems display the area of interest based on the treatment planning CT, have the capability of integrating MRI or PET images into the planning images, allow for designation of the intended target as well as critical surrounding normal structures, design radiation fields of the appropriate size and shape, and calculate intended

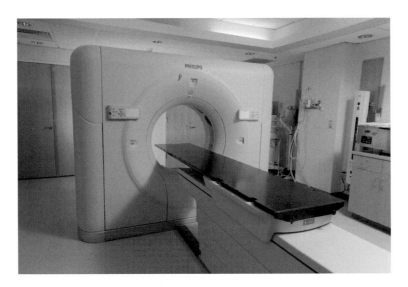

Figure 1. CT simulator.

Figure 2. Laser lines projected on patient's skin in radiation treatment room.

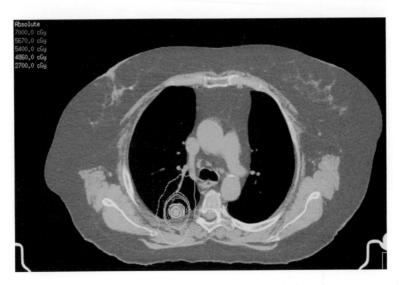

Figure 3. Isodose plan for treatment of lung tumor. (See insert for color representation.)

radiation dose. These systems are intuitive for the experienced user and are highly reliable with respect to performance. Radiation plans display doses of radiation, much like altitudes on a topographical map, overlaid on the structures as identified on the treatment planning CT images (Figure 3). The radiation oncologist and dosimetrist evaluate and alter the plan until the best plan with respect to radiation delivery to the tumor with minimal exposure to the surrounding structures is obtained.

Nearly all radiation treatment is delivered with a linear accelerator (Figure 4). Linear accelerators generate high energy X-rays by imparting energy on electrons as they pass through high-energy electromagnetic waves within the machine. These energized electrons can be aimed in the intended direction and ultimately at a metal target, usually made of tungsten. The interaction of these electrons with the metal target results in high energy X-rays aiming at the patient. The radiation beam can be shaped by inserting very dense metal devices into the path of the beam. Most commonly multiple, independently moving leaves of metal are positioned in the path generating highly customizable beam shapes using a device called a multileaf collimator (Figure 5).

Prior to treatment delivery, the location of the planned treatment within the patient must be confirmed. Linear accelerators have devices that allow for imaging of the beam as it passes through the patient that confirm the actual size and shape of the treatment beam used for treatment delivery. X-rays used for radiation treatment however are of such high energy that the differential absorption of these X-rays by bone, soft tissue, and air is minimal. Images obtained with the treatment beam have very low contrast so while they are a perfect representation of the size and shape of the radiation beam as it passes through the patient they provide images

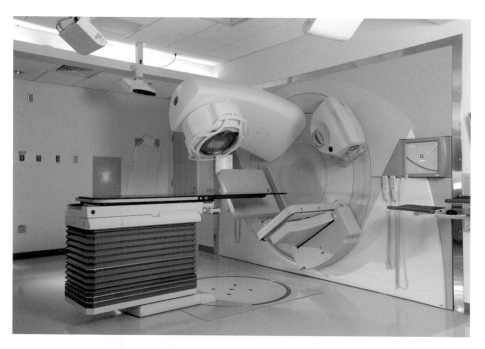

Figure 4. Linear accelerator.

Figure 5. Multileaf collimator.

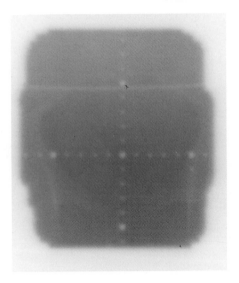

Figure 6. X-ray image of pelvis with treatment beam from linear accelerator.

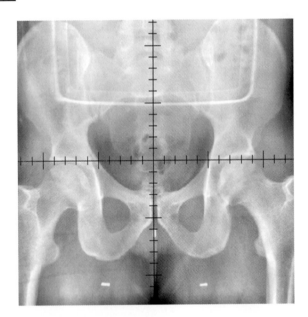

Figure 7. X-ray image of pelvis with dedicated imaging device on linear accelerator.

that are otherwise difficult to interpret (Figure 6). Diagnostic-quality, lower energy X-ray devices are now integrated into linear accelerators. These lower energy X-ray devices allow for high-quality imaging (Figure 7) and when rotated about the patient on the table can generate CT images (Figure 8). In addition to imaging devices integrated into the linear accelerator, systems that use ultrasound, implanted radiofrequency emitters, or the external surface of the patient in the treatment room may be employed.

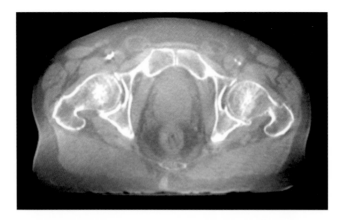

Figure 8. CT scan image of pelvis with dedicated imaging device on linear accelerator.

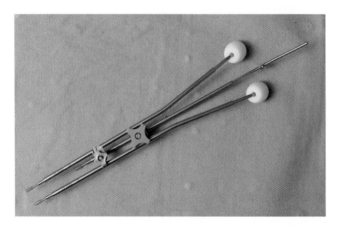

Figure 9. High-dose rate applicator for the treatment of cervix cancer.

While most patients who are treated with radiation are treated with beams of radiation generated by linear accelerators, some tumors are uniquely suited to the placement of radioactive sources within or immediately adjacent to the tumor. This treatment, termed brachytherapy (from the Greek *brachy* – "short") has been used for decades to deliver very high doses of radiation to tumors while minimizing radiation exposure to surrounding structures. Historically this was accomplished by placing sources in the appropriate location for days at a time. Currently, empty devices or applicators are inserted to the intended treatment site. Cancers of the cervix are frequently treated with specialized brachytherapy applicators that allow insertion of a radiation source into the uterus and cervix (Figure 9). These applicators or then loaded with a very small, highly radioactive source using a device called a high dose rate machine (Figure 10). This machine is essentially a computer-controlled spool of wire with the radiation source at the tip of the wire. Catheters from the HDR machine are attached to the applicator that had been previously placed within the patient. The machine then rotates the spool, so the source

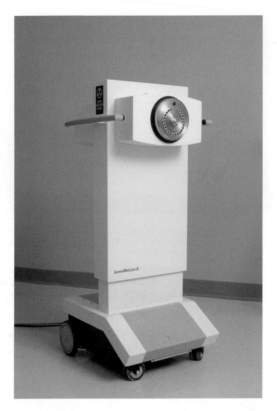

Figure 10. High-dose rate treatment machine.

is driven to various positions within the applicator for variable times. The radiation dose is determined by the amount of time the source is in each position.

Occasionally radiation sources are inserted directly into the tumor. This technique, interstitial implantation, is most commonly used for the treatment of prostate cancer. Transrectal ultrasound allows for good imaging of the prostate gland and can be used to direct these sources, frequently called seeds, into the gland at appropriate, predetermined positions. This approach allows for extremely highly conformal doses of radiation to be delivered to the prostate while minimizing the risk of radiation injury to the adjacent bladder and rectum (Figure 11).

THE EXPERIENCE

Individuals may be referred to radiation oncology for either curative or palliative treatment. All patients seen in a radiation oncology department are initially evaluated by a radiation oncologist. This appointment will include a review of the patient's history as well as an in-depth discussion about the potential role of radiation therapy given the specific circumstance of the disease type, stage, and other

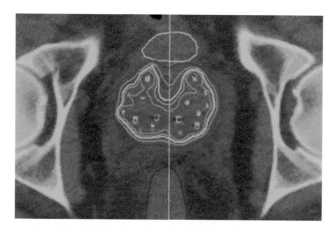

Figure 11. Prostate seed brachytherapy plan. (See insert for color representation.)

medical conditions that may have an effect on the proposed treatment. The consultation should include discussions about the goals of potential therapy, the expected treatment course, the expected result as well as a review of the potential short- and long-term risks of the recommended treatment. If radiation treatment is confirmed to be appropriate, most patients will need to go through the simulation process. As previously described most simulations are done on CT scanners however, unlike diagnostic CT scans, or other imaging procedures, the position of the patient is very important. The most appropriate position is determined by the radiation oncologist and radiation therapists and must account for patient comfort, reproducibility, and the location of the tumor with relation to other structures. Often a specific immobilization device is constructed to ensure that the patient's position can be easily reproduced from one day to the next. Patients who have tumors in the head or neck most frequently will have a plastic, form-fitting mask made at the time of simulation. Intravenous, oral, or rectal contrast may be utilized. After the CT images are obtained and the position of the tumor determined, tattoos, based on projection of laser lights in the simulator room on the patient's skin, will be placed.

The timing of the first radiation treatment following the simulation varies based on the clinical circumstance. Some tumors impact critical normal organ function or cause pain or bleeding and require urgent treatment. At other times, particularly when high doses of radiation are required to be given adjacent to highly sensitive critical structures the planning process will be complex and involve hours of work by the radiation oncologist, dosimetrist, and medical physicist. Planning times of over a week for these cases may be required.

Treatment begins after the plan is complete and approved by the radiation oncologist and medical physicist. The first day's treatment often takes longer than subsequent treatments. Time is spent with the radiation therapists orienting the patient to the treatment room and process. Great care is taken reproducing the position achieved in the simulator room, and X-ray images of the planned treatment field are obtained. While the therapists are not in the room while the treatment beam is on, video cameras within the room allow the therapists to see the patient and machine at all times and an intercom is always on so the therapists can hear the patient in the treatment room. The treatment itself is usually delivered in a series of treatment segments, with the machine aiming at the target area from many different angles. Depending on the circumstances and equipment, the beam may be on while the machine is rotating around the patient. While the patient may be in the treatment room for 15 minutes or longer, most treatment delivery takes less than 5 minutes. Treatments are painless, and other than the sound from the linear accelerator, most patients have no sensation while the beam is on. Some patients describe an unusual smell, either directly from the machine or possibly from ionization of the air in the room forming ozone with its distinct odor. Treatment very close to the eyes occasionally causes the patient to see flashes while the beam is on due to the interaction of energy on the retina [2].

The total number of treatments depends on the clinical circumstance. Patients who are being treated with palliative intent, that is treated solely to improve symptom control but without an expectation of a significant amount of added survival from the treatment, should be treated with abbreviated courses of therapy of two weeks or less [3]. These relatively short courses of treatment should be designed to minimize the potential side effects and inconvenience to the patient. Treatment schedules for definitive courses of treatment are dependent on the disease that is being treated. Some tumors including many lymphomas are typically quite sensitive to radiation and are usually treated over approximately three to four weeks while other cancers such as prostate cancer are most often treated over a period of up to nine weeks. Dividing the radiation doses up into small daily doses, termed "fractionation," takes advantage of the radiobiologic difference between normal tissues and tumors and maximizes the likelihood of tumor control while simultaneously limiting the risk of serious long-term toxicity.

SIDE EFFECTS

Radiation has many effects on normal tissue. Side effects that occur during and within 90 days of completion of treatment are considered "acute" and have fairly predictable clinical courses and underlying causes. While there are inherent differences in radiation sensitivity from cancer to normal organs, some tissues, particularly those that reproduce relatively rapidly (skin, lining of the mouth, throat, and gastrointestinal tract) in fact respond similarly to cancer in the short term. Acute effects of radiation on cancer and these rapidly responding tissues are caused by rapid induction of cell death and the body's attempt to ameliorate the damage. Some tissues have unique properties that cause them to manifest radiation damage roughly one to six months after radiation exposure. The most notable of these, the lung, can cause serious, and even life-threatening lung inflammation (radiation pneumonitis), that is caused by multiple interactions between normal lung and inflammatory cells [4]. An unfortunate consequence of radiation treatment is the potential for injury to treated organs or structures that does not manifest itself for many months or years after radiation exposure. This late injury is most commonly due to progressive proliferation of fibrous tissue within the inner linings of small blood vessels. This process can relentlessly progress until the blood supply cannot support normal organ architecture or function. This late effect can cause mild, minimally bothersome symptoms such as occasional rectal bleeding following the treatment of prostate cancer, but can be responsible for devastating cognitive changes in children and adults treated for brain tumors, or permanent injury to other central or peripheral nerves (spinal cord, brachial plexus), the coronary arteries (myocardial infarction), the lung (pulmonary fibrosis), the gastrointestinal tract (stricture, obstruction, poor absorption), the bladder (radiation cystitis), or soft tissues (fibrosis, lymphedema).

Acute radiation side effects are usually managed by the radiation oncologist and the radiation nurse. Generally, adequate nutrition and rest will help most get through treatment most easily. Some patients may be able continue normal life activities including work, while others may find the rigors of treatment excessive and need to focus all of their energy on getting through therapy.

Fatigue is one of the most commonly reported acute side effects of radiation treatment with patients receiving treatment to some sites more susceptible than others. The underlying cause for this fatigue is unclear [5]. There are a number of interventions that may be useful in helping a patient with fatigue [6]. There is an increasing interest in developing exercise programs for patients being treated with radiation therapy to help decrease radiation-induced fatigue [7]. Many patients do not find that rest is particularly helpful and simply carry on with their normal activities as well as they can. Cancer rehabilitation services are becoming widely available and may have a role in ameliorating some of the fatigue and other causes of disability that are inherent in the treatment of some cancers.

Skin reaction from radiation exposure is common to some specific disease sites. Modern linear accelerators are designed to minimize radiation effects on the skin, however some cancers including breast cancer may be present close to the skin surface and the radiation dose to the skin is relatively high by design. Skin reactions can range from mild redness and itching to peeling and blistering. Many radiation oncologists and radiation oncology nurses make specific recommendations to patients to prevent or treat skin reactions; however there is limited data to suggest that any specific approach is superior to others. Topical steroid creams appear to have some effect but have not become widely used in the prophylactic setting [8]. Most typically patients will be asked to have their skin clean and dry at the time of treatment and discuss with their nurse and physician any skin care product they consider using. Late radiation effects on the skin and the subcutaneous tissues were commonly seen prior to the use of high energy linear accelerators but are much less common now. Typical late changes may include thinning and loss of pigmentation of the skin and the development of superficial blood vessels under the skin.

The acute effects of radiation therapy on patients being treated for cancers of the head and neck region can be particularly difficult and challenging to manage. As previously noted some normal tissues are particularly susceptible to the early effect of radiation and these include the lining of the mouth and throat and salivary glands. Changes in the texture of the saliva may be noted as early as a few days after the start of treatment and near-total mouth dryness may occur within the first few weeks. Changes in taste also occur within the first few weeks, and frequently the taste of many foods becomes unpleasant. The lining of the mouth and throat can become very inflamed, painful, and frequently may produce copious amounts of thick, mucousy secretions. Pain management, coping with the excessive secretion production as well as maintaining adequate hydration and nutrition can be significant challenges for patients as they get through this treatment. Supportive measures provided by nursing and speech and language pathology can be invaluable during this treatment and recovery. Recovery from these effects can take months, and weight loss associated with this treatment often lasts for more than a year. Late effects of radiation treatment to the head and neck area may include skin edema and/or fibrosis, permanent decrease in salivary gland function, permanent change in taste, and the risk of injury to the jawbone (mandible) and can be treated with variable success.

Cancers that originate within the chest (lung, esophagus, lymphomas) require careful evaluation of radiation exposure to the normal lung tissue as these patients frequently have underlying lung disease and limited pulmonary reserve. While acute injury to the esophagus can be bothersome, it is usually short lived and reasonably well controlled. Not infrequently radiation plans will have to allow for some risk of acute radiation toxicity to the esophagus in an attempt to minimize the risk of more serious injury to the lung. Radiation esophagitis is usually managed with changing one's diet to softer food, pain medication, and some topical, swallowed medications with locally soothing, anesthetizing, or anti-inflammatory properties. Irritation of the large airways by radiation can cause local irritation and usually manifest as a dry cough. Over-the-counter cough medications and narcotics are frequently used to control these symptoms. Radiation pneumonitis, as previously described usually manifests as shortness of breath and or cough two to six months after the completion of radiation treatment. These symptoms are often self-limited, however can at times require treatment with steroid medications and oxygen. Long-term consequences of radiation to the chest include scarring of the esophagus, permanent fibrosis of the lung, and increased risk of heart disease. It is important for patients who require radiation to the chest to limit other risk factors for cardiac disease including maintaining control of hypertension, diabetes and hypercholesterolemia, and stopping smoking.

Radiation therapy to the abdomen is most commonly used for cancers of the upper gastrointestinal tract as well as some lymphomas. There are a number of critical organs within the abdomen that are quite sensitive to the effects of radiation; respecting the tolerances of these organs is critical in treatment planning. The most common side effects specific to irradiation of the abdomen include nausea and loss of appetite. Nausea can usually be controlled with medication and frequently patients are instructed to take anti-nausea medication prior to treatment. Late effects on the stomach may include ulceration, while late radiation injury to the small intestine may cause obstruction. Injury to the liver and kidneys is usually asymptomatic and detected with blood testing. The severity of this injury can vary from minimal and completely reversible to frank organ injury and failure.

Radiation therapy to the pelvis is most commonly used in the treatment of prostate cancer, but radiation is also commonly utilized in the treatment of rectal, anal, endometrial, and cervix cancers. Acute side effects of treatment of the pelvis may include increased bowel and bladder frequency and decreases in blood cell counts. Symptomatic treatment of bowel frequency and urgency may include changes in diet or the use of antispasmodic agents such as loperamide. Bladder irritative symptoms are frequently best managed with hydration to maintain dilute urine, but occasionally respond to medications such as phenazopyridine that provide some topical anesthesia to the bladder. Men with bladder outlet obstructive symptoms due to symptomatic prostate hypertrophy may note a worsening of these symptoms and alpha-1 antagonists can be of benefit. Clinically significant declines in blood counts are uncommon when radiation is used alone, however pelvic radiation is frequently combined with chemotherapy in the treatment of pelvic malignancies (rectal, anal canal, cervix), and clinically significant and even life-threatening declines in white blood cells can occur. Late effects of radiation to the bowel can lead to permanent alteration in bowel function and rarely bowel obstruction that may require surgical intervention. High doses of radiation treatment are frequently employed in the curative treatment of prostate and cervix cancer. While higher doses are thought to be more effective than lower doses in these disease sites, these higher doses are associated with risk of injury to blood vessels as well as organs including the rectum and bladder. Radiation proctitis most commonly causes bleeding that occasionally requires treatment with topical formalin or argon plasma coagulation. Radiation cystitis, while less common can be particularly bothersome and difficult to treat. All curative treatments for men with localized prostate cancer carry a significant risk of erectile dysfunction. Radiation treatment causes a gradual decrease in blood flow though the vessels responsible for erections. Medications may be helpful in this circumstance however, most men do develop irreversible erectile dysfunction in the years following treatment.

Postoperative radiation therapy following wide excision of soft tissue sarcomas of the extremities has been preferred over amputation for decades. As sarcomas can recur at the site of surgery including the scar, radiation fields are designed to include these areas. Significant skin reaction including redness, peeling, and moist desquamation are commonly seen and managed as discussed previously. Significant peripheral edema is the most common late effect of radiation to the extremities. Physical therapists expert in lymphedema management can provide a number of effective therapeutic modalities to alleviate the edema, compressive garments can help and there is interest in surgical approaches to remove the most affected area and transpose healthy lymph nodes into nodal regions that have been affected by postradiation fibrosis [9].

RESEARCH IN RADIATION ONCOLOGY

Many of the current recommendations for radiation treatment have been informed by successful completion of clinical trials over the past 50 years. Through well-designed and completed trials we have learned that most women with breast cancer can be successfully treated without removal of the breast [10], that most patients with cancer of the larynx can be treated without its removal [11], that sarcomas of the extremities can be treated without amputation [12], and that small lung cancers can be successfully treated without surgery [13]. There are many questions left unanswered with respect to the best use of radiation in the treatment of malignancy. While most clinical trials do not typically provide a direct advantage to the participant, successful completion of clinical trials allows for more effective and safer treatment for those who will need radiation in the future. Current questions that are of particular interest include determining which tumor-specific genetic mutations may be predictive of increased or decreased sensitivity to radiation, of specific patterns of tumor recurrence that could be helpful in radiation field design, and of possible mutations that predict enhanced sensitivity to combinations of systemic drugs and radiation. There is a great deal of interest in shortening radiation treatment schedules and some data confirms that shorter schedules may be successfully used in some circumstances however these shorter courses of larger daily radiation doses need to be appropriately tested for safety and efficacy. Patients are frequently asked to consider taking part in a clinical trial. While we have learned a great deal about the best way to treat cancer through well designed trials, there are many unanswered questions. Less than 3% of individuals with cancer participate in clinical trials [14] and improvement in accrual is required if we are to move our understanding of this disease and its best treatment forward. Research in radiation therapy may be developed and run by scientific investigators at individual institutions, through cooperative agreements with drug or machine manufactures or through large, national or international cancer cooperative study groups.

Modern radiation therapy is a safe and effective way to treat many cancers. Advances in our understanding of the effects of radiation on cancers and normal tissues and in ways we can integrate radiation therapy with other treatment modalities has improved our ability to provide cancer control with minimal side effects. Ongoing research on the fundamental principles of cancer biology will further improve our ability to take advantage of the unique properties radiation has in treating malignancy.

REFERENCES

1. Policy 17-B "Scope of Practice of Clinical Medical Physics." AAPM Position Statements, Policies and Procedures. American Association of Medical Physicists, 30 Nov. 2011. Web. 18 Apr. 2015.
2. Schardt, D., Kavatsyuk, O., Kramer, M., and Durante, M. (2013). Light flashes in cancer patients with heavy ions. *Brain Stimulation* 6 (3): 416–417.

3. Choosing Wisely. American Society for Radiation Oncology, 13 Sep 2013. Web. 18 Apr. 2015

4. Pelagia, G.T. and Koukourakis, M. (2006). Radiation Pneumonitis and fibrosis: Mechanisms underlying its pathogenesis and implications for future research. *Int. J. Rad. Onc. Biol. Phys.* 66: 1281–1293.

5. Ryan, J.I., Carroll, J.K., Ryan, E.P. et al. (2007). Mechanisms of cancer-related fatigue. *The Oncologist* 12 Suppl 1: 22–34.

6. Dy, S.M., Lorenz, K.A., Naeim, A. et al. (2008). Evidence-based recommendations for cancer fatigue, anorexia, depression and dyspnea. *J. Clin. Oncol.* 26: 3886–3895.

7. Golhofer SM, Wiskemann J, Schmidt ME et al. (2015) Factors influencing participation in a randomized controlled resistance exercise intervention study in breast cancer patients during radiotherapy. BMC Cancer 15 epub ahead of print.

8. Hindley, A., Zain, Z., Wood, L. et al. (2014). Mometasone furoate cream reduces acute radiation dermatitis in patients receiving breast radiation therapy: Results of a randomized trial. *Int. J. Rad. Onc. Biol. Phys.* 90: 748–755.

9. Patel, K.M., Lin, C.Y., and Cheng, M.H. (2015). From theory to evidence: Long-term evaluation of the mechanism of action and flap integration of distal vascularized lymph node transfers. *J. Reconstr. Microsurg.* 31: 26–30.

10. Fisher, B., Anderson, S., Redmond, C. et al. (1995). Reanalysis and results after 12 years of follow-up in a randomized clinical trial comparing total mastectomy with lumpectomy with or without irradiation in the treatment of breast cancer. *N. Engl. J. Med.* 333: 1456–1461.

11. The Department of Veterans Affairs Laryngeal Cancer Study Group (1991). Induction chemotherapy plus radiation compared with surgery plus radiation in patients with advanced laryngeal cancer. *N Engl J Med* 324: 1685–1690.

12. Rosenberg, S.A., Tepper, J., Glatstein, E. et al. (1982). The treatment of soft-tissue sarcomas of the extremities: Prospective randomized evaluations of (1) limb-sparing surgery plus radiation therapy compared with amputation and (2) the role of adjuvant chemotherapy. *Ann Surg* 196: 305–315.

13. Crabtree, T., Puri, V., Timmerman, R. et al. (2013). Treatment of stage I lung cancer in high-risk and inoperable patients: comparison of prospective trials using stereotactic body radiotherapy (RTOG 0236), sublobar resection (ACOSOG Z4032) and radiofrequency ablation (ACOSOG Z4033). *J Thorac Cardiovasc Surg* 145: 692–699.

14. Murthy, V.H., Krumholt, H.M., and Gross, C.P. (2004). Participation in cancer clinical trials: race-, sex-, and age-based disparities. *JAMA* 291: 2720–2726.

26

PAIN MANAGEMENT IN CANCER

Carlos A. Pino

The Robert Larner, MD College of Medicine, The University of Vermont, Burlington, VT, USA

INTRODUCTION

Cancer-related pain affects millions of people worldwide, causing fear and diminishing the quality of life in patients suffering from cancer. Despite increasing information on analgesics and interventional techniques, it is estimated that as many as 90% of patients with advance cancer still suffer from debilitating pain [1, 2]. While the reasons for inadequate treatment of cancer pain are multifaceted and vary depending on the development of health services in different countries, certain barriers to adequate pain relief have been identified. Patients and their family tend to fear the risk of addiction and underreport the intensity of pain for fear of giving up on the cancer treatment. Physicians are often limited by time, may lack the necessary knowledge, and may be restricted by excessive governmental regulations over opioid prescribing.

Pain from cancer can result from mechanical compression, inflammatory, and neuropathic and/or ischemic mechanisms. It can also emanate from direct tumor invasion of bone, soft tissues, or nerves. It is estimated that 60–80% of pain in patients with cancer originates from direct tumor involvement or metastasis [2, 3]. Pain can also present as a result of the primary antineoplastic treatment, such that

Cancer: Prevention, Early Detection, Treatment and Recovery, Second Edition. Edited by Gary S. Stein and Kimberly P. Luebbers.
© 2019 John Wiley & Sons, Inc. Published 2019 by John Wiley & Sons, Inc.

TABLE 1. Percentage of patients with pain by cancer site.

Site	%
Bone	85
Oral cavity	80
Genitourinary	75–78
Breast	52
Lung	45
GI	40
Lymphoma	20

20–25% of pain can present as a direct result of cancer therapy [2–4]. Surgeries (e.g., mastectomy, thoracotomy, amputations), especially when they are extensive, can be a frequent cause of pain in patients with cancer. This pain can be secondary to tumor resection itself, damage to surrounding structures or nerves, postamputation pain, etc. Radiation therapy has been associated with neural (plexopathy, myelopathy), myofascial, and osseous injuries that can present with invalidating pain. Chemotherapeutic agents have also been associated with pain in patients with cancer. Such is the case of alkaloids and platinum-based compounds that are known to cause peripheral neuropathy and refractory neuropathic pain. Neuropathic type pain is particularly difficult to treat and is present in about 40% of patients with cancer pain [5]. Finally, it is also important to keep in mind that patients with cancer may have pain completely unrelated to their cancer.

Pain varies depending on the affected site. Seventy-five to eighty-five percent of patients complain of pain when the cancer affects the genitourinary system, bone, or mouth. Bone is particularly painful and it is the site of secondary tumors in 90% of patients with advanced prostate, breast, or lung cancer [6]. Primary cancers of breast, lung, and gastrointestinal tract are painful in 40–50% of patients (Table 1).

Successful pain management of the patient with cancer will ultimately depend on the accurate assessment by the clinician, identification of the components that contribute to pain, and the formulation of a therapeutic plan that considers the patients' needs and goals.

PRINCIPLES OF PHARMACOTHERAPY

Comprehensive management of pain is an integral part of treating patients with cancer. Developing an efficacious and individualized pain control strategy requires collaboration and effective communication between the primary care provider, the oncologist, palliative care, and pain management specialists. This is particularly important when certain aspects of the cancer treatment can modify the need and intensity of the pain management strategy, if the disease interferes with routes of administration of medications or contraindicates invasive pain management procedures. Just as successful therapies can reduce the tumor burden and alleviate pain, certain antineoplastic treatment (i.e., chemotherapy) can also cause new and intractable pain. Likewise, pain therapy can interfere by increasing side effects such as, opioid induce bowel dysfunction, or decreasing side effects, as in the case of intrathecal pain therapies.

Life expectancy will also influence treatment decisions. The goals of patients whose life expectancy exceeds several months may include relieving pain and maintaining a certain lifestyle to allow for normal and uninterrupted activity. In patients whose life expectancy is reduced to days or weeks, the expectations will be focused on maintaining comfort and reducing side effects.

SYMPTOMATIC CANCER PAIN MANAGEMENT

Most patients will respond adequately to simple pharmaceutical strategies administered orally. When the decision is made to start the patient on analgesics, two basic principles should be kept in mind. Whenever possible, oral analgesics or other noninvasive routes of administration are preferable; and titrate opioids and adjuvant analgesics following the principles of the World Health Organization (WHO) analgesic ladder (Figure 1). The WHO analgesic ladder recommends a sequential approach to administering analgesics based on pain severity, initially with non-opioids for mild pain, progressing to mild, and strong opioids for more severe pain (Figure 1). It is recommended that opioids and adjuvant analgesics be carefully titrated to maximally effective doses or until dose-limiting side effects appear.

The WHO analgesic ladder is a simple and effective method to provide pain relief using relatively inexpensive drugs. It is estimated that 75–90% of cancer patients with pain obtain relief when the WHO analgesic principles are followed. Unfortunately, studies fail to predict the effectiveness of this strategy in any given patient, which underlines the importance of individualized treatment.

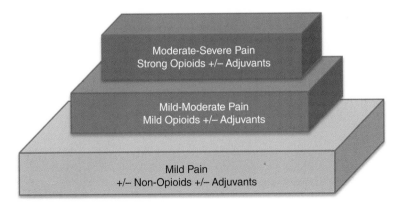

Figure 1. World Health Organization (WHO) analgesic ladder. (See insert for color representation.)

Non-opioids Analgesics

The most common non-opioid agents used to treat cancer pain include: acetaminophen; anti-inflammatory agents such as nonsteroidal anti-inflammatories (NSAIDs) and cyclooxygenase-2 (COX-2) inhibitors; corticosteroids; antineuropathic agents; and bisphosphonates. These agents have demonstrated their usefulness in the management of a wide variety of acute and chronic painful conditions and, in some cases, have contributed to reducing opioid requirements.

Acetaminophen is a common and inexpensive analgesic that is well tolerated and has good safety profile at therapeutic dose. Acetaminophen is sometimes classified as an NSAID, but it's mechanism of action has not been fully elucidated and appears to be different than traditional NSAIDs [7]. Acetaminophen is rapidly absorbed following oral administration and peak serum concentration is reached at two hours. Acetaminophen has few side effects in the usual dose range, without significant gastrointestinal effects or platelet dysfunction. Renal dysfunction can occur with acetaminophen, but less common than with NSAIDs. Hepatotoxicity and acute tubular necrosis can occur at doses higher than 4 g/day.

There is evidence supporting the use of acetaminophen in mild cancer pain [8, 9] but continuing acetaminophen once opioids have been started is controversial. A systematic review by Nabal et al. [10] found no overall benefit to adding acetaminophen to opioids in four of the five studies reviewed. In a small, randomized, blind, crossover trial, Israel et al. [11] found no additional benefit to adding acetaminophen to daily morphine in patients with cancer pain. The addition of acetaminophen and opioid combinations can, however, be helpful in patients already on opioids. Sima et al. found significant improvement in breakthrough pain of patients already on opioids, when prescribed additional combination oxycodone (5 mg) and acetaminophen (325 mg) compared to those receiving opioids plus placebo for pain from bone metastasis [12].

Nonsteroidal Anti-inflammatories (NSAIDS) and Cyclooxygenase-2 (COX-2) Inhibitors

NSAIDs and COX-2 inhibitors are a family of drugs whose mechanism of action is inhibition of the cyclooxygenase enzyme and reducing conversion of arachidonic acid to prostaglandins and thromboxane A2. There are two main forms of cyclooxygenase, COX-1 and COX-2. COX-1 is present in all tissues, while COX-2 is mainly expressed in inflamed tissue [13, 14]. Traditional NSAIDs are primarily nonselective inhibitors, although there is differential degree of affinity to each receptor.

NSAIDs help alleviate pain in patients with cancer when compared to placebo. In a meta-analysis looking at the efficacy of traditional NSAIDs, Eisenberg et al. [15] found benefit if NSAIDs were compared to placebo, but no benefit if they were compared to other NSAIDs or opioids. A Cochrane review [16] confirmed this finding. Seven of the eight studies reviewed revealed that NSAIDs are more effective than placebo for cancer pain, with no difference in side effects. A recent systematic review of the literature, with seven new articles and five studies previously included in the Cochrane review [10] supported the addition of NSAIDs to opioids in the treatment of moderate to severe pain from cancer [10]. Furthermore, two of the studies found that ketorolac or diclofenac reduced opioid consumption [10]. There does not appear to be any difference in analgesic efficacy between nonselective NSAIDs and COX-2 inhibitors.

Most studies of NSAIDs or COX-2 inhibitors in patients with cancer have looked at short-term use, so long-term efficacy and safety of NSAIDs and COX-2 inhibitors in this patient population remain unknown.

Corticosteroids

These agents may be quite useful in patients with cancer pain. Pro-inflammatory cytokines are involved in key steps of the pathophysiology of nociception, as well as in the development of inflammatory and neuropathic pain. Corticosteroids inhibit the inflammatory response to tissue injury by inhibiting phospholipase A2 that decreases the release of arachidonic acid from tissue phospholipids thereby limiting the formation of leukotrienes, thromboxanes, prostaglandins, and prostacyclin.

Additionally, corticosteroids can decrease pain by modulating tumor size or the inflammatory response to cancer therapy.

Systemic and prolonged used of corticosteroids can lead to a wide range of short- and long-term side effects. Steroids affect the stress and immune response as well as electrolyte regulation. They increase the conversion of amino acids to glucose in the liver and decrease peripheral utilization of glucose leading to hyperglycemia. Steroids also increase bone reabsorption leading to osteoporosis and may inhibit ACTH secretion causing adrenal insufficiency [17, 18].

Corticosteroids frequently used in the clinical setting are dexamethasone, methylprednisolone, betamethasone, prednisolone, and prednisone. Dexamethasone has minimal mineralocorticoid effect, causing less water retention. Corticosteroids can be used broadly when an anti-inflammatory effect is desired, such as spinal cord compression or brain metastasis. It is particularly useful in the management of pain secondary to bone infiltration and in neuropathic pain.

Despite its frequent use, there is very limited evidence for the efficacy of corticosteroids in cancer pain. Of four randomized, controlled trials, two study demonstrated efficacy, and two did not [19–22].

Anticonvulsants and Antidepressants

Opioid analgesia is often insufficient to control neuropathic pain and additional agents are needed, mainly antidepressants and anticonvulsants. The choice of agents is guided by multiple factors such as synergism with current medications, predominance of symptoms, avoidance of side effects, and financial costs.

Symptoms of neuropathic pain can originate from different pathophysiologic mechanism so trying different medications or combinations may benefit an individual patient.

Gabapentinoids are often used as adjuncts to opioids for treatment of neuropathic pain. Although they were design to be GABA agonist, neither gabapentin nor pregabalin bind GABA receptors. Gabapentinoids bind the alpha-2-delta-1 subunit of voltage gated calcium channels in the nervous system. It is speculated that by down-regulating calcium influx into neurons, gabapentinoids inhibit the release of excitatory neurotransmitters, specifically glutamate [23]. Gabapentin and pregabalin are well absorbed after oral administration and are not metabolized nor bind to plasma proteins. They are excreted unchanged, mainly in the urine [24].

Gabapentin has an effective dose range of 100 to 3600 mg/day, but should be started at low nightly doses and titrated up as tolerated. Pregabalin doses range from 25 to 600 mg/day, starting with 25–75 mg at night. Mishra et al. showed that all antineuropathic agents were effective for neuropathic cancer pain, and pregabalin provided an additional morphine-sparing effect [25]. In general, gabapentinoids are well tolerated and the most common adverse effects resolve within two to three weeks of onset. Common side effects include somnolence, dizziness, ataxia, and fatigue.

Tricyclic antidepressants (TCAs) are commonly used in the management of pain, specifically neuropathic pain. Tricyclics inhibit the reuptake of serotonin and norepinephrine in the descending modulating pathway of nociception [26]. A systematic review presenting numbers needed to treat (NNTs) found that TCAs were useful for neuropathic pain, with NNT of 3.1 (95% CI 2.7–3.7) [27]. Among the TCAs, amitriptyline and nortriptyline are commonly used. They may be started at 10 to 25 mg at night and increased up to doses of 150 mg/day. TCAs have several adverse effects that may be problematic, particularly in elderly patients, such as sedation, constipation, dry mouth, urinary retention, and orthostatic hypotension.

Duloxetine is a newer serotonin and norepinephrine reuptake inhibitor (SNRI) that is effective in neuropathic pain states and has gained popularity because of its low side-effect profile. In a randomized controlled trial, Smith et al. showed that 60 mg of duloxetine reduced chemotherapy induced neuropathic pain when compared to placebo after five weeks of treatment [28].

Bisphosphonates

Bisphosphonates are inhibitors of osteoclastic bone reabsorption and have demonstrated pain-relieving effects. Metastatic bone disease can present in up to 80% of patients with breast or prostate cancer and is a common cause of pain [29]. Bone targeting agents such as bisphosphonates have been shown to reduce complications (e.g., pathologic fractures, spinal cord compression) and improve pain associated with bone metastasis. Most of the studies involving bisphosphonates have had pain relief as a secondary outcome and several demonstrated improvement in bone pain [30–33]. Biphosphonates have also been shown to delay the progression of pain from moderate to severe. An analysis of two randomized studies looking at zoledronic acid in patients with bone metastasis showed that patients who had not developed pain prior to the start of therapy obtained greater benefit from therapy [34].

Treatment with denosumab, a human monoclonal antibody, has also demonstrated improvement in pain when compared to zoledronic acid [35, 36] and delayed the onset of moderate to severe pain [37].

Ketamine

In recent years, ketamine has gained renewed interest in pain management, primarily because of its antagonism of the N-methyl-D-aspartate (NMDA) receptor and its role in central sensitization. It is postulated that antagonism of NMDA receptor by ketamine is responsible for modulating the "wind up" phenomena and reducing the level of sensitization [38]. In the acute setting, ketamine has shown some efficacy in the treatment of neuropathic pain, as well as part of multimodal analgesia in the perioperative setting. It is beneficial providing analgesia, particularly in opioid-tolerant patients, and reducing opioid consumption [39]. It's use in cancer pain has been vindicated by several case series [40–42], however a randomized, double-blind, controlled trial revealed that patients receiving escalating doses of subcutaneous ketamine developed more side effects without significantly better pain control than placebo [43].

Cannabinoids

These are several compounds that act upon the cannabinoid receptors (CB). The CB1 receptors are found in the brain and peripheral nerve terminals. It is thought that CB1 receptors may be responsible for analgesia, euphoria and anticonvulsant effects. The CB2 receptors are found primarily in the immune system and are thought to have immunosuppressive and anti-inflammatory effects [44]. Three active compounds d-9-tetrahydrocannabinol (THC), cannabidiol (CBD), and cannabinol (CBN) are thought to be relevant in analgesia [45]. The discovery of endogenous compounds that bind these receptors has given rise to what is now known as the endogenous cannabinoid system [46]. The functions of this system are still under investigation and a number of agonists and antagonists are being developed for potential therapeutic use.

Studies on cannabinoids for pain have shown inconsistent results, with some studies failing to produce any significant analgesia [47]. In a study of cancer pain in 36 patients, Noyes et al. reported that 10 mg of THC produced analgesia comparable to 60 mg of codeine and 20 mg of THC had analgesia similar to 120 mg of codeine but the higher doses had more sedation than codeine [48]. In a placebo-controlled randomized trial in 360 cancer patients, nabiximol (a combination THC-cannabidiol) was well tolerated and improved analgesia after five weeks [49].

Opioids

Opioid analgesics are the most effective and commonly used agents for moderate to severe pain. Opioids can be classified by their interaction with mu, delta, and kappa receptors. Activation of these receptors by an opioid reduces neurotransmitter release presynaptically and hyperpolarizes the postsynaptic membrane, therefore interfering with pain signal transmission at the dorsal horn of the spinal cord. Opioid receptors are normally expressed throughout the nervous system, particularly in the dorsal horn of the spinal cord, thalamus, somatosensory cortex, and limbic system. Opioid receptors also appear in peripheral tissues, and their expression is activated by injury or inflammation.

Opioids can relieve moderate to severe pain and there is no easily measureable ceiling to its analgesic effects. However, dose-dependent side effects such as sedation, mental clouding, nausea, vomiting, or respiratory depression can limit their efficacy and use. Tolerance and physical dependence are predictable pharmacologic effects seen in animal studies and in the clinical setting in response to repeated administration of opioids. Other side effects common to opioids include constipation, pruritus, immunosuppression, hyperalgesia, and opioids-induced androgen deficiency. Addiction is a distinct phenomenon characterized by craving and a pattern of compulsive drug-seeking behavior.

Tramadol is a centrally acting analgesic that has properties as a weak opioid and weak activity as a norepinephrine and serotonin reuptake inhibitor. Tramadol has good oral bioavailability with standard doses between 50 and 400 mg/day. Doses of 100 mg four times per day may produce analgesia similar to that of oral morphine 5 mg, four times per day. Patients taking monoamine oxidase inhibitors are at risk of serotonin syndrome and tramadol is contraindicated.

Tapentadol is a new analgesic that relies on a dual mechanism of action. Tapentadol acts as an agonist on the central mu receptors and inhibits the reuptake of norepinephrine in the descending inhibitory pathway. These two mechanisms of action have been shown to be synergistic in animal models [50]. This explains that tapentadol has a significant lower affinity to the mu receptor than morphine, but its analgesic effect is only about three times less than morphine. Other advantages of tapentadol are that it has no active metabolite and has minimal serotonin activity reducing the potential for adverse effects.

Tapentadol is effective in the treatment of moderate pain that does not respond to conventional non-opioid medications. In a systematic review of 42 clinical trials, Riemsma et al. showed that tapentadol has comparable efficacy and reduced adverse effects when compared with hydromorphone, morphine, oxymorphone, oxycodone, and fentanyl [51].

Codeine is a commonly used mild opioid analgesic. It is a prodrug with variable oral bioavailability and relies on metabolism by cytochrome P450 2D6 to convert to morphine for effect. Codeine may be useful for mild pain, keeping in mind that 10% of the population may not have the enzyme needed to convert the prodrug to morphine. Codeine is also commonly used as an antitussive agent.

Morphine is a commonly use opioid with a wide variety of formulations, including immediate release tablets, liquid and slow release tablets, as well as parenteral and rectal. Morphine has a variable oral bioavailability and a plasma half-life of two hours, although analgesia can last four to six hours. Morphine has two major metabolites: morphine-3-glucuronide (M3G) and morphine-6-glucuronide (M6G). M6G contributes to the analgesia and it is eliminated by the kidney, such that reduced kidney function may lead to accumulation of this metabolite. M3G is the predominant metabolite in humans, also excreted by the kidney, and is responsible for the neuroexcitatory effects seen in patients with decreased renal function. The sustained release preparation of morphine provides analgesia for up to 12 hours and gives patients greater flexibility in dosing.

Hydromorphone, hydrocodone, and oxycodone are shorter acting opioids. Like morphine, hydromorphone can be given orally or parenterally. Hydrocodone, in combination with acetaminophen, is the most widely abused opioid in the United States. It must be kept in mind that the dose limit on the combination hydrocodone-acetaminophen is due to the acetaminophen maximum daily dose. Oxycodone is available orally and, like morphine, it is formulated as an immediate-acting or extended-release presentation. It is also available in combination with acetaminophen, aspirin, or ibuprofen.

Meperidine is rarely used because of its active metabolite, normeperidine. Normeperidine has a longer half-life (14–21 hours) than the parent compound and, when it accumulates, it can produce symptoms that range from increased irritability and myoclonus, to full-blown seizures.

Methadone has high oral bioavailability, it is relatively inexpensive, does not have toxic metabolites, and blocks the NMDA receptors, theoretically improving analgesia. Unfortunately, its highly variable plasma half-life ranges from 13 to 50 hours, which is much longer than the duration of analgesia, such that repetitive dosing can lead to drug accumulation. Sedation, overdose, and death have been reported in patients that are not carefully monitored while the dose is being adjusted. Additionally, methadone has the potential to prolong the QT interval and predispose patients to life-threatening arrhythmias. Finally, there is confusion about methadone equianalgesic dosing such that conversion to or from other opioids is unreliable (Table 2).

TABLE 2. Equianalgesic conversion of commonly prescribed opioids.

Drug	Equianalgesic Doses (mg)	
	Parenteral	Oral
Morphine [52]	10	30
Codeine	100	200
Fentanyl [53]	0.1	N/A
Hydrocodone [54]	N/A	30
Hydromorphone [55]	1.5	7.5
Oxycodone	N/A	20

Data presented in this table is approximate and based on commonly used tables. The clinician is urged to use good clinical judgment at all times.

INTERVENTIONAL PAIN THERAPIES

Intrathecal

Neuraxial infusions of opioids with or without local anesthetics are increasingly used in patients with cancer pain. In this patient population, both intrathecal and epidural analgesia significantly improves analgesia [56]. Enhanced technology of intrathecal devices along with the development of newer analgesic agents has transformed neuraxial analgesia into a major therapeutic option. The most commonly used modality is an implantable pump that delivers a constant dose of analgesics through a catheter into the subarachnoid space, directly into the cerebrospinal fluid (CSF) (Figure 2). The main advantages of intrathecal therapy are improved analgesia attributable to direct effect on the spinal receptors, lower total dose administered, and avoidance of systemic toxicity. The overall decrease in dose, along with the lack of systemic distribution, is responsible for the lower side effects observed with intrathecal therapy than with equivalent doses of oral or parenteral medications.

The most commonly used drugs in intrathecal therapy are opioids, local anesthetics, baclofen, and ziconotide. Neuraxial opioids act on receptors located in the substantia gelatinosa of the dorsal horn to provide analgesia, and the effect is determined by the opioids' specific affinity for these receptors. There are several factors to consider when selecting an opioid for intrathecal delivery, including area of pain, location of the catheter tip, and water solubility of the opioid. Highly hydrophilic intrathecal opioids like morphine exhibit greater rostral spread that may improve analgesia in certain clinical cases. There is abundant evidence that

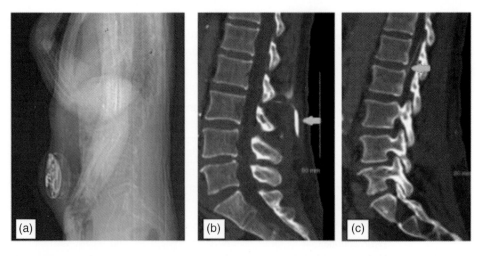

Figure 2. Intrathecal implantable pump. Intrathecal therapy with (a) reservoir in the left lower quadrant; (b) catheter anchored in the supraspinous ligament (red arrow); and (c) catheter in the intrathecal space (red arrow).

intrathecal opioids are superior to orally administered opioids in patients with cancer pain. In a randomized controlled trial, Smith et al. compared intrathecal drug delivery to optimized medical management in 200 cancer patients. Of patients with intrathecal therapy 67% had greater than a 20% improvement in pain, compared to only 36% of patients in the optimal medical management group. Visual analog scale pain score improved by 52% in the intrathecal group compared to 39% in those patients with optimized medical management. Also important, toxicity scores in the intrathecal pump group fell 50% versus 17% in the medical management group [57]. In a different prospective study of cancer patients, Rauck et al. also found significant reduction in pain and toxicity on patients with intrathecal therapy, but this difference disappeared at the 16-month follow up [58].

Side effects related to opioids are patient dependent, worse in opioid naïve patients. Pharmacologic side effects may include somnolence, nausea, vomiting, nightmares, pruritus, respiratory depression, hyperalgesia, constipation, myoclonus, pedal edema, and low testosterone and estrogen levels.

The mechanism of action of local anesthetics is through blockade of sodium channels inhibiting the depolarization of neurons. Numerous studies have documented the efficacy of local anesthetics in intrathecal therapy, mostly in combination with opioids or other agents. In a double-blind, randomized trial comparing intrathecal morphine alone to intrathecal morphine and bupivacaine, van Dongen et al. found that cancer patients receiving combination therapy developed less tolerance than the morphine only group. Even patients that have failed intrathecal opioids find improved analgesia with combination of local anesthetics and opioids, which suggests a synergistic analgesic effect [59]. Common side effects seen with intrathecal local anesthetics include numbness, weakness, paresthesias, and hypotension, all of which can be diminished by reducing the local anesthetic dose.

Ziconotide is a calcium channel blocker that is clinically used intrathecally to treat neuropathic pain. Ziconotide is a synthetic form of a peptide isolated from the venom produced by a marine snail and blocks the N-type calcium channel. In a multicenter, double-blind, placebo-controlled crossover study, Staats et al. found that patients in the treatment group had significantly better pain relief than those in the placebo group [60]. Unfortunately, a significant proportion of patients (12–33%) suffer side effects that include psychiatric (depression, confusion, cognitive impairment, hallucinations), neurologic (headaches, ataxia, abnormal gait, double vision), and other (somnolence, nausea, lightheadedness, and urinary retention) symptoms.

Complications associated to intrathecal therapy can be divided in those related to the device (catheter, pump) and those related to the medications. Device related complications include those affecting the catheter (migration, coiling, obstruction) or the pump (infection, battery failure, pain at the site). Pharmacologic side effects have been mentioned above.

One potentially devastating complications is the formation of a catheter tip granuloma. Depending on the location of the catheter tip, a granuloma may produce various neurologic symptoms including cord compression. Histopathologic analysis reveals that these are inflammatory masses and appear to be related to higher concentration rather than dose [61, 62]. Although most of the initial evidence implicated morphine, other opioids and baclofen have been associated with the formation of inflammatory catheter tip masses. Clinical suspicion is key for early diagnosis. One of the earliest symptoms may be an increase in pain or dose escalation, without clear progression of disease.

Vertebral Augmentation

Vertebroplasty and kyphoplasty are techniques of vertebral augmentation where acrylic cement is percutaneously injected into an affected vertebral body under fluoroscopic guidance. In vertebroplasty, cement is injected directly into the fractured vertebrae, while in balloon kyphoplasty, a balloon is inflated to create a cavity within the vertebral body and cement is injected into the newly created cavity. These procedures have been proposed as a way to reduce pain from vertebral fractures, as well as avoiding further collapse of the vertebral height. Metastatic spread of cancer to the vertebral bodies occurs more frequently in the thoracic spine. Bone metastatic disease is common in patients with advanced cancer, with up to 80% of patients with breast or prostate cancer [63]. In two systematic reviews of cancer patients, Chew et al. [64, 65] demonstrated reduction in pain ranging between 47 and 87%. Unfortunately, these procedures can have serious complications such as cement pulmonary embolism, neurologic damage from the instrumentation or cement leak, paralysis, and death.

Neurolytic Blocks

With the advent of newer technologies and intrathecal drug delivery systems, the use of neurolytic blocks, particularly chemical neurolysis, has been declining. Nevertheless, there are still some indications where it may be useful to consider chemical neurolysis. The two most common neurolytic agents are phenol and ethanol. Phenol is more viscous than ethanol, has a slower onset and it is not painful on injection. Ethanol needs to be concentrated over 50% to produce neurolysis. It may also cause pain on injection for which some practitioners add a long-acting local anesthetic. Both agents cause precipitation of endoneural lipoproteins, protein coagulation, and necrosis of neural tissue. We will only discuss celiac plexus, superior hypogastric plexus, and ganglion impar neurolysis.

Celiac Plexus Block

The celiac plexus is a network of nerve fibers deep in the retroperitoneum, located anterior to the aorta and over the celiac arterial trunk at the level of L1 vertebral body. The celiac plexus is formed predominantly by the preganglionic sympathetic efferent fibers derived from the greater, lesser, and least splanchnic nerves. Preganglionic parasympathetic efferent fibers from the vagus nerve also contribute to the plexus. The celiac plexus innervates the organs of the upper abdomen (e.g., pancreas, biliary, gastric) and the viscera from the distal esophagus to the transverse colon.

There are many approaches to the celiac plexus that have been described in the literature including a posterior approach guided by fluoroscopy or computed tomography (CT), ultrasonographic approach via the anterior route, and endoscopic ultrasound, all with advantages and disadvantages. Celiac plexus neurolysis has been shown to produce long-lasting improvement in pain in between 70 to 90% of patients, as well as reduced analgesic consumption [66–68]. It has been suggested that celiac plexus neurolysis can enhance the ability of patients to perform their activity of daily living [69] (Figure 3).

Superior Hypogastric Plexus Block

The superior hypogastric plexus is a retroperitoneal network of nerve fibers located at the level of the L5 and S1 vertebral bodies, anterior to the bifurcation of the abdominal aorta. The sympathetic fibers originate in the anterolateral column at the level of T11–T12 and it receives lumbar contributions, as well as from the intermesenteric plexus. The superior hypogastric plexus innervates the pelvic

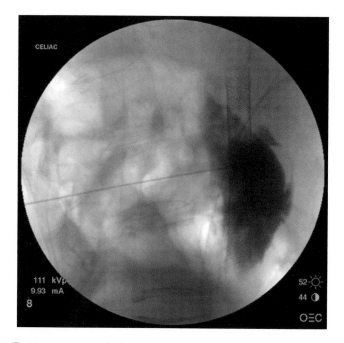

Figure 3. Fluoroscopy guided celiac plexus neurolysis from a posterior approach.

viscera and the primary indication is visceral pelvic pain, refractory to conventional medical treatment (gynecological, colorectal, or genitourinary).

This block is performed under fluoroscopic or CT guidance with bilateral needle placement or a single needle through the L5-S1 intervertebral disc. Several prospective case series have demonstrated good-to-excellent relief in 70% of patients and reduced opioid consumption, without significant adverse effects [70].

Ganglion Impar Block

The ganglion impar is a single, retroperitoneal, midline structure located at the level of the sacrococcygeal junction. This structure represents the termination of the paravertebral sympathetic chain and contains visceral afferent fibers that innervate the perineum, distal rectum, anus, distal urethra, distal vagina, and vulva.

Several techniques have been described to perform this block and, although it appears to be very safe, there are several theoretical complications including perforation of the rectum, sacral nerve root injury, and motor or sexual dysfunction [71].

REFERENCES

1. Paice, J.A. and Ferrell, B. (2011). The management of cancer pain. *CA Cancer J. Clin.* 61 (3): 157–182.
2. Sabiston, C.M., Brunet, J., and Burke, S. (2012). Pain, movement, and mind: does physical activity mediate the relationship between pain and mental health among survivors of breast cancer? *Clin. J. Pain* 28 (6): 489–495.
3. Foley, K.M. (1982). Clinical assessment of cancer pain. *Acta Anaesthesiol. Scand. Suppl.* 74: 91–96.
4. Twycross, R.G. and Fairfield, S. (1982). Pain in far-advanced cancer. *Pain* 14 (3): 303–310.
5. Ripamonti, C. and Longo, M. (2004). Cancer pain syndromes and pharmacotherapy of cancer pain. *Arch. Oncol.* 12 (3): 168–170.
6. Fallon, M.T. (2013). Neuropathic pain in cancer. *Br. J. Anaesth.* 111 (1): 105–111.
7. Anand, U. (2007). Mechanisms and management of cancer pain. In: *The Cancer Handbook*, 2e (ed. M.R. Alison). Chichester, UK: Wiley.
8. Doyle, D., Hanks, G.W.C., and MacDonald, N. (1998). *Oxford Textbook of Palliative Medicine*, 2e. New York, NY: Oxford University Press.
9. Caraceni, A., Hanks, G., Kaasa, S. et al. (2012). Use of opioid analgesics in the treatment of cancer pain: evidence-based recommendations from the EAPC. *Lancet Oncol.* 13: e58–e68.
10. Nabal, M., Librada, S., Redondo, M.J. et al. (2012). The role of paracetamol and nonsteroidal anti-inflammatory drugs in addition to WHO Step III opioids in the control of pain in advanced cancer: a systematic review of the literature. *Palliat. Med.* 26: 305–312.
11. Israel, F.J., Parker, G., Charles, M. et al. (2010). Lack of benefit from paracetamol (acetaminophen) for palliative cancer patients requiring high-dose strong opioids: a randomized, double-blind, placebo- controlled, crossover trial. *J. Pain Symptom Manag.* 39: 548–554.
12. Sima, L., Fang, W.X., Wu, X.M. et al. (2012). Efficacy of oxycodone/paracetamol for patients with bone- cancer pain: a multicenter, randomized, double- blinded, placebo-controlled trial. *J. Clin. Pharm. Ther.* 37: 27–31.

13. Radbruch, L. and Elsner, F. (2005). Emerging analgesics in cancer pain management. *Expert Opin. Emerg. Drugs* 10: 151–171.
14. Warner, T.D. and Mitchell, J.A. (2004). Cyclooxygenases: new forms, new inhibitors, and lessons from the clinic. *FASEB J.* 18: 790–804.
15. Eisenberg, E., Berkey, C.S., Carr, D.B. et al. (1994). Efficacy and safety of nonsteroidal anti-inflammatory drugs for cancer pain: a meta-analysis. *J. Clin. Oncol.* 12: 2756–2765.
16. McNicol, E.D., Strassels, S., Goudas, L. et al. (2005). NSAIDS or paracetamol, alone or combined with opioids, for cancer pain. *Cochrane Database Syst. Rev.* (2): Art. No.: CD005180. doi: https://doi.org/10.1002/14651858.CD005180.
17. Barrett, K.E., Boitano, S., Barman, S.M. et al. (2012). Chapter 20. The adrenal medulla & adrenal cortex. In: *Ganong's Review of Medical Physiology*, 24e (ed. K.E. Barrett, S. Boitano, S.M. Barman, et al.). New York, NY: McGraw-Hill http://accessmedicine.mhmedical.com/content. aspx?bookid=393&Sectionid=39736762 Accessed April 14, 2015.
18. Shih, A. (2007). Role of corticosteroids in palliative care. *J. Pain Palliat. Care Pharmacother* 21: 69–76.
19. Bruera, E., Roca, E., Cedaro, L. et al. (1985). Action of oral methylprednisolone in terminal cancer patients: a prospective randomized double-blind study. *Cancer Treat Rep.* 69: 751–754.
20. Bruera, E., Moyano, J.R., Sala, R. et al. (2004). Dexamethasone in addition to metoclopramide for chronic nausea in patients with advanced cancer: a randomized controlled trial. *J. Pain Symptom Manag.* 28: 381–388.
21. Della Cuna, G.R., Pellegrini, A., and Piazzi, M. (1989). Effect of methylprednisolone sodium succinate on quality of life in preterminal cancer patients: a placebo-controlled, multicenter study – The Methylprednisolone Preterminal Cancer Study Group. *Eur. J. Cancer Clin. Oncol.* 25: 1817–1821. 198.
22. Popiela, T., Lucchi, R., and Giongo, F. (1989). Methylprednisolone as palliative therapy for female terminal cancer patients: the Methylprednisolone Female Preterminal Cancer Study Group. *Eur. J. Cancer Clin. Oncol.* 25: 1823–1829.
23. Stahl, S.M., Porreca, F., Taylor, C.P. et al. (2013). The diverse therapeutic actions of pregabalin: is a single mechanism responsible for several pharmacological activities? *Trends Pharmacol. Sci.* 34: 332–339.
24. McNamara, J.O. (2011). Chapter 21. Pharmacotherapy of the epilepsies. In: *Goodman & Gilman's The Pharmacological Basis of Therapeutics*, 12e (ed. L.L. Brunton, B.A. Chabner and B.C. Knollmann). New York, NY: McGraw-Hill http://accessmedicine.mhmedical.com/content. aspx?bookid=374&Sectionid=41266227. Accessed May 05, 2015.
25. Mishra, S., Bhatnagar, S., Goyal, G.N. et al. (2012). A comparative efficacy of amitriptyline, gabapentin, and pregabalin in neuropathic cancer pain: a prospective randomized double-blind placebo-controlled study. *Am. J. Hosp. Palliat. Care* 29: 177–182.
26. O'Donnell, J.M. and Shelton, R.C. (2011). Chapter 15. Drug therapy of depression and anxiety disorders. In: *Goodman & Gilman's The Pharmacological Basis of Therapeutics*, 12e (ed. L.L. Brunton, B.A. Chabner and B.C. Knollmann). New York, NY: McGraw-Hill http://accessmedicine. mhmedical.com/content.aspx?bookid=374&Sectionid=41266221. Accessed May 06, 2015.
27. Finnerup, N.B., Otto, M., HJ, M.Q. et al. (2005). Algorithm for neuropathic pain treatment: an evidence based proposal. *Pain* 118 (3): 289–305.
28. Smith, E.M.L., Pang, H., Cirrincione, C. et al. (2013). Effect of duloxetine on pain, function, and quality of life among patients with chemotherapy-induced painful peripheral neuropathy: a randomized clinical trial. *JAMA* 309: 1359–1367.
29. Coleman, R.E. (2006). Clinical features of metastatic bone disease and risk of skeletal morbidity. *Clin. Cancer Res.* 12: 6243s–6249s.
30. Lipton, A., Theriault, R.L., Hortobagyi, G.N. et al. (2000). Pamidronate prevents skeletal complications and is effective palliative treatment in women with breast carcinoma and osteolytic bone metastases: long term follow-up of two randomized, placebo-controlled trials. *Cancer* 88: 1082–1090.

31. Body, J.J., Diel, I.J., Bell, R. et al. (2004). Oral ibandronate improves bone pain and preserves quality of life in patients with skeletal metastases due to breast cancer. *Pain* 111: 306–312.

32. Saad, F. and Eastham, J. (2010). Zoledronic acid improves clinical outcomes when administered before onset of bone pain in patients with prostate cancer. *Urology* 76: 1175–1181.

33. Cleeland, C.S., Body, J.J., Stopeck, A. et al. (2013). Pain outcomes in patients with advanced breast cancer and bone metastases: results from a randomized, double-blind study of denosumab and zoledronic acid. *Cancer* 119: 832–838.

34. Costa, L., Lipton, A., Hadji, P. et al. (2013). Treatment of bone metastases before the onset of pain. *Int. J. Clin. Oncol.* 18: 531–538.

35. Fizazi, K., Carducci, M., Smith, M. et al. (2011). Denosumab versus zoledronic acid for treatment of bone metastases in men with castration-resistant prostate cancer: a randomised, double blind study. *Lancet* 377: 813–822.

36. Stopeck, A.T., Lipton, A., Body, J.J. et al. (2010). Denosumab compared with zoledronic acid for the treatment of bone metastases in patients with advanced breast cancer: a randomized, double-blind study. *J. Clin. Oncol.* 128: 5132–5139.

37. von Moos, R., Body, J.J., Egerdie, B. et al. (2013). Pain and health-related quality of life in patients with advanced solid tumours and bone metastases: integrated results from three randomized, double-blind studies of denosumab and zoledronic acid. *Support Care Cancer* 21: 3497–3507.

38. Petrenko, A.B., Yamakura, T., Baba, H. et al. (2003). The role of N-methyl-D-aspartate (NMDA) receptors in pain: a review. *Anesth. Analg.* 97: 1108–1116.

39. Loftus, R.W., Yeager, M.P., Clark, J.A. et al. (2010). Intraoperative ketamine reduces perioperative opiate consumption in opiate-dependent patients with chronic back pain undergoing back surgery. *Anesthesiology* 113: 639–646.

40. Bell, R.F., Eccleston, C., and Kalso, E. (2003). Ketamine as adjuvant to opioids for cancer pain: a qualitative systematic review. *J. Pain Symptom Manag.* 26: 867–875.

41. Jackson, K., Ashby, M., Howell, D. et al. (2010). The effectiveness and adverse event profile of "burst" ketamine in refractory cancer pain: the VCOG PM1-00 study. *J. Palliat. Care* 26: 176–183.

42. Kerr, C., Holahan, T., and Milch, R. (2011). The use of ketamine in severe cases of refractory pain syndromes in the palliative care setting: a case series. *J. Palliat. Med.* 14: 1074–1077.

43. Hardy, J., Quinn, S., Fazekas, B. et al. (2012). Randomized, double-blind, placebo-controlled study to assess the efficacy and toxicity of subcutaneous ketamine in the management of cancer pain. *J. Clin. Oncol.* 30: 3611–3678.

44. Mackie, K. (2006). Cannabinoid receptors as therapeutic targets. *Annu. Rev. Pharmacol. Toxicol.* 46: 101–122.

45. Hill, L. and Schug, S.A. (2009). Recent advances in the pharmaceutical management of pain. *Expert. Rev. Clin. Pharmacol.* 2: 543–557.

46. Abrams, D. and Guzman, M. (2015). Cannabis in cancer care. *Clin. Pharmacol. Ther.* 97 (6): 575–586. https://doi.org/10.1002/cpt.108.

47. Frank, B., Serpell, M.G., Hughes, J. et al. (2008). Comparison of analgesic effects and patient tolerability of nabilone and dihydrocodeine for chronic neuropathic pain: randomized, crossover, double blind study. *BMJ* 336: 199–201.

48. Noyes, R., Brunk, S., Avery, D. et al. (1975). The analgesic properties of delta-9-tetrahydrocannabinol and codeine. *Clin. Pharmacol. Ther.* 18: 84–89.

49. Portenoy, R.K., Ganae-Motan, E.D., Allende, S. et al. (2012). Nabiximols for opioid-treated cancer patients with poorly-controlled chronic pain: a randomized, placebo-controlled, graded-dose trial. *J. Pain* 13 (5): 438–449.

50. Schröder, W., Vry, J.D., Tzschentke, T.M. et al. (2010). Differential contribution of opioid and noradrenergic mechanisms of tapentadol in rat models of nociceptive and neuropathic pain. *Eur. J. Pain* 14: 814–821.

51. Riemsma, R., Forbes, C., Harker, J. et al. (2011). Systematic review of tapentadol in chronic severe pain. *Curr. Med. Res. Opin.* 27: 1907–1930.

52. Kalso, E. and Vainio, A. (1990). Morphine and oxycodone hydrochloride in the management of cancer pain. *Clin. Pharmacol. Ther.* 47: 639–646.

53. Indelicato, R.A. and Portenoy, R.K. (2003). Opioid rotation in the management of refractory cancer pain. *JCO* 21: 87s–91s.

54. Hallenbeck, J.L. (2003). *Palliative Care Perspectives*, vol. 71. New York: Oxford University Press.

55. Vallner, J.J., Stewart, J.T., Kotzan, J.A. et al. (1981). Pharmacokinetics and bioavailability of hydromorphone following intravenous and oral administration to human subjects. *J. Clin. Pharmcol.* 21: 152–156.

56. Burton, A.W., Rajagopal, A., Shah, H.N. et al. (2004). Epidural and intrathecal analgesia is effective in treating refractory cancer pain. *Pain Med.* 5: 239–247.

57. Smith, T.J. and Coyne, P.J. (2005). Implantable drug delivery systems (IDDS) after failure of comprehensive medical management (CMM) can palliate symptoms in the most refractory cancer pain patients. *J. Palliat. Med.* 8 (4): 736–742.

58. Rauck, R.L., Cherry, D., Boyer, M.F. et al. (2003). Long-term intrathecal opioid therapy with a patient-activated, implanted delivery system for the treatment of refractory cancer pain. *J. Pain* 4 (8): 441–447.

59. van Dongen, R.T., Crul, B.J., and van Egmond, J. (1999). Intrathecal coadministration of bupivacaine diminishes morphine dose progression during long-term intrathecal infusion in cancer patients. *Clin. J. Pain* 15 (3): 166–172.

60. Staats, P.S., Yearwood, T., Charapata, S.G. et al. (2004). Intrathecal ziconotide in the treatment of refractory pain in patients with cancer or AIDS: a randomized controlled trial. *JAMA* 291 (1): 63–70.

61. Yaksh, T.L., Horais, K.A., Tozier, N.A. et al. (2003). Chronically infused intrathecal morphine in dogs. *Anesthesiology* 99: 174–187.

62. Allen, J.W., Horais, K.A., Tozier, N.A. et al. (2006). Time course and role of morphine dose and concentration in intrathecal granuloma formation in dogs: a combined magnetic resonance imaging and histopathology investigation. *Anesthesiology* 105: 581–589.

63. Coleman, R.E. (2006). Clinical features of metastatic bone disease and risk of skeletal morbidity. *Clin. Cancer Res.* 12: 6243s–6249s.

64. Chew, C., Craig, L., Edwards, R. et al. (2011). Safety and efficacy of percutaneous vertebroplasty in malignancy: a systematic review. *Clin. Radiol.* 66: 63–72.

65. Chew, C., Ritchie, M., O'Dwyer, P.J. et al. (2011). A prospective study of percutaneous vertebroplasty in patients with myeloma and spinal metastases. *Clin. Radiol.* 66: 1193–1196.

66. Eisenberg, E., Carr, D.B., and Chalmers, T.C. (1995). Neurolytic celiac plexus block for treatment of cancer pain: a meta-analysis. *Anesth. Analg.* 80 (2): 290–295.

67. Ischia, S., Polati, E., Finco, G. et al. (1998). 1998 Labat lecture: the role of the neurolytic celiac plexus block in pancreatic cancer pain management: do we have the answers? *Reg. Anesth. Pain Med.* 23 (6): 611–614.

68. Ischia, S., Ischia, A., Polati, E. et al. (1992). Three posterior percutaneous celiac plexus block techniques. A prospective, randomized study in 61 patients with pancreatic cancer pain. *Anesthesiology* 76 (4): 534–540.

69. Staats, P.S., Hekmat, H., Sauter, P. et al. (2001). The effects of alcohol celiac plexus block, pain, and mood on longevity in patients with unresectable pancreatic cancer: a double-blind, randomized, placebo-controlled study. *Pain Med.* 2 (1): 28–34.

70. Plancarte, R., de Leon-Casasola, O.A., El-Helaly, M. et al. (1997). Neurolytic superior hypogastric plexus block for chronic pelvic pain associated with cancer. *Reg. Anesth.* 22 (6): 562–568.

71. Plancarte, R.A., Amescua, C., Patt, R.B. et al. (1990). Presacral blockade of the ganglion of Walther (ganglion im- par). *Anesthesiology* 73: A751.

27

COMPLEMENTARY AND ALTERNATIVE MEDICINE

Phil Trabulsy

University of Vermont, Burlington, VT, USA

Prior to the 1990s, nonconventional medical disciplines and therapies outside mainstream medicine were termed "alternative." Academic health centers rarely used alternative therapies and allopathic medical schools did not teach this material. In 1990, Dr. David Eisenberg and colleagues at Harvard Medical College published a comprehensive review on the use of complementary and alternative medicine (CAM) in the United States. This pioneering study reported that the percentage of US individuals using some form of CAM reached nearly 45%, and less than 30% of these patients disclosed this to their conventional medical provider [1]. To the mainstream medical community this fact was an important statistic and deserved attention. Follow-up studies on CAM usage over the past 30 years, have reported CAM use by the public ranging from 28 to 60%, with incidence of use increasing yearly [2]. In addition, studies revealed a significant proportion of patients were using CAM in parallel with their conventional healthcare.

In 1991, the United States Congress established the Office of Alternative Medicine. In 1998, the National Center for Complementary and Alternative Medicine (NCCAM) became a division of the National Institute of Health (NIH). In December 2014, NCCAM was again renamed the National Center for Complementary and Integrative Health (NCCIH), by mandate from the Obama Healthcare Reform Bill. The reported mission of NCCIH is "to define, through

Cancer: Prevention, Early Detection, Treatment and Recovery, Second Edition. Edited by Gary S. Stein and Kimberly P. Luebbers.
© 2019 John Wiley & Sons, Inc. Published 2019 by John Wiley & Sons, Inc.

rigorous scientific investigation, the usefulness and safety of complementary and integrative health interventions and their roles in improving health and health care" [3]. In 2005, the Institute of Medicine (IOM) defined CAM as "a broad domain of health-related resources that encompasses health systems, modalities, and practices … other than those intrinsic to the dominant healthcare system of a particular society or culture and in a given historical period." NCCIH subcategorizes the CAM disciplines to specific domains based on therapeutic methodology and distinctive medical philosophies. See Table 1 for an expanded view of the NCCAM categories.

Increasing collaboration between CAM and conventional medical practitioners, along with the use of complementary approaches within mainstream medicine, have given rise to the term Integrative Medicine [4]. Over the past 15 years, over 60 medical schools from the United States and abroad have joined to form the Academic Consortium for Integrative Health. The Consortium defines Integrative Medicine as "the practice of medicine that reaffirms the importance of the relationship between practitioner and patient, focuses on the whole person, is informed by evidence, and makes use of all appropriate therapeutic approaches, healthcare professionals, and healthcare disciplines to achieve optimal health and healing." Concisely, integrative medicine is the use of evidence-based complementary therapies in collaboration and coordination with evidence-based conventional healthcare [5]. With increasing emphasis placed on the bio-psychosocial and spiritual components of patient care, the use of integrative medicine, and integrative oncology, fits well with the multidimensional cancer care model. An approach to the cancer patient that fosters physical, mental, and spiritual well-being, in conjunction with use of evidence-based complementary therapies, is the core concept of Integrative Oncology [5]. The patient-centric perspectives found in the Integrative Oncology paradigm are also evident within other patient-centric areas of healthcare, including palliative care, quality of life improvement initiatives, spiritual well-being, and narrative-based medical encounters. Table 2 is an adaptation of the NCCAM categories specific for integrative oncology.

TABLE 1. NCCAM Categories.

Category	Examples
Nonconventional traditional medical systems	TCM, Ayurvedic, naturopathy, anthroposophic, Native/Indigenous
Mind–Body Therapies	Meditation, mindfulness, yoga, Tai Chi, psychology, biofeedback, relaxation, art, music, nature
Biologically based therapies	Dietary supplements, herbs, vitamins, minerals, botanicals, medicinal mushrooms
Manipulative and Body-Based therapies	Massage therapy, chiropractic, osteopathic
Energy Medicine, Biofield- therapies	Healing Touch, Therapeutic touch, Reiki

TABLE 2. Integrative oncology categories.

Category	Examples
Nutritional Support	Counseling, weight management, dietary supplements
Lifestyle Medicine	Physical fitness, behavioral changes, sleep hygiene, social support
Mind–body medicine	Stress management, mindfulness practice, spirituality, religious practice, prayer, expressive arts
Movement/Manual Therapies	Massage, yoga, Tai Chi, Feldenkreis, dance, nature
Physical Therapies	Acupuncture, physical therapy and rehabilitation
Traditional Healing Systems	TCM, ayurveda, homeopathy, osteopathy, naturopathy

THE COMPLEMENTARY AND INTEGRATIVE APPROACH

The interest in and use of complementary and integrative therapies (CIT) by cancer patients has increased over the past 30 years with estimates of use ranging from 33 to 47% depending on where in the cancer trajectory the patient resides [6]. Women with breast cancer are among the highest users of CIT averaging 45–50% of patients across differing treatment stages [7]. It is interesting that use of CIT has increased alongside the substantial improvements and technical advancements in mainstream cancer treatment, including increased survivorship. Issues related to psychological support, personal empowerment, improved emotional well-being, enhancing the body's ability to fight disease, diminishing risk of recurrence, reducing the adverse effects of the cancer and treatment side effects, and reconnecting with purpose and meaning in life are the reasons for use of CIT consistently reported by patients [6]. Additionally, patients turn to complementary therapies due to a paucity of conventional medical options for symptom relief and a desire for less invasive and non-pharmacologic therapies. A study examining the use of integrative therapies among patients with various cancer diagnoses at a comprehensive cancer center show that the most common therapies used were nutritional supplements (91%), dietary changes (27%), exercises (23%), acupuncture (20%) and, movement therapy such as yoga and tai chi (16%), massage (10%), energy therapies (7%), prayer and spirituality practices (6%) [8].

Oncologists concur that treatment compliance is a critical component in promoting cancer control, and recent reports draw attention to the capacity of complementary interventions to improve patients' tolerance for conventional medicine. Theoretically, this increased ability to complete conventional cancer treatment without treatment breaks allows patients to comply with proven protocols more fully [9]. With many questions still unanswered, there is a continued need for rigorous research to support use of these interventions. The recommended research priorities are aimed at facilitating the creation of comprehensive evidence-based clinical practice guidelines on the use of integrative therapies across all populations of cancer patients and survivors. With standardized guidelines established, credible outcome studies and comparative effectiveness research can follow.

THE INTEGRATIVE ONCOLOGY ENCOUNTER

For most patients, receiving a cancer diagnosis is an anxiety provoking experience. During this overwhelming time, many of our patients who are interested or curious find it difficult to make informed decisions about the use of integrative therapies. An integrative health professional, knowledgeable about cancer and integrative therapies, can be an invaluable resource. The main objective of the integrative oncology encounter is to enhance the exchange of ideas between patients and clinicians regarding CAM decisions, and to direct the patient to beneficial therapies. The integrative oncology approach aspires to increase the efficacy of conventional cancer treatments, reduce treatment-related side effects and symptoms, and improve quality of life for the cancer patient, while making certain that safety and efficacy remain a priority [10] (see Table 3). Integrative oncology incorporates the best of rational data-based complementary therapies with the best of mainstream cancer care. To paraphrase a quote from integrative oncologist, Dr. Sagar, "the cancer care team strives for a 'one medicine' ideal that synergistically combines therapies and treatments respectfully and collaboratively, with the goal that the collaborative sum is greater than the individual parts in isolation" [11]. Recent reports from cancer centers throughout the United States and abroad, affirm that this integrative medicine approach is in high demand. Research confirms that the combination of relationship-centered care and interdisciplinary collaboration have a high probability of improving patients' quality of life, reducing distress, increasing self-empowerment, enhancing cooperation with care, and possibly improving clinical outcomes [8].

PARTNERSHIP, COMMUNICATION, DECISION-MAKING

US studies report that a significant percentage of adult and pediatric patients with cancer use at least one complementary therapy. Without credible information, patients may be unknowingly engaging in practices that could be unsafe and ineffective, while potentially diminishing the efficacy of the conventional treatments that they may be using for cancer control. Many of these patients do not disclose

TABLE 3. Characteristics of effective patient–practitioner communication.

Characteristics
a. Respect patient's preferences
b. Empathy
c. Trust
d. Open-minded information exchange
e. Address psychosocial and emotional issues
f. Collaborative clinical decision-making

their complementary therapy use with their oncologist nor do they obtain quality information from the physician or healthcare team to make an educated decision. Both physicians and patients report that discussion of complementary and integrative approaches significantly enhances the physician–patient relationship [12]. Experts stress that the complementary and integrative therapy discussion occurs as early as possible, and hopefully during one of the initial consultative visits. Highlighting the potential value of early discussions, a recent study evaluated dietary supplement use in 650 cancer patients during all stages of their care, and found that greater than 28% of these adult cancer patients were at increased risk for drug-herb interactions with their conventional medical treatments. In addition, these patients did not disclose their use to the treating cancer team [13]. It is important to note that most patients who do report the use of CIT during cancer treatment view it as supporting quality of life and complementary to their conventional treatments, rather than it being an alternative or for a "cure" [14]. An intriguing study demonstrated that the percentage of patient's who would discuss and disclose their use of complementary therapies would increase by more than 90% if the patient's physician would simply inquire, in a nonjudgmental way, about past or present use of any complementary therapies. Understanding the reasons patients do not disclose their use of complementary therapies may be insightful for the practitioner (see Table 4).

In an effort to improve service to the patient's requesting integrative therapy consultations, a number of major cancer centers have developed Decision Support Programs. These programs facilitate and promote open dialog about the use of CIT within cancer care, direct patients toward high-quality information resources, and educate patients and practitioners on the safe integration of these therapies within standard care. In addition, offering an integrated decision support program may be an invaluable asset to an institution's patient-centered care initiatives [15]. This particular approach to care requires a partnership between practitioner and patient. The shared and practiced responsibilities of both partners will provide this relationship with integrity and strength. A framework of several key responsibilities to consider for developing an effective patient-practitioner relationship can be seen in Tables 5 and 6.

TABLE 4. Patient's report the following reasons for nondisclosure of CAM practices.

Reasons for nondisclosure
1. Physician noninquiry
2. Patient anticipation of the physicians nonapproval
3. Physician perceived disinterest
4. Perceived inability to be able to provide useful information on complementary modalities
5. Patient's perception that their use of a complementary modality is irrelevant to their conventional cancer treatment
6. Patient physician consult time constraint
7. Patient's desire to maintain autonomy and control.

TABLE 5. Health partnership guidelines: patient responsibilities.

Patient's Responsibilities
Share with Cancer Team previous, present, and possible desire for use of complementary therapies
Use trusted, reliable sources for information regarding complementary and integrative therapies during cancer treatment.
Always share with cancer team any possible benefits or side-effects you may experience with a complementary therapy, as well as the benefits
Always disclose your use of therapies or interventions that have the potential to interact with your treatments (vitamins, special diets, herbs, supplements etc.)

TABLE 6. Health partnership guidelines: clinician responsibilities.

Clinician's Responsibilities
Most cancer patients are very receptive to a shared decision-making approach
Most cancer patients view CIM as complementary, NOT alternative
Discuss "nonconventional therapies" as specific as possible by categorizing, i.e., herbs, supplements, yoga, massage acupuncture
Avoid the label "alternative" when discussing patient's use of CIM
Have easily accessible up-to-date literature on common integrative therapies for patient discussion and dissemination
Consider creating a standard decision-making program in your Center with a Lead CIM navigator and ambassador
Always strive for offering balanced evidence-based advice based on the known risk/benefit ratio of a therapy
Respectfully accept the use of a CIM therapy when there is no evidence of harm and the patient cost and commitment is manageable
Always discourage CIM treatments that delay conventional care, treatments by non-credentialed or unlicensed professionals, or when a biologic therapy is not approved by a regulatory body

Clinical Decision-Making

The integrative oncologist understands that it is important to apply the same standard of clinical efficacy in integrative medical clinical decision making as with conventional medical decision-making. Similar to the conventional clinician, the integrative oncologist relies on sound evidence acquired from randomized control trials, meta-analyses, and systematic reviews to grade the safety, efficacy, and effectiveness of the intervention whenever possible. Cohen and Eisenberg [16] developed a helpful resource for evaluating evidence of CAM research studies and their use in clinical decision-making for specific CAM therapies [16]. It is important for the clinician discussing evidence and safety to bear in mind those high-quality studies on the effectiveness of a particular complementary therapy may be limited [17]. Therefore, it is understandable that discussions of uncertainty may occur and can be difficult for both the healthcare practitioner and patient in supporting their decision. Studies suggest that when a high probability of benefit over risk exists, even with limited safety data, the clinician can consider use of the therapy, while providing caution and close monitoring for safety. An example of this clinical dilemma is in the patient's desire for use of mindfulness practices for anxiety and depression. Here, the data on effectiveness, though trending in a positive direction, is limited; there is essentially no concern about safety issues. The practitioner could support this decision comfortably. Similar example would be the patient's desire for use of biofield energy therapy for pain management. Here, the research evidence for efficacy is in the early stages, yet safety is essentially of no concern. This is an area for the clinician where respectively supporting the patient's desire seems appropriate [18].

Education

As the oncologist is the lead coordinator (navigator) of patient care, it is essential that they be aware of all care and interventions that the patient receives. Clinicians are increasingly required to provide reliable up-to-date information regarding the use of CIT therapies throughout the cancer trajectory. Unfortunately, more than 75% of conventional oncologists report lack of sufficient knowledge base to discuss specific CIT therapies with their patients. Likewise, studies evaluating how patients obtain this information find that their healthcare practitioner was the least frequently used resource (with M.D. 7%, nurses 9%, CAM practitioners 18%, family 43%, media 55%, and friends 75%) [19].

In a response to the apparent lack of understanding among practitioners on the appropriate use of CIT therapies, the IOM and the World Health Organization (WHO) now recommend that conventional medical schools educate students in traditional and complementary medical disciplines at both the undergraduate and postgraduate level. A major goal of these efforts is to improve communication and remove cultural barriers that commonly exist between conventional and complementary practitioners. These IOM learning objective are highlighted in Table 7.

A number of unique topic specific educational workshops are available. Their aim is to educate and guide oncologists and oncology practitioners in methods to appropriately query and advise patients regarding safe and effective use of CIT. Educational opportunities are available for oncology health professionals to

TABLE 7. Learning objectives and competencies put forth by the Institute of Medicine.

1. *Knowing and learning how to ask patients about their use of complementary medicine*
2. *Become familiar with the most commonly used forms of traditional and complementary medicine*
3. *Being confident and comfortable in referring interested patient's toward reliable sources of information and practitioner*
4. *Learn how to obtain reliable information about the safety and efficacy of complementary medical approaches.*

Source: Adapted from IOM report Ref. [20].

acquire awareness, knowledge, and access to evidence-based information about complementary, integrative, and alternative medicine and its interactions with conventional treatments [21]. Workshops for oncologists may include reviewing the high prevalence use of herbs and supplements by cancer patients highlighting the most commonly used herbs and supplements, and potential drug-herb interactions [22]. Similarly, specialty training is available to CIT practitioners such as massage therapists and licensed acupuncturists engaged in the care of cancer patients, and are often a prerequisite for employment in larger cancer centers.

OVERVIEW OF COMMONLY USED COMPLEMENTARY AND INTEGRATIVE THERAPIES

Mind Body Medicine

Use of mind body therapies in supportive cancer care has become commonplace throughout cancer centers in the United States and abroad. Up to 50% of breast cancer survivors report using one or more mind body therapies within the past 12 months [23]. In 2014, the NCCIH designated mind–body medicine a research and grant funding priority. Previous research has documented connections between the workings of the conscious mind, the emotions, one's feelings, and many of the physiologic processes of the body [24]. Studies reveal that the mind body connection can play a role in enhancing immune function, improving hormonal balance, and in the healthy modulation of crucial brain neurotransmitters, even affecting gene expression. These positive physiologic effects result in part from their effect on the autonomic nervous system and modulation of the stress response [25].

The diagnosis of cancer is frequently associated with emotional and physiologic distress experienced as disbelief, anxiety and depression, sleep disturbance, appetite imbalances, relationship struggles, and interruption of daily routine activities. It has become increasingly clear that the successful care of the cancer patient must include the recognition and awareness of the physiologic impact of distress and stress [26]. Especially given the fact there exists an extensive body of research demonstrating chronic stress's ability to promote cancer growth and progression [27]. With the prevalence of cancer-related stress reported to be between 25 and 67%, it is critically important that clinicians and patients alike make every effort to address this issue. Awareness and effective management of life stress related to cancer diagnosis and treatment may be important for improving outcomes [28]. In 2010, the National Comprehensive Cancer Network (NCCN) made cancer-related stress a clinical practice priority. The NCCN guidelines stress the screening of all cancer patients for clinically significant stress with treatments employed based upon the results of screening [29].

The use of complementary therapies among cancer patients is highest when patients report high levels of personal distress and perceive that conventional medicine is not meeting their psychosocial needs. Many patients seek mind–body modalities to help manage stress through improved resilience, coping skills, quality of life, and reconnecting with meaning and purpose in their lives (see Table 8 for examples). Research supports that mind body practices have a positive effect on numerous physiologic systems by reversing and stabilizing the harmful effects of stress and facilitating positive functional changes within the brain [30]. Many experts now suggest, based on the extensive positive research in this field, that introduction of mind–body practices and techniques to patients occur early in the course of their illness to help them manage their stress as effectively as possible. Although research has yet to link these approaches to an improvement in survival, a comprehensive approach to stress management may still have value for improving outcomes, quality of life, and general well-being.

Following is an in-depth description of the more commonly employed mind body practices used by cancer patients.

Imagery/Relaxation Techniques

Imagery has gained significant popularity as a mind–body technique used by cancer patients throughout the course of treatment and into survivorship. Imagery is a form of mental picturing, which can include all the senses related to envisioning a positive outcome. Purposeful intentions include the following: slowing the heart rate, strengthening the immune system, ameliorating pain, and reducing stress symptoms [31]. Dr. Jeanne Achterberg, an early pioneer in imagery research, defined imagery as "the thought process that evokes and uses the senses, as well as the communication mechanism between perception, emotion, and bodily change" [32]. Imagery stimulates specific areas of the brain as effectively as if we were actually seeing, hearing, tasting, or otherwise sensing [33]. The most effective images are individualized and vary from person to person.

Mind–body techniques elicit a physiologic effect called the relaxation response [34]. A number of **relaxation** techniques have been successfully studied and are

TABLE 8. Stress reduction.

Cognitive Behavioral Therapy, Psychotherapy
Support Groups:
Bereavement, Survivorship, Cancer Support
Aromatherapy
Relaxation Techniques:
Meditation, Imagery, Music, yoga, breath work
Biofeedback, Autogenic Training
Journaling, Self- Expression, Humor
Biofield Therapies:
Therapeutic Touch, Healing Touch, Reiki
Massage Therapy

presently employed by cancer patients including intentional breathing techniques (soft belly technique, deep-slow inhalation exhalation technique), muscle relaxation techniques (progressive, passive), autogenic suggestions, and the use of focused repetitive phrases (prayer words, mantras). Research has demonstrated that these techniques combined with guided imagery are highly effective in reducing anxiety before, during, and after surgery in adult and pediatric patients. Radiation oncology patients who undertook relaxation training demonstrated significant reductions in stress, depression, anger, and fatigue in comparison to those who did not receive training [35]. Imagery and relaxation techniques have been successfully integrated into existing patient support programs. There are reports of improved pain control and recovery from cancer surgery [36]. They have shown promise for decreased chemotherapy-related nausea and vomiting, and improved emotional resilience during radiotherapy, surgical biopsies, and bone marrow aspirations.

Another well-studied and popular mind body technique is that of **progressive muscle relaxation (PMR)**. American physician Edmund Jacobson developed this technique in the early 1920s [37]. The technique involves deliberately inducing tension, followed by release and relaxation, in a series of specific muscle groups throughout the body. One randomized study in breast cancer patients demonstrated improved quality-of-life measures and significant reductions in anxiety, depression and anger with PMR and guided imagery compared to controls. An additional finding was a significant reduction in anticipatory and post-chemotherapy-induced nausea and vomiting [38]. A recent comprehensive review of relaxation techniques including muscle relaxation, guided imagery, and cognitive distraction found that these techniques effectively control anticipatory nausea and vomiting from chemotherapy, lessen the anxiety and distress associated with the cancer diagnosis and treatment, and can be effective tools in the self-management of cancer-related pain [39].

Hypnotherapy

Hypnosis is a state of calm focused attention that utilizes enhanced mental imagery and directed suggestions for therapeutic purposes in patients. Clinical hypnosis is used to alter perception, sensation, emotion, thought, or behavior. For decades, studies have demonstrated improved pain and distress in hospitalized patients using clinical hypnosis [40]. A recent review of randomized control trials showed reduced pain and distress with clinical hypnosis administered prior to diagnostic breast biopsy, as well as reduced pain, distress, fatigue, and nausea when used prior to breast cancer surgery [41]. Another study found the combination of hypnosis and cognitive behavioral therapy (CBT) significantly decreased radiotherapy-related distress and fatigue. Hypnosis may reduce hot flashes in women with and without a history of breast cancer [42]. One study from a systematic review found that metastatic breast cancer patients who combined hypnosis and supportive group therapy experienced significant reductions in pain and distress [43]. For many cancer-related symptoms authors give a strong consideration for the use of hypnosis as an ancillary intervention from the time of breast biopsy, throughout the course of cancer care [41]. A number of studies also show that hypnotherapy is highly effective and quite acceptable in the pediatric cancer patient population [44].

Meditation/Mindfulness

Meditation is an ancient healing discipline dating back many centuries. The practice of meditation draws attention to the vital role of the mind and its influence on an individual's mental and physical health. Simply speaking, it is a practice of intentionally focusing one's attention. With practice, there is the potential of increasing one's mental awareness, clarity, and acuity as well as fostering a more relaxed and resilient state of consciousness. The core component of meditation is the concept of mindfulness. It has been described as a nonjudgmental moment-to-moment awareness that when practiced, can be brought into an individual's everyday life, outside the formal meditation practice [45].

Although meditation practices vary, the form most extensively studied in cancer patients has been the multicomponent program created by Dr. Jon Kabat-Zinn at the University of Massachusetts in the 1970s. This **Mindfulness-based Stress Reduction** (MBSR) program involves the exploration and use of different mindfulness practices within an eight weeks structured group format. The mainstay of the MBSR program involves sitting meditation, walking meditation, hatha yoga, and body-scan meditation. The body-scan technique sequentially draws the patient's awareness, or attention, to consecutive parts of the body, including those areas of the body that may be experiencing pain, tension, or distress [45]. Several systematic reviews and meta-analyses have reported positive outcomes from mindfulness-based programs in cancer care. The studies have routinely reported that these mindfulness programs can significantly improve symptoms of pain, stress, anxiety, depression, and fatigue in cancer patients of any type [27, 46]. At the basic science level, research on mindfulness interventions has focused on biologic markers and physiologic changes. Findings from these studies are quite encouraging, with reports demonstrating the ability of mindfulness interventions to enhance the cancer patients immune function as well as positively affect patients' anti-inflammatory systems associated with inhibiting cancer growth and progression [47].

Table 8 provides a concise framework for available stress-reduction techniques and practices.

MANUAL/MOVEMENT THERAPIES

Massage Therapy

Massage therapy is a manual, hands-on form of healing dating as far back as Hippocrates in 400 BCE. At present, 40% of patients undergoing cancer treatment report using massage therapy [48]. Massage therapy research, involving cohorts of cancer and chronic pain patients has expanded to an all-time high through funding from the NCCIH and NIH. Research is underway with particular focus on determining appropriate massage "dosages," mechanisms, and differing techniques, as well as performance of comparative effectiveness studies. The NCCIH defines massage therapy as "an assortment of techniques involving the manipulation of the soft tissue of the body through manual or instrumental pressure and movement" [3]. Based on a patient's symptoms and preference, a variety of massage techniques are available including reflexology, deep tissue massage, Swedish massage, Thai massage, neuromuscular massage, and myofascial release.

Recent theories put forth to explain massage therapy's mechanism of action include the gate-control theory, increased vagal nerve activity, increased serotonin levels, improved blood and lymphatic flow, and the therapeutic effect of the relationship between practitioner and patient [49]. A vast majority of the randomized controlled trials regarding massage have found massage therapy to be effective for relieving many cancer related side effects including pain, nausea, anxiety, depression, anger, stress, and fatigue.

With 40–90% of patients reporting cancer-related pain, many cancer centers, hospitals, and hospices recognize the value of a multidisciplinary approach to pain management. Generally, this includes therapeutic massage. Clinical research has found significant positive effects with massage therapy for cancer related pain, mood enhancement, and improved sleep [50].

Although massage is generally safe for both adults and children, it is important for cancer patients, especially those undergoing treatment, to work with therapists who have experience in therapeutic medical massage or advanced training in oncology massage [51]. Therapists trained to work with cancer patients avoid intense pressure massage especially in close proximity to lesions, enlarged lymph nodes, medical devices, postoperative wound sites, or near the field of radiation therapy. They exercise caution and modify the depth of touch when working with patients with bleeding tendencies or those receiving anticoagulant medication [52]. The Society for Oncology Massage has published standards of practice and safety guidelines for therapists working with cancer patients.

Yoga Therapy

Yoga therapy is a mind–body discipline that combines physical poses (asanas) with breathing and meditation. Several studies have demonstrated the health benefits of yoga in the noncancer population [53]. These benefits include improved motor skills and proprioception, improved cardiac and lung function, increased muscle strength and flexibility, and decreased stiffness and joint pain. Recent studies report that yoga was safe and feasible for most patients with cancer and at any time in the individual patient's cancer treatment course [54, 55]. A comprehensive review of the literature examining the role of yoga in cancer care found that yoga resulted in significant reductions in distress, anxiety, and depression. There was mild to moderate reduction in cancer related fatigue, and moderate increases in health-related quality of life metrics [56]. This meta-analysis also found yoga to have a moderate effect on fatigue, which is reported by more than 50% of patients to be their most concerning symptom [56]. There were small but significant effects on sleep disturbances noted within one study of a yoga intervention in lymphoma patients, reporting improvement in sleep quality. Although there was a trend toward more positive outcomes on mood, fatigue, and symptom management with increased class attendance; benefits were also demonstrated for improved quality of life and self-reported physical function in those attending class less frequently [56].

Although the literature is robust on the physical function and fitness benefits of a number of exercise programs, many cancer patients find that barriers exist that preclude the use of traditional exercise programs [57]. Two recent studies found yoga to be especially suited to cancer patients who perceived barriers to other forms of exercise [46, 58]. A frequently expressed concern for cancer patients is the safety of yoga. Five studies in Buff art's systematic review (>250 patients) evaluated the safety of yoga. Four studies reported no adverse events and one study reported one episode of mild back spasm that resolved with rest. To improve outcomes the authors hypothesized that an increase in intervention duration from the usual 6–12 week period might significantly improve long-term physical function and functional well-being [56]. With studies supporting yoga as a successful intervention for improving both physical and psychosocial functioning, experts in the field suggest earlier yoga interventions may impart greater benefit throughout the patient's cancer care trajectory. A number of important questions remain, including optimal duration and frequency of yoga interventions, and long-term outcomes with consistent practice.

Tai Chi

Tai Chi is a form of martial arts, which involves a slow sequence of continuous movements performed with awareness and mindfulness. Two studies of breast cancer survivors, including one randomized study, found significant improvements in functional capacity, aerobic capacity, muscular strength, and flexibility in patients participating in Tai Chi [59, 60]. A 2010 randomized controlled pilot study of a Tai Chi intervention similarly reported significant improvement in self-esteem, bone health, immune function, and quality-of-life measures in breast cancer survivors [61]. Tai Chi may have wide-ranging applications in cancer care as a movement therapy, due to its relatively gentle controlled fluid sequences. It may be especially suited for those deconditioned from cancer treatment, as well as aging cancer survivors. With nearly 70% of cancer survivors failing to achieve the recommended level of aerobic and resistance training from traditional exercise, Tai Chi may offer health benefits in a more accessible manner [62].

Biofield Therapies/Energy Medicine

A significant percentage of cancer patients utilize biofield or energy therapies, including Healing Touch, Therapeutic Touch, Reiki, and Qi Gong. At present, research regarding the efficacy, effectiveness, and mechanisms of action of the various therapies is limited. However, high patient demand and satisfaction with these modalities has resulted in cancer care programs and hospitals widely offering them. Energy therapies conceptualize the human body as an energy field comprised of circuits or channels through which subtle energy moves [63]. This subtle energy has many names depending on the tradition or culture studied, including "prana" in Ayurveda, "Qi" in Chinese medicine, and "spirit" in Native American traditions. A prevailing hypothesis states that disturbances or "blockages" to the free flow of energy are associated with the development of physical symptoms and illness [64]. The goal of treatment is to restore energy flow and balance to the patient.

Although the number of well-designed studies is limited, there are smaller studies that demonstrate significant improvements in mood, anxiety, and quality-of-life measure in cancer patients receiving these treatments. These effects relate to the enhancement of the relaxation response, which was demonstrated with the use of biofield therapy in cancer patients [65, 66]. A recent systematic review of the clinical benefits of biofield therapy for cancer patients reported a moderate yet significant improvement in acute cancer pain. These same studies reported conflicting evidence for longer-term pain, cancer-related fatigue, and quality-of-life measures [67]. These early research efforts have raised a number of unique clinical questions regarding biofield energy interventions. In response to these equivocal data, the NIH has funded several ongoing preclinical and clinical research studies in this field.

Acupuncture and Traditional Chinese Medicine

Acupuncture is an ancient discipline embedded in Traditional Chinese Medicine that dates back more than 4000 years. Present day acupuncture has become a specialty in its own right and in cancer patient care; it represents a foundational component on the list of integrative oncology therapies. The technique of acupuncture involves the insertion of tiny, filamentous needles into specific body points, occasionally accompanied by the application of heat or electric pulses through the needles. A licensed acupuncturist or certified professional provides the treatment, which lasts on average 20–40 minutes [68]. In 1997, the NIH, in response to increasing public use and interest in acupuncture, convened an expert consensus panel to investigate its use and effectiveness in clinical medicine. They concluded there was significant evidence to support acupuncture as a credible medical discipline.

The popularity of acupuncture for medical use has continued to grow both nationally and internationally. Estimates range anywhere from 17 to 31% of conventionally treated cancer patients use acupuncture as adjunctive therapy. Classical Chinese texts, such as the Ne Ching, claim acupuncture exerts its effect by regulating the flow of energy along meridians in the body Recent research findings suggest that mediation of acupunctures effect occurs by modulation of the nervous system activity [69]. According to the Society for Acupuncture Research, acupuncture has shown a range of reproducible effects on the peripheral, central, and autonomic nervous systems, which correlate with reductions in pain and inflammation, and regulation of endocrine function. There is a release of a wide array of endogenous neurotransmitters during acupuncture treatments. Coupled with MRI scan findings of specific brain changes and reorganization it suggests that chronic pain reduction with acupuncture has the potential to affect structural changes positively and facilitate creation of new healthier pathways within the brain [70]. Likewise, findings from animal models using electro-acupuncture show the release of a full spectrum of endogenous opioid neuropeptides known to suppress pain [71]. These neuropeptides mediate the analgesic effects of acupuncture [25]. Anatomically, the traditionally described acupuncture points (acupoints) and meridians (energy pathways) appear to occur frequently at locations where connective tissue planes merge. Current models suggest that stretching of the connective tissue takes place during acupuncture needling, which results in active fibroblast-induced cytoskeletal remodeling. Although these responses may lead to physiologically relevant effects, their relationships to clinical outcomes remain unknown and investigation continues [72].

The use of acupuncture for patients with cancer has become more prevalent in recent years. Leading cancer centers such as Dana Farber, Memorial Sloan Kettering, MD Anderson, the Cleveland Clinic, and the Mayo Clinic offer adjunctive acupuncture for cancer patients. It is important to note that all reports make the point that the role of acupuncture in oncology is for cancer-related symptom management and not for direct cancer treatment. Recent published clinical trials have added evidence to support the use of acupuncture for management of cancer-related symptoms including pain, nausea and vomiting, fatigue, dry mouth (xerostomia), decreased appetite, hot flashes, insomnia, anxiety, and depression. The NIH has recognized acupuncture as a safe and effective treatment for chemotherapy related nausea and vomiting since 1997. The practice has been widely used as an alternative or adjunctive therapy to standard antiemetics including 5-hydroxytryptamine-3 (5-HT$_3$) receptor antagonists. In 2013, a randomized sham control trial reported that true acupuncture was no more effective than sham acupuncture for post-chemotherapy chronic nausea and vomiting. Interestingly, both the true and the sham acupuncture demonstrated positive effects at nausea reduction. Careful evaluation by reviewers of this paper cited significant limitations to this study including a short treatment period (only six treatments) and no usual care comparison [73].

Cancer-related fatigue is comprised of a complex group of physical and emotional symptoms, which affects more than 60% of patients. In 2012, a randomized controlled trial of 300 breast cancer patients, given a six-week series of weekly acupuncture treatments, demonstrated significant improvement in the physical and mental aspects of their fatigue. Subsequently, the authors randomized these same study groups to a second trial in which one of the groups self-administered weekly acupuncture treatments. There was a nonsignificant trend toward further improvement in both the therapist-delivered and self-administered acupuncture groups, suggesting a role for patient-directed maintenance therapy [74]. Most recently, a study from Asia on cancer related fatigue has hypothesized that effective treatment for severe fatigue, requires a multipronged therapeutic approach [75]. Although early findings of acupuncture for fatigue are quite promising, there is a need for well-designed, randomized control trials with large sample sizes.

Chemotherapy-induced peripheral neuropathy (CIPN) is a concerning symptom that can be extremely painful and can significantly impair function. It has a reported incidence of 30–40% in patients who receive neurotoxic chemotherapy agents [76]. Acupuncture has shown promise in treating CIPN in a number of small studies. Two recent pilot studies, with relatively small sample sizes (27 and 19 patients), reported a reduction in pain, numbness, and tingling, with minimal changes in nerve conduction studies [77, 78]. One study reported that the greatest improvement occurred after the first month of acupuncture usually requiring 8–12 treatments. This highlights the importance of further well-designed research studies aimed at establishing appropriate acupuncture treatment dosages and optimal treatment durations.

Xerostomia, or dry mouth, is a distressing side-effect of radiation therapy for head and neck cancer patients. Concomitant symptoms are loss of taste and difficulty in speaking and swallowing. In a number of well-designed clinical trials, there was a significant clinical benefit to patients during radiation treatment and more than six months posttreatment [79]. Recently a small trial found that patients who received preventive acupuncture before and during radiotherapy had higher salivary flow rates and less severe xerostomia-related symptoms than those who did not receive the intervention [80]. A randomized multicenter trial found eight sessions of weekly group acupuncture provided significantly greater relief of symptoms from radiation-induced xerostomia than the control education group [81, 82].

Anti-estrogen therapy and aromatase inhibition are mainstays of treatment in hormone receptor positive breast cancer. Hot flashes may be a distressing treatment side effect of these medications in nearly 70% of those patients undergoing anti-estrogen therapy. This side effect is responsible for over 30% of patients discontinuing their anticancer medication. Similarly, 50% of patients treated with aromatase-inhibitors report troubling joint pain and stiffness. A recent systematic review of acupuncture for vasomotor symptoms found an average reduction of 43.2% in hot flashes for both breast cancer and prostate cancer patients [83]. A smaller randomized controlled trial (N = 23) in breast cancer patients reported 52% of patients in the acupuncture groups experienced a significant reduction in hot flashes compared with 24% of patients in the sham acupuncture group [84]. In 2014, the largest randomized controlled trial to date of acupuncture for aromatase inhibitor related pain, found significant reductions in pain severity scores for both the electro-acupuncture and sham-acupuncture groups as compared to a wait list control [85]. Acupuncture has become a well-established technique for cancer-related side effects. More than 60% of National Cancer Institute designated comprehensive cancer centers in the United States now recommend acupuncture as a viable treatment option for cancer-related symptom management of chemotherapy-induced nausea and vomiting, radiation-induced xerostomia, and others. There is compelling evidence for use of acupuncture for treatment of other cancer-related symptoms including peripheral neuropathy, musculoskeletal symptoms, hot flashes, anxiety and depression, pain, fatigue, diarrhea, and leucopenia among others. Acupuncture offers an impressive safety profile in the hands of well-trained, credentialed practitioners, following well-established safety guidelines [86]. The effective use of acupuncture for integrative cancer care requires open communication between the patient, oncology team, and qualified acupuncture practitioner [87]. As encouraging evidence continues to emerge, the support on the use of acupuncture as a complement to first-line treatment of patients with cancer-related symptomatology is likely to increase.

Spirituality

In order to foster patient-centered, whole-person cancer care, assessment of the patient's spiritual concerns is vitally important. Performance of a thorough spirituality assessment has been shown to improve clinician–patient relationships, allow physicians to accommodate religious beliefs into cancer care, and facilitate appropriate referrals. Recently, a consensus panel of the American Association of Medical Colleges put forth the following description: "Spirituality is recognized as a factor that contributes to health in many persons and found in all cultures and societies. Spirituality is expressed in an individual's search for ultimate meaning through participation in religion, ceremony, and rituals, with beliefs in higher powers, ancestry, nature, rationalism, humanism, and the arts. All of these factors can influence how patients and healthcare professionals perceive health and illness and how they interact with one another" [88]. Although debate continues regarding the definition of spirituality, literature confirms the significant role spirituality occupies in the fields of oncology, palliative care, and at end-of-life care [89]. Common elements of most definitions include a personal spiritual dimension, varied cultural preferences, a broader domain than religion alone, multidimensional aspects with transcendent components, wholeness, a connectedness to self, others, nature, and Higher Power, as well as the individual's desire for meaning and purpose in life [90].

Hospital surveys report that the act of addressing the patient's spirituality and spiritual concerns correlates highly with patient satisfaction. Cancer patients routinely report that attention to this aspect of their health is a critical component of the cancer care provided by their physicians and nurses [91]. Most spiritual care organizations recommend placing an increased emphasis on spirituality in healthcare, especially regarding the cancer patient and their family. Likewise, the Joint Commission, the National Consensus Project Guidelines for Quality Palliative Care, the Association of Palliative Care Practitioners, as well as numerous health-care delivery systems have aligned efforts to create educational platforms to advance communication skills and enhance the knowledge bases regarding spirituality in patient care. Presently, a number of organizations offer spiritual care training programs for oncology professionals, and there exist a number of easily accessible web-based practitioner resources. Performance of spiritual assessments and or spiritual screening is highly recommended for all cancer patients. Studies reveal that a physician, nurse, spiritual care provider, or other health professional involved in the patient's care can successfully conduct this assessment [92]. Tools for the purpose of spiritual assessment frequently include the FICA interview guide and the HOPE guide [93].

Although evidence is limited in confirming the healing power of therapeutic spiritual practices, clinical reports have validated a variety of related interventions as helpful and supportive. These include support groups, prayer, meditation, music, art therapy, nature-based therapy, journaling, and spiritual counseling. There is strong evidence that these modalities have significant psychological value in conjunction with improvement in spiritual well-being and quality of life [94]. Patients frequently receive spiritual support from a number of sources including friends, family, and when appropriate their religious community, yet more often expect to be referred to a spiritual counselor or chaplain, especially if problems exist. Studies report that there is an institutional need to increase education and training of staff and physicians in an effort to satisfy the demand for spiritual assessments in cancer care [95]. Authors in the United States and abroad have put forward novel models of spiritual care that emphasize the use of multiple healthcare team members. Research efforts are underway to illuminate the meaning and vastness of spiritual care, with the goal of discovering specific, individualized interventions that demonstrate accessibility and effectiveness for the cancer patient, their family, and the caregiver. Helpful resources on this topic can be found in Table 9.

TABLE 9. Spirituality resources.

The Association for Clinical Pastoral Education www.spiritualityandhealth.ufl.edu
George Washington Institute for Spirituality and Health smhs.gwu.edu/gwish
Center for Spirituality and Healing at the University of Minnesota www.csh.umn.edu
The Healthcare Chaplaincy www.healthcarechaplaincy.org
The Tannenbaum Center for Interreligious Understanding https://tanenbaum.org

Dietary Supplements and Biologically Based Approaches

The use of biologically based substances, including dietary supplements and herbs, is common practice among cancer patients and survivors [96]. Astonishingly, the dietary supplement industry, which includes herbs, vitamins, and minerals, is estimated to be a 19 billion dollar a year industry with upwards of 50% of Americans taking daily supplements [97]. The prevalence and reported use of dietary supplements ranges from 26 to 75% following cancer diagnosis, with some variability depending on cancer type [7, 98, 99]. Cancer patients report using dietary supplements to strengthen the immune system, increase resilience, improve the chance of cure, increase sense of control, and improve their quality of life [100].

Due to the bioactive nature of dietary supplements, herbs, and natural compounds there is concern for potential adverse drug-supplement interactions, especially in those undergoing cancer treatment. These interactions may decrease effectiveness of conventional therapy, or otherwise adversely affect patient health. In addition studies report that 31–68% of cancer patients and survivors who use dietary supplements do not discuss their use with their physicians, increasing the risk of unintended effects [101]. Fortunately, there are an increasing number of specialty trained conventional and naturopathic physicians, who understand the role of supplements in oncology. These practitioners serve as important resources for cancer patients regarding the safe and appropriate use of dietary supplements throughout the cancer trajectory. They view conventional treatments with chemotherapy, radiation, and surgery as first-line cancer therapy, and the use of dietary supplements as complementary and supportive in practice.

Potential Drug-Supplement Interactions

Many herbs and biologically based supplements show significant promise as anticancer and immune enhancing agents. However the concomitant use of supplements, especially high-dose antioxidants or complex botanical agents, during chemotherapy or radiation therapy is of particular concern to oncologists due to the potential for therapy–supplement interactions [102]. There are thousands of supplements readily available on the market today with little oversight. Many health claims made by manufacturers are the result of laboratory studies and not confirmed by clinical trials. This can lead to overzealous claims of clinical benefits by the manufacturers. Since the FDA does not regulate these products, quality control of dietary supplements can be substandard. At the same time many reputable products are available and the manufacturers producing them have taken additional steps to ensure quality and safety.

EXAMPLES OF POTENTIAL INTERACTIONS

The concomitant use of biologically based supplements with radiation and chemotherapy has some merits yet clinicians should proceed with caution [103, 104]. Preclinical studies indicate that certain herbal medicines can "increase the sensitivity of cancer cells to chemotherapeutic agents, improve survival rates, enhance tumor response to drugs, and reduce side-effects" [104]. For instance a number of small clinical studies found milk thistle effective in protecting and improving patient's liver function while they received chemotherapy agents known to cause liver injury [105].

Some natural supplements have been found to have both positive and negative effects on traditional chemotherapy. Studies confirm that the use of green tea, ashwaghanda, and astragalus assist in a synergistic manner the effectiveness of some chemotherapies and are reportedly safe. Another report found evidence astragalus can potentiate the cancer cell's ability to destroy cisplatin in patients with non-small cell lung cancer [106]. The botanical supplement turmeric, whose active constituent is curcumin, exhibits both positive and negative interaction with chemotherapy. Curcumin negatively interferes with the metabolism of taxane-based chemotherapeutic agents yet has been shown to improve effectiveness of the chemotherapeutic agent, 5-fluorouracil.

A number of herbs have been found to interact negatively with chemotherapy and radiation therapy, primarily in laboratory and preclinical trials. Only a few human studies have directly assessed the effect of any herb on a chemotherapeutic agent. Notwithstanding, these concerns are still worrisome [107]. Given the sheer number of natural products on the market, and the ever-increasing number of chemotherapeutic agents, it is unlikely that the majority of specific drug–herb interactions will ever be studied. It is therefore crucial to be aware of a number of categories of potential drug–herb interactions that are most concerning.

Drug metabolism: Some herbal supplements may either increase or decrease chemotherapy concentrations by affecting the rate of drug clearance in the liver. The majority of these interactions occur in the liver's cytochrome P450 enzyme system [108] (see Table 10 for an overview of herb interactions). St. John's Wort is a classic example: it stimulates the P450 system, thus accelerating the clearance of other drugs metabolized by this system. This results in significantly lower drug concentrations, which may reach subtherapeutic levels. Although it is one of the top selling herbal remedies for mild depressive symptoms, it is strongly contraindicated for use with all chemotherapeutic regimens [109].

Immune stimulation: A number of botanicals and herbs are known to have strong immune-stimulating properties; they are traditionally used for this purpose. An example is Echinacea, a commonly sought over-the-counter remedy used to stimulate the immune response and shorten the duration of colds and flus. While Echinacea may be safely used in many instances, it is contraindicated in patients with hematologic cancers such as leukemia and lymphoma. These cancers are characterized by the abnormal growth and proliferation of cells of the immune system. Therefore dietary supplements such as Echinacea that potentially stimulate white blood cell production are contraindicated.

Hormonal effects: Dietary supplements that may have an estrogen-like or progesterone-like action can be a concern in hormone sensitive tumors including Estrogen receptor (ER) and progesterone receptor (PR) positive breast, uterine, and ovarian cancers. These include red clover, hops, and soy isoflavones. However literature supports the use of foods that are rich in some of these constituents, such as whole soy foods that may provide a protective effect in women with breast cancer. It is critical that cancer patients discuss specific options with a knowledgeable physician prior to the use of any hormonally active dietary supplements or herbal preparation (see Table 10).

Anesthetic effects: Special attention must be paid to those oncology patients undergoing surgery or other invasive procedures. A number of commonly used dietary supplements have the potential to interact with anesthetic agents, thereby increasing the associated risk. Most surgeons require patients to stop all dietary supplements for one to two weeks prior to surgery to reduce the risk of these and other potential drug–herb interactions.

Blood thinning effects: a number of dietary supplements are known to have anticoagulant effects at various doses. These include garlic, ginger, ginkgo, dong guai, and high dose (>4 g daily) omega three fatty acids found in fish oil. These supplements have the potential to interact with prescription blood thinners or to impair normal blood clotting, conferring an increased risk of a bleeding complication during any invasive procedure. In general these supplements should be stopped one to two weeks prior to planned procedures. Table 11 provides a review of the supplements that may cause adverse reactions for patients undergoing surgery.

Antioxidant effects: The use of high dose antioxidant supplements during cancer treatment continues to be controversial. Theoretically, dietary antioxidants may alter the effectiveness of chemotherapy and radiation therapy by reducing the

TABLE 10. Common supplements to avoid during chemotherapy.

Herbal Supplement	System Affected
Echinacea	Taxanes, vinca alkaloids, cyclophosphamide (CYP 3A4 inducer)
Gingko	Taxanes, vinca alkaloids, TK inhibitors, cyclophosphamide (CYP 3A4 inhibitor, CYP2C19 inducer)
Ginseng	Possible stimulator ER positive cancer growth
Curcumin	Taxanes (weak JNK enzyme induction)
Goldenseal	Strong inhibitor CYP3A4
St. John's wort	Avoid with all concurrent chemotherapy
Kava kava	cyclophosphamide, TK inhibitors, taxanes vinca alkaloids (strong inhibitor CYP2E1)
Grape seed	camptothecins, cyclophosphamibe, TK inhibitors, taxanes vinca alkaloids (CYP 3A4 inducer)

TABLE 11. Supplements to be avoided by the surgical patient.

Risk	Supplement
Anesthesia	Kava, yohimbe, valerian, licorice, milk thistle, quercetin, N-actyl cysteine
Bleeding	Gingko, ginseng, ginger, St. John's wort, red clover, green tea, Bromelain, Garlic, feverfew, omega 3's, Dong Quai

desired oxidative damage stimulated by the therapy. Although this is of concern to many oncologists, a systematic review of randomized trials found patients receiving antioxidants during chemotherapy did not experience differences in treatment response. In addition patients were more likely to experience fewer side effects and tolerate a greater number of chemotherapy cycles [110]. Indeed, Mesna and amifostine are examples of prescription antioxidants now used to improve treatment efficacy and protect patients from toxicity during cancer therapy.

Conversely, several well-known studies have demonstrated increased risk associated with specific antioxidants. A randomized trial in 540 head and neck cancer patients, found that vitamin E and beta-carotene supplementation during and after radiation therapy was associated with a higher incidence of recurrence and second primary cancers. Nonsmokers demonstrated no increased risk of those reported outcomes with antioxidant supplementation [111]. Due to the prevalence of contradictory studies, most experts agree there is need for additional well-designed randomized controlled trials to determine the safety and efficacy of high dose antioxidant supplementation during cancer treatment. It is advised that as healthy a diet as possible be adopted through the cancer trajectory to supply antioxidants through phytonutrients in food.

COMMON PATIENT USES OF DIETARY SUPPLEMENTS IN CANCER CARE

As previously noted, patients use dietary supplements during cancer care for improved quality of life, immune support, recurrence reduction, and relief from side effects of conventional treatment. There are many reports in the literature that supplemental CoQ10 reduces cardiotoxic effects when used in concert with Adriamycin, Danorubicin, and Doxil chemotherapy [112]. Several studies report a significant reduction in severity of CIPN with the coadministration of the amino acid L-glutamine in large gram doses initiated at the onset of chemotherapy [113]. L-glutamine also has a considerable body of evidence for its protective role in reducing mucositis, esophagitis, and colitis caused by radiation therapy to these areas, or from the chemotherapy regimen. An informed discussion should be held with patients considering the use of glutamine in these high doses, as there is concern that this amino acid may also support cancer cell growth. Table 12 contains a list of common uses of supplements.

TABLE 12. Examples of effective agents in cancer support.

Support	Agent
Immune support	PSK mushroom extract, Coriola, Agaricus, Grifola mushrooms, Wheat germ extract, astragulus
Stomatitis/ mucositis	Zinc, *Arnica Montana*, *Aloe Vera*
Neuropathy	L-Glutamine, vitamin E,
Nausea	Ginger
Radiation dermatitis	Topical Calendula, curcumin
Cachexia	Omega-3 fatty acids
Lymphedema	Selenium selenite, ginkgo
Intestinal toxicity	Probiotics, glutamine
Cardiac toxicity	Coenzyme Q10
Fatigue	Carnitine
Hot flashes	Black cohosh (Klimdynon)

Source: Adapted from [114].

PROMISING BIOLOGICALLY BASED ADJUVANT THERAPIES

It is compelling to note that over 60% of anticancer drugs approved internationally from 1940 to 2006 had their origins in a wide array of plants. Many scientific reports from China and Japan demonstrate the effectiveness of plant-based botanical preparations for symptoms related to cancer and its treatment. Randomized controlled trials have demonstrated that use of herbal medicines containing chemoprotective phytochemicals were effective in sensitizing cancer cells to treatment and minimizing the side effects of conventional chemotherapy. Although questions have arisen related to the scientific methodology used in these studies, the preponderance of positive clinical data from abroad has resulted in renewed interest of the research community in the United States. The NIH and the NCCIH have prioritized this work by funding US academic research centers to corroborate these findings [115].

An example of a promising adjuvant therapy is the product Krestin (polysaccharide-Kureha; PSK), which has been used as an immunotherapy agent during chemotherapy in Asia for over 30 years. PSK is derived from strains of the Coriolos versicolor mushroom, and has documented anticancer activity *in vitro*, *in vivo*, and in human clinical trials. Several randomized clinical trials have demonstrated PSK's potential as an adjuvant to conventional cancer treatments. Studies are underway to determine the specific anticancer mechanisms of the protein-bound polysaccharide substances found in mushrooms [116, 117].

Curcumin, the active component of turmeric, is a botanical powerhouse with numerous targeted actions. It has demonstrated therapeutic activity against breast cancer cells through a variety of metabolic pathways. There have been hundreds of published studies on cur cumin's anti-cancer activities. The majority of these studies demonstrate cur cumin's effectiveness in inducing cell death (or apoptosis) both *in vitro* and *in vivo* [118]. As more evidence accrues, curcumin may prove to be a powerful intervention in the chemoprevention, as well as an adjuvant treatment for a wide variety of cancer types. There is need for additional well designed Phase I, II, and III studies to corroborate these earlier findings regarding its efficacy as an anticancer agent.

Fermented wheat-germ extract is a dietary supplement that shows promise as an adjuvant agent in cancer care. Clinical studies in patients with colorectal cancer and melanoma have demonstrated wheat germ extract's effectiveness in reducing recurrence rates, increasing quality of life, and in one study increasing survival when administered in combination with conventional care. As with other promising therapies, additional large-scale randomized, controlled, clinical studies are required to clarify the value of fermented wheat-germ extract as an adjuvant to future cancer care regimens [119].

Finally, there is increasing evidence that vitamin D may exert anti-proliferative, pro-differentiating, and immune modulatory effects on tumor cells *in vitro*, and delay tumor growth *in vivo* [120]. The hypothesis that vitamin D offers a protective effect against cancer was initially based on geographic observations in which increasing latitude was associated with an increased incidence and mortality from cancer [121]. Follow-up controlled and randomized trials on vitamin D supplementation and cancer have been contradictory. Further data on cancer outcomes will materialize over the next few years as results emerge from large randomized vitamin D prevention trials currently underway in the United States and Europe. Table 13 reviews some of the supplements found to be most promising from a research perspective. Understandably, there is need for further research to confirm these early reports.

TABLE 13. Promising botanical supplements for chemoprevention.

Supplement	Potential use
Melatonin	Most solid tumors, lung, colorectal
Medicinal Mushroom	Breast, colorectal, gastric
Curcumin	Breast, prostate, lung, head/neck, melanoma, ovarian, glioblastoma
Cruciferous (indole-3C)	Breast, prostate
Green tea	Breast, prostate, colorectal, head/neck, multiple myeloma, lung
Astragulus	Lung, lymphoma, leukemia
Resveratrol	Glioblastoma multiforme, skin cancers
Soy	Ovarian, breast, prostate
Milk Thistle	Hepatocellular, prostate
Gingko bilboa	Ovarian
Vitamin D	Breast, colorectal, prostate, probably most cancers
Antioxidants: A, C, E, Selenium	Multiple cancer types
Multivitamin Supplementation	Multiple cancer types
Wheat germ extract	Most solid tumors, melanoma

Source: Adapted from Standish et al. [122].

THE CRITICAL IMPORTANCE OF NUTRITION

There is a wealth of data regarding the impact of diet in the prevention and promotion of cancer. Although investigation into promising dietary supplements and biologically based therapies for cancer treatment continues, most experts agree that dietary consumption of these phytonutrients through a wholefoods approach affords patients with the optimal source of most anticancer nutrients. The "biologic action package" theory suggests that whole foods such as fruits and vegetables contain a diversity of dietary compounds that work synergistically through their complex interconnected biologic effects. This interconnectedness produces the positive health effects consistently seen with wholefood diets, in contrast to the frequently contradictory studies of single-agent supplementation. Data does support the use of supplements to provide nutritional support not obtained with diet alone, or for patients with documented nutrient deficiencies related to the cancer treatment. Deficiencies of omega-3 essential fatty acids, vitamin D, and healthy gut flora are commonplace in cancer treatment and may warrant use of omega 3's from fish oil or flaxseed oils, vitamin D, and probiotics in these patients [123]. Table 14 highlights examples of some of the foods scientifically shown to exhibit anticancer properties. Studies confirm that consumption of these types of foods may have beneficial effects in relation to cancer prevention and cancer control [122].

CONCLUSION

Demand and interest for CIT continues to increase in the United States and abroad. In response to this challenge, cancer centers worldwide are developing and implementing integrative oncology programs and services. Although desire for patient satisfaction is essential, the primary goal of these integrative oncology programs is to help cancer patients obtain and maintain optimal health throughout their treatment course and beyond. Given that cancer care is complex and multifaceted, a

TABLE 14. Examples of foods with anticancer profiles.

Food	Profile
Green tea polyphenols	Aromatase inhibitory activity
Citrus fruits	Aromatase inhibitory activity
Cruciferous vegetables	Indoles: reduce the incidence of cancer occurrence and recurrence.
Dark berries and purple fruit	Ellagic acids and Proanthocyanidines: chemoprotective
Garlic	Allinin, Allicin: support cellular differentiation
Legumes	High in protease inhibitors
Greens and colorful vegetables	Bioflavonoids: anti-inflammatory, antioxidant. Inhibit tumor promotion

new paradigm seems appropriate. Each patient's treatment plan should be comprehensive, multidisciplinary, integrative, personalized, evidence-based, holistic, and safe. The combination of mainstream cancer care with safe and effective complementary therapies, aimed at reducing symptoms and improving quality of life, constitutes a major objective of the integrative movement. These collaborative efforts empower patients to gain tools to help them navigate their cancer trajectory while interacting openly with their clinicians, managing symptoms, improving self-care skills, enhancing physical, emotional, spiritual, and overall well-being.

In an effort to understand the wide array of integrative approaches used in the clinical setting, new research methodologies have emerged. The integrative oncology community is excited about the initiatives set forth by the NCCIH division of the NIH. This new research framework is tasked with evaluating comprehensive multidisciplinary treatment approaches that better reflect the realities of day-to-day clinical care. The primary methodologies championed by the integrative health research community are comparative effectiveness research and patient-centered outcome studies.

These research endeavors may hold the answer to some of the difficult questions raised over the past decade regarding efficacy and effectiveness of the various integrative therapies employed in cancer care. Prioritizing and supporting these types of research efforts will aid in the development of comprehensive evidence-based integrative health recommendations. Efforts are presently underway by organizations like the Society for Integrative Oncology (SIO) in the United States, and the National Information Center for Complementary and Alternative Medicine at the University of Tromsø, Norway, to facilitate the creation of evidence-based standards, norms, and benchmarks in the field of integrative medicine and oncology. In time, this will enable the development of clinical guidelines, medical treatment recommendations, and quality assurance metrics that will lead to a wider use of integrative medicine within private practice and academic centers. The outcome of these efforts has the potential to yield valuable and credible evidence for key stakeholders and decision-makers within the healthcare arena, which will ultimately provide greater safety and better quality for our patients, which they deserve.

REFERENCES

1. Eisenberg, D.M., Davis, R.B., Ettner, S.L. et al. (1998). Trends in alternative medicine use in the U.S. 1990–1997. Results of a follow-up national survey. *JAMA* 280: 1569–1575.
2. Barnes, P.M., Powell-Griner, E., McFann, K. et al. (2004 May 27). Complementary and alternative medicine use among adults: United States. *Advance Data* 343: 1–19.
3. NCCIH, 2014 http://nccih.nih.gov
4. Dalen, J. (1999). Is integrative medicine the medicine of the future? A debate between Arnold S. Relman, MD, and Andrew Weil, MD. *Archives of Internal Medicine* 159 (18): 2122–2126.
5. Kligler, B., Maizes, V., Schachter, S. et al. (2004 Jun). Core competencies in integrative medicine for medical school curricula: a proposal. *Academic Medicine* 79 (6): 521–531.

6. Horneber, M., Bueschel, G., Dennert, G. et al. (2012). How many cancer patients use complementary and alternative medicine: a systematic review and meta-analysis. *Integrative Cancer Therapies* 11 (3): 187–203.

7. Greenlee, H., Kwan, M.L., Ergas, I.J. et al. (2009a). Complementary and alternative therapy use before and after breast cancer diagnosis: the Pathways Study. *Breast Cancer Research and Treatment* 117 (3): 653–665.

8. Frenkel, M., Cohen, L., Peterson, N. et al. (2010). Integrative medicine consultation service in a comprehensive cancer center: findings and outcome. *Integrative Cancer Therapies* 9: 276–283.

9. von Gruenigen, V.E., Huang, H.Q., Gil, K.M. et al. (2012). The association between quality of life domains and overall survival and ovarian cancer patients during adjuvant chemotherapy: a Gynecologic Oncology Group Study. *Gynecologic Oncology* 124: 379–382.

10. Abrams, D. and Weil, A. (2009). *Integrative Oncology*. NY: Oxford University Press.

11. Sagar, S.M. (2006). Integrative oncology in North America. *Journal of the Society for Integrative Oncology* 4 (1): 27–39.

12. Richardson, M.A., Sanders, T., Palmer, J.L. et al. (2000). Complementary/alternative medicine use in a comprehensive cancer center and the implications for oncology. *Journal of Clinical Oncology* 18: 2505–2514.

13. McCune, J.S., Hatfield, A.J., Blackburn, A.A. et al. (2004). Potential of chemotherapy herbal interactions and adult cancer patients. *Supportive Care in Cancer* 12 (454–462): 58.

14. Verhoef, M.J., Balneaves, L.G., Boon, H.S. et al. (2005 Dec). Reasons for and characteristics associated with complementary and alternative medicine use among adult cancer patients: a systematic review. *Integrative Cancer Therapies* 4 (4): 274–286.

15. Weeks, L., Balneaves, L.G., Paterson, C. et al. (2014). Decision-making about complementary and alternative medicine by cancer patients: integrative literature review. *Opened Medicine* 8 (2): 54–66.

16. Cohen, M.H. and Eisenberg, D.M. (2002). Potential physician malpractice liability associated with complementary and integrative medical therapies. *Annals of Internal Medicine* 136: 596.

17. Verhoef, M.J., Mulkins, A., Carlson, L.E. et al. (2007). Assessing the role of evidence in patients evaluation of complementary therapies: a quality study. *Integrative Cancer Therapies* 6 (4): 345–353.

18. Weiger, W.A., Smith, M., Boon, H. et al. (2002 Dec 3). Advising patients who seek complementary and alternative medical therapies for cancer. *Annals of Internal Medicine* 137 (11): 889–903.

19. Swisher, E.M., Cohn, D.E., Goff, B.A. et al. (2002 Mar). Use of complementary and alternative medicine among women with gynecologic cancers. *Gynecologic Oncology* 84 (3): 363–367.

20. IOM (2003). *Health Professions Education: A Bridge to Quality*. Washington, DC: The National Academies Press.

21. Seely, D. and Oneschuk, D. (2008). Interactions of natural health products with biomedical cancer treatments. *Current Oncology* 15: 81–86.

22. Lee, R.T., Barbo, A., Lopez, G. et al. (2014). National survey of the US oncologists knowledge, attitudes, and practices patterns regarding herb and supplement use by patients with cancer. *Journal of Clinical Oncology* 32 (36): 4095–4101.

23. Ganz, P.A. (2009). Survivorship: adult cancer survivors. *Primary Care* 36: 721–741.

24. Dusek, J.A., Hibberd, P.L., Buczynski, B. et al. (2008 Mar). Stress management versus lifestyle modification on systolic hypertension and medication elimination: a randomized trial. *Journal of Alternative and Complementary Medicine* 14 (2): 129–138.

25. O'Regan, D. and Filshie, J. (2010). Acupuncture and cancer. *Autonomic Neuroscience: Basic and Clinical* 157: 96–100.

26. Spiegel, D. (2011). Mind matters in cancer survival. *JAMA* 305: 502–503.

27. Cramer, H., Lauche, R., Paul, A. et al. (2012). Mindfulness-based stress reduction for breast cancer: a systematic review and meta-analysis. *Current Oncology* 19 (5): 343–352.

28. Chaoul, A., Milbury, K., Sood, A.K. et al. (2014 Dec). Mind-body practices in cancer care. *Current Oncology Reports* 16 (12): 417.

29. Molassiotis, A.L., Yung, H.P., Yam, B.M. et al. (2002). The effectiveness of progressive muscle laxation training and managing chemotherapy-induced nausea and vomiting and Chinese breast cancer patients: a randomized control trial. *Supportive Care in Cancer* 10: 237.

30. Carlson, L.E., Doll, R., Stephen, J. et al. (2013 Sep 1). Randomized controlled trial of mindfulness-based cancer recovery versus supportive expressive group therapy for distressed survivors of breast cancer. *Journal of Clinical Oncology* 31 (25): 3119–3126.

31. Gruzelier, J.H. (2002). A review of the impact of hypnosis, relaxation, guided imagery in individual differences on aspects of community in health. *Stress* 5 (2): 147–163.

32. Achterberg, J. (1985). *Imagery in healing:shamanism and modern medicine*, 188–189. Boston: New Science Library, Shambal.

33. Jeannerod, M. and Frak, V. (1999). Mental imaging of motor activity in humans. *Current Opinion in Neurobiology* 9: 735–739.

34. Benson, H. and Klipper, M.Z. (1975). *The relaxation response*. New York: William Morrow and Company.

35. Nuness, D.F., Rodriguez, A.L., da Silva, H.F. et al. (2007). Relaxation and guided imagery program in patients with breast cancer undergoing radiotherapy is not associated with neuroimmunomodulatory effects. *Journal of Psychosomatic Research* 63: 647–655.

36. Syrjala, K.L., Donaldson, G.W., Davis, M.W. et al. (1995). Relaxation and imagery and cognitive-behavioral training reduce pain during cancer treatment: a controlled clinical trial. *Pain* 63: 1189–1198.

37. Jacobson, E. (1938). *Progressive Muscle Relaxation*. Chicago: University of Chicago Press.

38. Yoo, H.J., Ahn, S.H., Kim, S.B. et al. (2005 Oct). Efficacy of progressive muscle relaxation training and guided imagery in reducing chemotherapy side effects in patients with breast cancer and in improving their quality of life. *Supportive Care in Cancer* 13 (10): 826–833.

39. Mayden, K.D. (2012 Nov). Mind-body therapies: evidence and implications in advanced oncology practice. *Journal of the Advanced Practitioner in Oncology* 3 (6): 357–373.

40. Montgomery, G.H., Schnur, J.B., and Kravits, K. (2013 Jan). Hypnosis for cancer care: over 200 years young. *CA: A Cancer Journal for Clinicians* 63 (1): 31–44.

41. Cramer, H., Lauche, R., Paul, A. et al. (2015). Hypnosis in breast cancer care: a systematic review of randomized controlled trials. *Integrative Cancer Therapies* 14 (1): 5–15.

42. Elkins, G.R., Fisher, W.I., and Johnson, A.K. (2011 Oct 11). Hypnosis for hot flashes among postmenopausal women study: a study protocol of an ongoing randomized clinical trial. *BMC Complementary and Alternative Medicine* 11: 92.

43. Spiegel, D. and Bloom, J.R. (1983 Aug). Group therapy and hypnosis reduce metastatic breast carcinoma pain. *Psychosomatic Medicine* 45 (4): 333–339.

44. Rheingans, J. (2007). A systematic review of non-pharmocologic adjunctive therapies for symptom management in children with cancer. *Journal of Pediatric Oncology Nursing* 24: 81–94.

45. Kabat-Zinn, J. (1990). *Catastrophe Living: Using the Wisdom of Your Body and Mind to Face Stress, Pain, and Illness*. New York: Delta trade paperback/Bantam Dell.

46. Johns, S.A., Brown, L.F., Beck-coon, K. et al. (2015). Randomized controlled pilot study of mindfulness-based stress reduction for persistently fatigued cancer survivors. *Psycho-Oncology* 24: 885–893.

47. Carlson, L.E., Speca, M., Faris, P. et al. (2007). One-year pre-post intervention follow-up of psychological, immune, endocrine, and blood pressure outcomes of mindfulness-based stress reduction breast and prostate cancer L patients. *Brain, Behavior, and Immunity* 21 (8): 1038–1049.

48. Lee, S.H., Kim, J.Y., Yeo, S. et al. (2015 Jul). Meta-analysis of massage therapy on cancer pain. *Integrative Cancer Therapies* 14 (4): 297–304.

49. Moyer, C.A., Rounds, J., and Hannum, J.W. (2004 Jan). A meta-analysis of massage therapy research. *Psychological Bulletin* 130 (1): 3–18.

50. Pan, Y.Q., Yang, K.H., Wang, Y.L. et al. (2014 Oct). Massage interventions and treatment-related side effects of breast cancer: a systematic review and meta-analysis. *International Journal of Clinical Oncology* 19 (5): 829–841.

51. Furlan, A.D., Brosseau, L., Imamura, M. et al. (2002 Sep 1). Massage for low-back pain: a systematic review within the framework of the Cochrane Collaboration Back Review Group. *Spine (Phila Pa 1976)* 27 (17): 1896–1910.

52. Lee, M.S., Lee, E.N., and Ernst, E. (2011). Massage therapy for breast cancer patients: a systematic review. *Annals of Oncology* 22: 1459–1461.

53. Ross, A., Friedmann, E., Bevans, M. et al. (2013 Aug). National survey of yoga practitioners: mental and physical health benefits. *Complement Theoretical Medicine* 21 (4): 313–323.

54. Lin, K.Y., Hu, Y.T., Chang, K.J. et al. (2011). Effects of yoga on psychological health, quality-of-life, and physical health of patients with cancer: a meta-analysis. *Evidence-Based Complementary and Alternative Medicine* 2011.

55. Raub, J.A. (2002). Psychophysiologic effects of Hatha yoga on musculoskeletal and cardiopulmonary function: a literature review. *Journal of Alternative and Complementary Medicine (New York, N.Y.)* 8: 797–812.

56. Buffart, L.M., van Uffelen, J.G., Riphagen, I.I. et al. (2012 Nov 27). Physical and psychosocial benefits of yoga in cancer patients and survivors, a systematic review and meta-analysis of randomized controlled trials. *BMC Cancer* 12: 559.

57. Appling, S.E., Scarvalone, S., MacDonald, R. et al. (2012 May 1). Fatigue in breast cancer survivors: the impact of a mind-body medicine intervention. *Oncology Nursing Forum* 39 (3): 278–286.

58. Sprod, L.K., Fernandez, I.D., Janelsins, M.C. et al. (2015 Jan). Effects of yoga on cancer-related fatigue and global side-effect burden in older cancer survivors. *Journal of Geriatric Oncology.* 6 (1): 8–14.

59. Mustian, K.M., Katula, J.A., Gill, D.L. et al. (2004 Dec). Tai Chi Chuan, health-related quality of life and self-esteem: a randomized trial with breast cancer survivors. *Support Care Cancer* 12 (12): 871–876.

60. Mustian, K.M., Katula, J.A., and Zhao, H. (2006 Mar). A pilot study to assess the influence of Tai Chi Chuan on functional capacity among breast cancer survivors. *The Journal of Supportive Oncology* 4 (3): 139–145.

61. Peppone, L.J., Mustian, K.M., Janelsins, M.C. et al. (2010). Effects of a structured weightbearing exercise program on bone metabolism among breast cancer survivors: a feasibility study. *Clinical Breast Cancer* 10: 224–229.

62. Wei, G.X., Dong, H.M., Yang, Z. et al. (2014 Apr 17). Tai Chi Chuan optimizes the functional organization of the intrinsic human brain architecture in older adults. *Frontiers in Aging Neuroscience* 6: 74.

63. Oschman, J.H.L. (2000). *Energy Medicine: The Scientific Basis*. New York, NY: Churchill Livingstone.

64. Mentgen, J.L. (2001 Mar). Healing touch. *The Nursing Clinics of North America* 36 (1): 143–158.

65. Jain, S. and Mills, P.J. (2010). Biofield therapies: helpful or full of hype? A best evidence synthesis. *International Journal of Behavioral Medicine* 17: 1–16.

66. Post-White, J., Kinney, M.E., Savik, K. et al. (2003). Therapeutic massage and healing touch improve symptoms incancer. *Integrative Cancer Therapies* 2 (4): 332–344.

67. Hart, L.K., Freel, M.I., Haylock, P.J. et al. (2011 Oct). The use of healing touch in integrative oncology. *Clinical Journal of Oncology Nursing* 15 (5): 519–525.

68. Kapchuk, T.J. (2002). Acupuncture: very efficacy and practices. *Annals of Internal Medicine* 136: 374–383.

69. Stone, J.A. and Johnstone, P.A. (2010). Mechanisms of action for acupuncture and the oncology setting. *Current Treatment Options in Oncology* 11: 118–127.

70. Yan, B., Li, K., Xu, J.Y. et al. (August 2005). Acupoint-specific FMRI patterns in human brain. *Neuroscience Letters* 383 (3): 236–240.

71. Han, J.S. (January 2003). Acupuncture: neuropeptide release produced by electrical stimulation of different frequencies. *Trends in Neurosciences* 26 (1): 17–22.

72. Langevin, H.M. and Yandow, J.A. (2002). Relationship of acupuncture points and meridians to connective tissue planes. *The Anatomical Record* 269: 257–265.

73. Deng, G., Chan, Y., Sjoberg, D. et al. (2013). Acupuncture for the treatment of post-chemotherapy chronic fatigue: a randomized, blinded, sham controlled trial. *Supportive Care in Cancer* 21 (17): 35–41.

74. Molassiotis, A., Bardy, J., Finnegan-John, J. et al. (2013). A randomized, control trial of acupuncture self needling as maintenance therapy for cancer related fatigue after therapist delivered acupuncture. *Annals of Oncology* 24 (16): 45–52.

75. Ling, W.M., Lui, L.Y., So, W.K. et al. (2014 Nov 1). Effects of acupuncture and acupressure on cancer-related fatigue: a systematic review. *Oncology Nursing Forum* 41 (6): 581–592.

76. Windebank, A.J. and Grisold, W. (2008). Chemotherapy-induced neuropathy. *Journal Peripheral Nerve System* 13 (1): 27–46.

77. Garcia, M.K., McQuade, J., Haddad, R. et al. (2013 Mar 1). Systematic review of acupuncture in cancer care: a synthesis of the evidence. *Journal of Clinical Oncology* 31 (7): 952–960.

78. Bao, T., Goloubeva, O., Pelser, C. et al. (2014). A pilot study of acupuncture in treating bortezomib-induced peripheral neuropathy in patients with multiple myeloma. *Integrative Cancer Therapies* 13: 396–404.

79. Meng, Z., Garcia, M.K., Hu, C. et al. (2012). Sham controlled randomized, feasibility trial of acupuncture for prevention of radiation-induced xerostomia among patient's with nasopharyngeal carcinoma. *European Journal of Cancer* 48: 1692–1699.

80. Braga, F.P., Lemos Junior, C.A., Alves, F.A. et al. (2012). Acupuncture for the prevention of radiation-induced xerostomia in patients with head and neck cancer. *Brazilian Oral Research* 25: 180–185.

81. Garcia, M.K., Driver, L., Haddad, R. et al. (2014a Mar). Acupuncture for treatment of uncontrolled pain in cancer patients: a pragmatic pilot study. *Integrative Cancer Therapies* 13 (2): 133–140. https://doi.org/10.1177/1534735413510558.

82. Garcia, M.K., McQuade, J., Lee, R. et al. (2014b Dec). Acupuncture for symptom management in cancer care: an update. *Current Oncology Reports* 16 (12): 418.

83. Garcia, M.K., Graham-Getty, L., Haddad, R. et al. (2015 Nov 15). Systematic review of acupuncture to control hot flashes in cancer patients. *Cancer* 121 (22): 3948–3958.

84. Bokmand, S. and Flyger, H. (2013). Acupuncture relieves menopausal discomfort in breast cancer patients: prospective double blinded randomized study. *Breast* 22 (3): 20–23.

85. Mao, J.J., Xie, S.X., Farrar, J.T. et al. (2014). A randomized trial of electro-acupuncture for arthralgia related to aromatase inhibitor use. *European Journal of Cancer* 50 (2): 67–76.

86. Ernst, G., Strzyz, H., and Hagmeister, H. (2003). Incidence of adverse effects during acupuncture therapy–a multicenter survey. *Complementary Therapy Medicine* 11: 93–97.

87. Tsai, W.H., Yang, C.C., Li, P.C. et al. (2013 Jul). Therapeutic potential of traditional Chinese medicine on inflammatory diseases. *Journal of Traditional and Complementary Medicine* 3 (3): 142–151.

88. AAMC task force report: Spirituality, cultural issues, and end-of-life care. In: Association of American medical colleges: Report III. Contemporary issues and medicine: Communication medicine. Washington, DC: Association of American medical colleges, 1999, pages 24–9

89. Whitford, H.S., Olver, I.N., and Peterson, M.J. (2008). Spirituality as a core domain in the assessment of quality of life in oncology. *Psycho-Oncology* 17 (11): 1121–1128.

90. Sulmasy, D.P. (2006). *The Re-birth of the Clinic: Introduction to Spirituality in Healthcare.* Washington: Georgetown University pressed.

91. Kristeller, J.L., Rhodes, M., Cripe, L.D. et al. (2005). Oncologists assisted spiritual intervention study (OASIS): patient acceptability and initial evidence of effects. *International Journal of Psychiatry in Medicine* 35 (4): 329–347.

92. Puchalski, C.M. and Larson, D.B. (1998). Developing curriculum and spirituality and medicine. *Academic Medicine* 73 (9): 970–974.

93. Spirituality in Cancer Care (PDQ), Health Professional Version. PDQ Supportive and Palliative Care Editorial Board, July 17th, 2015

94. Garland, S.N., Carlson, L.E., Cook, S. et al. (2007). A non-randomized comparison of mindfulness-based stress reduction and healing arts programs for facilitating post-traumatic growth and spirituality in cancer out patients. *Supportive Care in Cancer* 15 (8): 949–961.

95. Sinclair, S., Raffin, S., Pereira, J. et al. (2006). Collective soul: spirituality of an interdisciplinary palliative care team. *Palliative & Supportive Care* 4 (1): 13–24.

96. Yates, J.S., Mustian, K.M., Morrow, G.R. et al. (2005). Prevalence of complementary and alternative medicine use in cancer patients during treatment. *Support Care Cancer* 13 (10): 806.

97. Jiang, T. (2009 Jun). Re-thinking the dietary supplement laws and regulations 14 years after the Dietary Supplement Health and Education Act implementation. *International Journal of Food Sciences and Nutrition* 60 (4): 293–301.

98. Velicer, C.M. and Ulrich, C.M. (2008). Vitamin and supplement use among US adults after cancer diagnosis: a systematic review. *Journal of Clinical Oncology* 26: 665–673.

99. Greenlee, H., Hershman, D.L., and Jacobson, J.S. (2009b Jun). Use of antioxidant supplements during breast cancer treatment: a comprehensive review. *Breast Cancer Research and Treatment* 115 (3): 437–452.

100. Hulbocky, F.J., Ratain, M.J., Wen, M. et al. (2007). Complementary and alternative medicine among advanced cancer patients enrolled on phase I trials: a study of prognosis, quality of life, and preferences for decision making. *Journal of Clinical Oncology* 25 (5): 548–554.

101. Swarup, A.B., Barrett, W., and Jazieh, A.R. (2006). The use of complementary and alternative medicine by cancer patients undergoing radiation therapy. *American Journal of Clinical Oncology* 29 (5): 468–473.

102. Labriola, D. and Livingston, R. (1999). Possible interactions between dietary antioxidants and chemotherapy. *Oncology* 13: 1003–1008.

103. Mann, J. (2002). Natural products in cancer chemotherapy: past, present, and future. *Nature Reviews. Cancer* 2: 143–148.

104. Murphy, R.A., Mourtzakis, M., Chu, Q.S. et al. (2011 Aug 15). Supplementation with fish oil increases first-line chemotherapy efficacy in patients with advanced nonsmall cell lung cancer. *Cancer* 117 (16): 3774–3780.

105. Ladas, E.J., Kroll, D.J., Oberlies, N.H. et al. (2010). A randomized, controlled, double-blind, pilot study of milk thistle for the treatment of palatal toxicity in childhood acute lymphoblastic leukemia (ALL). *Cancer* 116: 506–513.

106. He, H., Zhou, X., Wang, Q. et al. (2013). Does the course of astragalus-containing Chinese herbal prescriptions and radiotherapy benefit to non-small-cell lung cancer treatment: a meta-analysis of randomized trials. *Evidence-based Complementary and Alternative Medicine* 2013: 426207.

107. Mills, E., Wu, P., Johnston, B.C. et al. (2005 Oct). Natural health product–drug interactions: a systematic review of clinical trials. *Therapeutic Drug Monitoring* 27 (5): 549–557.

108. Meijerman, I., Beijnen, J.H., and Schellens, J.H. (2006 Jul-Aug). Herb–drug interactions in oncology: focus on mechanisms of induction. *The Oncologist* 11 (7): 742–752.

109. Zhou, S., Gao, Y., Jiang, W. et al. (2003 Feb). Interactions of herbs with cytochrome P450. *Drug Metabolism Reviews* 35 (1): 35.

110. Block, K.I., Koch, A.C., Mead, M.N. et al. (2007). Impact of antioxidant supplementation on chemotherapeutic efficacy: a systematic review of the evidence from randomized control trials. *Cancer Treatment Reviews* 33 (5): 407–418.

111. Bairati, I., Meyer, F., Gelinas, M. et al. (2005). A randomized trial of antibiotics and vitamins to prevent secondary primary cancers in head and neck cancer patients. *Journal of the National Cancer Institute* 97: 481–488.

112. Greenlee, H., Shaw, J., Lau, Y.K. et al. (2012 Sep). Lack of effect of coenzyme q10 on doxoru-bicin cytotoxicity in breast cancer cell cultures. *Integrative Cancer Therapies* 11 (3): 243–250.

113. Wang, W.S., Lin, J.K., Lin, T.C. et al. (2007 Mar). Oral glutamine is effective for preventing oxaliplatin-induced neuropathy in colorectal cancer patients. *The Oncologist* 12 (3): 312–319.

114. Hardy, M.L. (2008). Dietary supplement use in cancer care: help or harm. *Hematology/Oncology Clinics of North America* 22: 581–617.

115. Yang, C., Chien, L., and Tai, C. (2008). Use of complementary and alternative medicine among patient's with cancer receiving outpatient chemotherapy in Taiwan. *Journal of Alternative and Complementary Medicine* 14 (4): 413–416.

116. Fisher, M. and Yang, L.X. (2002 May–June). Anti-cancer effects and mechanisms of polysaccha-ride-K (PSK) implications of cancer immunotherapy. *Anticancer Research* 22 (3): 1737–1754.

117. Wasser, S.P. (2014 Nov-Dec). Medicinal mushroom science: current perspectives, advances, evidences, and challenges. *Biomedical Journal* 37 (6): 345–356.

118. Fridlender, M., Kapulnik, Y., and Koltai, H. (2015). Plant derived substances with anticancer activity: from folklore to practice. *Frontiers in Plant Science* 6: 799.

119. Mueller, T. and Voigt, W. (2011 Sep 5). Fermented wheat germ extract – nutritional supplement or anticancer drug? *Nutrition Journal* 10 (89): 1–6.

120. Giammanco, M., Di Majo, D., La Guardia, M. et al. (2015 Oct). Vitamin D in cancer chemopre-vention. *Pharmaceutical Biology* 53 (10): 1399–1434.

121. International agency for research on cancer. IARC Working groups Report: Vitamin D and cancer 2008.

122. Standish, L.J., Alschuler, L.N., Ready, A.B. et al. (2009). Botanical medicine in integrative oncology. In: *Integrative Oncology* (ed. D.I. Abrams and A. Weil), 104–146. Oxford Press.

123. Bazzan, A.J., Newberg, A.B., Cho, W.C. et al. (2013). Diet and nutrition in cancer survivorship and palliative care. *Evidence-Based Complementary and Alternative Medicine* 2013: 1–12.

28

GENE THERAPY

Mai K. ElMallah[1,2], Michael Kalfopolous[1], and Terence R. Flotte[1,2,3]

[1] *Gene Therapy Center, University of Massachusetts*
Medical School, Worcester, MA, USA
[2] *Department of Pediatrics, University of Massachusetts*
Medical School, Worcester, MA, USA
[3] *Microbiology & Physiologic Systems, University of Massachusetts*
Medical School, Worcester, MA, USA

CANCER GENE THERAPY

Cancer gene therapy is an emerging and exciting field in cancer therapeutics that offers promising treatments for both hematological and solid tumors. Recombinant viral vectors emerged in the 1980s and the first retroviral mediated gene transduction into lymphocytes of cancer patients occurred in 1990 [1]. Since that time there has been a rapid progression to several human clinical trials. Gene therapy involves the delivery of a therapeutic gene that can function in a variety of ways to modify the target cells. As such, the goals of cancer gene therapy are to overexpress a desired protein, to restore normal cellular phenotype, to knockdown expression of a macromolecule, directly induce cancer cell death, or decrease the blood supply to a tumor. Delivery of the gene can be through viral or non-viral vectors. Viral gene delivery involves using a virus as a vector to deliver a gene to a specific target cell, whereas, nonviral gene delivery is when plasmid is directly delivered to the target cell. Over 60% of current gene therapy

Cancer: Prevention, Early Detection, Treatment and Recovery, Second Edition. Edited by Gary S. Stein and Kimberly P. Luebbers.
© 2019 John Wiley & Sons, Inc. Published 2019 by John Wiley & Sons, Inc.

clinical trials are designed to target cancer treatment [2]. However, one challenge of gene therapy for cancer is finding a strategy to specifically deliver a vector to the target tumor tissue without inducing gene expression in normal healthy tissue, thereby causing significant side effects. Another challenge is maintaining adequate levels of gene expression in the target organ to destroy and maintain suppression of cancer cells. Different forms of gene manipulation and gene therapy are currently being explored in clinical trials with variable success rates ([2], http://clinicaltrials.gov). The area of cancer gene therapy is extensive; therefore discussion in this chapter will be limited to a review of genetic modification of vector mediated immunotherapy, specifically CAR T cells, therapy targeting tumor destruction, and anti-angiogenic therapy.

VECTOR-MEDIATED IMMUNOTHERAPY

CAR T Cells

The immune system plays a vital role in the recognition and destruction of cancer cells. Cancer immunosurveillance involves a healthy immune system that recognizes and destroys precursors of cancer. T lymphocytes, or T cells, are a vital component of cell mediated immunity and help protect us from invading organisms by activating a variety of cell mediated adaptive responses. This activation occurs following the binding of T cells to specific antigens presented by major histocompatibility complex (MHC). MHC antigens are usually expressed by diseased cells and are necessary for the activation of the T cell mediated immune response. Unfortunately, many cancer cells effectively escape recognition and destruction by T cells by downregulating MHC, thus T cells become unaware of their presence. In addition, since the tumor is derived from self-tissue, tumor cells sometimes express tumor associated antigens (TAA) not recognized by T cells cancer cells. Thus, when cancer occurs, tumor cells replicate without limits and avoid destruction by the immune system through reduced immunogenicity and immune subversion, and possibly even active suppression of the immune response [3].

Gene therapy and genetic engineering aim to refine and enhance the function of T cells and the immune response against cancer. This genetic modification of cellular immunity, or adoptive cell therapy, has been an ongoing goal of cancer therapy therapeutics and the focus of many recent clinical trials of both hematological and solid tumors. One area in cancer gene therapy focuses on engineering T cells to encode cell surface receptors that can recognize the TAAs. An example of this is the chimeric antigen receptors (CARs) system used to genetically modify a patient's own T cell to express an artificial T cell receptor. This CAR system is a promising genetically engineered molecular biology technique whereby tumor antigen binding regions are cloned and expressed as a single chain fragment variable (scFv) antigen binding region.

"CAR T cells" were first reported in the late 1980s by Zelig Eshhar [4]. CARs are composed of a number of components (Figure 1): (i) an extracellular scFv of a

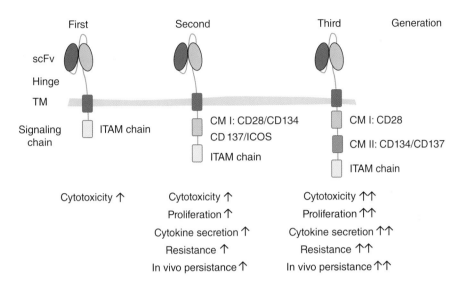

Figure 1. First, second, and third generation CAR T cells. *Source*: Cartellieri et al. 2010. Reproduced with permission of *Journal of Biomed Biotechnol* [5].

monocolonal antibody (mAb) to target a specific tumor cell surface molecule, (ii) a spacer domain for flexibility that also aids with T cell and target cell engagement, (iii) a transmembrane domain, and (iv) signaling modules that trigger T cell effector functions. Since they are derived from antibodies, CAR T cells do not need MHC to recognized TAAs, so they can target cancer cells that have downregulated MHC. Thus, CARs combine antibody type specificity with effector T cell function in order to enable T cells to target tumor cells independent of native receptor activation. Once tumor cells are recognized, CAR T cells mediate specific IL-2 secretion and induce target cell lysis. However, CARS are only able to detect antigens on the cell surface (about 20% of antigens expressed by cancer cells) so if the tumors lose their cell surface antigens, they can escape CAR detection [6].

CARs are "modular by design" thus any antibody and multiple combinations of signaling and cosignaling parts can cause T cells to be redirected against any desired target antigen. CARs are constantly evolving and have been altered significantly over the last 15 years. Three generations of CAR T cells have been generated and studied in preclinical and clinical trials [6–9] (Figure 1). The major distinctions between each of the generations are the intracellular signaling modules that trigger the T cell effector function. First generation CAR T cells were comprised of only CD3ζ as an intracellular signaling domain without any costimulatory components. However, clinical trials with these first generation CAR T cells resulted in suboptimal antitumor activity, short-term T cell expansion and a lack of persistence of CAR T cells [10, 11]. Thus, second and third generation CARs were developed to enhance cytokine production by incorporating one or more costimulatory domains fused to CD3ζ (Figure 1). Examples of costimulatory receptors include CD28, 4-1BB, OX-40, and DAP10. The ideal costimulatory molecule is still unknown since comparisons between second generation CARs have not been extensive [12, 13]. However, second generation CAR T cells were found to be superior to first generation CARs when their function was compared in patients with B cell non-Hodgkin Lymphoma (NHL). The second generation CARs induced better cytokine production and T cell proliferation upon repeated antigen exposure. In addition, comparison of expansion and survival of both T cell populations in the same recipient showed that second generation CAR T cells expand more efficiently and survive longer [14–17]. They also mediate tumor regression of ALL in xenograft models [12].

The most promising findings with CAR T cells have been in hematological tumors. In contrast, the antitumor activity in solid tumors of second and third generation CAR T cells is limited. This is presumably because the solid tumor microenvironment suppresses the T cell response with immunosuppressive cytokines, regulator modulators and co-inhibitory receptors [18, 19]. CARs have been generated against a large number of cell surface molecules including CD19, HER2, GD2, prostate-specific membrane antigen (PSMA), and mesothelin. There are currently many clinical trials ongoing to test these CARs with over 30 Phase I clinical trials listed on clinicaltrials.gov. However, the most impressive and dramatic outcomes with CAR T cells have been in patients treated with autologous CAR T cells targeting the CD19 receptor in B cell leukemia and lymphoma [14, 16, 17, 20, 21].

CAR T Cells in End-Stage Hematological Cancers

Initial clinical trials using CD19 CAR T cells for patients with relapsed B cell acute lymphoblastic leukemia (B-ALL) are very promising and may result in a dramatic shift of the standard of care for B-ALL. CD19 is an ideal target for B cell malignancies because it is expressed on the B lineage from the early pro-B to mature B cell stages [22]. Thus, it is also expressed on most B cell malignancies including >95% of B cell ALL, B cell NHL and chronic lymphocytic leukemia (CLL). Since it is not expressed on other hematological cell lines or non-hematological cells, it does not result in bone marrow toxicity or destruction of non-B cells. However, because it is expressed on all B cells, including normal B cells, targeting this antigen results in B cell aplasia and hypogammaglobulinemia. Although it is an adverse effect, B cell aplasia is a marker for functioning CD19 CAR T cells and is followed as an indicator for persistent CD19 CAR T cells.

CAR T cells targeting CD19 are second generation CAR T cells with costimulatory receptors. An initial study of first generation CD19 CAR T cells in patients with refractory follicular lymphoma showed no response [23]. The patients underwent lymph depletion and administration of IL2, but T cell persistence was limited to less than one week and there was no tumor response or toxicity [23]. This poor outcome was thought to be due to the lack of costimulatory domains. As previously mentioned, these costimulatory domains are incorporated into second generation CAR T cells that results in enhanced T cells function and destruction of

CD19-expressing cancer cells [24]. In addition, these second generation CD19 CAR T cells also show greater cytokine proliferative responses and greater persistence in the target [13]. Third generation CD19 CAR T cells have also been examined, and as mentioned previously these have additional costimulatory domains. However, for B cell malignancies, this combination of costimulatory domains is not superior to the second generation CD19 CAR T cells, which have only one costimulatory domain. In fact, second generation CD19 CAR T Cells containing a 4-1BB domain were compared in a preclinical trial to third generation CD19 CAR T cells containing both a 4-1BB and CD28 domains and the cytotoxic function of the second generation CARs were found to be superior to the third generation effects [25].

Several clinical trials using CAR T cells to target CD19 have recently been reported, many with variable success [14, 16, 21, 26–29]. The first promising clinical trial was reported by Kochenderfer et al. [21] in a patient with relapsed advanced follicular lymphoma involving all major lymph node areas. In addition, he had bilateral pleural effusions, night sweats, and weight loss. He was treated with lymphodepletion using cyclophosphamide for two days followed by five days of fludarabine. The day following completion of fludarabine, he received CD19 CAR T cells (T cells transduced with retrovirus encoding the anti-CD19 CAR), followed by IL-2 administration. There was persistence of CD19 CAR transduced T cells measured by quantitative polymerase chain reaction (qPCR) for up to 27 weeks. In addition, the patient achieved a partial response lasting 32 weeks with a similar duration of B-cell aplasia [21].

In the past few years, the same group reported outcomes after treating 8 and 15 patients, respectively, with second generation anti-CD19 CAR T cells transduced by gamma-retroviral vectors and containing the CD28 costimulatory domain [29]. In the first study [15] they treated eight patients with advanced, progressive B cell malignancies. One patient had complete remission at 15 months, six had partial remission, and one had stable disease. All patients received pretreatment with cyclophosphamide and fludarabine, followed by a single infusion of CD19 CAR T cells. However, IL-2 following the CD19 CAR T cells administration was not given in the study reported in 2015. In the second study, 15 patients were treated but one patient died soon after treatment and another was lost to follow up secondary to noncompliance. Nine patients had diffuse large B-cell lymphoma, eight of which had chemotherapy refractory lymphoma. However, only seven of these were evaluable – of these four achieved complete remission, two had partial remission and one had stable disease. Of the other six patients in this report, two had indolent lymphomas and four had CLL. Three of the four treated patients were in ongoing complete remission at the time of the report. Thus out of the 13 evaluable patients, eight achieved a complete response and four others had a partial response. There was persistence of CAR T cells seen up to 75 days with the best response duration on going at 23 months [29]. However, these responses were associated with significant toxicity due to cytokine release syndrome (CRS) and a variety of neurological toxicities including confusion, obtundation, aphasia, apraxia, gait disturbances, and facial paresis [29].

In 2013 and 2014, the first studies using CD19 CAR T cells in ALL were published [8, 26]. In one study, five adults with relapsed B cell acute lymphocytic leukemia (B-ALL) were treated with second generation CD19-targeted CAR T cells using the CD28 costimulatory domain [26]. This was used as a bridge therapy for four of the five patients prior to allogenic hematopoietic stem cell transplantation. The fifth patient, who was not eligible for either stem cell transplantation or further CAR T cell therapy (secondary to long-standing comorbidities), relapsed 90 days after therapy (while he was receiving high dose systemic steroids for CRS) [26].

Maude et al. treated a total of 30 patients (25 of which were pediatric patients) with second generation anti-CD19 CAR T cells (transduced using lentivirus and renamed CTL019). Twenty-six patients had B cell ALL in a first to fourth relapse, three had primary refractory B cell ALL and one had relapsed T-cell ALL that expressed CD19. All patients underwent chemotherapy aimed at depleting T cells one week before administration of CTL019. Ninety percent of the patients achieved complete remission one month following CTL019. Of the 27 patients who had complete remission, 19 remained in remission, 15 received no further therapy, and 4 patients withdrew from the study to receive other therapy. Two patients had CNS disease which cleared. There were no deaths related to the treatment but seven patients died after disease progression or relapse. At six months, event free survival rate was 67% and overall survival was 78% [8]. Durable response correlated with higher peak levels of circulating CAR T cells and with increased duration of B cell aplasia. CRS occurred and eight patients required admission to the intensive care unit and vasopressor support for hypotension. An IL6 receptor blocking antibody as well as steroids successfully treated the CRS. Thirteen patients developed neurological toxicity. However, full recovery of both CRS and neurological toxicity was reported in these patients [8].

The variability of response in different patients depends on several factors. Several of the clinical trials examining the effects of anti-CD19 CAR T cells on B cell malignancies seem to indicate that a large tumor burden is inversely correlated with the response to CD19 CAR T cell therapy. However, while the overall responses are greater when tumor burdens are smaller, tumor burden is not the sole predictor of response and should not be used to exclude patients from trials [22]. CD19 CAR T cell function is also enhanced with pre-infusion chemotherapy. When a direct comparison of outcomes of CAR T cell between three patients treated without prior conditioning chemotherapy and another treated with cyclophosphamide, conditioning chemotherapy enhanced both T-cell persistence and disease outcome [14]. However, the authors did highlight that the patients who received cyclophosphamide had had it previously with unsuccessful chemotherapy so, it may have mediated the lymphodepleting effects but did not have any activity on the cyclophoshopamide refractory tumor cells [22]. Interestingly, the T cell dose does not appear to influence the outcome of the CD19 CAR T cell therapy [22].

T cell subsets used in CD19 CAR T cells have been highly variable. Usually there is a transfer of bulk transduced T cells with different CD4:CD8 ratios, different proportions of regulatory T cells (Treg) and memory T cell subsets. The ideal T cell type seems to be unknown but efficacy has been shown to improve with cotransfer of CD4 and CD8 T cells [30]. The adoptive transfer of CD4 and CD8 central memory cells transduced separately and infused at a defined ratio is currently being studied [31].

Response rates to CAR T cells in ALL seem to be higher than in NHL or CLL. This may reflect the inhibitory nature of the tumor microenvironment in NHL [32] or the defect in T cell function in CLL [33]. It may also be due to the affinity of CD19 CAR T cells to the bone marrow compared with lymph nodes. In addition, it seems that second generation CARs differ based on the costimulating domain. Figure 2 illustrates the different CARS used to treat CLL patients at the Memorial Sloan Kettering Cancer Center (MSKCC), the National Cancer Institute (NCI), and at the University of Pennsylvania (UPenn) [22]. It seems that those CARs that contain the 4-1BB CAR persistently transduced T cells for up to two years [8]. In parallel, this construct resulted in B cell aplasia that lasted much longer than those patients which had the CAR constructs with the CD28 domain. The CD28 domain constructs lasted approximately one to three months [28]. In addition, CRS seems to occur earlier in CD28 containing CARS than 4-1BB. These data suggest that a 50–90% response rate can be achieved in patients with ALL, with lower responses in NHL and CLL. No clear relationship between T cell dose and efficacy exists because low dose CAR T cells can expand *in vivo*.

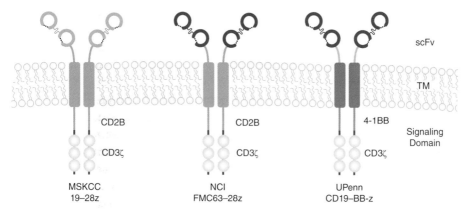

Figure 2. Schematic diagram of chimeric antigen receptor (CARs) used to treat chronic lymphocytic leukemia (CLL) patients at MSKCC, NCI, and UPenn. *Source*: Davila et al. 2012. Reproduced with permission of *Oncoimmunology* [22].

Adverse Effects of CAR T Cells

As with any cancer therapy, the on target, off tumor effects can result in severe side effects. B cell aplasia occurs in CD19 CAR T cells in patients because these genetically engineered T cells target CD19 that is on B cell malignancies as well as normal B cells. This results in an increased risk of infection. The duration of B cell aplasia depends on the construct used and it is often an indicator of the persistence of CD19 CAR T cells. Immune globulin replacement is sometimes given to decrease the risk of infections in patients with persistent B cell aplasia [7]. CRS is associated with elevated serum cytokine levels, mainly IL6, IL10, and interferon gamma. It is common with CD19 CAR T cell therapy and typically occurs 4–21 days after CAR T cell infusion. The severity of CRS seems to be proportional to tumor load [8, 26, 29]. It is unclear whether the CAR T cells, the dying tumor cells or the locally activated macrophages are responsible for cytokine release. However, the effects of cytokine release range from mild flu like symptoms to multiorgan failure and respiratory failure. Therapies for this syndrome include corticosteroids (but these might inhibit T cell function) or the IL6 receptor antibody tocilizumab [8, 29]. The IL6 receptor antibody may not prevent T cell expansion, but it is typically held until established organ dysfunction because the cytokine storm may be vital for the expanding CAR T cell population [7, 28]. Another adverse effect is neurotoxicity, which is normally transient and resolves spontaneously without therapy. Neurotoxicity varies from aphasia to delirium, seizures, confusion, obtundation, aphasia, apraxia, gait disturbances, and facial paresis [29] This may be due to a substance secreted from the CAR T cells, systemic cytokines crossing the blood brain barrier or cross reactivity of CAR T cells with a target in the CNS [22, 28, 29].

Nonviral vs Viral CAR T Cells

Both viral and nonviral gene therapy have been used in order to introduce CAR constructs into T cells. Plasmid, or nonviral DNA, transfection is thought to offer some advantages over viral-based therapy because DNA plasmids are considerably cheaper, are non-immunogenic and are not limited by DNA size. However, this method is not efficient because of silencing of transgene expression and can sometimes result in integration of multiple copies. In addition, nonviral inoculation requires long-term cultures and antibiotic selection due to the relative inefficiency of gene transfer. The long-term culture may result in decreased activity and persistence of infused T cells. Expression of the transgene is brief due to intracellular breakdown and delivery into the cells of interest is inefficient. Thus, this method has been abandoned as a means to produce CAR T cells because it may undermine the biological quality of the final cell product. Another nonviral gene transfer method is the transposon/transposase system – the Sleeping Beauty system. The sleeping beauty transposon system consists of a transposon containing a gene expression cassette and DNA or RNA, which codes for the transposase enzyme. Thus, the expression cassette can be transposed from a plasmid into the genome by recognition of specific terminal repeated flanking the CAR expression cassette by the transposases and the cassette is inserted in the T-cell's genome. This results in sustained transcription of a transgene [34]. Although the sleeping beauty was shown to provide efficient stable gene transfer with sustained transgene expression, the culture time is still relatively long. In addition these protocols result in considerable cell death and typically require prolonged culture for cellular recuperation. These systems are currently being studied clinically and if they are found to be equivalent to the vector approaches, the cost and complexity of CD19 CAR T cell production will be considerably decreased [7].

Viral gene therapy transduction using retroviruses has been the popular method for anti- CD19 CAR T cells. One of two retroviruses has been used: lentiviral or gamma-retroviral. Both these retroviruses can permanently insert DNA into the genome. Split packaging, packaging viral components separately, renders these viruses safe and eliminates the concern that they will generate replication competent viruses. They are highly efficient in transducing T cells after *in vitro* activation with cytokines. However, there is concern for potential oncogenes caused by randomly inserting transgenes into the genomes. For example, when 10 patients were treated by gene transfer to restore missing interleukin-2 (IL-2) receptor gene to hematopoietic stem cells using gamma-retroviral vectors, nine of the patients were successfully treated. However, four of these nine patients developed T cell leukemia several years later. Likewise, when a lentiviral vector was used to treat beta-thalassemia patients successfully in one human patient, a clonal expansion was observed raising concerns about the long-term safety of this approach. Thus, the safety of gene transfer with retroviral vectors was questionable. However, when retroviral therapy was specifically used on gene modified T cells there was an absence of significant adverse events. The proposed reason for the lack of adverse events was that mature human T cells are resistant to insertional genotoxicity. Thus, this safety and long-term persistence of engineered T cells using retroviral vectors is reassuring [35].

When comparing transduction efficiency of gamma-retroviral vectors and lentiviral vectors, it appears that transduction of gene transfer efficiency is higher with gamma-retroviral vectors than with lentiviral vectors [14–16, 20, 21]. Lentiviral vectors can transduce quiescent cells whereas gamma retroviruses require cells in mitosis. However, this is not really a concern because CAR T cell protocols require powerful mitogenic stimuli. Lentiviruses may be safer because they are self-inactivating so that after insertion the viral promoters are truncated. On the other hand gamma retroviruses generate a stable producer cell line and allow the production of indefinite quantities of vector. Thus, although CAR T cell therapy is very promising, more optimal delivery methods and the long term impact of the different viral vectors and non-viral gene transfer methods still remain unknown [35].

TUMOR DESTRUCTION

Oncolytic Gene Therapy Vectors

Oncolytic viruses are replication competent viruses that provide promising therapeutic agents for cancer treatment. These viruses are unique because they selectively replicate and destroy cancer cells without damaging normal healthy cells. They can infiltrate throughout a tumor and effectively destroy the replicating cancer cells [36]. Oncolytic viruses are not a recent phenomenon. In fact, the idea of an oncolytic virus first emerged in the mid-1800s when case reports described resolution of leukemia following coincidental infections with naturally occurring viruses, namely influenza like viral illnesses [37, 38]. However, since the advent of genetic engineering in the late 1980s, oncolytic viruses have been extensively studied. In particular, there have been rapid advancements in this field in the last two decades, following the report by Bischoff et al., which describes the possible use of adenoviruses in the treatment of cervical carcinoma [39].

Cytotoxic Activity

Oncolytic viruses work by a variety of different mechanisms but the main mechanism involves direct cytotoxic effects on the tumor cells [36]. In order to be safe and effective, oncolytic viruses must have specificity for the target cancer. In this way, there will be minimal damage and destruction to normal tissue. This specificity occurs naturally in several viruses but can also be achieved by genetically engineering the viruses to enhance their tumor specificity. The genetic engineering includes transcriptional, translational, transductional, and pro-apoptotic targeting mechanisms (Figure 3).

In translational targeting, viruses are modified to destroy their interferon suppressor mechanisms and thus interferon is released from non-tumor cells infected with the virus signaling neighbor cells to shut down protein synthesis so that no more viruses are produced. However, in tumor cells there is a defective interferon signaling pathway, thus the tumor cells remain susceptible to these viruses with the interferon suppressor mechanism. For example, vesicular stomatitis virus (VSV) normally blocks the synthesis and release of interferons, but genetically modifying VSV by mutating the M protein results in an inability of this virus to inhibit interferon release [41]. This renders VSV oncolytic but not pathogenic to non-tumor cells.

Transcriptional gene modification of viruses involves inserting tissue specific or tumor specific promotors in their genomes to regulate expression of viral genes – the products of these genes are essential for virus propagation. Examples of viruses which have been genetically modified to become tumor selective are adenoviruses, herpes viruses, and retroviruses [42–45].

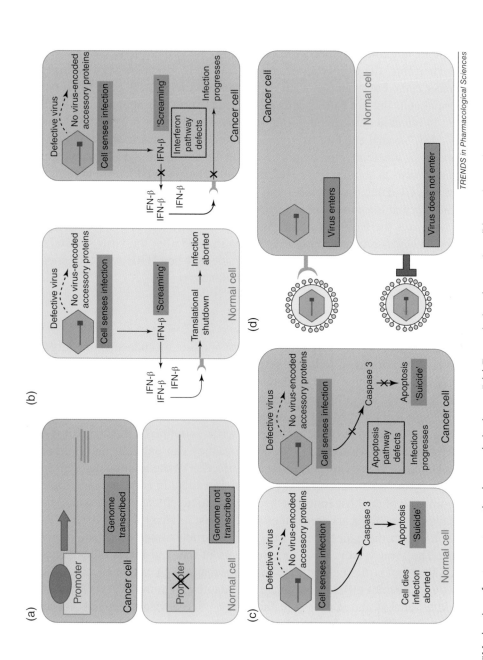

Figure 3. "Mechanisms for tumor targeting by oncolytic viruses." (a) Transcriptional targeting. (b) Translational targeting. (c) Pro-apoptotic targeting. (d) Transductional targeting. *Source:* Russell and Peng 2007. Reproduced with permission of Elsevier [40].

Transductional targeting utilizes gene therapy to modify viral attachment to receptors that are expressed on tumor cell surfaces. Recombinant measles viruses are an example of transductional modification. The measles virus has a hemagglutinin (H) attachment protein and the fusion (F) protein. The H or F proteins have been modified to target tumor cells that express the targeted receptors [46, 47].

Pro-apoptotic targeting engineers the virus to disable viral proteins that prevent apoptosis. Therefore, the oncolytic virus can generate more viruses and multiply only in cancer cells. For example, adenovirus depends on their E1 gene to propagate in normal cells but not cancer cells. The E1 gene prevents apoptosis and cell cycle arrest in infected cells by inhibiting the tumor suppressor proteins: p53 and Rb [40, 44, 45]. Normal cells undergo rapid Rb induced cell cycle arrest and p53 induced apoptosis when infected by oncolytic adenovirus lacking E1 function and propagation of virus is aborted. However, in tumor cells, p53 and Rb are not functional therefore E1 deleted adenovirus can still propagate efficiently without fear of triggering premature apoptosis. The first marketed approval for this type of oncolytic virus was the H101 type 5 adenovirus: a therapy that targeted head and neck cancer. This was granted approval in China in 2005 after the virus showed a superior response rate in head and neck cancer patients when H101 was combined with chemotherapy, compared to chemotherapy alone [40, 48].

Effects on Tumor Microenvironment

Oncolytic viruses can also destroy cancer cells by indirect mechanisms that target the cancer microenvironment [49, 50]. For example, the oncolytic vaccinia virus and the VSV target the tumor neovasculature and initiate an inflammatory reaction, which results in clots within the vasculature. This finding is unique to the tumor vessels with no effect on normal vasculature. This blood loss to the tumor results in the synergistic destruction of the tumor cells when combined with the cytotoxic effects of the oncolytic viruses [51]. In addition, they can enhance the immune system by inducing cytokine release following infection of the tumor cells. This in turn results in cytotoxic effects on surrounding tumor cells. In addition, innate and adaptive immunity of the host contributes to the cytotoxic efficacy of the oncolytic virus, and thus their effects are sometimes more robust *in vivo* [52].

In order to infect the cancer cells, many viral particles must reach the tumor vasculature. The viral vectors must then replicate in order to infect other tumor cells. Once they infect the cells, the primary mechanism of action of oncolytic viruses is immunogenic cell death by innate and adaptive immune effects. When an oncolytic virus attacks the cancer cell, the tumor microenvironment triggers an immune response. The virus disrupts the tumor cells as well as the tumor microenvironment: the tissue architecture and immune homeostasis. The first line of defense following the attack of the oncolytic virus on the tumor cell is the innate immune system. The innate immune system is activated by inflammatory cytokines that are released once the macrophages contact the viral structures. Furthermore, other innate immune cells invade the damaged tumor tissue and induce an acute inflammatory response against the virus. Then, neutrophils invade the oncolytic tumor and result in the immediate antitumor cytotoxic effects. This neutrophil activating signal has been a target for oncolytic virotherapy. For example, the measles virus can induce a robust neutrophil activation. On the other hand, others have shown that the innate immune response can be suppressed to enhance oncolytic virus spread throughout the tumor. Therefore, it may be necessary to selectively suppress innate immune cell subpopulations to propagate the oncolytic virus [53].

Barriers to Effective Oncolytic Viruses

Initial results using adenovirus as an oncolytic agent only showed modest effects but provided the proof of concept that these agents could be safely administered to patients. There are many factors that interfere with the effectiveness of viral oncolytic therapy. For example, the immune system can neutralize the virus before it reaches the tumor cells. In addition, the virus may not be able to penetrate the tumor tissue. Furthermore, if it does penetrate the tumor, the stroma within the tumor may inhibit virus spread. Finally, within the tumor tissue, viral replication may be inhibited by hypoxic conditions and elimination by the immune system [36, 54].

To avoid neutralization and barriers to reaching the tumor, the oncolytic virus can be delivered directly to the tumor. However, systemic (intravenous) delivery would be preferable particularly in the setting of metastatic cancer. In addition, the virus is more likely to reach more of the tumor if delivered systemically, specifically, it will reach the tumor near the vessels from which the virus was delivered and it may also promote inflammatory changes resulting in disruption of the vasculature feeding the tumor [49]. However, vascular delivery of oncolytic viruses can be absorbed by the liver and spleen thus it does not reach the tumor. Alternatively these viruses can be deactivated by antibodies. Several ways to block antibody recognition have been explored. For example, serotype switching has been considered where a different viral serotype is given with each treatment cycle, e.g., Adenovirus, or the viruses are genetically engineered so that they are no longer neutralized by antibodies which recognize the original virus [55].

In addition, high interstitial fluid pressure affects intravenous delivery of viral factors because it results in a flux of fluid to the periphery of tumors limiting access of vectors into the tumor core. These high pressures also limit the movement of structures out of the vasculature and into the tumors [54].

Viral targeting efficiency can also be increased by reducing the concentration of antiviral antibodies prior to treatment with the genetically modified oncolytic virus. The innate immune response of the host is very active following intratumor viral infection. This response plays a crucial role in limiting viral oncolysis. Immunosuppression can counter this effect and can reduce the concentration of antiviral antibodies. Antibody evasion is another method to suppress the antibody response. This can be done by using cells infected with virus to transport the virus to the target tumor or by delivering the genome as a non-immunogenic infectious nucleic acid. Although non-immunogenic nucleic acids evade phagocytosis, there is a problem with specific tissue targeting with this method [56]. In addition, once the oncolytic virus reaches the tumor, it must propagate throughout the tissue to effectively destroy all the cancer cells. This can be aided by suppressing antiviral cytotoxic T and B cell response using cyclophosphamide. Cyclophosphamide can increase the efficiency of oncolytic viral spread and limit the increase in the antiviral antibody titers between each oncogenic viral treatment dose [57]. For example, this immunosuppressive medication renders HSV oncolytic viruses more effective in the fight against intracerebral tumors [57].

Suicide Genes

Suicide gene therapy utilizes genes that result in an active cytotoxic compound. The herpes simplex virus type 1 thymidine kinase/ganciclovir system is an example of oncolytic gene therapy. It involves the delivery of the gene (thymidine kinase) to the tumor which causes ganciclovir to be converted to an activated chemotherapeutic drug (ganciclover triphosphate). However, suicide gene therapy has had variable clinical response. For example, in glioblastoma multiforme (GBM), suicidal genes resulted in no significant differences in survival between the groups. Despite these setbacks, there continue to be studies focused on enhancing delivery of these "suicide genes" and promoting their synergy with chemotherapy [58, 59].

Tumor Suppressor Genes

Tumor suppressor genes prevent oncogenesis, but are often mutated or absent in cancer. For example, in GBM there is at least one tumor suppressor gene that is mutated or deleted in all patients. In 91% of patients two or more of the genes are inactivated. Therefore, gene therapy to deliver tumor suppressor genes has been the focus of much research over the last decade. The genes encode functional tumor suppressors and are incorporated in the site of neoplasia in order to restore their function and stop unregulated growth. Examples of these include p53, cyclin inhibitors, and miRNAs. They have been shown to increase survival significantly in animal models. However, clinical trials have not shown the same results as those found in animals. Adenovirus was used to transfer p53 into 12 patients with recurrent glioma but there was not widespread distribution of the vector and p53 [60].

Tumor suppressor genes can also be targeted by RNA guided delivery of chimeric nucleases. These nucleases disrupt, remove, or replace mutated DNA in target cells and include zinc-finger nucleases (ZFNs), transcription activator like effector nucleases (TALENs), and clustered regularly interspaced short palindromic repeats (CRISPR). CRISPRs are commonly associated with CRISPR-associated (Cas) genes. The chimeric nucleases typically cause double stranded breaks (DSB) in DNA, increasing the frequency of homology-directed repair (HDR).

HDR is a template dependent pathway for DSB repair that can be utilized by target nucleases to make small edits in the genome. ZFNs typically consist of 30 amino acids that recognize DNA sequences 9–18 bp in length and have been used to correct numerous disease-causing mutations associated with X-linked severe combined immune deficiency, alpha-1 anti-trypsin deficiency, hemophilia B, and sickle-cell disease. TALENs are created from naturally occurring TALE proteins and are composed of a series of 33–35 amino acid repeat domains that recognize a single base pair. TALENs differ from ZFNs because they do not require the recreation of linked repeats to construct long arrays. Additionally, TALEs have greater design flexibility due to its single base recognition, unlike ZFNs triplet recognition, but this can be problematic when widespread identical repeat sequences are present. TALENs, along with ZFNs, have been used to introduce targeted alterations in plants to confer invaluable traits such as disease and herbicide resistance [61].

In comparison to ZFNs and TALENs, which use endonuclease catalytic domains bound to DNA-binding proteins to induce DSBs at specific regions, CRISPR-Cas makes DSBs using nucleases guided by small RNAs through normal base-pairing mechanisms. The CRISPR-Cas method of genome editing is often preferred over ZFNs and TALENs because of its ease of design, high specificity, and efficiency for multiplexed gene editing [62].

CRISPR-Cas9 is a promising technique for gene therapy. Viral infections associated with carcinogenesis, such as human papillomavirus (HPV) and Epstein-Barr virus (EBV) are targeted by CRISPR-Cas9 to inactivate or clear the oncogenic virus and provide a cost-effective option for treatment [63]. Utilization of this treatment is reported in EBV-positive Burkitt's lymphoma cells, where the result was a greatly reduced viral load and cell proliferation. Similarly, in HPV-positive cervical carcinoma cells the CRISPR-Cas9 system restored cellular p53 and Rb levels when E6 and E7 genes were targeted [64]. As previously mentioned, CRISPR-Cas9 also has the potential for therapeutic genome manipulations to activate onco-suppressor genes or inactivate oncogenes. In a study by Liu et al. [65], therapeutic genome editing was done by "AND" gate circuits based on a constructed CRISPR-Cas9 bladder cancer cell-specific genome. An "AND" circuit refers to two inputs that must coexist to generate an output. In the study, bladder cancer cell onco-suppressor genes were activated by inhibiting the LacI protein by Cas9, resulting in growth inhibition, migration suppression, and apoptosis. However, this method of cancer therapy has limitations, such as the genetic heterogeneity of cancerous tumors and the high edit efficiency needed to adequately eliminate unedited cells and reduce their growth advantage [63, 64]. Targeting epigenetics for genome editing by Cas9 is another attractive candidate for therapy because epigenetic regulators are essential for a cell's transformation into a cancer cell and when tumor related, can promote carcinogenesis [66].

For delivery methods of a Cas9 vector, adenovirus and lentivirus are capable of delivery ex vivo and have achieved high levels of gene disruption. However, a major concern is immunogenicity. Adeno-associated virus (AAV) is a much more attractive target for Cas9 delivery that is being explored because of its small size and minimal immune response [66]. However, the Cas9 gene that is most commonly used is too large to transduce into wild-type AAV, so an AAV engineered to carry a larger genetic load must be developed or an ortholog of Cas9 that is smaller must be utilized [63].

Outside gene therapy, the CRISPR-Cas9 system is used to create human cell lines that exhibit identical chromosomal translocations to cancers such as acute myeloid leukemia and lung cancer [67]. The advantage of these engineered oncogenic chromosomal rearrangements is to study human cancers in mice [68]. These animal models will further our understanding of the development and progression of cancers and ideally will lead to new therapies in the future. However, limitations to the CRISPR-Cas9 system are genome instability, disruption to normal gene function, and unintended epigenetic alterations due to unintended off-target effects. Off-target effects can be either off-target binding or off-target cleavage. Methods for reducing off-target effects are currently in development [64]. Additionally, more efficient delivery vectors, more potent Cas9 and an increased rate of HDR over NMEJ is desired for the CRSIPR-Cas9 system [63].

ANTI-ANGIOGENIC GENE THERAPY

In order to grow beyond a small multicellular cluster, a solid tumor requires new vessel formation from the surrounding vasculature (angiogenesis). Vascular endothelial growth factor – A (VEGF) drives angiogenesis in cancer and it is over-expressed in many solid tumors [69, 70]. Thus, blocking VEGF signaling has been the target of many solid cancer gene therapy studies. Tyrosine kinase inhibitors (TKI) are a group of agents designed to inhibit VEGF receptor signaling. TKI's improved overall survival in patients with metastatic renal cell carcinoma and have now been FDA approved [71–77]. They are also effective as single agents and prolong survival in hepatocellular carcinoma and advanced pancreatic neuroendocrine tumors [78, 79]. In metastatic colorectal cancer, the combination of chemotherapy with a monoclonal antibody that specifically binds to VEGF-A, Bevacizumab, was found to be superior to chemotherapy alone [80]. In addition, aflibercept, which binds to VEGF-A, VEGF-B, and placental growth factor (PLGF) has been recently approved as an adjunct to chemotherapy for the treatment of colorectal cancer [77]. Bevacizumab has minimally improved survival in non-small cell lung cancer – there was an improvement in overall survival of 4% at one year [81]. Anti-angiogenic therapy in combination with chemotherapy also improved progression free survival by a little over a month in relapsed ovarian cancer [82, 83]. However, anti-angiogenic therapy seems to be cancer specific and does not seem to impact survival in breast, melanoma, pancreatic, or prostatic cancer – the reason for this variability in effect is still unclear.

SUMMARY AND FUTURE DIRECTIONS

As described here, cancer gene therapy has evolved along a number of promising avenues as biologically sound strategies for antitumor effect have been matched with suitable gene transfer vectors. After decades of fruitless efforts, strategies such as CAR-T cells and oncolytic viruses are beginning to show clinically significant benefits. The future of cancer gene therapy will likely rest on this continuing co-evolution of "Precision Medicine"-based antitumor strategies and improved vector platforms.

REFERENCES

1. Rosenberg, S.A., Aebersold, P., Cornetta, K. et al. (1990). Gene transfer into humans – immunotherapy of patients with advanced melanoma, using tumor-infiltrating lymphocytes modified by retroviral gene transduction. *N. Engl. J. Med.* 323 (9): 570–578.
2. Ginn, S.L., Alexander, I.E., Edelstein, M.L. et al. (2013). Gene therapy clinical trials worldwide to 2012 – an update. *J. Gene Med.* 15 (2): 65–77.
3. Zitvogel, L., Tesniere, A., and Kroemer, G. (2006). Cancer despite immunosurveillance: immunoselection and immunosubversion. *Nat. Rev. Immunol.* 6 (10): 715–727.
4. Gross, G., Waks, T., and Eshhar, Z. (1989). Expression of immunoglobulin-T-cell receptor chimeric molecules as functional receptors with antibody-type specificity. *Proc. Natl. Acad. Sci. U. S. A.* 86 (24): 10024–10028.
5. Cartellieri, M., Bachmann, M., Feldmann, A. et al. (2010). Chimeric antigen receptor-engineered T cells for immunotherapy of cancer. *J Biomed Biotechnol* 2010: 956304.
6. Jensen, M.C. and Riddell, S.R. (2014). Design and implementation of adoptive therapy with chimeric antigen receptor-modified T cells. *Immunol. Rev.* 257 (1): 127–144.
7. Ghorashian, S., Pule, M., and Amrolia, P. (2015). CD19 chimeric antigen receptor T cell therapy for haematological malignancies. *Br. J. Haematol.* 169 (4): 463–478.
8. Maude, S.L., Frey, N., Shaw, P.A. et al. (2014). Chimeric antigen receptor T cells for sustained remissions in leukemia. *N. Engl. J. Med.* 371 (16): 1507–1517.
9. Maus, M.V., Grupp, S.A., Porter, D.L. et al. (2014). Antibody-modified T cells: CARs take the front seat for hematologic malignancies. *Blood* 123 (17): 2625–2635.
10. Park, J.R., Digiusto, D.L., Slovak, M. et al. (2007). Adoptive transfer of chimeric antigen receptor re-directed cytolytic T lymphocyte clones in patients with neuroblastoma. *Mol. Ther.* 15 (4): 825–833.
11. Till, B.G., Jensen, M.C., Wang, J. et al. (2008). Adoptive immunotherapy for indolent non-Hodgkin lymphoma and mantle cell lymphoma using genetically modified autologous CD20-specific T cells. *Blood* 112 (6): 2261–2271.
12. Brentjens, R.J., Santos, E., Nikhamin, Y. et al. (2007). Genetically targeted T cells eradicate systemic acute lymphoblastic leukemia xenografts. *Clin. Cancer Res.* 13 (18 Pt 1): 5426–5435.
13. Milone, M.C., Fish, J.D., Carpenito, C. et al. (2009). Chimeric receptors containing CD137 signal transduction domains mediate enhanced survival of T cells and increased antileukemic efficacy in vivo. *Mol. Ther.* 17 (8): 1453–1464.
14. Brentjens, R.J., Riviere, I., Park, J.H. et al. (2011). Safety and persistence of adoptively transferred autologous CD19-targeted T cells in patients with relapsed or chemotherapy refractory B-cell leukemias. *Blood* 118 (18): 4817–4828.

15. Kochenderfer, J.N., Dudley, M.E., Feldman, S.A. et al. (2012). B-cell depletion and remissions of malignancy along with cytokine-associated toxicity in a clinical trial of anti-CD19 chimeric-antigen-receptor-transduced T cells. *Blood* 119 (12): 2709–2720.

16. Porter, D.L., Levine, B.L., Kalos, M. et al. (2011). Chimeric antigen receptor-modified T cells in chronic lymphoid leukemia. *N. Engl. J. Med.* 365 (8): 725–733.

17. Savoldo, B., Ramos, C.A., Liu, E. et al. (2011). CD28 costimulation improves expansion and persistence of chimeric antigen receptor-modified T cells in lymphoma patients. *J. Clin. Invest.* 121 (5): 1822–1826.

18. Gajewski, T.F., Meng, Y., Blank, C. et al. (2006). Immune resistance orchestrated by the tumor microenvironment. *Immunol. Rev.* 213: 131–145.

19. Kakarla, S. and Gottschalk, S. (2014). CAR T cells for solid tumors: armed and ready to go? *Cancer J.* 20 (2): 151–155.

20. Kalos, M., Levine, B.L., Porter, D.L. et al. (2011). T cells with chimeric antigen receptors have potent antitumor effects and can establish memory in patients with advanced leukemia. *Sci. Transl. Med.* 3 (95): 95ra73.

21. Kochenderfer, J.N., Wilson, W.H., Janik, J.E. et al. (2010). Eradication of B-lineage cells and regression of lymphoma in a patient treated with autologous T cells genetically engineered to recognize CD19. *Blood* 116 (20): 4099–4102.

22. Davila, M.L., Brentjens, R., Wang, X. et al. (2012). How do CARs work?: early insights from recent clinical studies targeting CD19. *Oncoimmunology* 1 (9): 1577–1583.

23. Jensen, M.C., Popplewell, L., Cooper, L.J. et al. (2010). Antitransgene rejection responses contribute to attenuated persistence of adoptively transferred CD20/CD19-specific chimeric antigen receptor redirected T cells in humans. *Biol. Blood Marrow Transplant.* 16 (9): 1245–1256.

24. Campana, D., Schwarz, H., and Imai, C. (2014). 4-1BB chimeric antigen receptors. *Cancer J.* 20 (2): 134–140.

25. Kochenderfer, J.N., Feldman, S.A., Zhao, Y. et al. (2009). Construction and preclinical evaluation of an anti-CD19 chimeric antigen receptor. *J. Immunother.* 32 (7): 689–702.

26. Brentjens, R.J., Davila, M.L., Riviere, I. et al. (2013). CD19-targeted T cells rapidly induce molecular remissions in adults with chemotherapy-refractory acute lymphoblastic leukemia. *Sci. Transl. Med.* 5 (177): 177ra138.

27. Cruz, C.R., Micklethwaite, K.P., Savoldo, B. et al. (2013). Infusion of donor-derived CD19-redirected virus-specific T cells for B-cell malignancies relapsed after allogeneic stem cell transplant: a phase 1 study. *Blood* 122 (17): 2965–2973.

28. Davila, M.L., Riviere, I., Wang, X. et al. (2014). Efficacy and toxicity management of 19-28z CAR T cell therapy in B cell acute lymphoblastic leukemia. *Sci. Transl. Med.* 6 (224): 224ra225.

29. Kochenderfer, J.N., Dudley, M.E., Kassim, S.H. et al. (2015). Chemotherapy-refractory diffuse large B-cell lymphoma and indolent B-cell malignancies can be effectively treated with autologous T cells expressing an anti-CD19 chimeric antigen receptor. *J. Clin. Oncol.* 33 (6): 540–549.

30. Adusumilli, P.S., Cherkassky, L., Villena-Vargas, J. et al. (2014). Regional delivery of mesothelin-targeted CAR T cell therapy generates potent and long-lasting CD4-dependent tumor immunity. *Sci. Transl. Med.* 6 (261): 261ra151.

31. Riddell, S.R., Sommermeyer, D., Berger, C. et al. (2014). Adoptive therapy with chimeric antigen receptor-modified T cells of defined subset composition. *Cancer J.* 20 (2): 141–144.

32. Yang, Z.Z. and Ansell, S.M. (2012). The tumor microenvironment in follicular lymphoma. *Clin. Adv. Hematol. Oncol.* 10 (12): 810–818.

33. Christopoulos, P., Pfeifer, D., Bartholome, K. et al. (2011). Definition and characterization of the systemic T-cell dysregulation in untreated indolent B-cell lymphoma and very early CLL. *Blood* 117 (14): 3836–3846.

34. Aronovich, E.L., McIvor, R.S., and Hackett, P.B. (2011). The Sleeping Beauty transposon system: a non-viral vector for gene therapy. *Hum. Mol. Genet.* 20 (R1): R14–R20.

35. Scholler, J., Brady, T.L., Binder-Scholl, G. et al. (2012). Decade-long safety and function of retroviral-modified chimeric antigen receptor T cells. *Sci. Transl. Med.* 4 (132): 132ra153.

36. Woller, N., Gurlevik, E., Ureche, C.I. et al. (2014). Oncolytic viruses as anticancer vaccines. *Front. Oncol.* 4: 188.

37. Ferguson, M.S., Lemoine, N.R., and Wang, Y. (2012). Systemic delivery of oncolytic viruses: hopes and hurdles. *Adv. Virol.* 2012: 805629.

38. Kelly, E. and Russell, S.J. (2007). History of oncolytic viruses: genesis to genetic engineering. *Mol. Ther.* 15 (4): 651–659.

39. Bischoff, J.R., Kirn, D.H., Williams, A. et al. (1996). An adenovirus mutant that replicates selectively in p53-deficient human tumor cells. *Science* 274 (5286): 373–376.

40. Russell, S.J. and Peng, K.W. (2007). Viruses as anticancer drugs. *Trends Pharmacol. Sci.* 28 (7): 326–333.

41. Stojdl, D.F., Lichty, B.D., tenOever, B.R. et al. (2003). VSV strains with defects in their ability to shutdown innate immunity are potent systemic anti-cancer agents. *Cancer Cell* 4 (4): 263–275.

42. Chung, R.Y., Saeki, Y., and Chiocca, E.A. (1999). B-myb promoter retargeting of herpes simplex virus gamma34.5 gene-mediated virulence toward tumor and cycling cells. *J. Virol.* 73 (9): 7556–7564.

43. Dalba, C., Klatzmann, D., Logg, C.R. et al. (2005). Beyond oncolytic virotherapy: replication-competent retrovirus vectors for selective and stable transduction of tumors. *Curr. Gene Ther.* 5 (6): 655–667.

44. Dobbelstein, M. (2004). Replicating adenoviruses in cancer therapy. *Curr. Top. Microbiol. Immunol.* 273: 291–334.

45. Ko, D., Hawkins, L., and Yu, D.C. (2005). Development of transcriptionally regulated oncolytic adenoviruses. *Oncogene* 24 (52): 7763–7774.

46. Nakamura, T., Peng, K.W., Harvey, M. et al. (2005). Rescue and propagation of fully retargeted oncolytic measles viruses. *Nat. Biotechnol.* 23 (2): 209–214.

47. Springfeld, C., von Messling, V., Frenzke, M. et al. (2006). Oncolytic efficacy and enhanced safety of measles virus activated by tumor-secreted matrix metalloproteinases. *Cancer Res.* 66 (15): 7694–7700.

48. Garber, K. (2006). China approves world's first oncolytic virus therapy for cancer treatment. *J. Natl. Cancer Inst.* 98 (5): 298–300.

49. Breitbach, C.J., De Silva, N.S., Falls, T.J. et al. (2011). Targeting tumor vasculature with an oncolytic virus. *Mol. Ther.* 19 (5): 886–894.

50. Coffin, R.S. (2015). From virotherapy to oncolytic immunotherapy: where are we now? *Curr. Opin. Virol.* 13: 93–100.

51. Breitbach, C.J., Arulanandam, R., De Silva, N. et al. (2013). Oncolytic vaccinia virus disrupts tumor-associated vasculature in humans. *Cancer Res.* 73 (4): 1265–1275.

52. Sobol, P.T., Boudreau, J.E., Stephenson, K. et al. (2011). Adaptive antiviral immunity is a determinant of the therapeutic success of oncolytic virotherapy. *Mol. Ther.* 19 (2): 335–344.

53. Prestwich, R.J., Errington, F., Diaz, R.M. et al. (2009). The case of oncolytic viruses versus the immune system: waiting on the judgment of Solomon. *Hum. Gene Ther.* 20 (10): 1119–1132.

54. Smith, E., Breznik, J., and Lichty, B.D. (2011). Strategies to enhance viral penetration of solid tumors. *Hum. Gene Ther.* 22 (9): 1053–1060.

55. Novella, I.S., Gilbertson, D.L., Borrego, B. et al. (2005). Adaptability costs in immune escape variants of vesicular stomatitis virus. *Virus Res.* 107 (1): 27–34.

56. Carlisle, R.C., Briggs, S.S., Hale, A.B. et al. (2006). Use of synthetic vectors for neutralising antibody resistant delivery of replicating adenovirus DNA. *Gene Ther.* 13 (22): 1579–1586.

57. Fulci, G., Breymann, L., Gianni, D. et al. (2006). Cyclophosphamide enhances glioma virotherapy by inhibiting innate immune responses. *Proc. Natl. Acad. Sci. U. S. A.* 103 (34): 12873–12878.

58. Fischer, U., Janssen, K., and Schulze-Osthoff, K. (2007). Cutting-edge apoptosis-based thera-peutics: a panacea for cancer? *BioDrugs* 21 (5): 273–297.

59. Fischer, U., Steffens, S., Frank, S. et al. (2005). Mechanisms of thymidine kinase/ganciclovir and cytosine deaminase/ 5-fluorocytosine suicide gene therapy-induced cell death in glioma cells. *Oncogene* 24 (7): 1231–1243.

60. Lang, F.F., Bruner, J.M., Fuller, G.N. et al. (2003). Phase I trial of adenovirus-mediated p53 gene therapy for recurrent glioma: biological and clinical results. *J. Clin. Oncol.* 21 (13): 2508–2518.

61. Gaj, T., Gersbach, C.A., and Barbas, C.F. 3rd (2013). ZFN, TALEN, and CRISPR/Cas-based methods for genome engineering. *Trends Biotechnol.* 31 (7): 397–405.

62. Ran, F.A., Hsu, P.D., Wright, J. et al. (2013). Genome engineering using the CRISPR-Cas9 sys-tem. *Nat. Protoc.* 8 (11): 2281–2308.

63. Xiao-Jie, L., Hui-Ying, X., Zun-Ping, K. et al. (2015). CRISPR-Cas9: a new and promising player in gene therapy. *J. Med. Genet.* 52 (5): 289–296.

64. Wen, W.S., Yuan, Z.M., Ma, S.J. et al. (2015). CRISPR-Cas9 systems: versatile cancer modelling platforms and promising therapeutic strategies. *Int. J. Cancer* 138: 1328–1336.

65. Liu, Y., Zeng, Y., Liu, L. et al. (2014). Synthesizing AND gate genetic circuits based on CRISPR-Cas9 for identification of bladder cancer cells. *Nat. Commun.* 5: 5393.

66. Yao, S. and Chen, C. (2015). CRISPR/Cas9-mediated genome editing of epigenetic factors for cancer therapy. *Hum. Gene Ther.* 26: 463–471.

67. Doudna, J.A. and Charpentier, E. (2014). Genome editing. The new frontier of genome engineer-ing with CRISPR-Cas9. *Science* 346 (6213): 1258096.

68. Maddalo, D., Manchado, E., Concepcion, C.P. et al. (2014). In vivo engineering of oncogenic chromosomal rearrangements with the CRISPR/Cas9 system. *Nature* 516 (7531): 423–427.

69. Carmeliet, P. and Jain, R.K. (2011). Molecular mechanisms and clinical applications of angio-genesis. *Nature* 473 (7347): 298–307.

70. Ellis, L.M. and Hicklin, D.J. (2008). VEGF-targeted therapy: mechanisms of anti-tumour activ-ity. *Nat. Rev. Cancer* 8 (8): 579–591.

71. Motzer, R.J., Hutson, T.E., Cella, D. et al. (2013). Pazopanib versus sunitinib in metastatic renal-cell carcinoma. *N. Engl. J. Med.* 369 (8): 722–731.

72. Motzer, R.J., Hutson, T.E., Tomczak, P. et al. (2009). Overall survival and updated results for sunitinib compared with interferon alfa in patients with metastatic renal cell carcinoma. *J. Clin. Oncol.* 27 (22): 3584–3590.

73. Motzer, R.J., Hutson, T.E., Tomczak, P. et al. (2007). Sunitinib versus interferon alfa in meta-static renal-cell carcinoma. *N. Engl. J. Med.* 356 (2): 115–124.

74. Rini, B.I., Escudier, B., Tomczak, P. et al. (2011). Comparative effectiveness of axitinib versus sorafenib in advanced renal cell carcinoma (AXIS): a randomised phase 3 trial. *Lancet* 378 (9807): 1931–1939.

75. Sternberg, C.N., Davis, I.D., Mardiak, J. et al. (2010). Pazopanib in locally advanced or meta-static renal cell carcinoma: results of a randomized phase III trial. *J. Clin. Oncol.* 28 (6): 1061–1068.

76. Sternberg, C.N., Hawkins, R.E., Wagstaff, J. et al. (2013). A randomised, double-blind phase III study of pazopanib in patients with advanced and/or metastatic renal cell carcinoma: final overall survival results and safety update. *Eur. J. Cancer* 49 (6): 1287–1296.

77. Vasudev, N.S. and Reynolds, A.R. (2014). Anti-angiogenic therapy for cancer: current progress, unresolved questions and future directions. *Angiogenesis* 17 (3): 471–494.

78. Llovet, J.M., Ricci, S., Mazzaferro, V. et al. (2008). Sorafenib in advanced hepatocellular carci-noma. *N. Engl. J. Med.* 359 (4): 378–390.

79. Raymond, E., Dahan, L., Raoul, J.L. et al. (2011). Sunitinib malate for the treatment of pancre-atic neuroendocrine tumors. *N. Engl. J. Med.* 364 (6): 501–513.

80. Hurwitz, H., Fehrenbacher, L., Novotny, W. et al. (2004). Bevacizumab plus irinotecan, fluoro-uracil, and leucovorin for metastatic colorectal cancer. *N. Engl. J. Med.* 350 (23): 2335–2342.

81. Soria, J.C., Mauguen, A., Reck, M. et al. (2013). Systematic review and meta-analysis of ran-domised, phase II/III trials adding bevacizumab to platinum-based chemotherapy as first-line treatment in patients with advanced non-small-cell lung cancer. *Ann. Oncol.* 24 (1): 20–30.

82. Burger, R.A., Brady, M.F., Bookman, M.A. et al. (2011). Incorporation of bevacizumab in the primary treatment of ovarian cancer. *N. Engl. J. Med.* 365 (26): 2473–2483.

83. Perren, T.J., Swart, A.M., Pfisterer, J. et al. (2011). A phase 3 trial of bevacizumab in ovarian cancer. *N. Engl. J. Med.* 365 (26): 2484–2496.

VIII

PATIENT RESOURCES

29

CANCER-CENTER-BASED CARE

George J. Weiner

Holden Comprehensive Cancer Center,
The University of Iowa, Iowa City, IA, USA
Department of Internal Medicine, Carver College
of Medicine, Iowa City, IA, USA
Department of Pharmaceutical Sciences and Experimental
Therapeutics, College of Pharmacy, Iowa City, IA, USA

The term "cancer center" is used to refer to everything from a small medical office where a single community oncologist sees patients and provides chemotherapy through major, internationally renowned academic cancer centers that employ hundreds of highly trained oncologists, thousands of other staff members, and provide highly subspecialized care, conduct cutting-edge research, and train the next generation of cancer care givers and researchers. For the sake of consistency, the term "cancer center" in this discussion will be used to refer to academic cancer centers that support delivery of interdisciplinary cancer care by subspecialists in a setting that also supports research. Using this definition, the distinguishing factors of cancer-center-based care are the ability of such centers to (i) provide state-of-the-art care based on rapidly changing science; (ii) provide care from clinical oncologists with focused expertise in specific cancer types; (iii) provide interdisciplinary care;

Cancer: Prevention, Early Detection, Treatment and Recovery, Second Edition. Edited by Gary S. Stein and Kimberly P. Luebbers.
© 2019 John Wiley & Sons, Inc. Published 2019 by John Wiley & Sons, Inc.

(iv) participate in, and play a leadership role in, clinical cancer research; and (v) adopt new technologies quickly [1, 2]. Each of these will be discussed individually.

  a. Changing science – progress in cancer medicine is moving faster than ever before, and this is resulting in major and frequent changes in patient-care options and standards of care. For example, two broad areas of therapy where this change is ongoing and incredibly rapid are targeted therapy and immunotherapy. Many of these advances in targeted therapy and immuno- therapy emerged from the laboratories of cancer centers that provide cancer care. Physicians at such centers are often aware of such advances and pro- vide them to their patients well before they are adopted by community oncology. Thus, physicians practicing at such cancer centers are well posi- tioned to provide cutting-edge care based upon the latest scientific advances. This includes not only advances in new approaches to cancer therapy but also awareness of the results of population-based research exploring cancer outcomes in various populations including advances in cancer prevention, early detection, and quality of life.

  b. Subspecialization – it is extremely difficult for even the most outstanding, well- read general oncologist to keep up with all of the advances taking place in the field. Cancer-center-based care is distinguished by the subspecializa- tion of oncologists who work in such centers. Such oncologists often focus their practices, and their continuing medical education effort, on a single type or subset of malignancies (breast cancer, leukemia, GI cancers, etc.). This facilitates their ability to keep up with the latest advances.

  c. Interdisciplinary care – an additional distinguishing characteristic of cancer- center-based care is that it is interdisciplinary. Teams of physicians with different areas of expertise (medical oncology, surgical oncology, radiation oncology, pathology, etc.) are organized around specific cancer types. These groups are referred to by various names such as Multidisciplinary Oncology Groups, Disease Oriented Groups, or Multidisciplinary Oncology Teams. Coordinators are often assigned to disease-focused interdiscipli- nary teams to assist in triage of patients to the correct practitioner based on the needed primary modality of care (surgical oncology, medical oncology, radiation oncology). Such coordinators also facilitate health record, patho- logic, and radiologic review so the interdisciplinary team has the appropri- ate data available to make a recommendation. Interdisciplinary teams usually have a dedicated disease-specific tumor board where management of patients with a particular cancer is discussed by all the experts in that cancer. A discussion of clinical trial eligibility for each patient is often a key component of such tumor board deliberations. This is in contrast to smaller hospitals and community practices that most commonly have a single, general tumor board attended by general oncologists. In some cancer cent- ers, interdisciplinary teams have a joint clinic, which further facilitates interaction. When such teams are operating optimally, patients receive efficient, high-quality care provided by an expert team with cutting-edge knowledge in a given specific cancer type.

d. Holistic, patient-centered care – academic cancer centers are historically viewed as being impersonal and focused solely on the disease as opposed to the patient. This is now changing. Cancer centers are increasingly working to address the needs of the whole patient through Palliative and Supportive Care, Psychosocial Oncology, and Adolescent/Young Adults Programs. These efforts are difficult to justify financially as independent entities if their return on investment is viewed independently. Fortunately, most cancer centers are able to justify the investment based on the recognition that caring for the whole patient enhances the quality of care and patient satisfaction with an ultimate positive impact on market share, costs, and profit.

e. Clinical cancer research – oncologists who practice at cancer centers and focus on specific cancer types are often extensively involved in clinical research. This includes not only participation in national trials supported by the NCI and industry (which can also be available to patients receiving their care from community oncologists) but also investigator-initiated trials based on concepts developed by the cancer center oncologist, other members of the interdisciplinary team or other academic collaborators. This provides additional therapeutic options for patients.

f. New technologies – the cancer-specific focus of the clinical oncologists, the interdisciplinary approach to cancer care, and the involvement in innovative clinical research results in early adoption of new technologies. This includes hardware such as imaging equipment, approaches to molecular analysis, surgical techniques, and systemic therapies such as cancer immunotherapy.

g. Outcomes at cancer centers – many of those who provide cancer-center-based care claim that such care results in superior outcomes. There is growing evidence this may be the case, at least in specific circumstances. Wolfson et al. found that patients between the ages of 22 and 65 years residing in Los Angeles County with newly diagnosed adult-onset cancer treated at NCI-designated comprehensive cancer centers experienced superior survival compared with those treated at other facilities. It is important to point out that the populations treated at the cancer centers were distinct from those treated in the community, and that race/ethnicity, insurance, socioeconomic status, and distance to a cancer center may contribute to differences in outcome. Institutions with cancer centers generally have large patient volumes. Patient volume has also been shown to correlate with better outcomes [3–6].

h. Cost and value – there is evidence that cancer-center-based care may be more expensive than care provided in the community. The costs of individual treatments were found to be more expensive at academic cancer centers. However, this difference is not seen when the entire cost of care for a patient is examined. This may be due to reduced use of expensive treatment near the end of life at academic cancer centers when compared to the community. Value in healthcare, at its most simple, is patient health outcomes achieved per dollar spent [7]. The discussion of value, and in particular how to measure it with respect to cost and outcomes is more complex [8, 9]. Considering the value of cancer-center-based care needs to go beyond comparing the cost of individual healthcare items (e.g., cost of a CT) and should include assessment of the economic impact on overall outcome. In an era of unsustainable increases in the cost of cancer care [10], cancer centers need to strike the balance between cost and outcomes if they are to survive economically [11].

i. Free standing cancer centers – free-standing cancer centers are independent entities. These cancer centers may be affiliated with a university institution at some level, but are administratively and financially separate. Examples of such institutions are the Memorial Sloan-Kettering Cancer Center in New York, NY, the St. Jude Children's Research Hospital in Memphis, TN, and the University of Texas M.D. Anderson Cancer Center in Houston, TX. In terms of financing, these centers control the revenue they generate and are responsible for their expenses. A number of free-standing cancer centers are exempt from the Department of Health and Human Services Centers for Medicare & Medicaid Services Prospective Payment System (PPS). "PPS-exempt" cancer centers benefit significantly from this alternative payment model. Many also have substantial fundraising capabilities that allow them to supplement the quality of care they provide. As freestanding organizations, they have greater flexibility in organizing themselves internally (Departments, Divisions, etc.) in a manner that facilitates their ability to provide interdisciplinary, subspecialty-focused cancer care. However, their free-standing nature has some limitations. Many cannot sponsor graduate students directly. They often need to hire non-oncology specialists to help manage their patients' noncancer health problems. Finally, they may not fully benefit from the research advantages provided by a truly comprehensive research intensive medical center.

j. Matrix cancer centers – matrix cancer centers comprise the majority of NCI-designated cancer centers. These centers are part of a university and usually reside in an academic hospital, hence making them dependent on the university structure. Financing for these centers is highly variable with each matrix cancer center establishing specific arrangements between itself and the medical center. Some cancer centers have clear service line organizations where the cancer center has access to the profit margin created by the cancer service line. Even with a service line model, a portion of the resources generated by the cancer service line are often shared among the different units within the medical center. Other matrix cancer centers operate more on a cost basis. Given how quickly cancer medicine is changing, for a matrix cancer center to be successful, a mechanism needs to be in place for that cancer center to have resources to invest in new approaches to cancer care. Matrix cancer centers benefit from various strengths of their home institutions including availability of noncancer experts, shared research cores and a breadth and depth of research that is often greater than can be found at a free-standing cancer center. From the point of view of the patient, both free-standing and matrix cancer centers provide the advantages discussed above including an ability to respond to rapidly changing science, clinical oncologists with expertise in specific cancer types, interdisciplinary care, innovative clinical cancer research, and new technologies.

k. Patient education and care coordination – patient education and care coordination are particularly important for patients receiving their care at cancer centers. In an interdisciplinary setting where multiple experts are involved in the care of a patient, that patient can easily get confused about the identity of their primary physician, their diagnosis, options, and next steps. Cancer centers have developed programs to inform and educate patients regarding the diagnosis, treatment options, and expected outcomes and to assist in their navigation through the system. Patient navigators were initially pioneered by Dr. Harold Freeman and have been found to be particularly vital in underserved communities where patients and their families can easily get confused by, and lost in, a complex system [12]. It is becoming increasingly clear that educating patients, providing care coordination and navigation, and assuring patients do not get lost in the complex systems of a cancer center improve outcomes and increase patient satisfaction.

l. Efficiency – academic cancer centers are constantly challenged to improve efficiency. This creates tension between the caregivers, who in large cancer centers often understand little about the costs of providing quality cancer care, and cancer center administrators who are held accountable for cancer center finances, need to pay attention to the bottom line and strive to improve the operational methods of their facility to minimize costs. Successful management of the unavoidable tension between delivery of state-of-the-art, high-quality care, and cost, is vital for both free-standing and matrix cancer centers if they are to remain viable and provide top-quality care to patients. The most successful centers are transparent and collaborative in how they manage the relationship between mission and money.

m. Networks – the line between cancer-center-based care and community-based care is increasingly complicated. Cancer centers that previously were found at a specific location are evolving into cancer center systems by expanding their services to networks that extend across larger geographic areas. In some cases, oncologists practicing at network sites are well integrated into the culture and interdisciplinary activities of the cancer center. In other cases, the network is based more on financial factors (e.g., the community practice is owned by the cancer center or medical center) or the cancer center supports monitoring of the quality of care in the practice, but there is minimal true integration. In still others, the affiliation is in name only. An ongoing initiative of the Association of American Cancer Institutes, led by Dr. Stanton Gerson, is focusing on developing a more accurate understanding of such models and the value they bring to the cancer center, and more importantly, the patients they serve (see http://www.aaci-cancer. org/networkcare.asp).

n. Cancer center/community collaboration – increasingly, networks are being established that provide the best of cancer-center-based and community-based care. When such networks work well, patients are able to receive the benefits of cancer center care including subspecialty, interdisciplinary opinions, and access to innovative clinical trials, while also receiving the actual therapy closer to home. Unfortunately, there are a number of barriers to such successful networks including restrictions imposed by payers, physician preferences, and patient lack of information on options.

CONCLUSIONS

Cancer-center-based care is increasingly important as the pace of progress advances and cancer medicine becomes more complex. Cancer centers are designed to provide state-of-the-art care based on rapidly changing science, care from clinical oncologists with focused expertise in specific cancer types in an interdisciplinary setting, play a leadership role in clinical cancer research, and adopt new technologies quickly. Preliminary data suggests outcomes in some cases may be superior at cancer centers, although perhaps at a higher cost. Various administrative structures, including free-standing and matrix cancer centers, have proven successful in providing top-quality cancer-center-based care. Patient education and navigation is particularly important for patients receiving care at a cancer center so they do not get lost in the system. Balancing quality care and cost remains a challenge. Ultimately, development of networks that link cancer centers and community oncology may provide the best approach to providing the best possible care to patients close to home.

REFERENCES

1. Simone, J.V. (1999). Understanding academic medical centers: Simone's Maxims. *Clin. Cancer Res.* 5 (9): 2281–2285.
2. Simone, J.V. (2002). Understanding cancer centers. *J. Clin. Oncol.* 20 (23): 4503–4507.
3. Sosa, J.A., Bowman, H.M., Gordon, T.A. et al. (1998). Importance of hospital volume in the overall management of pancreatic cancer. *Ann. Surg.* 228 (3): 429–438.

4. Wolfson, J., Sun, C.L., Kang, T. et al. (2014). Impact of treatment site in adolescents and young adults with central nervous system tumors. *J. Natl. Cancer Inst.* 106 (8): dju166.

5. Wolfson, J., Sun, C.L., Wyatt, L. et al. (2017a). Adolescents and young adults with acute lymphoblastic leukemia and acute myeloid leukemia: impact of care at specialized cancer centers on survival outcome. *Cancer Epidemiol. Biomark. Prev.* 26 (3): 312–320.

6. Wolfson, J., Sun, C.L., Wyatt, L. et al. (2017b). Impact of treatment site on disparities in outcome among adolescent and young adults with Hodgkin Lymphoma. *Leukemia* 31 (6): 1450–1453.

7. Porter, M.E. (2010). What is value in health care? *N. Engl. J. Med.* 363 (26): 2477–2481.

8. Ramsey, S. and Schickedanz, A. (2010). How should we define value in cancer care? *Oncologist* 15 (Suppl 1): 1–4.

9. Schickedanz, A. (2010). Of value: a discussion of cost, communication, and evidence to improve cancer care. *Oncologist* 15 (Suppl 1): 73–79.

10. Mariotto, A.B., Yabroff, K.R., Shao, Y. et al. (2011). Projections of the cost of cancer care in the United States: 2010–2020. *J. Natl. Cancer Inst.* 103 (2): 117–128.

11. Alvarnas, J., Majkowski, G.R., and Levine, A.M. (2015). Moving toward economically sustainable value-based cancer care in the academic setting. *JAMA Oncol.* 1 (9): 1221–1222.

12. Freeman, H.P. and Rodriguez, R.L. (2011). History and principles of patient navigation. *Cancer* 117 (15 Suppl): 3539–3542.

30

SUPPORTIVE CARE

Kathleen McBeth

University of Vermont Medical Center, Burlington, VT, USA

The goal of supportive care is to prevent or treat, as early as possible, the symptoms of a disease, the side effects caused by the treatment of that disease, and the psychological, social and spiritual problems related to the disease or its treatment [1].

WHAT DOES SUPPORT REALLY MEAN?

Emotional support during a cancer diagnosis can vary a great deal from person to person. What works for one patient may be less effective for another – even another patient working through the same diagnosis. I have spent the past 10 years as a clinical psychologist in an oncology clinic, and what I have learned is that support actually means creating a strong foundation. Just as a building does not sit on one brick, someone dealing with a cancer diagnosis will often need a foundation built on multiple supports. Without a strong base, the patient won't have what is essential to supportive care.

Cancer: Prevention, Early Detection, Treatment and Recovery, Second Edition. Edited by Gary S. Stein and Kimberly P. Luebbers.
© 2019 John Wiley & Sons, Inc. Published 2019 by John Wiley & Sons, Inc.

INDIVIDUAL THERAPY

The focus of therapy is to establish supports and build a foundation. We discuss the individual needs, desires, and concerns and, once we identify the most pressing issues, we look for acceptable solutions as well as possible barriers. During the course of our lifetime we develop coping strategies that help us adjust to difficult circumstances and address traumas. One of principal goals of supportive therapy is to uncover and reactivate those coping strategies to help individuals reestablish what's been beneficial in the past.

An example of this is when a woman came to therapy to discuss the anxiety she was experiencing about an upcoming medical procedure. She explained that she had recently learned about mindfulness practices and was now trying to meditate as a way to relax and quiet her mind. However, this new technique was not working. During our session we noticed that this woman was wearing a religious cross. We asked about her faith. She explained to us how important her belief system is to her and went on to explain how saying the Rosary provided her with feelings of support and strength. We immediately began talking with her about incorporating prayer – as opposed to trying to simultaneously learn and apply the practice of meditation – would be most beneficial. In this case, the woman already had a support system and a foundation: her beliefs. It was not only easier for her to activate this system, it was enormously beneficial. Later, this woman reported that her procedure went very well and, once she started drawing on the supports and foundation she already had in place, she was able to greatly reduce her anxiety. Drawing on existing coping skills and strategies – and the process of discovering what exactly ours are as individuals – is how we establish our inner support and the basis of a strong foundation.

Joanne would describe herself as a little on the rough side. As she would often say, she "just cannot tolerate all that pink stuff." Not surprisingly, she had a hard time participating in events, rallies, or support groups – even when invited by other breast cancer survivors. She described herself as "angry" the majority of the time she was in treatment. She felt her diagnosis was unfair and that her body had "betrayed her." She came to therapy because she could not control her tears. Even this, however, was difficult for Joanne to admit. And yet, on more than one occasion, Joanne's oncologist felt that she needed additional support.

She came into my room and told me, "I've been sent to the shrink." Of course, this did not feel like a positive start to our relationship. However, by allowing her to continue to express her feelings of anger, frustration, and hurt, Joanne began to have some compassion for herself and eventually became less judgmental of her response to her own diagnosis. Joanne was even able to reach out to friends and family members who were going through difficult time themselves. One day Joanne returned to therapy and stated that she'd read a book that referred to breast cancer as a "silver lining." She was surprised that her reaction to this statement was not negative and instead she was able to see how understanding her own feelings of anger and fear helped her to be more compassionate toward herself and others. Joanne was able to establish a group of friends who went on to start programs that would support other cancer patients.

EMOTIONAL AND SOCIAL SUPPORTS

One of the difficulties of a cancer diagnosis is that the illness can feel like throwing a pebble in a pond. The ripple that forms is the challenges that everyone in the group will experience as a result of the diagnosis. What's difficult to understand, as an individual with cancer, is that you are not the rock that causes the ripple. Cancer is the rock that causes this disturbance. You are caught in it just as much as your family members and the people who care about you. Therefore, the individual who has the cancer diagnosis is not the burden; they are also burdened by this diagnosis. What can be supportive to them can be supportive to the entire group.

Emotional supports can come from family and friends however, more often than not; they want to "protect" their loved one. This is an emotional response that can undermine their ability to provide real support. Instead, they worry that they are a "burden" among other things. For this reason and myriad others, discussing emotional concerns with family members is often difficult; therefore, having the support of someone whose had a cancer diagnosis – or a professional trained in psychosocial oncology – can be very useful. The bonds we have seen formed between patients and oncologists, nurses, social workers, and therapists can be profound – and profoundly helpful.

Many people feel it is important to remain hopeful at all costs. The challenge here is that all emotions are important and if we do not allow for anger, sadness, or disappointment we run the risk of shutting off a significant way to cope. Sometimes people find that their distress is a great motivator for change. In many cases, people see their previous behavior, ignoring their physical and/or emotional health, as a sign that they need to change their thinking. Lawrence LaShann wrote a book titled *Cancer as a Turning Point* [2], in which he explained that, through his experiences working as a therapist with hundreds of cancer patients, he observed that individuals who saw their cancer as a reason to look at their lifestyle choices and challenge themselves to "do better" were often able to see their cancer as a "wake-up call."

Many times an individual will come to therapy to "vent" because they worry that their family "cannot handle" the true depth of their grief or fear. The most important part is that we do not stay "alone with the problem" emotionally. George had been married to his wife for 63 years. He candidly told us that, while their marriage was not always a picnic, together the two had worked through many struggles throughout their lives. The most difficult struggle they faced was when George's wife was diagnosed with cancer. George felt it was unfair of him to "be upset." He stated that, "after all, she has the cancer, not me. So what do I have to complain about?" This was a wonderful opening for support. It was important for George to understand that, during the course of her diagnosis, many aspects of his life had changed too. His wife may be the one with the illness; however a cancer diagnosis is an ordeal that's felt by every member of the family. Understanding that allowed George to become more open about his feelings of distress and he was able to talk openly to his adult sons about his concerns for the future.

Another area of support within your social connections can be family members, neighbors, work colleagues, and often fellow cancer patients. Donald is a constant learner. He spent his career organizing efforts in the community, building teams of individuals to network across agencies. As such, he has seen firsthand the strength of programs that support one another. Therefore, when Donald was diagnosed with lymphoma he intuitively understood the value of his treatment team. He worked to establish healthy connections across disciplines, which meant that he became involved in his oncology team as well as his psychosocial supports and alternative treatment programs such as acupuncture and massage. When Donald came to therapy, he wanted to understand the uncertainty of his diagnosis and the distress familiar to most individuals during early treatment. He was greatly comforted to learn that the majority of individuals will have some level of distress during treatment and that much of this can be alleviated through supports and by developing individual coping strategies. Donald continued to engage in supportive therapy and a number of healing modalities such as yoga and reiki. Post cancer treatment, Donald also sought out oncology rehabilitation and he participates in an ongoing effort to reestablish his physical health and well-being.

PRACTICAL SUPPORT

Another area of support is physical or practical matters. For some, the needs of daily living – grocery shopping, house cleaning, child care, and work-related issues – are their largest stressor. In this case, we are often able to identify individuals who can be of assistance including family members, community members, social workers, and volunteers. There are also a number of websites that help to organize people's efforts to provide concrete supports. These sites include www. meal–train.com and www.lotsofhelpinghands.com. Both use calendars that family members or individuals with cancer can fill in to inform their supports of what would be most helpful when – such as rides to the hospital, bringing meals to your home, light housework, or grocery shopping. These tasks may seem easy but when you are going through treatment, they can be overwhelming.

It would not be an over statement to say that cancer treatment can feel like a full-time job. As such, practical supports can be a tremendous help. Susan had always been very independent. She was a single mother of two and she worked in law enforcement. When she was diagnosed with breast cancer for the second time, she realized that she needed to consider letting her friends provide some necessary supports. During her first diagnosis, she was largely stoic and unable to accept support of any kind. However, she realized that she and her family both suffered a great deal as a result of her inability to accept help from others. From that experience, Susan learned that the people who care about her felt helpless because she would not allow them to help. By accepting their help, she would not only be alleviating her own distress, she'd be alleviating the distress of those who were helping her, too. Therefore, Susan said, "When you let people help you, you are giving them a gift." For Susan, this was a tremendous realization. In fact, she feels that it's the most important message she could give to someone with a cancer diagnosis.

MEDICAL SUPPORT

Medical support is more than having a team of doctors to take care of the disease, treat the body, or help us understand medical terminology and reports. One oncologist explained to her patient that, "she was learning a new language." This simple statement helped the patient to ask questions and not assume that she "ought to know" what everything meant. This patient was able to say to her doctor, "I've never been good with new languages, so please bear with my questions." This was the start of a useful and meaningful relationship, one that proved beneficial throughout and even after her treatment.

Often the main concern of medical support is treatment-related when, in fact, we also need to identify the concerns of the individual. In many cases, the patient needs education about their cancer treatment. This education may include an explanation of chemotherapy or a discussion aimed at helping the patient understand upcoming surgical procedures and options. Nurses and support staff can help familiarize patients with tests like CT scans, blood work, and biopsies. Simply knowing and understanding what exactly is in your treatment plan can be hugely supportive and empowering. Knowing that you can choose when and how you receive information is also supportive. For example, you may choose to have a family member or friend present when meeting with your oncologist. You can even tell your treatment team what, or how much, you want to know. Many individuals want all their test results and the latest treatment options but, for others, this can be overwhelming. They only want to understand each treatment's effects and what they have to do to stay well during treatment. In either case, your doctor will need to understand your wishes and you'll want to establish helpful communication as soon as possible.

Eric made a therapy appointment to discuss end-of-life concerns. Because this was our first meeting, we discussed the history of Eric's cancer. During the conversation, we became confused about what Eric had told us regarding his diagnosis and treatment. Because we needed to understand his prognosis to guide him toward the best end-of-life care, we asked for Eric's permission to talk with his oncologist. When we did that, it immediately became clear that Eric had misunderstood what his oncologist told him at their last meeting. He was not, in fact, told that his cancer was terminal. Imagine his relief! To be sure, Eric's case is unique. But it does illustrate how beneficial it can be to have another person present whenever your oncologist is sharing information about your prognosis or treatment. In Eric's case, doing so would've spared him incredible emotional stress. In almost every case, it will prove helpful.

While being hopeful is useful in addressing the full experience of emotions, many individuals believe that, if they are not positive, they will negatively affect their cancer treatment. A woman came to therapy with the statement that, "Everyone around me has to have positivity. I will not accept any negativity in my life because of my cancer and the need for a positive outcome." She was convinced that if she accepted any negative thinking that she would destroy her ability to create a positive treatment outcome. This emphatic response to the people on her treatment team and her loved ones actually worked against their communication. They felt that they would somehow precipitate her physical decline if they were less than 100% positive. They began avoiding her whenever treatment did not work. Ultimately, the medical team needed to discuss a different approach. The point is this: believing that a positive attitude will increase your chances of survival can be a tremendous burden, not only on the individual with cancer but also on the people who support them. All emotions are essential to open communication. Understanding that emotions like fear and hope need to be heard and expressed in a useful way will ensure that communication is not hampered by forced beliefs or the potential to hurt feelings.

WHEN DO WE NEED SUPPORT?

The timing of support is very personal. For many, it is essential to meet their support team at the time of diagnosis. But in some cases, the time of diagnosis is a period of shock [3]. Over the years, we have seen that when individuals first learn that they have cancer, it takes a while for this information to be absorbed. This "numbing" or disbelief is often expressed by an almost robotic response to the news. Patients will learn about upcoming appointments, meet their medical team, and schedule tests or biopsies almost completely devoid of any emotion. Because the initial diagnosis period is a very busy time for most people, they struggle to find the time to process their feelings. Instead, the patient feels that he/she needs to focus on schedules and appointments. Logistics take precedence over their real need for support. However, this response is usually brief [4]. During this time, it is useful to understand the role of supports within the organization including psychologists, social workers, and/or support groups.

Once treatment is established, we often find that our emotions can get in the way of our daily functioning. Many patients experience these emotions in the form of constant fatigue or sleeplessness. A person may find themselves angry most of the time, or in a state of constant worry or fear. These are common manifestations and they can be effectively addressed with emotional support. In cancer, we often use the term "distress" rather than depression or anxiety; for many this diagnosis brings on emotional responses that can be outside our normal behavior. These emotions are often specific to the cancer diagnosis and the treatment issues. By understanding that distress is common, we feel less alone with the problem and are better positioned to utilize our supports to address these concerns.

In recent years there's been a strong effort to include the assessment of distress as the sixth vital sign in patient care [5]. The American College of Surgeons Commission on Cancer included this as a standard practice for accreditation starting in 2015. Distress screening focuses on areas such as practical concerns, family problems, emotional issues, and spiritual concerns. Distress covers a wide range of emotions such as feeling powerless, sad, fearful, and hopeless. It can also manifest itself in feelings of depression, anxiety, guilt, or panic. These emotions may interfere with areas of your life including your interactions with others in the form of intrusive thoughts or behaviors like shutting out family members or loved ones. The primary reason that the medical community has begun to focus on distress in cancer is because it is normal to have these feelings. Nonetheless, they can be debilitating and can severely interfere with a person's well-being, both emotionally and physically. Assessment for these issues is a start, but assessment alone is not enough. We need to help people access the supports they need, which may require involving social workers, psychologists, and spiritual advisors.

As stated, for some people early diagnosis may not be the right time to seek supports. Posttreatment can be a difficult time as well. This has been described by some as "posttreatment depression." It's important to address these feelings [6]. During treatment, we are surrounded by a medical support team, our family, and social supports. However, as one patient stated, "sympathy has an expiration date." It is at this time that we may feel most alone. We can help ourselves by accessing supports in the form of groups or individual therapy, both of which can be enormously beneficial. We may also find ways to "give back" by participating in survivor programs and community events like rallies, 5-K walks or day-long celebrations of cancer survivorship. Learning more about the events in your area can be a useful way to network with others who have been touched by cancer.

We have seen many people engage in posttreatment oncology rehabilitation to improve their physical functioning. They have reached out to nutritionists to address dietary concerns and, in many cases, have reduced their weight. As therapists, we have seen many people address their resentments and failed relationships. They have worked to achieve better communication with their loved ones and, often, their goal is to remove any regrets they may be harboring from their emotional concerns.

WHO CAN PROVIDE SUPPORT?

Building your support foundation often starts as a conversation with your oncologist in which you express your need to speak to someone "whose been there." In many communities, there are peer supports or individuals who, following their own experience with a cancer diagnosis, volunteer their time to offer support based on their experiences. Often there is a desire to speak to a professional therapist. In this case, it is important to find an individual trained in the unique needs of oncology patients. This person has a focus on health psychology or psycho-oncology and understands the adjustment to illness concerns that accompany this disease [7]. This type of support focuses on coping strategies and the communication issues that may naturally arise during this stressful time.

As a psychology intern at a cancer center, I was introduced to a number of individuals during their chemotherapy treatments. One morning I went to say hello to a gentleman who was waiting to receive treatment for lung cancer. This gentleman read my badge and noticed I was from the psychology department. His immediate response was, "I have cancer, I'm not nuts." This was a perfect opportunity to open the conversation to the distress and adjustment associated with a cancer diagnosis. We formed an excellent relationship and this gentleman quickly began looking forward to our visits – particularly when he learned his cancer had progressed and he needed to speak to his young family about his prognosis. Being able to process his end-of-life needs, including how to conduct a healthy communication with his family was extremely important and gave him and his family the tools and the ability to face the next phase of his life.

HOW DOES CANCER DIFFER FROM OTHER DIAGNOSES?

Uncertainty begins at the very first moment of a cancer diagnosis. This stems from the fact that cancer is often curable and, in some cases, fatal. As a result, our feelings can range from optimism to despair in the span of a single hour. They can vary wildly from one hour to the next. These feelings are often influenced and informed by our own history with cancer – whatever that may be. In particular, if we have had a family member who has had a cancer diagnosis, that family member's outcome will tend to strongly influence or color our feelings about our own diagnosis.

Carol was a 35-year-old girl whose mother died of cancer when she was 17 years old. She understood that her mother had a gene mutation called BRCA and therefore, there was a strong likelihood that she would have cancer herself. Carol spent her young adulthood waiting for her own cancer diagnosis and, at age 34, she had a positive biopsy. She stated that she had worried about the "worst thing that could happen in my life" for so long that, when she was diagnosed, it almost felt like a letdown. The good news for Carol was that her cancer was detected early and she immediately joined a young survivors group and formed supports. She states now that her cancer experience was nothing like what her mother had gone through – although, at the outset, she expected her outcome would be the same as her mother's. Her anticipation was more challenging than her actual experience. She now reports that she's somewhat thankful she does not have to continue to worry about how she would manage this diagnosis because she feels that her response was positive, productive, and supportive.

WHAT DOES IT FEEL LIKE TO GIVE BACK?

Be a role model. Pay it forward. They are not just sayings. In fact, studies show that altruism is a key to resilience [8]. When we help others, we are able to make sense of a cancer diagnosis. By working with and for others, we help to develop our own coping strategies and problem-solving abilities. This can give us a larger sense of purpose and, rather than letting cancer define us, we learn to see it as an opportunity to help others based on our own experiences.

Charles was diagnosed with leukemia. He suffered through three years of difficult treatment and needed to accept early retirement. Afterwards he began to look back on his experience being in the hospital and he realized that he could have used supports for himself and his wife during that time. Charles has since become a volunteer at the desk in the ICU (intensive care unit). He greets family members when they come to see loved ones and is always ready to hand them a tissue when things get rough. His empathetic ear and deep sense of compassion is a function of the fact that he's been in that circumstance himself. Before his own diagnosis, Charles would have described himself as a "curmudgeon." But his experience opened his awareness to others and helped him realize that he has something important and worthwhile to share with other people experiencing a similarly serious diagnosis.

By telling our stories we help to normalize and honor another person's cancer experience. By listening to others, we give them an opportunity to express their feelings of fear and strength. The goal is to let another individual feel less alone and connected with another person who has had the experience of cancer. Individuals often look for opportunities to share their experience, strength, and hope with one another. In many cases, support groups can provide this opportunity. When you participate in the support group you are able to learn how others manage, cope with, and communicate their illness and difficult times with family members. By sharing your own experience, you are often able to see some of the resilience and insights that you have learned through your diagnosis.

The role in Supportive Oncology care is to help a person identify and build a foundation of support – to provide the tools and expertise needed to go through this challenging diagnosis and see that the person is not alone. Learning that you have multiple ways to access support can be essential to our emotional well-being. To be sure, cancer can be a life-altering event. But the more understanding we have about what's helpful to us, the less disturbing the process will be. Using the coping skills that've been helpful in the past, allowing others to be supportive, and giving ourselves the time to feel and process this part of our life can be enough. Understanding that by accessing support and looking for ways to cope we are building our foundation of support and helping to improve our quality of life.

REFERENCES

1. Definition of Supportive Care – NCI Dictionary of Cancer Terms. https://www.cancer.gov/ publications/dictionaries/cancer-terms?cdrid=46609
2. LaShann, L. (1994). *Cancer as a turning point*. London: Penguin.
3. Weisman, A.D. and Worden, J.W. (1976–77). The existential plight in cancer: significance of the first 100 days. *Int. J. Psychiatry Med.* 7 (1): 1–15.
4. Adjustment to Cancer: Anxiety and Distress – National Cancer Institute https://www.cancer.gov/ about-cancer/coping/feelings/anxiety-distress-pdq
5. Bultz, B.D. and Carlson, L.E. (2005). Emotional distress: the sixth vital sign in cancer care. *J. Clin. Oncol.* 23 (26): 6440–6441.
6. Jaouad S Lost in Transition After Cancer. New York Times.com. 3/16/2015
7. Holland, J.C. (2002). History of psycho-oncology: overcoming attitudinal and conceptual barriers. *Psychosom. Med.* 64: 206–221.
8. Southwick, S.M. and Charney, D.S. (2012). *Resilience the Science of Mastering Life's Greatest Challenges*. Cambridge, UK: Cambridge University Press.

31

INVESTIGATING CANCER SENTIMENTS ACROSS SOCIAL MEDIA

Eric M. Clark, W. Christian Crannell, Jesse Moore,
Ted James, Mitchell Norotsky, and Chris A. Jones

Department of Surgery, University of Vermont, Burlington, VT, USA

INTRODUCTION

Twitter is a popular online social media platform that encompasses 288 million monthly active members. Users interact via short, 140 character messages called tweets. Currently, Twitter accumulates approximately 500 million tweets per day [1]. Notably, the Pew Research Center reported that Twitter usage has increased from 18 to 23% over the past year, with a significant increase of 5–10% in users older than 65 [2], which indicates a broad range of sentiments to analyze [2]. Researchers have performed studies with the social media platform to understand how socioeconomic factors underscore happiness [3].

Recent healthcare studies have integrated Twitter analytics to understand trending infectious disease [4–6], obesity, and allergies [7]. Diez et al. sought to qualitatively characterize the content of chronic diseases on Facebook and Twitter [8]. Twitter has also been studied to investigate the social implications of breast cancer awareness month [9], and to qualitatively categorize cervical and breast cancer screening patient dialog [10].

Researchers have also sought to understand the interconnectedness of cancer patients on Twitter and have tried to characterize those relationships with data

Cancer: Prevention, Early Detection, Treatment and Recovery, Second Edition. Edited by Gary S. Stein and Kimberly P. Luebbers.

analytics [11]. Social-media analytics could be beneficial for uncovering potential patient reported outcomes regarding surgery/treatment as well as providing hope and online support groups. In this chapter, the analytic tools and potential pitfalls for conducting research with Twitter data are outlined for those interested in performing their own studies with social media analytics.

NAVIGATING TWITTER DATA

Twitter shares a free, small (1%), random sample of Twitter data downloadable publicly through an Application Programming Interface (API). Although 1% seems meager, at Twitter's current (2015) daily traffic, this sample can comprise as many as five million tweets per day, which is a sizeable data mine for running all sorts of interesting analyses. (For more information visit Twitter's streaming API support site [https://dev.twitter.com/streaming/overview].)

After building a feed to download a stream from Twitter's API, the data (each tweet) will be returned in a JavaScript Object Notation (JSON) data structure. This type of structure is unpackable with almost any programming language (Perl, Python, etc.).

Specific modules have been created to unpack the data into an easily navigable key-value pair. The digital copy of a tweet is much more than just a 140 character text string. Along with the raw text, Twitter also relays the meta-data concerning the authorship of the tweet, including the time of creation (also called a timestamp), the time zone local to where the tweet was posted, the language, geographical coordinates (geo-stamp), and user information (screen-name, id). (For a detailed list of all of the fields associated with a JSON encoded tweet see- https://dev.twitter.com/overview/api/tweets.)

In general, an analysis involving tweets will be too massive to curate by hand. There are many types of software readily available for textual analyses in almost every programming language. The next step is to filter tweets for terms that are relevant to a specific area of interest. Key phrasal pattern matching relevant terms is a natural first step in filtering tweets for an exploratory analysis. For example, a study interested cancer-related social-media posts may choose the search terms "cancer," "chemotherapy," "chemo," etc.

Pattern-matching software for key terms exists universally in every programming language. In general, these algorithms use tokenization methods to isolate each specific word from a string of text, and then match a string of interest case insensitively. Sometimes it is useful to remove punctuation, so the tokenizer can match terms on the boundary of a sentence (i.e., cancer!, #cancer, cancer). It is also preferable that the match case task is performed insensitively, which can be done by forcing all characters to be lowercase (i.e., Cancer, cancer).

CLEANING YOUR DATASET: (SAMPLING RESULTS)

When choosing your key terms to filter, it's important to consider possible false positives, i.e., nonrelevant data. After you have a corpus of pattern-matched relevant tweets, it is important to try to understand the ratio of true positives and false positives. Observing a random sample of the data, and reporting the relevant percentage of tweets may help strengthen the integrity of the results. Try to examine a few false positives to identify the theme of these nonrelevant data points. Then try to notice underlying sentence structure (i.e., key phrases) that could be used to identify and exclude these types of data points. Removing false positives from the data will help dampen a noisy dataset and is a crucial step to avoid misreporting trends in the data.

For example, a common false positive for a search involving key terms related to "cancer," the medical condition, would be tweets about the astrological sign for cancer. There are also automated Twitter accounts that tweet frequently about daily horoscope readings, which could account for a nontrivial portion of the captured data. These types of accounts generally have a limited vocabulary, since they are consistently tweeting similar messages [12, 13].

Observing the sentence structure of these automaton's tweets could help identify key phrases to include for removal from the data. For example, we list below a couple of actual tweets that mention the key word Cancer.

- *My horoscope is on point #Cancer*
- *Cancer horoscope for Jun 24 2014 by Daily Horoscope URL*
- *I have never really gotten into horoscopes, I mean I am a cancer but who really cares lol*
- *Cancer Today: Love: 51% | Sex: 56% | Relationship: 67%. Read Full Love Horoscope Here: URL*

None of these tweets have any relevance to the medical condition. A key trait within each of this set of false positives is the word "horoscope." In general, a tweet mentioning the word "horoscope" and "cancer" is most likely not relevant to the disease. Using this knowledge, all tweets mentioning the word "horoscope" can then be removed from the dataset. This process is called "cleaning the data" and is extremely important in dampening potential noise from a dataset.

After cleaning the data it is important to evaluate the relevance of the collected data to area of interest. One method to benchmark the relevance of the data is to pull a random sample of tweets and evaluate their relevance. It is useful to utilize at least two separate human evaluators in order to minimize error or potential bias. Discrepancies between reviewers can then be reevaluated for a more accurate depiction of the sample's relevance. This evaluation is extremely important in strengthening the integrity of results that are too massive to curate.

DATA VISUALIZATION: SAMPLE PRELIMINARY ANALYSES

Lung cancer brought on by the inhalation of combustible tobacco products is the single most preventable form of cancer [14]. Now we introduce a sample preliminary analysis, performed by the Department of Surgery at the University of Vermont, of tweets that are relevant to cigarettes. Social media provides an outlet to observe trends, sentiments, and organic opinions regarding tobacco products, their use, and societal stigma.

The first step was to create a key-phrasal list of all cigarette related key words (e.g., cigarette, cancer stick) and then troll a sample of Twitter's streaming API for tweets containing case insensitive mentions of these terms. We did not include the search term "smoke" for this sample due to its potential in producing nonrelevant data (like a tweet about "smoked cheese" for example).

Using these key phrases, Twitter's firehose feed (a 10%) random sample was trolled spanning 2012–2013, as well as a 1% sample of Twitter containing tweets that contain geo-stamps, i.e., latitude and longitudinal coordinates. It has been reported that approximately 1% of all Tweets contain geo-coordinates, so this second data set can provide the means for interesting geo-spatial analyses, even with an order of magnitude less data.

As part of the cleaning process, it is also important to consider all forms of potential noise in the data. For studies looking at broad societal trends, it's important to consider the distribution of individuals and their contribution to the data. In Figure 1 the user tweet distribution of this data set is plotted on

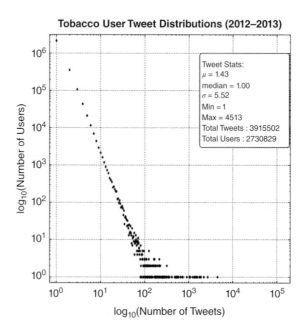

Figure 1. The user tweet distribution of the tobacco tweet dataset is plotted on logarithmic axes.

logarithmic axes. Individuals are binned by their number of tweets appearing in the dataset. We see the vast majority of individuals only contributed one tweet to the dataset, which is preferable for this type of study. However, in the tail of the distribution we see some users are responsible for 100s to even 1000s of tweets. In this particular data set, these were representative of automated advertisement Twitter accounts that were promoting electronic cigarettes, also reported in Clark et al. [13]. In particular, automated accounts can pollute the dataset with a high density of irrelevant tweets. Identifying and removing these types of noisy components from the data can help strengthen results and help avoid misreporting trends.

After the data has been collected, and cleaned for false positives, the next step is to perform exploratory analyses to find any trends (if any) that are hiding within the data. A natural first step is to observe the frequency of the collected tweets as a function of time – a time series analysis.

Notably, the time stamp of each tweet could differ considerably depending on the time-zone of its authorship. The time displayed by each tweet is in Greenwich Mean Time (GMT), which is not necessarily the actual time at which the tweet was authored. To account for this, it is useful to transform the tweet into its "local time" by using the time zone metadata provided by Twitter. Many packages exist in most programming languages to easily transform time data using their corresponding time zones (which are provided by Twitter's streaming API).

The next step is to decide how to group intervals of tweets – also known as "binning the data." A bin is a collection of data points. The goal is to create a visualization that is informative without overcrowding, or with many empty bins. Some finesse is usually required for this step. In the scope of Twitter data, one may choose been tweets at the monthly, daily, or even hourly level.

Sentiment analysis is becoming a useful tool for exploring the types of emotionally charged words appearing in written expression. Hedonometrics, developed by Dodds et al. [15, 16] explore how the positive or negative words differ between subsets of text. This group created frequency word distributions in several languages of the most frequently appearing 10000 words compiled from multiple sources (including Twitter, and the Google Web Crawl). Each of the words was then evaluated by 50 native speakers on a "happiness scale" that was numerically scaled between one and nine. The scores were averaged and compiled into a happiness database known as LabMT. Positive words had an average happiness score (h_{avg}), greater than six, like "love" (8.42), "pancakes" (6.96), "haha" (7.64). Negative words are words that have scores lower than four like "sad" (2.38), "bad" (2.64), "pain" (2.10).

Using the happiness scores from LabMT, the average emotional rating of a corpus is calculated by tallying the appearance of words found in the intersection of the wordlist and a given corpus, in this case subsets of tweets. A weighted arithmetic mean of each word's frequency, f_w, and corresponding happiness score, h_w for each of the N words in a text yields the average happiness score for the corpus, $\bar{h}_{text}$:

$$\bar{h}_{text} = \frac{\sum\limits_{W=1}^{N} f_w \cdot h_w}{\sum\limits_{W=1}^{N} f_w}$$

For exploratory analyses, a time series of frequency counts, computed happiness values, or other statistics can be useful in identifying potential trends in the data. Extreme spikes or dips in the time series, or outliers, can be indicative of a significant event or even a potential source of noise, i.e., misrepresented data. In Figure 2 the tobacco tweet dataset that was previously mentioned visualizes frequency and happiness both as a function of time. Notably there are two distinct mini spikes on the frequency time-series on June 29 in both 2012 and 2013. Although these two dates had the highest daily frequency of tweets (5319 in 2012 and 5326 in 2013), these values are not too substantially different from the daily fluctuation. It is however suspicious that this peak has reoccurred at the same date over both years. Trends like these could warrant further investigation.

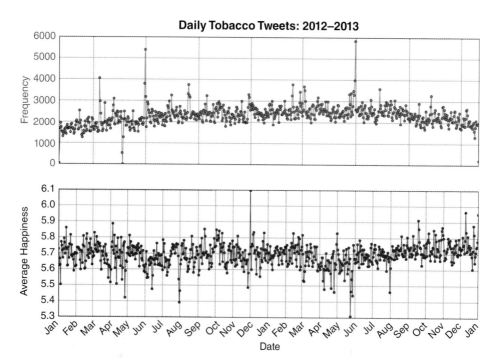

Figure 2. Tweets spanning 2012 through 2013 with tobacco-related keywords were collected and plotted by their daily frequencies and happiness averages. Tobacco-related tweets are slightly more negative when compared to the entire twitter sphere.

We see the "average happiness time-series" (blue) has relatively smooth fluctuations. High-frequency neutral words, like articles and prepositions (of, the, etc.) can significantly distort potential trends in the time series. It is also important to remove the key words from the initial search from the calculation, in order to minimize selection bias. For instance, the word cigarette has a negative score (3.09) in the happiness database, so if this word were included it would be a significant contributor to a negative shift in computed word happiness. For this happiness calculation, neutral words (also known as stop-words) were removed from the analysis by excluding words with average happiness scores between four and six. This bolsters the signal of the extremely emotional content of the tweets. There is an extreme peak in December, which can be attributed to positive tweets mentioning the holiday season. A notable dip in the frequency time series occurs in May 2012, where zero relevant tweets were collected for one day. This is most likely a data collection error (perhaps a power outage). Hence a time-series analysis could help understand potential errors in the data collection process as well as a guiding step in generating new hypotheses to investigate.

Now we've visualized how the frequency and computed average happiness are changing as a function of time. A background trend that is influencing the previous time series is the general weekly cycle of human emotion and interaction with Twitter [17].

Now, we may want to understand the daily cycle of sentiments and tobacco related frequencies that are hidden in the data. At the hourly resolution, converting the tweet to its local time of authorship is extremely important. In Figures 3a and b,

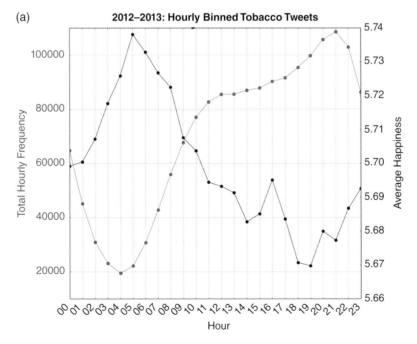

Figure 3. (a) (Left) Average number of tobacco related tweets captured per local hour of occurrence spanning 2012–2013. (b) (right) Wordshift graph illustrating the words influencing a shift in positivity from 10:00 in comparison to 3:00. (See insert for color representation.)

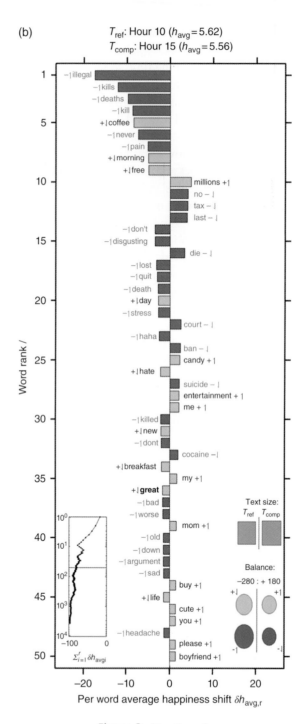

Figure 3. (*Continued*)

tobacco-related tweets are rebinned based upon their local hour of authorship (0–23), spanning 2012 through 2013, along with the average computed happiness of each bin plotted on a duel axis. Now we can see a negative correlation developing between the number of tobacco related tweets and their average computed word happiness. The next step is to try to investigate what is potentially causing this relationship.

Wordshift graphs, developed by Dodds et al. [18, 19], are a tool for quantitatively visualizing the specific words that are influencing the computed average happiness score between two subsets of text. A reference period T_{ref.}, creates a basis of the emotional words being used to compare with another period, T_{comp}. The top 50 words responsible for a happiness shift between the two periods are displayed, along with their contribution to shifting the average happiness of the tweet set. The arrows next to a word indicate an increase or decrease, respectively, of the word's frequency during the comparison period with respect to the reference period. The addition and subtraction signs indicate if the word contributes positively or negatively, respectively, to the average happiness score.

In this scope, wordshift graphs are created to compare the happiness scores of each hour of the day and illustrate the emotionally charged words contributing to the change in happiness. Here, 14:00 ($h_{avg} = 5.69$) is compared to 8:00 ($h_{avg} = 5.73$). An increase in the negative words "hate," "disgusting," "never," "gross," and a decrease in the positive words "coffee" and "morning" (among others) are the culprits for the negative shift. This happiness relationship may be more indicative of the natural "daily cycle" of written emotion as described by Dodds et al. [20] The negative tobacco related lingo may also drive this shift, but it is important to recognize this natural fluctuation of computed sentiments throughout the day. Prior trends in data could lead to misleading results, hence it is extremely important to heavily analyze these phenomena in order to avoid a misinterpretation of an apparent relationship.

We've investigated the temporal for tweets spanning two years of Twitter data from the larger 10% sample of Twitter's streaming API. Now, we turn our focus to the geo-spatial component of this analysis. Approximately 1% of all tweets report the geo-location to within 10 meters of accuracy of the user. Using the Geo-tagged Database of tweets, we can investigate geographical trends in the data set. For this analysis, we focus on English tweets authored in the United States, and compare the differences in frequency and happiness at the state level. The first step is to visualize the number of relevant tweets authored per state. Comparing a state with minimal Twitter activity to others could lead to results that misrepresentative of the true population.

In Figure 4, the number of tobacco-related tweets collected per state is visualized with a histogram. Here we see significant variation of frequencies per state. The Center of Disease Control (CDC) 2012 estimates of smoking rates per state were then collected and correlated to the tobacco-tweet density of states with a substantial Twitter representation. There is notably a significant correlation (0.580,

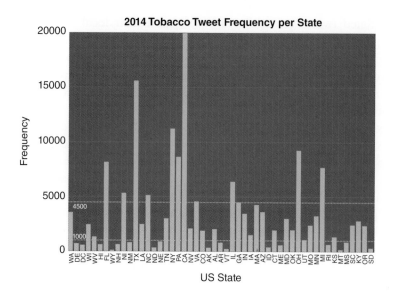

Tobacco Tweet Density vs 2012 Smoking Rate

Type	Rho	P value	# States	Min Tweets
Adult	0.580	0.048	12	4500
Youth	Insig.	NA	NA	NA

2012 Smoking Rate vs. Tweet Happiness

Type	Rho	P value	# States	Min Tweets
Adult	−0.348	0.044	34	1000
Youth	−0.465	0.015	27	1000

Figure 4. (Above) Tables show the correlations between the happiness of tweets, relative tobacco tweet density, and smoking rate per state (provided by the CDC 2012 estimates). There is a high Spearman Correlation between the adult smoking rate and density of tobacco related tweets in states with at least 4500 tweets (N = 12). Using states with a large number of tweets opens up the possibility to infer rates of tobacco use through the relative twitter activity. There is a high negative Spearman Correlation (second table) between both the adult and youth smoking rates versus the tobacco tweet average happiness for states with at least 1000 tweets (N = 27).

$p < 0.05$) of states mentioning higher densities of tobacco key terms and that have at least 4500 tweets in our dataset, which comprised 12 US states. A significant negative correlation is found between the adult and youth smoking rates versus the average happiness of the state's collected tobacco tweets.

Heat maps are graphical tools that use color coding to help visualize the flow of numerical data in two dimensions. Geographical heat maps are created by binning values into regions and coloring the regions based on a numerical scale. You have the freedom to determine an appropriate color scale for the situation, but it's

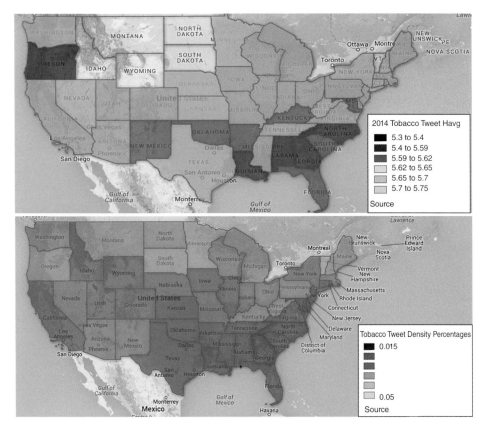

Figure 5. Approximately 40 000 geo-tagged tobacco related tweets spanning March 2014–May 2014. (Top) The density of tobacco tweets is computed by finding the ratio of tobacco related tweets to the total number of geo-located tweets per state appearing in the geo-stamped subset of twitter (~1% of all tweets). (Bottom) The states are recolored based on their average happiness scores of their tobacco-related tweets, uncolored states have less than 500 tweets. (See insert for color representation.)

important to create a scale that gives a truthful representation of the flow of data and how it changes at the regional level. The geographical bins are another variable that should be manipulated appropriately.

In Figure 5, a happiness heat map of the 50 States helps visualize the happiness and frequencies of tobacco related tweets at the regional level. The heat maps were generated with Google Fusion Tables, which are free to use and have a user-friendly interface. Custom geographical heat maps can also be created in many programming languages like, Python, R, and Matlab. States with less than 500 tweets were removed from the hedonometric analysis (clear states on the map). The first map displays the density (i.e., #TobaccoTweets/#StateTweets) of tobacco-related tweets per state. These densities are quite varied, however, it is notable that high population areas are not necessarily the driving factor for a high density of relevant tweets.

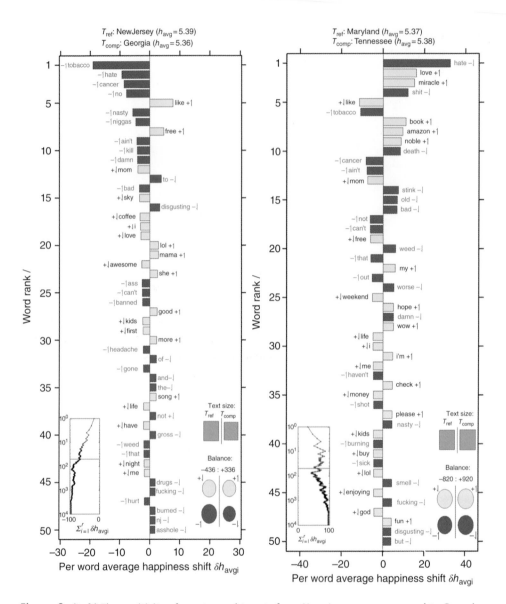

Figure 6. (Left) The positivity of geo-tagged tweets from New Jersey are compared to Georgia. A slightly negative shift in positivity is due to an increase in words like "smoke," "headache," and "cancer," as indicated by the up arrow (more frequent) and minus sign (negative sentiment). (Right) Maryland Tobacco Tweets are compared to Tennessee. A slightly positive shift occurs due to an increase of "love" and decrease of "hate." (See insert for color representation.)

Now, we can use wordshift graphs to compare the words influencing positive or negative shifts between states. We conclude our analysis with a geographically binned wordshift graphs in Figure 6. On the left, tobacco tweets from Georgia are compared to tweets from New Jersey, and on the right tweets from Tennessee are compared to Maryland. Using these tools we can better understand the sentiments surrounding tobacco use and how they compare at both the regional and temporal scopes.

ANALYSIS OF GEO-TAGGED TWEETS MENTIONING CANCER

In a similar analysis to tobacco-related tweets, a large sample of cancer tweets authored in English from March 2014 through December 2014 were obtained from Twitter's API. Pattern matching using "cancer" as a keyword returned 186 406 tweets. As already mentioned, the tweets were cleaned to remove unwanted content, and a total of 25 103 tweets were removed from the data set, resulting in a clean data set of 146 357 tweets from the United States. Using the National Cancer Institute (NCI) data for the highest incidence cancers [20], search terms were used to extract tweets pertinent for each unique cancer diagnosis. The most frequently tweeted cancers were breast (n = 15 421), lung (n = 2928), prostate (n = 1036), and colon/rectal (n = 773) (Table 1).

The fact that breast cancer was the top-tweeted cancer was not surprising, considering breast cancer is one of the most prevalent cancer types [20], the large public awareness surrounding the disease, and the highly publicized and endorsed October breast cancer awareness month. The national incidence of lung, prostate, and colorectal cancer is most likely reflected in the relative frequency of tweets pertaining to these diseases. One can also note that patients were more likely to be identified depending on the cancer type. For instance, only 1% of all breast cancer tweets were patient tweets, most likely representing the fact that nondiagnosed

TABLE 1. For each National Cancer Institute (NCI) cancer diagnosis, filter terms were used to extract cancer-specific tweets. These tweets were then manually filtered to select patients.

NCI Diagnosis	Filter terms	Tweets (n)/fraction of total[a] (%)	Patients (n)/fraction of diagnosis tweets (%)
Breast	"breast cancer"	15 421/70.9	161/1.0
Lung	"lung cancer"	2928/13.5	37/1.3
Prostate	"prostate cancer"	1036/4.8	15/1.4
Colon, colorectal, rectal	"colon cancer," "colorectal", "rectal cancer"	773/3.6	24/3.1
Pancreas	"pancreas cancer," "pancreatic cancer"	673/3.1	6/0.9
Thyroid	"thyroid cancer"	195/0.9	23/11.8
Lymphoma	"lymphoma"	180/0.8	22/12.2
Leukemia	"leukemia"	177/0.8	10/5.6
Melanoma	"melanoma"	141/0.6	14/9.9
Bladder	"bladder cancer"	93/0.4	3/3.2
Kidney	"clear cell," "kidney cancer," "renal cell carcinoma," "renal cancer"	83/0.4	8/9.6
Endometrial	"uterine cancer," "endometrial cancer"	43/0.2	10/23.3

[a]Total number of diagnosis specific tweets = 21 743.

individuals are tweeting about the disease for fundraising purposes or sharing feelings about a loved one with the disease. This is in contrast to endometrial cancer, where 23% of the disease tweets were patients, potentially representing that either patients with endometrial cancer are more likely to tweet than other cancer patients, or there is less public awareness surrounding the diagnosis and thus there are fewer non-patient tweets. Overall, non-patient tweets for all cancer diagnosis were largely individuals sharing about a loved one with the diagnosis. Further investigation into this area of patient tweets with respect to cancer type is warranted, particularly with cancer types that are more likely to affect younger patients, who may be more likely to tweet about their disease. One can analyze the cancer-tweet data in a similar fashion to tobacco-tweets with a hedonometric analysis. Here, the anonymous Twitter ID number associated with each extracted patient tweet allows for collection of all tweets (not just cancer-related tweets) for each user over the same time period. This "patient tweet-set" was then quantitatively categorized with the hedonometric analysis. The average computed happiness value was greatest for thyroid (h_{avg} =6.1625), breast (h_{avg} =6.1485), and lymphoma (h_{avg} =6.0977) cancers, whereas the average computed sentiment score was lowest for pancreatic (h_{avg} =5.8766), lung (h_{avg} =5.8733), and kidney (h_{avg} =5.8464) cancers (Table 2). The happiness value for each respective cancer diagnosis was driven by specific wordshifts inherent to that diagnosis, when that cancer diagnosis was compared to all others (Table 2 and Figure 7).

Since the happiness values were computed using all tweets from the patients, not just cancer-related tweets, it is apparent that the cancer diagnosis permeates through patients' lives and reinforces the disease-illness dichotomy [21]. This idea represents the notion that the "disease" is the pathological diagnosis, whereas the "illness" refers to how each individual responds, copes, and manages the disease in daily life. The fact that variations are observed between cancer types, with thyroid cancer patients having the highest average happiness value and kidney cancer patients having the lowest happiness value (Table 2), demonstrates that the natural history inherent to the cancer type may affect patients in unique ways. The differences in happiness may perhaps be explained by the prognosis of the cancer type, as thyroid cancer (97.8% five-year survival) and breast cancer (89.2% five-year survival) are generally amenable to treatment with favorable outcomes, whereas pancreatic (6.7% five-year survival) and lung (16.8% five-year survival) cancer carry a worse prognosis. However, the low happiness value of kidney cancer does not conform to the trend (72% five-year survival) [20] and the value may represent morbidity associated with the treatment or the natural history of the disease. For instance, kidney cancer patients more often use the negative words "surgery" and "hospital." Likewise, endometrial cancer patients use the terms "surgery" and "pain" more frequently. Through the differences in computed happiness values, it appears that the specific cancer diagnosis, not just the mere fact of having "cancer," affects patients throughout their lives and patients manifest their illness through social media. At the moment, these results are correlative, and whether certain cancer patients are indeed happier than others requires more thorough investigation.

TABLE 2. Computed average word happiness value (h_{avg}) for each cancer diagnosis and summary of major word shifts responsible for sentiment value.

Cancer type	Tweetset (n)	h_{avg}	Increased frequency words	Decreased frequency words
Thyroid	5673	6.1625	**"blessed," "thank," "Christmas," "love"**	*expletives, "not," "no," "lost," "die"*
Breast	72528	6.1485	**"happy," "love," "welcome,"**	*expletives, "hate," "never"*
Lymphoma	5143	6.0977	**"god," "win," "photo," "proud," "miss"**	*"not," "don't,"* **"happy"**
Endometrial	4939	6.0913	**"love,"** *"sorry," "surgery," "pain"*	*expletives, "hate," "don't"*
Bladder	1579	6.0843	**"good," "great," "win"**	**"love,"** *"don't," "hate"*
Melanoma	13418	6.0611	**"love,"** *"bloody," "hell"*	**"happy," "great," "good"**
Prostate	16161	6.0223	**"good," "great," "nice,"** *expletives*	**"love," "happy"**
Colorectal	9682	6.0149	**"lol," "good,"** *"not," "no," "hell"*	**"happy," "love," "beautiful," "welcome"**
Leukemia	6042	5.9730	*"smoke," "hate," "bored,"* **"haha"**	**"happy," "beautiful"**
Pancreas	5117	5.8766	*expletives, "don't," "bad"*	**"happy," "great," "thanks"**
Lung	38379	5.8733	*expletives, "don't," "hate," "mean"*	**"love," "happy," "great," "thanks"**
Kidney	7245	5.8464	*expletives, "don't," "hospital," "surgery"*	**"happy," "lol," "thank"**

Positive sentiment words are displayed in **bold**, whereas negative sentiment words are displayed in *italics*. In general, as the h_{avg} increases, the data set contains increased frequency of positive words and decreased frequency of negative words.

587

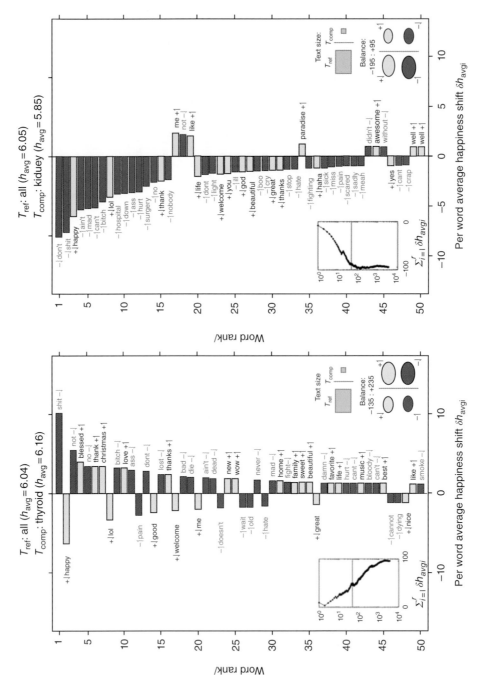

Figure 7. Wordshift graphs for the happiest cancer patient tweet-set and the saddest cancer patient tweet-set. (See insert for color representation.)

FUTURE SOCIETAL IMPLICATIONS

It is imperative, from a societal perspective, that we continue to study the implications and value of sentiments that are inferred across publicly accessible networks. As these sentiments can eventually lead to treatment changes, so they may also lead to changes in reimbursement policy and practice. Could social sentiments sway medical consensus? Could a person be held accountable for receiving the treatment they revealed to prefer on social media, if that treatment is more expensive than the standard of care? Could insurance companies discriminate against such persons as holding a revealed potential for being "high-cost" patients? These and numerous other issues will continue to need to be debated in open dialog with patient advocacy groups, academics, industry voices, and policy makers. As digital health solutions that are newly regulated by the US Food and Drug Administration (US-FDA) are predicted to save up to $100 billion by 2018 [22], an economic outlook would caution against moral hazards such as increased awareness of available tools combined with required coverage, and increased fear around pathology that could lead to more, rather than less, testing and treatments. The future of leveraging social media to infer sentiments seems to offer a unique opportunity to evaluate treatments and conditions in small snippets of revealed language. Much like the revelations from RNA splicing, Twitter sentiments contain a lot of useful and useless words that are variously truncated, misspelled, used as humor, offered as invitation for further dialog, or used in proper context to express social sentiments around important health conditions, among a network of followers. With the right combination of rapidity and relevance, actioning upon the correct combination of health sentiments with more efficient treatment and healing could help bend our ever rising cost curves.

REFERENCES

1. About. Twitter. [Online] 2015. Accessed March 2, 2015. https://about.twitter.com/company.
2. Duggan M, Ellison NB, Lampe C, et al. "Social Media Update 2014." Pew Research Center, January 2015.
3. Mitchell, L., Frank, M.R., Harris, K.D. et al. (2013). The geography of happiness: connecting Twitter sentiment and expression, demographics, and objective characteristics of place. *PLoS One* 8 (5): e64417.
4. Lampos V, N Cristianini. "Tracking the flu pandemic by monitoring the social web." *Cognitive Information Processing (CIP), 2010 2nd International Workshop on*. IEEE, 2010.
5. Signorini, A., Segre, A.M., and Polgreen, P.M. (2011). The use of Twitter to track levels of disease activity and public concern in the US during the influenza A H1N1 pandemic. *PLoS One* 6 (5): e19467.
6. Goff, D.A., Kullar, R., and Newland, J.G. (2015). Review of Twitter for infectious diseases clinicians: useful or a waste of time? *Clinical Infectious Diseases* 60 (10): 1533–1540.
7. Paul MJ, M Dredze. "You are what you Tweet: analyzing Twitter for public health." *ICWSM*. 2011. https://www.aaai.org/ocs/index.php/ICWSM/ICWSM11/paper/view/2880/3264

8. De la Torre-Díez, I., Díaz-Pernas, F.J., and Antón-Rodríguez, M. (2012). A content analysis of chronic diseases social groups on Facebook and Twitter. *Telemedicine and e-Health* 18 (6): 404–408.

9. Thackeray, R., Burton, S.H., Giraud-Carrier, C. et al. (2013). Using Twitter for breast cancer prevention: an analysis of breast cancer awareness month. *BMC Cancer* 13 (1): 508.

10. Lyles, C.R., López, A., Pasick, R. et al. (2013). 5 mins of uncomfyness is better than dealing with cancer 4 a lifetime: an exploratory qualitative analysis of cervical and breast cancer screening dialogue on Twitter. *Journal of Cancer Education* 28 (1): 127–133.

11. Sugawara, Y., Narimatsu, H., Hozawa, A. et al. (2012). Cancer patients on Twitter: a novel patient community on social media. *BMC Research Notes* 5 (1): 699.

12. Clark EM, Williams JR, Jones CA, et al. "Sifting Robotic from Organic Text: A Natueral Language Approach for Detecting Automation on Twitter." arXiv preprint arXiv:1505.04342 (2015).

13. Clark EM, Jones CA, Williams JR, et al. (2015). Vaporous Marketing: Uncovering Pervasive Electronic Cigarette Advertisements on Twitter. *arXiv preprint arXiv:1508.01843*.

14. Minna, J.D., Roth, J.A., and Gazdar, A.F. (2002). Focus on lung cancer. *Cancer Cell* 1 (1): 49–52.

15. Dodds, P.S., Harris, K.D., Kloumann, I.M. et al. (2011). Temporal patterns of happiness and information in a global social network: hedonometrics and Twitter. *PLoS One* 6 (12): e26752.

16. Dodds, P.S., Clark, E.M., Desu, S. et al. (2015). Human language reveals a universal positivity bias. *Proceedings of the National Academy of Sciences of the United States of America* 112 (8): 2389–2394.

17. Dodds, P.S., Harris, K.D., Kloumann, I.M. et al. (2011). Temporal patterns of happiness and information in a global social network: hedonometrics and Twitter. *PLoS One* 6 (12): e26752.

18. Dodds, P.S., Harris, K.D., Kloumann, I.M. et al. (2011). Temporal patterns of happiness and information in a global social network: Hedonometrics and twitter. *PloS one* 6 (12): e26752.

19. Dodds, P.S., Clark, E.M., Desu, S. et al. (2015). Human language reveals a universal positivity bias. *Proceedings of the National Academy of Sciences* 112 (8): 2389–2394.

20. American Cancer Society (2015). *Cancer Facts & Figures 2015*. Atlanta: American Cancer Society.

21. Helman, C.G. (1981). Disease versus illness in general practice. *The Journal of the Royal College of General Practitioners* 31 (230): 548–552.

22. US-FDA Accenture Report, June 2015.

IX

THE PATH TO REDUCING THE CANCER BURDEN: PREVENTION, EARLY DETECTION, RECOVERY, SURVIVORSHIP

32

CANCER GENETICS: RISK ASSESSMENT

Wendy McKinnon

University of Vermont Cancer Center, Burlington, VT, USA

INTRODUCTION

Approximately one in two men and one in three women will develop cancer in their lifetime. It is estimated that 5–10% of most common cancers are due to inherited mutations in cancer susceptibility genes. Inheritance of such mutations increases the lifetime risk for developing cancer dramatically over the general population risk. Recognizing which individuals and families may be in this higher risk group is imperative in reducing cancer burden. The goal of cancer risk assessment is to identify individuals who are at elevated risk based on personal and/or family history or the results of genetic testing and to provide screening and prevention strategies to identify cancer early or prevent cancer altogether.

Evidence that there is benefit to identifying individuals with inherited cancer risk and reducing cancer burden is emerging. For example, the Evaluation of Genomic Applications in Practice and Prevention (EGAPP) committee found sufficient evidence that universal screening of all individuals with newly diagnosed colorectal cancer for the most common form of inherited colon cancer (Lynch syndrome) reduces morbidity and mortality in relatives [1]. Many institutions now perform universal Lynch screening for colorectal cancers. Identifying at risk

Cancer: Prevention, Early Detection, Treatment and Recovery, Second Edition. Edited by Gary S. Stein and Kimberly P. Luebbers.
© 2019 John Wiley & Sons, Inc. Published 2019 by John Wiley & Sons, Inc.

relatives allows early detection and prevention, as it has been shown that frequent colonoscopic examinations improve mortality in individuals with Lynch syndrome [2]. Similarly, it has been shown that women who carry a BRCA1 or BRCA2 gene alteration who undergo prophylactic surgery (removal of at-risk breast and/or ovarian tissue) have a survival advantage [3]. These are just a couple of examples of the importance of identifying individuals at high risk for cancer to reduce cancer burden.

This chapter will walk through the process of cancer risk assessment from beginning to end. For additional references regarding cancer risk assessment, the National Society of Genetic Counselors (NSGC) has published guidelines for cancer risk assessment, which includes the essential elements of this process [4]. Guidelines also exist specifically for hereditary breast and ovarian cancer syndrome and Lynch syndrome [5, 6].

REFERRAL – WHO IS A GOOD CANDIDATE FOR CANCER RISK ASSESSMENT?

The first step in cancer risk assessment is identifying who is a candidate to undergo such an evaluation. The features of inherited cancer include earlier age of onset than seen in the general population, multiple primary cancers in the same individual, rare cancers, specific histologies, and a positive family history of the same or related cancers on the same side of the family (see Table 1).

Not knowing the criteria for who is a candidate for a cancer risk assessment is a major barrier to referral of appropriate individuals. The American College of Medical Genetics (ACMG) recently published an excellent practice guideline outlining referral indications for cancer predisposition assessment [7]. The purpose of the guideline is to offer a comprehensive list of tumors that, when seen in a single individual or in family members, can be associated with an inherited cancer syndrome thus, facilitating referral for cancer genetic counseling. For example, some clinicians may recognize that a patient with a history of colon cancer or breast cancer diagnosed under that age of 50 would be a candidate to consider genetic counseling and testing. However, many clinicians may not realize

TABLE 1. Features of inherited cancers.

Features of inherited cancers
• Early age of diagnosis relative to general population (i.e., under age 50 for breast or colon cancer)
• Positive family history of same or related cancers (i.e., breast and ovarian cancer, colon and uterine cancer on the same side of the family)
• Individuals with multiple primary cancers
• Rare cancers (i.e., male breast cancer, paraganglioma)
• Certain histologies (i.e., triple negative breast cancer, high grade serous ovarian cancer)

that a woman with breast cancer of lobular histology and a relative with diffuse gastric cancer is a candidate for such an evaluation. This combination of features could be associated with a mutation in the CDH1 gene, predisposing to both diffuse gastric cancer and lobular breast cancer. The ACMG guideline is a quick and easy reference for clinicians when faced with a patient with a personal or family history of cancer to determine if a referral to a cancer genetics risk assessment program is indicated.

COLLECTION OF FAMILY HISTORY INFORMATION

To provide a cancer risk assessment, accurate and thorough collection of family history including at least three generations, including maternal and paternal lineages, is essential in accurately assessing risk. Having multiple generations helps ascertain a pattern of tumor development and likelihood of an inherited syndrome. Providers taking family histories should attempt to use standardized pedigree nomenclature [8], so that the family history information is understandable to any provider. The American Society of Clinical Oncology has published a statement regarding their recommendations for what constitutes the minimum family history that should be gathered in the oncology setting [9].

Sending questionnaires ahead of time to gather information about referred individuals' personal health history as well as the health of his/her relatives, focusing on tumor history (both benign and malignant) can help facilitate the intake process. Many cancer genetic programs use paper questionnaires, but some programs are transitioning to web-based questionnaires, which may also translate this information directly into a pedigree. Some of these programs will also provide a cancer risk assessment based on the information entered. This information can then be reviewed together with the individual at their appointment.

There are also web-based family history programs available that individuals can access themselves that facilitate collection of family history information. The surgeon general also has a family history collection tool referred to as My Family Health Portrait (https://familyhistory.hhs.gov). Using My Family Health Portrait, you can enter your family health history, learn about your risk for conditions that can run in families, print your family health history to share with family or your healthcare provider. You can also save your family health history and update it over time. Invitae is a commercial testing company and they include a family history tool on their website (www.invitae.com).

While some families have an obvious pattern, others are not as clear. Factors that can impede assessment of family history include adoption, small family size, gender distribution (especially if assessing for gender-related cancers), early deaths, and lack of knowledge of relatives' health history due to variety of reasons [10].

GENETIC COUNSELING AND TESTING – DISCUSSION OF RISKS/BENEFITS/LIMITATIONS OF GENETIC TECHNOLOGIES

When the history is suggestive of an inherited cancer syndrome, referral to a trained cancer genetics provider for genetic counseling and discussion of testing options is recommended. Genetic counseling is a critical part of the genetic testing process to help individuals make informed decisions and that they are provided with appropriate supports when needed. When genetic testing is performed by providers who are not trained in genetics, there is potential for harm [11]. The authors identified three themes in their case series of genetic testing performed without involving a certified genetic counselor. These themes include: wrong genetic tests ordered, genetic test results misinterpreted, and inadequate genetic counseling. Patient morbidity and mortality were an issue in several of these cases.

There are several ways to provide genetic counseling to at risk individuals. The standard, traditional mode is face-to-face genetic counseling. However, with increasing demand and lack of access to a trained genetics professional, other modes are being instituted. Telemedicine consults are increasingly being used in rural areas to provide genetic counseling without having to travel to the medical center. Research has shown high satisfaction from individuals receiving care in this format. Some states have passed legislation mandating insurance companies to reimburse for telemedicine consults at the same rate as face-to-face consultations. Telephone counseling is also on the rise and there are companies hiring genetic counselors to provide telephone genetic counseling to individuals who do not have access to a local genetics providers or the telemedicine option. A recent study comparing standard face-to-face genetic counseling to telephone genetic counseling for women considering BRCA testing showed that genetic counseling can be effectively and efficiently delivered via telephone and increases access and decreases costs [12]. Insurance companies are beginning to utilize telephone counseling services to increase appropriate referrals for genetic testing.

Key Principles in Genetic Testing

It is important, whenever possible, to initiate genetic testing with an individual affected with malignancy. In particular, testing the person diagnosed at the youngest age, or with the most severe phenotype based on pathology, will provide the most informative results. If the affected relative is found to carry a specific gene alteration, then other relatives can have site-specific testing for that alteration and it's a "yes" or "no" result. If the affected relative is found to NOT carry any genetic alterations, then no genetic testing is recommended for the unaffected relatives. Their cancer risk should be based on the family history, not on the results of genetic testing. While it is recognized that it is not always possible to test affected relatives (i.e., deceased, cannot participate due to illness, dementia, or lack of interest), it is imperative that the unaffected individual proceeding with testing understand the limitations of a negative result. Testing unaffected individuals can lead to uninformative results. This is not a true negative result, but an inconclusive result, and cancer risk is based on family history. A major concern with testing unaffected individuals, without testing an affected relative first, is that a negative result will be falsely reassuring.

Possible Results from Genetic Testing

- **Positive** for a pathogenic (disease causing) mutation.
- **Negative** for any identifiable mutations:
  - If the first person in the family to undergo testing is negative, this is an inconclusive result and risk for cancer is based on family history;
  - If the first person in the family is found to carry a pathogenic mutation and a relative tests negative for this mutation, this is a true negative result.
- **Variant of uncertain significance** (VUS) refers to a result in which a DNA change is identified but there is not enough data/evidence to know if this variant increases cancer risk or is a benign polymorphism.
- Testing of relatives should only be performed if a pathogenic variant is identified in another relative. Testing relatives for a VUS is not recommended unless this is done as part of a research endeavor.

Many insurance companies cover the cost of genetic testing based on clinical criteria and medical necessity. The National Comprehensive Cancer Network (NCCN), as well as Medicare has established criteria for genetic testing for inherited breast and colon cancer. Many insurance companies have adopted these criteria for coverage (http://www.nccn.org/professionals/physician_gls/pdf/genetics_screening.pdf; http://www.cms.gov/medicare-coverage-database/details/lcd-details.aspx?LCDId=24308&ContrId=348&ver=76).

PROTECTIONS AGAINST GENETIC DISCRIMINATION

In May 2008, the President signed into law the Genetic Information Nondiscrimination Act (GINA) that protects Americans against discrimination based on their genetic information when it comes to health insurance and employment. The bill had passed the Senate unanimously and the House by a vote of 414 to 1. The long-awaited measure, which had been debated in Congress for 13 years, paved the way for people to take full advantage of the promise of personalized medicine without fear of discrimination. A fact sheet regarding GINA is available at: http://www.genome.gov/Pages/PolicyEthics/GeneticDiscrimination/GINAInfoDoc.pdf

It is important to recognize that there is no legislation to protect against discrimination with regards to life insurance, disability insurance, or long-term care.

NEW TECHNOLOGIES

Prior to 2012, genetic testing was typically performed gene by gene. For individuals and families with overlapping phenotypes, this process was expensive and time consuming. With the advent of next generation sequencing, also known as massive parallel sequencing, came the possibility to test for many genes simultaneously, with less cost, and less time involved. There are pros and cons to consider when offering multigene panel testing. These include possible identification of variants in genes that are not well described in terms of cancer risk and/or lack medical management guidelines. Additionally, the incidence of finding variants of uncertain significance increases significantly with this technology, as many of the genes being tested are rare and not well studied. Multigene panel testing raises the potential for greater complexity and uncertainty. It is imperative that such testing be done in collaboration with a cancer genetics specialist (see Table 2) [13].

The PROMPT (Prospective Registry of Multi-Plex Testing) registry is a collaborative effort across academic institutions and commercial labs to answer important questions about new genetic tests. It is a patient-driven registry and individuals who receive results from panel testing are encouraged to enter their results and their family history information into the registry to increase data collection and to help increase our knowledge about these rarer cancer susceptibility genes.

MEDICAL MANAGEMENT

If, after analyzing an individual's personal and family history, there is not an indication to proceed with genetic testing (i.e., low likelihood of finding a gene alteration), cancer risk can be calculated using empiric data and/or available models. For example, a woman, whose mother was diagnosed with breast cancer at age 47 who has since passed away and has no other family history to suggest an inherited syndrome, has a risk that is slightly above that of women in the general population. Models to help calculate her risk include the Gail Model, the Claus Model, the Tyer-Cusik Model, and a model called Boadicea. No model is perfect in assessing risk, but can be helpful in providing an estimate and can help guide appropriate screening. Women who score greater than 20–25% on the later three of these models are candidates for having breast screening MRI annually in addition to

TABLE 2. How to find a cancer genetics specialist.

Cancer genetics specialists
National Society of Genetic Counselors (NSGC): http://www.nsgc.org
National Cancer Institute (NCI): http://www.cancer.gov/cancertopics/genetics/directory
American College of Medical Genetics (ACMG): http://www.ACMG.net
Informed DNA: http://informeddna.com

mammography according to the American Cancer Society [14]. These models can also be used for women after receipt of a negative genetic test result, assuming a specific gene alteration has not been identified in another relative.

For those individuals found to carry a specific pathogenic alteration screening based on published guidelines is recommended. If no guidelines exist, then the genetics provider will use expert opinion, in addition to the family history to guide screening. While evidence-based guidelines exist for many cancer susceptibility genes, there remains a lack of guidelines for many of the genes now included on multigene panels. The NCCN have published guidelines for many inherited cancer syndrome (http://www.nccn.org/professionals/physician_gls/f_guidelines.asp#detection).

PSYCHOSOCIAL ISSUES

Psychosocial assessment is critical and should be part of both the pretest and post-test genetic counseling process. During the course of gathering personal and family history information, it is important to inquire about psychosocial issues as well.

Psychosocial issues to address include cancer worry, anxiety, intrusive thoughts (how often do you think about cancer), depression, anger, fear, guilt, experiences with cancer in self, family and friends, perception of risk for self and others, competence for giving informed consent, social stressors and supports, family communication, and readiness for testing. This information may provide clues about how an individual and their family may interpret and cope with genetic information.

Studies have shown that individuals can derive significant psychological benefits from genetic testing information. Living with uncertainty regarding one's own cancer risk in a family full of relatives with cancer can be overwhelming. Learning one's risk more definitely can be empowering and allow one to take action rather than living with the unknown. While minimal adverse effects have been observed among most people with positive test results, a select subset of people appear to be more vulnerable to testing distress and require more professional assistance. Many cancer risk assessment programs have access to psychological services if additional counseling is needed to adjust to genetic risk. Additionally, if available, referral to support groups may be helpful. Other sources of support may include peer support, internet-based support organizations, and patient-focused gatherings focused on hereditary cancer (see Table 3).

T A B L E 3. Examples of support organizations for inherited cancer.

Support organizations
FORCE (Facing our Risk of Cancer Empowered) – http://www.facingourrisk.org
Bright Pink – http://www.brightpink.org
Lynch Syndrome International – www. http://lynchcancers.com
Familial Adenomatous Polyposis (FAP) – http://www.fapgene.com
Li-Fraumeni Association – http://www.lfsassociation.org

At the University of Vermont Cancer Center we have offered a series of retreats for individuals and their families identified as carrying a pathogenic mutation in either the BRCA genes associated with hereditary breast and ovarian cancer syndrome or the mismatch repair genes associated with Lynch syndrome. With grant money, we have offered a full-day event with invited expert speakers, to update families on medical management recommendations, as well as address psychosocial issues of being at increased cancer risk. We published our experience in 2007 in the *Journal of Genetic Counseling*. We showed that individuals were more likely to adhere to recommended screening and prevention strategies and feel more empowered about their situation after attendance [15].

REPRODUCTIVE OPTIONS

The majority of inherited cancer syndromes are inherited in an autosomal dominant pattern. This means that any child of an individual who carries a pathogenic change in a cancer susceptibility gene has a 50% chance of having inherited this gene change. For some individuals, if given the opportunity, might chose to avoid the possibility of passing on a gene change associated with a significant risk for cancer development to their future children.

Pre-implantation genetic diagnosis (PGD) is a technique used to identify genetic alterations in embryos created through *in vitro* fertilization (IVF) before pregnancy. PGD refers specifically to when one parent has a known genetic abnormality and testing is performed on an embryo to determine if it also carries the genetic abnormality. Because only unaffected embryos are transferred to the uterus for implantation, PGD provides an alternative to current postconception diagnostic procedures such as amniocentesis or chorionic villus sampling, which are followed by the difficult decision of whether to terminate the pregnancy if the results are unfavorable. PGD is presently the only option available for avoiding the risk of having a child affected with a genetic disease prior to implantation.

It is understandable that not all couples at-risk for passing an inherited cancer susceptibility gene onto a child would be interested in such technology. Assisted reproductive technologies including sperm and egg donation, as well as adoption, are other ways to reduce the risk of passing an inherited risk to future offspring. For individuals of reproductive age who are carriers of an inherited cancer risk, it is recommended that they be counseled about their reproductive options so they can make an informed choice about their future families.

ROLE OF LIFESTYLE AND ENVIRONMENT IN CANCER RISK

Genetic susceptibility can play an important role in determining cancer risk, but it is not the whole story. Can one impact or overcome genetic susceptibility by modifying lifestyle and/or environment? There is evidence in the literature to suggest that exercise can reduce cancer risk and reduce risk of recurrence in breast cancer survivors.

There is a growing emphasis on wellness and incorporating a healthy lifestyle to impact overall health, in particular with regards to exercise, diet, and stress reduction. In the Women's Health Initiative, those women who followed healthy lifestyle behaviors for nutrition and physical activity was associated with lower risk of cancer and death in postmenopausal women. Those women who had the best adherence to guidelines for weight control, diet, alcohol consumption, and exercise had a 22% lower rate of breast cancer than women with the worst adherence to guidelines [16]. The NCCN has recommended counseling regarding the importance of weight control, exercise, moderating alcohol consumption, and avoidance of combination hormone replacement therapy to reduce risk (see http://NCCN.org).

SUMMARY

In conclusion, it is hoped that undergoing cancer genetic risk assessment with a trained genetic provider will help reduce distress, improve the accuracy of the perceived risk of, and increase knowledge about cancer and genetics.

REFERENCES

1. Evaluation of Genomic Applications in Practice and Prevention (EGAPP) Working Group (2009 Jan). Recommendations from the EGAPP working group: genetic testing strategies in newly diagnosed individuals with colorectal cancer aimed at reducing morbidity and mortality from Lynch syndrome in relatives. *Genet. Med.* 11 (1): 35–41.
2. Lindor, N.M., Petersen, G.M., Hadley, D.W. et al. (2006 Sep 27). Recommendations for the care of individuals with an inherited predisposition to Lynch Syndrome: a systematic review. *JAMA* 296 (12): 1507–1517.
3. Domchek, S.M., Friebel, T.M., Singer, C.F. et al. (2010 Sep 1). Association of risk-reducing surgery in BRCA1 or BRCA2 mutation carriers with cancer risk and mortality. *JAMA* 304 (9): 967–975.
4. Riley, B.D., Culver, J.O., Skrzynia, C. et al. (2012). Essential elements of genetic cancer risk assessment, counseling, and testing: updated recommendations of the National Society of Genetic Counselors. *J. Genet. Couns.* 21: 151–161.
5. Berliner, J.L., Fay, A.M., Cummings, S.A. et al. (2013). NSGC practice guideline: risk assessment and genetic counseling for hereditary breast and ovarian cancer. *J. Genet. Couns.* 22: 155–163.
6. Weissman, S.M., Burt, R., Church, J. et al. (2012 Aug). Identification of individuals at risk for Lynch syndrome using targeted evaluations and genetic testing: National Society of Genetic Counselors and the Collaborative Group of the Americas on Inherited Colorectal Cancer joint practice guideline. *J. Genet. Couns.* 21 (4): 484–493.
7. Hampel, H., Bennett, R.L., Buchanan, A. et al. (2015 Jan). A practice guideline from the American College of Medical Genetics and Genomics and the National Society of Genetic Counselors: referral indications for cancer predisposition assessment. *Genet. Med.* 17 (1): 70–87.
8. Bennett, R.L., French, K.S., Resta, R.G. et al. (2008 Oct). Standardized human pedigree nomenclature: update and assessment of the recommendations of the National Society of Genetic Counselors. *J. Genet. Couns.* 17 (5): 424–433.

9. Lu, K.H., Wood, M.E., Daniels, M. et al. (2014). American Society of Clinical Oncology Expert Statement: collection and use of a cancer family history for oncology providers. *J. Clin. Oncol.* 32 (8): 833–841.1.

10. Weitzel, J.N., Lagos, V.I., Cullinane, C.A. et al. (2007 Jun 20). Limited family structure and BRCA gene mutation status in single cases of breast cancer. *JAMA* 297 (23): 2587–2595.

11. Bonadies, D.C., Brierley, K.L., Barnett, R.E. et al. (2014 Jul-Aug). Adverse events in cancer genetic testing: the third case series. *Cancer J.* 20 (4): 246–253.

12. Schwartz, M.D., Valdimarsdottir, H.B., Peshkin, B.N. et al. (2014 Mar 1). Randomized noninfe-riority trial of telephone versus in-person genetic counseling for hereditary breast and ovarian cancer. *J. Clin. Oncol.* 32 (7): 618–626.

13. Fecteau, H., Vogel, K.J., Hanson, K. et al. (2014). The evolution of cancer risk assessment in the era of next generation sequencing. *J. Genet. Couns.* 23 (4): 633–639.

14. Saslow, D., Boetes, C., Burke, W. et al. (2007 Mar-Apr). American Cancer Society Breast Cancer Advisory Group. American Cancer Society guidelines for breast screening with MRI as an adjunct to mammography. *CA Cancer J. Clin.* 57 (2): 75–89.

15. McKinnon, W., Naud, S., Ashikaga, T. et al. (2007 Aug). Results of an intervention for individu-als and families with BRCA mutations: a model for providing medical updates and psychosocial support following genetic testing. *J. Genet. Couns.* 16 (4): 433–456.

16. Thompson, C.A., ML, M.C., Wertheim, B.C. et al. (2014). Nutrition and physical activity cancer prevention guidelines, cancer risk, and mortality in women's health initiative. *Cancer Prev. Res.* 7 (1): 42–53.

33

CANCER PREVENTION: DIET, NUTRITION, AND LIFESTYLE

Jennifer L. May[1] and Jean R. Harvey[2]

[1]*Outpatient Oncology Dietitian, The University of Vermont Cancer Center, VT, USA*

[2]*Associate Dean for Research, College of Agriculture and Life Sciences, Robert L. Bickford, Jr. Endowed Professor, Chair, Department of Nutrition and Food Sciences, University of Vermont, Vermont, VT, USA*

INTRODUCTION

We know there are some things we can't control. We know that even people who live healthy lifestyles still get cancer. However, while we can't change risk factors like our family history (which is predicted to contribute to about 5–10% of all cancers), we can modify our health behaviors to lower our cancer risk. In fact, the American Institute for Cancer Research (AICR) suggests that that lifestyle-related factors account for about 30% of all cancers in Western countries. Therefore, it is no surprise that controllable lifestyle choices are receiving a lot of attention in aiding the fight against cancer diagnosis and recurrence.

However, this issue is complex. Despite what you hear on the news or read in popular magazines, no single miracle food or wonder diet alone can prevent you from getting cancer. There can be a lot of conflicting information out there. Survivorship and cancer research is still in its early stages. Dependable, science-based advice can be hard to come by and may seem contradictory at times.

Cancer: Prevention, Early Detection, Treatment and Recovery, Second Edition. Edited by Gary S. Stein and Kimberly P. Luebbers.
© 2019 John Wiley & Sons, Inc. Published 2019 by John Wiley & Sons, Inc.

Fortunately, even though study results can seem equivocal, we do know that there are dietary choices that are more likely to keep your body healthy, boost your immune system, and help keep your cancer risk as low as possible. We also know that research over the past 25 years has shown that there are three controllable lifestyle choices that can decrease risk factors for diagnosis or recurrence. These are: (i) maintaining a healthy body weight; (ii) eating a healthy diet; and (iii) getting regular physical activity.

According to the AICR, one-third of cancers in the 1.5 million cancers that occur every year in the United States could be prevented by following these three controllable lifestyle choices. "Follow any one, and you will lower your risk. Follow all three, and you will receive the most protection."

HOW DO WE KNOW THIS?

In 1997, the AICR and its international affiliate, the World Cancer Research Fund (WCRF) in the United Kingdom, produced the first landmark report based on worldwide scientific evidence linking food, nutrition, physical activity, and cancer prevention. Since that time, thousands of additional studies were conducted. That is why in 2001, AICR and WCRF began producing a second expert report [1] that could consider the new research while using new methods to analyze large numbers of studies at once.

The result was the publication of "Food, Nutrition, Physical Activity, and the Prevention of Cancer: a Global Perspective" in 2007. For this report, nine independent research teams collected all the research on the relationships between diet, physical activity, and body size that was relevant to cancer risk. Together they identified 500 000 studies, narrowed them down to 22 000 that were reputable, and then identified 7000 that met their strict standards. An expert panel of 21 world-renowned scientists compared and evaluated these studies. Their final analysis and the resulting publication reported on how diet, physical activity, and body fat affect cancer occurrence at 17 sites. Cancer survivorship and obesity were also evaluated.

The panel formulated recommendations to help individuals lower their cancer risk for both cancer occurrence and recurrence. In the remainder of this chapter, the recommendations will be reviewed as they relate to the three main AICR Guidelines.

As stated previously, the three controllable AICR Guidelines are related to:

- Weight: How Much We Weigh
- Physical activity: How Much We Move
- Diet: What We Eat

Additionally, since "Food, Nutrition, Physical Activity, and the Prevention of Cancer: a Global Perspective" was published, The Continuous Update Project (CUP) [2], was established. The CUP came about because the field of research in cancer prevention related to food, nutrition, and physical activity is growing rapidly.

The CUP is now the world's largest resource of scientific literature on food, nutrition, physical activity, and cancer. The process of updating the research on a rolling basis is underway, to provide comprehensive cancer research reports. It is important to note that even with the ongoing research from the CUP, the three controllable factors and 10 recommendations established in the 2007 report have remained constant.

Weight: How much we weigh

The link between excess body fat and cancer was one of the strongest findings from the AICR's report and its continuous updates.

AICR estimates that, "approximately 122,200 cases of cancer occurring in the United States every year are attributable to excess body fat. The figure has increased steadily, from the 100,000 preventable cases of cancer cited in 2009 to almost 117,000 estimated in 2013. More obesity-related cancers and increases in incidences have led to the increased estimates" [3].

According to the Centers for Disease Control and Prevention, most recent statistics show that one-third of US adults is obese. Furthermore, the percentage of children aged 6–11 years in the United States who are obese increased from 7% in 1980 to nearly 18% in 2012. As described by the AICR, there are several ways in which excess body fat may increase cancer risk. For starters, fat tissue produces proteins called cytokines that can cause chronic inflammation, which increases cancer risk. Being overweight and obese also increases blood levels of insulin and related hormones that can spur the growth of cancer cells.

In order to best prevent cancer and promote survivorship, the AICR recommends that individuals try to stay as lean as possible within the healthy range of body weight. This means not only maintaining a healthy body weight but also avoiding major weight gain throughout adulthood. Table 1 shows the percent of cancers that can be prevented if body weight remains at a healthy level.

T A B L E 1. Percentage of cancers that could be prevented by staying lean.

Cancer Site	% link to excessbody fat: MEN	% link to excessbody fat: WOMEN	Cases Prevented Annually
Prostate (advanced)	11	–	1794
Ovarian	–	5	1099
Breast (postmenopausal)	–	17	39 554
Endometrium	–	50	26 315
Kidney	20	28	14 766
Gallbladder	11	28	2139
Esophagus	32	38	6025
Pancreas	17	20	8578
Colorectal	17	15	21 961
TOTAL			**122 231**

Source: Policy and Action for Cancer Prevention 2009. Adapted from AICR/WRCF [4, 5].

Body Mass Index (BMI)

Body Mass Index is a measurement that expresses the relationship of your weight (in kilograms) to your height (in meters). It is only a proxy measure of body fat but is generally highly correlated, or related to, the amount of body fat you have. Individuals with a BMI in the healthy range (18.5–24.9) have a lower cancer risk. A BMI of 25.0–29.9 is considered Overweight and a BMI of 30.0 and above is considered Obese. If you are unsure of what your BMI is, access the BMI calculator from the Centers for Disease Control and Prevention and find out: http://www. cdc.gov/healthyweight/assessing/bmi/index.html.

Waist Circumference

Waist circumference is another method of assessing risk from excess body weight. A waist circumference measure will give you an estimate of the accumulation of visceral or belly fat that you have. Visceral fat accumulates in the abdominal area and surrounds vital organs. People with too much visceral fat have been shown to be at greater risk of developing obesity-related diseases and cancer. A waist measurement of 31.5 inches or greater for women or 37 inches or greater for men indicates increased health risk. In order to measure your waist circumference, use these guidelines:

- Place a tape measure around your bare abdomen, above the tip of the hipbone.
- Be sure that the tape is snug, but does not compress your skin, and is parallel to the floor.
- Relax, and measure your waist after exhaling.

How to Achieve and Maintain a Healthy Body Weight

It is important to note at the outset that weight is not infinitely malleable. You cannot just "decide" what you want to weigh. No amount of caloric restriction and power yoga is going to get you to a size 2 if you've always been a size 12. However, losing just 5–10% of your body weight has substantial health benefits. So how do you manage this?

How Do Calories Affect Your Weight?

Your weight is all about energy balance: calories in = calories out. If the number of calories you eat equals the number of calories you burn, your weight stays the same. Consume more calories than you burn, and you gain weight. If you are looking to lose weight, you must create an energy imbalance by decreasing calories in (eating less) or increasing calories out (exercising more). The best method is to do both.

There are approximately 3500 calories in one pound of body weight. To lose one pound, you need to create a negative balance of 3500 calories. A negative balance of 500 calories per day will result in a weight loss of about one pound per week. A negative balance of 1000 calories per day will produce a weight loss of about two pounds per week.

How Many Calories Do You Need?

In other words, what do you subtract 500 or 1000 calories from? The amount of calories you need to maintain your weight depends on your size, metabolic rate, and activity levels. If you want to know how many calories you are eating right now to maintain the weight you are, multiply your weight in pounds times 12. (Example: If you weigh 150 pounds, you need 150×12 or 1800 calories per day to remain at that weight. A person who weighs 200 pounds requires about 2400 calories to maintain that weight.)

If you want to lose weight subtract 1000 calories from your weight maintenance number (your weight $\times 12$). This will create a weight loss of approximately one to two pounds per week. You should try not to eat below 1200 calories/day as it is really hard to get enough nutrients if you regularly try to eat less than this. In weight management, a calorie is a calorie. In terms of gaining or losing weight, your body does not care if the potato chips you just ate were baked or fried (your heart does though … but that is not what we are focused on at the moment!) – it just adds the 150, 250, 350 calories… whatever you ate … into your daily total. And the more you eat, the more you need to burn. Choosing carbs and proteins – which, gram for gram, have fewer calories than fat, may allow you to eat more food, but not more calories. For example, 1200 calories equals 17 apples (high in fiber, lots of volume); it also equals 24 Oreos (which probably will not fill you up much or provide your body with any of the nutrients it needs).

What Should You Eat?

Get the most out of your food. As stated previously, fat contains twice as many calories per gram than protein and carbs, making fat an easy target for eliminating excess calories. And research supports monitoring your fat intake as well as your calorie intake for successful weight management. Cutting back on fat intake and making smart choices with carbs, proteins, and fats will give you more or your calorie budget.

So how much fat do you need? Your fat gram budget can be between 20 and 25% of your calorie intake goal. For example, if you have a calorie intake goal of 1400 kcal/day, you would have a fat gram goal of 31–38 g/day. To determine the fat gram goal, multiply the calorie goal by 0.20 or 0.25 and then divide that number by 9 (there are 9 calories in a gram of fat) (for example, $1400 \times 0.25 = 350$ calories; 350 calories $/9 = 38$ g of fat).

While there has been a lot of research on the type of foods you should select while trying to lose weight (e.g., low carb, high fat, high protein), in the end, it appears that being able to adhere to a dietary pattern for a long time is really key. You need to manage your calories. How you do that, and what food choices you make within a specified calorie range are not as important as you might think. Of course, choosing a healthy diet is always a goal!

How Do You Know If You're on Track?

You know that you are on track by keeping track of what you are eating. While this may sound tedious (OK, it is tedious), keeping track of your food intake is the best predictor of weight loss success. Write down everything you eat and drink. This can tip you off to behaviors (and calories/fat!) that you might not be aware of. Follow these guidelines:

- Write it right after you bite it. That way, you'll avoid the forgot-it factor.
- Write it right after you sip it. Beverage calories count, too!
- Be specific. Record the type and amount of food in as much detail as possible in a journal or on line tracker using the My Plate SuperTracker (https://www. supertracker.usda.gov). You can also note where you were and how you were feeling, if you think it had an influence on what or how much you ate.
- Record calories. Refer to food labels, the My Plate [6, 7] SuperTracker database, or any other reliable source of nutrition information.
- Record fat. Again, use all your references to determine the grams of fat in what you ate. Food labels, or other reliable sources of nutrition information can help you with this.
- Subtotal. Add up your calories after each entry. That way, you will know if you have any calories in your budget for that after-dinner snack – before you eat it.

It does not matter whether you use an app, a website, or a good old fashion paper and pencil. "Keeping track" is a behavior and you just need to do it. Sometimes you may find yourself away from your measuring spoons or scale. In that case, Table 2 has some useful portion size estimates to help you stay on track even when you may be eating away from home.

PHYSICAL ACTIVITY: HOW MUCH WE MOVE

As well as helping us avoid weight gain, activity itself can help to prevent cancer. The AICR report identifies that regular activity can help to keep hormone levels healthy. This is important as having high levels of some hormones can increase your cancer risk. Physical activity may also strengthen your immune system, help keep your digestive system healthy, and allow you to consume a greater quantity of healthy foods with cancer-protective nutrients – without the consequence of weight gain.

T A B L E 2. Portion size estimates of common foods.

FOODS	PORTION SIZE ESTIMATE
GRAINS	
1 cup cereal flakes	Baseball
1 pancake	Compact disc
½ cup cooked pasta or rice	Light bulb
1 slice bread	Cassette tape
1 bagel	6 oz can tuna or Hockey puck
3 cups popcorn	3 baseballs
DAIRY and CHEESE	
1 oz cheese	4 stacked dice
1 cup yogurt	Baseball
½ cup frozen yogurt or ice cream	Light bulb
FRUITS & VEGETABLES	
1 medium fruit	Baseball
1 cup grapes	~32 grapes
1 cup strawberries	~ 12 berries
1 cup lettuce	4 lettuce leaves or baseball
1 cup carrots	~12 baby carrots
1 cup cooked vegetables	Baseball
1 medium potato	Computer mouse
FATS and OILS	
1 tbsp butter or spread	Poker chip
1 tbsp mayonnaise or salad dressing	Poker chip
1 tbsp oil	Poker chip
MEATS, FISH, and NUTS	
3 oz lean meat or poultry	Deck of cards
3 oz fish	Checkbook
3 oz tofu	Deck of cards
2 tbsps peanut butter	Golf ball
2 tbsps hummus	Golf ball
¼ cup almonds	12 almonds
¼ cup pistachios	24 pistachios
SWEETS and TREATS	
1 piece chocolate or 1 brownie	Dental floss package
1 slice cake	Deck of cards
1 cookie	~ 2 poker chips

Source: Based on data from AICR/WRCF.

As suggested by multiple studies, physical activity may help to reduce the risk of some secondary cancers and recurrence of certain cancers. Moreover, physical activity may help improve tolerance of cancer treatment and quality of life during and after cancer treatment. The advice of "take it easy" during cancer treatment may no longer be accurate! An expert panel assembled by the American College of Sports Medicine (ACSM) concluded that exercise is safe and beneficial for cancer survivors after, and even during, treatment [8]. Studies have demonstrated that exercise, when carefully monitored by the healthcare team, is a powerful tool to improve endurance, and boost sense of well-being and self-esteem, while decreasing fatigue and depression. Additionally, it can help increase appetite in a healthy way, plus aid in digestion. Most experts now recommend that people with cancer become and stay as physically active as they safely can.

Physical activity can help improve the following areas:

- Quality of life
- Maximum walking distance
- Muscle mass
- Muscle strength and power
- Aerobic Fitness
- Flexibility

How Much Activity Do You Need?

For cancer prevention and survivorship AICR recommends being physically active for **at least 30 min/day**. Physical activity in any form helps to lower cancer risk. Aim to build more activity, like brisk walking, into your daily routine. Try also to limit sedentary habits, such as watching television.

Research shows that to avoid weight gain, doing more activity is beneficial. For maximum health benefits, scientists recommend that we aim for **60 minutes or more of moderate activity every day, or 30 minutes or more of vigorous activity.**

What Is Moderate vs. Vigorous Activity?

Moderate activity is anything that gets your heart beating a bit faster and makes you breathe more deeply. Vigorous activity raises your heart rate so that you warm up, start to sweat and feel out of breath. Consider the examples in Table 3.

How Do You Get Started?

If you're not used to doing much activity, start by working toward 30 minutes of moderate activity each day. Don't forget, something is always better than nothing! You can build up slowly until you reach your target. Shorter bouts of activity are

TABLE 3. Examples of moderate and vigorous physical activity.

MODERATE ACTIVITY	VIGOROUS ACTIVITY
Brisk walking	Race walking, jogging, running
Water aerobics	Swimming fast or doing laps
Tennis (doubles)	Tennis (singles)
Ballroom or line dancing	Fast dancing or aerobic dancing
Biking on level ground or with a few hills	Biking faster than 10 miles per hour

Source: Based on data from AICR/WRCF.

also just as beneficial (i.e., three 10-minute intervals). It's the total time that's most important. Select exercise that you enjoy and will be able to maintain on a regular basis. For weight management, any exercise that requires continuous movement of large muscle groups – such as walking and biking – is good. Simply walking is a great exercise. Walking can be done alone, with others, at any time of day, indoors or out, at home or while traveling, and it requires no special equipment. Mile for mile, it burns as many calories as running does and certainly "counts" from a cancer prevention perspective.

It's also helpful if you establish a routine; schedule exercise just as you work in any other appointment. Try to set aside time each day. Look over your schedule and find a time that you think will work and make yourself "an appointment with health." Don't forget to record your exercise just like you record your food intake. Many of the dietary intake websites and apps will also have an option that allows you to record your exercise as well.

You can also start thinking of ways to add a bit more activity into your daily routine. Here's a quick list to get you started:

- Take the stairs. Do not be that person who takes the elevator to the second floor!
- Always park farther away. Make the "miles away" spot your space every time.
- Change classes. Remember the four minutes between classes when you were in high school. Bring it back. Every hour, take four minutes to move around: go get a drink of water, use the restroom, walk something down to someone's office. Set an alarm to remind you to move, if that's what you need.
- Fire the lawn guy. Mow your own lawn; rake your own leaves.
- Do not sit still. Watching television? Do sit-ups during commercials. Ride an exercise bicycle or walk in place.
- Switch your social agenda. Instead of going to a movie and sharing popcorn with a friend; go for a walk and a coffee. Instead of throwing a barbecue, suggest a group hike.

We have found so many ways to be efficient and save time (think: dishwashers, electric toothbrushes, garage door openers ...) that we now need to engineer activity back into our lives.

DIET: WHAT WE EAT

A healthy, balanced diet – one filled with a variety of vegetables, fruits, whole, grains, nuts, seeds, and legumes (beans), and low in red and especially processed meat, has been shown from multiple studies to decrease one's risk factors for developing cancer. This way of eating provides vitamins, minerals, antioxidants, and phytonutrients – naturally-occurring plant substances. Unlike vitamins and minerals, phytonutrients are not essential for keeping you alive, however, when you eat or drink phytonutrients, they may help prevent disease.

Researchers have identified a plethora of naturally occurring substances in plant foods, which have the ability to defuse potential carcinogens. "Heal Well – A Cancer Nutrition Guide," a joint project of the AICR, the LIVESTRONG Foundation, and Meals to Heal, describes these nutrients and phytonutrients as actually seeking out toxins and eliminating them from the body before they can cause cell damage that could lead to cancer. Other nutrients and phytonutrients can make it easier for the body to make repairs at a cellular level and others may help stop cancer cells from reproducing. So even after a cell begins to experience damage that can lead to cancer, your lifestyle choices can still help put a halt to the cancer process.

So Exactly What Should You Eat?

For cancer prevention and survivorship AICR recommends:

- Eat more of a variety of vegetables, fruits, whole grains, and legumes such as beans.
- Limit consumption of red and processed meats.
- Limit consumption of salty foods and foods processed with salt (sodium).
- Limit alcoholic beverages.
- Aim to meet nutritional needs through diet alone.

Each of these recommendations are discussed in more detail below.

Vegetables, Fruits, Whole Grains, and Legumes

The recommendation specifies eating at least five portions/servings of a variety of non-starchy vegetables and fruits every day. (Example of a serving: one cup raw or cooked vegetables or one medium apple.) Additionally, eat whole grains and/or legumes (beans and lentils) with every meal.

A practical way to do this is to make a habit of filling at least two-thirds of your plate with vegetables, fruits, whole grains, legumes, and nuts, while allotting one-third or less of your plate to poultry, fish, lean meats, and low-fat dairy and plant-based proteins.

Some other tips for adding this many servings of fruits and vegetables into your diet include:

- Keep a bowl of fruit within easy reach on the kitchen counter or your desk at work so that you can grab a piece of fruit when you are hungry.
- Buy packaged, ready-to-eat fresh vegetables, and fruits. These cut down on preparation time.
- Keep dried fruit on hand for a snack that is easy to take with you when you are away from home.
- Use the microwave to quickly cook vegetables.
- Dip raw vegetables in low-fat salad dressing, hummus, or peanut butter.
- Toss raw or cooked broccoli and cauliflower with low-fat Italian dressing and parmesan cheese to make a flavorful side dish.
- Roast vegetables and fruits to bring out their flavor. Just drizzle them with a small amount of olive oil, and bake them in the oven until they are tender.
- Season cooked vegetables with lemon juice and a small amount of olive oil. For extra flavor, add fresh herbs such as basil, tarragon, and sage.
- Try baked apples or pears topped with cinnamon and honey for a delicious dessert.
- Mix sliced fruit or frozen berries with yogurt or cereal.
- Add apple chunks, pineapple, grapes, or raisins to tuna or chicken salad.
- Make fruit smoothies by blending together fresh or frozen fruit, fruit juice, and yogurt.
- Add dried or fresh fruit to oatmeal, pancakes, and waffles.
- Add lots of colorful vegetables, such as red cabbage, carrots, and bell peppers, to green salads.
- Top salads with dried cranberries or raisins, or with sliced pears, oranges, nectarines, strawberries, or grapefruit.
- Add extra vegetables, such as grated zucchini or carrots, spinach, kale, and bell peppers, to pasta sauces and soups.
- Add lots of vegetables and fruit to sandwiches. Lettuce, tomatoes, cucumbers, bell peppers, and avocado slices are flavorful choices.

Again, keeping track of how many fruits and vegetables you eat each day is one way to know if you are meeting the target. You are more likely to eat more fruits and vegetables if you write down how many servings you get. Additionally, have a goal. Start with small goals you can achieve easily (this by the way, is good advice for setting any goal!). Then set larger goals as you go. For example, you might want to start by eating one extra serving of fruit or vegetables a day. When you have achieved that goal, your next goal could be to include an extra serving of fruit or vegetables at most meals.

Red and Processed Meats

Limit your intake of cooked red meat (i.e, beef, pork, lamb, and game) to 18 oz. or less per week. Try to avoid processed meats (i.e., cold cuts, bacon, sausage, hot dogs, and ham) to decrease the risk of colorectal cancer. Opt for meatless meals a few times a week, such as, vegetable stir-fry, bean soups, eggplant parmesan, or black bean burritos.

Salty Foods

Consuming too much salt can be harmful to your health, including, increasing your risk for stomach cancer, and high blood pressure. Most salt in American diets comes from processed foods, such as boxed, canned, and frozen prepared items, in addition to fast foods and other restaurant foods. It's recommended that we try to limit our sodium intake to 1500 mg/day.

Alcohol

Despite some evidence linking moderate alcohol consumption to lower risk for heart disease, this protective effect does not apply to some cancers. AICR recommends avoiding even small amounts of alcohol. Alcohol increases risk for cancers of the colon and rectum, breast, esophagus, mouth, and liver. If you choose to drink, limit intake to one drink per day for women and two for men.

One drink is defined as 12 fl oz of beer, 1.5 fl oz of 80-proof distilled spirits, or 5 fl oz of wine.

Foods Not Supplements

To reduce your risk of cancer, choose a balanced diet with a variety of foods rather than taking supplements. In general, the best source of nourishment is through food and drinks, not from dietary supplements. Nutrient-rich whole foods contain substances that are necessary for good health, such as, fiber, vitamins, minerals, and phytonutrients. Plant-based foods are the source of many cancer-fighting compounds.

The Dietary Supplement Health and Education Act (DSHEA) of 1994 defines dietary supplements as "products taken by mouth that contain 'dietary ingredients' used to supplement the diet. Dietary ingredients are vitamins, minerals, herbs, botanicals, and substances like amino acids, enzymes, metabolites, or organ tissues. These supplements come in many forms, including, tablets, capsules, softgels, gelcaps, liquids, tinctures, teas, extracts, concentrates, or powders."

Dietary supplementation may be recommended and prescribed for you by your healthcare team for specific medical conditions, such as osteoporosis and iron-deficiency anemia. If you are not currently taking supplements but would like to, it's important to discuss this with your healthcare provider.

SUMMARY

While there are still some questions that remain concerning weight management, how much we move, and what we eat, the previous recommendations, don't appear to be changing anytime soon.

Unfortunately, we know there are some things like genetics and the environment that we don't have control over. However, we also know that we do have control over the lifestyle choices we choose to make. And, we know healthy choices can help protect against serious illnesses like heart disease, stroke, and adult onset diabetes. And now we know they can also potentially decrease our risk of development or recurrence of cancer. These changes can only enhance your health and most importantly, add pleasure to your life. Choose wisely, choose foods you enjoy, and choose to eat them in good company.

REFERENCES

1. Food, Nutrition, Physical Activity and the Prevention of Cancer: A Global Report http://www.aicr.org/assets/docs/pdf/reports/Second_Expert_Report.pdf
2. The Continuous Update Project. http://www.dietandcancerreport.org/cup
3. The American Institute for Cancer Research. www.aicr.org
4. AICR/WRCF, *Policy and Action for Cancer Prevention* 2009 available at: https://www.wcrf.org/int/policy/evidence-policy/our-policy-report
5. Continuous Update Project reports (updated March 2014).
6. National Institutes of Health Weight-Control Information Network; WIN. http://www.niddk.nih.gov/health-information/health-communication-programs/win/Pages/default.aspx
7. United States Department of Agriculture My Plate http://www.choosemyplate.gov
8. American College of Sports Medicine Guidelines https://www.acsm.org/about-acsm/media-room/acsm-in-the-news/2011/08/01/new-guidelines-strongly-recommend-exercise-for-cancer-patients-survivors

34

CANCER SURVIVORSHIP

David Cranmer

Vermont Cancer Survivor Network, Montpelier, VT, USA

Increasing innovations in medical technology have led to improved treatment resulting in more Vermonters surviving with cancer each year. It is estimated that over 3600 Vermonters will be diagnosed with cancer in 2004. Of those, 64% will live at least five years. These people will join over 36 000 Vermonters currently living with cancer [1].

The term "cancer survivor" refers to an individual from the time of diagnosis with cancer through the remaining years of life. It is important to also recognize the needs, concerns, and issues of "caregivers." These are the people in a cancer survivor's life who are affected by the diagnosis, including family, friends, coworkers, and others.

Cancer survivors face numerous physical, psychological, social, spiritual, and financial issues throughout their diagnosis and treatment and for the remaining years of their lives. With this in mind, we are sharing ways to improve the quality of life for Vermonters living with cancer, as well as those who support them.

DIAGNOSIS

Learning you have cancer is a shock. You may feel numb, frightened, angry, or all of the above. Disbelief sets in. Most people feel completely overwhelmed. These feelings are normal. The shock of diagnosis wears off quickly. Feelings can change

Cancer: Prevention, Early Detection, Treatment and Recovery, Second Edition. Edited by Gary S. Stein and Kimberly P. Luebbers.
© 2019 John Wiley & Sons, Inc. Published 2019 by John Wiley & Sons, Inc.

rapidly from hour to hour. Stress and anxiety are joined by fear and depression. Welcome to the roller coaster ride of cancer survivorship.

How we deal with diagnosis can evolve through nine stages: denial, anger, fear, anxiety, loss of control, depression, guilt, loneliness, and hope. You may not progress through each stage, or in this order, but this emotional evolution is normal.

Most people cope by initially refusing to believe the diagnosis. You need to get the facts about your cancer diagnosis. Once you have accepted the diagnosis, it's normal to express your anger. Having cancer is scary and feeling angry would be expected. A cancer diagnosis is what people fear the most. Fear is a natural human response – it prepares you mentally and physically to react – fight or flight. When you think about it, most fear is based on the unknown – and surviving cancer is nothing but a series of unknowns. Asking questions – lots of them – of your healthcare team, helps you gain the knowledge and ease your fears.

Anxiety, loss of control, and depression. Let your healthcare team know how you're feeling, not only physically, but emotionally, there are medications that can help. Also talking with a social worker or mental health professional can help. Many folks have said that talking to other cancer survivors has helped.

Guilt and loneliness. Support groups have proven to be a great way to deal with the guilt of cancer.

Hope. After going through this series of negative feelings and emotions, most people develop a sense of hope. Hope is the "magic ingredient" in motivating yourself. You'll need to motivate yourself to get through these treatments and to get on with your life. Try to avoid getting caught up in the statistics of cancer – they are based on thousands of other people – you are you. No one can accurately predict the outcome of your treatment. Hope may be inspired by religious or spiritual beliefs. How your level of hope can affect the outcome of your treatment has been debated, but there is little doubt that a positive attitude and a hopeful outlook can improve the quality of your life.

Choosing to live after a cancer diagnosis is the most important step. Then you must be prepared to make choices at every step of your treatment. This means talking to your healthcare team – asking questions – at every visit.

GETTING INFORMATION

Using the internet is a popular way to obtain information. Most of us have internet access at home, work, or our local library. Although the web is full of information, there is no system in place to help one determine the accuracy of the information. It is easy for false and misleading information to multiply online.

How do you know if the source of information is accurate? Consider the source. The website should include basic information on who runs the site, where they are located, how to get in touch with them, and why they have established the site. This is usually a page called "About This Site" or "About Us."

Also be aware of the site's address. Sites addresses ending in "gov." are government sponsored sites (The National Cancer Institute – part of the National Institute of Health has as its address – cancer. gov). An address ending in ".edu" indicates an educational institution, such as Harvard.edu. Those addresses ending in ".org" usually indicate a non-for-profit organization, such as the American Cancer Society at http://cancer.org. Addresses ending in ".com" usually denote a commercial site. These sites may be sponsored by drug companies or organizations wishing to sell you something. The source of funding can impact what information is listed and how it is presented.

The site should not be asking for information from you. Many sites ask you to sign up for their mailing list – or to access their bulletin boards or discussion sites. If a site wants your name and address or email address before letting you in, it may be a commercial site more interested in selling you something. Never give out your social security or credit card number to any site asking for one to allow you to enter.

The last thing to look for is the "Links" page on the site. Reliable sites usually have a policy about who they will allow as links.

In addition to websites, the internet offers bulletin boards, chat rooms, blogs, and lists. These will contain information generated by those visiting the site. While it is good to hear about the experiences of other survivors, remember, there is no way to verify what is being said.

A great online source for cancer information is the American Society of Clinical Oncology at www.cancer.net.

In no way is the internet going to be a replacement for talking with your medical team. They know you, your health, and what options are available.

YOUR TREATMENT HAS ENDED – NOW WHAT?

"Your tests are negative … that's good!" This means you are in remission, the end of one phase of your cancer survivorship and start of the next phase.

While in active treatment you are doing everything possible to beat your cancer. You have been focused on your target – getting better. Now what?

In 2010, the University of Vermont surveyed nearly 1600 Vermont cancer survivors to gather information about survivors' needs and identify unmet needs. The purpose is to use the research to improve services and support for Vermont cancer survivors.

The top cancer survivor needs (whether they were met or unmet) were:

- To feel like you were managing your health together with the medical team.
- More information about after-effects.
- Information provided in a way you can understand.
- Assurance your doctors talked to each other to coordinate your care.
- More information about your cancer.

The top cancer survivor unmet needs were:

1. Help in reducing stress in your life.
2. More information about after-effects.
3. Help finding ways to reduce your worrying.
4. Help managing your concerns about your cancer coming back.
5. Help dealing with the impact of cancer on your relationship with your partner.

A closer look found a need for greater coordination of care, and more information about side effects and after-effects of treatment, as well as better access to complementary and alternative medicine. There were also unmet needs in the areas of sexual function/fertility issues: adjusting to body changes and dealing with partner.

SURVIVORSHIP CARE PLANS

After completing active treatment, coordinated follow-up care is required to address the needs listed above. Your cancer team will create a Survivorship Care Plan that details what treatment you received as well as follow-up care recommendations. This plan is to be used by, you, your caregiver, and your primary care provider to ensure continuous care.

Your plan should include resources for information on your cancer, as well as health promotion activities and support programs. This information is always changing, so bring your plan with you to your cancer follow appointments to get any updates.

Meeting your unmet needs is as individual as you are. Use the resources in your survivorship care plan to find the help you are looking for. Your cancer center has access to the programs that will best work for you. You need to ask for the help.

CAREGIVERS

The cancer caregiver goes through many of the same stages and feelings as the cancer patient. A caregiver said, "we both had cancer, it just wasn't in my body."

Caregivers need to be sure to take care of their own emotional and physical needs in order to be a better caregiver. Stress is an important issue to address. Talk to your doctor if you are having trouble coping. Your cancer center has the resources to help you, you will need to ask. To be a good caregiver, you need to be good to yourself.

There are online resources for caregivers, such as Cancer Care's My Cancer Circle™.

REFERENCE

1. Kachajian, J. and Verbyla, A. 2017 *Cancer Incidence in Rutland.* http://www.healthvermont.gov/sites/default/files/documents/pdf/stat_RRMCForum102815Final.pdf

35

THE TOOLS OF ADVOCACY IN CANCER PREVENTION, EARLY DETECTION, RECOVERY, AND TREATMENT

Jill Sudhoff-Guerin

The Vermont Medical Society, Montpelier, VT, USA

The ultimate conquest of cancer is as much a matter of public policy as it is a scientific and medical challenge.
—*John R. Seffrin, PhD, Former Chief Executive Officer for the American Cancer Society*

Over the past century, America has made incredible advances in the science of cancer prevention, early detection, recovery, and treatment. In fact, the CDC released a Morbidity and Mortality Weekly report in March, 2015 verifying that cancer is no longer a death sentence. Innovations in early detection and treatment of cancer are paying off; as the CDC report found that 65% of patients diagnosed with invasive cancers between 2003 and 2010 lived past the five-year survival mark, whereas just a few decades ago only 50% of cancer patients lived beyond five years [1] Dr. Julie Vose, president of the American Society of Clinical Oncology, credited the improvement in the five-year survival rates to the increase in access and efficacy of screening programs and diagnostic analysis. "The screening done for many of the malignancies has improved over the last few years, and thanks to the Affordable Care Act, it is now covered for many people," said Vose, of the University of Nebraska Medical Center in Omaha, "[t]his will definitely help in years to come to decrease the 5-year mortality, as many of the common

Cancer: Prevention, Early Detection, Treatment and Recovery, Second Edition. Edited by Gary S. Stein and Kimberly P. Luebbers.
© 2019 John Wiley & Sons, Inc. Published 2019 by John Wiley & Sons, Inc.

cancers can be caught with early screening and therefore detected at an earlier stage with resultant better outcomes" [2].

These better outcomes are not only the direct result of the continued targeted efforts of researchers, oncologists, and medical professionals, but also from the concentrated effort of advocacy. Without the persistent work of cancer advocates and other public health advocacy organizations the Affordable Care Act (ACA) would not have become law and would not include critical chronic disease prevention tenets.

THE ACA: ADVOCACY AT WORK

As Dr. Vose pointed out, at its most basic level, the ACA has the potential to dramatically decrease the likelihood of cancer and other chronic disease among the majority of Americans, primarily because of the broad expansion of health insurance coverage for the estimated 47 million Americans living without insurance in 2010 when the law was passed. Another critical game-changer, for population health, is the fundamental requirement for all health insurance plans to provide coverage with zero-cost sharing for 45 preventive services, including smoking cessation, vaccinations, and cancer screenings. According to the American Public Health Association, in 2014, when the law was implemented, 1 in 6 Americans were uninsured and 7 out of 10 deaths were linked to preventable chronic diseases, such as cancer, obesity, diabetes, heart disease and high blood pressure [3].

This vital health reform law has also directly changed access to health insurance for those living with a cancer diagnosis by eliminating the ability for health insurance plans to deny coverage for preexisting conditions and by abolishing annual or lifetime caps on benefits. Also, of crucial benefit to cancer patients, and others living with chronic disease, are stipulations that prohibit healthcare plans from charging more for sick patients and put in place maximum thresholds for out-of-pocket costs including prescription drug costs.

Undoubtedly, these provisions will serve to reduce the rate of chronic disease and make a significant impact in the lives of those living with a cancer diagnosis, but these policy faculties were not accidental. Countless public health advocacy organizations and cancer advocacy volunteers not only helped to pass the ACA, but worked tirelessly to ensure that these protections were included in final version of the law.

THE POWER OF COALITION

The protections were included due in part to the efforts of an aggressive advocacy alliance called the *Health Coverage Coalition for the Uninsured* (HCCU), which in January 2007 began pushing Congress to adopt a federal health reform proposal that would provide health insurance coverage for at least half of the estimated 47 million uninsured Americans [4]. Early in the process their proposal advocated for three core initiatives that were eventually included in the final version of the ACA. The HCCU wanted Congress to provide additional funding for the State Children's Health Insurance Program (SCHIP), to cover the then-estimated nine million uninsured children, to give states the option to expand Medicaid eligibility to cover families whose household income is up to 138% of the federal poverty level (FPL) and to offer tax credits to adults with incomes between 100 and 300% FPL to help them purchase private coverage.

The success of the HCCU was due in part to the broad interest of the participating organizations, including: AARP, American Academy of Family Physicians, American Hospital Association, American Medical Association, American Public Health Association, America's Health Insurance Plans, Blue Cross Blue Shield Association, Catholic Health Association, Families USA, Federation of American Hospitals, Healthcare Leadership Council, Johnson & Johnson, Kaiser Permanente, Pfizer Inc., United Health Foundation, and the US Chamber of Commerce.

Although, some would assume that groups like the American Public Health Association; which is made up of 25,000 of the most credible health organizations in the world, vs. the US Chamber of Commerce or Pfizer, the pharmaceutical company, would represent an unholy alliance because of entirely different operating principles and presumably, completely different objectives. But the burden of the ever-increasing cost of healthcare had become a trend too difficult to ignore. These businesses and organizations knew that as long as the United States was carrying a large number of uninsured lives, premiums for private insurance plans would continue to rise in order to cover soaring emergency and uncompensated healthcare costs. They also knew that businesses offering employer-sponsored insurance would continue to flounder economically, as long as they were unable to predict and plan for wild-card annual increases in the price tag for healthcare benefits. According to a Kaiser Family Foundation report from 2014, average premium increases have risen more slowly in the last five years, but in the late 1990s and early 2000s American employers faced double-digit healthcare premium increases annually [5]. The historical evidence of the rapid variation in healthcare premiums, versus the static incline in business revenue and wages indicated that unless the broader economic disadvantage created by the impact of the sheer numbers of uninsured lives was addressed, as a nation, we would never get ahead of the steep curve in healthcare costs. So, despite obvious differences, this unlikely group of businesses and organizations were able to put aside any existing hostilities in order to align on the singular policy mission of providing health insurance coverage for as many uninsured Americans as quickly as possible.

The HCCU became an extremely powerful advocacy alliance. The diversity of the group played to their favor, in their ability to broaden their reach to potential allies by having such an expansive range of organizations providing representation of the business side, the medical provider side, the pharmaceutical company side, the insurance side and the public health advocate side of this significant health reform dilemma; this advocacy soon became the primary undertaking of the ACA.

PUBLIC HEALTH AND ADVOCACY

In 2011, the CDC reported that during the twentieth century, population health among US residents improved at an unprecedented rate, proven by the dramatic shift in life expectancy from 47.3 years in 1900 to 76.8 years in 2000 [6]. According to an *American Journal of Public Health* article published in September 2009, public health policy changes have substantially influenced this outcome; all of the ten great public health achievements of the last century have been due in part to evidence-based legislation, regulation, and guidelines including seat belt laws, vaccination requirements, and breast and cervical cancer screening recommendations [7].

Along with population health gains, achievement in public health policy has largely been incremental, usually taking decades if not longer to fully implement. These compromises are due in part to the fact that whether it is taxation, prohibition, or regulation, each policy initiative generally requires the direct confrontation of established cultural, social, political, and financial norms and conventions. Therefore, advocates pushing for policies are continually asked to make concessions to lessen the impact of their intervention on vested interest groups. These interest groups can either become allies, obstacles, or "the opposition," depending on how the policy intervention is framed. Far-reaching public health goals, no matter how rational, can often be portrayed as overly ambitious or punitive to established interest groups. When the public policy intervention faces hefty political and financial opposition, advocates are generally told to take the "long view" and often the policy campaign timelines are extended, sometimes over decades.

SURGEON GENERAL'S REPORT: CASE IN "LONG-VIEW" PUBLIC POLICY

One of the greatest incremental public health campaigns of the twentieth century was launched by the release of the landmark Surgeon General's report, "Smoking and Health: Report of the Advisory Committee to the Surgeon General of the United States" by the then Surgeon General Luther Terry in 1964. Although there had been many public debates about the potential negative health impacts of smoking, and a growing body of evidence mounting in the 15 years prior to the report, the report itself was one of the first declarations by a US official that announced a direct, causal relationship between cigarette smoking and lung cancer, bronchitis, emphysema, and for pregnant women who smoke a high likelihood of an underweight newborn. The report called for "appropriate remedial action," but in the face of the behemoth power of the tobacco industry and the outright popularity of tobacco use in every facet of US society, the Advisory Committee responsible for the report did not go so far as suggest specific interventions in the report, as they were wary of the weight of the pushback. Even the alliance of prominent public health organizations, such as the American Cancer Society

(ACS), the American Heart Association, and the American Public Health Association, who in 1961 were responsible for calling on President John F. Kennedy to create a national commission on smoking, cited the need to consider the cultural and financial forces that would be directly impacted by the potential negative health findings related to tobacco use. The letter they addressed to President Kennedy requested that the commission be dedicated to "seeking a solution to this health problem that would interfere least with the freedom of industry or the happiness of individuals" [8].

In 1964, half of all American men and one-third of American women smoked [9], second-hand smoke was relatively unavoidable in public places like restaurants, airplanes, and doctor's offices, and the US media depicted cigarettes as an attractive and integral accessory in major movies and television and print advertising. And even though the causal relationship between lung cancer and cigarette smoking was becoming indisputable, these cultural, social, and financial factors represented a formidable obstacle in combatting this public health crisis. A Gallup Survey conducted in 1958 found only 44% of Americans believed smoking caused cancer, but with overwhelming empirical evidence and a deliberate public awareness campaign, a second study conducted only ten years later (1968) found that 78% believed in the connection [8]. This dramatic shift in consciousness happened relatively quickly, but the groundwork had been laid by policy interventions pushed by public health advocates far before the release of the Surgeon General's report and by two significant legislative actions Congress implemented after the release of the report.

By the end of the 1950s, because of mounting evidence that smoking caused lung cancer and chronic disease, most states had laws prohibiting the sale of cigarettes to minors. And in relatively direct response to the Surgeon General's report, Congress passed the *Federal Cigarette Labeling and Advertising Act of 1966*, which required every cigarette pack sold in the United States to display a health warning saying, "Caution: Cigarette smoking may be hazardous to your health [10]." One of the sponsors of the legislation, Sen. Warren G. Magnuson, D-Washington, praised the new law, calling it "a forthright, historic step towards the responsible protection of the health of this nation's citizens" and yet public health advocates did not celebrate this action [11].

Rather, tobacco industry lobbyists and cigarette manufacturers were the ones high-fiving this seeming public health victory, because it actually significantly delayed efforts to pass a more meaningful law that would prohibit cigarette advertising on television and radio. The tobacco industry feared advertising regulations far more than labeling laws, as they were the number one product advertiser on television in the 1960s and spent approximately $297 million on cigarette advertising and promotion in 1966 alone [12]. The industry not only used advertisements to develop new markets for their targets but also ran advertisements discrediting scientific health claims against cigarettes. And this is why they began developing their own team of well-connected political insiders, lawyers, and lobbyists before the release of the report, who pushed for a strategy using the labeling law as a smoke screen to protect their real interests. When passed, the labeling law included a key provision that blocked the Federal Trade Commission from regulating tobacco advertising for four years and prohibited states from developing cigarette advertising regulations as well [11].

So, while the smoking rates have been more than halved since the first Surgeon General's report, and undeniable progress has been made in the field of tobacco control and prevention, tobacco use is still the number one preventable cause of disease in the nation and reducing tobacco use to meet the CDC's *Healthy People 2020* goals remains elusive. It remains elusive not only because of the addictive nature of smoking and nicotine, but also due to the incredible influence and political power of what's been known as "Big Tobacco." Big Tobacco is not only made up of the three major tobacco companies in the world, but also armies of lobbyists and lawyers with deep pockets, public relations companies, "independent researchers," and most importantly, lawmakers who in exchange for campaign donations have helped protect the industry from regulation.

Policy is catching up with Big Tobacco though, as the public health community has years of scientific-based evidence showing that strong smoke-free laws, high prices for tobacco products, and properly funded tobacco control programs work together to effectively drive state smoking rates down. The public health community is also making headway against Big Tobacco because public opinion in the United States has swung strongly against cigarette smoking due to ongoing public awareness campaigns and targeted media tactics.

These media tactics in some ways mimicked or used tobacco industry advertisements to make the US public aware of the dangers of smoking. After the law regulating cigarette television and radio advertisements was eventually implemented in 1970, the tobacco industry was forced to throw its advertising weight into print. One of the most iconic and effective tobacco advertising campaigns was the "Marlboro Man," in which supposedly real "cowboys" were hired to embody the rugged, individualism of the American smoker. The ACS ended up also using this campaign to spread the message that cigarette smoking causes chronic disease, when former Marlboro Man actor Eric Lawson appeared in a 1997 anti-smoking advert after being diagnosed with chronic obstructive pulmonary disease, or COPD [13]. Eric Lawson starred in Marlboro Man advertisements from 1978 to 1981 and died at age 72 from complications due to the disease.

THE PRIMARY TOOLS OF ADVOCACY

When we examine the advocacy successes of the ACA and the advancement of US tobacco control policies, four core principles emerge as the basis for effective public health policy campaigns. The first is the integral role the researcher plays in building the body of evidence for the policy intervention, the second is the critical importance strong coalitions play in building allies and diffusing opposition, the third is the value of public awareness campaigns and the fourth is the overwhelming impact the personal story has to create the emotional rationale to take action on policies that often confront social and political norms in order to achieve a better health outcome.

RESEARCH AND POLICY: HAND IN HAND FOR PUBLIC HEALTH

As said previously, public health gains have historically been reliant on public policy, but there is an interdependence on the epidemiology that ultimately informs successful policy implementation. The scientific data is the critical foundation of formulating any evidence-based policy intervention, while multidisciplinary research and evaluation results of the implemented program serve to drive the advancement of policies that will ultimately achieve the greatest health benefits. Unfortunately, policymakers often have more to consider than just whether there is evidence to support a public health program. Often they are juggling multiple interest groups, budgetary constraints, accepted cultural behaviors, community biases, and various other political and economic forces when considering a proposed policy. In the face of these obstacles, the intrinsic value that public health experts and researchers offer both the policymaker and the public health advocate is objectivity. Rather than voicing personal opinions, strong, objective research speaks for itself in making the case for the best policy derived from the careful evaluation of a broad field of perspectives and inquiries.

According to a 2013, *CA: Cancer Journal for Clinicians* article published by the ACS, the 1964 Surgeon General's report, *Smoking and Health*, was not only a landmark public health publication because of its conclusions, but also because it set a precedent on how to interpret controversial medical and scientific data and how to build an objective review team [14]. Amidst the overwhelming acceptance of smoking in all areas of US society and the incredible economic might and influence of the tobacco industry, it was of significant importance for the Advisory Committee to create a bullet-proof methodology for evaluating the data. While the tobacco industry scrambled to poke holes in the data, the Committee based their life-saving, yet indisputable conclusions on the review of over 7000 white papers, 27 case control and cohort studies, and 7 other broader cohort studies.

HOW COALITIONS AND GRASSROOTS CAPACITY INFLUENCE

Unlike the concrete accuracy of research data, advocacy has to operate in the unpredictable and uncontrollable realm of politics, in which evidence is largely anecdotal, information the only true currency, and relationships paramount. Like the impact that the *HCCU* had on influencing the ACA, broad-based coalitions have the potential to change the game for public health advocates if they are comprised of a wide range of respected community, business, and political leaders, grassroots activists, trade associations, and other relevant third-party organizations, beyond just public health advocates and experts. Coalitions can be made up of "grasstops" members, or community leaders who are already connected to policymakers, media sources and local business groups, but also must have representation from the "grassroots" constituents, who can appeal directly to their lawmaker.

Once the coalition has been built, key, consistent messaging is developed and the coalition is mobilized into directing their energy into specific advocacy tactics that will impact the campaign targets. Whether initiating high-level grasstops meetings with key decision makers, flooding a public hearing with supporters, or generating numerous letters from grassroots activists to elected officials, broad-based coalitions have the ability to garner support and generate real influence through the power of their collective message and the diversity of their individual affiliations and relationships.

MASS MEDIA, MEDIA ADVOCACY, AND CANCER PREVENTION

Every day Americans are bombarded with a plethora of information, some of this information is pursued, but most of it is passive, consumer-motivated advertisements seeking to market products and services, establish social norms, and influence cultural behaviors. Over the past several decades, public health advocates have adopted similar mass media strategies in order to reach broad audiences relatively quickly with counter messages focused on public health epidemic education, prevention, and outreach. These media strategies have been extremely powerful when it comes to cancer in terms of cancer awareness, cancer-related stigma, and cancer prevention and early detection because the majority of cancer diagnoses are preventable. According to a 2008 article, only 5–10% of cancer cases are related to genetic defects, which means that 90–95% of cancer diagnoses are attributable to lifestyle choices and preventable behaviors, such as: tobacco use, diets high in sugar and fat, alcohol use, sun exposure, environmental pollutants, stress, obesity, and lack of physical activity [15]. Public health professionals have used persuasive, targeted media campaigns like the Centers for Disease Control and Prevention's "Tips from Former Smokers" (TIPS) campaign, which uses real life stories from former smokers living with smoking-related diseases to develop graphic, hard-hitting, emotional profiles. According to an article printed in *The Lancet* medical journal, after the first TIPS advertisements ran for three months in 2012 1.6 million additional smokers made a quit attempt and 100,000 of these remain nonsmokers as a result of the advertisements [16]. Advocates were then able to demonstrate the cost-effectiveness of the media campaign to lawmakers and in December 2014 former US Senator Tom Harkin announced as part of the FY2015 omnibus appropriations bill, a $6 million increase in funding for the TIPs campaign [17].

Whereas, these mass media advertisements focus on personal responsibility and influencing individual behaviors, "media advocacy" targets decision-makers in order to implement specific policy recommendations. Lawrence Wallack, in his 1993 book, *Media Advocacy and Public Health: Power for Prevention*, calls media advocacy a "blend of science, politics and activism" [18]. Media advocacy essentially uses the news media to spread important social and public health policy concern by framing the message, identifying specific policy solutions and finding experts, celebrities and constituents to relay this message in order to attract press coverage. The ACS used this strategy in 1974 when they named actress Eva Gabor as the National Women's "IQ" Chair of the ACS anti-smoking campaign. "IQ" stood for I Quit and although ACS was also running paid advertisements with the former smoker, they knew with such a high profile actress and newsworthy Gabor quotes, such as "Smoking is bad for glamor," and "Life is so beautiful, why cut it shorter?" they would also gain vital press coverage for their campaign [19]. They were also successful in framing the message, as they then used Gabor's press opportunities to push back against the cigarette industry's campaign to promote smoking as part of the 1970s women's liberation movement.

OUR STORIES ARE UNDEBATABLE TRUTHS

Critical to media advocacy and driving both public awareness and public policy is the power of personal experience to drive the message. A 2009 report, released by the Kaiser Family Foundation and the ACS before the ACA was passed, stated that cancer, as one of the most costly medical conditions in the United States [20], is unaffordable for most patients living with a cancer diagnosis because of high insurance premiums, cost-sharing requirements, and the potential inability to work while going through treatment [21]. The report itself was compiled of 20 different patient stories, representative of the more than 20,000 Americans who have called the ACS Health Insurance Assistance Service because they could not find adequate and affordable health insurance or were struggling to pay for healthcare despite being insured. These stories served to highlight the unfair access to quality care through many insurance plans and the limitations of employer-sponsored insurance prior to the implementation of the ACA. But more importantly, these stories served to personalize and humanize the insurance policy gaps related to a cancer diagnosis and are an integral tool in advocating for better access to care for cancer patients.

CONNECTION AS THE ROOT OF ACTION

"Coach" Dave Koch took the call on Christmas Eve, 1996 and he fell on the floor. At the time of the call, the energetic postal worker and father of four felt "great" and never expected the persistent lump in his neck would be diagnosed as non-Hodgkin lymphoma. From that moment, he has been on what he calls, "an incredible journey." With three of his children in college and one in high school, he struggled not only to comprehend how his cancer treatments were affecting him, but also how to help his family and friends understand what he was going through. He channeled this newfound awareness into running a weekly cancer treatment support group and into volunteering for the ACS Cancer Action Network, where he uses his story as a means to advocate for evidence-based policies aimed at eliminating cancer as a major health problem.

Advocacy is the art of focused persuasion to bring positive change at the local, state and national levels. Almost every decision-maker in the United States has been touched by cancer in some way. Most lawmakers have an intimate understanding of the emotional and physical struggles of cancer diagnoses, cancer treatment, and recovery through a family member, a friend, or their own cancer journey. Therefore, along with the science of cancer; the epidemiological data and critical medical research, public awareness campaigns and the outreach of credible public health organizations, it is undoubtedly the cancer patient's story that connects the lawmaker to the urgency of the need to pass policies to prevent, treat, control, and cure cancer.

When Coach met with a key decision-maker to ask her to support policies to improve cancer patient's lives, his story worked as a great equalizer. That law-maker may not remember the details of the data or the demographics discussed in their meeting, but she will remember Coach's story and how it was similar to her grandmother's battle against cancer or maybe her own. His vulnerability and the undebatable truth of his story, along with the collective cancer experience of our culture creates the emotional will to drive policymakers to follow through on supporting critical life-saving policies. As further evidence of this point, the Wharton School conducted a study that asked participants to read three stories and contribute five dollars to alleviate hunger in Africa. In one version, their donation would go to a particular seven-year-old girl in Mali named Rokia; in the second, to millions of suffering Africans; in the third, to Rokia – but in this version, she was presented within the larger context of world hunger: "Rokia is just one of millions suffering from hunger." The study found that people were more likely to give directly to the first story of Rokia – not to anonymous millions and not to Rokia when presented as part of a larger scenario [22].

To summarize, through careful data collection we can determine evidence-based solutions to public health and chronic disease dilemmas, through coalition-building and media advocacy we can develop the public awareness to build support for the solution, and through meaningful storytelling we are able to educate lawmak-ers on the personal impact of the disease and the impetus to prevent others from going through the same experience. After fostering this human connection, the decision-maker will likely remember Coach's story and be more amenable to be part of the solution when it comes time to vote on the issue.

BREAKING DOWN THE STIGMA

Although many nonprofits employ the use of the personal story in order to garner support, traditionally talking about cancer has been taboo. A critical component of cancer control advocacy is simply generating public awareness of the disease and breaking down the cancer-related stigma and the culture of silence that has historically been part and parcel of the cancer diagnosis. Until quite recently in the United States and currently in other countries around the globe, myths around cancer diagnosis and treatment have encouraged patients and their families not to talk about their cancer for fear of being ostracized and not to become educated on prevention, proper diagnosis, and the best forms of treatment.

As a child, I remember my grandmother referring to her neighbor and mouthing "she has the cancer." My grandmother didn't even want to say the word cancer out loud, as she, along with most Americans at the time, believed cancer was contagious, that it could be caused by mental trauma or injury and if people knew you had it your chances of attaining gainful employment or a marriage proposal could be significantly reduced. Although, we have made substantial progress in reducing stigma related to cancer in this country, in many countries around the globe, deep shame is still associated with a cancer diagnosis, not only in the financial burden it places on families, but also in the origins of the disease. In a 2011 survey conducted by the Lance Armstrong Foundation, 25% of respondents from the nine countries polled, including Japan, Mexico, Russia, Argentina, China, France, India, Italy, and South Africa, believe cancer patients brought the disease on themselves [23]. The shame and lack of meaningful information surrounding cancer can lead to the disease going undetected at the precancerous stages, undiagnosed at the early, treatable stages and untreated until the consequences of the disease in terms of pain and prognosis are dire.

My grandmother who was a forward-thinking "educated" woman from suburban New Jersey, who despite having three children, always worked outside the home and taught her daughter to put absolute faith in her intellectual abilities. She was also a devout Irish Catholic, extremely superstitious and completely unconnected to her health and her own body. The family was shocked when she was diagnosed with Stage 4 Ovarian cancer when only 64 years old, yet she had ignored all the signs of the disease because of embarrassment of diagnosis and fear of treatment. Unfortunately, I never got to know my grandmother as she died soon after her diagnosis, but her legacy for my family is one of awareness and proactive prevention. My mother has always been completely open with my sister and I that we face a higher ovarian cancer risk because of my grandmother's history. She has taught us to routinely get our screenings, to educate ourselves on the risk, and to proactively seek healthcare to avoid all cancers and I plan to do the same with my daughters.

TURNING STIGMA INTO AWARENESS AND AWARENESS INTO ADVOCACY

With only one generation between my grandmother and I, I have a completely different outlook on cancer and my chances of survival of a diagnosis. But for my grandmother, who was born in 1920, the words, "you have cancer" almost certainly meant death. Through the 1950s and 1960s, my grandmother's experience was not unusual, as most Americans were afraid to speak openly about cancer and therefore knew very little about the causes and possible treatments of the disease. Shortly before my grandmother was born in 1913, a concerned group of physicians and businessman organized the American Society of the Control of Cancer (ACCU), with the primary goal of raising public awareness of cancer in order to fill the extreme information void and to ultimately, achieve better cancer survival and prevention [24]. Through the leadership of the ACCU, the group began engendering critical cancer awareness by writing educational articles for popular magazines and professional journals, publishing *Campaign Notes,* a monthly bulletin with cancer prevention information and recruiting doctors around the country to share cancer information with the general public. In 1945, the ACCU was renamed the ACS, after a powerful businesswoman named Mary Woodard Lasker offered to help fundraise for the ACCU if they agreed to devote 25% of what she raised toward cancer research [25]. After the Society agreed, Lasker went to work and with the breadth of her connections raised over $4 million in research funds and eventually convinced Congress to pass the National Cancer Act in 1971 [24]. The Cancer Act was a major political victory that resulted in dramatic increases in National Institutes of Health funding, specifically in the field of cancer.

Lasker was a masterful networker, relentless in her bid to pass the National Cancer Act, who employed the primary advocacy tools and then some in her "War on Cancer." She brought top medical researchers to Washington to testify on the medical, societal, and personal benefits of enhanced research opportunities and to promote the idea of a "cure" for cancer [26]. She used her husband's advertising connections to generate mass media around the cause and even convinced RCA president David Sarnoff that the word "cancer" could be said on the radio [27]. She used her social prowess to build an incredibly influential coalition of the most powerful, wealthy businessmen and women of the time who not only significantly increased her fundraising potential, but who were also formidable allies in the halls of Congress. She was quoted often in telling lawmakers, "If you think research is expensive, try disease!" [27].

In December 1969, a coalition she founded called the Citizens Committee for the Conquest of Cancer placed full-page advertisements in *the New York Times* and *the Washington Post* that said: "Mr. Nixon: You Can Cure Cancer" [28]. Whereas, in April 1971, right before the Cancer Act bill was up for a vote, she mobilized 500,000 Americans to send letters to Congress after asking Ann Landers to solicit the support in her national syndicated column [26]. This strategic media advocacy and outpouring from constituents was largely credited with the passage of the law.

Although some called Lasker's tactics "ruthless," she was extremely successful and according to a 2015 article in The Hill, was "widely credited for steering more public dollars into health-related research than any other individual in U.S. history" [26]. She was successful because she had the long-view in pursuing her policy objectives and because she used every advocacy tool at her disposal.

WHAT IS THE FUTURE OF CANCER ADVOCACY?

We have made the case that in this country we have changed a cancer diagnosis from a death sentence to a manageable disease, due in part to the collective voice of advocacy organizations demanding better treatment, better screenings, and more funding for research. In fact, there are more than 345 organizations around the globe focused on the treatment, research, and support of cancer [29]. Hopefully, these organizations can begin to make headway on the next horizon of the "war on cancer," which is in the disparity of cancer stigma, cancer education and prevention, and cancer treatment in economically developing countries like those in sub-Saharan Africa and India. The roughly 650,000 Africans who develop cancer annually face an 80% mortality rate, as they currently lack basic access to adequate cancer care and treatment [30]. Also, one-third of these cancers are preventable or easily treated if detected early. In order to combat this epidemic, we must build strong partnerships focused on eliminating cancer stigma through critical information sharing, we must pass stronger tobacco policies, as Big Tobacco is targeting the continent with flashy ad campaigns and cheap products and we must aid in creating the infrastructure to properly screen for cancer and treat those with cancer.

REFERENCES

1. Henley, S.J., Singh, S.D., King, J. et al. (2015). Invasive cancer incidence and survival – United States, 2011. *Morbidity and Mortality Weekly Report* 64: 238–242.
2. Bankhead, Charles, "More Cancer Patients Living Longer: Two-thirds of patients with diagnoses in 2003 to 2010 lived 5 years or longer." *Medpage Today: Oncology/Hematology*, March 13, 2015. Accessed at: https://www.medpagetoday.com/hematologyoncology/othercancers/50476
3. APHA. *Health Reform*. Accessed December 28, 2018 at: https://www.apha.org/topics-and-issues/health-reform
4. American Hospital Association (2007). *Expanding Coverage*. Chicago: AHA.
5. Claxton, G., Rae, M., Panchal, N., et al. Health Benefits In 2015: Stable Trends In The Employer Market. *Health Affairs*, available at: https://www.healthaffairs.org/doi/10.1377/hlthaff.2015.0885
6. National Center for Health Statistics (2011). *Health, United States, 2010: With Special Feature on Death and Dying*. Hyattsville, MD: CDC, National Center for Health Statistics.
7. Brownson, C., Chriqui, J., and Stamatakis, K. (2009 September). Understanding evidence-based public health policy. *American Journal of Public Health* 99 (9): 1576–1583. https://doi.org/10.2105/AJPH.2008.156224.

8. Surgeon General's Advisory Committee on Smoking and Health, and United States. Public Health Service. Office of the Surgeon General. 1964. Smoking and Health. Official Report. *Profiles in Science* from the National Library of Medicine. Accessed December 28, 2018 at https://profiles.nlm.nih.gov/ps/display/About

9. U.S. Department of Health and Human Services (2014). *The Health Consequences of Smoking—50 Years of Progress: A Report of the Surgeon General*. Atlanta, GA: U.S. Department of Health and Human Services, Centers for Disease Control and Prevention, National Center for Chronic Disease Prevention and Health Promotion, Office on Smoking and Health.

10. FTC. *Statutes Enforced or Administered by the Commission*. Accessed December 28, 2018 at: https://www.ftc.gov/enforcement/statutes/federal-cigarette-labeling-advertising-act-1966

11. Drew, Elizabeth. "The Quiet Victory of the Cigarette Lobby: How It Found the Best Filter Yet—Congress" *Atlantic Monthly,* September, 1965. Accessed November 18, 2015 at http://www.theatlantic.com/magazine/archive/1965/09/the-quiet-victory-of-the-cigarette-lobby-how-it-found-the-best-filter-yet-congress/304762

12. National Center for Chronic Disease Prevention and Health Promotion (US) Office on Smoking and Health (2012). *Preventing Tobacco Use Among Youth and Young Adults: A Report of the Surgeon General*. Atlanta: Centers for Disease Control and Prevention.

13. "Actor who played Marlboro Man in ads dies from smoking-related disease." *Associated Press*. 2014-01-27.

14. Brawley, O.W., Glynn, T.J., Khuri, F.R. et al. (2013). The first surgeon general's report on smoking and health: the 50th anniversary. *CA: A Cancer Journal for Clinicians* https://doi.org/10.3322/caac.21210.

15. Anand, P., Kunnumakara, A.B., Sundaram, C. et al. (2008). Cancer is a preventable disease that requires major lifestyle changes. *Pharmaceutical Research* 25 (9): 2097–2116. https://doi.org/10.1007/s11095-008-9661-9.

16. McAfee, T., Davis, K.C., Alexander, R.L. Jr. et al. (2013). Effect of the first federally funded US antismoking national media campaign. *The Lancet* 382 (9909): 2003–2011.

17. Rita Rubin, M.A. (2015). Successful CDC campaign "Tips from former smokers" to be expanded. *JAMA* 313 (6): 558. https://doi.org/10.1001/jama.2014.18588.

18. Wallack, L., Dorfman, L., Jernigan, D. et al. (1993). *Media Advocacy: The Power of Prevention*. Newbury park, CA: Sage Publications.

19. Desert Sun, April 2, 1974. "Smoking Not Sexy." Accessed at http://cdnc.ucr.edu/cgi-bin/cdnc?a=d&d=DS19740402.2.54

20. National Cancer Institute at the National Institutes of Health. 2011 Cancer costs projected to reach at least $158 billion in 2020. Accessed at http://www.cancer.gov/newscenter/newsfromnci/2011/CostCancer2020

21. Schwartz, K., Claxton, G., Martin, K. et al. (February, 2009). *Spending to Survive: Cancer Patients Confront Holes in the Health Insurance System*. Kaiser Family Foundation and American Cancer Society https://www.kff.org/health-costs/event/spending-to-survive-cancer-patients-confront-holes.

22. Capecci, J. and Cage, T. (2012). *Living Proof: Telling Your Story to Make a Difference*. Minneapolis, MN: Granville Circle Press.

23. Beaulieu, N., Bloom, D., Bloom, L.R. et al "Breakaway: The global burden of cancer – challenges and opportunities." *Economist Intelligence Unit Limited* for LIVESTRONG, August 2009.

24. American Cancer Society. (2015) Our History: the Early Years. Accessed at http://www.cancer.org/aboutus/whoweare/our-history

25. Strickland, S.P. (1972). *Politics, Science, and Dread Disease. A Short History of United States Medical Research Policy*. Harvard University Press. Accessed at http://www.smokershistory.com/ASCC.htm.

26. Pomeroy, Claire (March 20, 2015). The Empress of All Maladies: Mary Lasker." *The Hill*. Retrieved from http://thehill.com/blogs/congress-blog/healthcare/236121-empress-of-all-maladies-mary-lasker

27. Hunt, N. (December 13, 2007). "Mary Woodard Lasker: First Lady of Medical Research." Retrieved from http://www.laskerfoundation.org/media/firstlady.htm
28. The Mary Lasker Papers Cancer Wars. *Profiles in Science* from the National Library of Medicine. Accessed at https://profiles.nlm.nih.gov/ps/retrieve/Narrative/TL/p-nid/201
29. National and International Cancer Organizations. Cancer Index, July, 2015. Accessed at http://www.cancerindex.org/clinks7.htm.
30. American Cancer Society. (2015) Sub-Saharan Africa. Accessed at http://www.cancer.org/aboutus/globalhealth/globalinitiativesbyregion/sub-saharan-africa

INDEX
